AF877790

MOLECULAR RECOGNITION AND INCLUSION

MOLECULAR RECOGNITION AND INCLUSION

Proceedings of the Ninth International Symposium on Molecular Recognition and Inclusion, held at Lyon, 7-12 September 1996

Edited by

A. W. COLEMAN
Institut de Biologie et Chimie des Protéines,
CNRS,
Lyon, France

KLUWER ACADEMIC PUBLISHERS
DORDRECHT / BOSTON / LONDON

A C.I.P. Catalogue record for this book is available from the Library of Congress.

ISBN 0-7923-5330-7

Published by Kluwer Academic Publishers,
P.O. Box 17, 3300 AA Dordrecht, The Netherlands.

Sold and distributed in North, Central and South America
by Kluwer Academic Publishers,
101 Philip Drive, Norwell, MA 02061, U.S.A.

In all other countries, sold and distributed
by Kluwer Academic Publishers,
P.O. Box 322, 3300 AH Dordrecht, The Netherlands.

Printed on acid-free paper

All Rights Reserved
© 1998 Kluwer Academic Publishers
No part of the material protected by this copyright notice may be reproduced or
utilized in any form or by any means, electronic or mechanical,
including photocopying, recording or by any information storage and
retrieval system, without written permission from the copyright owner.

Printed in the Netherlands

TABLE OF CONTENTS

INTRODUCTION

This volume contains the Proceedings of the Ninth International Symposium on Molecular Recognition and Inclusion, ISMRI 9 which was held in Lyon, France during 7 to 12 September 1996. The articles reflect the over 50 oral presentations and 140 posters which were presnted at ISMRI 9, both in the range of topics and also in the layout of the volume which comprises five sections, Plenary, Invited, Oral and Emerging Lectures and the four poster sessions. Some words should be said about the Emerging lectures, these were a means of allowing young scientists, often doctoral students to present short 15 minute talks on their work and were one of the great scientific successes of ISMRI 9. I would again like to thank the presenters of these lectures for their contributions.

The scientific content of ISMRI 9 reflected the logo of the conference showing the symbiotic interactions between Chemistry , Physics and Biology which contribute so strongly to the inter- and pluridisciplinary nature of Supramolecular Science. The topics ranged from Glycobiology through Membrane Systems through Synthetic Organic and Inorganic Chemistry to the construction of Complex Edifices in solution and the Solid-State to arrive at the Physics of Molecular Interactions via the understanding of Water and Gas-Clathrates. Once more to all the speakers who us the breadth of the subjects, thank you.

Finally my thanks to the International Organising Committe of ISMRI for allowing me the chance to organise ISMRI 9, to the Scientific Advisory Board for their help in setting up the programme, to the Local Committee for running the Symposium so smoothly and a special thanks To Professeur Hèlene Parrot-Lopez, and Drs Marc Munoz and Mohamed Eddaoudi without whom ISMRI 9 would not have been the success it was.

Dr. A.W.COLEMAN
LYON
1998

BIOLOGICAL IMPORTANCE OF GLYCOSYLATION

Glycobiology

RAYMOND. A. DWEK
University of Oxford
Department of Biochemistry
The Glycobiology Institute
The Rodney Porter Building
South Parks Road
Oxford OX1 3QU UK

The chemistry of simple sugars was worked out in the late nineteenth century by Emil Fischer, and the ring structures determined in the inter-war years by Haworth and colleagues. Simple polysaccharides such as starch, glycogen and cellulose, as well as more complex molecules such as chitin and hyaluronic acid had also received attention and their component sugars identified by classical means. By the 1960's, especially through work on blood-group determinants, it had become clear that besides simple mono-and polysaccharides, naturally occurring carbohydrates were commonly conjugated to proteins and lipids (as glycoproteins and glycolipids).

Little progress could be made to determine the structure or function of these complex molecules until sensitive and sophisticated techniques became available to analyse the component sugars and the order and structural details of their attachment to protein. Today automatic techniques are available for analysis of glycoproteins (in picomole amounts) and the progress in technology has advanced considerably our understanding of carbohydrate structures attached to proteins.

Protein glycosylation is influenced by three main factors: the overall protein conformation, the effect of local conformation, and the available repertoire of glycosylation-processing enzymes for the particular cell type. In general, the pattern of glycoforms is protein-specific, site-specific, and tissue- or cell-specific.

Glycobiology deals with the nature and role of carbohydrates in biological events. Glycoproteins are now known to be fundamental to many important biological processes including fertilisation, immune defence, viral replication, parasitic infection, cell growth, cell-cell adhesion, degradation of blood clots and inflammation. They are major components of the outer surface of mammalian cells. Over half the biologically important proteins are glycosylated (Figure 1).

A. W. Coleman (ed.), Molecular Recognition and Inclusion, 1–6.
© 1998 *Kluwer Academic Publishers. Printed in the Netherlands.*

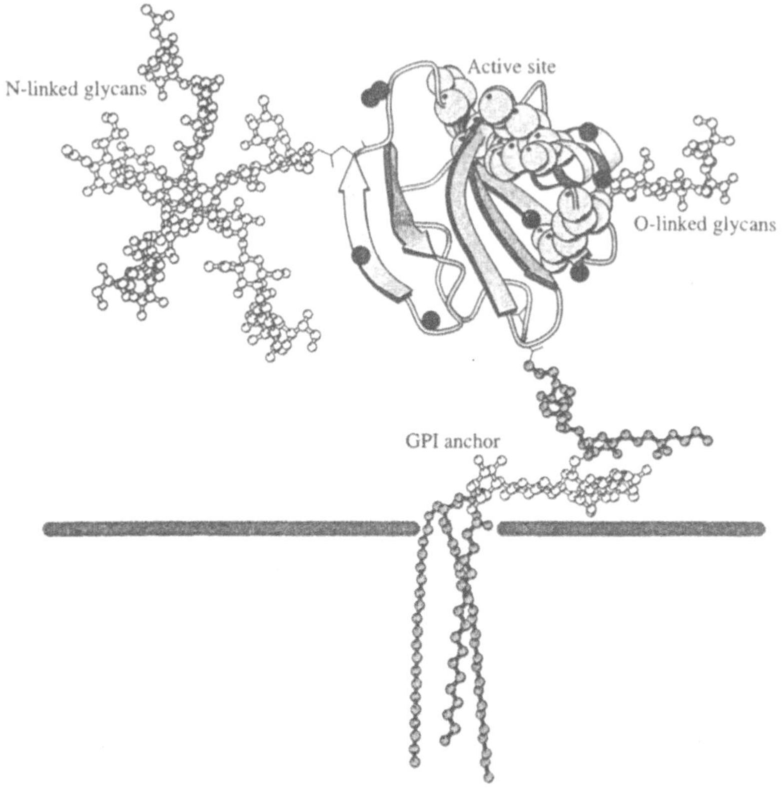

Figure 1

Molecular model of CD59 showing the relative sizes of the N-linked, O-linked and GPI anchor to the protein

Oligosaccharide structures change dramatically during development and it has been shown that specific sets (i.e. specific sequences) of oligosaccharides are expressed at distinct stages of differentiation. Further, alterations in cell surface oligosaccharides are associated with various pathological conditions including malignant transformation.

The finding that glycosylation may vary with disease also leads to the concept that its manipulation might alter the properties of glycoproteins and result in beneficial therapeutic results. The ability to manipulate and modify sugar structures also provides an important approach in understanding the different functions of oligosaccharides.

The elegant biosynthetic glycan-processing pathway in the cell allows, in principle, the same oligosaccharide to be attached to quite different proteins without having to code the information into the DNA of the individual proteins. However, the orientation of the attached oligosaccharide with respect to the polypeptide may markedly affect the properties of the glycoproteins. Further, different glycoforms of a protein may display quite different orientations of the oligosaccharides with respect to the protein, thus conferring different properties. A striking example is the structure of the Fc fragment of IgG (Figure 2).

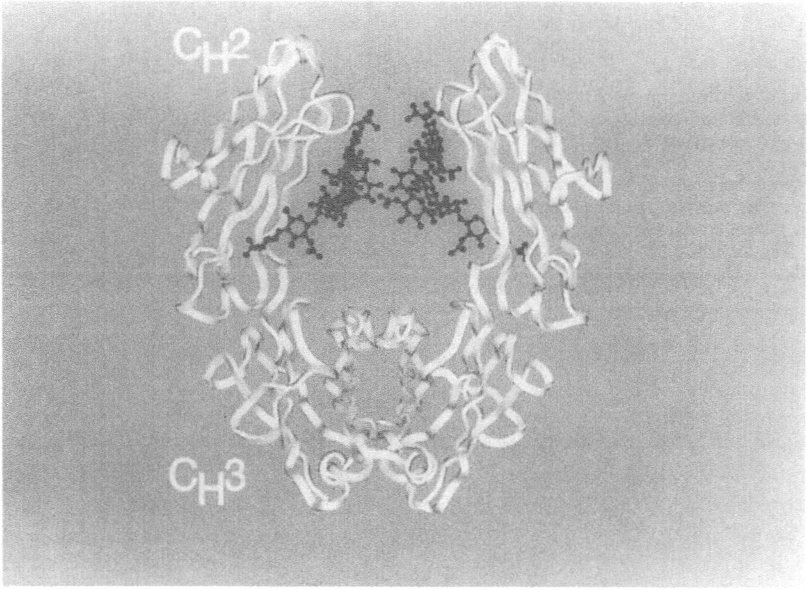

Figure 2

The Fc part of the IgG molecule showing the intrinsic oligosaccharides filling the interstitial space between the two C_H2 domains.

The conserved N-linked complex oligosaccharides at Asn^{297} on each heavy chain of the C_H2 domain occupy the interstitial space between the domains and also interact with the domain surface. Loss of the two terminal galactoses from the oligosaccharide as in the Fc fragment from patients with rheumatoid arthritis, results in a loss of interaction between the domain surface and the oligosaccharide. This permits displacement and consequent exposure of the oligosaccharides, giving them the potential to be recognised by endogenous receptors lectins such as the Mannose Binding Protein.

The recognition of oligosaccharides (lectins) is influenced by their accessibility, the number of copies of the oligosaccharides and their precise geometry of presentation. These factors introduce a high degree of specificity and control as to whether the recognition is physiologically relevant or not.

That one set of structures on different proteins can result in quite dramatic variations in properties of glycoproteins or that different glycoproteins may have different properties emphasises that there is no single unifying function for oligosaccharides.

Clearly, a major function is to serve as recognition markers. Additionally, oligosaccharides can modify the intrinsic properties of proteins to which they are attached by altering the stability, protease resistance or quaternary structure. The large size of oligosaccharides may allow them to cover functionally important areas of proteins (Figure 3), to modulate the interactions of glycoconjugates with other molecules and to affect the rate of processes which involve conformational changes.

Figure 3

The effect of flexibility of the Asn-34 chain on the orientation of the oligosaccharide attached to ribonuclease B. The Man-9 glycoform of RNase B based on the 2.5 Å X-ray crystal structure with an overlay of 10 oligosaccharide conformations from a 500 psec molecular dynamic trajectory of Man-9. The total Van der Waals surfaces of the oligosaccharides are shown.

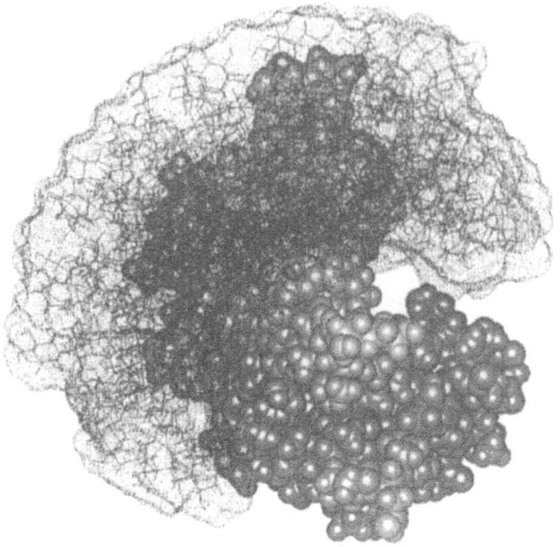

Glycosylation is highly sensitive to alterations in cellular function, and abnormal glycosylation is diagnostic of a number of disease states including rheumatoid arthritis and cancer. The control of glycosylation by the cell affords, in principle, a means of putting the same recognition markers on quite different proteins without having to code the information into the DNA of that protein. Site-specific glycosylation of a protein also suggests that the 3-D structure of the protein plays a role in determining the extent and type of its own glycosylation.

Glycosylation is also a highly sensitive probe of the correct functioning of a cell. This makes it necessary to define in detail the glycosylation of recombinant products which have possible industrial or pharmaceutical applications, since altering the glycosylation of a glycoprotein may significantly affect its properties. The major factors affecting the intrinsic properties of the glycoprotein would seem to be the size of the attached oligosaccharide which may affect intermolecular interactions or intramolecular rearrangements, site specific glycosylation and variable glycosylation site occupancy (Figure 4a and b).

Figure 4 a and b

HPLC sugar prints of normal IgG from a healthy individual and a patient with rheumatoid arthritis. The sugars are labelled with a fluorescent probe at their reducing termini.

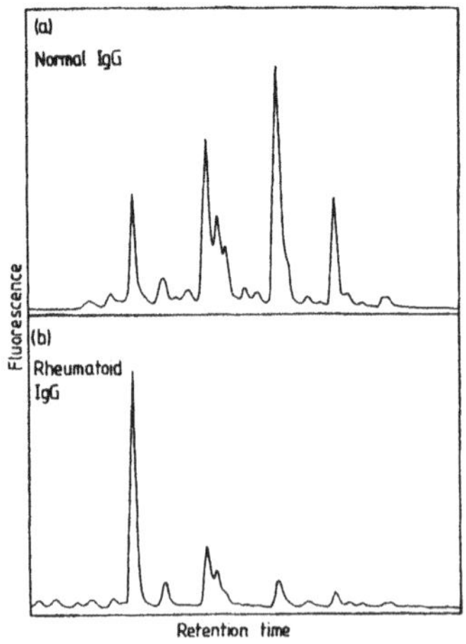

For example, the rate of fibrin dependent activation of plasminogen by tissue plasminogen activator (tPA) depends on the occupancy of the glycosylation sites on Kringle 2 in tPA (site 184) (Figure 5a,b and c) and on Kringle 3 in plasminogen (site 288). The combinatorial effect of glycoforms of both the tPA and plasminogen molecules results in a 4-fold range of activity. In a cascade process, such as extra cellular matrix re-modelling, which involves a number of glycosylated enzymes including tPA and plasminogen, such variations in activity may allow a high degree of control.

Figure 5a,

(a) A schematic molecular model of plasminogen type 1 and type 2. Plasminogen consists of five kringle regions and a serine protease domain. Type 1 plasminogen (left) has two occupied glycosylation sites - at Asn289ArgThr in kringle and at Thr 345 in kringle 4. Type 2 lacks the N-linked sugar at Asn 289.

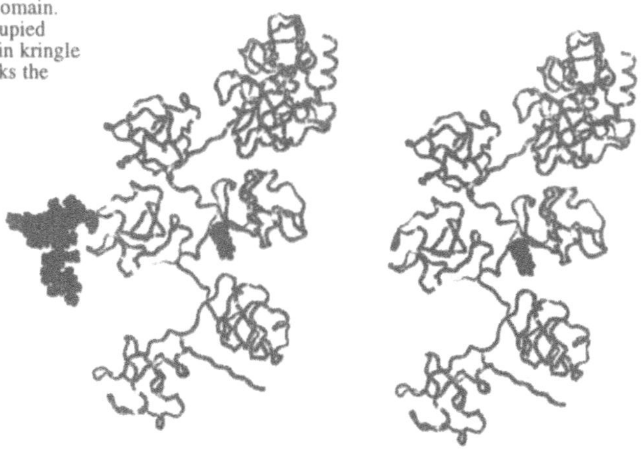

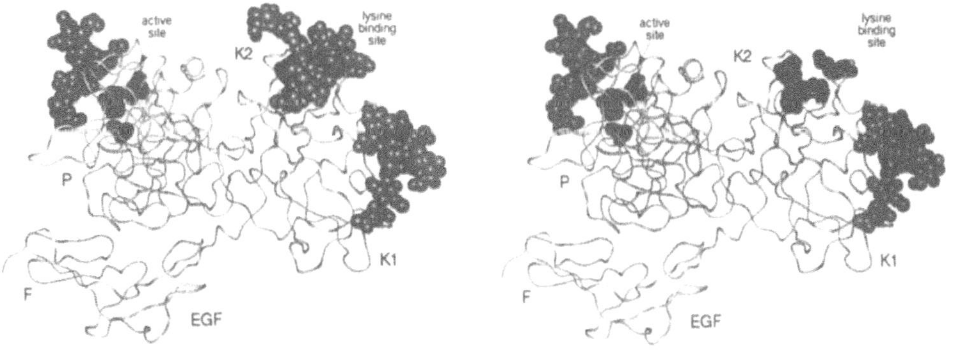

Figure 5b and c

(b,c) Schematic model of tissue-type plasminogen activator types I and II. tPA is composed of five domains: a fibronectin type 1 finger module, and EGF-like module, two kringles and a serine protease domain. This model was constructed using the co-ordinates of the finger growth factor pair (Smith, B.O., Downing, A.K. & Campbell, I.D.) and kringle 2 (93) from human tPA. Kringle 1 and the serine protease domains were modelled by homology. The high-mannose carbohydrate at position 117 and the complex sugars at sites 184 and 448 are shown.

Control of glycosylation can also be influenced by imino sugars. N-butyldeoxynojirimycin (NB-DNJ) inhibits the processing enzymes α -glucosidases I and II. Treatment with this compound (at a concentration which exhibits anti-HIV activity *in vitro*) results in glycoproteins with uniform glycosylation, where immature endo H sensitive oligosaccharides are retained. This has been demonstrated for recombinant gp 120 expressed in CHO cells, as well as for gp120 derived from H9 cells, acutely infected with the HIV-1 III_b strain. Two consequences of treatment with NB-DNJ are the inhibition of syncytia formation in cells infected with HIV-1, and the reduction in infectivity of released virus. Although the exact mechanism of action still has to be established, alteration of the glycans of the HIV envelope by NB-DNJ is a possible candidate for forming the basis for this activity. In contrast to the HIV envelope glycoproteins, which contains about 30 glycosylation sites, the hepatitis B virus envelope proteins contain only one or two glycosylation sites. *In vitro* treatment of this virus with NB-DNJ results in a high proportion of virus particles being retained inside the cell. Preliminary data show that these viruses contain a large proportion of endo H sensitive oligosaccharides. This suggests that correct glycosylation is necessary for the processes involving transport of the hepatitis B virus out of the cell. Comparison of the effects of NB-DNJ on these two viruses emphasise that oligosaccharides attached to proteins can have very different functions.

In summary:

- Glycosylation is the primary cause of microheterogeneity in proteins (Glycoforms). These reflect complexity at both molecular and cellular levels.

- Protein sugar prints are conserved and not random under normal physiological conditions.

- There are many potential functions of glycosylation. For instance, physical properties include: folding, trafficking, packing, stabilisation, protease protection, quaternary structure and organisation of water structure. Properties relating to recognition and biological triggering are characterised by: weak interactions, multiple presentation and precise geometry.

- Many of the properties may only operate in a specific biological context.

- Changes in sugar prints may both reflect and results in physiological changes, e.g. cancer and rheumatoid arthritis.

For a general reference see:

Dwek, R.A. **Glycobiology: Toward Understanding the Function of Sugars.** (1996) *Chemical Reviews* **96** Number 2, 683-720

Hydration Interactions: Their Role in Recognition and Bioassembly Phenomena

Felix Franks
BioUpdate Foundation
7 Wootton Way
Cambridge CB3 9LX
UK

The unique and eccentric physical and physicochemical properties of water, particularly in its liquid state, originate from its molecular structure which can be represented by a tetrahedron with sp^3 hybridized orbitals directed toward the four corners, as shown in Fig. 1. The molecules interact weakly by hydrogen bonding, giving liquid water a three-dimensional network structure, the ideal of which is found in hexagonal ice. The chemistry of life processes is sensitively attuned to this structure and to the energy of the hydrogen bond in ^1H_2O. Even the minor (?) isotopic modification to ^2H_2O produces a physiologically toxic environment. The physical and biophysical chemistry of water and aqueous solutions has been discussed in detail in the series Water - A Comprehensive Treatise [1] and, more recently, in Water Science Reviews [2].

Because of its tetrahedral quadrupolar structure, the interactions of water with other chemical species are expected to be relatively weak, highly cooperative and orientation-specific. So-called hydration interactions can be divided into three distinct classes:

1. Ion-water interactions, mainly of an electrostatic nature, relatively strong

2. Direct molecule-water hydrogen bonding, e.g. with polar groups in organic molecules and in which water can act either as proton donor or proton acceptor

3. So-called hydrophobic or apolar hydration in which water appears to interact with molecules (e.g. hydrocarbons) or molecular moieties that cannot participate in hydrogen bonding. This type of "interaction" is unique to water as solvent.

Although hydrogen bonds between molecules are very weak, this is not necessarily the case for ion-molecule hydrogen bonds. Ion-water interactions in solution are thus of a relatively long range, compared to purely molecular hydration effects. It was not until the advent of neutron diffraction, that definitive evidence of ionic hydration structures in solution has been obtained. Figure 2 shows the average dispositions of water molecules surrounding Ca^{++} and Cl^- ions. In each case the hydration shell consists of six water molecules, forming an

A. W. Coleman (ed.), Molecular Recognition and Inclusion, 7–18.
© 1998 *Kluwer Academic Publishers. Printed in the Netherlands.*

octahedron about the central ion [3]. The diffraction data also provide evidence for second layers of less-well oriented water molecules. The dynamics of ion hydration in solution has been studied mainly by n.m.r. Life times of water molecules in the ion hydration shell range from several picoseconds to microseconds, i.e. long-lived hydration shells.

A mystery which has puzzled scientists for more than a century concerns the manner in which salts direct many chemical processes in aqueous solutions. Hofmeister, while studying the effects of salts on protein solubility, found that ions could be divided into two groups: those that enhance and those that reduce the protein solubility [4]. He reported that the order (although not the magnitudes) in which the ions affect the solubility was identical, and independent of the nature of the protein. The "Hofmeister Series" in an abridged form is shown below:

$$SO_{42-} > HPO_{42-} > F^- > OAc^- > Cl^- < Br^- < I^- < NO_{3-} < HClO_{4-}$$

salting-out salting-in

In later years it was found that the same sequence applies to the effect of ions on the solubility of argon, the stability of proteins (see below), the critical micelle concentration of amphiphiles and biological membrane phase transition temperatures. Although during the past century the ionic series has been "rediscovered" on many occasions, its origin is still quite uncertain and subject to continuing speculation.

Molecular interactions by direct solute-water hydrogen bonding are termed hydrophilic hydration. Molecular groups capable of participating in hydrogen bonding include -O-, -OH, NH-, NH_2, C=O etc. Of the various types of hydration, this is the least well-defined. Because of the complexity of many of the molecules involved, it is hardly amenable to study by diffraction. The characterisation of hydrophilic hydration has been based mainly on n.m.r. [5,6], dielectric [7] and thermodynamic measurements [8]. It has long been clear that solute-water interactions play a major role in directing the stereochemistry of polar molecules, such as carbohydrates. Thus, a combination of n.m.r. and Molecular Dynamics (MD) simulation studies on the stereoisomers mannitol and sorbitol in water and in pyridine has established that the two isomers take up different configurations from one another in solution and that these configurations differ for the two solvent environments and from those characteristic of the crystalline states of the two molecules [9]. The configurations of the two isomers in aqueous solution is shown in Fig. 3. This solvent sensitivity is also reflected in the solution properties of saccharides, e.g. anomeric ratios, tautomerism and glycosidic bond flexibility, all of which are extremely sensitive to their molecular hydration geometry [10]. This observation leads naturally to the speculation that hydration effects may well play a role in immunochemical and other glycobiological phenomena.

The third type of hydration arises from the introduction of apolar molecules or apolar residues into water. Its simplest manifestation is in aqueous solutions of rare gases and hydrocarbons [11]. It involves the reorientation of water molecules so as to create cavities able to accommodate the apolar guest molecule. Water cannot directly interact with the inert solute; it therefore attempts to maintain its intermolecular hydrogen bond network by performing a series of reorientations, as a result of which the empty volume within the structure is redistributed [12]. In ice and liquid water, pairs of H_2O molecules are arranged in *gauche* configurations. The introduction of an apolar group produces cooperative rotations about the hydrogen bonds to produce *cis* arrangements of water molecules. As shown in Fig. 4, this allows the formation of cavities of various dimensions, able to encase the apolar guest moecule, but without net breakage of water-water hydrogen bonds. It is thus apparent, that so-called apolar or hydrophobic hydration is largely confined to interactions between water molecules. Crystalline analogues of such cage structures are well known in the gas hydrates of the lower alkanes and other small molecules [13]. That similar structures also exist in solution has been convincingly established by neutron diffraction studies of aqueous argon solutions which show that each argon atom has 16 nearest H_2O neighbours, placed at a range of 0.28 - 0.54 nm from the centre of the cage [14].

The water-water hydrogen bond lengths and energies in crystalline gas hydrates are almost identical to those in ice. The low solubility (positive excess free energy of mixing) of rare gases and hydrocarbons in water does not therefore arise from unfavourable, repulsive energetic interactions between the solutes and water. It is due to constraints on the configurational degrees of freedom placed on the water molecules forming the cage, because -OH vectors must only be oriented along the edges of, or away from the cavity. In other words, the positive excess free energy results from the decrease in the entropy upon mixing the substances. This is a unique type of "interaction" which plays an important role in the formation and stabilisation of biological structures. The recent neutron diffraction results on aqueous argon solutions have confirmed computer simulation studies, pioneered by Stillinger and Rahman [15] and earlier n.m.r. relaxation results [16] which had suggested a general rotational slowing down of water molecules in the neighbourhood of apolar residues. Apolar hydration is usually referred to as "structure making" and continues to be the subject of great interest .

Since the transfer of an apolar residue R from hydrocarbon (or gas) is seen to be thermodynamically unfavourable, then the converse, i.e., the association of R residues in water, should be accompanied by a negative free energy change. At the simplest level, two hydrocarbon molecules, each with its associated hydration cage, would gain in stability by their association, because this would "release" water molecules from the cages which could then relax into their more stable, unperturbed configurational states. The process

$$2R(\text{hydrated}) \rightarrow R_2(\text{hydrated}) + \text{water}$$

would therefore be expected to take place spontaneously. The driving force for such an association does *not*, however, derive from an attraction (e.g. by van der Waals forces) between apolar molecules or groups, but from an extrusion of alkyl groups by water for configurational reasons [17]. The process is said to be entropy-driven, in the sense that $T\Delta S_\phi > 0$ and $T|\Delta S_\phi| > |\Delta H_\phi|$, where the subscript ϕ describes the association process in the above equation. Thus, what *appears* to be an attraction between two apolar residues or molecules (negative free energy) is actually the sum of several water-solute repulsions. The term apolar (or hydrophobic) interaction which is commonly used to describe the process is really a misnomer.

Irrespective of the origin of nature's molecular and supermolecular building materials, a given biological function is generally associated with a specific three-dimensional structure, maintained largely by weak, noncovalent forces, the formation and stability of which require the involvement of water.

Considerable progress has been made in the elucidation of specific water molecule coordinates in protein crystals [18]. It must however be remembered that water molecules can interact with the peptide chain only by hydrogen bonding and are labile, subject to more or less rapid exchange. Even in crystal diffraction studies, therefore, one is not observing actual water electron densities or neutron intensities, but *probability* densities with life times governed by exchange rates. The situation in liquid or *in vivo* environments is even more complicated. Despite a vast, and rapidly growing literature devoted to protein folding, there is as yet little real understanding about the molecular and energetic details of hydration interactions and their essential role in determining the conformational or functional attributes of proteins. Incorporation of such factors into calculation and computer simulation procedures presents severe challenges, but protein folding and stability results, arrived at without attempts at including hydration effects, must, at best, be misleading and of questionable value. Direct protein hydration studies should be based on dynamic methods of measurement, usually n.m.r. relaxation. Such measurements are informative, because they provide life-times and exchange rates.They also require a high level of expertise and are laborious to perform [19].

Even under the most favourable conditions, the conformational stability (free energy) margins of native proteins hardly ever exceed 50 kJ mol^{-1}, corresponding to no more than three hydrogen bonds. It is obvious, therefore, that whatever may be the stabilising influences, they are almost cancelled by destabilising effects, leaving only a marginal net free energy of stabilisation. The physical properties of water are sensitive to the same factors that influence protein stability, so that some connection is likely. Probably the three types of hydration discussed above play the major role. Accordingly, amino acids are classified into ionogenic, apolar and polar groups. To maintain a stable globular structure, a peptide chain must contain at least ca. 50 per cent of apolar residues. These residues also tend to be more highly conserved than the polar residues. They form the structural core of globular proteins, whereas the polar and ionogenic residues tend to be located on the periphery or flexible loops and are associated with the biochemical function of the particular molecule.

Figure 5 illustrates a typical protein thermal stability profile, together with the associated thermodynamic functions describing the reversible inactivation/reactivation processes [20]. Two important points are apparent: 1) the strongly curved $\Delta G(T)$ profile is indicative of a large specific heat change, and 2) the small free energy is due to an almost cancelling out of enthalpic and entropic contributions, neither of which needs to be small. This latter effect is again one of the mysteries of aqueous solutions. Examination of Fig. 5, which is quite typical of proteins in general, reveals that, even under optimum pH conditions, the stability of the biologically active state is limited to a relatively narrow temperature range. In other words, an ordered, folded structure can be destroyed by heating *and* by cooling. However, the delicate stability balance of proteins is perturbed in distinctly different ways at the two temperature extremes [21]. These differences have as their common basis the temperature sensitivity of the physical properties of the common solvent: water. The major stabilising contributors to $\Delta G(T)$ are probably apolar hydration/aggregation and intrapeptide effects, whereas configurational entropy and hydrophilic/ionic hydration provide the driving force for destabilisation. Thus, interactions for which $d(\Delta G)/dT > 0$ weaken at low temperature, and vice versa [22]. The net effect of temperature changes is to perturb the delicate balance between large stabilising and destabilising contributions which, under physiological conditions, maintains the marginal stability of active proteins.

At the molecular level the causes of high- and low-temperature inactivation are seen to be quite different. Probably the main cause of cold inactivation is the weakness of the collective water-apolar group repulsions which provide the main driving force for maintaining the folded structure under *in vitro* conditions at physiological temperatures. A subsidiary drive for cold inactivation is due to the increasing affinity of ionic and polar groups for water. In the language of polymer science, water becomes a "good" solvent (by direct hydrogen bonding) at low temperatures.

Saccharide shapes and conformations are even more sensitive to solvent effects than are those of peptides, lipids or nucleotides. This sensitivity probably arises from the fact that, like water, crystalline (and fused ?) sugars exist as three-dimensional hydrogen-bonded networks, the bonding details depending on the topological details of the -OH groups, already referred to above.

Chemical processes occurring in nature are characterised by the economy with which energy is utilised. It follows that in living organisms which might consist of up to 97 per cent of water, this liquid fulfils a function other than that of an inert substrate. It is much harder to elucidate the exact role(s) of water in life processes. Apart from acting as a proton exchange medium, water moves through the organism, carrying nutrients and removing waste products and also functions as lubricant in the form of surface film and viscous juices, e.g. dilute secretions of mucopolysaccharides.

Water participates in four major types of biochemical reactions: oxidation, reduction, condensation and hydrolysis. There are many other biochemical reactions in which water splitting or synthesis form important stages but where the exact mechanisms are still a mystery. To biochemists the chemical transformation of organic molecules in metabolism and synthesis takes precedence over second order effects (?) and "proton book keeping", as related to the oxidation/reduction of the common solvent medium. It is unlikely, however, that correct mechanisms for complex metabolic reaction sequences can ever be established without taking such effects into account. It is no exaggeration to claim that biochemistry is primarily the chemistry of water.

The production of water through the combustion of carbohydrates in the mitochondrion forms a good example. The normal human adult has a daily water turnover of approx. 4 per cent of the body weight: 2.5 kg, of which 300 g is produced endogenously by the oxidation of carbohydrate; the remainder is absorbed by the intake of food and drink, while the loss is accounted for by perspiration, transpiration and excretion. Glucose is oxidised according to the equation

$$C_6H_{12}O_6 + 6O_2 = 6CO_2 + 6H_2O$$

The synthesis of 300 g of water by this mechanism is accompanied by the liberation of energy to the amount of 7,600 kJ, enough to raise the body temperature by 26°C. Actually the energy is converted into chemical energy which is stored in the form of ATP. A more correct form of the above equation is

$$C_6H_{12}O_6 + 6O_2 + 38ADP + 38P_i = 6CO_2 + 6H_2O + 38ATP$$

The 300 g of water produced by this reaction are therefore accompanied by the synthesis of about 100 mol ATP which is stored and provides the energy requirements of the many physiological functions of the body. Even written in the above form, the equation is a gross oversimplification of the real reaction sequence. The oxidation of glucose and the simultaneous synthesis of ATP (and water) takes place in a cascade of 14 reactions, each controlled by an enzyme. Water participates in each step. The *in vivo* mechanisms and rates of all such coupled reactions have, in the course of evolution, become sensitively attuned to the properties of water, such as its ionisation equilibrium and its hydrogen bonding pattern. Even small changes in any of these properties can cause chaos to the coupling between biochemical reactions, and hence to the viability of the organism.

The production of 300 g of water also requires 185 litres of oxygen (approximately 40 per cent of the total daily oxygen requirement) which the lung extracts from air with an efficiency of 14 per cent. Since air contains 21 per cent of oxygen, the lung must process some 6,300 litres of air daily, in order to generate the necessary supply of oxygen. By extending such calculations, it can be shown that, just to supply the cells with enough oxygen for the daily combustion of glucose, the heart must pump 7,000 litres of blood around the vascular system.

All other biosynthetic processes, e.g. protein and nucleotide synthesis and hydrolysis, are similarly coupled to the chemistry of water. If, however, the combustion of proteins, lipids and carbohydrates to yield ATP could proceed in an unbalanced manner, then animal life could only continue for a few days. The replacement of the basic nutrients is performed by plants which, by means of photosynthesis, generate the energy in the form of those chemicals that are utilised by the animals, mainly carbohydrates, but also proteins and lipids, to a limited extent. The key reaction is the oxidation of water, *ie* the converse of the reaction used by the glucose consumers.

Essentially, oxidation is the removal of two electrons from O^{2-} to form O^- and eventually gaseous oxygen. The whole photosynthetic process occurs in three stages: a photochemical excitation of the photosynthetic pigments, causing a release of electrons, the electron transfer reactions, leading to the reduction of NADP, and the "biochemistry", involving the conversion of CO_2 to carbohydrate. The second stage is least well understood; it includes the oxidation of water and the synthesis of ATP. The reaction is believed to be

$$4H_2O = O_2 + 4H^+ + 2H_2O + 4e$$

Little is known about the detailed mechansim whereby water is oxidised to gaseous oxygen during photosynthesis, except that enzymes containing clusters of four manganese atoms are involved. So far these enzymes have resisted attempts at their isolation in a functional state.

The short and superficial survey of the physical properties of water illustrates how the unique molecular geometry of the water molecule and the energy of the O-H...O hydrogen bond, as it exists in liquid water, are basically responsible for directing the many complex assembly and kinetic processes involved in the Chemistry of Life.

References

1. Franks, F. (ed). Water - A Comprehensive Treatise, Vols. 1 - 7, Plenum Press, New York, 1972-1982.

2. Franks, F. (ed.) Water Science Reviews, Vols. 1-5, Cambridge University Press, Cambridge, 1990-1995.

3. Enderby, J.E. and Neilson, G.W. (1979). X-ray and neutron scattering by aqueous solutions of electrolytes. In Ref. 1, Vol. 6, pp. 1-46.

4. Hofmeister, F. (1888). Nunyn-Schmiedebergs Archiv für experimentelle Pathologie und Pharmakologie 24, 247-260.

5. Suggett, A., Ablett, S. and Lillford, P.J. (1976). Journal of Solution Chemistry 5, 17-31.

6. Girlich, D. Multikernresonanzuntersuchungen zur molekularen Dynamik waessriger Saccharidloesungen. Ph.D. Thesis, Regenburg University, 1991.

7. Suggett, A. and Clark, A.H. (1976). Journal of Solution Chemistry 5, 1-15.

8. Goldberg, R.N. and Tewari, Y.B. (1989). Journal of Physical and Chemical Reference Data 18, 809-880.

9. Franks, F., Dadok, J., Ying, S., Kay, R.L. and Grigera, J.R. (1991). Journal of the Chemical Society, Faraday Transactions 87, 579-585.

10. Franks, F. and Grigera, J.R. (1990). Solution properties of low molecular weight polyhydroxy compounds. In ref. 2, Vol 5, 187-289.

11. Ben Naim, A. (1980). Hydrophobic Interactions. Plenum Press, New York.

12. Stillinger, F.H. (1980). Science 209, 451-457.

13. Davidson, D.W. (1973). Clathrate hydrates. In ref. 1, Vol. 2, pp. 115-234.

14. Broadbent, R.D. and Neilson, G.W. (1994). Journal of Chemical Physics 100, 7543-7547.

15. Stillinger, F.H. and Rahman, A. (1974). Journal of Chemical Physics 60, 1545-1557.

16. Zeidler, M.D (1973). NMR specroscopic studies. In ref. 1, Vol. 2, pp.529-584.

17. Franks, F. (1975). The hydrophobic interaction. In ref. 1, Vol. 4, pp. 1-94.

18. Finney, J.L. and Poole, P.L. (1985). Journal of Biosciences 8, 25-35.

19. Denisov, V.P. and Halle, B. (1995). Journal of Molecular Biology 245, 682-697.

20. Hatley, R.H.M. and Franks, F. (1989). FEBS Letters 257, 171-173.

21. Franks, F. (1995). Advances in Protein Chemistry 46, 105-140.

22. Franks, F. (1993). Protein hydration. In Protein Biotechnology (ed. F. Franks). Humana Press, Totowa, NJ, pp.437-465.

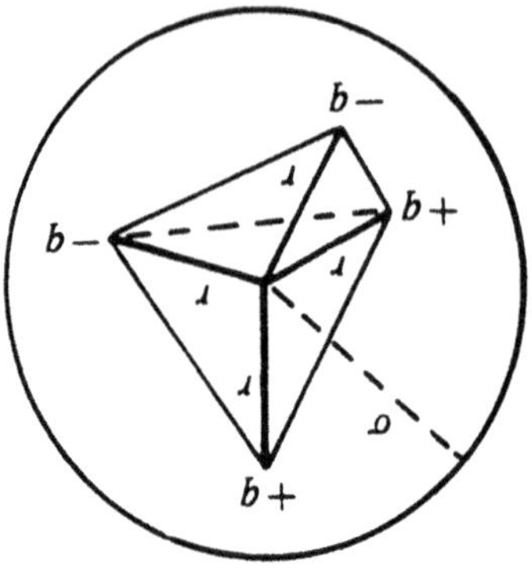

Figure 1: The four-point charge model for the water molecule. Positive charges correspond to the positions of hydrogen atoms and negative charges to those of lone pair electrons. The van der Waals radius (σ) is fitted to the O-----O distance in hexagonal ice.

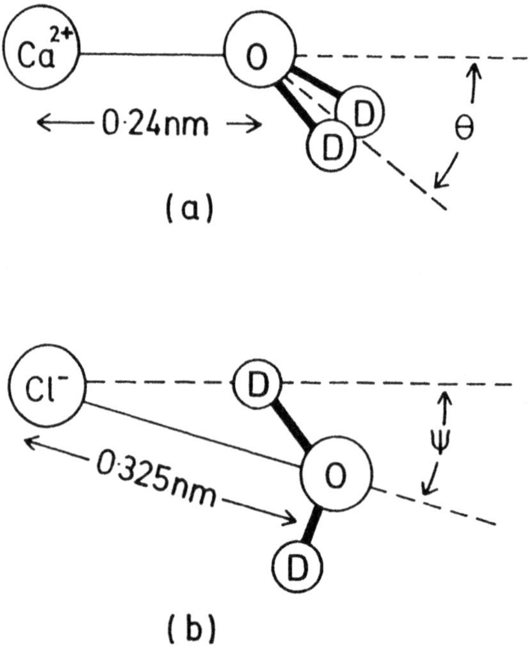

Figure 2: Cation- and anion-water conformations consistent with the experimental neutron scattering data on aqueous solutions. The tilt angles θ and ϕ increase with increasing concentration and the average hydration number is 6, indicating an octahedral configuration of water molecules about the ions. After Enderby & Neilson, ref. 3.

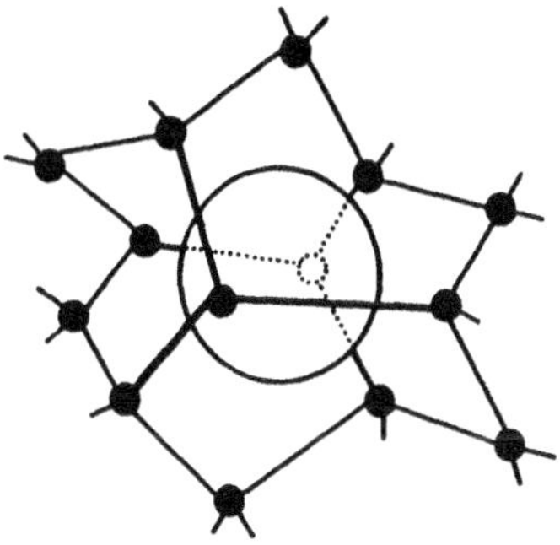

Figure 3. One of several possible cage arrangements of water molecules that can be produced in the proximity of an apolar species (large circle) by the reorientation mechanism shown in Figure 5. The net effect in water is the redistribution of its free volume. In the representation shown, each guest molecule has eight nearest water neighbours and four next-nearest neighbours.

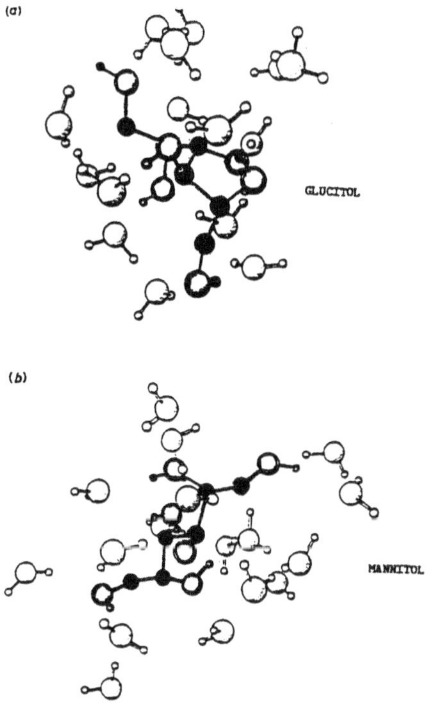

Figure 4. The aqueous solution configurations of the stereoisomeric alditols (a) glucitol (sorbitol) and (b) mannitol, as obtained from a combination of n.m.r. and Molecular Dynamics studies [9]. Carbon atoms are shown as small black circles and sugar oxygen atoms as large open shaded circles. For details see text.

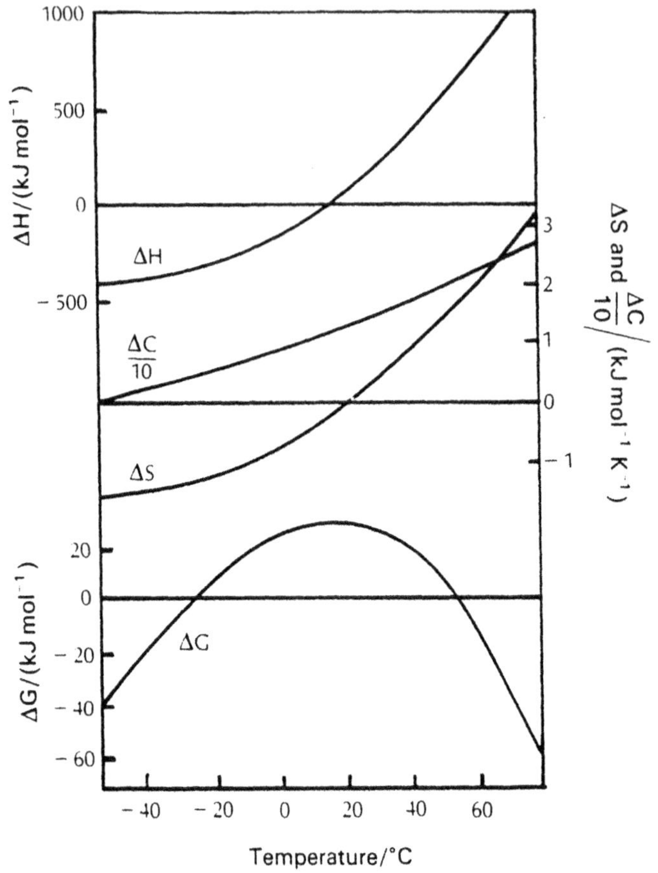

Figure 5. Thermodynamic functions describing the thermal stability profile of lactate dehydrogenase, illustrating heat and cold denaturation phenomena, obtained from measurements at elevated and unfrozen, subzero temperatures (supercooled water). The maximum stability margin does not exceed 30 KJ mol^{-1}, despite the large values of the enthalpic and entropic contributions. Redrawn from Franks, ref. 21.

TRIS(MACROCYCLES) AS MODELS FOR TRANSMEMBRANE, CATION-CONDUCTING CHANNELS

GEORGE W. GOKEL, ERNESTO ABEL, STEPHEN L. DEWALL, JOHN P. EVANS, TAKASHI JIN,[1] GLENN E. M. MAGUIRE, ERIC S. MEADOWS, OSCAR MURILLO, AKIO NAKANO,[2] MAYUR R. SHAH, IWAO SUZUKI,[3] GREGORY P. TOCHTROP, AND SHIGERU WATANABE[4]
Bioorganic Chemistry Program and
Dept. of Molecular Biology & Pharmacology
Washington University School of Medicine
660 South Euclid Ave., Campus Box 8103
St. Louis, MO 63110 U.S.A.

1. Introduction

All living systems are bounded by such structures as cell walls or membranes which protect them from their environment and which prevent their interior contents from being lost. Cells cannot be completely isolated from their environment because nutrients must be acquired and waste products must be expelled. The flow of such species as cations through membranes is controlled by cation-conducting, transmembrane proteins. These proteins are currently receiving enormous attention from the biochemical, biophysical, and molecular biological communities.[5] In addition to the study of naturally occurring protein channels, a number of chemical research groups have designed, synthesized, and studied model channel systems.[6]

2. Flexibility *vs.* Rigidity

One of the fundamental issues that must be addressed in the design of any receptor or biological model is whether the system will be flexible or rigid. A rigid system possesses several advantages. For example, the equilibrium constant for binding can be expressed as K. It is known that $\Delta G = -RT\ln K$. This implies that as the free energy of binding is enhanced by 1.36 kcal/mole, the equilibrium binding constant will increase tenfold. Thus, an increase in ΔG of less than 3 kcal/mol will enhance binding by 100-fold.

A rigid model system or receptor also has the advantage from the experimenter's point of view that the positions of all interactive elements incorporated into the design are in positions known with respect to each other. There are two disadvantages of the rigid design philosophy. The first is simply that an incorrect guess may be made about the requirements of the system. If a donor, for example, is positioned incorrectly its utility is lost and it may even become a liability in the overall design. Second, the rigid system cannot make conformational adjustments *in situ* that might correct design

A. W. Coleman (ed.), Molecular Recognition and Inclusion, 19–29.
© 1998 *Kluwer Academic Publishers. Printed in the Netherlands.*

flaws.

Our design philosophy has long been to mimic nature in the use of relatively flexible molecular arrays that can adjust conformation depending upon the environment. The lariat ethers,[7] for example, were designed to have the flexibility of simple macrocycles such as 18-crown-6 but to have the cation-enveloping architecture of the cryptands. In the figure below, 18-crown-6 is shown on the left and [2.2.2]-cryptand is illustrated at the right. The lariat ether (center) is illustrated to show its similarity to the cryptand although it does not assume such a conformation in the absence of a cation.

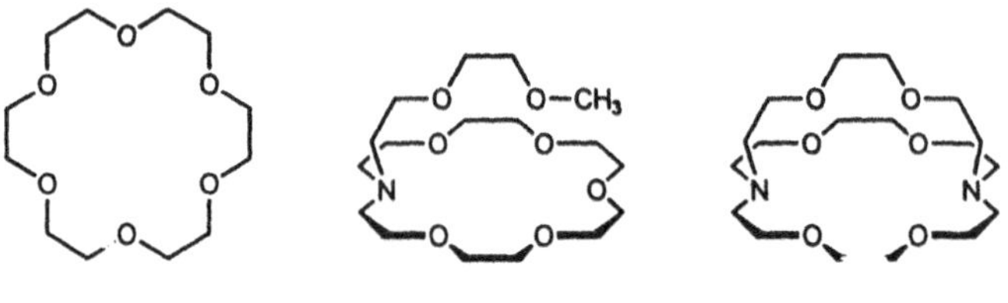

The importance of flexibility is apparent in binding rates rather than in the binding constant. The complexation, binding, or stability constant, K_S, for any of the three compounds shown is given by $K_S = k_1/k_{-1}$ or $k_{complex}/k_{release}$. For 18-crown-6, both the complexation and release rates are fast. For cryptands, complexation is a rapid process but the cation is so well-encapsulated that release poses a problem. For binding, these rates are favorable but for cation transport, they are an issue to consider. Thus, cation transport by the lariat ethers is more efficacious than by cryptands precisely because cation release occurs more readily.

The issue of flexibility *vs.* rigidity enters the design considerations in several ways as noted below. The ability of the system to adjust conformationally was thought to be of critical importance since, at the outset of this effort, no synthetic, non-peptidic, channel-like transporter of alkali metal cations had ever been reported.

3. Design of a Model, Cation-conducting Synthetic Channel

Our original notion for a cation channel system can be represented schematically using a circle and line cartoon. The circles represent crown ethers and the wavy line represents alkyl chains. The compound contains three macrocyclic rings and four alkyl chains. Two of the alkyl chains covalently link the distal macrocycles to the central one. The two distal (non-covalent) chains are mobile and, it was hoped, would adjust to constitute the opposite "wall" of the channel structure. Obviously, the entire assembly would be inserted into the bilayer such that the distal crowns would constitute headgroups (presumed to be favorable) and the central macroring would function as a cation relay. The latter position is less favorable for a subunit that contains polar residues but the presence of such a relay was thought to be essential for cation transport.

A number of issues were considered in the design of our cation channel model system. First, if the channel is to span a lipid bilayer membrane, it is important to know the thickness (width) of the membrane. This can vary considerably depending upon the glycerol headgroups, the lengths and unsaturation of the fatty acids, and the level of interdigitation. It was our intention to study the model channel system in a phosphatidyl choline/phosphatidyl glycerol liposome system that approximates a natural bilayer. In biological studies, a phospholipid bilayer membrane is thought to average about 30Å. It is also known from solid state studies that gramicidin, a dimeric, membrane-spanning peptide, has a distance from one end of the coil to the other of 26Å.[8] The solid state structure is also known for the cation-channel-former bacteriorhodopsin which has α-helical, transmembrane segments of about 30Å.[9] Designing for a transmembrane distance of about 30Å was adopted as the target.

It was important to decide on the structural elements that would serve as entry and exit points for cations. Our interest in crown ethers as cation complexing agents made the choice a simple one for us.[10] There are at least two issues that are critical for a consideration of crown ethers in a synthetic channel. First, can they serve effectively as "headgroups" to stabilize the position of the molecule in the bilayer? This is essential since a mobile system might trap the cation within the bilayer and serve, at best, as a cation carrier. Second, it was important not to use the crown ether having the strongest cation complexing power. A strong complexing agent will bind and hold a cation. This is a favorable outcome in the sense of complexation and selectivity. It is an unfavorable outcome in terms of transport, whether the system under consideration is a carrier or a channel. In other words, if the crown binds the cation of interest too strongly, it will not pass through the pore but be held within in it.

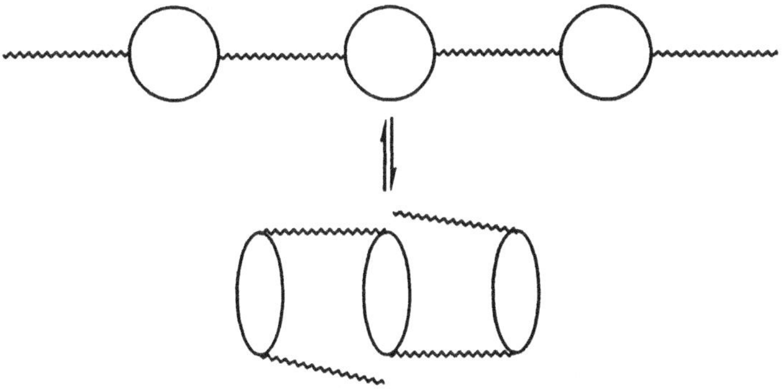

The first question was addressed by considering whether the crown ether residue could serve as headgroup for an amphiphile. The groups of Okahara and of Kuwamura had prepared a variety of crown-ether-derived amphiphiles.[11] Their studies suggested that azacrown ethers having a single, alkyl tail readily formed micelles. Our hope was that crown ethers could serve as headgroups in amphiphilic molecules that form stable vesicles. Our strategy was to use the steroidal sidechain to add order to the aggregates. Cholesterol is well known to be an "organizing agent" for membranes. Indeed, we found that whereas alkylaza-15-crown-5 derivatives formed micelles in water, the steroidal derivative shown forms stable vesicles.[12] It was felt that the crown residue could be relied upon to serve as an entry portal that stabilized the transmembrane channel-former.

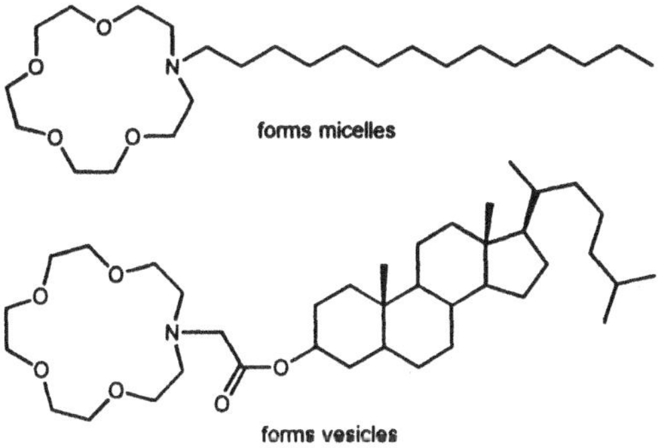

An ancillary question arose during this work. We wondered whether there was any particular amino acid sidechain that could serve to stabilize α-helical, transmembrane spans in natural protein channels. A worthy candidate seemed to us to be tryptophan since the indole sidechain is known to organize water about its >N-H bond. We thus prepared a family of *N*-substituted alkylindoles with a plan to assess their ability to form vesicles. The indole residue is not charged and only a single alkyl tail was present. Both of these facts suggested that if any aggregate formed, it would be micellar. Notwithstanding these expectations, when ten or more carbons were present in the alkyl chains of *N*-alkylindoles, stable vesicles were formed.[13] Further studies in this area are being pursued separately.

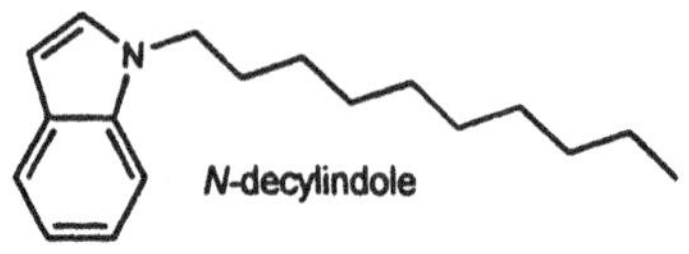

The crown ether chosen for inclusion in the channel prototype was diaza-18-crown-6. Our experience in the synthesis of this compound's derivatives is extensive and we have studied its binding properties with a variety of cations.[14] It was thought that the diazacrown could function as both entry and exit portals as well as a central relay unit. The presence of a crown ether at the midplane of the bilayer would permit the cation to "jump" from one side of the membrane to the middle and then to the opposite side of the bilayer. The presence of the third crown was expected to add a modicum of cation selectivity to the system.

A particular advantage of the diazacrown system is that the nitrogen atoms can invert. Thus the conformation of *N*-alkyl-substituted diazacrown ethers remains flexible. This would not be the case if the sidearm was attached at a carbon atom. Of course, in the latter case, the stereochemistry could be clearly defined but the system could not adjust in case the "fit" was not optimal.

Dodecyl chains were selected to connect the three macrorings. A variety of 1,12-disubstituted dodecyl derivatives are available commercially and it was thought that this would facilitate synthesis. Moreover, the approximate linear distance spanned by a carbon-carbon bond is 1Å. Of course, this assumes an extended hydrocarbon chain in which each link is in a *gauche* conformation. The two covalent links would provide ~24Å of span. Adding the thickness of the three crown ethers, this meets the 26-30Å design criterion specified above.

Taken together, these design features lead to a cation channel model that was expected to function as a "tunnel" or "pore" within the bilayer. This is illustrated below. It was also anticipated that the pore would be filled with water molecules which would be associated transiently with both cation and channel. The precise number and position of such water molecules is unknown.

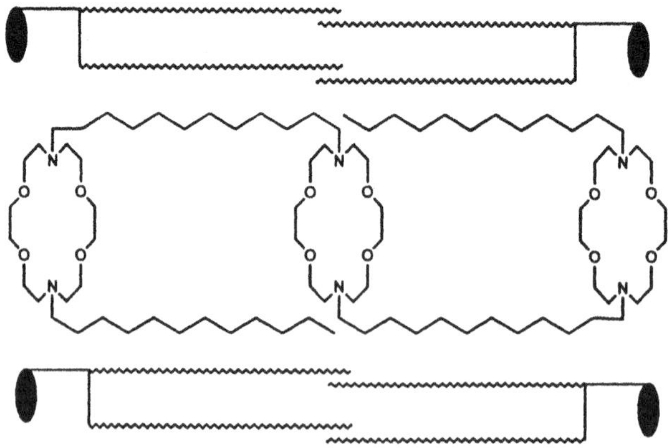

4. Synthesis of the Channel Model

The synthesis of the compound above, called in our group "channel 1," poses some interesting challenges. The compound is almost symmetrical but not quite. Thus, two of the dodecyl chains must be attached at both ends and two only at one end. Likewise, one of the crowns must be attached to two others and two of the crowns only to one crown. A number of attempts were made to prepare the compound and one of the successful approaches is illustrated below. As the project continues, methodology is evolving and some of the onerous purification steps have gradually been made more facile.[15]

Diaza-18-crown-6 was prepared as previously reported.[16] Reaction with a sub-stoichiometric (80%) amount of 1-bromododecane, Na_2CO_3, and KI (catalytic amount) in refluxing butyronitrile for 4 d gave *N*-dodecyl-4,13-diaza-18-crown-6 [C_{12}<N18N>H] in 27% yield along with 14% of *N,N'*-bis(dodecyl)-4,13-diaza-18-crown-6. The resulting monosubstituted diazacrown was treated with 1,12-dibromododecane (0.3 equiv) approximately as described above but reflux time was kept to only 1.5 h. The bromododecyl crown $CH_3(CH_2)_{11}$<N18N>$(CH_2)_{12}Br$ [C_{12}<N18N>$C_{12}Br$] was obtained as a pale, yellow oil in 66% yield. The two precursors, diaza-18-crown-6 (H<N18N>H), C_{12}<N18N>$C_{12}Br$ (2 equiv), Na_2CO_3 (20 eq) and KI (0.3 eq) were heated in 2/3 v/v $CH_3CN/CH_3CH_2CH_2CN$ for 4 d. Channel 1 was obtained as in 23% yield a white solid (mp 61-63 °C).

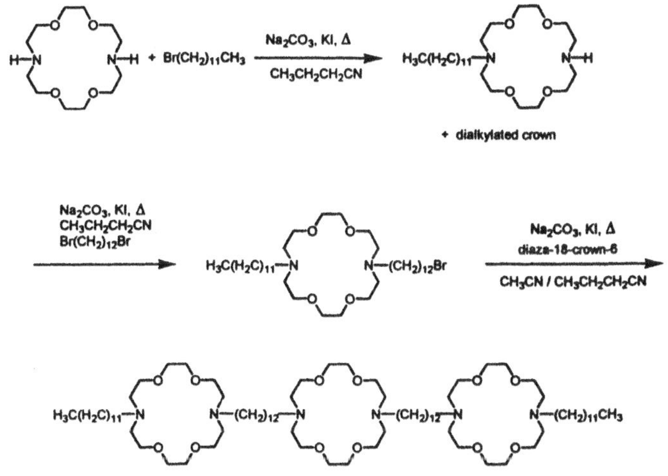

5. Establishing Efficacy

There are many means by which cation conduction can be established. One approach is to use the method devised by Fendler and Kano[17] which relies on differences in the fluorescence of trapped pyranine as the pH changes due to proton flux. We favored the ^{23}Na-NMR method devised by Riddell and developed by Hinton and their coworkers.[18] In brief, vesicles are prepared from a mixture of phosphatidyl choline and phosphatidyl glycerol. These are prepared in the presence of NaCl, some of which is included within the vesicles. The channel-former is inserted into the bilayer by brief incubation of the two. If the ionophore conducts cations, the sodium inside and outside of the vesicles will be in equilibrium. The cation flux cannot be detected since the $^{23}Na^+$ signal is the same whether the salt is in the liposome or in the surrounding medium. A dysprosium polyphosphate shift reagent is therefore added to the external medium. The non-included Na^+ signal is observed at a different chemical shift so that the two environments can be distinguished.

Exchange of the sodium in the two different environments can be evaluated by changes in the sodium linewidths. In the slow exchange region, the rate constant, $k = 1/\tau = \pi(\Delta\nu - \Delta\nu_0)$ is directly proportional to the line broadening observed where $\Delta\nu$ is the linewidth at half-height of the observed resonance line in the presence of the ionophore and $\Delta\nu_0$ is the corresponding value in its absence. In our studies, the ionophore gramicidin is always run simultaneous with other compounds of interest. Thus, all rate data can be normalized to a value of 100 ($k \approx 174\ s^{-1}$). Data are recorded in the table below for a number of compounds that have been studied by this method.

Table 1. Cation flux rates in egg lecithin vesicles assessed by ^{23}Na-NMR spectroscopy

Compound investigated	*Relative rate*
gramicidin	100
$C_{12}<N18N>C_{12}<N18N>C_{12}<N18N>C_{12}$[a]	28
$C_{12}<N18N>EOEOEOE<N18N>EOEOEOE<N18N>C_{12}$[b]	3
$C_{12}<N18N>C_{12}<N15N>C_{12}<N18N>C_{12}$	25
$C_{12}<N18N>C_{12}OEOEOEOC_{12}<N18N>C_{12}$	14
$PhCH_2<N18N>C_{12}<N18N>C_{12}<N18N>CH_2Ph$	39
$H<N18N>C_{12}<N18N>C_{12}<N18N>H$	28
$<18N>C_{12}<N18N>C_{12}<N18>$	<2
$St\text{-}E<N18N>C_{12}<N18N>C_{12}<N18N>E\text{-}St$[c]	<2
$St\text{-}OCOM<N18N>C_{12}<N18N>C_{12}<N18N>MCOO\text{-}St$[d]	5
$C_{12}<N18N>C_{12}$	<2
$C_6H_5CH_2<N18N>CH_2C_6H_5$	<2
$C_{12}<N18N>C_{12}<N18N>C_{12}$	<2

a. $<N18N>$ represents diaza-18-crown-6, $<N15N>$ represents diaza-15-crown-5, and $<N18>$ represents aza-18-crown-6. b. E represents ethylene, CH_2CH_2. c. St represents 3-β-cholestanyl. d. M represents CH_2.

Several interesting features are apparent. First, the sodium cation is conducted with considerable efficacy by several of these completely synthetic structures. The best of the structures, $PhCH_2<N18N>C_{12}<N18N>C_{12}<N18N>CH_2Ph$, is almost half as good as gramicidin. We also noted that changing the central crown unit from an 18- to a 15-membered ring did not alter Na^+ flux beyond the experimental margin for these measurements. This suggests that the central macroring is not required to be in the "tunnel" or "pore" conformation. The implication of this is that the cation is not required to pass through the macrocycle. If this is so, then the crown is probably extended between the spacer chains and more or less perpendicular to the two distal crowns. There are two primary consequences of this. First, the polarity at the mid-plane of the bilayer is reduced relative to the "tunnel" conformation. Second, if the central macrocycle is extended, the overall length of the channel-former is longer than anticipated. The conformation suggested by these studies is shown in Figure 1.

Three facts are of interest concerning the sidearms present in these systems. First, we note that when the dodecyl sidechain is eliminated but the nitrogen atom remains, *i.e.*, $>N\text{-}R \rightarrow >N\text{-}H$, there is no change in efficacy. When the sidearm is removed but the heteroatom is changed to oxygen, *i.e.*, $>$N-R $\rightarrow$ $>$O, ionophoretic activity is lost. Although our explanation for this is speculative, it appears that protonation may play a role in stabilizing the channel within the bilayer.

Second, replacement of the dodecyl sidechain by a benzyl group leads to a substantial enhancement (~40%) of cation flux. This may be due to stabilization of the channel within the bilayer by aromatic ring interactions on the surface of the bilayer. Such interactions are known[19] and could stabilize the extended conformation of 7. In such a case, cation flux would be enhanced by organization within the bilayer, *i.e.*, by forging a defined conduit.

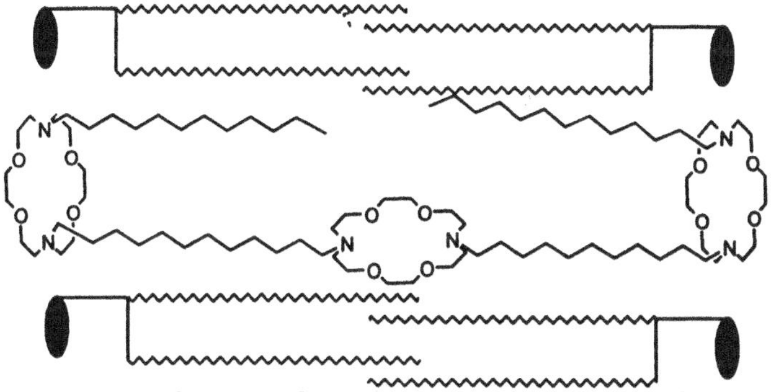

Figure 1. Inferred conformation for "channel 1" in a bilayer

Third, steroids do not seem to be effective sidearms for this particular channel model. This lack of activity may be due to stabilizing contact between the dodecyl spacer chains and the nearly planar α-surface of cholestanol. The latter is essentially the only flat aliphatic hydrocarbon surface known and the extent of contact should be considerable. Such cross-channel contacts would obviously impede cation flux. When the cholestanyl sidearm is attached by a more rigid spacer unit, contact of these surfaces is inhibited and cation flux is measurably, if only slightly, higher.

6. Control Experiments

A number of control experiments were done to check for the presence of non-channel mechanisms of cation conduction. The first controls are apparent in the final three compounds shown in the table. The compounds in question are $C_{12}<N18N>C_{12}$, $C_6H_5CH_2<N18N>CH_2C_6H_5$, and $C_{12}<N18N>C_{12}<N18N>C_{12}$. The first two are known carrier molecules which constitute subunits of the channel-formers and they do not conduct cations within our ability to detect it in this system. The final structure is "two-thirds" of the channel 1 molecule and is likewise inactive.

An additional possibility, no matter how remote, is that these structures function as simple detergents that render the lipid bilayer "leaky." Of course, in a sense, this is what occurs when a protein inserts in a bilayer. The fact that a protein is inherently more complicated than the compounds studied here does not alter the fact that both compounds conduct cations at low concentrations in a reproducible fashion. Even so, the following studies were undertaken. In each study the egg lecithin vesicles described above were prepared and incremental amounts (0, 5, 10, 15, and 20 μM) of the known detergents sodium dodecylsulfate and Triton X-100 were added. No line broadening was observed for any of these individually studied cases. In the most concentrated system, microtiter additions of detergent solution were made until a final concentration of 190 μM was reached. This most concentrated sample did not show vesicular lysis nor did it show any line broadening. Vesicular lysis was observed, however, when either sodium dodecylsulfate or Triton X-100 was present at a concentration of 2 mM.

An additional control was to test that the tris(macrocycles) are soluble in the membrane rather than in the aqueous medium surrounding the liposomes. The issue in this case is whether differences in activity can be attributed simply to differences in solubility of the ionophore in the membrane. We thus calculated log P (octanol-water partition coefficient) values for several of the candidate structures.

18-Crown-6 has a log P value of 0.21. *N,N'*-Dibenzyldiaza-18-crown-6 prefers octanol by an experimentally determined log P value of 4.21.[20] It was not possible to determine log P values for C_{12}<N18N>C_{12}<N18N>C_{12}<N18N>C_{12} by experiment so the HINT module of the Sybyl molecular modeling package was used. The result was a log P value of 18.5. Even if this value is in error by 100%, there is no water solubility for this compound. Values calculated for several other channel-formers all gave values of >10 which clearly resolves this issue.

An additional concern is whether all of these compounds function as carriers and, for some unknown reason, function better than any other known carriers. Evidence on this question could be obtained by conducting U-tube-type transport experiments in which sodium picrate is transported through a bulk $CHCl_3$ membrane by the compound in question. We have previously studied a series of lariat ether derivatives using a concentric tube transport apparatus[21] and found that transport rates correlated well with both picrate extraction constants and with log K_S values determined in anhydrous methanol solution.[22] In this case, gramicidin was used as the standard for the bilayer and valinomycin was used as standard for the bulk $CHCl_3$ phase. Data obtained for several compounds are shown in Table 2.

Table 2. Cation transport by selected ionophores.

Ionophore	Relative Rate ($CHCl_3$)	Rel. rate (bilayer)
Valinomycin	1.0	0.14
Gramicidin	0.02	1.00
$PhCH_2$<N18N>CH_2Ph	0.53	0.01
C_{12}<N18N>C_{12}	0.48	0.01
$HOOC(CH_2)_{11}$<N18N>$(CH_2)_{11}COOH$	0	0.01
<18N>C_{12}<N18N>C_{12}<N18>	0.58	0.02
H<N18N>C_{12}<N18N>C_{12}<N18N>H	0.27	0.28
C_{12}<N18N>C_{12}<N18N>C_{12}<N18N>C_{12}	0.26	0.28
$PhCH_2$<N18N>C_{12}<N18N>C_{12}<N18N>CH_2Ph	0.46	0.38

An analysis of the data reveals that there is little, if any, correlation between the transport efficacies of the various ionophores in the bilayer and in the $CHCl_3$ membrane. The remarkable mitochondrial potassium cation carrier is the best sodium transporter in $CHCl_3$ but a poor carrier in the bilayer. Gramicidin is nearly inactive in $CHCl_3$ but very active in the phospholipid membrane. Among the synthetic carriers, no correlation in activity can be gleaned from the data. This is not, of course, proof of a difference in mechanism, but it is certainly suggestive.

7. A Structure-activity Relationship

We postulated that the enhanced transport ability of the benzyl-substituted channel, $PhCH_2$<N18N>C_{12}<N18N>C_{12}<N18N>CH_2Ph, was due to an interaction between the benzyl group arenes and the phospholipid headgroups. If so, not only should Na^+ pass through the distal crowns, the transport should be affected by any substituent present on the aromatic ring. A time-honored means for evaluating this possibility is to apply the Hammett equation and to correlate cation transport with the substituent constant. In the present case, we used the Taft σ^0 values developed for use in substituted phenylacetic acid derivatives.[23] Three channel-formers were prepared for this study: $XC_6H_4CH_2$<N18N>C_{12}<N18N>C_{12}<N18N>$CH_2C_6H_4X$ in which X is H, NO_2, or OCH_3. The transport rate for the *para*-H compound was found previously to be 38% that of gramicidin. Presumably, the transport rate would be retarded by nitro and accelerated by methoxy. The experimentally determined values were, respectively, 30% and 43% that of gramicidin. When plotted against the Taft σ^0 values, the three points gave a straight line with a correlation coefficient of >0.96. This leads us to speculate that the channel former has a structure in the bilayer as shown below.

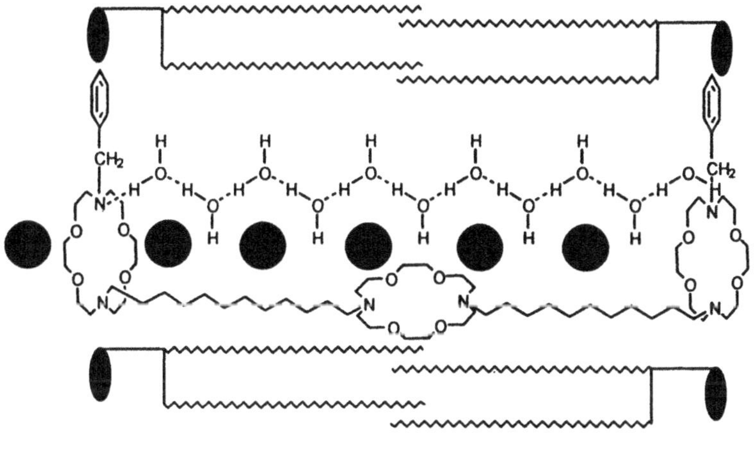

We presume that the channel is filled with hydrated sodium cations. The black spheres represent sodium cations. The water chain is drawn in such a way that it is highly ordered. There is currently no evidence that this is so. Further, water may occupy other sites either within the channel or near the crown ethers.

8. Conclusions

We have described the *de novo* design and synthesis of a family of tris(macrocyclic) compounds that function as transmembrane channels in a phospholipid bilayer. The ability of this family of structures to transport Na^+ does not depend upon differential membrane solubility. It is also not attributable to a simple detergent effect. The formation of a transmembrane pore does not involve all three macrocyclic rings parallel to each other. The distal crown ethers appear to serve as headgroups for the ionophore and also entry portals into the membrane. The flexible sidearms may play a critical role in the conformation of the macrocycle. Changes from alkyl to steroidal to benzylic give significiant changes in cation flux. Studies with structural fragments of the tris(macrocyclic) system and use of a concentric tube bulk membrane apparatus both suggest that the carrier mechanism does not account for the sodium transport. The central macroring may serve as a cation relay unit but is not required to be parallel to the other two crowns. Passage of the sodium cation through the distal macorings is established by use of a Hammett correlation which shows that transport rate varies with substituent as expected.

9. Acknowledgment

We thank the NIH for a grant (GM 36262) that has supported the development of the synthetic, cation-conducting transmembrane channels described here.

10. Notes and References

1 Research Institute for Electronic Science, Hokkaido University, Sapporo, Japan.
2 Department of Chemistry, Toa University, Shimonoseki, Japan.
3 Pharmaceutical Institute, Tohoku University, Sendai, Japan.
4 Department of Chemistry, Kochi University, Kochi, Japan.
5 (a) Nicholls, D.G.; *Proteins, Transmitters, and Synapses*, Blackwell Science, Oxford, 1994. (b) Hille, B.; *Ionic Channels of Excitable Membranes*, Sinauer Press, Sunderland, MA, 1992. (c) Stein, W.D.; *Channels, Carriers, and Pumps*, Academic Press, New York, 1990.
6 (a) Tabushi, I., Kuroda, Y., Yokota, K. *Tetrahedron Lett.* **1982,** 23(44), 4601-4604. (b) Menger, F. M.; Davis, D. S.; Persichetti, R. A.; Lee, J. J. *J. Am. Chem. Soc.* **1990**, *112*, 2451-2452. (c) Kobuke, Y.; Ueda, K.; Sokabe, M. *J. Am. Chem. Soc.* **1992**, *114*, 7618-7620. (d) Neevel, J. G.; Nolte, R. *Tetrahedron Lett.* **1984**, 25(21), 2263-2266. (e) Nolte, R. J. M.; Beijnen A. J. M.; Neevel, J. G.; Zwikker, J. W.; Verkley, A. J.; Drenth, W. *Israel. J. Chem.* **1984**, *24*, 297-301. (f) Kragten, U. F.; Roks, M. F. M.; Nolte, R. J. M. *J. Chem. Soc. Chem. Comm.* **1985**, 1275-1276. (g) Fuhrhop, J.-H.; Liman, U.; David, H.H.; *Angew. Chem. Int. Ed. Engl.* **1985**, *24*, 339-340. (h) Jullien, L.; Lehn, J. *Tetrahedron Lett.* **1988**, *29*, 3803-3806. (i) Jullien, L.; Lehn, J. M. *J. Inclusion Phenom.* **1992**, *12*, 55-74. (j) Canceill, J.; Jullien, L.; Lacombe, L.; Lehn, J. M. *Helv. Chim. Acta* **1992**, *75*, 791-811. (k) Pregel, M.; Jullien, L.; Lehn, J. M. *Angew. Chem. Int. Ed. Engl.* **1992**, 31, 1637-1639. (l) Pregel, M.; Jullien, L.; Canceill, J.; Lacombe, L.; Lehn, J. M. *J. Chem. Soc. Perkin Trans.* 2, **1995**, 417-426. (m) Carmichael, V.E.; Dutton, P.; Fyles, T.; James, T.; Swan, J.; Zojaji, M. *J. Am. Chem. Soc.* **1989**, *111*, 767-769. (n) Fyles, T.; James, T.; Kaye, K. *Can. J. Chem.* **1990**, *68*, 976-978. (o) Fyles, T.; Kaye, K.; James, T.; Smiley, D. *Tetrahedron Lett.* **1990**, 1233. (p) Kaye, K.; Fyles, T. *J. Am. Chem. Soc.* **1993**, *115*, 12315-12321. (q) Fyles, T.; James, T.; Pryhitka, A.; Zojaji, M.; *J. Org. Chem.* **1993**, *58*, 7456-7468. (r) Ghadiri, M.R., Granja, J.R., Buehler, L.K. *Nature* **1994**, *369*, 301-304. (s) Khazanovich, N., Granja, J. R., McRee, D.E., Milligan, R.A., Ghadiri, M. R. *J. Am. Chem. Soc.* **1994**, *116*, 6011-6012. (t) Stadler, E.; Dedek, P.; Yamashita, K.; Regen, S. *J. Am. Chem. Soc.* **1994**, *116*, 6677-6682. (u) Voyer, N.; Robitaille, M. *J. Am. Chem. Soc.* **1995**, *117*, 6599-6600. (v) Tanaka, Y.; Kobuke, Y.; Sokabe, M. *Angew. Chem. Int. Ed. Engl.* **1995**, *34*(6), 693-694. (w) Deng, G.; Merritt, M.; Yamashita, K.; Janout, V.; Sadownik, A.; Regen, S.L.; *J. Am. Chem. Soc.* **1996**, *118*, 3308.
7 (a) Gokel, G.W.; Dishong, D. M.; Diamond, C. J.; *J. C. S. Chem. Commun.*, **1980**, 1053. (b) Dishong, D. M.; Diamond, C. J.; Gokel, G.W.; *Tetrahedron Letters*, **1981**, 1663. (c) Schultz, R. A.; Dishong, D. M.; Gokel, G.W.; "Lariat Ethers. 3. Macrocyclic Polyethers Bearing Donor Groups on Flexible Arms Attached at a Nitrogen Pivot Point," *Tetrahedron Letters*, **1981**, 2623.
8 (a) Wallace, B.A., Ravikumar, K. Science **1988**, *241*, 182-187. (b) Langs, D. *Science* **1988**, *241*, 188-191.
9 (a) Henderson, R.; Baldwin, J.M.; Ceska, T.A.; Zemlin, F.; Beckmann, E.; Downing, K.H. *J. Mol. Biol.* **1990**, *213*, 899-929. (b) Deisenhofer, J.; Epp, O.; Miki, K.; Huber, R.; Michel, H. *Nature* **1985**, *318*, 618-624.

10 (a) G.W. Gokel, *Crown Ethers and Cryptands*, Volume 3 in J. F. Stoddart (Ed.), Monographs in Supramolecular Chemistry, The Royal Society of Chemistry, London, England, 1991, 190 pp. (b) Gokel, G.W.; *Chem. Soc. Reviews*, **1992**, *21*, 39-47.

11 (a) Okahara, M.; Kuo, P.L.; Yamamura, S.; Ikeda, I.; *J. Chem. Soc. Chem. Commun.* **1980**, 586. (b) Ikeda, I.; Iwaisako, K.; Nakatsuji, Y.; Okahara, M.; *Yukagaku* **1986**, *35*, 1001 (*Chem. Abstr.* 106:86714). (c) Kuwamura, T.; Kawachi, T.; *Yukagaku* **1979**, *28*, 195 (*Chem. Abstr.* 90:206248). (c) Kuwamura, T.; Akimaru, M.; Takahashi, H.L.; Arai, M.; Kenkyu Hokoku-Asahi Garasu Kogyo Gijutsu Shorekai **1979**, *35*, 45 (*Chem. Abstr.* 95:61394q). (e) Kuwamura, T.; Yoshida, S.; *Nippon Kagaku Kaishi* **1980**, 427 (*Chem. Abstr.* 93:28168e).

12 (a) Echegoyen, L.E.; Hernandez, J.C.; Kaifer, A.; Gokel, G.W.; Echegoyen, L.; *J. Chem. Soc. Chem. Commun.* **1988**, 836-837. (b) Nakano, A.; Hernandez, J.C.; DeWall, S.L.; Wang, K.; Berger, D.R.; Gokel, G.W.; *Supramol. Chem.* **1997**, *in press.*

13 Abel, E.; Fedders, M.F.; Gokel, G.W.; *J. Amer. Chem. Soc.* **1995**, *117*, 1265-1270.

14 (a) Gatto; V.J.; Gokel, G.W.; *J. Am. Chem. Soc.* **1984**, *106*, 8240-8244. (b) Gandour, R. D.; Fronczek, F.R.; Gatto, V.J.; Minganti, C.; Schultz, R.A.; White, B.D.; Arnold, K.A.; Mazzocchi, D.; Miller, S. R.; Gokel, G.W.; *J. Am. Chem. Soc.*, **1986**, *108*, 4078-4088. (c) Gatto, V.J.; Arnold, K.A.; Viscariello, A.M.; Miller, S.R.; Morgan, C.R.; Gokel, G.W.; *J. Org. Chem.* **1986**, *51*, 5373-5384. (d) Schultz, R.A.; White, B.D · Dishong, D.M.; Arnold, K.A.; Gokel, G.W.; *J. Am. Chem. Soc.* **1985**, *107*, 6659-6668.

15 (a) Nakano, A.; Xie, Q.; Mallen, J.; Echegoyen, L.; Gokel, G.W. *J. Am. Chem. Soc.* **1990**, *112*, 1287. (b) Murillo, O.; Watanabe, S.; Nakano, A.; Gokel, G.W.; *J. Amer. Chem. Soc.* **1995**, *117*, 7665-7679.

16 Gatto, V.J.; Miller, S. R.; Gokel, G.W.; *Organic Syntheses* **1989**, *68*, 227.

17 Kano, K.; Fendler, J.H.; *Biochim. Biophys. Acta* **1978**, *509*, 289.

18 (a) Riddell, F.G.; Hayer, M.K.; *Biochim. Biophys. Acta* **817**, 313-317 (1985). (b) Buster, D.C.; Hinton, J.F.; Millett, F.S.; Shungu, D.C.; *Biophys. J.* **53**, 145-152 (1988). (c) Riddell, F.G.; Arumugam, S.; Brophy, P.J.; Cox, B.G.; Payne, M.C.H.; Southon, T.E.; *J. Amer. Chem. Soc.* **110**, 734-738 (1988). (d) Riddell, F.G.; Arumugam, S.; *Biochim. Biophys. Acta* **984**, 6-10 (1989). (e) Riddell, F.G.; Tompsett, S.J.; *Biochim. Biophys. Acta* **1024**, 193-197 (1990).

19 (a) Petti, M.A.; Shepodd, T.J.; Barrans, Jr.; R.E.; Dougherty, D.A.; *J. Amer. Chem. Soc.* **1988**, *110*, 6825. (b) Kearney, P.C.; Mizoue, L.S.; Kumpf, R.A.; Forman, J.E.; McCurdy, A.; Dougherty, D.A.; *J. Amer. Chem. Soc.* **1993**, *115*, 9907.

20 Murillo, O.; Suzuki, I.; Abel, E.; Gokel, G.W.; *J. Am. Chem. Soc.* **1996**, *118*, 7628-7629.

21 Lamb, J. D.; Christensen, J.J.; Izatt, S.R.; Bedke, K.; Astin, J.S.; Izatt, R.M. *J. Am.Chem. Soc.* **1980**, *102* (10), 3399. (b) Lamb, J. D.; Izatt, R. M.; Garrick, D. G.; Bradshaw, J. S.; Christensen, J. J. *J. Membr. Sci.* **1981**, *9*, 83-107.

22 Hernandez, J.C.; Trafton, J.E.; Gokel, G.W.; *Tetrahedron Lett.* **1991**, 6269-6272.

23 Taft, R.W.; *J. Phys. Chem.* **1960**, *64*, 1805.

CONSTRUCTION OF THE INTERFACES POSSESSING BOTH FUNCTIONALITIES OF MOLECULAR RECOGNITION AND ELECTRON TRANSFER

T. OSA

Pharmaceutical Institute, Tohoku University

Aobayama, Aoba-ku, Sendai 980-77 Japan

Supramolecular structures possessing useful functionalities have been paid to much attention by many investigators. My group have been interested in modified electrode interfaces which play not only molecular recognition to specific substrate but also electron transfer or charge transfer between electrode and specific substrate. Focussing my group's recent results, I would present 1) selective electroorganic reactions on poly(acrylic acid) thin layer coated graphite felt (PAA-GF) electrode in which mediator(s), enzyme(s) and co-factor(s) are immobilized, and 2) enzyme sensors based on alternate deposition of avidin and biotinylated enzyme on electrode surface.

ENANTIOSELECTIVE ELECTROORGANIC REACTIONS ON PAA-GF ELECTRODES IMMOBILIZED BY MEDIATOR, ENZYMES OR/AND CO-FACTOR

Selective electroorganic reactions can be enhanced by modified electrodes, especially, by mediator-immobilized electrode. For preparative electrosynthesis, much loading of mediator substituents is necessary and polymer-coated electrodes can be powerful candidates to achieve this purpose [1]. After examining a various kind of polymers, we have attained to use poly(acrylic acid) [2]. PAA has many favorable properties to prepare a polymer-coated electrode as described:

a) Easy control of the layer thickness by dip-coating of electrode from PAA/methanol solution.
b) Strong physical adsorption of PAA to electrode surface, particularly, to GF surface.
c) Easy chemical immobilization of mediation substituents to PAA layer by the reactions of carboxy group of PAA.
d) Control of hydro- or lipo-philicity by the presence or absence of free carboxy groups of PAA.

A.W. Coleman (ed.), Molecular Recognition and Inclusion, 31-38.
© 1998 *Kluwer Academic Publishers. Printed in the Netherlands.*

e) In the case of no-free carboxy groups, organic compounds can be condensed to PAA layer.
f) A suitable size of domain can be formed by crosslinking of carboxy group of PAA which would be very important to construct a supramolecular structure.
g) Chemical stability of PAA layer.

2,2,6,6-Tetramethylpiperidinyl-N-oxy (TEMPO)-modified PAA-GF was prepared as shown in Scheme 1 [3].

GF → (0.25% PAA in CH_3OH) → PAA-coated GF → (4-amino-TEMPO in DMF, 72 h, r.t. (with DCC*)) → TEMPO-modified PAA-coated GF → ($H_2N(CH_2)_6NH_2$ in DMF, 48 h, r.t. (with DCC*)) →

cross-linked and TEMPO-modified PAA-coated GF → ($(C_4H_9O)_2SO_2$, DBU**) → carboxyl-free, cross-linked and TEMPO-modified PAA-coated GF

*DCC : dicyclohexyl-carbodiimide
**DBU : 1,8-diazabicyclo-[5.4.0]-7-undecene

Scheme 1. Preparation of TEMPO-modified GF electrode for macroelectrosynthesis.

The TEMPO-modified PAA-GF succeeded in enantioselective binaphthyl formation of 2-naphthol and 2-methoxynaphthalene in the presence of (-)-sparteine as deprotonating agent in acetonitrile[4]. Table 1 shows the results of 2-naphthol coupling reaction on the TEMPO-modified PAA-GF electrode together with those on bare GF electrode and chemical oxidation in the presence of TEMPO and (-)-sparteine in solution. Highly enantioselective binaphthyl formation was achieved only by TEMPO-modified GF electrode which means the advantage of PAA-GF electrode comparing with bare electrode or chemical reagent. A plausible oxidative coupling reaction of 2-naphthol by the use

TABLE 1. Oxidative coupling reactions of 2-naphthol to 1,1'-binaphthol in the presence of (-)-sparteine [a)]

Method	Charge passed / C	Current efficiency / %	Isolated yield / %	$[\alpha]_D^{20}$	$R:S$ [e)]	ee [e)] / %	Turnover [f)] number
electrocatalytic on TEMPO-modifiedGF [b)]	513.2	88.8	93.6	-33.6°	0.3 : 99.7	99.5	410
electrocatalytic on bare GF [c)]	519.9	92.4	93.4	-3.4°	44.5 : 55.5	11.1	10
reagent oxidation [d)]	—	—	90.2	-2.0°	46.7 : 53.3	6.7	1

a) the presence of 5 mmol 2-naphthol and 5 mmol (-)-sparteine in each reaction, b) in 5 ml of 0.2 M $NaClO_4$ / CH_3CN, c) 0.5 mmol 4-acetylamino-TEMPO in 5 ml of 0.2 M $NaClO_4$ / CH_3CN, d) 5 mmol of oxoammonium tosylate of 4-acetylamino-TEMPO in 5 ml of CH_3CN, 24 hr, e) obtained by HPLC, f) calculated from 1,1'-binaphthyl (mol) x 2 / TEMPO (mol).

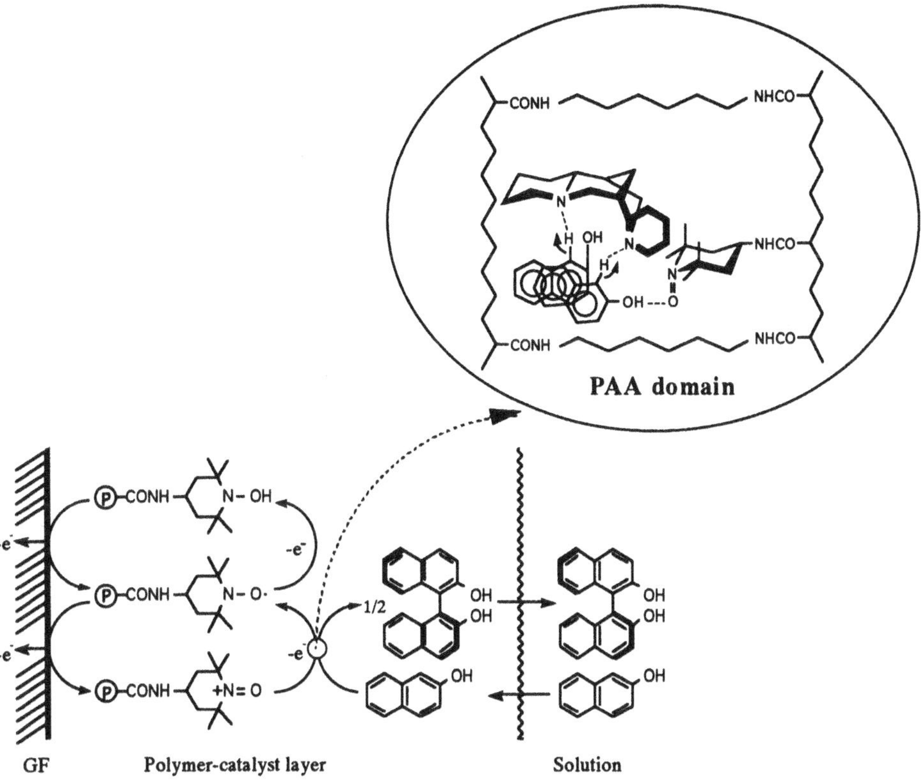

Scheme 2. A plausible mechanism of TEMPO-modified GF electrode for the oxidation of 2-naphthol.

of TEMPO-modified PAA-GF is depicted in Scheme 2. An intermediate of 2-naphthol cation is first formed with nitrosonium ion of TEMPO by one electron oxidation. The enantioselectivity of binaphthyl formation will be simultaneously controlled by hydrogen bonds between the intermediate species and (-)-sparteine. Judging from the low enantioselective binaphthyl formation on the bare GF in the presence of TEMPO and (-)-sparteine in solution, we conclude that the domain of PAA layer is important. The spatial size of domain can accommodate 2-naphthol and (-)-sparteine very tightly and can enhance the hydrogen bond interaction more strongly than that in solution.

Enantioselective oxidation of racemic *sec*-alcohols was also achieved on TEMPO-modified GF electrode (Table 2) [5]. Since another optical isomer of racemic alcohols is remained unreacted, the reaction can be used to a new type of electrochemical resolution process for racemic alcohols.

Preparative electroenzymatic oxidations and reductions have been carried out on a PAA-GF immobilizing ferrocene, diaphorase, NAD^+, and alcohol dehydrogenase[6] and on a PAA-GF immobilizing methylviologen, diaphorase, NAD^+, and alcohol dehydrogenase [7], respectively.

TABLE 2. Enantioselective oxidation of racemic alcohols on a TEMPO-modified GF electrode by use of (-)-sparteine

Racemic alcohol	Product	Main remained alcohol	Charge passed (C)	Oxidized alcohol (%)	Selectivity of product (%)	Current efficiency (%)	Turnover number	Unreacted alcohol (%)	$[\alpha]_D^{20}$	*R* : *S*	ee (%)
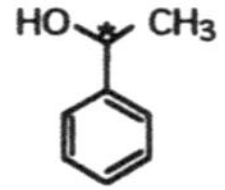	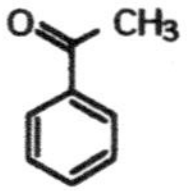	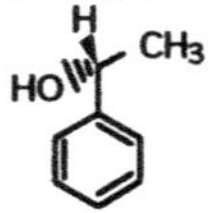	53.7	52.9 a)	92.4	95.1	87	46.2 a)	+44° c)	99.8 : 0.2	99.6
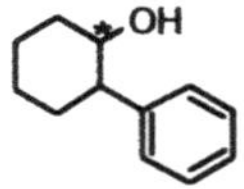	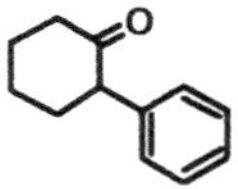	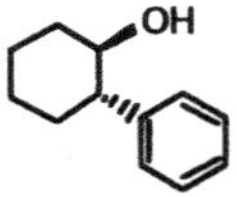	56.5	53.8 a)	90.2	91.8	88	45.1 a)	-58° d)	99.7 : 0.3	99.4
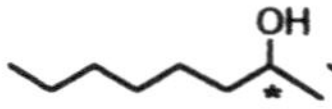	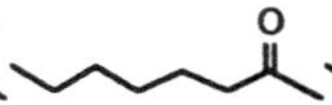	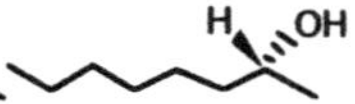	52.7	52.1 b)	93.0	95.4	85	46.5 b)	-9.5° e)	99.8 : 0.2	99.6
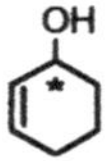	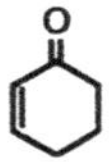	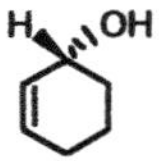	49.4	50.3 b)	96.4	97.7	82	48.2 b)	+109° f)	99.9 : 0.1	99.8
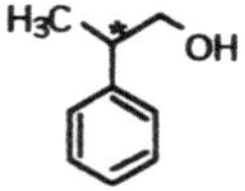	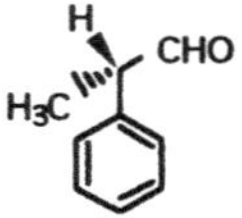		101.8	48.9 a)	97.8	92.7	160	0 a)	0°	50.1 : 49.9	0.2
	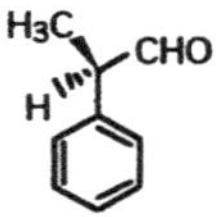			48.9 a)	97.8			0 a)			

a) Mesured by HLPC (column : CHIRALCEL-OD, 0.46 cm ϕ x 25 cm, solvent : hexane-isopropanol = 95 : 5, flow speed : 0.5 ml·min $^{-1}$). b) Mesured by GC (Cp-Cyclodextrin-B-2,3,6-M-19, 0.25 mm ϕ x 25 m/raising temp 3°C·min $^{-1}$ from 80 to 150°C, inj. temp 200°C, detc. temp 240°C). c) Compared with authentic sample (Fulka Chemica-Biochemica) : $[\alpha]_D^{20}$ -45±1° (*c* 5, CH_3OH); *R* : *S* > 99.5 : 0.5 (GC). d) Compared with authentic sample (Fulka Chemica-Biochemica) : $[\alpha]_D^{20}$ -59±1° (*c* 10, CH_3OH); (1*R*,2*S*) : (1*S*,2*R*) > 99 : 1 (GC). e)Lit. $[\alpha]_D^{25}$ -9.9° (neat). f)Lit. $[\alpha]_D^{20}$ +112.0° (*c* 0.60, CH_3Cl).

ENZYME SENSOR BASED ON ALTERNATE DEPOSITION OF AVIDIN AND BIOTINYLATED ENZYME ON ELECTRODE SURFACE

We have studied enzyme sensors by the use of albumin [8,9] and Langmuir-Blodgett membrane [10,11] supports last two decades. Since 1993, we have used avidin and biotinylated enzyme for biosensor fabrication [12]. Avidin of a highly stable glycoprotein in egg-white binds biotin and its derivatives very strongly and highly specifically (Figure 1) [13]. Many kinds of biotin- or avidin-labelled reagents are not only commercially available, but also enzymes are generally biotinylated with easiness [14].

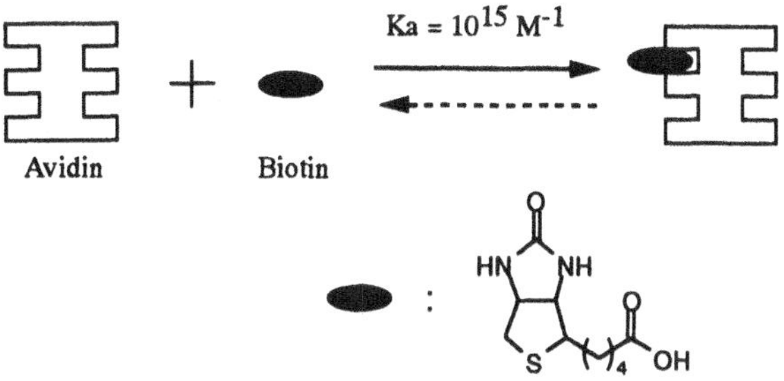

Figure 1. Complex formation between avidin and biotin.

At the beginning of our studies, a surface of metal oxide was converted to the hydrophobic surface via fatty acid LB monolayer to adsorb avidin [15] or to introduce reactive groups to bind avidin chemically [16]. However, it became clear that the surface of platinum, gold and glassy carbon adsorbs avidin directly from an avidin solution (10 mg/ml phosphate buffered saline, PBS), being stable to be used for biosensors [17]. After washed with PBS to remove weakly adsorbed avidin, the avidin-modified electrode was immersed in a biotinylated enzyme solution (20-100 mg/ml PBS). The same procedure to deposit avidin and biotinylated enzyme alternately could to carried out till 20 times (20 enzyme layer deposition). This stepwised deposition is demonstrated in Figure 2 for an example of avidin and biotinylated glucose oxidase (B-GOx) on a platinum electrode surface.

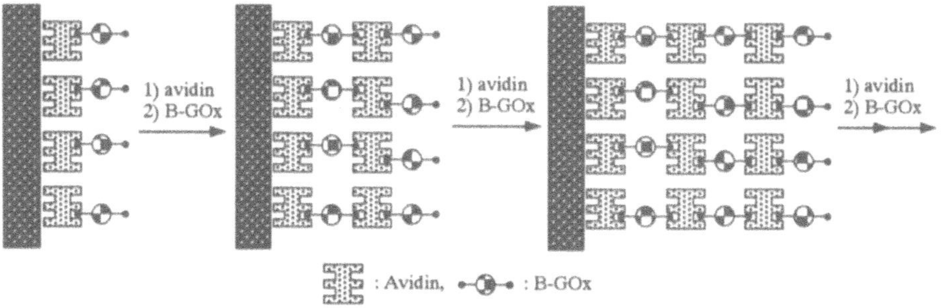

Figure 2. Stepwise deposition of avidin and biotinylated GOx (B-GOx) on a Pt electrode surface.

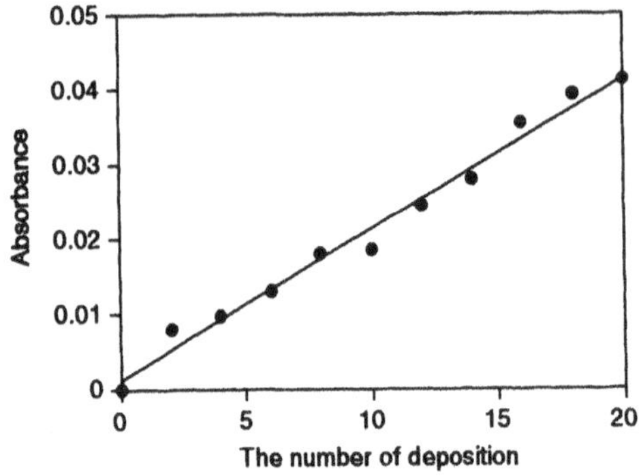

Figure 3. A successive deposition of biotinylated GOx and FITC-avidin on quartz slide monitored by absorbance at 495nm.

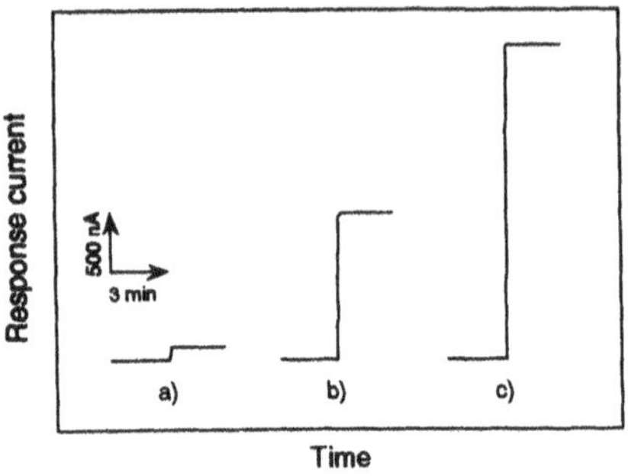

Figure 4. Typical response curves of glucose sensors with monolayer (a),10-layers (b) and 20-layers (c) of biotinylated GOx to 1mM glucose.

A linear increase of enzyme loading by alternate deposition was confirmed by a linear increase of response current and absorbance (use of fluorescein-5-isothiocyanate (FITC)-conjugated avidin) till 15 times deposition (Figure 3) [18]. The response time of the sensors was satisfactorily fast (ca. 20 s), irrespective of the number of GOx layers (Figure 4). The results mean that the multilayers of avidin and B-GOx do not interfere the diffusion of substrate and enzyme-reaction product. The response current extended the dynamic range of detection, especially at a lower glucose concentration.

A physiological level of ascorbate in human blood (ca. 0.1 mM) often disturbs the accurate determination of glucose by electrochemical method, because ascorbate can be oxidized directly on the surface of base electrode of device. The interference of ascorbate for glucose sensor was almost removed by the multi-deposition of inner 10 GOx layers and outer 10 ascorbate oxidase (AOx) layers (Figure 5) [19].

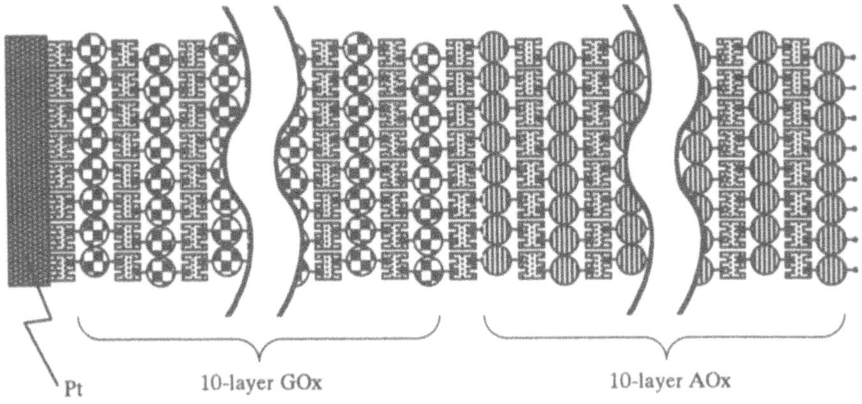

Figure 5. A schematic representation of the GOx/AOx multilayer-modified electrode.

$$\text{Ascorbic acid} + 1/2\ O_2 \xrightarrow{\text{AOx}} \text{Dehydroascorbic acid} + H_2O$$

The lactate [17,20] and choline [20] sensors fabricated by the similer method indicated also the same effective abilities as the glucose sensor.

The above method is not only characterized by a molecular-level modification of electrode surfaces but also applicable to the construction of protein architectures, accompaning both functionalities of molecular recognition and electron-tarnsfer mediator.

This research was supported in part by Grant-in-Aids for Scientific Research on Priority Areas (No. 05235102).

References

1. Osa, T., Kashiwagi, Y., Bobbitt, J.M., and Ma, Z.: Electroorganic Synthesis on Catalyst Coated Electrodes, in R. D.Little and N. L. Weinberg (eds.), *Electroorganic Synthesis*, Marcel Dekker, Inc., New York, 1991, pp. 343-354.
2. Osa, T., Akiba, U., Segawa, I., and Bobbitt, J.M.: Electrocatalytic Oxidation of Nerol with Nitroxyl Radical Covalently Immobilized to Poly (acrylic acid) Coated on Carbon Electrodes, *Chem. Lett.* (1988), 1423-1426.
3. Kashiwagi, Y., Ono, H., and Osa, T.: Electrocatalytic Oxidative Coupling of Methylquinolines on TEMPO-modified Graphite Felt Electrodes, *Chem. Lett.* (1993), 257-280.
4. Osa, T., Kashiwagi, Y., Yanagisawa, Y., and Bobbitt, J.M.: Enantioselective, Electrocatalytic Oxidative Coupling of Naphthol, Naphthyl Ether and Phenanthrol on a TEMPO-modified Graphite Felt Electrode in the Presence of (-)-Sparteine (TEMPO=2,2,6,6-tetramethylpiperidin-1-yloxyl), *J. Chem. Soc., Chem. Commun.* (1994), 2535-2537.
5. Kashiwagi, Y., Yanagisawa, Y., Kurashima, F., Anzai, J., Osa, T., and Bobbitt, J.M.: Enantioselective, Electrocatalytic Oxidation of Racemic Alcohols on a TEMPO-modified Graphite Felt Electrode by Use of Chiral Base (TEMPO=2,2,6,6-tetramethylpiperidin-1-yloxyl), *J. Chem. Soc., Chem. Commun*. (1996: in press).
6. Osa, T., Kashiwagi, Y., and Yanagisawa, Y.: Electroenzymatic Oxidation of Alcohols on a Poly(acrylic acid)-coated Graphite Felt Electrode Terimmobilizing Ferrocene, Diaphorase and Alcohol Dehydrogenase, *Chem. Lett.* (1994), 367-370.
7. Kashiwagi, Y., Yanagisawa, Y., Shibayama, N., Nakahara, K., Kurashima, F., Anzai, J., and Osa, T.: Preparative, Electroenzymatic Reduction of Ketones on an All Components-immobilized Graphite Felt Electrode, *Electrochim. Acta*. (1996: in press).
8. Anzai, J., Kusano, T., Osa, T., Nakajima, H., and Matsuo, T.: Urea Sensor Based on Ion Sensitive Field Effect Transistor Coated with Cross-Linked Urease-Albumin Membrane, *Bunseki Kagaku* **E 33** (1984), 131-136.
9. Anzai, J., Tezuka, S., Osa, T., Nakajima, H., and Matsuo, T.: Urea Sensor Based on Ion-Sensitive Field Effect Transistor. IV. Determination of Urea in Human Blood, *Chem. Pharm. Bull.* **35** (1987), 693-698.
10. Anzai, J., Lee, S., and Osa, T.: Reactive Langmuir-Blodgett Membrane for Biosensor Applications. Use of Succinimidyl Behenoate-Based Membranes as Support for Covalently Immobilizing α–Chymotrypsin, *Bull. Chem. Soc. Jpn.* **62** (1989), 3018-3020.
11. Anzai, J. and Osa, T.: Langmuir-Blodgett Membranes in Chemical Sensor Applications, *Selective Electrode Rev.* **12** (1990), 3-33.

12. Anzai, J., Hoshi, T., and Osa, T.: Electrochemical Preparation of Active Avidin Films for Enzyme Sensor Applications, *Chem. Lett.* (1993), 1231-1234.
13. Wilchek, M. and Bayer, E.A.: The Avidin-Biotin Complex in Bioanalytical Applications, *Anal. Biochem.* **171** (1988), 1-32.
14. Bayer, E.A and Wilchek, M.: Applications of Avidin-Biotin Technology, *Method Enzymol.* **184** (1990), 14-45.
15. Anzai, J., Hoshi, T., Lee, S., and Osa, T.: Use of the Avidin-Biotin System for Immobilization of an Enzyme on the Electrode Surface, *Sensors and Actuators B* **13-14** (1993), 73-75.
16. He, P.-G., Takeshita, H., Hoshi, T., Anzai, J., Suzuki, Y., and Osa, T.: Preparation of Enzyme Multilayers on Electrode Surface by Use of Avidin and Bio-labeled Enzyme for Biosensor Applications, *Materials Science & Engineering* **C2** (1994), 103-106.
17. Anzai, J., Takeshita, H., Hoshi, H., and Osa, T.: Regulation of Output Current of *L*-Lactate Sensors Based on Alternate Deposition of Avidin and Biotinylated Lactate Oxidase on Electrode Surface through Avidin/Biotin Complexation, *Chem. Pharm. Bull.* **43** (1995), 520-522.
18. Hoshi, T., Anzai, J., and Osa, T.: Controlled Deposition of Glucose Oxidase on Platinum Electrode Based on an Avidin/Biotin System for the Regulation of Output Current of Glucose Sensors, *Anal. Chem.* **67** (1995), 770-774.
19. Anzai, J., Takeshita, H., Hoshi, T., and Osa, T.: Elimination of Ascorbate Interference of Glucose Biosensors by Use of Enzyme Multilayers Composed of Avidin and Biotin-labeled Glucose Oxidase and Ascorbate Oxidase, *Denki Kagaku* **63** (1995), 1141-1142.
20. Hoshi, T., Takeshita, H., Anzai, J., and Osa, T.: Use of Electrodeposited Avidin Film for the Preparation of Lactate and Choline Sensors, *Anal. Sci.* **11** (1995), 311-312.

ELECTRICAL SIEVES FOR MOLECULE RECOGNITION

C.L. BOWES, T. JIANG, A.J. LOUGH, G.A. OZIN,*
S. PETROV, A. VERMA, G. VOVK, and D. YOUNG
Materials Chemistry Research Group
Lash Miller Chemical Laboratories, University of Toronto,
80 St. George Street, Toronto, Ontario, Canada, M5S 3H6

R.L. BEDARD
Universal Oil Products,
25 E. Algonquin Rd., Des Plaines, Illinois, USA, 60017.

1. Abstract

Nanoporous materials with crystallographically defined pores represent an attractive class of materials for molecule discriminating chemical sensor applications. Recently, zeolite-modified surface acoustic wave (SAW), zeolite-loaded piezoelectric quartz crystal microgravimetric (QCM), and aluminophosphate molecular sieve coated capacitance devices have shown promise in selectively detecting organic molecules. As zeolites are electrical insulators it would be interesting to find electrically conducting open-framework materials whose charge-transport response is sensitive to the size and shape of adsorbed molecules. Recently, nanoporous tin(IV) sulfides have been synthesised and structurally characterized. Considering the semiconductor character of tin(IV) disulfide, the nanoporous tin(IV) sulfide versions are potentially interesting electronic molecular sieves for chemical sensing. In this paper synthetic, structural, ion-exchange, thermochemical and vibrational information for various members of the $R_2Sn_3S_7$ structure class of nanoporous tin(IV) sulfides, where R represents an alklyammonium template cation, will be briefly highlighted. The adsorption, optical and electrical responses of these materials are found to be sensitive to the identity and concentration of a range of molecular guests. Furthermore, their adsorption processes are fast and reversible. These are attributes of a molecule recognition material for sensing applications.

2. Introduction

Molecular recognition drives research and development in zeolite science. The unique molecule size and shape dependent adsorption properties of this class of oxide-based open-framework materials makes them the "inorganic enzymes" of industrial catalysis

A. W. Coleman (ed.), Molecular Recognition and Inclusion, 39–58.
© 1998 *Kluwer Academic Publishers. Printed in the Netherlands.*

and separation processes. Molecule recognition is also a pre-requisite for chemoselective sensing and the ability of zeolites to selectively detect molecules has been under active investigation since the first report of this phenomenon. [1] Although zeolites are electrical insulators, their molecule specificity has been demonstrated by a number of groups, through the measurement of, for example, mass and capacitance transducer action. [2]

The recent discovery of microporous metal sulfides, [3] raises the intriguing possibility of building electrical sieving devices that are capable of recognizing molecules through optical and electrical conductivity changes. [4] The x-ray structures of open-framework metal sulfides, [5] their potential for molecular recognition, and the concept of constructing an "inorganic electronic nose" have recently been described. [6] In this invited paper, we will describe the synthesis-structure-adsorption-optical-electrical conductivity relationships for some of these materials. [7] Information on the speed and reversibility of the electrical responses observed for these materials when exposed to a range of probe molecules, will be discussed. This allows a preliminary charge-transport mechanism for the host and host-guest systems to be proposed. The results suggest the use of sensor arrays of these materials in an artificial neural network, to obtain the characteristic electrical fingerprint of a mixture of molecules.

3. Synthesis and Structure

3.1 SYNTHESIS

All of the $R_2Sn_3S_7$ materials, discussed in this paper were prepared under hydrothermal reaction conditions from mixtures of $1SnS_2/1ROH/33H_2O$ or $1Sn/2S/1ROH/30H_2O$, where R represents an alklyammonium template cation. [8] The reaction mixtures were usually heated at 150°C for one to four days. To obtain pure $R_2Sn_3S_7$, static crystallisation at 150°C followed by slow cooling of the reaction mixture to room temperature over a period of 1-2 days can be employed. A tumbling reactor was sometimes used in conjuction with alternative mineralizers, such as HCO_3^- or SH^-, to enhance the synthesis of phase pure $R_2Sn_3S_7$ materials.

3.2 STRUCTURE

$R_2Sn_3S_7$ represents one member of a growing family of tin(IV) sulfides, with a nanoporous layer topology. The general formula of this class of materials is $A_{2+2x}Sn_nS_{2n+1+x}$. The structure of the sheets of three of these are illustrated in Figure 1. They have been established by SC-XRD structure analysis. [5,9,10,11] The porous layers of $A_{2+2x}Sn_nS_{2n+1+x}$ are negatively charged and the A^+ counter cations (*i.e.*, organic or inorganic) reside in the void spaces between and/or within the porous layers. One interesting structural feature of $A_{2+2x}Sn_nS_{2n+1+x}$ is the presence of periodic arrays of Sn_3S_4 broken-cube clusters. In the case of $R_2Sn_3S_7$, six of these cubes are linked together by six double sulfide $(\mu\text{-}S)_2$ bridge-bonds to form layers containing hexagonal-shaped 24-atom rings, Figure 1A.

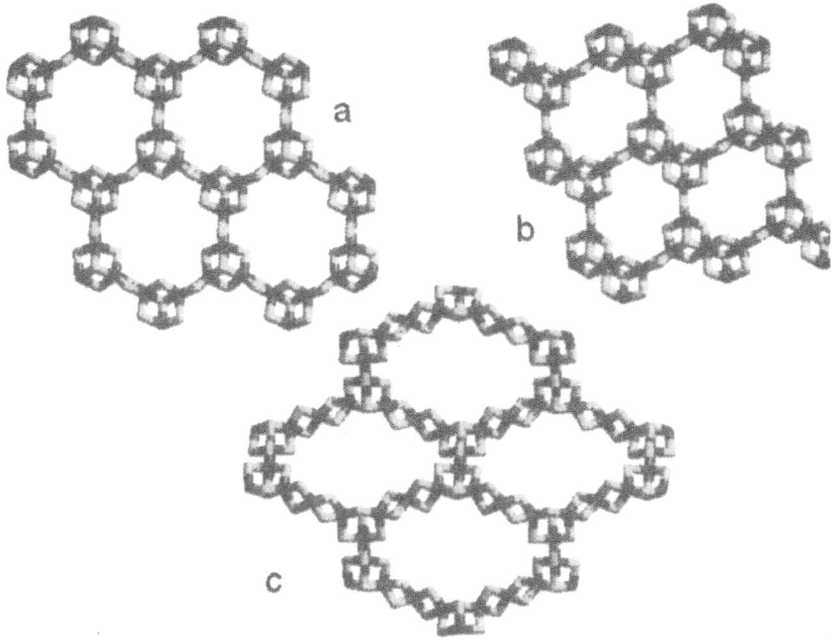

Figure 1. The structure of the nanoporous sheets of three layered $A_{2+2x}Sn_nS_{2n+1+x}$ materials, A) $Sn_3S_7^{2-}$ B) $Sn_5S_{12}^{4-}$ C) $Sn_4S_9^{2-}$.

The local symmetry around the five-co-ordinated tin(IV) center is pseudo-trigonal bipyramidal, SnS_5, with respect to sulfur(-II). Three edge-sharing SnS_5 units constitute one broken-cube cluster.

3.3 ASSEMBLY PATHWAY

Our current understanding of the mode of formation of $R_2Sn_3S_7$ open-frameworks is encapsulated graphically in Figure 2. [8,12] In brief, solution phase $Sn_2S_6^{4-}$ building units snap together, as donor-acceptor pairs, to create Sn_3S_4 broken-cube clusters. These assemble around template cations R^+ to form porous anionic $Sn_3S_7^{2-}$ sheets. Template cations organise the sheets into stacks which move into registry in the porous lamellar $R_2Sn_3S_7$ product phase.

The symmetry of the porous layers and their stacking sequence depends sensitively on the nature of the template cation and the water content of the unit cell. This is the origin of the polytypism found in the $R_2Sn_3S_7$ class of materials and the main cause of the difficulty of growing single crystals suitable for structure-property measurements.

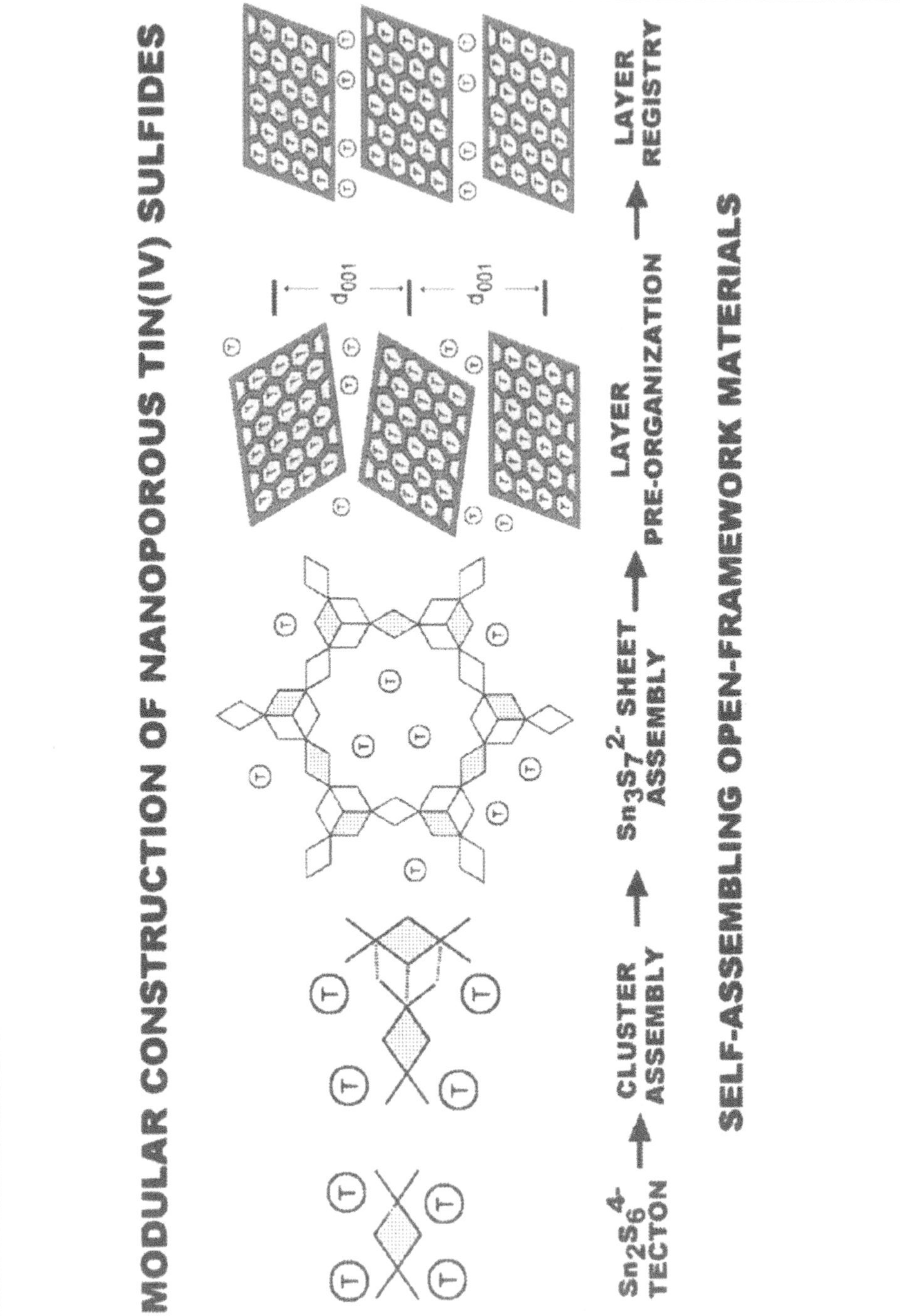

Figure 2. Graphical summary of the current understanding of the mode of formation of $R_2Sn_3S_7$ open-frameworks.

4. $R_2Sn_3S_7$ as a Molecular Sieve Or Intercalate

$R_2Sn_3S_7$ is like a hybrid of a layered metal disulfide and a molecular sieve. Where will its properties fit into this range of behaviours? Will it display differences in stacking sequences and polytypism like the dense phase berndtite, SnS_2? Will it sustain removal of the organic space-filler and retain the structural integrity of the open-framework like a zeolite-type material? Will it undergo intercalation or sieving behaviour or both? What type of adsorption isotherm will it display? Will its electronic band and charge-transport properties resemble those of the bulk semiconductor SnS_2 or the electrically insulating zeolites, or rather will they be unique? Will it display molecule discriminating transducer action? These are some of the questions that we will address briefly in this study.

4.1 FRAMEWORK FLEXIBILITY

The diffraction, Figure 3, and spectroscopic characterisation of a substantial group of $R_2Sn_3S_7$ materials leads one to a number of conclusions. There is structural similarity among the various types of $R_2Sn_3S_7$, which present three double-edge-sharing distorted trigonal bipyramidal tin(IV) centres in the broken-cube cluster building units that constitute the porous sheets. The remarkable flexibility of the structure [10] is seen,

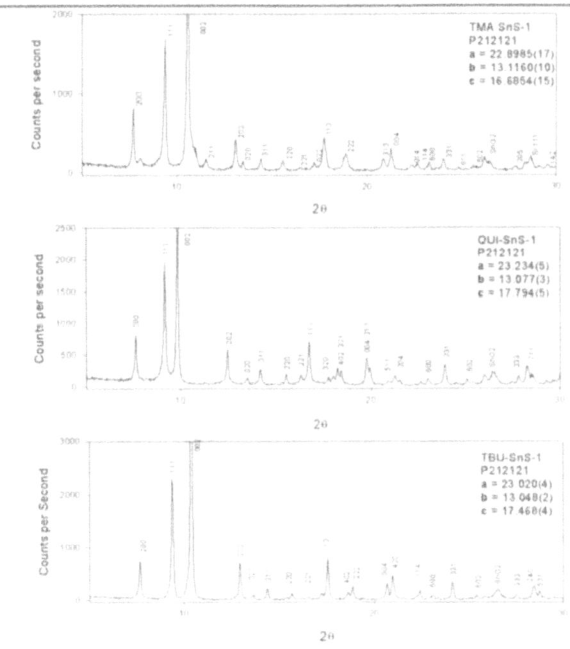

Figure 3. The PXRD diffraction patterns, indexing and unit cell dimensions of three isostructural orthorhombic space group $P2_12_12_1$ $R_2Sn_3S_7$ materials, where from top to bottom, R = tetramethylammonium, quinuclidinium and t-butylammonium cations.

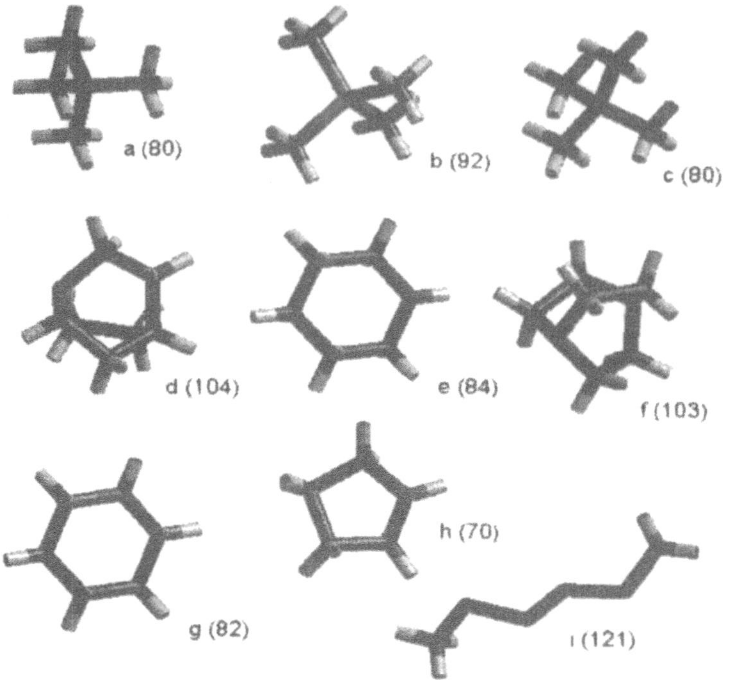

Figure 4. Illustration of some of the distinct void fillers that can form the same or different polytypes of the $R_2Sn_3S_7$ structure.

both through the number of distinct void fillers that can form the same or different polytypes, as well as the number of distinct polytypes that can be obtained using the same void filler, Figure 4. As mentioned above, this makes crystallisation of phase pure materials with sufficient crystal size and quality for SC-XRD structure determinations a challenge. For example, both $(Et_4N)_2Sn_3S_7$ and $(NH_4)_{0.5}(Et_4N)_{1.5}Sn_3S_7$ have been synthesised and structurally characterized. [6,7,13] They have related structures with two charge-compensating cations residing within and between the layers. In $(NH_4)_{0.5}(Et_4N)_{1.5}Sn_3S_7$, one-quarter of the Et_4N^+ cations in $(Et_4N)_2Sn_3S_7$ are replaced by NH_4^+ cations. Small polar molecules can cyclically adsorb and desorb from these two materials with concomitant reversible flexing of the framework structure. For example, when a dry $(Et_4N)_2Sn_3S_7$ sample was saturated with water, its structure transforms from monoclinic ($\beta = 95.26°$) to pseudo-orthorhombic ($\beta = 90.11°$), accompanied by an increase in the unit cell volume of 5.8 %, Figure 5, Table 3. Reversible adsorption of water in $(NH_4)_{0.5}(Et_4N)_{1.5}Sn_3S_7$ leads to a 26 nm blue-shift of the optical absorption edge, and a million-fold drop of its electrical resistance, Figures 14, 15, see later.

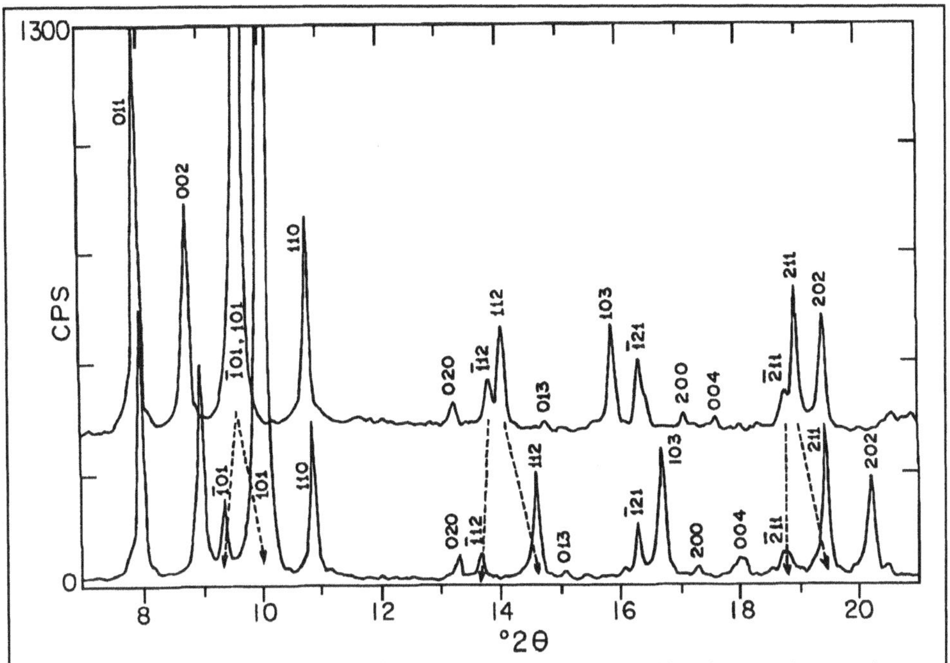

Figure 5. PXRD diffraction patterns of a dry and water saturated sample of $(Et_4N)_2Sn_3S_7$, depicting the water induced transformation from a monoclinic (β = 95.26°) to a pseudo-orthorhombic (β = 90.11°) unit cell, accompanied by an increase in the unit cell volume of 5.8 %.

Studies of this kind for a range of $R_2Sn_3S_7$ materials have illustrated the remarkable flexibility of their structures. This elasticity appears to originate from co-operative interactions between the template cations, guest species and the sulfide framework, and can be traced to angular distortions centred mainly on the broken-cube clusters which appear to mechanically function as hinge-points. This responsiveness gives rise to changes in unit cell dimensions and space groups, intra- and interlayer restructuring, pore size and shape distortions, changes in layer stacking arrangement and spacing, and polytypism. The following will show how this interesting behaviour will combine with the thermochemical, adsorption, optical and electrical properties to unveil a materials class attractive for chemical sensing applications.

4.2 STABILITY OF $TMA_2Sn_3S_7$

Knowledge of the thermochemical properties of $R_2Sn_3S_7$ materials is central to any planned application. [14] Here we present *in situ* VT-PXRD, TGA and MS data for the representative member $TMA_2Sn_3S_7$, Figures 6, 7. The VT-PXRD traces, recorded under vacuum, show that there is a small peak shift to lower d-spacing because of the removal of physisorbed/chemisorbed water from the framework, as seen in the TGA. This is followed by a slight increase in d-spacings, corresponding to the thermal expansion of the materials. $TMA_2Sn_3S_7$ retained crystallinity up to approximately 300°C under vacuum and 200°C under nitrogen. $TMA_2Sn_3S_7$ displays its greatest TGA recorded weight loss at approximately 300°C, in reasonable agreement with the VT-PXRD data. At temperatures below about 175°C, only ions corresponding to water are observed by MS, indicating that the weight loss observed in the TGA below this temperature is most likely water.

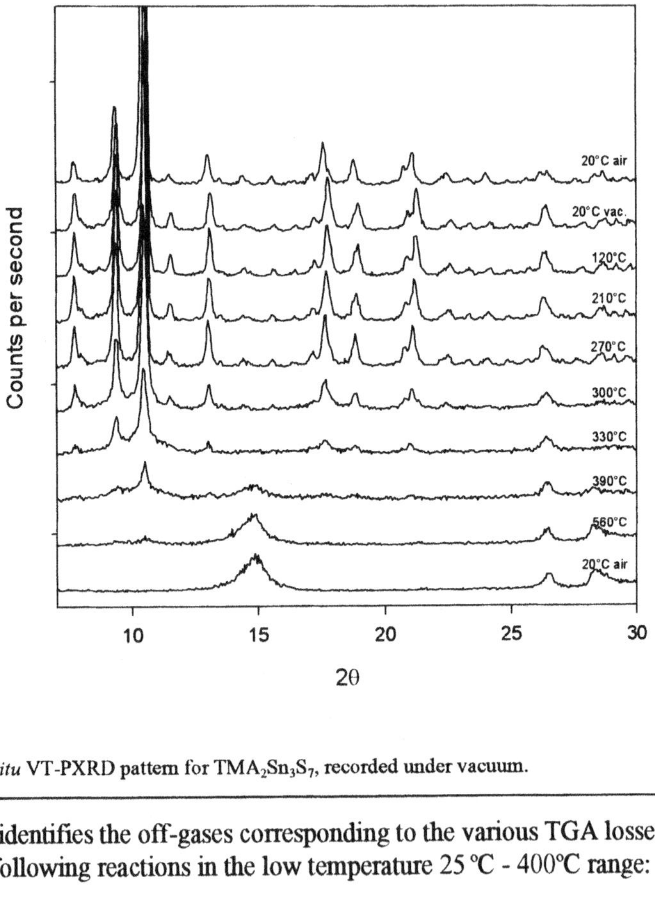

Figure 6. *In situ* VT-PXRD pattern for $TMA_2Sn_3S_7$, recorded under vacuum.

The MS identifies the off-gases corresponding to the various TGA losses and suggest the following reactions in the low temperature 25 °C - 400°C range:

Room Temperature - 150°C Thermal Transitions

$$[(CH_3)_4N]_2Sn_3S_7 \cdot xH_2O \longrightarrow [(CH_3)_4N]_2Sn_3S_7 + x\,H_2O$$

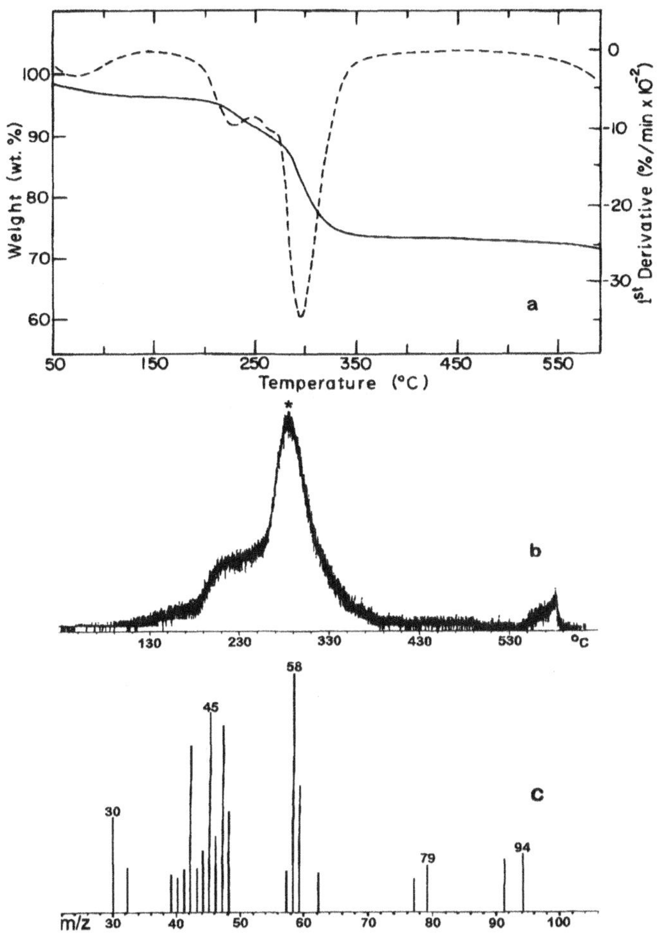

Figure 7. *In situ* TGA data (a), total ion chromatograph, (b), and MS data at the peak of the TIC (indicated with an asterisk), (c), for $TMA_2Sn_3S_7$.

200 - 400°C Thermal Transitions

$$[(CH_3)_4N]_2Sn_3S_7 \longrightarrow 2\ (CH_3)_3N + (CH_3)_2S + 3\ SnS_2$$

The first loss is attributed to water and represents 2% by mass of the remaining material, which corresponds to 1.6 molecules per pore of $TMA_2Sn_3S_7$. The TGA-MS and HT-PXRD suggest that the process above 400°C involves the reduction of tin(IV) and oxidation of sulfur(-II) to give an overall thermal loss equation [14]:

$$[(CH_3)_4N]_2Sn_3S_7 \longrightarrow 2\ (CH_3)_3N + 1/2\ (CH_3)_2S + 1/2\ (CH_3)_2S_2 + 3\ SnS_x$$

where x = 1, 1.5, 2.

These results may be summarised by comparing, once again, the $R_2Sn_3S_7$ materials to the dichalcogenide phases and the zeolites. Unsurprisingly, they are less thermally stable than the dense SnS_2 phases, as is expected for open-framework materials, and less stable than the three-dimensional, oxide-based zeolites. Nevertheless, they are sufficiently robust to undergo thermal activation for mild template-removal experiments which are the subject of the following section on ion exchange-adsorption-optical-conductivity relationships.

4.3 ION-EXCHANGE AND SIEVING BEHAVIOUR

It is expected that any microporous material should have some sieving ability, although it is not necessarily so for a layered material. The layered metal disulfides, for example, simply swell to accomodate the guest. Ion-exchange of charge-balancing ions is also expected for zeolite-like materials, while the intercalation of guests to the neutral layered transition metal disulfides is usually accompanied by some reduction in oxidation state of the metal centres. In this section the aqueous and vapour phase amine-exchange and sieving properties of $(RH)_2Sn_3S_7$, where R = t-butylamine, are briefly explored.

The exposure of $(RH)_2Sn_3S_7$ materials to guest amines of differing size and basicity was evaluated by changes in the PXRD patterns coupled with VT-MS identification of the off-gases, as described above:

$$(RH)_2Sn_3S_7 + x\,R' \longrightarrow (RH)_{2-x}(R'H)_x\,Sn_3S_7 + x\,R$$

$$(RH)_{2-x}(R'H)_x\,Sn_3S_7 \longrightarrow (2-x)\,R + x\,R' + H_2S + 3\,SnS_2$$

The results of the vapour phase amine exchange experiments are summarized below. In addition, significant ion-exchange of $(t\text{-}BuNH_3)_2Sn_3S_7$ was observed with Me_4NCl under aqueous conditions.

TABLE 1. Vapour Phase Amine Exchange in $(t\text{-}BuNH_3)_2Sn_3S_7$

amine	pKb	kinetic diameter, Å	exchange result
$t\text{-}BuNH_2$	3.17	6.5	
Me_3N	4.19	6.5	minimal
Et_3N	2.77	7.5	significant
Pr_3N	3.45	8.5	none
$s\text{-}Bu_2NH$	2.90	8.0	none

One can deduce from these observations that $(t\text{-}BuNH_3)_2Sn_3S_7$ has attributes of a molecular sieve, exhanging small alkylammonium cations like Me_4N^+, but can also behave like an intercalation material, swelling to accomodate larger more basic amines, such as Et_3N. The results imply essentially a middle ground in which, within a certain range, the layers are able to swell to allow entry of a species larger than the roughly 6-7

Å diameter pores. Outside this range, molecules may be excluded showing a sieving effect. This will be further elucidated in the next section in which the adsorption properties of $R_2Sn_3S_7$ are investigated.

4.4 ADSORPTION AND MOLECULE DISCRIMINATION

The adsorption isotherms of $TMA_2Sn_3S_7$ with several gases, H_2O, CO_2, N_2, and Ar, have been recorded at the temperatures indicated in Figure 8. The as-synthesised sample was activated at 120°C to remove water. These adsorption data provide considerable insight into the intercalation/sieving issue. The adsorption of CO_2 follows a regular, Type I isotherm. This is typical of a microporous solid having very little external surface relative to the internal surface. [15] A Type IV isotherm, characteristic of micropore filling and interlayer swelling, is not observed for either CO_2 or H_2O, indicating that $TMA_2Sn_3S_7$ is displaying sieving rather than intercalation behaviour towards these molecular guests.

A second outcome of these data is that the lattice very clearly excludes N_2 and Ar while allowing CO_2 and H_2O to enter and produce the isotherms obtained. This is quite engaging, given the difference in kinetic diameter between CO_2 (3.3 Å) *versus* Ar (3.4 Å). This appears to be obvious evidence for the sub-angstrom precision molecular sieving ability of $TMA_2Sn_3S_7$. It was possible to calculate the pore volume from the CO_2 adsorption data, using a Dubinin plot, Figure 9. [16] The usual form of the Dubinin equation is given below:

$$\log_{10}x = \log_{10}(W_o\rho) - D(\log_{10}p_o/p)^2 \qquad (1)$$

where W_o is the pore volume, ρ is the density of the liquid adsorbate, taken to be 1.56 g/cm^3, and x is the adsorption in grams. D is an empirical constant, so that plotting $\log_{10}x$ against $(\log_{10}p_o/p)^2$ gives an intercept equal to $W_o\rho$, from which the pore volume can be extracted. The results of this analysis are summarised in Table 2.

The 0.04 cm^3/g pore void volume determined for $TMA_2Sn_3S_7$ is small by comparison to the dehydrated zeolite A and X cases, but reasonable for a material containing all of its void filling organic. [17] The value also compares quite well with the calculated free volume accessible to CO_2, 0.07 cm^3/g, obtained using the Connelly surface package of Molecular Simulations. [18] The void volume of the hypothetical template-removed $TMA_2Sn_3S_7$ material is calculated to be 0.40 cm^3/g, similar to the zeolites. The following section will focus attention on the optical and electrical properties of $TMA_2Sn_3S_7$ to unveil a material with chemical sensing potential.

TABLE 2. Analysis of Observed and Calculated Pore Volumes of $TMA_2Sn_3S_7$ As Compared to Zeolites

$TMA_2Sn_3S_7$	Template in, observed	0.04 cm^3/g
$TMA_2Sn_3S_7$	Template in, calculated	0.07 cm^3/g
$TMA_2Sn_3S_7$	Template out, calculated	0.40 cm^3/g
NaA	Template out, observed	0.26 cm^3/g
NaX	Template out, observed	0.31 cm^3/g

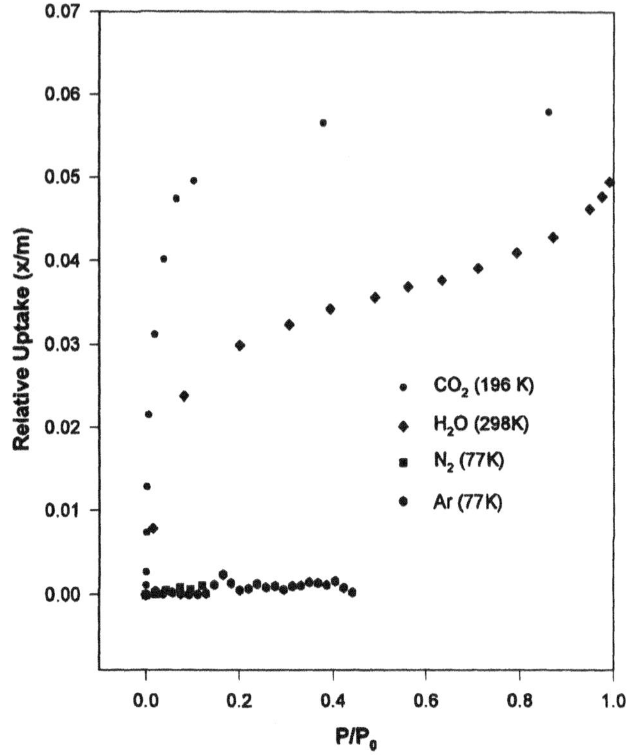

Figure 8. The adsorption isotherms of $TMA_2Sn_3S_7$ with several gases, H_2O, CO_2, N_2, and Ar, recorded at the indicated temperatures.

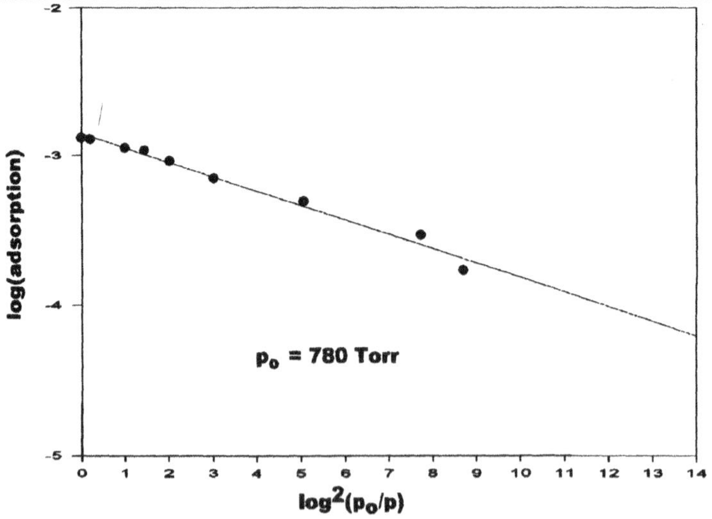

Figure 9. Dubinin plot used to calculate the pore volume of $TMA_2Sn_3S_7$ from the CO_2 adsorption data.

5. Electronic Structure

Based on the above results, $TMA_2Sn_3S_7$ can now be classified as an open-framework material with adsorption properties characteristic of a non-oxide crystalline molecular sieve. In what follows, we investigate the optical and electrical properties of this material and its responses to the inclusion of guest species. The room temperature optical absorption data for $TMA_2Sn_3S_7$ is shown in Figure 10. Reflectance data were converted to absorbance using the Kubelka Munk transformation. The best calculated fit of the optical absorption edge was to the analytical expression for an indirect-forbidden band gap transition: [19]

$$\alpha_{if} = C_{if}(h\omega - Eg - h\omega_c)^3/(1 - \exp(-h\omega_c/k_BT)) \quad (2)$$

In this analysis, the absorption coefficient α_{if} was fit to a function varying with $\omega^{1/3}$ and the resulting linear form was extrapolated to zero absorption to obtain the value of the band gap Eg, as has been done for SnS_2. [20] For indirect systems, the phonon ω_c usually appears as a linear deviation at the onset of the absorption edge. The phonon was obtained from the difference in intercept between the major slope and the deviation, Figure 10. This is inexact, as there may be several phonons of various sorts contributing to the absorption edge. Using this procedure, the room temperature band gap was estimated to be 2.44 ± 0.05 eV, and the associated phonons at 4500 cm^{-1}.

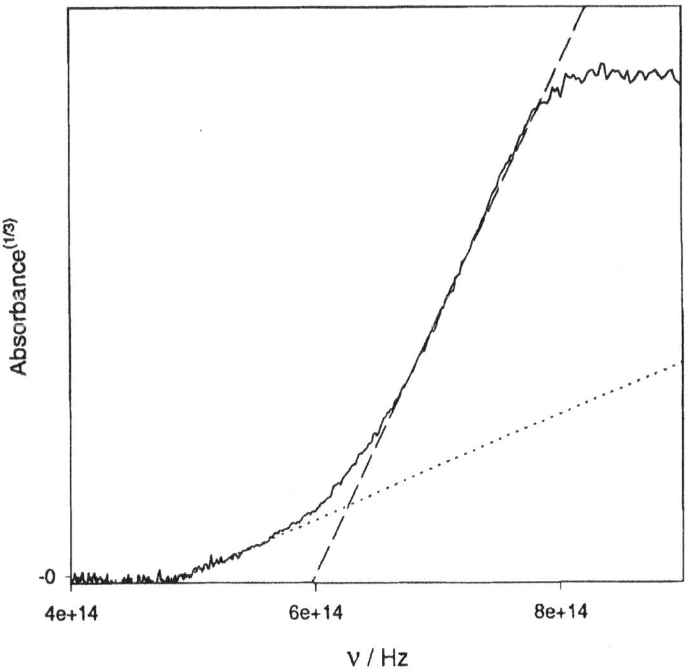

Figure 10. The room temperature optical absorption data for $TMA_2Sn_3S_7$ and the best calculated fit of the absorption edge to the analytical expression for an indirect-forbidden band gap electronic excitation.

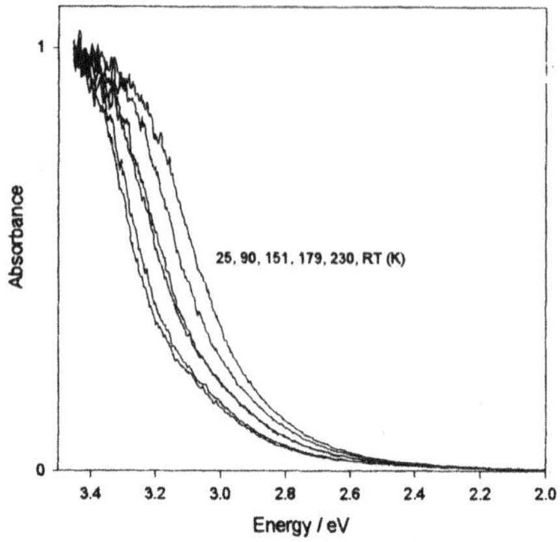

Figure 11. Temperature dependence of the band gap, $E_g(T)$, of $TMA_2Sn_3S_7$ over the range of 25 - 300K.

The band gap also displayed a clear temperature dependence, over the recorded range of 25 - 300K, Figure 11. The contribution of lattice dilation and electron-lattice interactions to the band-to-band excitation energy Eg has a temperature dependence that is expressed in the commonly used empirical formula: [21]

$$E_g = E_o - \alpha T^2/(T + \beta) \quad (3)$$

where E_o is the band gap energy at absolute zero, α and β are empirical constants, and β is taken, on the basis of theory, to be the Debye temperature given approximately by $\theta_D = h\nu_{max}/k$.

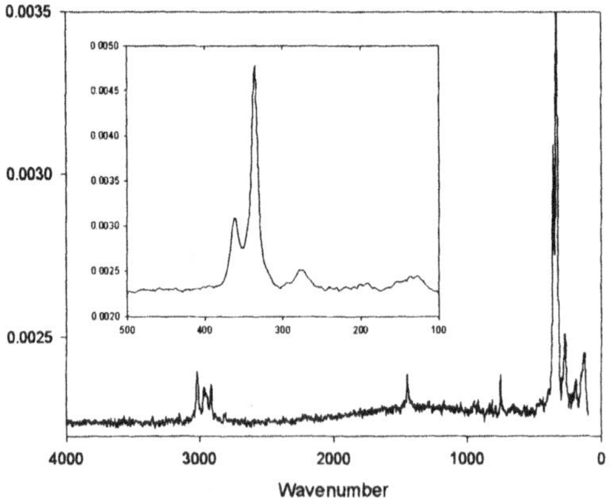

Figure 12. The FT-Raman spectrum of $TMA_2Sn_3S_7$. Inset shows low frequency framework phonon modes

From the FT-Raman spectrum of $TMA_2Sn_3S_7$, Figure 12, the highest νSnS mode was observed around 400 cm^{-1} and is taken to be the phonon cut-off frequency, ν_{max}, of the nanoporous tin sulfide framework, which corresponds to a θ_D of about 500K. Thus the Debye temperature lies well above the 25 - 300K range used for the measurement of Eg(T). This is consistent with the improved linear fit obtained when Eg was plotted as a function of T^2 rather than T. An analysis of these data yield an insightful comparison between the electronic band parameters for $TMA_2Sn_3S_7$, relative to those of bulk SnS_2:

$TMA_2Sn_3S_7$	SnS_2
Estimated Band gap 2.44 eV	Band gap 2.07 eV
Indirect forbidden (best fit)	Indirect allowed
dEg/dT = -5.6 x 10^{-4} eV/K	dEg/dT = -7 x 10^{-4} eV/K

Interestingly, they are both wide band gap, indirect semiconductors with comparable sensitivity of Eg to temperature. Like bulk SnS_2, the optical band gap in these $R_2Sn_3S_7$ materials is believed to correspond to an intralayer 3p(S-II) → 5d(SnIV) valence to conduction band electronic excitation, Figure 12. [22] The extent of interlayer electronic coupling in these 2-D anisotropic layered materials remains to be elucidated.

5.1 SENSOR RESPONSE, OPTICAL

The molecular recognition ability of these materials was first discovered by the observation of different PXRD patterns for $(Et_4N)_2Sn_3S_7$ at different seasons of the year. The structure of this material changes reversibly with the change of season. In the humid summer of Toronto, $(Et_4N)_2Sn_3S_7$ has a pseudo-orthorhombic unit cell in which for example, the (1 0 1) and ($\bar{1}$ 0 1) reflections are degenerate. By contrast, in the dry winter $(Et_4N)_2Sn_3S_7$ has a monoclinic unit cell in which the (1 0 1) and ($\bar{1}$ 0 1) reflections are non-degenerate. The reversible change of unit cell parameters is believed to result from differences in atmospheric humidity. This is confirmed by an *in situ* PXRD water adsorption study, Figure 5. Under vacuum, the unit cell of $(Et_4N)_2Sn_3S_7$ has the largest monoclinic β angle (95.26°) and the smallest unit cell volume (2667 Å^3), Table 3. By comparison, a water-saturated $(Et_4N)_2Sn_3S_7$ sample exhibits the largest unit cell volume (2823 Å^3) and the smallest β angle (90.11°), that is, essentially a pseudo-orthorhombic unit cell. Depending upon the amount of the physisorbed water in the sample, the unit cell dimensions and β angle of $(Et_4N)_2Sn_3S_7$ can adopt any intermediate value between that of the above monoclinic and pseudo-orthorhombic end-members. An inspection of the unit cell parameters, Table 3, reveals that the expansion of the unit cell occurs mainly along the **c** and **a** directions, the **b** parameter remains almost unperturbed. The nanoporous layer of $(Et_4N)_2Sn_3S_7$ resides on the [2 0 2] Miller plane. The expansion of the **a** and **c** parameters creates a larger interlamellar spacing and also changes the registry of the pores in the layers. Therefore the adsorption of water into $(Et_4N)_2Sn_3S_7$ actually creates space between the nanoporous layers. Note that the 156 Å^3 water-induced expansion of the unit cell volume actually corresponds to 16 water molecules, or 4 water molecules per $(Et_4N)_2Sn_3S_7$ formula unit (based on the reported kinetic diameter of 2.65 Å for water). Recall that there are four crystallographically

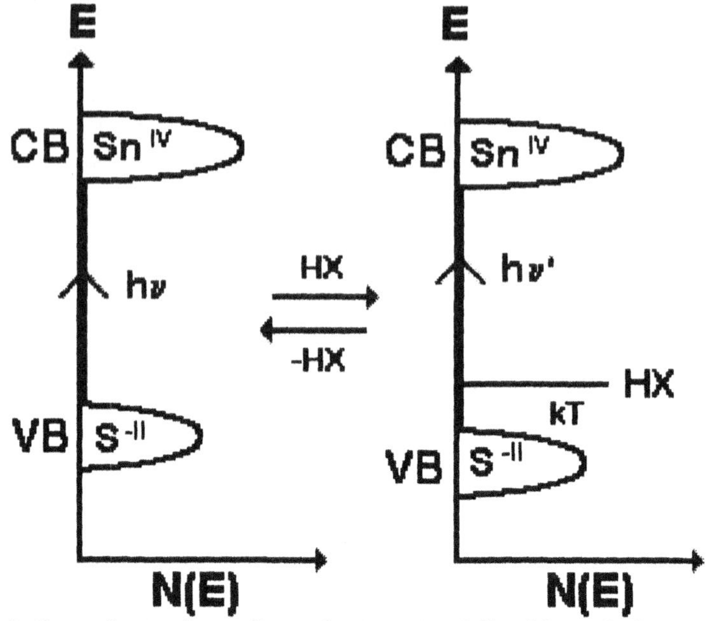

Figure 13. Qualitative density-of-states diagram for $TMA_2Sn_3S_7$ (left) and the result of its perturbation by a hydrogen-bonding HX guest, to the sulfur of the open-framework (right).

equivalent positions for the P21/n space group. It seems that the adsorption of water into $(Et_4N)_2Sn_3S_7$ results in a co-operative reconstruction of both the framework and the positions of the template molecules. The single crystal XRD structures of the orthorhombic and monoclinic polytypes of $(Me_4N)_2Sn_3Se_7$ support this proposal. [10] Co-operative reconstructions of this type are believed to be responsible for the observed reversible optical absorption and electrical conductivity changes of these materials on adsorption-desorption of water molecules, as described below. The materials are behaving like "inorganic enzymes".

The unit cell of $(NH_4)_{0.5}(Et_4N)_{1.5}Sn_3S_7$ also expands upon adsorption of water, but without an accompanying change of unit cell symmetry, Table 3. The volume expansion is found to be 36 Å^3, which approximately corresponds to 4 water molecules.

The removal of the adsorbed water from $(NH_4)_{0.5}(Et_4N)_{1.5}Sn_3S_7$ induces a red-shift of the optical absorption edge of about 26 nm, Figure 14A. Re-exposure of this sample to water causes a reversible blue-shift of the optical absorption edge. By controlling the water content, a monotonic blue-tuning of the absorption edge was observed as illustrated in Figure 14B. The adsorbed water can co-ordinate to the framework Sn(IV), hydrogen-bond to the framework S(-II), and interact with the charge-balancing ammonium cations. The adsorption-desorption of water is expected to introduce perturbations to the optical band gap and charge-transport behaviour, Figure 13, see below.

TABLE 3. Unit Cell Parameters for the Two End-Members, $(Et_4N)_2Sn_3S_7$ and $(NH_4)_{0.5}(Et_4N)_{1.5}Sn_3S_7$.

	$(Et_4N)_2Sn_3S_7$ (dry)	$(Et_4N)_2Sn_3S_7$ (wet)
Crystal System	Monoclinic	Pseudo-orthorhombic
Space Group	$P2_1/n$	$P2_1/n$
a (Å)	10.182 (2)	10.386 (4)
b (Å)	13.296 (3)	13.315 (3)
c (Å)	19.779 (5)	20.414 (8)
β (°)	95.26 (3)	90.11 (5)
V (Å^3)	2667 (1)	2823 (2)
	$(NH_4)_{0.5}(Et_4N)_{1.5}Sn_3S_7$(dry)	$(NH_4)_{0.5}(Et_4N)_{1.5}Sn_3S_7$(wet)
Crystal System	Primitive	Primitive
Space Group	$P\bar{3}m1$	$P\bar{3}m1$
a (Å)	11.477 (3)	11.509 (1)
b (Å)	6.620 (1)	6.645 (1)
c (Å)	26.856 (8)	27.717 (3)
V (Å^3)	2040 (1)	2076 (2)

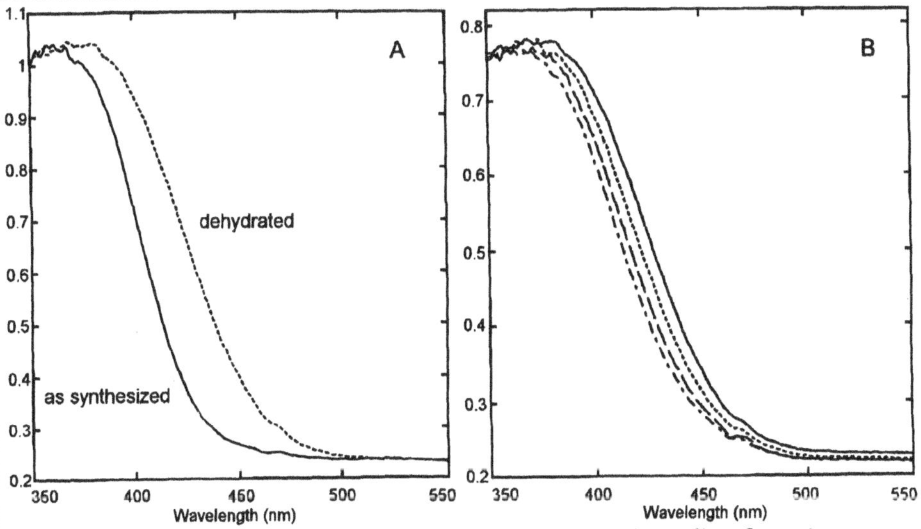

Figure 14. (A) Shifts of the optical absorption edge on adsorption-desorption cycling of water in $(NH_4)_{0.5}(Et_4N)_{1.5}Sn_3S_7$. (B) Monotonic blue-tuning of the optical absorption edge by controlling the water content of $(NH_4)_{0.5}(Et_4N)_{1.5}Sn_3S_7$.

5.2 SENSOR RESPONSE, ELECTRICAL

The effect of exposure to various gaseous molecules on the DC conductivity of $TMA_2Sn_3S_7$, measured with a four-probe technique on a polycrystalline thick film, has been evaluated, Figure 15. [22] Gold wire ohmic contacts have been employed. [23] Measurements were made on $TMA_2Sn_3S_7$, evacuated at room temperature, to correspond to the adsorption isotherms. Ar, which was excluded, had essentially no effect on the conductivity. Neither did the adsorption of CO_2, which even at room temperature was significantly adsorbed. Only H_2O had a significant effect on conductivity, raising it by four orders of magnitude, Figure 15. Similar results have been observed for a variety of $R_2Sn_3S_7$ materials, showing responses to water of up to seven orders of magnitude, bringing the conductivity into a reasonable range for a semiconductor. Recent experiments have resulted in equally large conductivity responses to D_2O, NH_3, CH_3OH, C_2H_5OH, and iso-C_3H_7OH, while for Ar, CO_2, N_2, CH_3F and H_2, no change was observed. The response time measured for adsorption-desorption of H_2O in $(NH_4)_{0.5}(Et_4N)_{1.5}Sn_3S_7$ is better than one second.

The commonality between the first group is the opportunity for hydrogen-bonding. This suggests a model in which hydrogen-bonding to framework sulfur allows charge-transfer from the valence band, predominantly composed of S(-II) states, into the hydrogen-bonding acceptor orbital, Figure 13. This implies an empty non-bonding orbital present at an energy higher than the valence band by about 1.5 kcal/mol, the approximate energy of a S···H hydrogen-bond. [25] At room temperature, such acceptor states might be significantly populated, leaving an equal population of holes in the valence band, similar to a p-doped semiconductor. It remains that even given this model, conductivity may occur via any of a number of mechanisms, for example, hole conductivity in the valence band and/or hole hopping between the non-bonding states of the guest. There may also be some contribution from ionic conduction, probably through proton mobility, however, the observed three-fold increase in the conductivity

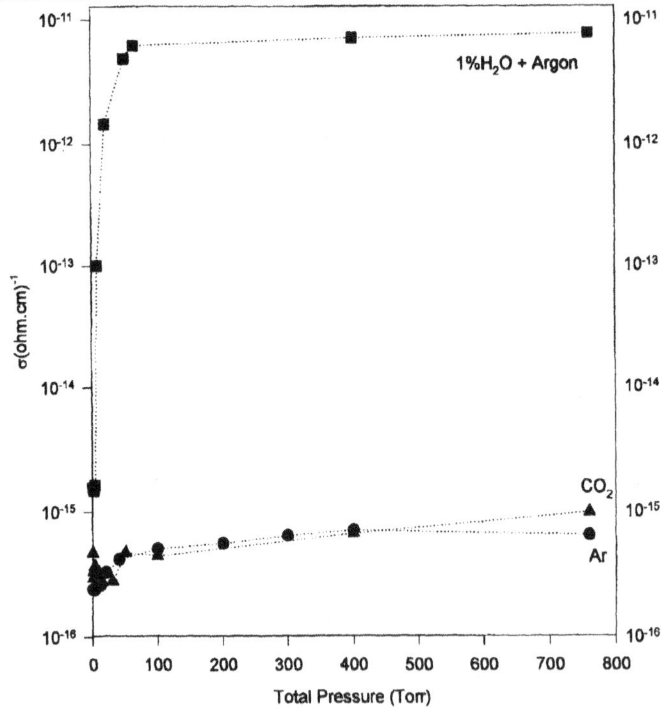

Figure 15. The effect of exposure to various gaseous molecules on the DC conductivity of $TMA_2Sn_3S_7$ measured with a four-probe technique on a polycrystalline thick film sample.

with D_2O versus H_2O and the similar conductivity values for H_2O/Ar versus $H_2O/Ar/H_2$ (note that more active platinum wire contacts were used for this experiment), suggests that protonic transport is not a significant contributor, but has not yet been ruled out. Ongoing studies, including Seebeck measurements, will be required to elucidate details of the mechanism of adsorbate-induced charge-transport.

6. Conclusion

The results of this study have important ramifications regarding the way in which $R_2Sn_3S_7$ materials are classified. They show the molecular sieving aspect of microporous materials like zeolites, but also have the ability to incorporate molecules far larger than the excluded small molecules. These characteristics are more reminiscent of an intercalation compound. Consequently, $R_2Sn_3S_7$ must be placed somewhere between these two extrema, a layered chalcogenide with holes in the sheets giving it molecular sieving capabilities. $R_2Sn_3S_7$ materials are seen as wide band gap solids, with a room temperature band gap in the region of 500 nm and a temperature dependence which is comparable to bulk SnS_2. Finally, evidence of a reversible and rapid chemical sensor response to small molecules has been observed in both the optical and electrical conductivity measurements. While no final conclusion has yet been made regarding the mechanism of this adsorbate-induced charge-transport, it augers well for the use of materials of this type for molecule size and shape selective chemical sensing.

7. Acknowledgements

We would like to acknowledge the Natural Sciences and Engineering Research Council of Canada (NSERC), Universal Oil Products (UOP) and the Canadian Space Agency (CSA) for generous financial support of this work. CB expresses her appreciation to NSERC for graduate scholarship support, and TJ and CB are grateful for graduate scholarships from the University of Toronto.

8. References

1. G.A. Ozin, M.D. Baker, M.M. Olken, *J. Am. Chem. Soc.* **110**, 5709 (1988).
2. T.E. Mallouk, D.J. Harrison, *in Interfacial Design and Chemical Sensing*, ACS Symposium Series, **561**, (1993); K. Brown, T. Bein, G.C. Frye and C.J. Brinker, *J. Am. Chem. Soc.*, **111**, 7640, (1989); Y. Yan, T. Bein, *Mater. Res. Soc. Symp. Proc.*, **233**, 175, (1991); Y. Yan, T. Bein, *J. Phys. Chem.* **96**, 9387, (1992); Y. Yan, T. Bein, *Chem. Mater.*, **4**, 975 (1992); K.J. Balkus, S.J. Riley, B.E. Gnade, *Thin Solid Films*, **260**, 4, (1995); K.J. Balkus, L.J. Sottile, H. Nguyen, S.J. Riley, B.E. Gnade, *Mater. Res. Soc. Symp. Proc.*, **371**, 33, (1995); T. Bein, *Chem. Mater.*, **8**, 1636, (1996).
3. R.L. Bedard, L.D. Vail, S.T. Wilson, E.M. Flanigen, U.S. Pat. 4,880,761, (1989).
4. G.A. Ozin, *Adv. Mater.* **4**, 612, (1992).
5. C.L. Bowes, G.A. Ozin, *Adv. Mater.*, **8**, 13, (1996).
6. G.A. Ozin, *Supramolec. Chem.*, **6**, 125, (1995).
7. C.L. Bowes, G.A. Ozin, A. Verma, G. Vovk, submitted.

8. T. Jiang, G.A. Ozin, D. Young, *Adv. Mater.*, **6**, 860, (1994).
9. W.S. Sheldrick, *Z. Anorg. Allg. Chem.*, **562**, 23, (1988); W.S. Sheldrick, H-G. Braunbeck, *Z. Naturforsch.*, **45b**, 1643, (1990); J.B. Parise, Y. Ko, J. Rijssenbeek, D.M. Nellis, K. Tan, S. Koch, *J. C. S. Chem. Commun.*, 527, (1994).
10. H. Ahari, R.L. Bedard, C.L. Bowes, T. Jiang, A.J. Lough, G.A. Ozin, S. Petrov, D. Young, *Adv. Mater.*, **7**, 375, (1995).
11. T. Jiang, R.L. Bedard, A.J. Lough, G.A. Ozin, S. Petrov, D. Young, *Chem. Mater.*, **7**, 245, (1995).
12. R.J. Francis, S.J. Price, J.S.O. Evans, S. O'Brien, D. O'Hare, S.M. Clark, *Chem. Mater.*, **8**, 2102, (1996).
13. T. Jiang, Ph.D. Thesis, University of Toronto
14. T. Jiang, R.L. Bedard, G.A. Ozin, *Adv. Mater.*, **7**, 166, (1995).
15. K.S.W. Sing, D.H. Everett, R.A.W. Haul, L. Moscou, R.F. Pierotti, J. Rouquerol, T. Siemeiniewska, *Pure and Appl. Chem.*, **57**, 603, (1985).
16. .J. Gregg, K.S.W. Sing, *Adsorption, Surface Area and Porosity*, Academic Press, New York, 1967.
17. D.W. Breck, *Zeolite Molecular Sieves*, Wiley, Toronto, 1974.
18. M.L. Connolly, *Science*, **709**, 221, (1983); M.L. Connolly, *J. Appl. Crystal.*, **16**, 548, (1983).
19. G. Burns, *Solid State Physics*, pg. 502, Academic Press, Toronto, 1985.
20. J. George, C.K. Valsala Khumari, K.S. Joseph, *J. Appl. Phys.*, **54**, 5347, (1983).
21. Y.P. Varshini, *Physica*, **34**, 149, (1967).
22. J. Robertson, *J. Phys. C: Solid State Phys.*, **12**, 47, (1979); E. Lifshitz, Z. Chen, L. Bykov, *J. Phys. Chem.*, **97**, 238, (1993).
23. A. Hamnett in *Solid State Chemistry: Techniques*, A.K. Cheetham, P. Day, eds., Clarendon Press, Oxford, 1987.
24. G.F. Yousefi, M.K. Agarwal, *J. Mat. Sci. Lett.*, **12**, 1447, (1993).
25. D.F. Shriver, P. Atkins, C.H. Langford, *Inorganic Chemistry*, pg. 407, W.H. Freeman, N.Y., 1994.

SUPRAMOLECULAR COMPLEXATION OF FULLERENES AND 1,2-DICARBADODECA-BORANE(12)

COLIN L. RASTON
Department of Chemistry, Monash University, Clayton, Melbourne, Victoria 3168 Australia

Abstract

Calix[n]arenes, n = 5, 6 and 8, and cyclotriveratrylene (= CTV) form host-guest complexes with C_{60} in toluene, and in toluene in mesoporous silica with 5.4 nm pores. Host-guest complexation is the primary process and the associated polarisation of the fullerene can lead to micelle like aggregation of the fullerenes. For CTV complexation the final aggregates are ≥ 5.4 nm in diameter whereas in the solid state the fullerenes form polymeric arrays of fullerenes at the van der Waals limit with each CTV molecule forming a perched structure with a fullerene. *p*-Bu^t-calix[8]arene can be used for purification of C_{60}, the structure of the complex based on an aggregate of three fullerenes in a triangle with each calixarene in a double cone conformation and spanning an edge of the triangle. Fullerene C_{70} forms complexes with various calixarenes. *o*-$C_2B_{10}H_{12}$ [= 1,2-dicarbadodeca-borane(12)] which like C_{60} is icosohedral, also forms a perched complex with CTV with the H-atoms attached to carbon residing over two of the aromatic rings; theory for $CH_{carborane}$ ....π-benzene predicts an interaction energy of 2.717 kcal mol^{-1}; 6-31G*.

1.1 Introduction

Supramolecular complexes of fullerenes have been prepared for C_{60} and C_{70} for a variety of host systems including calixarenes,[1-6] cyclotriveratrylene (= CTV),[7,8] and γ-cyclodextrin,[9] azacrown ethers,[10] and a porhyrazine.[11] Complexation in hydrocarbon solvents is of interest, in part, in the use of macromolecules as a method for purification of the fullerenes. Indeed, in 1994 we[1] and Shinaki *et al.*[2] reported the use of *p*-Bu^t-calix[8]arene for the retrieval of C_{60} from toluene solutions of fullerite derived from carbon soot, involving the recrystallisation of the precipitate from toluene followed by decompostion of the complex using chlorinated hydrocarbons. An independent study on the solid state ^{13}C NMR of the 1:1 complex has been reported,[3] and since the original foregoing reports significant advances have been made in understanding the nature of the complexes involved and the work has been extended to other host molecules, CTV and a substituted analogue.[8,12] Any complexation of electron deficient, neutral C_{60} and C_{70} will be inherently weak and predetermined in part by the dimensions and shape of the cavity and solvation effects, the nature of the interactions involving fullerenes within the supermolecules, solvation effects, and fullerene-fullerene interactions.[8] For C_{60} complexation is thwarted by the spherical and non-polarised nature of the fullerene.

None of the C_{60} or C_{70} calix[n]arene complexes, n = 5, 6 or 8, have been structurally substantiated, with solid state NMR spectroscopy being inconclusive,[3,13] although deduction of one of them has been possible combining physical data with molecular mechanics.[6] In contrast, for CTV two discrete complexes of composition $(C_{60})_n(CTV)$, n = 1 or 1.5, have been structurally authenticated.[7,8]

Carboranes, *o*-, *m*-, and *p*-$C_2B_{10}H_{12}$ [= 1,2-, 1,7-, 1,12- dicarbadodeca-borane(12)] have been known for over 30 years but their supramolecular chemistry is limited to α, β, or γ-cyclodextrin inclusion complexation of *o*-$C_2B_{10}H_{12}$ as 2:1 and 1:1 complexes in aqueous medium.[14] This is surprising given the nexus between these clusters and C_{60}. The carboranes and C_{60} are all remarkably thermally stable, icosahedral clusters with not too

A. W. Coleman (ed.), Molecular Recognition and Inclusion, 59–66.
© 1998 *Kluwer Academic Publishers. Printed in the Netherlands.*

dissimilar diameters of ~ 8 and 10.0 Å respectively. However, only the *p*-carborane is like C_{60} in being non-polarised in the ground state, the other carboranes having significantly different dipole moments.

This article is devoted to the supramolecular chemistry of fullerenes and *o*-$C_2B_{10}H_{12}$ for polycyclic calixarenes and CTV hosts bearing aromatic rings, as developed in our laboratory and in collaboration with J. L. Atwood and R. J. Blanch.

1.2 Calixarenes

Isolated discrete calixarene complexes of C_{60} and C_{70} are shown in Table 1. For calix[4]arene only the *p*-phenyl system was effective, although in the three dimensional structure the fullerene forms intimate contact with the π-cloud of the aromatic rings, *exo*- to the calixarene cavity (low precision X-ray structure).[15] Molecular mechanics shows the pitch of the fullerene cavity is too high for favourable docking of the fullerene in the calixarene cavity. The same is also the case for calix[4]arene itself and also for *p*-Bu^t-calix[4]arene. For the other complexes molecular mechanics shows that docking of the fullerenes within the calixarene cavity is favourable and the complexes are assumed to be based on association of such supermolecules in the solid state, possibly with fullerene-fullerene interactions.

Table 1. Isolated stoichiometric calixarene complexes of C_{60} or C_{70}.

L = *p*-X-Calix[4]arenes	*p*-X-Calix[5]arenes	*p*-X-Calix[6]arenes	*p*-X-Calix[8]arenes
$(C_{60})(L)$, X=Ph[14]	$(C_{60})(L)$, X=H[16]	$(C_{60})_2(L)$, X=H[1]	$(C_{60})(L)$, X=Bu^t[1-3]
	$(C_{60})(L)_{1.5}$, X=Bu^t[16]	$(C_{60})_2(L)$, X=Bu^t[16]	$(C_{60})(L)$, X=Pr^i[4]
	$(C_{60})(L)$, X=CH_2Ph[17]		
	$(C_{70})(L)_{1.5}$, X=H[16]	$(C_{70})_2(L)$, X=H[16]	$(C_{70})_2(L)$, X=Bu^t[4]
	$(C_{70})(L)$, X=Bu^t[16]	$(C_{70})_2(L)$, X=Bu^t[16]	

p-Bu^t-calix[8]arene forms a 1:1 complex with C_{60} which crystallises essentially quantitatively from solution in the presence of *ca* 10% of the calixarene. In toluene above *ca* 80°C the complex decomposes, reforming the complex on cooling. Retrieval of the fullerene from the complex is effected by the addition of either chloroform or methylene chloride in which the fullerene is only sparingly soluble and crystallises from solution. Formation of non-classical C-H....$\pi_{aromatic}$ hydrogen bonds between the chlorinated solvent and aromatic rings of the calixarene may take precedence over the host-fullerene interactions resulting in decomposition of the complex. This type of interaction is found in methylene chloride inclusion complexes of calix[4]arenes,[18] and is favoured energetically by 3.94 kcal mol^{-1}.[19] Applying the ability of *p*-Bu^t-calix[8]arene to form a complex with C_{60} forms the basis of obtaining >99.5% purity C_{60} from fullerite after the initially formed precipitate, which contains up to 11% C_{70}, is recrystallised twice from toluene. It is noteworthy that pure C_{70} does not form a complex with the same calixarene but does so in benzene, crystallising as the 2:1 complex, $(C_{70})_2$(*p*-calix[8]arene).[4]

The *p*-Bu^t-calix[8]arene complex with C_{60} has an absorption band at 470 nm which is assigned as a resonant inter-fullerene molecular transition,[6] similar to that observed in thin films of C_{60},[10] rather than a charge transfer transition between the host system and the electron deficient fullerene. Formation of the complex possibly involves a monomeric 1:1 transient intermediate, Scheme 1, which distorts the electron cloud of the fullerene[6] resulting in micelle like formation featuring fullerene-fullerene interactions in the interior core with the fullerenes encapsulated by the host calixarene molecules. ^{13}C CP-MAS NMR data[13] suggests that no well ordered structure of the complex exists and a variety of modifications are possible in the solid state. Nevertheless there is a common underlying ratio of *ca* 2:6[3] for two different types of Bu^t-phenyl groups. A structure which is consistent with this ratio has each calixarene in the double cone conformation spanning two fullerenes such that three of these Bu^t-phenyl groups wrap around each fullerene with a fourth Bu^t-phenyl group also associated with each fullerene, yet astride both fullerenes. This double cone conformation of the calixarene would require minimal perturbation of the H-bonded phenol network relative to uncomplexed *p*-Bu^t-calix[8]arene,[20] as concluded from earlier IR studies.[1,2] Complexation of C_{70} by *p*-Bu^t-calix[8]arene only in the presence of C_{60} is consistent with an aggregate of fullerenes with at least one of the C_{60}'s replaced by C_{70}, the maximum molar incorporation of C_{70} in the complex being 15±1.0% (HPLC).[6]

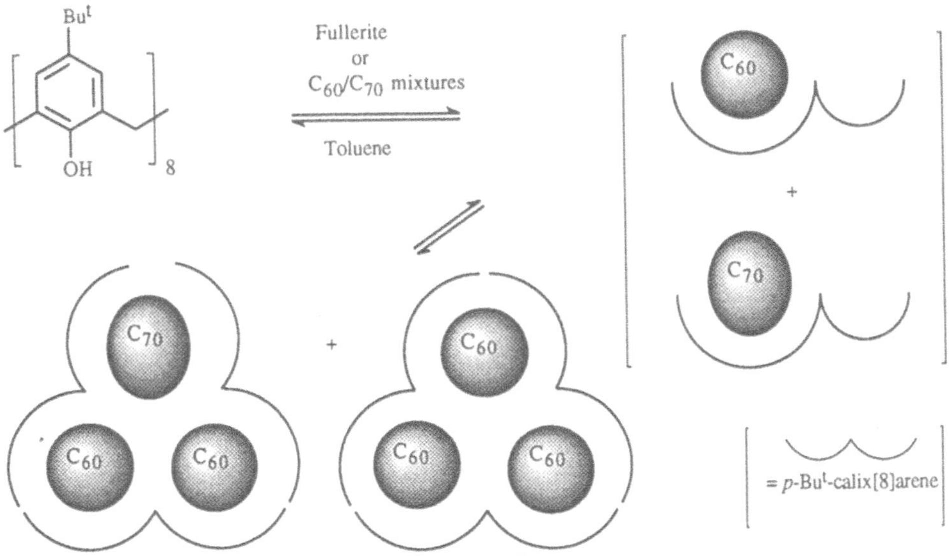

Scheme 1

The determination of the number of fullerenes in the cluster as three rests on consideration of molecular mechanics using BIOSYM[6] with the minimised structure for [(*p*-Bu^t-calix[8]arene)$_3$(C_{60})$_3$] shown in Figure 1. Moreover, comparison of the X-ray powder diffraction patterns of the C_{60} complex with that containing some C_{70}[13] is consistent with some isomorphous replacement of C_{60} by C_{70}. Incorporating more than one molecule of C_{70} per trimer of fullerenes may be too disruptive to the tight fullerene/calixarene meshing as suggested by molecular mechanics. The presence of aggregates of fullerene complexes rather than monomeric species explains why the so called size specific nature of the calixarene complexing C_{60} is associated with some complexation of C_{70}, since pure C_{70} fails to form a complex under the same conditions.

A dimeric structure has been ruled out on several grounds including that the structure with each of the two calixarenes in a similar double cone conformation has considerable surface area of the fullerenes exposed at that ends of the dimer such that additional fullerene-fullerene interactions are possible. Aggregates based on more than three fullerenes while maintaining maximum π-π interaction is not possible as they would have insufficient surface to accommodate the calixarenes while maintaining a 1:1 ratio of fullerene to calixarene. Molecular mechanics on the monomeric 1:1 (*p*-Bu^t-calix[8]arene)(C_{60}) (and the C_{70} analogue), Scheme 1, gives a minimum corresponding to five of the aromatic rings of the calixarene forming a shallow cavity associated with docking of the fullerene, the other three aromatic rings in a smaller cone with the phenolic protons forming a closed hydrogen bonded loop.[6]

Fullerene C_{70} forms a 2:1 fullerene rich complex in benzene and not toluene and the solid state NMR data,[4] which indicates the fullerenes are symmetrical, can be interpreted as being representative of a complex with two C_{70} molecules at the van der Waals limit residing in each cavity of the calixarene in a double cone conformation. This is a reasonable structure using molecular mechanics, with the overall structure dominated by fullerene-fullerene interactions. Similarly the structures of the calix[6]arene fullerene rich complexes, 2:1 C_{60} or C_{70}: calix[6]arenes, Table 1, can be similarly rationalised with a double cone conformation associated with two fullerenes, and the structure dominated by fullerene - fullerene interactions. The 1:1 complexes for the calix[5]arenes is as expected on consideration of molecular mechanics; a single cone shape for the calixarene allows maximum overlap of the π-cloud of the calixarenes with the C_{60}, especially with alignment of a C_5 axis of the fullerene with that of the calixarene. Overall symmetry matching may be important in encouraging host - guest interactions. This theme is further apparent in the CTV complexes discussed below where in principle a C_3 axis of the fullerene can align with the C_3 axis of the CTV.

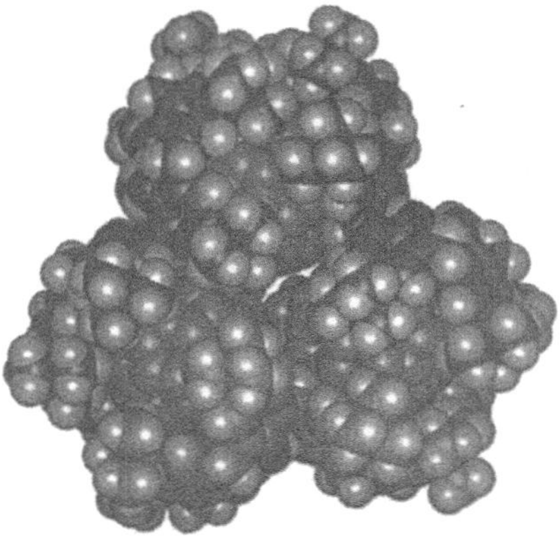

Figure 1. Space filling molecular mechanics structures of $[(C_{60})_3(p\text{-Bu}^t\text{-calix[8]arene})_3]$ (standard CVFF forcefields; structure minimised using the conjugate gradients algorithm with a convergence criterion of the average derivative being less than 0.001 kcal mol^{-1}.

1.3 Cyclotriveratrylenes

Complexation by toluene solutions of CTV, **I**, R = Me, results in polymeric structures in the solid, either a fullerene rich phase $(C_{60})_{1.5}$(CTV), or a 1:1 phase (C_{60})(CTV) for saturated solutions of CTV. The structures are dominated by fullerene-fullerene interactions, and each CTV has a C_{60} associated within the cavity of the CTV as a 'ball and socket' nano-structure, Figure 2. For $(C_{60})_{1.5}$(CTV) the fullerenes collectively comprise a two dimensional closed pack array[15] with alternating disposition of the CTV molecules in a hexagonal motif with the central fullerene devoid of CTV. In (C_{60})(CTV) each fullerene has a very short contact to two other symmetry related fullerenes, centroid-centroid distances being 9.87 Å, subtending an angle of 120.2°, the fullerenes thus form a zigzag array.

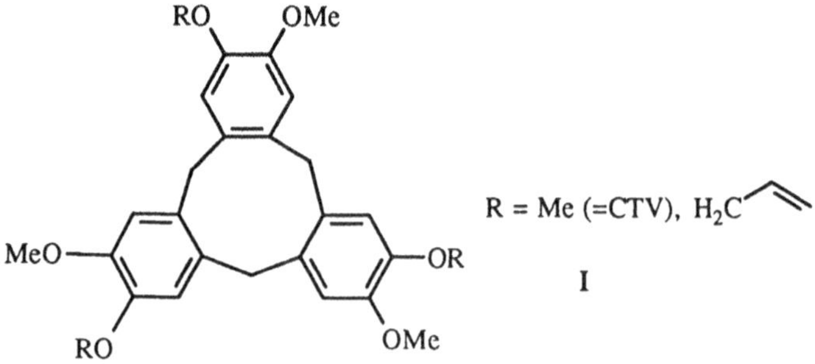

I

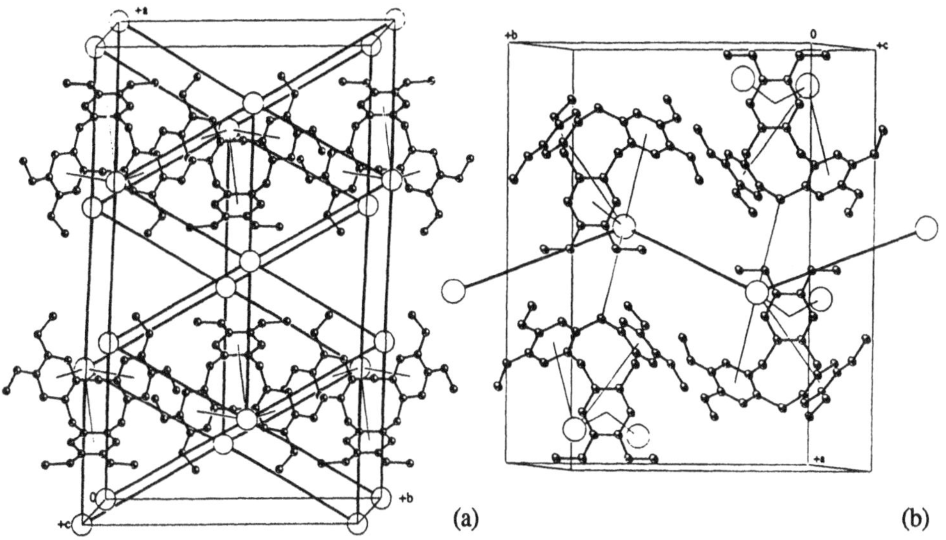

Figure 2. Solid state structures of (a) $(C_{60})_{1.5}$(CTV) and (b) (C_{60})(CTV); for simplicity a small arbitrary radius for C_{60} with lines representing nearest fullerene-fullerene and only 'ball and socket' guest host contacts are shown.

Solvated host-guest species, $[(CTV)(C_{60})]_{solvated}$, are formed first in solution. This results in polarisation of the fullerene promoting aggregation which is evident by the appearance of a broad at 475 nm band assigned to a resonant inter-fullerene transition. The allyl substituted analogue **I**, R = allyl (=CTVallyl$_3$), with C_{60} shows only solvated host-guest species. Presumably for CTVallyl$_3$ the extended allyl groups on the CTV and the deeper cavity disfavours aggregation of the host-guest complexes of the type observed for CTV; any fullerene-fullerene interaction is seemingly at the expense of host-guest interactions with pure C_{60} crystallising from toluene as the unsolvated cubic close packed modification[21] at concentrations near to the threshold of solubility of the fullerene in pure toluene.

1.4 Qualitative Test for Supramolecular Complexation of Fullerenes

In addressing the difficulties in identifying and encouraging host-guest interactions, a new amorphous mesoporous silica (= MPSiO$_2$) with sufficient pore size to allow mobilisation of C_{60} has been developed.[22] The material has a narrow range of pores, *ca* 5.4 nm, which are large enough to accommodate small aggregates of fullerenes and associated host molecules, at least for small aggregates where conventional particle size analysis would be difficult. When this material is loaded with C_{60} it offers a qualitative test for supramolecular complexation of the fullerene, and the silica devoid of C_{60} is a material for uptake of preformed micelles $\leq$ 5.4 nm in diameter.

Contact of the magenta coloured toluene solution of C_{60} with MPSiO$_2$ results in magenta colouration of the silica which is retained after air drying, but on heating at 320 K to remove the toluene the material turns yellow. A broad new band at 450 nm in the reflectance spectrum, ascribed to inter-fullerene resonant transitions and is consistent with disruption of toluene solvated C_{60} molecules,[23] affording aggregates of C_{60} within the pores of MPSiO$_2$. The band is similar to that found in thin films and aggregates of colloidal C_{60},[10,23] and that for the complexation of C_{60} with *p*-Bu^t-calix[8]arene and CTV (see above). Contact of the yellow material with toluene restored the magenta colour of toluene solvated C_{60} in the pores.

Addition of toluene solutions of a variety of calixarenes and CTV to magenta MPSiO$_2$ solid yields brown/yellow material which show a new broad reflectance band at *ca* 450 nm. This is consistent with the formation of aggregates of C_{60}, most likely as part of micelle-like structures with the fullerene cores surrounded by host molecules in the pores of MPSiO$_2$. The new band is indicative of supramolecular complexation of the fullerene in the pores of the silica and is a qualitative test for host-guest interaction. It is noteworthy that some calixarenes, e.g. *p*-Bu^t-calix[4]arene and its permethylated ether analogue, failed to yield isolable host-guest complexes, and failed to show any host-guest interaction in solution, unlike in the pores of $MPSiO_2$. However, host-guest interaction is implied in solution by the distinctly different crystal morphology of pure solid C_{60} obtained on evaporation of toluene solutions of the fullerene and host.[16]

Addition of mixtures of C_{60} with a selection of host molecules to $MPSiO_2$ also yielded brown material to varying degrees. A 1:1 mixture of C_{60} to CTV gave a similar colour change and spectrum to that of adding the same host to C_{60} already in the pores. However, for a 1:10 ratio there was little uptake of fullerene, which is consistent with the above studies that micelle-like species are being formed under such conditions. The bulk of these must be stable micelles > 5.4 nm in diameter.

1.5 Carboranes

Both CTV and $CTVallyl_3$ complex with *o*-$C_2B_{10}H_{12}$, affording in the solid the 1:2 and 1:1 adducts, (*o*-$C_2B_{10}H_{12}$)$(L)_n$, L = CTV, n = 2, L = $CTVallyl_3$, n = 1.[24] The 1:2 ratio occurs despite using equimolar amounts of the two compounds and while the ratio can be indicative of two host molecules encapsulating the cluster, the X-ray structure revealed only one of the ligands to be bound to the cluster, the other seemingly acting as a space filler, forming a zigzag π-stacked column. The IR of both compounds shows a significant shift for $\nu_{C\text{-}H}$ of the carborane of 13.5 and 12.5 cm^{-1} respectively to lower energy and a shift to higher energy for $\nu_{B\text{-}H}$, by 15.3 and 3.2 cm^{-1}. This is consistent with the presence of non-classical hydrogen bonding for the C-H bonds with dispersal of the extra charge on the carbons resulting in strengthening of the B-H bonds. ^1H NMR solution spectra for the 2:1 complex show a rapid exchange process between a 1:1 species and solvated CTV with one resonance for the CTV.

In the solid state three hydrogens of a triangular face of the icosahedron are directed towards the aromatic rings of the CTV albeit unsymmetrically, Figure 3. The origin of this most likely arises from the energetically favoured non-classical hydrogen bonds between each of the two acidic hydrogens attached to carbon cluster atoms and two aromatic rings of the CTV versus repulsion between the hydrogen attached to a boron atom of the cluster and the third aromatic ring of the CTV. This is supported by theoretical studies[24] which gives the protons on carbon a residual charge of +0.299 cf 0.068 - 0.091 for the hydrogen atoms on boron, with a single C-H......π interaction for benzene, **II**, and the carborane favoured by 2.717 kcal mol^{-1} (CH......benzene ring centroid 2.694 Å).

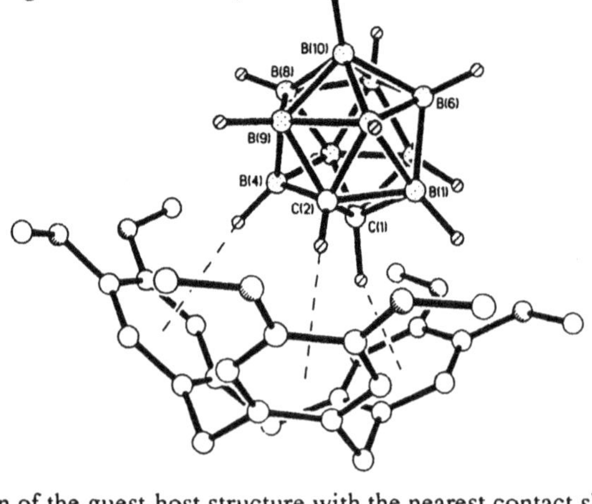

Figure 1. Projection of the guest-host structure with the nearest contact shown as broken line; distances from the centroid of the aromatic rings to the H atoms attached to C(1), C(2), and B(4) are 2.184, 2.556, and 2.765 Å, respectively (corresponding distances from C(1), C(2), and B(4) to the centroids, 3.299, 3.503, 3.701 Å).

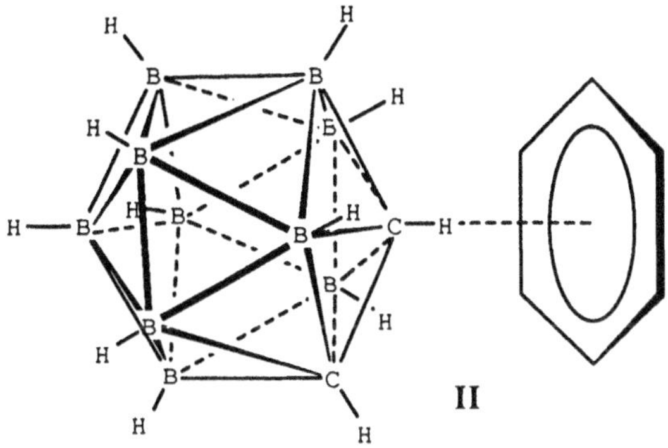

II

1.6 Conclusions

The ability to purify C_{60} by complexation, and with some progress on C_{70} using the same strategy is noteworthy, as is the novelty of the structures of the complexes. The primary process in supramolecular complexation of C_{60} is formation of a host - guest species with polarisation of the fullerene resulting in aggregation of the fullerenes. The fate of this is complete encapsulation of a single fullerene (not yet established in non-aqueous systems), the formation of a stable micelle - like structure (in mesoporous silica as well as in bulk solution), or the breakdown of the latter resulting in crystallisation of pure C_{60}. Complexation of higher fullerenes is a synthetic goal as is the complexation of reduced fullerenes, the associated complexes likely to possess unusual properties. Also the non-classical hydrogen bond interaction as a means of forming host-guest nano-structures based on carboranes offers scope for purifying the isomeric *o*- *m*, and *p*-isomers, and well as assembling complex structures.

Acknowledgments

Contributions to this work by several associates whose names appear in the references are gratefully appreciated, along with financial support from the Australian Research Council, the Department of Industry, Science and Tourism (Australia) and the National Science Foundation (USA).

References

[1] J. L. Atwood, G. A. Koutsantonis, C. L. Raston, *Nature*, **1994**, *368*, 229.
[2] T. Suzuki, K. Nakashima, S. Shinkai, *Chem. Lett.*, **1994**, 699.
[3] R. M. Williams, J. M. Zwier, G. H. Nachtegaal, A. P. M. Kentgens, *J. Am. Chem. Soc.*, **1994**, *116*, 6965.
[4] T. Suzuki, K. Nakashima, S. Shinkai, *Tetrahedron Lett.*, **1995**, *36*, 249.
[5] R. M. Williams, J. W. Verhoeven, *Recl. Trav. Chim. Pays-Bas*, **1992**, *111*, 531.
[6] C. L. Raston, J. L. Atwood, P. J. Nichols, I. B. N. Sudria, submitted to *Angew. Chem.*
[7] J. L. Atwood, M. J. Barnes, R. S. Burkhalter, P. C. Junk, J. W. Steed, C. L. Raston, *J. Am. Chem. Soc.*, **1994**, *116*, 10346.
[8] J. L. Atwood, M. J. Barnes, M. G. Gardiner, and C. L. Raston, *Chem. Commun.*, **1996**, 1449.
[9] T. Anderson, G. Westman, G. Stenhagen, M. Sundahi, O. Wennerstrom, *Tetrahedron Lett.*, **1995**, *36*, 597, and references therein.
[10] F. Diederich, J. Effing, U. Jonas, L. Jullien, T. Plesnivy, H. Ringsdorf, C. Thilgen, D. Weinstien, *Angew. Chem., Int. Ed. Engl.*, **1992**, *31*, 1599.
[11] D. M. Eichhorn, S. Yang, W. Jarrell, T. F. Baumann, L. S. Beall, A. J. P. White, D. J. Williams, A. J. Barrett, B. M. Hoffman, *J. Chem. Soc., Chem. Commun.*, **1995**, 1703.
[12] J. Canceill, A. Collet, G. Gottarelli, *J. Am. Chem. Soc.*, **1984**, *106*, 5997.
[13] J. L. Atwood, C. L. Raston, J. A. Ripmeester, unpublished results.
[14] A. Harada and S. Takahashi, *J. Chem. Soc., Chem. Commun.*, **1988**, 1352.
[15] J. L. Atwood and C. L. Raston, unpublished results.
[16] J. L. Atwood, C. L. Raston, I. B. N. Sudria, unpublished results.
[17] C. L. Raston and C. Sandoval, unpublished results.
[18] J. L. Atwood, S. G. Bott, C. Jones, C. L. Raston, *J. Chem. Soc., Chem. Commun.*, **1992**, 1349.
[19] W. Jorgenson, D. Steverance, *J. Am. Chem. Soc.*, **1990**, *112*, 4768.

[20] C. D. Gutsche, A. E. Gutsche, A. I. Karaulov, *J. Incl. Phenom*, **1985**, *3*, 447.
[21] H.-B. Burgi, R. Restori, D. Schwarzenbach, *Acta Crystallogr. Sect. B.*, **1993**, *49*, 832, and references therein.
[22] A. Drljaca, C. Kepert, L. Spiccia, C. L. Raston, and T. D. Smith, submitted to *Chem. Commun.*
[23] A. Beedy, J. Eatoe, and R. K. Heetan, *J. Chem. Soc., Chem. Commun.*, **1994**, 173.
[24] R. J. Blanch, M. Williams, G. D. Fallon, M. G. Gardiner, R. Kaddour, and C. L. Raston, submitted to *Angew Chemie.*

MOLECULAR SWITCHES BASED ON MOLECULAR INCLUSION

David N. Reinhoudt*, André M. A. van Wageningen, Bart-Hendrik Huisman
Laboratories of Supramolecular Chemistry and Technology
University of Twente
P.O. Box 217, 7500 AE Enschede, The Netherlands

Abstract

The synthesis of calix[4]arene-based carceplexes via two different methods is described, as well as a method for modifying the behavior of guests after incarceration. A new approach for the development of a molecular switch is presented.

1. Introduction

The developments in microelectronics and dataprocessing during the past twenty years have raised the demand for increasingly larger data storage capacity. The smallest "storage device" is one molecule, a so-called molecular switch. This device can be defined as a molecule which under external influence may switch to a meta-stable state. The lifetime relevant to qualify as meta-stable is to be defined by the application. The molecular switch is required to have distinct (and different) properties in the two states

A. W. Coleman (ed.), Molecular Recognition and Inclusion, 67–75.
© 1998 Kluwer Academic Publishers. Printed in the Netherlands.

permitting a "read" operation to determine the state of the switch. Not only this property is important but the switch must be organized in an array of such devices in time and in position. We are approaching such molecular switches on the basis of our current work on molecular inclusion properties of calix[4]arene-based carceplexes and on our work on self-assembly of resorcin- and calixarenes on gold surfaces. The results of both lines of research will be discussed. First, the synthesis of calix[4]arenes carceplexes and their stereo-chemical properties is discussed. Secondly, the self-assembly of such molecules on gold will be treated. In particular the relation between the headgroup and the supporting alkyl chains and the characterization of these monolayers will be reported. Finally, a method to address these arrays of molecular switches individually will be discussed.

2, Calix[4]arene-based Carcerands

Cram *et al.* [1] have shown that resorcin[4]arene-based carcerands can permanently incarcerate guest molecules. Although incarcerated guests can adopt different orientations this does not lead to different stereoisomers due to the symmetry of the carcerand. Combination of calix[4]arenes with resorcin[4]arenes leads to calix[4]arene-based carcerands **2** which possess a non-symmetric cavity [2]. Therefore different orientations of incarcerated guests lead to different diastereoisomers. This makes these molecules of interest because of their potential use as molecular switches. Due to the flexibility of the calix[4]arene skeleton direct coupling between a tetra-functionalized calix[4]arene and a tetrol cavitand does not result in the formation of a calix[4]arene-based carcerand. Therefore, a new method for the introduction of amino groups from iodo-substituted calix[4]arenes was developed [3]. Reaction of 1,2-bis(chloroacetamido)-3,4-dinitrocalix[4]arene with tetrol-resorcin[4]arene predominantly leads to an 1:1 *endo* coupled product. This preference for the *endo* orientation is probably a result of electrostatic interactions between the nitro groups on the calix[4]arene and the hydroxyl groups on the resorcin[4]arene. The 1:1 coupled product is converted into **1** via reduction of the remaining nitro groups and reaction with chloroacetyl chloride. This compound can be used for the synthesis of calix[4]arene-based carceplexes either via *solvent* or *doped* inclusion (see Scheme 1).

2.1. SYNTHESIS VIA SOLVENT INCLUSION

One method for the synthesis of calix[4]arene-based carceplexes comprises the formation of the final two bridges in an appropriate solvent (method A). During this reaction one solvent molecule is permanently incarcerated. Only highly polar solvents such as amides and sulfoxides can be used, *e.g.* DMF and ethyl methyl sulfoxide, respectively. Increasing the size of the solvent going from DMF to 1,5-dimethyl-2-pyrrolidinone leads to a dramatic decrease of the yield from quantitative to only 5%, respectively [4].

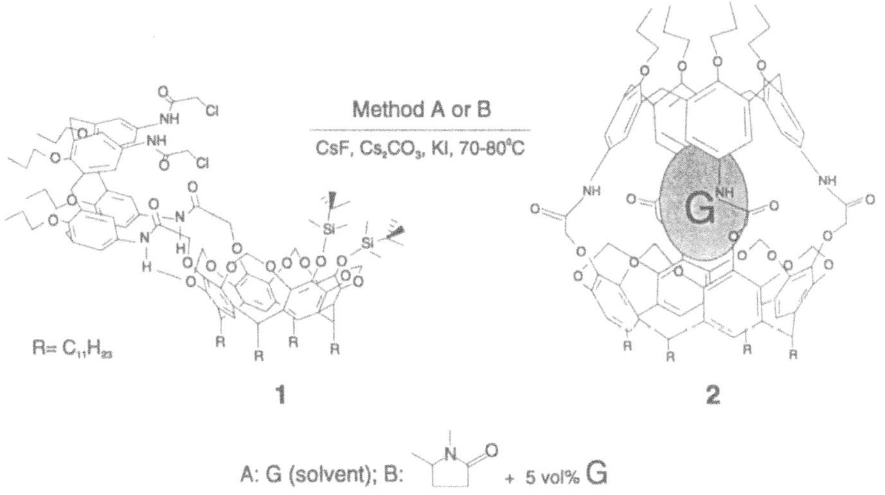

Scheme 1

2.2. SYNTHESIS VIA DOPED INCLUSION

The possibility to use only highly polar solvents for the synthesis of calix[4]arene-based carceplexes limits the number of potential guests. Therefore we developed a synthesis method called *doped* inclusion (method B) [4]. By using 1,5-dimethyl-2-pyrrolidone as a solvent, which itself is a bad template for the closure reaction, and adding potential guests, *e.g.* 2-butanone or 3-sulfolene (5-10 vol%) the carceplex with the added guest is exclusively formed. The ability to synthesize carceplexes via doped inclusion enlarges the scope of potential guests to solids, *e.g.* 3-sulfolene, and deuterated guests, *e.g.* DMF-d_7 and DMSO-d_6. The incarceration of deuterated guests allows investigation of the rotational behavior of guests inside calix[4]arene-based based carcerands using solid state ^{2}H NMR spectroscopy.

2,3. MODIFICATION OF THE BRIDGES AFTER INCARCERATION OF GUESTS

The orientation of the guests inside the carcerand was studied by 2D NOESY and 2D ROESY NMR spectroscopy. With exception of *N,N*-dimethyl acetamide and *N*-methyl-2-pyrrolidone the guests either rotate fast on the NMR chemical shift timescale (DMF, DMSO) or preferably adopt one orientation in a temperature range of -40° to 100° C (ethyl methyl sulfoxide, 2-butanone). Therefore it is interesting to have a method for changing the behavior of guests after incarceration. An apparent position for modification are the amide bridges between the resorcin[4]- and calix[4]arene moiety. The amide bridges in carceplexes **2** could be converted into thioamides in essentially quantitative yields using Lawessons reagent in refluxing xylene giving thiacarceplexes **3** [4]. The incarcerated guests *do not react* which means that they are not reactive under the reaction conditions. This indicates that the guests are very well shielded. The energy barriers ($\Delta G^{\neq}$) for the interconversion between the different diastereoisomers were determined by 2D EXSY NMR spectroscopy. The results are summarized in table 1.

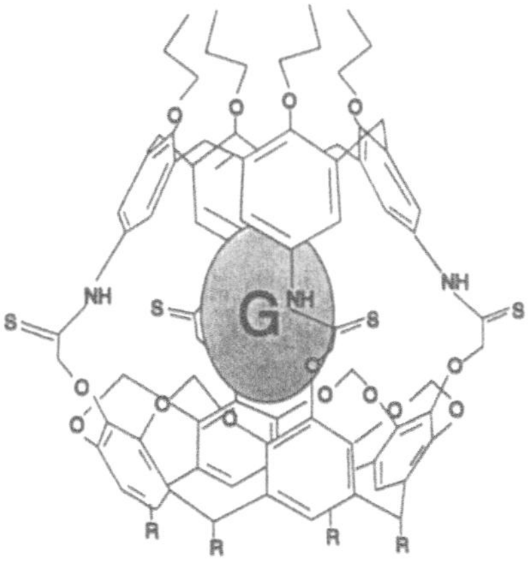

3

Chart 1

TABLE 1. Energy barriers ($\Delta G^{\neq}_{273}$) for interconversion between different diastereoisomers of guests inside calix[4]arene-based carcerands determined by 2D EXSY NMR spectroscopy in $CDCl_3$ at 273 K and distance between diametrical N*H* atoms determined using molecular mechanics calculations.

Guest	Bridge	$\Delta G^{\neq}_{273}$ (kcal/mol)	d (NH..HN) (Å)
N,N-Dimethyl acetamide	Amide	12.7 ± 0.5	9.29 ± 073
	Thioamide	15.2 ± 0.5	8.56 ± 0.24
N-Methyl-2-pyrrolidinone	Amide	15.7 ± 0.5	9.60 ± 0.69
	Thioamide	17.5 ± 0.5	8.85 ± 0.51

It is clear that the conversion of the amide bridges into thioamides results in an increase of the energy barriers for interconversion by 1.8-2.5 kcal mol^{-1}. An explanation for this behavior may be the stronger hydrogen bond donating character of thioamides compared to amides. Furthermore, molecular modelling studies revealed that in the calix[4]arene-based carceplexes with thioamide bridges the distance between the diametrical N*H* atoms (across the cavity) is significantly smaller [5].

3. Self-Assembled Monolayers on Gold

Self-assembled monolayers have been investigated extensively during the past decade [6], especially in view of the potential applications, like in piezoelectrical devices, nonlinear optics, and microsensors [7]. Previously, highly-ordered monolayers were obtained of resorcin[4]arene [8] and calix[4]arene [9] derivatives, substituted with four dialkylsulfides for the adsorption to gold as depicted in Chart 2. The driving force for the formation of a monolayer is the Au-S interaction, while van der Waals forces between the alkyl chains provide a high degree of order. Essential in the design of the adsorbates is that the alkyl chains cover a space which is comparable in size to the area covered by the 'arene' unit [10].

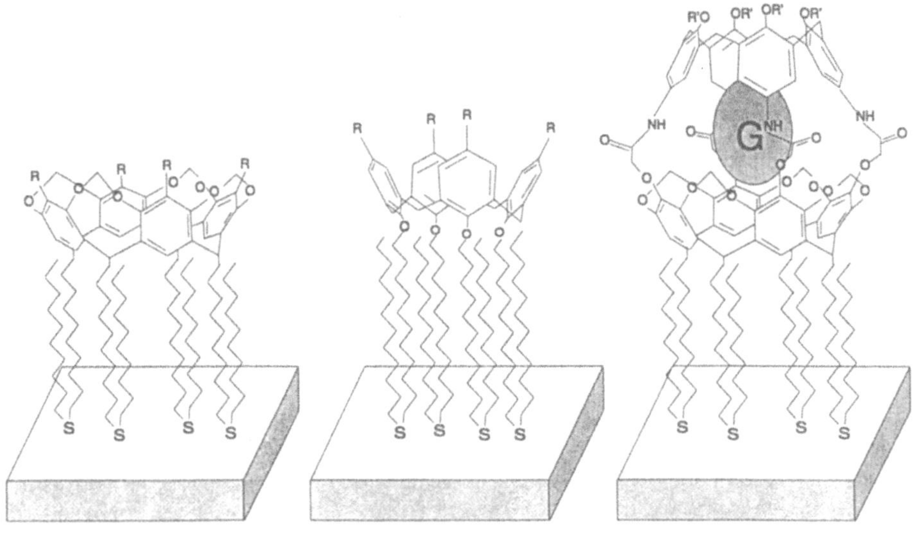

4 5 6

Chart 2

3.1. MONOLAYERS OF CARCEPLEXES ON GOLD

The experience with self-assembled monolayers was recently combined with the calix[4]arene-based carcerands to form highly-ordered thin films. The carcerand can be oriented in space in a controlled manner by assembling this compound in a monolayer on gold, similar to adsorbates **4** and **5**. Thus the orientation of the carcerand is fixed, while the guest remains free to be switched independently between two different orientations. As appeared from the systematic study of guests described before, different energy barriers between the two states can be obtained, depending on the size of the cavity and of the incarcerated guest. Self-assembled monolayers of carceplex **6** (synthesized via method A with DMF as guest as shown in Scheme 1, R = $C_{10}H_{20}$-S-$C_{10}H_{21}$) were prepared from a 0.1 mM solution in ethanol on gold [11]. The monolayers were characterized by a number of surface sensitive techniques (like FTIR, ellipsometry, contact angle measurement, XPS, and SIMS). Well-ordered monolayers were obtained, for which a structure is proposed as depicted in Chart 2.

Convincing data about the order and orientation of the carceplex **6.DMF** is obtained from the monolayer infrared spectrum (Table 2). The methylene C-H stretches have the same absorption maxima in the monolayer as in the bulk, which indicates a high degree

TABLE 2. Selected infrared peak positions for carceplex 2 in the bulk and the monolayer.

vibration	bulk [cm^{-1}]	monolayer [cm^{-1}]
-CH_2-, symmetrical	2853	2853
-CH_2-, asymmetrical	2924	2926
Amide I	1696	n.o.
Amide II	1526	1537

of order. The amides in the carceplex appear in the bulk infrared spectrum as amide I (predominantly C=O stretch) and amide II. In the monolayer spectrum, however, only amide II appears at 1537 cm^{-1} as a clear absorption, while amide I is very weak or absent. The orientation of the transition dipole of amide I is in the proposed monolayer structure parallel to the gold, and therefore is not absorbed by the (p-polarized) infrared beam. This confirms that the orientation of the carceplex **6.DMF** is 'upright', similar to the previously reported adsorbates.

3.2. SWITCHING OF CARCEPLEXES IN A MONOLAYER

The two positions of the incarcerated guest provide in principle the possibility to switch between these two states. Currently, the possibilities to determine and affect these orientations are under invesitgation. Under influence of an external electrical field the dipoles of the guest will be alligned. This will result in a change of the local electrical field, that can be detected using the extremely sensitive SQUID (super-conducting quantum interference device). Thus, the lateral resolution for switching will be limited to the abillity to control the size of the measuring device, which is estimated to be in the submicron range now. An even higher resolution may be obtained using scanning probe techniques, like AFM and STM, which ultimately leads to molecular resolution.

4. Summary

It has been shown that calix[4]arene-based carceplexes can be obtained *via* solvent or doped inclusion. Furthermore, the behavior of incarcerated guests can be altered after incarceration by modification of the bridges between the resorcin[4]- and calix[4]arene moiety. Self-assembled monolayers are an elegant tool to control molecular orientation. The calix[4]arene-based carceplex was confined in space by means of self-assembly to gold. This approach is a step toward a new type of molecular switch.

References and notes

1. D. J. Cram and J. M. Cram, *Container Molecules and their Guests* in *Monographs in Supramolecular Chemistry Vol. 4*, ed. J. F. Stoddart, Royal Society of Chemistry, Cambridge (1994).

2. Timmerman, P.; Verboom, W.; van Veggel, F. C. J. M.; van Duynhoven, J. P. M.; Reinhoudt, D.N. *Angew. Chem. Int. Ed. Engl.* **1994**, *33*, 2345-2348.

3. Timmerman, P.; Verboom, W.; Reinhoudt, D. N.; Arduini, A.; Grandi, S.; Sicuri, A. R.; Pochini, A.; Ungaro, R. *Synthesis* **1994**, 185-189.

4. van Wageningen, A. M. A.; van Duynhoven, J. P. M.; Verboom, W.; Reinhoudt, D. N. *J. Chem. Soc., Chem. Commun.* **1995**, 1941-1942.

5. Molecular mechanics calculations were carried out using the CHARMm force field incorporated in Quanta.

6. For a recent review see: Dubois, L. H. ; Nuzzo, R. G. *Annu. Rev. Phys. Chem.* **1992**, *43*, 437-463.

7. Ulman, A. *An Introduction to Ultrathin Organic Films;* Academic Press: San Diego, CA, 1991.

8. Thoden van Velzen, E. U.; Engbersen, J. F. J.; Reinhoudt, D. N. *J. Am. Chem. Soc.* **1994**, *116*, 3597-3598.

9. Huisman, B.-H.; Thoden van Velzen, E. U.; van Veggel, F. C. J. M.; Engbersen, J. F. J.; Reinhoudt, D. N. *Tetrahedron Lett.* **1995**, *36*, 3273-3276.

10. Thoden van Velzen, E. U.; Engbersen, J. F. J.; de Lange, P. J.; Mahy, J. W. G.; Reinhoudt, D. N. *J. Am. Chem. Soc.* **1995**, *117*, 6853-6862.

11. Huisman, B.-H.; Rudkevich, D. M.; van Veggel, F. C. J. M.; Reinhoudt, D. N. *J. Am. Chem. Soc.* **1996**, *118*, 3523-3524.

FLUORESCENT CYCLODEXTRINS AS CHEMOSENSORS FOR MOLECULAR RECOGNITION

AKIHIKO UENO
Department of Bioengineering, Faculty of Bioscience and Biotechnology, Tokyo Institute of Technology, 4259 Nagatsuta-cho, Midori-ku, Yokohama 226, Japan

1. Introduction

Cyclodextrins (CDs) with one or two chromophores can act as chemosensors for molecular recognition [1-5]. In most cases, they form self-inclusion forms and the chromophore included in the cavity is excluded upon accommodation of a guest molecule, thus leading to environmental change around the chromophore from the hydrophobic cavity to water solution. When the chromophore is a fluorescent moiety, this induced-fit type of complexation results in variation of the fluorescence spectra. We have constructed various guest-responsive CD systems which have naphthalene, pyrene, dansyl, or 4-dimethylaminobenzoyl moiety as a fluorophore [1,2]. Here much attention is paid to the effects of environmental unit incorporated into these sensor systems (Figure 1). One example is [CD]-[L- or D-leucine]- [fluorophore (dansyl)] combined system where leucine is the amino acid residue with a hydrophobic isobutyl side chain and acts as a chiral hydrophobic environment. We observed different sensitivities and binding constants between these L- and D-leucine-incorporated systems. We also constructed [CD]-[fluorophore(dansyl)]-[monensin] system where monensin is an antibiotic hydrophobic residue and changes its structure to cyclic form in the presence of sodium ion. The sensitivity and binding abilities of this system changed by sodium ion. The other example is protein-bound system of [CD]-[fluorophore]-[biotin] where avidin is bound to biotin residue and the fluorescence of 4-N,N-dimethylaminobenzoyl moiety was greatly enhanced by the presence of avidin.

A. W. Coleman (ed.), Molecular Recognition and Inclusion, 77–84.
© 1998 *Kluwer Academic Publishers. Printed in the Netherlands.*

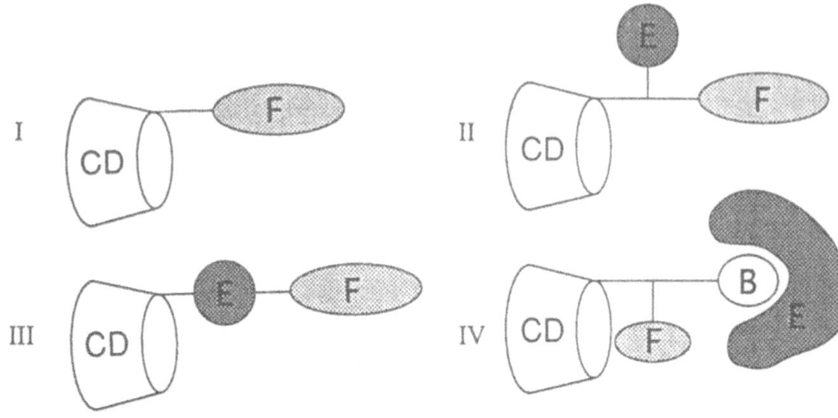

Figure 1. Various CD-based fluorescent sensors. F: fluorophore; E: environmental unit; B: protein binding unit. The system I represents simple fluorescent sensors. The systems II and III represent leucine- and monensin-incorporated ones, respectively. The system IV represents the protein-bound sensor in which E is protein.

2. Dansylleucine-modified CDs

Dansyl unit is a good fluorescence probe because its fluorescence intensity is strong in hydrophobic environment whereas it is weak in polar water environment. On this basis, we have reported some simple CD systems with a dansyl moiety as sensors of molecular recognition. These systems have a dansyl- or dansylglycine moiety and actually worked as sensors for molecule detection. In many cases, the dansyl moiety forms self-inclusion complexes in which the moiety is included in its own CD cavity, showing the strong fluorescence emission. Upon addition of guest species, the dansyl moiety is excluded from the CD cavity to polar water medium associated with accommodation of guest species, decreasing the fluorescence intensity (Figure 2, (1)).

We have recently prepared L- and D-dansylleucine-modified β-CDs (**1**,**2**) and γ-CDs (**3**,**4**) and examined the effects of the hydrophobic leucine unit and its chirality difference on the binding constant as well as on the sensitivity as sensors, using dansylglycine-modified β- and γ-CD (**5**,**6**) as reference hosts [6]. We used N-dansyl-L-

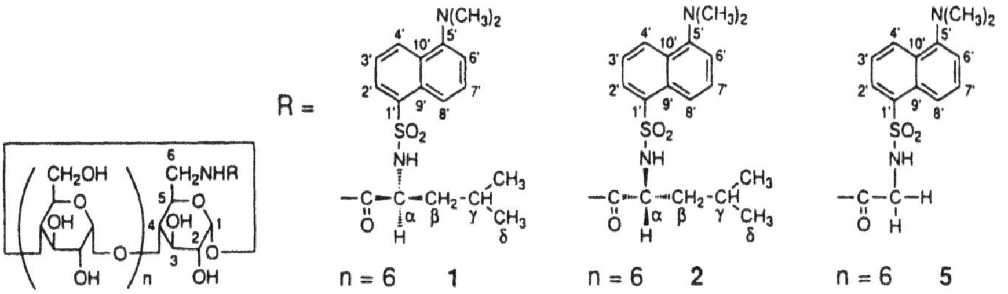

n = 6 **1** n = 6 **2** n = 6 **5**

n = 7 **3** n = 7 **4** n = 7 **6**

leucine (**7**) as the reference compound for various spectroscopic measurements of **1**-**6**. The maximum wavelengths of the fluorescence spectra of **1**-**7** are 543, 540, 550, 553, 546, 556, and 571 nm, respectively, and the order of the intensities is **2**>**1**>**5**>**4**>**3**>**6**>**7**, reflecting the order of the hydrophobic nature around the dansyl moiety. Table 1 shows binding constants of **1**-**6** for various guests.

TABLE 1. Binding constants of various guests of **1-6** in aqueous solution at 25° a)

	K / M^{-1}					
Guest	**1**	**2**	**3**	**4**	**5**	**6**
8	77100	4100	6250	24600	37600	4110
9	21900	11800	5620	16100	10200	2660
10	2660	1520	5600	9420	1030	658
11	1650	588	320	1960	60.4	b)
12	59300	20400	b)	b)	9630	b)
13	54200	19900	116	370	8910	b)
14	1740	860	27.1	43.5	540	29.3
15	6110	2450	b)	b)	1700	76.2

a) In phosphate buffer (pH 7.0) for **8-12** and in aqueous solution for other guests.

b) The values are too small to be accurately determined.

Chart 1

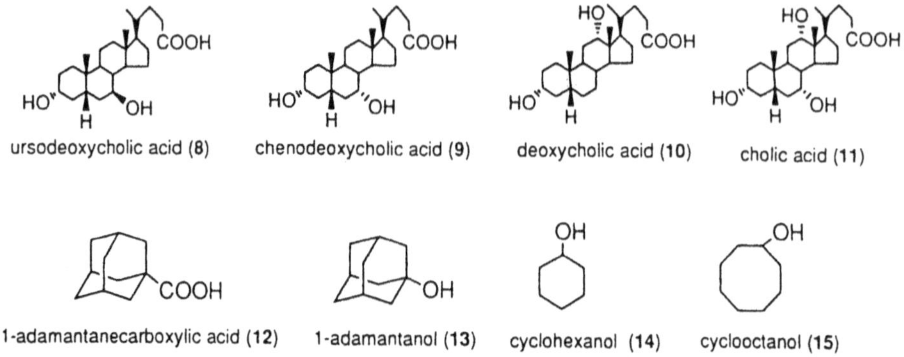

The binding constants were obtained from the guest-induced variations of the fluorescence intensity. The binding constants of the β-CD derivatives are in the order of **1**>**2**>**5**, suggesting that the hydrophobic leucine moiety enhances the binding abilities of **1** and **2** and the L-leucine moiety is more preferable for stronger guest binding than the D-leucine one. The analysis of NMR measurements indicates that the leucine moiety of **2** is more deeply included in the CD cavity than **1**, forming a stable self-inclusion complex which is unfavorable for accommodation of a guest molecule into the cavity. In contrast to the larger binding ability of L-leucine-incorporated β-CD derivative (**1**) for **1** and **2**, γ-CD derivatives **3** and **4** exhibit larger binding constants for D-leucine-incorporated one (**4**). All these dansyl-modified CDs, including **5** and **6** exhibit remarkable molecular recognition abilities as shown by the order of binding constants of steroidal guests **8**>**9**>**10**>**11**. It is noted that γ-CD derivatives exhibit guest-induced enhancements due to co-inclusion of the dansyl moiety and the guest in the large γ-CD cavity (Figure 2, (2)) when small compound such as cyclohexanol is used as a guest species.

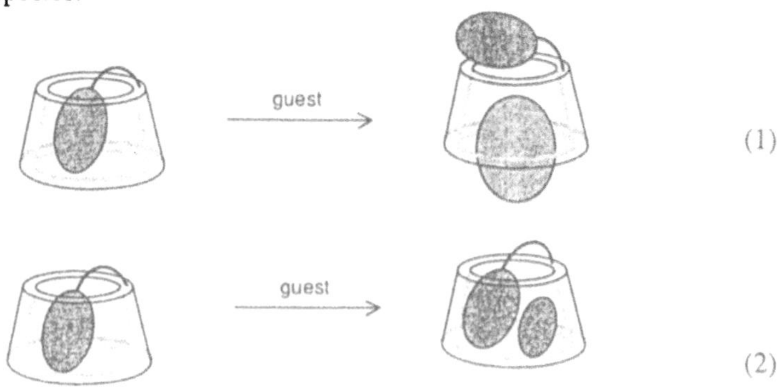

Figure 2. Induced-fit guest binding of dansyl-modified CDs.

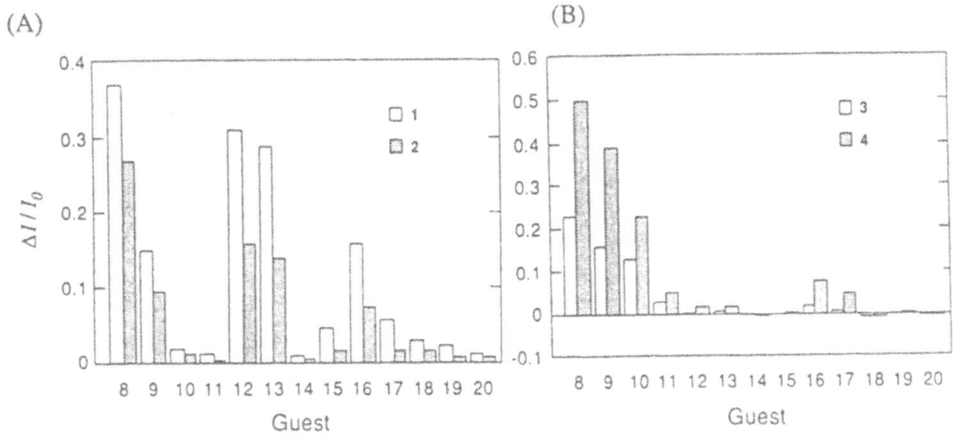

Figure 3. Sensitivity parameters ($\Delta I/I^{\circ}$) of 1 and 2 (A) and 3 and 4 (B). Guest compounds 8-15 are shown in Chart 1. 16, (-)-borneol; 17, (+)-camphor; 18, (-)-menthol; 19, nerol; 20, geraniol. [1-4]=2 x 10^{-6} M, [guest]=(A) 10^{-5} M, (B) 10^{-4} M. λ_{max}=370 nm.

Figure 3 shows the sensitivity parameters $\Delta I/I_0$, where $\Delta I=I_0 - I$ and I and I_0 are fluorescence intensities in the presence and absence of guest species. The sensitivities are almost parallel to the binding constants.

3. Dansyl-modified CDs with a distorted cavity

All these have a dansyl fluorophore at the primary hydroxyl side of CD, however, we have recently prepared CD derivatives which have a dansyl fluorophore at the secondary hydroxyl side of β- and γ-CD, by the reaction between 3-deoxy-3-amino-CD and dansyl chloride or dansylglycine. Since the synthesis of the amino-CD includes the tosylation of the C-2 hydroxyl group followed by the conversion to 2,3-mannoepoxide which is then opened by ammonia through the attack to C-3, the stereochemistry of the amino-CD should be different from the usual one, with one altrose unit being incorporated into the CD ring. We employed molecular dynamic approach to get insight into the structural features of the altrose-incorporated CDs and the results are shown in Figure 4. The obtained structures are all distorted and the cavity is extremely narrowed for α-CD and β-CD cases, and only γ-CD still has the cavity to accommodate a guest species. We examined variations of fluorescence intensity of C-3 dansyl (or dansylgycine)-modified β- and γ-CDs and found that the β-CD derivatives are unfavorable to accommodate guest species and γ-CD derivatives bind guest species weakly, being in consistent with the results of molecular dynamics [7].

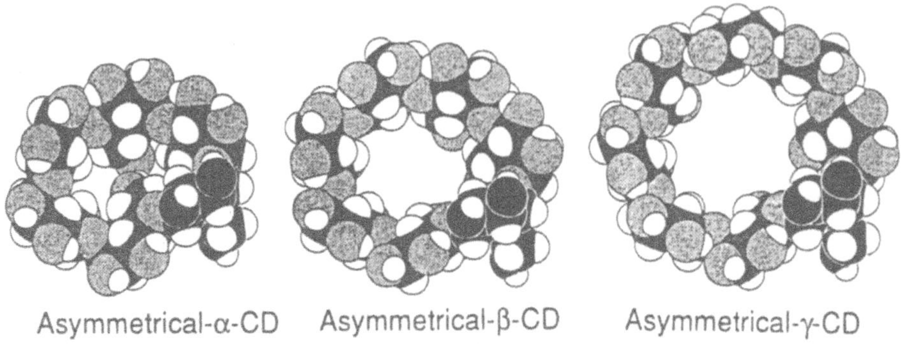

Figure 4. The structures of altrose-incorporated CDs obtained by molecular dynamics. The carbons of the altrose part is shown in black.

4. Monensin-Modified Fluorescent CD

Monensin is antibiotic compound and it has the ability of transporting sodium ion through biomembranes, forming cyclic structure with a sodium cation in the center. We have prepared dansyl-modified β-CD **21**, which has a monensin residue connected with ω-amino group of incorporated lysine as a hydrophobic and metal ion-responsive environmental moiety [8].

21:

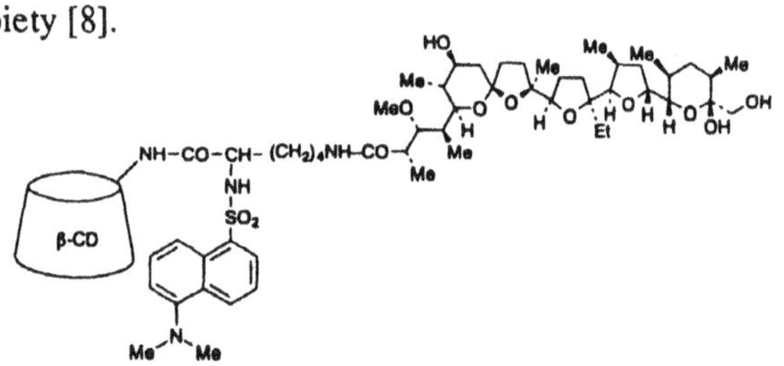

Table 2 shows binding constants of monesin-dansyl-modified β-CD **21** in the absence and presence of sodium ion and dansylglycine-modified β-CD (**22**). The guest alcohols gave larger binding constants for **21** than for **22**, particularly in the cases of acyclic monoterpenes such as nerol (64-fold) and geraniol (92-fold). The results indicate that the monensin acts as an effective hydrophobic cap, enhancing the binding ability of the host. The effect of sodium cation (1 M) is also remarkable, giving 2-fold or much larger binding constants.

TABLE 2. Binding constants of monensin-dansyl-modified β-CD **21** and dansylglycine-modified β-CD **22** in aqueous solution

	K / M^{-1}		
Guest	**21**	**21** + Na^{+}(1M)	**22**
Nerol	5370	10300	83.3
Geraniol	6320	11900	68.4
Cyclohexanol	2250	1580	540
Cyclooctanol	3310	9010	1700

The sensitivities of these systems of these system indicated as $\Delta I/I_0$ are also enhanced by the presence of sodium ion, for example, showing 2.3-fold larger value of $\Delta I/I_0$ for cyclohexanol.

5. Fluorescent CD with protein as an the environmental factor.

We have already shown that the modified CDs with a fluorophore which exhibits twisted intramolecular charge transfer (TICT) fluorescence can be used as effective fluorescent chemosensors [7].

On this basis, we have recently prepared a CD system **23** which has both the fluorophore (4-N,N-dimethylaminobenzoyl unit) and protein (avidin) -binding site (biotin) [9].

23:

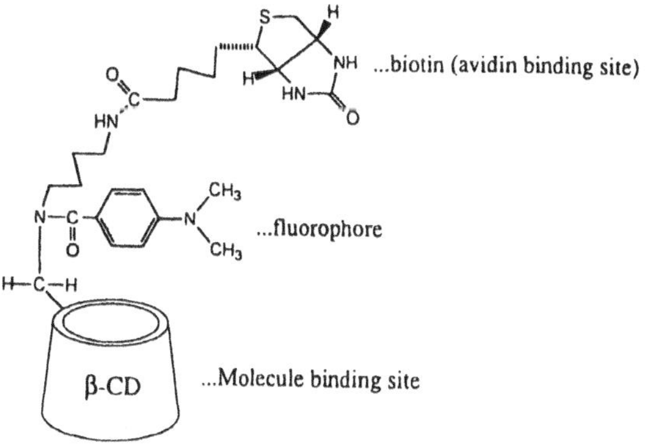

The CD derivative predominantly exhibits normal fluorescence from nonpolar planar (NP) excited state around 375 nm with weak TICT emission around 500 nm. We found that the NP fluorescence is markedly enhanced (8.4-fold) by the presence of equimolar

avidin. In the absence of avidin, in many cases, the intensity of the NP fluorescence decreases with increasing concentration of the guest species. However, we found that the intensity increases in the presence of avidin. The binding constants of various guests also increased in the presence of avidin. These results suggest that proteins can be very useful as the environment to construct chemosensors highly responsive to guest molecules.

References

1. Ueno, A. (993) Fluorescent sensors and color-change indicators for molecules, *Adv. Mat.* **5**, 132-134.
2. Ueno, A. (1996) Fluorescent cyclodextrins for molecule sensing, *Supramol. Sci.* **3**, 31-36.
3. Ueno, A., Suzuki, I., and Osa, T. (1990) Detecting organic compounds by pyrene excimer fluorescence, *Anal. Chem.* **62**, 2461-2466.
4. Ueno, A., Kuwabara, T., Nakamura, A., and Toda, F. (1992) Modified cyclodextrins as a guest-responsive color-change indicator, *Nature* **356**, 136-137.
5. Hamasaki, K., Ikeda, H., Nakamura, A., Ueno, A., Toda, F., Suzuki, I., and Osa, T. Fluorescent sensors of molecular recognition. Modified cyclodextrins capable of exhibiting guest-responsive twisted intramolecular charge transfer fluorescence, *J. Am. Chem. Soc.* **115**, 5035-5040.
6. Ikeda, H., Nakamura, M., Ise, N., Oguma, N., Nakamura, A., Ikeda, T., Toda, F., and Ueno, A. Fluorescent cyclodextrins for moelcule sensing: Fluorescent properties, NMR characterization, and inclusion phenomena of N-dansyl-leucine-modified cyclodextrins, *J. Am. Chem. Soc.* in press.
7. Hamasaki, K., Usui, S., Ikeda, H., Ikeda, T., and Ueno, A. Dansyl-modified cyclodextrins as fluorescent chemosensors for molecular recognition, *Supramol. Chem.*, in press.
8. Nakamura, M., Ikeda, A., Ise, N., Ikeda, T., Toda, F., and Ueno, A. (1995) Dansyl-modified β-cyclodextrin with a monensin residue as a hydrophobic, metal responsive cap, *J. Chem. Soc., Chem. Commun.* 721-723.
9. Wang, J., Nakamura, A., Hamasaki, K., Ikeda, H., Ikeda, T., and Ueno, A. (1966) A fluorescent molecule-recognition sensor with a protein as an environmental factor, *Chem. Lett.* 303-304.

TETRATHIAFULVALENES IN MACROCYCLIC AND SUPRAMOLECULAR CHEMISTRY:

SELF ASSEMBLY WITH TETRATHIAFULVALENES

JAN BECHER*, ZHAN-TING LI, PHILIPPE BLANCHARD, NIELS SVENSTRUP, JESPER LAU, MOGENS BRØNDSTED NIELSEN AND KLAUS B. SIMONSEN.

Department of Chemistry, Odense University, 5230 Odense M, Denmark

Abstract: We describe that use of stepvise selective protection-deprotection of tetrathiafulvalene (TTF)-thiolates give rise to either two or three-dimensional macrobicyclic tetrathiafulvalenophanes in high yields by convergent syntheses. Configurationally selective self-assembly of cyclic tetrathiafulvalenes using the cyclic π-acceptor cyclo-bis(paraquat-*p*-phenylene) yields [2]- or [3]pseudocatenanes. A simple method for the preparation of donor-acceptor tetrathiafulvalenophanes is reported.

1. Tetrathiafulvalenes

Tetrathiafulvalenes (TTF's) and related heterocycles have been of high interest for more than two decades, due to their unique π-donor properties[1].

The unsubstituted TTF-molecule **1** is a planar non-aromatic 14-π system. Oxidation takes place sequentially in two discrete reversible steps with formation of ; *i*: a radical cation **2** and *ii*: the 2 x 6π dication **3**:

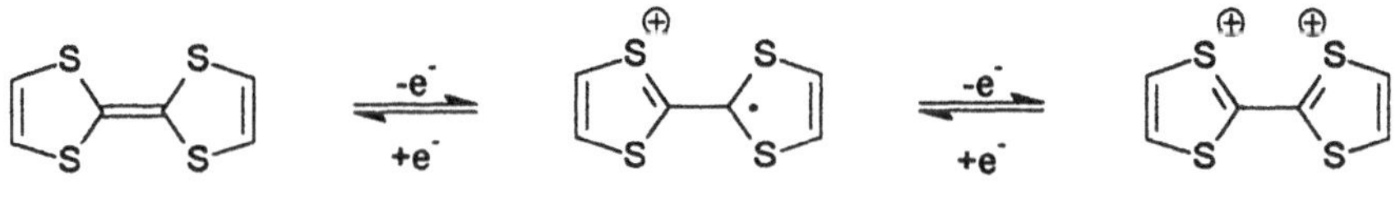

1 TTF **2** radical cation **3** dication

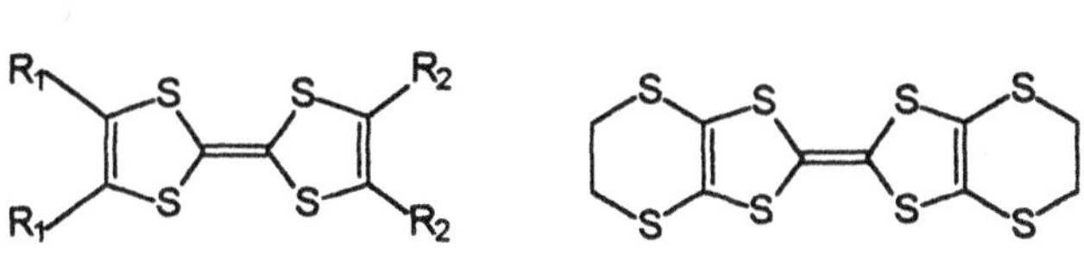

4 **5** (BEDT-TTF)

A. W. Coleman (ed.), Molecular Recognition and Inclusion, 85–95.
© 1998 *Kluwer Academic Publishers. Printed in the Netherlands.*

The redox potential for TTF is relative low: $E^1_{1/2} = 0.35$ and $E^2_{1/2} = 0.71V$ (sv Ag/AgCl, MeCN/TBAHFP). Change of substituents (R^1 or R^2) at the TTF system **4**, allows "fine tuning" of the redox properties. Among the large number of tetrathiafulvalenes which have been prepared, BEDT-TTF **5** is relative unique, it has given rise to more superconducting salts than any other TTF-derivative so far.

2. Conducting materials

In order to develop new conducting charge transfer complexes and salts based on tetrathiafulvalenes it is necessary to control both the redox potential of the donor and the electron affinity of the acceptor simultaneously. It is a fact that the control of the stoichiometry resulting in formation of partly ionized states between the donor and acceptor is very serendipitous. Therefore control of the 2- and 3-dimensional molecular architecture of macrocyclic and oligo tetrathiafulvalenes is of great current interest, because it can be expected that their derived charge transfer salts may reveal new types of molecular interactions.

3. Macrocyclic and supramolecular chemistry

It is exactly the facile and reversible formation of a stable radical cation that makes the tetrathiafulvalene system such an interesting building block, not only for molecular conductors but also as a π-donor in macrocyclic and supramolecular chemistry[2]. From the beginning of our reseach on tetrathiafulvalene chemistry we were well aware that new TTF-building blocks might be useful in many areas of materials chemistry, some examples where TTFs have been used are shown in the scheme.

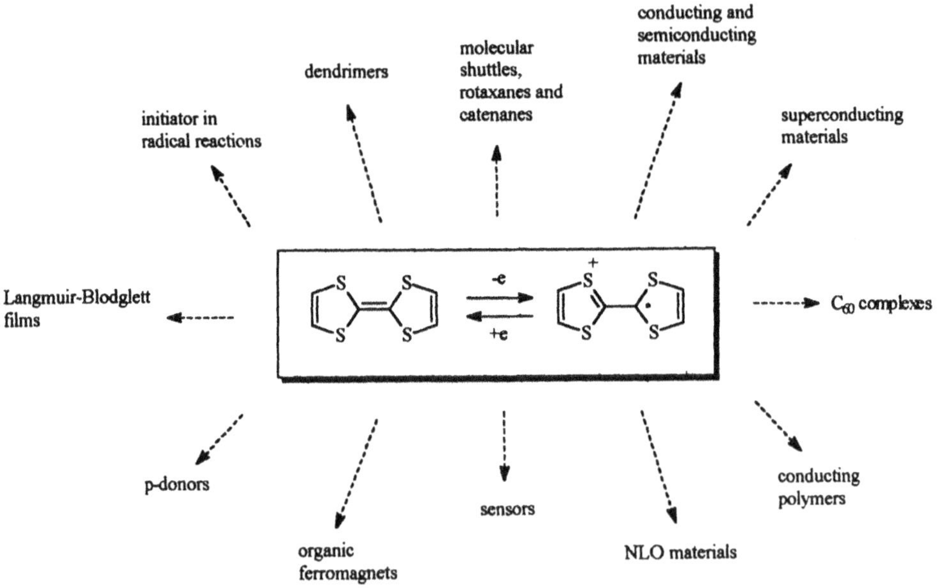

4. New TTF-building blocks

There are two standard routes to TTF-derivatives; *i*: coupling between two 1,3-dithiole derivatives and *ii*: functionalization of a preformed TTF-system according to the retrosynthetic scheme:

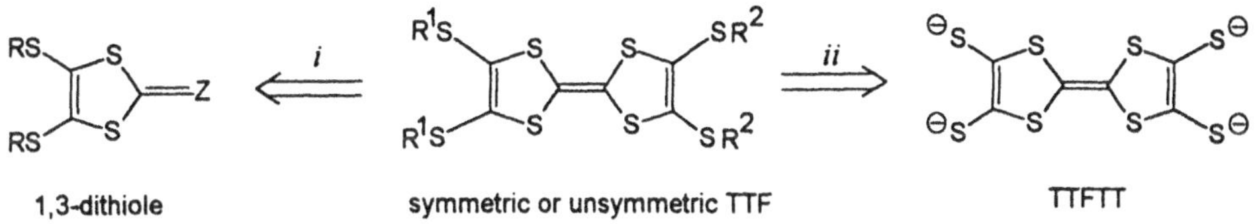

6 **7** **8**

It is a fact that coupling of 1,3-dithioles to tetrathiafulvalenes often gives rise to synthetic problems and furthermore that bis-1,3-dithioles show a strong propensity to give intramolecular coupling reactions rather than bis-TTF macrocycles. We therefore realised, that the most useful route to macrocyclic and oligo-TTFs should be based on a preformed TTF-group *i.e.* route *ii*. This clearly would require a stable and versatile protecting group R.

5. Protection and deprotection of TTFTT

In a number of papers we have reported an effective protection/deprotection methodology for preparation of TTFTT derivatives using the cyanoethyl group as a versatile protecting group for TTF thiolate groups[3]. The cyanoethyl group almost univesally gives soluble and well crystalline products, both in the case of the 1,3-dithioles **6** as well as for the TTFs **7**. Not only can the tetrathiafulvalene tetrathiolate (TTFTT) **8** be prepared and subsequently realkylated in almost quantitative yield, but it is also possible to carry out stepvise protection/deprotection reactions using caesium hydroxide as base in the cleavage step. Although TTF-thiolates must be handled under inert atmosphere when for example sodium methoxide/methanol is used for dealkylation, the corresponding TTF-caesium thiolates are normally stable in the air and can be isolated. As noted above deprotection takes place stepvise and use of either 1, 2 or in the case of TTFTT, 4 equivalents of caesium hydroxide (in dry DMF under N_2) followed by realkylation allows the preparation and subsequent construction of multible-TTF systems with both 2- and 3-dimensional structures, overall yields deprotection/realkylation are in the range 50 - 100%. The following schemes gives an overview of the various TTF-building blocks:

6. Unsymmetric TTFs

The unsymmetrical TTFs are prepared by cross coupling of a suitable 1,3-dithiole-2-thione with 4,5-bis(2'-cyanoethylthio)-1,3-dithiole-2-one.

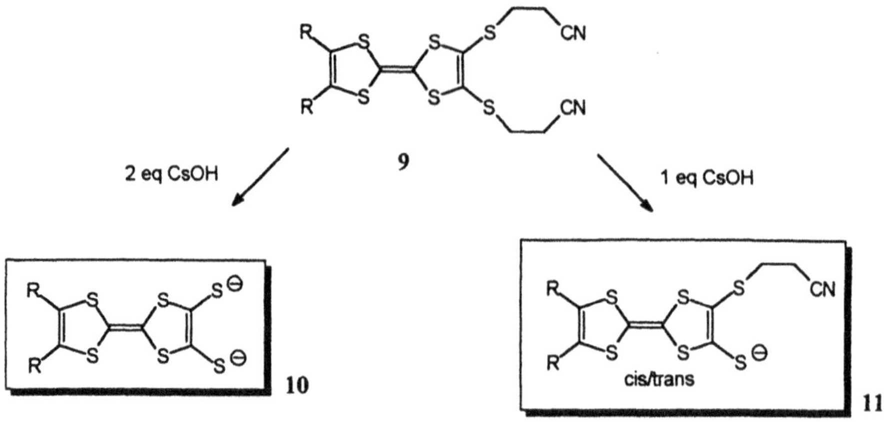

7. Mono- and bis-deprotection

Using the key starting material 2,3,5,7-tetrakis(2'-cyanoethylthio)tetrathiafulvalene **12** and a suitable amount of caesium hydroxide gives access to either TTFTT as well as bis thiolate **13** and the monothiolate **14.**

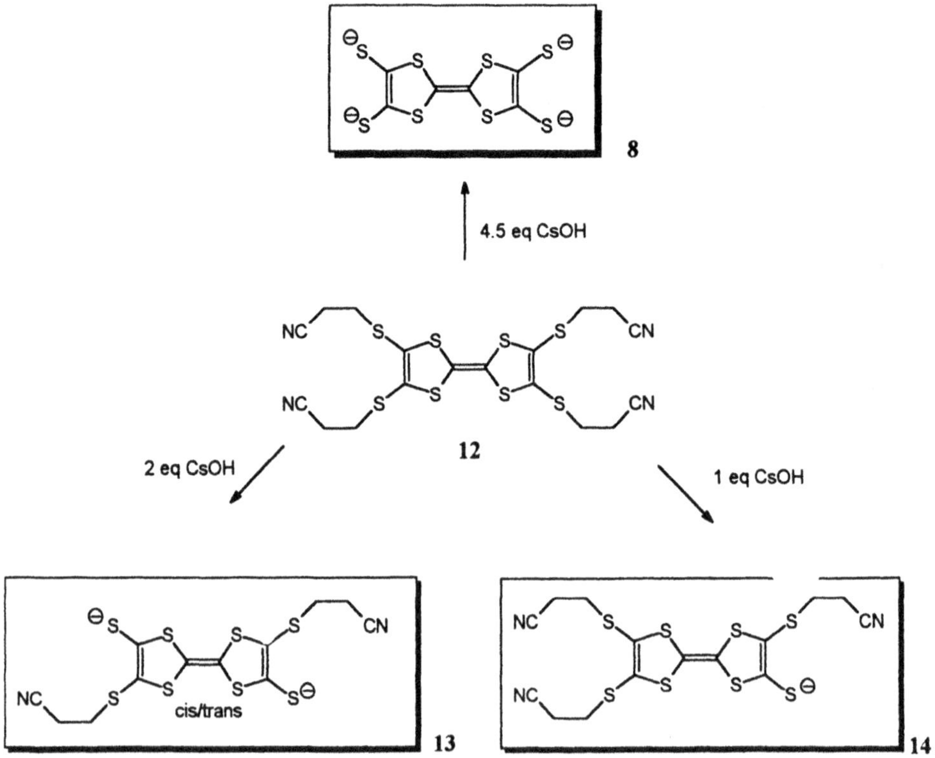

By suitable combination of these building blocks it has been possible to prepare a range of multible-TTF systems as described below.

8. Cyclic TTF-oligomers

The *charge-transfer* salt of the multi-donor derivatives will be of great interest because of the possibility of obtaining materials with higher dimensionality as mentioned previously. CPK models of the cyclic TTF-tetramer **16** showed that one possible conformation had a calix-like geometry that perhaps would be appropriate for the interaction with a variety of acceptors. The synthetic strategy using the TTF-building blocks described above enabled us to prepare compound **16** as well as a number of related oligomers. The yield are high in all steps as this synthetic strategy allows the sequential assembling of each synthon ensuring an almost complete control of the oligo-purity. The methodology employed could be extended to a variety of different spacer groups as well as different substituents. Exchange of the propylene units for example with small aromatic systems, methylene or acetylene units would give rise to more rigid macrocycles with different properties

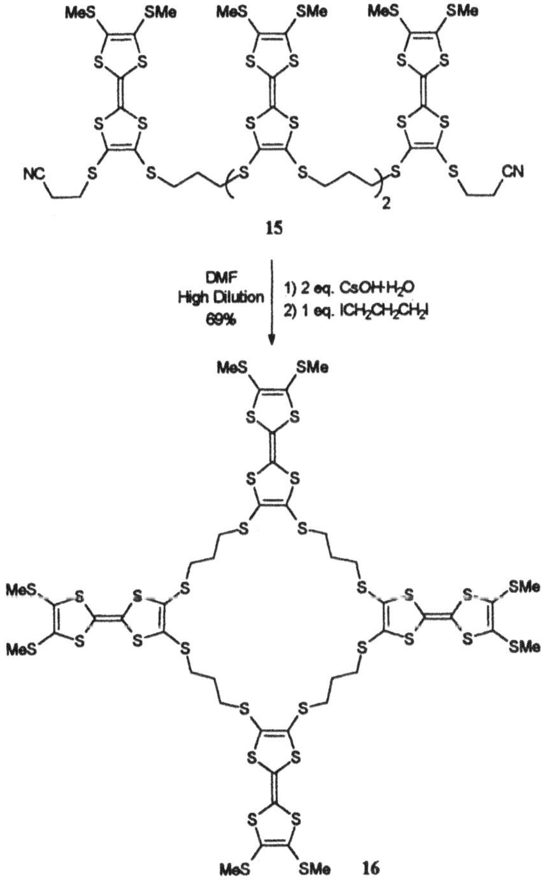

9. Catenanes

The formation of catenanes by self-assembly is controlled by the information stored in the "preprogramed" starting materials. The information, locked within the catenane molecules, also results in catenanes with properties quite different from those of their individual molecular components. Based on the donor-acceptor interactions, catenanes of different structures have been prepared from a variety of starting materials. By incorporating two dioxyphenylene units into a TTF-based dimacrocyclic system, we succeeded in obtaining both *cis* and *trans* [3]pseudocatenanes, whereas replacing the dioxyphenylenes with 9,10-dioxyanthrylenes resulted in the selective formation of the corresponding *trans* [3]pseudocatenane[4]. The example shown in the following scheme illustrate the selective self-assembling synthesis of a *cis* [3]pseudocatenane, using the dimacrocycle precursor **20**, in which the central TTF unit constitutes the bridge of the dimacrocycle while the peripheral TTF units control the configuration of the resulting catenane[5].

Preparation of compound **20** was chosen because the investigation of a CPK model suggested that only when the central TTF unit in [3]catenane **21** adopts a *cis* configuration, then the pheripheral TTF units will be able to form π-π donor-acceptor interactions with the two bipyridinium units. Such interaction are essential for the configurational control during the possible self-assembling process. The reaction scheme illustrate well the sequential assembly of the macrocyclic TTFs.

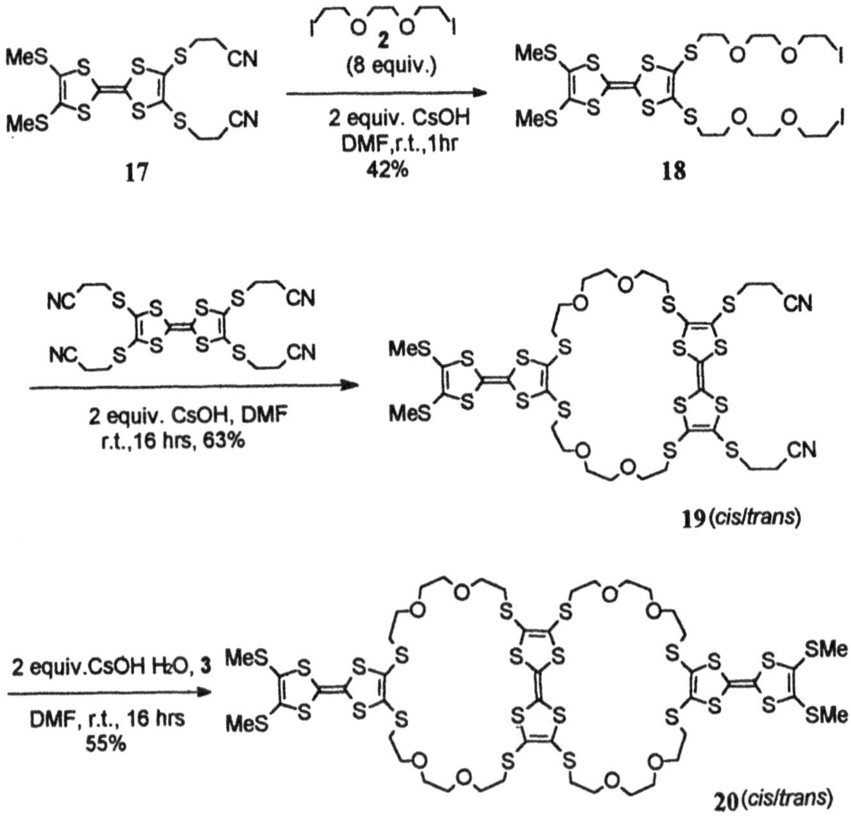

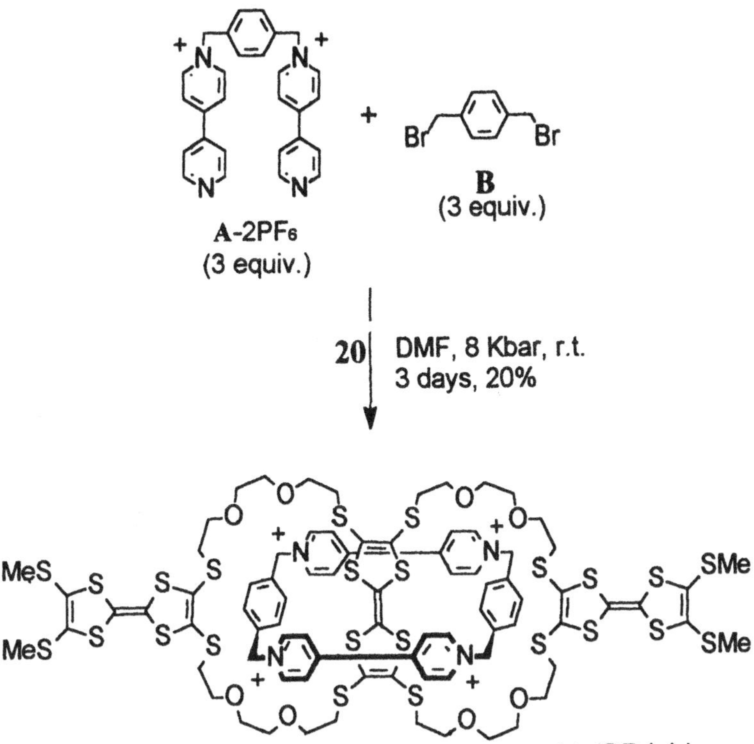

21-4PF6(*cis*)

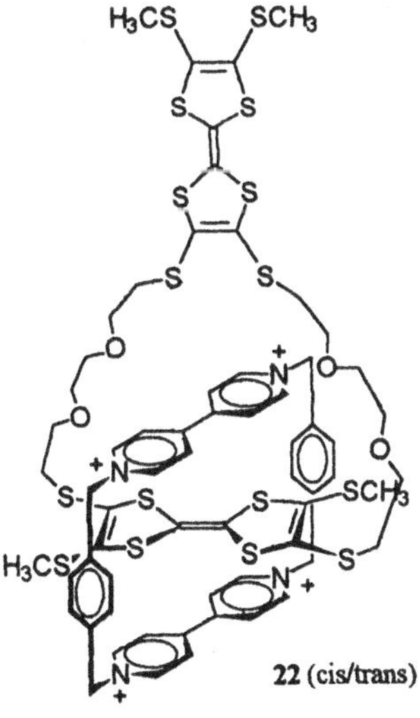

22 (cis/trans)

The two-step macrocyclizations using high dilution conditions (perfusor pump) afforded **20** via the intermediate **19** in a 35% overall yield. No catenane products were detected after stirring the solution of **20** in DMF or acetonitrile in the presence of **A**-$2PF_6$[2a] and **B** at room temperature for three weeks at atmospheric pressure. However under ultra-high pressure (8 Kbar), **21**-$4PF_6$ was obtained in 20% yield from the self-assembly of **20**, **A**-$2PF_6$, and **B** at room temperature in three days. Catenane **21**-$4PF_6$ was characterized by electrospray mass spectrometry, which exhibited ion peaks at m/z = 1160, 725, and 508, corresponding to $[M-2PF_6]^{2+}$, $[M-3PF_6]^{3+}$, and $[M-4PF_6]^{4+}$, respectively and CPK-models, as well as NMR studies indicate that tris-TTF catenane **21** adopt a structure in which the three TTF-groups effectively sandwiches the cyclic acceptor around the central *cis*-TTF group. Two peripheral TTF are necessary for formation of the cis-isomer, thus the donor composed of only two TTF-groups gives rise to an inseparable mixture of the corresponding cis/trans isomers of the following [2]catenane **22**.

10. TTF-macrocycles and cages

The construction of macrocycles containing tetrathiafulvalene (TTF) units has received a growing attention. Apart from the preparation of novel organic conducting materials, such molecules may act as hosts in host-guest chemistry. Moreover, since they contain one binding site close to an electroactive recognition site, host-guest interactions can be investigated electrochemically.

The properties of the TTF moiety as a reversible, stable two electron donor makes it a good candidate for the incorporation into macrocyclic systems that might act as receptors for electron poor compounds.

The molecular design of three dimensionally bridged macropolyclic compounds is currently a challenge in synthetic chemistry and incorporation of redox active groups into such molecules is of interest for the preparation of macropolycyclic receptor molecules, such as the Cram cacerand synthesis. Our TTF-methodology is well suited to prepare macrobicyclic cyclotetrathiafulvalenophanes with three TTF-bridges, via tris-TTF intermediates as shown in the following scheme[6].

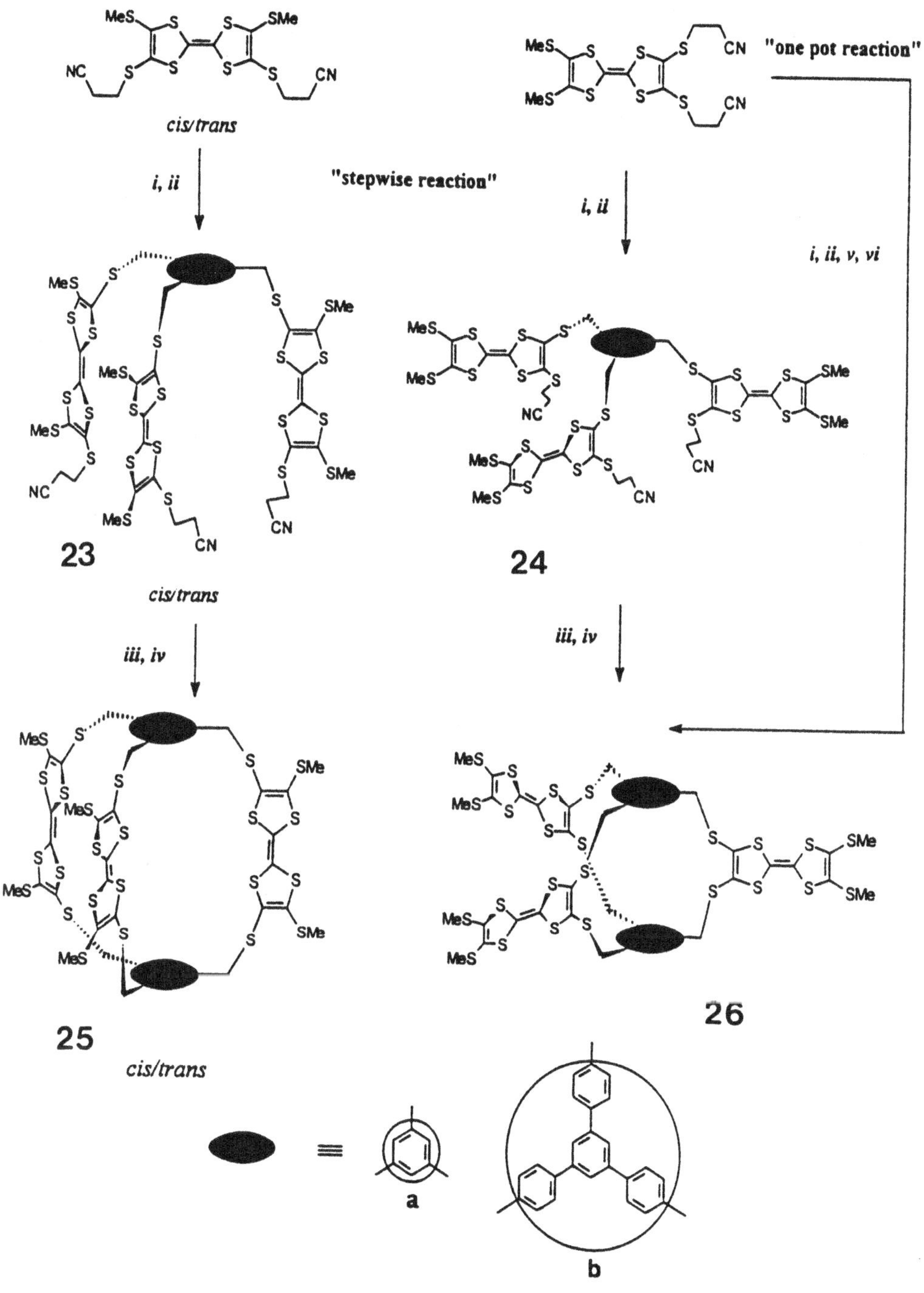

The reagents in this scheme are the following: i) CsOH, H_2O (1.05 eq), MeOH/DMF; ii) tribromide (0.33 eq), DMF; iii) CsOH,H_2O, DMF-high dilution; iv) tribromide (1 eq), MeOH/DMF; v) CsOH,H_2O (1.05 eq), MeOH/DMF; vi) tribromide (0.33 eq), DMF, (under N_2 at room temp.).

11. Donor-acceptor cyclophanes

The relative effective self-assembly of the macrocyclic tetrathiafulvalenes using the π-acceptor cyclo-bis(paraquat-*p*-phenylene) gave us the idea to a direct synthesis of donor-acceptor cyclophanes, using suitable TTF-building blocks. Thus reaction of a 2,6-bisalkylthiotetrathiafulvalene gives for n = 4 the cyclophane directly in up to 48% yield.
We are currently trying to determine the scope and limitations of these types of reactions, the surprisingly good yields observed in some cases, when the right linker is used, may be a result of self assembly during the formation of the donor-acceptor pair.

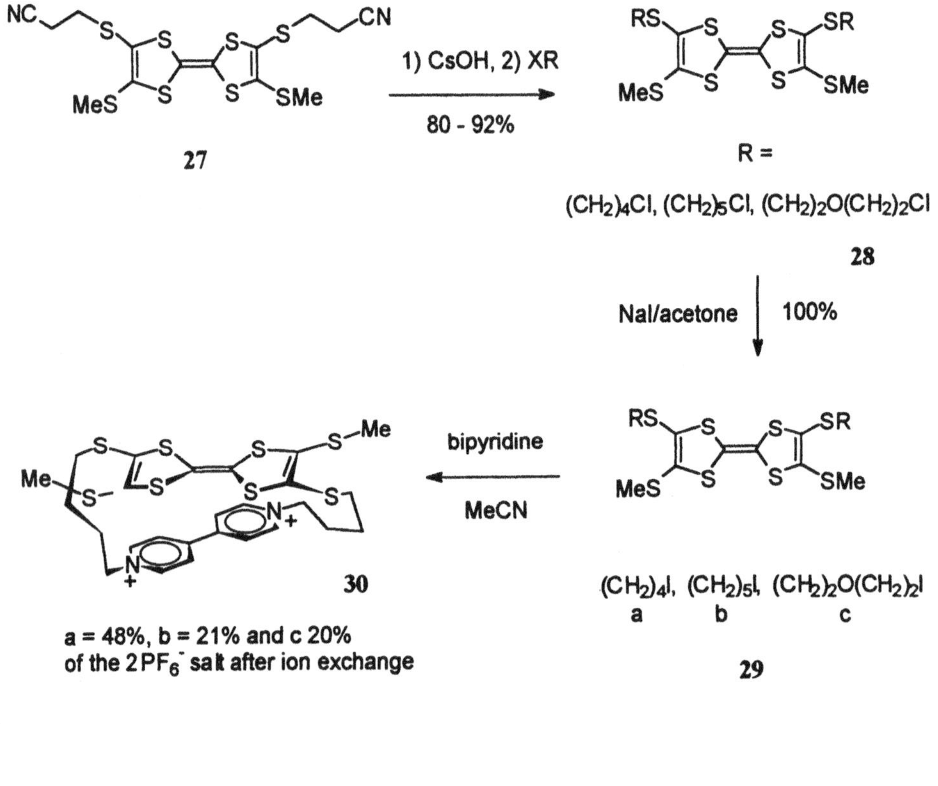

12. Dendrimers

The syntheses and investigations of dendrimeric macromolecules is currently a most active research topic at the interface of supramolecular and polymer chemistry[7].
Dendrimers are sometimes called starburst polymers or cascade molecules and beceause of their hyperbranched structure, with a branching point at each monomer unit, they have a high level of three-dimensional order. Using specific linkers it is possible to control molecular weight, interior cavity size, topology, surface functionality etc.. An important

aspect of the interest in dendrimers are due to the multiplication of functional components, and in the case of TTF-dendrimers, their sequential and reversible oxidation to radical cations at relative low potentials. It can be anticipated that such functional cascade molecules will be usefull for a wide area of practical applications. We have prepared stable pentakis-TTFs in a convergent way using our cyanoethyl protection/deprotection methodology[8]. Cyclic voltammetry CV of pentamer **31** showed 4 reversible well resolved redox waves, consistent with a multidonor system with no intramolecular electron transfer. TTF-pentamers of this type are soluble in unpolar solvents such as methylenchloride etc., confirming the expectation that oligo-TTFs having the right linkers are soluble and therefore processable for example for incorporation in polymeric materials.

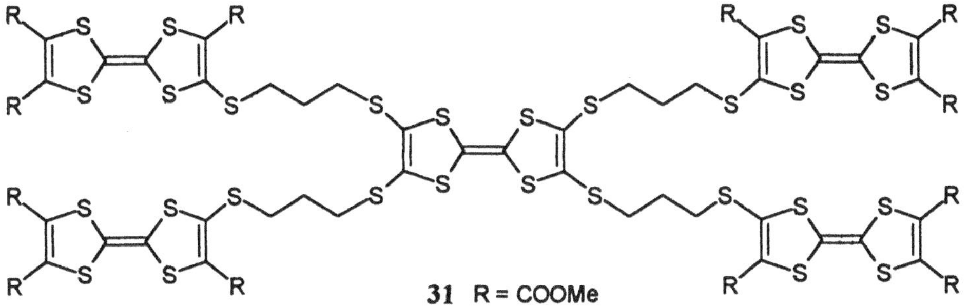

31 R = COOMe

References

1. G. Schukat, A. M. Richter, E. Fanghänel, *Sulfur Rep.* **1987**, *7*, 155; b) G. Schukat, E. Fanghänel, *ibid.* **1993**, *14*, 245.
2. For reviews on TTF supramolecular chemistry, see a) T. Jørgensen, T. K. Hansen, J. Becher, *Chem. Soc. Rev.* **1994**, *23*, 41; b) M. Adam, K. Müllen, *Adv. Mater.* **1994**, *6*, 439.
3. Use of the protection/deprotection methodology is described in the following papers: a) N Svenstrup, K M Rasmussen, T. K. Hansen, J. Becher, *Synthesis* **1994**, 809. b) J. Becher, J. Lau, P. Leriche, P. Mørk, N. Svenstrup, J. *Chem. Soc. Chem. Commun.* **1994**, 2715. c) K. B. Simonsen, N. Svenstrup, J. Lau, O. Simonsen, P. Mørk, G. J. Kristensen and J. Becher, *Synthesis*, **1996**, 407.
4. Z.-T. Li, P. C. Stein, N. Svenstrup, K. H. Lund, J. Becher, *Angew. Chem.* **1995**, *107*, 2719; *Angew. Chem. Int. Ed. Engl.* **1995**, *34*, 2524.
5. Zhan-Tinh Li and J. Becher, *Chem. Commun.*, **1996**, 639.
6. P. Blanchard, N. Svenstrup and J. Becher, *Chem. Commun.*, **1996**, 615.
7. See for example, J. Issberner, R. Moors and F. Vögtle, *Angew. Chem. Int. Ed. Engl.* **1994**, *33*, 2413.
8. Lau, O. Simonsen and J. Becher, *Synthesis*, **1996**, 521.

ANION SELECTIVE RECOGNITION AND SENSING BY NOVEL TRANSITION METAL RECEPTOR SYSTEMS

PAUL. D. BEER
Inorganic Chemistry Laboratory, University of Oxford, South Parks Road, Oxford OX1 3QR UK.

1. Introduction

The syntheses of positively charged or neutral electron-deficient abiotic receptor molecules designed to non-covalently bind anionic guest species is an area of ever increasing research activity[1]. This is because anions are known to play ubiquitous roles in biological and chemical processes[2], for example nature utilises anions for numerous biochemically important pathways and polyanions for the storage and transmission of genetic information. In addition many anions act as nucleophiles, bases, redox agents and phase-transfer catalysts. In recent years the importance of being able to detect and/or extract particular environmental anionic pollutants[3] such as nitrate, phosphate and radioactive pertechnetate, has also been realised. Surprisingly in view of this the fabrication of specific ligands that have the capability of sensing anions via optical[4] and/or electrochemical[5] methodologies has not been fully exploited. We have initiated a research programme aimed at the design and construction of innovative spectral and electrochemical sensory reagents for anions based on novel transition metal organometallic and coordination receptor systems[6]. This article reviews our current progress in this area.

2. Acyclic and Macrocyclic Cobaltocenium Receptor Systems

The first redox-responsive class of anion receptor based on the redox-active, pH independent positively charged cobaltocenium moiety was reported by our group in 1989[7]. The macrocyclic receptor (1) bound and electrochemically detected bromide guest anionic species in acetonitrile solution. As illustrated by the natural phosphate and sulphate binding proteins[8] the importance of hydrogen bonding to the anion recognition process by these organometallic systems was soon realised and various acyclic amide (CO-NH) linked cobaltocenium receptors were synthesised (Fig. 1)[9,10].
The addition of tetrabutylammonium salts NBu_4X (X = Cl, Br, NO_3, HSO_4, H_2PO_4) to deuteriated acetonitrile or dimethyl sulphoxide ^1H NMR solutions of these receptors resulted in remarkable downfield shifts of the respective receptor's protons, especially the amide protons. These results suggest a significant -CO-NH--X^- hydrogen bonding interaction is contributing to the overall anion complexation process. The resulting titration curves suggested 1 : 1 receptor : anion stoichiometry in all cases. The importance of hydrogen bonding in anion recognition by the cobaltocenium class of receptor is further highlighted by receptors containing the tertiary amide ($CONR_2$) linkage being unable to form solution halide anion complexes, as evidenced from ^1H NMR

A. W. Coleman (ed.), Molecular Recognition and Inclusion, 97–102.
© 1998 *Kluwer Academic Publishers. Printed in the Netherlands.*

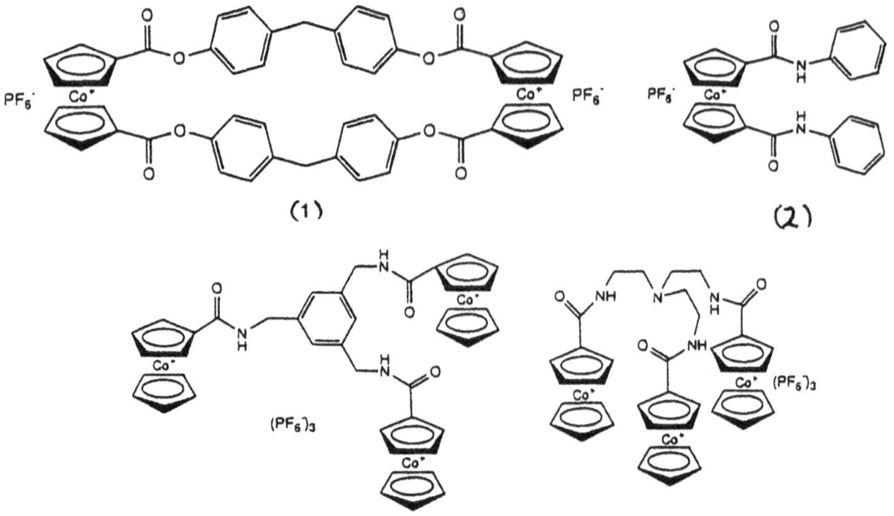

Fig. 1

titration experiments. The single-crystal structure of the bromide complex of (2) underlines this, revealing Br^- hydrogen bonding to the amide proton and to aryl and cyclopentadienyl protons of the receptor[10].
Electrochemical anion recognition investigations showed all the (CO-NH) containing cobaltocenium receptors to undergo significant one-wave cathodic shifts of the respective cobaltocenium-cobaltocene redox couples. The complexed anion effectively stabilises the positive charge of the cobaltocenium unit causing the redox couple to be perturbed to a more negative potential. It is noteworthy that very large cathodic perturbations are observed with $H_2PO_4^-$, an observation which complements the stability constant data in which the highest K values are obtained with this anion.
In an effort to impart selectivity and enhance complex stability for this class of anion receptor we have prepared a series of ditopic bis(cobaltocenium) receptor molecules in which two positively charged metallocene centres, linked via various alkyl, aryl and calix[4]arene spacers, may cooperate in the molecular recognition of mono- or di-anionic guest substrates[11] (Fig. 2). Proton NMR halide anion coordination studies revealed these alkyl-linked derivatives (3)-(5) form 1 : 1 stoichiometric complexes in acetonitrile solution. Stability constant evaluations suggest that as the length of the alkyl chain increased the selectivity preference for chloride and the general stability of the halide complex decreased. This latter experimental observation provides experimental evidence for the existence of an anionic chelate effect.

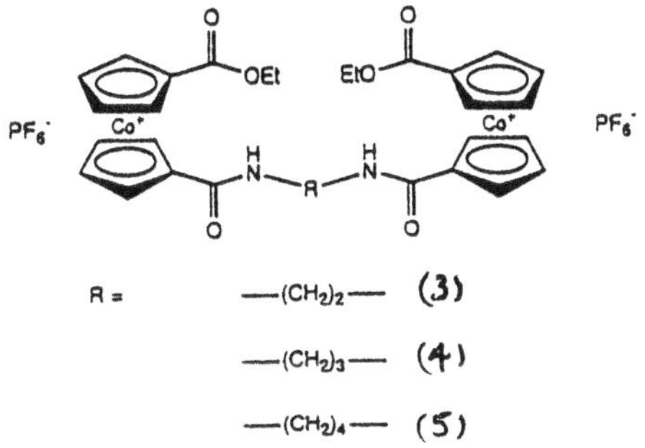

Fig. 2

3. Halide and Nitrate Selective Cobaltocenium Porphyrin Receptors

A series of new cobaltocenium porphyrin receptors have been prepared and shown to spectrally and electrochemically sense anions[12,13] (Fig. 3). Notable anion selectivity differences are displayed by the various atropisomers which highlights the importance of the relative positions of the cobaltocenium amide moieties in the anion recognition process. For example, the cis-α,α,α,α-atropisomer (6) exhibits the selectivity trend Cl^- > Br^- >> NO_3^- in which all four cobaltocenium moieties cooperatively form a cavity complementary to the spherical halide anion guest[12]. In contrast the α,α,- (7) atropisomer displays the rare selectivity sequence NO_3^- > Br^- > Cl^- indicating a complementary trigonal host cavity exists for nitrate[13].

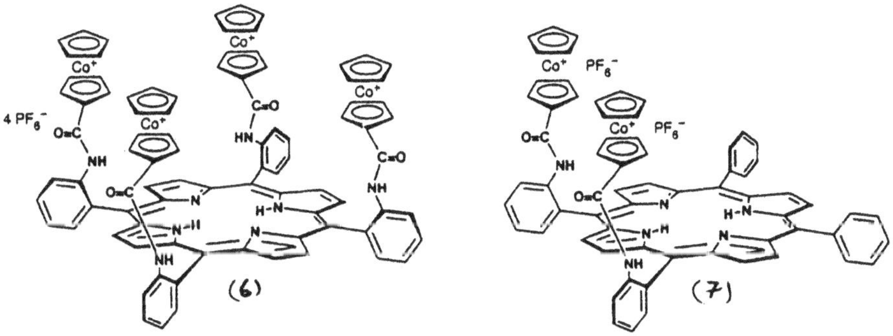

Fig. 3

4. Fluorescence Emission Spectral and Electrochemical Sensing of Anions by Acyclic, Macrocyclic and Calix[4]arene Ruthenium(II) Bipyridyl Receptor Molecules

We have recently incorporated the Lewis-acidic redox- and photo-active ruthenium(II) bipyridyl moiety, in combination with amide (CO-NH) groups, into acyclic, macrocyclic and lower-rim calix[4]arene structural frameworks to produce a new class of anion receptor with the dual capability of sensing anionic guest species *via* electrochemical and optical methodologies[14,15] (Fig. 4). Single-crystal X-ray structures of (8).Cl^- and (9).$H_2PO_4^-$

display again the importance of hydrogen bonding to the overall anion complexation process. Stability constant determinations in dimethyl sulfoxide demonstrated these

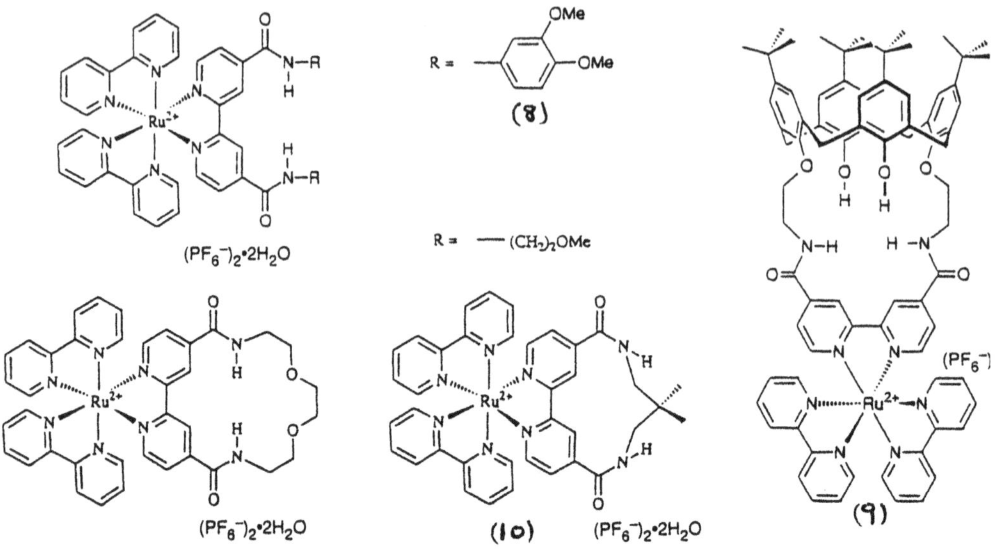

Fig. 4

receptors form strong, and in the case of the macrocyclic (10) and calix[4]arene containing receptor (9), highly selective complexes with $H_2PO_4^-$. Substantial anion-induced cathodic perturbations of the respective ligand-centred amide substituted bipyridyl reduction redox couple were detected in electrochemical anion recognition experiments with (9), in agreement with stability-constant values, able to sense $H_2PO_4^-$ in the presence of tenfold excess amounts of HSO_4^- and Cl^-.

Fluorescence emission spectroscopic measurements were also undertaken to probe binding[15,16]. All receptors exhibited significant blue shifts in the respective MLCT λ_{max} emission band on addition of Cl^- and $H_2PO_4^-$, not observed with $[Ru(bipy)_3]^{2+}$, with (9) displaying the largest perturbation of 16nm. These shifts were accompanied by large increases in emission intensity (higher quantum yields) which may be a consequence of the bound anion rigidifying the receptor, inhibiting vibrational and rotational relaxation modes of non-radioactive decay.

5. Remarkable Chloride Selective Macrocyclic Ruthenium(II)-Bipyridyl-Metallocene Receptors

The majority of the various anion receptors already discussed in this article exhibit pronounced selectivity for the dihydrogen phosphate anion in preference to halide anions. The chloride anion is crucial for a large number of biological processes[17]. For example, the relatively common hereditary disease cystic fibrosis is known to result from a genetically caused misregulation of chloride anion channels[18]. Thus there is a real need for selective detection as established methods for chloride determination based on titrimetric analysis lack selectivity and are not suitable for biological applications[19].

We have prepared the novel macrocyclic bis[ruthenium(II) bipyridyl] and ruthenium(II)-bipyridyl-metallocene receptors (11)-(13) (Fig. 5) and ^1H NMR titration studies suggested each receptor formed an extremely stable 1 : 1 stoichiometric complex with chloride in $(CD_3)_2SO$ solutions[20]. In fact the magnitudes of the stability constants up to 4 x $10^4 M^{-1}$ are amongst the largest known for any anion-abiotic amide receptor complex and are *ca.* two orders of magnitude greater than the stability constant obtained for the acyclic receptor (14). Analogous ^1H and ^{31}P NMR titration experiments with $H_2PO_4^-$ gave *no* evidence of binding this anion by these macrocyclic receptors in $(CD_3)_2SO$ solution. This contrasts with acyclic receptor (14) which forms a stronger complex with $H_2PO_4^-$ than Cl^-. This remarkable Cl^- over $H_2PO_4^-$ selectivity preference exhibited by the macrocyclic systems may be attributed to their inherently rigid structures. Molecular modelling calculations (MM2) and C.K.P. models suggest the minimised structure of (11) has all the amide and 3,3′-bipyridyl protons lying in a coplanar arrangement which creates a host cavity of similar dimensions to the chloride anion (r = 1.81 Å) capable of forming eight hydrogen bonds with this spherical anionic guest species. The larger size and tetrahedral shape of $H_2PO_4^-$ is non-complementary to the macrocyclic receptor's host cavity and consequently complex formation with this anion is not favoured.

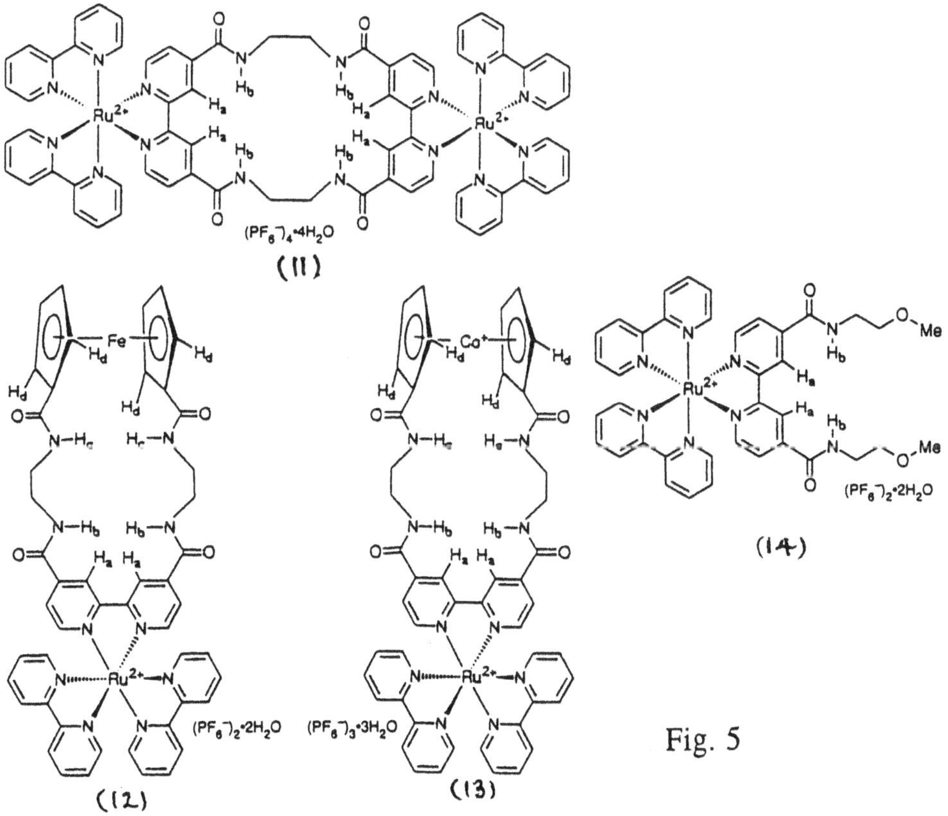

Fig. 5

Fluorescence emission spectroscopic measurements corroborated the NMR findings. Although excess amounts of $H_2PO_4^-$ had no effects on the emission spectra of the macrocyclic receptors the addition of chloride produced substantial blue shifts ($\Delta\lambda_{max}$ =

6nm) with significant intensity increases demonstrating (11)-(13) are first-generation prototype chloride selective sensory reagents.

6. Conclusions

We have seen in this article a variety of transition-metal organometallic and coordination receptor systems which are capable of recognising and sensing anionic guest species both by electrochemical and spectroscopic means. The respective topological positioning of redox-/photo-active transition-metal Lewis-acidic centres and hydrogen-bonding functionalities such as amide (CO-NH) groups in particular, dictates not only the receptor-anion complex thermodynamic stability and sensing proficiency, but also crucially the distinct anion selectivity trend the receptor displays. No doubt the fabrication of these types of systems into membranes, electronically conducting polymeric supports and optical fibres will produce novel prototype molecular sensory devices of the future.

7. Acknowledgements

I thank my postgraduate and postdoctoral coworkers for all their efforts in this multidisciplinary research field; their names appear in the references. Special thanks go to Dr Michael G.B. Drew for his many structure determinations and molecular modelling expertise. I gratefully acknowledge the financial support of EPSRC, The Royal Society, Kodak Limited, MediSense and Serpentix.

8. References

1. Schmidtchen, F.P. (1988), *Nachr. Chem. Tech. Lab.* **36**, 8; Katz, H.E. (1991), in *Inclusion Compounds*, ed. J.L. Atwood, J.E.D. Davies and D.D. MacNicol, Oxford University Press, New York. vol. 4, p. 391; Dietrich, B. (1993), *Pure Appl. Chem.*, **65**, 1457; Hawthorne, M.F., Yang, X. and Knubler, C. (1991), *Angew. Chem., Int. Ed. Engl.*, **30**, 1507; Sessler, J.L. and Burrell, A.K. (1992), *Top. Curr. Chem.*, **161**, 177; Reinhoudt, D.N., Rudkevich, D.M., Stauthamer, W.R.P.V., Verboom, W., Engersen, J.F.J. and Harkenna, S.J. (1994), *J. Am. Chem. Soc.*, **116**, 4341; Holman, K.T., Halihan, M.M., Steed, J.W., Jurisson, S.S. and Atwood, J.L., (1995), *J. Am. Chem. Soc.*, **117**, 7848; Xu, W., Vittal, J.J. and Puddephatt, R.J., (1995), *J. Am. Chem. Soc.*, **117**, 8362.
2. Frausto da Silva, J.J.R. and Williams, R.J.P. (1976), *Struct. Bonding (Berlin)*, **29**, 67.
3. Mason, C.F. (1991), *Biology of Freshwater Pollution*, 2nd edn. Longman, Harlow.
4. Czarnik, A.W. (1994), *Acc. Chem. Res.*, **27**, **302.**
5. Beer, P.D. (1994), *Adv. Mater.*, **6**, 607.
6. Beer, P.D. (1996), *Chem. Comm.*, 689.
7. Beer, P.D. and Keefe, A.D. (1989), *J. Organometal. Chem.*, **375**, C40.
8. Pflugrath, J.W. and Quiocho, F.A. (1985), *Nature*, **314**, 257; He, J.J. and Quicho, F.A. (1991), *Science*, **251**, 1479.
9. Beer, P.D., Hesek, D., Hodacova, J. and Stokes, S.E. (1992), *J. Chem. Soc., Chem. Commun.*, 270; Beer, P.D., Hazlewood, C., Hesek, D., Hodacova, J. and Stokes, S.E. (1993), *J. Chem. Soc., Dalton Trans.*, 1327.
10. Beer, P.D., Drew, M.G.B., Graydon, A.R., Smith, D.K. and Stokes, S.E. (1995), *J. Chem. Soc., Dalton Trans.*, 403.
11. Beer, P.D., Hesek, D., Kingston, J.E., Smith, D.K., Stokes, S.E. and Drew, M.G.B. (1995), *Organometallics*, 3288.
12. Beer, P.D., Drew, M.G.B., Hesek, D. and Jagessar, R. (1995), *J. Chem. Soc. Chem. Commun.*, 1187.
13. Beer, P.D., Hesek, D. and Jagessar, R. unpublished work.
14. Beer, P.D., Chen, Z., Goulden, A.J., Grieve, A., Hesek, D., Szemes, F. and Wear, T. (1994), *J. Chem. Soc., Chem. Commun.*, 1269.
15. Szemes, F., Hesek, D., Chen, Z., Dent, S.W., Drew, M.G.B., Goulden, A.J., Graydon, A.R., Grieve, A., Mortimer, R.J., Wear, T., Weightman, J.S. and Beer, P.D. (1996) *Inorg. Chem.*, **35**, 5868.
16. Beer, P.D., Mortimer, R.J., Stradiotto, N.R., Szemes, F. and Weightman, J.S. (1995), *Anal. Commun.*,**32**, 419.
17. Kirk, K.L. (1991), *Biochemistry of the Elemental Halides*, Plenum, New York.
18. Kairn, W. and Schwederski, B. (1991), *Bioinorganic Chemistry: Inorganic Elements in the Chemistry of Life*, Wiley, New York, p. 283.
19. *Official and Standardized Methods of Analysis*, ed. C.A. Watson. The Royal Society of Chemistry, Cambridge, 3rd edn., 1994.
20. Beer, P.D. and Szemes, F. (1995), *J. Chem. Soc., Chem. Commun.*, 2245.

MACROCYCLIC SUGAR THIOUREAS: CYCLOOLIGOSACCHARIDES MIMICKING CYCLOPEPTIDES

J.M. GARCÍA FERNÁNDEZ,[a]* C. ORTIZ MELLET,[b] J.L. JIMÉNEZ BLANCO,[b] J. FUENTES,[b] M. MARTÍN-PASTOR[c] and J. JIMÉNEZ-BARBERO[c]

[a]*Instituto de Investigaciones Químicas, C.S.I.C., Américo Vespucio s/n, Isla de la Cartuja, E-41092 Sevilla, Spain*

[b]*Departamento de Química Orgánica, Facultad de Química, Universidad de Sevilla, E-41071 Sevilla, Spain*

[c]*Instituto de Química Orgánica General, C.S.I.C., Juan de la Cierva 3, E-28006 Madrid, Spain*

1. Introduction

Interactions between carbohydrates and other organic molecules, specially proteins, play a prominent role in many biological recognition processes. The complexity of such phenomena has stimulated the use of model systems to get information about their nature (lipophilic, polar, hydrogen bonds, etc.) and to evaluate their contribution to the stability and specificity of the carbohydrate-containing supramolecular entity. Among model sugar-derived hosts, the natural cyclodextrins (CDs) have been by far the most widely used for this purpose, due to their commercial availability and to the presence of a hydrophobic cavity that can accommodate a guest molecule surrounded by a hydrophilic environment.[1] Contributory reasons to the popularity of CDs are the high symmetry and rigidity of the structure that facilitate the study of inclusion complexes by NMR techniques. However, this lack of conformational flexibility is a limitation regarding efficiency of complexation, since the host cannot adapt its geometry to the guest for an optimal interaction. Such ability is, in contrast, a main structural feature of polyamide receptors, from which cyclopeptides are paradigmatic examples. We have now imported this faculty into cyclic oligosaccharides by inserting thiourea spacers that would act as symmetric amide surrogates. The possibility of switching between *Z* and *E*

A. W. Coleman (ed.), Molecular Recognition and Inclusion, 103–108.
© 1998 *Kluwer Academic Publishers. Printed in the Netherlands.*

configurations at the pseudoamide linkages allows then a restricted number of rotameric forms whose relative populations are conditioned by the strong hydrogen bond donnor character of the NH thiourea protons.[2,3] In addition, these sugar-thiourea coronands may be considered as comformationally restricted glycopeptide analogues, useful as model compounds for the study of specific carbohydrate-peptide interactions, e.g. interactions involving the pseudoamide NH protons and the carbohydrate oxygen atoms.

2. Results and Discussion

Our first target molecules have been mixed saccharide-thiourea coronands incorporating two α,α'-trehalose units bridged through their primary C-6 positions by two thiourea segments. The use of trehalose derived building blocks is advantageous for several reasons: it is a rigid, commercially available disaccharide with a concave shape that fits well in macrocyclic structures and it is symmetric, thus facilitating the NMR studies. The synthesis of these new pseudocyclooligosaccharide receptors (**3-7**) was achieved in satisfactory yield (40-60%) by coupling of 6,6'-diamino-6,6'-dideoxy-α,α'-trehaloses[4] (**1a-c**) with the related *O*-protected diisothiocyanates[5] (**2b,c**) in pyridine (Scheme 1). The starting materials were obtained from the native disaccharide by replacement of the primary hydroxyl group by iodine, nucleophilic displacement by azide, Staudinger reduction and isothiocyanation reaction, in combination with standard hydroxyl protection methodologies, as previously the reported for monosaccharide models.[6,7] Attempts at the direct condensation of **1a** and **2a** failed, formation of intramolecular cyclic thiocarbamates being favoured under the reaction conditions.

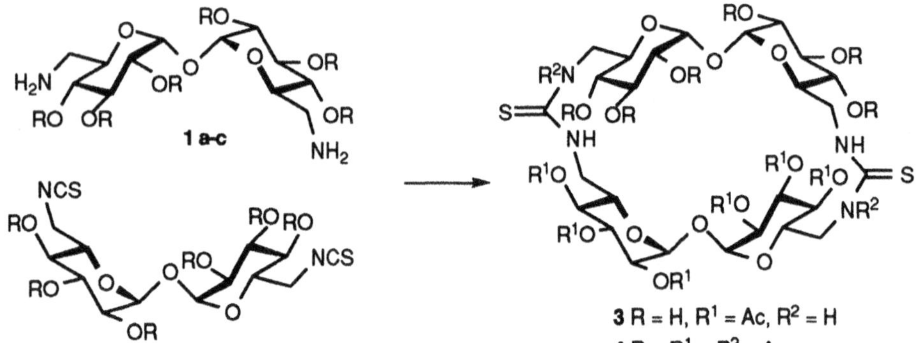

2 a-c

a R = H; **b** R = Ac; **c** R = $SiMe_3$

3 R = H, R^1 = Ac, R^2 = H
4 R = R^1 = R^2 = Ac
5 R = R^1 = Ac, R^2 = H
6 R = R^1 = R^2 = $SiMe_3$
7 R = R^1 = R^2 = H

Scheme 1. Synthesis of macrocyclic sugar thioureas

The conformational properties of these macrocyclic sugar thioureas are mainly governed by the presence of four slow-rotating NH—C(=S) bonds. Examination of molecular models reveals that only two configurational patterns are compatible with the macrocyclic structure, namely the *Z,E:E,Z* (cross conformation) and the *Z,E:Z,E* (parallel conformation), both having C_2 symmetry. Shifting from one to the other offers a possibility for conformational adaptability similar to that of a small cyclopeptide (Figure 1).

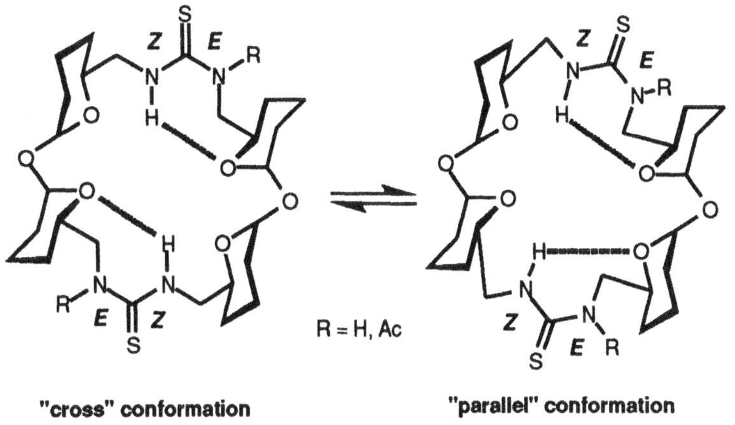

Figure 1. Possible conformations in α,α'-trehalose—thiourea macrocycles showing the intramolecular hydrogen bonds. The oxygen substituents at the pyranose rings are omitted.

In compound **4**, obtained as the sole reaction product by *in situ* treatment of **3** with Ac_2O-pyridine, the cross conformation is necessarily fixed after *N*-acetylation. The acetyl groups are directed to the outside of the internal cavity, whereas the NH protons are inside directed and participate in intramolecular hydrogen bonding with the opposite pyranoid oxygen atoms, as supported by ^1H chemical shift temperature coefficient measurements and computational calculations. The selectively *O*-acetylated coronand **5** also existed as a monoconformational compound having the cross disposition (Figure 2). In contrast, the per-*O*-trimethylsilyl derivative **6** showed in the low temperature range NMR spectra signals for the two possible rotameric forms (cross and parallel) in similar relative proportions, suggesting that the capability of conformational adaptability of these receptors can be modulated by acting on the hydroxyl protecting groups. In all cases the tridimensional structure is stabilised by the presence of two seven-membered NH···O hydrogen bonds (figure 1). It is noteworthy that seven membered hydrogen bonds also play a decissive role in the secondary structure of polypeptide chains, where they fix the so-called γ-turn loops.

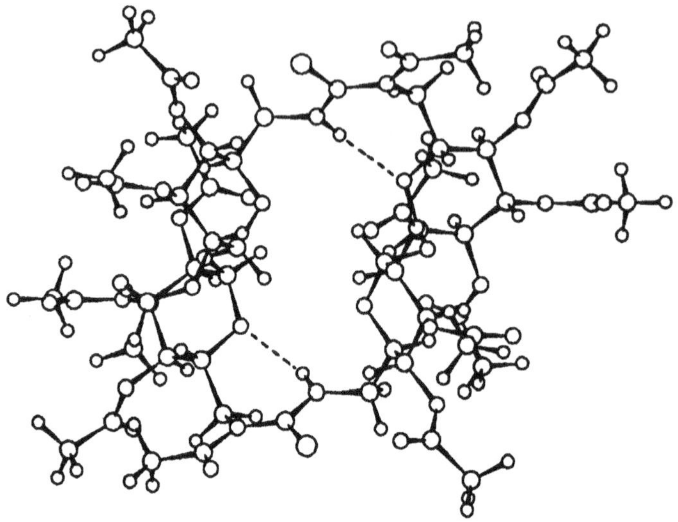

Figure 2. Three dimensional model of the more stable conformer (MM2*) of macrocycle 5 showing the seven-membered NH···O hydrogen bonds. Dimensions of the cavity: 6.04 x 3.06 Å.

The formation of the seven-membered intramolecular hydrogen bond probably explains the high efficiency of the macrocyclization reaction: after the first thiourea bridge has been stablished, the reactive ends are broght close toguether by formation of a pseudo γ-turn, favouring cyclization towards oligomerization.

The fully unprotected target macrocyle **7** was obtained in quantitative yield by acid hydrolysis of the TMS groups of **6**. Preliminary cation-binding studies in water, using thin-layer ligand-exchange chromatography,[8] showed that **7** forms strong complexes with soft cations that interact specifically with sulfur, such as Cu^{2+} and Ag^{+}. Other mono and divalent cations (Cs^{+}, K^{+}, Na^{+}, Zn^{2+}, Ba^{2+}, Ca^{2+}) formed only weak complexes, indicative of absence of efficient coordination by oxygen. The three dimentional model built after computational calculations agrees with these results. The cavity of the macrocycle is essentially apolar, similar to that of cyclodextrins. Only the two interglycosidic oxygen atoms are at the inside, whereas the hydroxyl groups are outside directed. On the other hand in the more stable conformation, both thiocarbonyl sulfur atoms are directed towards the same side of the main plain of the molecule, providing efficient corrdination sites for soft cations. The dimensions of the cavity (6.14 x 5.26 Å), similar to that recently reported for cyclomaltopentaose,[9] let suppose that inclusion of small molecules is feasable, with lipophilic interactions and hydrogen bonding, in competition with the intramolecular ones, being the expected driving forces, These aspects are currently being sought in our laboratory.

3. Experimental

3.1. GENERAL METHODS

The ^{1}H and ^{13}C NMR spectra were recorded on Bruker AMX 300 and 500 apparatus. The FABMS spectra were obtained in the positive mode using a Kratos MS-80 RFA instrument. NaI was generally used as cationizing agent. Thin-layer ligand-exchange chromatography was performed on Polygram® Ionex-25 SA-Na plates. All new compounds gave satisfactory microanalysis (C, H, N, S, Si).

3.2. MACROCYCLE **4**

A solution of 6,6'-diamino-6,6'-dideoxy-α,α'-trehalose (**1a**, 0.22 g, 0.65 mmol) and the corresponding per-*O*-acetyl diisothiocyanate (**2b**, 0.36 g, 0.53 mmol) in pyridine (5 mL) was stirred at 40 °C for 12 h. Evaporation of the solvent and preparative TLC (30:1 CH_2Cl_2-MeOH) afforded **4** (0.43 g, 60%): $[\alpha]_D$ +89 (*c* 1.4, $CHCl_3$). ^{1}H NMR (500 MHz, $CDCl_3$, 328 K): δ 8.50 (bs, 2 H, NH), 5.53 (d, 1 H, H-1 *E*), 5.29 (d, 1H, H-1 *Z*). ^{13}C NMR (75.5 MHz, $CDCl_3$, 323 K): δ 91.4 (2 C), 71.1, 70.7, 69.9, 69.7, 69.5, 69.4, 68.4, 66.9, 49.2, 45.6. FABMS: *m*/*z* 1357 ([M + Na]$^+$).

3.3. MACROCYCLE **5**

A solution of 2,3,4,2',3',4'-hexa-*O*-acetyl-6,6'-diamino-6,6'-dideoxy-α,α'-trehalose dihydrochloride (**1b**, 0.31 g, 0.45 mmol) and **2b** in pyridine (5 mL) was stirred at 40 °C for 48 h. The reaction mixture was poored into ice-water (80 mL) and the solid was filtered and purified by column chromatography to give **5** (0.22 g, 40%) as a syrup: $[\alpha]_D$ +195 (*c* 1, $CHCl_3$). ^{1}H NMR (300 MHz, $CDCl_3$, 323 K): δ 6.46 (t, 1 H, NH), 5.38 (d, 1 H, H-1); (248 K): δ 6.91, 6.65 (2 t, each 1 H, NH), 5.73 (d, 1 H, H-1 *Z*), 5.39 (d, 1 H, H-1 *E*). ^{13}C NMR (75.5 MHz, $CDCl_3$, 323 K): δ 91.4, 70.2, 69.8, 69.4, 69.0, 44.4. FABMS: *m*/*z* 1292 ([M + Na]$^+$).

3.4. MACROCYCLE **6**

A solution of the hexa-*O*-trimethylsilyl derivatives **1c** (0.26 g, 0.31 mmol) and **2c** (0.24 g, 0.31 mmol) in pyridine was stirred at 40 °C for 12 h. Evaporation of the solvent and column chromatography of the residue afforded **6** (0.28 g, 60%) as an amorphous solid: $[\alpha]_D$ +139 (*c* 0.8, $CHCl_3$). ^{1}H NMR (500 MHz, $CDCl_3$, 328 K): δ 6.50 (bs, 1 H, NH), 4.83 (d, 1 H, H-1); (248 K): cross rotamer, δ 7.11, 6.12 (d and t, respectively, each 1 H, NH), 4.83 (d, 1 H, H-1 Z), 4.65 (d, 1H, H-1 *E*); parallel rotamer, δ 7.01, 6.04 (bs and t, respectively, each 1 H, NH), 4.72 (d, 1 H, H-1 *Z*), 4.65 (d, 1H, H-1 *E*). ^{13}C NMR (125.7 MHz, $CDCl_3$, 323 K): δ 91.4, 72.6, 72.2, 71.7, 70.3, 42.5. FABMS: *m/z* 1615 ([M - Me + H]$^+$).

3.5. MACROCYCLE **7**

Treatment of **6** (53 mg, 0.03 mmol) in 4:2:1 $CHCl_3$-MeOH-H_2O (2mL), with 10% aqueous AcOH (0.6 mL) at 60 °C for 3 h gave, after elimination of the solvents and freeze-drying, pure **7** (24 mg, 98%) as an amorphous solid: $[\alpha]_D$ +135 (*c* 0.6, H_2O). ^1H NMR (500 MHz, D_2O, 363 K): δ 5.61 (d, 1 H, H-1).^{13}C NMR (75.5 MHz, D_2O, 353 K): δ 96.3, 75.7, 74.2, 73.9, 73.8, 47.9.

4. Acknowledgements

We thank the Dirección General de Investigación Científica y Técnica for finantial support (grant no. PB 94/1440-C02-01 and PB 93/0127) and the Junta de Andalucía for a scholarship award to J.L.J.B.

5. References

1. Wenz, G. (1994) Cyclodextrins as building blocks for supramolecular structures and functional units, *Angew. Chem., Int. Ed. Engl.*, **32**, 336-358.
2. Ortiz Mellet, C., Moreno Marín, A., Jiménez Blanco, J.L., García Fernández, J.M. and Fuentes, J. (1994) Influence of intramolecular hydrogen bonding on the conformational properties of sugar thioureas, *Tetrahedron: Asymmetry*, **5**, 2325-2334.
3. García Fernández, J.M., Ortiz Mellet, C., Jiménez Blanco, J.L., Fuentes, J., Diánez, M.J., Estrada, M.D., López-Castro, A. and Pérez-Garrido, S. (1996) Influence of intramolecular hydrogen bonding on the conformation of 3-deoxy-3-thioureido sugars, *Carbohydr. Res.*, **286**, 55-65.
4. Ortiz Mellet, C. Jiménez Blanco, J.L., García Fernández, J.M. and Fuentes, J. (1995) *O*-Acetyl protection of 6-aminoaldopyranosides and 1-aminoalditols, *J. Carbohydr. Chem.*, **8**, 1133-1152.
5. García Fernández, J.M., Ortiz Mellet, C., Jiménez Blanco, J.L., Fuentes, J., Gadelle, A., Coste-Sarguet, A. and Defaye, J. (1995) Isothiocyanates and cyclic thiocarbamates of α,α'-trehalose, sucrose and cyclomaltooligosaccharides, *Carbohydr. Res.*, **268**, 57-71.
6. García Fernández, J.M., Ortiz Mellet, C. and Fuentes, J. (1993) Chiral 2-thioxotetrahydro-1,3-O,N-heterocycles from carbohydrates. 2. Synthesis of oxazolidine pseudo *C*-nucleosides and bicyclic oxazine-2-thione, *J. Org. Chem.*, **58**, 5192-5199.
7. García Fernández, J.M., Ortiz Mellet, C., Díaz Pérez, V.M., Jiménez Blanco, J.L. and Fuentes, J. (1996) Conformational energetics of sugar thioureas and synthesis of glycosyl thioureido sugars, *Tetrahedron*, **52**, 12947-12970.
8. García Fernández, J.M., Jiménez Blanco, J.L., Ortiz Mellet, C. and Fuentes, J. (1995) Synthesis, conformational flexibility and preliminary complexation behaviour of α,α'-trehalose-based macrocycles containing thiourea spacers *J. Chem. Soc., Chem. Commun.*, 57-58.
9. Immel, S., Brickmann, J. and Lichtenthaler, F.W. (1995) Small-ring cyclodextrins: their geometries and hydrophobic topographies, *Liebigs Ann.*, 924-942.

MOLECULAR CLEFTS DERIVED FROM KAGAN'S ETHER. SYNTHESIS AND SOLID STATE INCLUSION COMPLEXES OF A CHIRAL MOLECULAR TWEEZER

MICHAEL HARMATA*,† MEHMET KAHRAMAN,† SRIRAM TYAGARAJAN,† CHARLES L. BARNES† AND CHRISTOPHER J. WELCH§
†*Department of Chemistry*
University of Missouri-Columbia
Columbia, Missouri 65211 USA
§*REGIS Technologies, Inc.*
8210 Austin Avenue
P.O. Box 519
Morton Grove, Illinois 60053 USA

Research in the synthesis and study of molecular tweezers is an established and growing area of interest in the field of molecular recognition and inclusion. The work of Zimmermann has been of particular impact in this area. His contributions focus on a series of molecular tweezers which are achiral.[1]

We have been interested in the preparation of chiral molecular tweezers which are congeners of Kagan's ether **1**. This molecule possesses chirality and is shaped appropriately for incorporation into tweezer-like molecules.

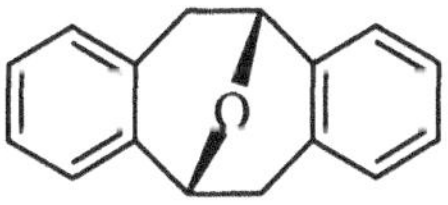

1: Kagan's Ether

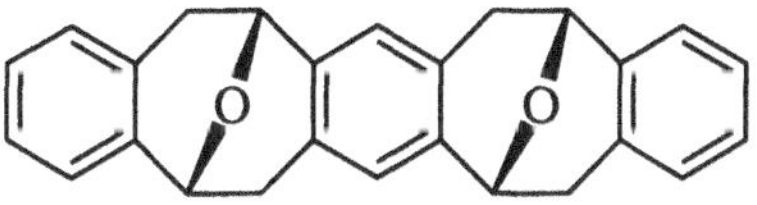

2

The first chiral molecular tweezer we prepared was **2**. This racemic compound exhibited interesting solid-state clathrate properties but did not show significant binding to pi-acidic guests in solution, as indicated by ^1H NMR experiments.[2] We conjectured that better binding properties could be afforded by a tweezer which had a deeper cleft. To that end, we chose to synthesize a molecular tweezer with naphthalene pincers, compound **3**.

The synthesis of **3** parallels our synthesis of **2** (Scheme 1).[2] Thus, metal-halogen exchange of the dibromide **4** followed by quenching of the resultant organodilithium species with 1-naphthylacetaldehyde gave the diol **5** in 78% yield as an approximately 1:1 mixture of d,l and meso isomers. Treatment

A. W. Coleman (ed.), Molecular Recognition and Inclusion, 109–116.
© 1998 *Kluwer Academic Publishers. Printed in the Netherlands.*

of the d,l diol with catalytic tosic acid in dichloromethane gave the

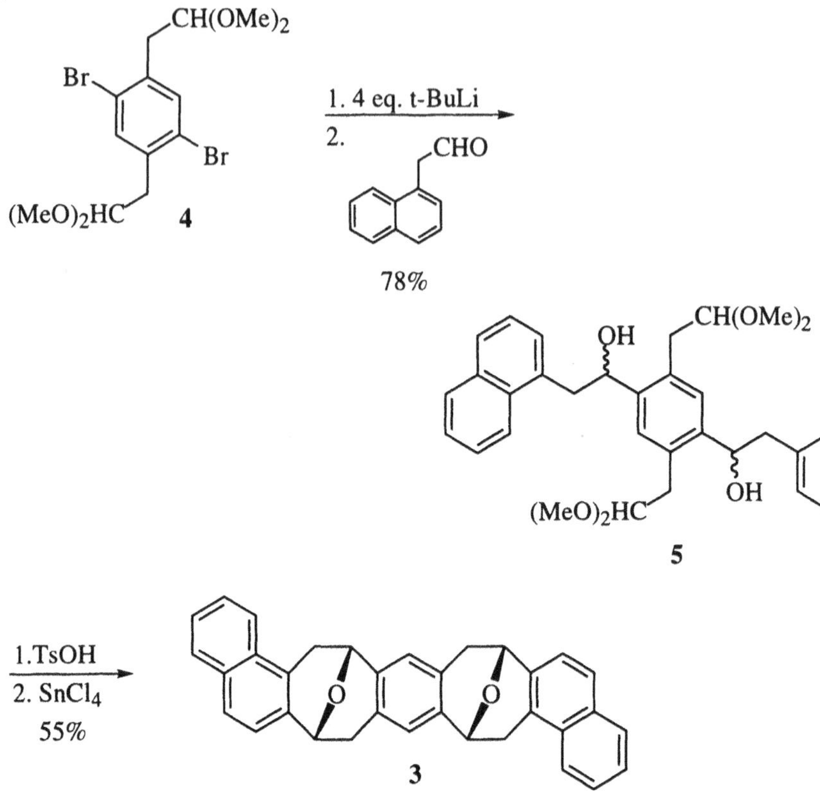

Scheme 1

corresponding cyclic acetal in high yield. This was transformed to the molecular tweezer **3** by treatment with tin tetrachloride in an overall yield of 55% from **5**.[3]

We initially examined some of the solid state inclusion properties of **3**. Crystal structures for the host-guest complexes of **3** and three pi-acidic guests were determined. A number of structural features of these complexes remain relatively constant in the series and suggest a pattern which may be the basis for prediction of structures of as yet unknown host-guest complexes.

Slow evaporation of a 1:1 mixture of **3** and maleic anhydride in chloroform produced crystals suitable for X-ray analysis.[4] Two views of the structure are shown in Figure 1. Various structural features of the complex are noteworthy and those cited are represented diagrammatically to the left in Figure 1. Thus, the distance (a) between the naphthalene pincers is 6.81Å. The distance (c) between the centroid of the guest and the centroid of the central benzene ring of the host is 5.62Å and the distance (b) between the closest hydrogen of the guest and the centroid of the same benzene ring is

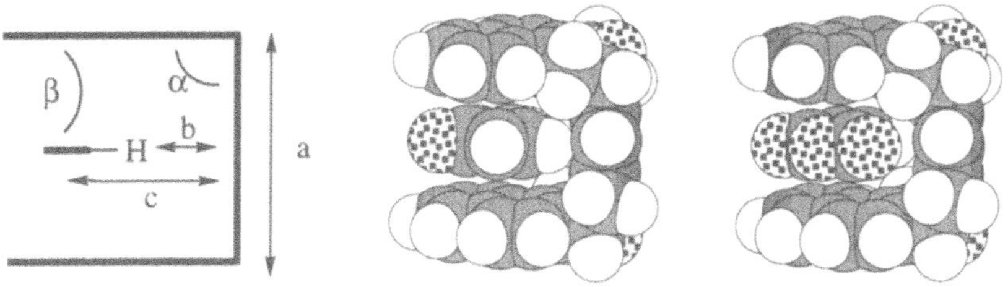

Figure 1. Inclusion Complex of maleic anhydride and 3.

2.52Å The angles (α) between the least squares planes of the naphthalene pincers and the central benzene ring of the host are 87.2° and 92.2°. The angles (β) between the pincers and the guest are 6.7° and 5.9°. The naphthalene pincers and the guest are thus nearly parallel and are presumably experiencing a pi-acid/pi base stacking interaction.[5] A hydrogen atom of the maleic anhydride is directed toward the central

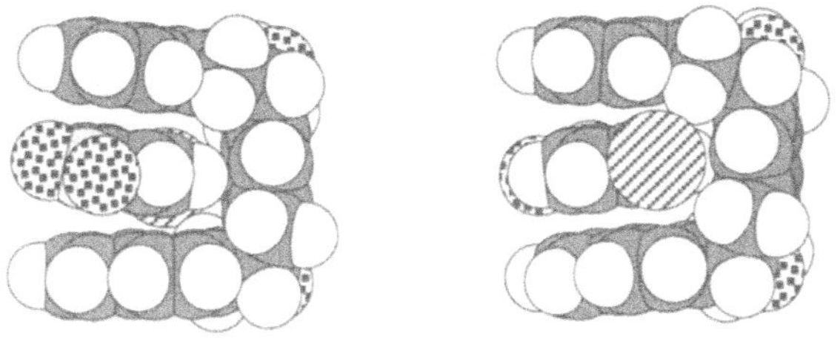

Figure 2. Inclusion Complex of p-chloronitrobenzene and 3.

benzene ring of **3** and this appears to be an apex-face interaction: a hydrogen bond between the guest and the host.[6] Both the stacking interactions and the hydrogen bond could be responsible for directionality in the solid-state inclusion complex, though packing forces may also play a role.

Crystals grown in the same way as described above but using p-chloronitrobenzene as a guest showed similar characteristics.[7] Two views of the inclusion complex are shown in Figure 2. The measured distances and angles corresponding to those reported for the maleic anhydride inclusion complex are as follows: (a): 7.07Å; (b): 2.47Å; (c) 4.98Å; (α): 84.7°, 92.7°; (β): 1.3°, 7.8°. There was some question as to which of the two hydrogens available for hydrogen bonding to the benzene ring of the host would engage in such an interaction. That those hydrogens ortho to the chlorine atom in the guest do so suggests that a steric effect may be operative. The length of the carbon-chlorine bond allows the relatively electropositive ortho-hydrogen to approach the central benzene ring of **3** more closely than might be possible for the hydrogen ortho to the nitro group of the guest.

Some support for this conclusion comes from the X-ray analysis of the host-guest complex of **3** and trinitrobenzene (TNB).[8] The inclusion complex looks as might be predicted based on the general pattern of pi-stacking and hydrogen bonding that appears to be the mode of guest-binding to this molecular tweezer. However, the calculated distance from the

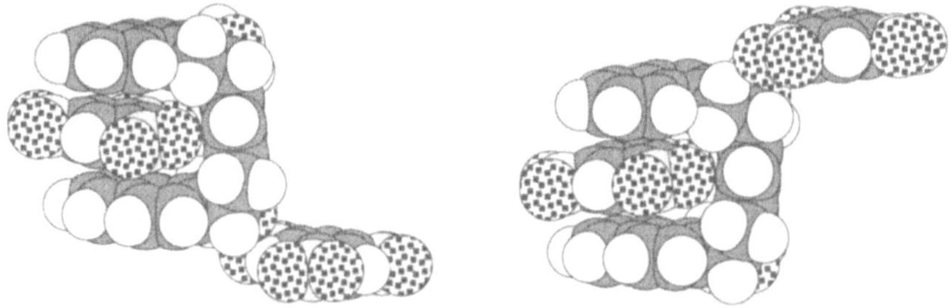

Figure 3. Inclusion Complex of 1,2,5-trinitrobenzene and 3.

hydrogen on the trinitrobenzene to the centroid of the benzene ring in **3** is 2.99 Å, considerably longer that that for either of the other guests. We surmise that this is due to a steric effect created by the nitro groups which prohibits a closer approach of the guest to the benzene ring of **3**. Other features of this host-guest inclusion complex are similar to the others as the following data indicate: (a): 6.84Å; (b): 2.99Å; (c): 5.46Å; (α): 95.8°, 92.7°; (β): 3.0°, 6.6°.

Interestingly, the crystals of the inclusion complex of **3** and TNB were obtained in a stoichiometry of 2:1/TNB:**3**. While one TNB molecule occupies the site expected within the cleft of **3**, the other is located outside. There appears to be a hydrogen-bonding interaction between the outside TNB and a tweezer molecule. Further, a host-guest unit is stacked on top of the outside TNB and the included TNB of this latter unit experiences

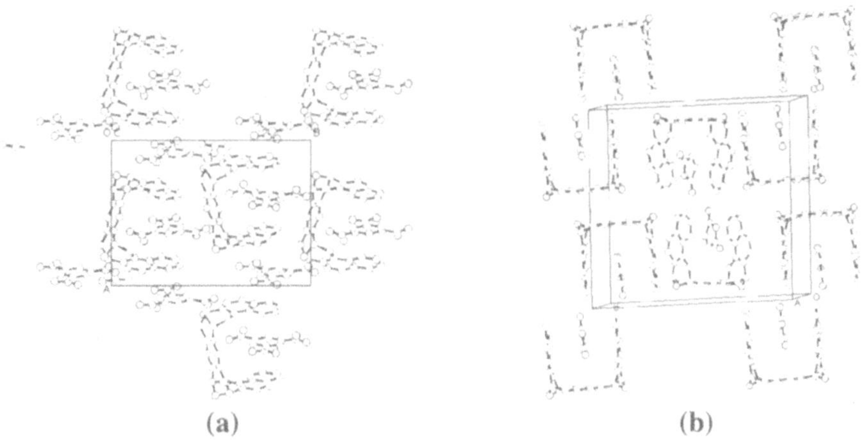

(a) (b)

Figure 4. Packing diagrams of 3/TNB inclusion complex (a) and 3/maleic anhydride inclusion complex (b).

an apex-face interaction with the back of the benzene ring of the tweezer to which the outside TNB is hydrogen-bonded. Thus, two host-guest units are joined via an TNB molecule through hydrogen-bonding and stacking interactions.

A packing diagram in Figure 4(a) helps to illustrate this. The packing of the host-guest units of the other two inclusion complexes is quite different. In the latter, the host-guest units are packed in a herringbone-type arrangement such that the complexes experience edge-face interactions between their naphthalene units.

To date, only a few solution binding studies with **3** have been undertaken. The data obtained via ^1H NMR titration experiments indicated that in chloroform at room temperature, **3** binds TNB with an association constant of K = 45-120 M^{-1}.[9]

These data can be compared to binding studies conducted with molecular tweezer **4**. By virture of its dibenzofuran pincers, this tweezer has a deeper

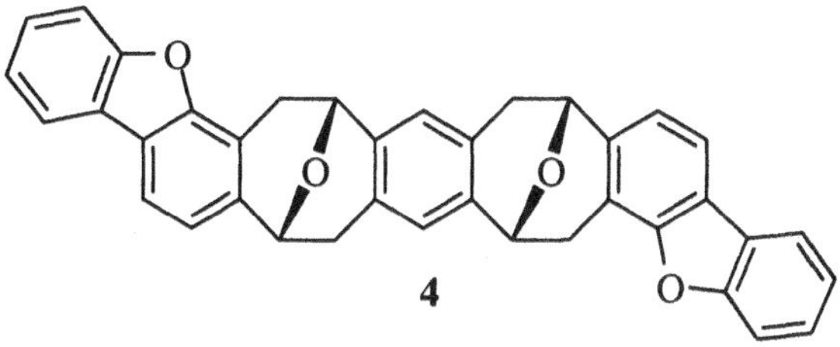

4

cavity than **3** and also differs from it electronically. The influence of each of these variables on binding has not been separately evaluated. In any case, solution binding studies with and TNB give an association constant of 2000 M^{-1}.[11] Clearly, one or both of the aforementioned factors is beneficial to the binding process. It is interesting to examine the structural feactures of the solid state complex between **4** and TNB. Details of this aspect of our work have appeared elsewhere.[12] The relevant data are as follows: (a): 6.66Å; (b): 3.22Å; (c): 5.58Å; (α): 89.3°, 89.4°; (β): 5.27°, 0.22°. It is interesting to note that the distance from the hydrogen of the TNB ring to the central benzene ring of **4** is longer than the corresponding distance in the **3**/TNB complex. One possible interpretation is that the increased volume of the cleft in **4** allows the TNB to engage in more energetically favorable pi-stacking interactions, sacrificing the apex-edge interaction in the process.

We have recently begun a study of enantioselective recognition of chiral guests by molecular tweezers. The approach involves the use of a pi-acidic stationary phase, specifically a (3R, 4S) Whelk-O 1 column. This chiral stationary phase (CSP) incorporates a ligand which should be a good guest for our tweezers. In addition, the column has been used for the separation of a number of racemates, and the mechanism generally used for enantiodiscrimination is fairly well understood.[13] Furthermore, the use of chromatography to study molecular recognition offers a number of conveniences. For example, only a small amount (≤ 1mg) of a candidate tweezer is required for analysis. This compound can be analyzed as a racemate and need not be chemically pure.

The structure of the (3R, 4S)-Whelk-O 1 chiral stationary phase (CSP) is shown as structure **5**. Columns based on this structure with the stereochemistry shown were used exclusively for our studies.

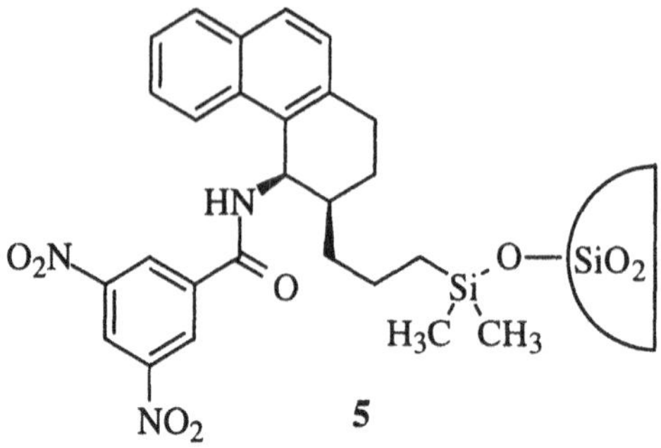

5

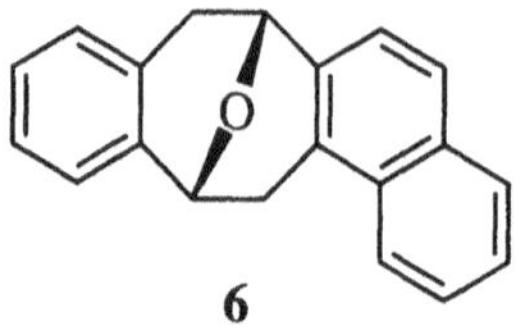

6

Data for the chromatographic behavior of **4** on the (3S, 4R)-Whelk-O column are summarized in Table 1. Interestingly, the nature of the alcohol modifier has a considerable influence on enantioselectivity, with the bulky 2-propanol affording the best separation factor (α=1.75) and methanol the worst (α=1.57). This general trend is frequently seen, but seems quite pronounced in this case. Ethyl acetate, which cannot serve as a hydrogen bond donor, gave the worst separation (α=1.36). These results point to the importance of solvation in enantioselective guest inclusion by these tweezers.

A comparison of Kagan's ether **1**, tweezer **2** and tweezer **4** in 20% isopropanol in dichloromethane (Table 2) shows an increase in both retention factor (k') and enantioselectivity (α) as one moves from the relatively shallow cleft of **1** to the deeper cleft of **4**.

TABLE 1. Chromatographic data for **4** on a (R,R)-Whelk-O column.

Solvent	**α value**	**k_1 value**
20% IPA in CH_2Cl_2	1.75	0.92
20% EtOH in CH_2Cl_2	1.65	0.85
20% MeOH in CH_2Cl_2	1.57	0.89
100% EtOAc	1.36	1.60

TABLE 2. Chromatographic data for substrates on a (R,R)-Whelk-O column using 20% IPA in CH_2Cl_2 as eluent.

Compound	**α value**	**k_1 value**
1	1.21	0.096
2	1.40	0.12
4	1.75	0.92

A preliminary comparison of **1**, **2** and **6** in two different solvent systems underscores consideration of solvation in binding studies and demonstrates that increasing the surface area of the pi systems interacting with the CSP affects retention times as well as separation factors. These data are summarized in Table 3.

TABLE 3. Chromatographic data for substrates on a (R,R)-Whelk-O column using two solvent systems.

Solvent	**k'_1/α-1**	**k'_1/α-2**	**k'_1/α-6**
20% IPA in hexane	1.13/1.22	12.22/1.25	2.69/1.33
CH_2Cl_2	0.19/1.30	0.67/1.40	0.26/1.37

With a reasonable amount of chromatographic data in hand, we conducted a variable temperature study in order to elucidate the thermodynamic parameters associated with the binding and hence resolution process. Using **4** as a substrate, chromatograms were run over a temperature range of 0-70 °C with 10% IPA in methylene chloride serving as eluent. The results of the van't Hoff analysis (Table 4) show the enthalpy and entropy of binding between the CSP and either of the enantiomers of **4** and a model compound, dibenzofuran (DBF). Interestingly, the ΔH values for the tweezer enantiomers are roughly the same as that for dibenzofuran despite the fact that the tweezers are capable of a number of interactions unavailable to DBF. However, the entropy lost on adsorption is much greater for dibenzofuran than for the tweezers. The net result is that the entropy lost for a given enthalpy gain (ΔS/ΔH) is greatest for the dibenzofuran and lower for the enantiomers of **4**. The strong entropic bias in the current example is interesting, but may simply reflect an effect of solvation.

TABLE 4. Thermodynamics of Adsorption of **4** on a (R,R)-Whelk-O column using 10% IPA in CH_2Cl_2 as eluent.

Compound	ΔH (kcal/mol)	ΔS (cal/mol °K)
Dibenzofuran	-1.76	-9.9
Enantiomer 1 of **4**	-1.40	-5.2
Enantiomer 2 of **4**	-1.89	-5.5

The absolute configuration of the two enantiomers of **4** has not yet been established. However, inspection of molecular models and application of the chiral discrimination model developed for the interaction of the Whelk-O CSP with naproxen[14] suggest that the enantiomer of **4** which should be more strongly bound has the (S,S,S,S) configuration. On the other hand, models based on the crytallographic data obtained with host/guest complexes **4** and **3** do not necessarily support this conclusion. Further work is necessary to define the specific interations responsible for the resolution of **4** and related molecular tweezers on the Whelk-O 1 column.

In conclusion, we have developed a synthesis of chiral molecular tweezers which produces the target compounds in generally good overall yield. The solid state inclusion complexes of these tweezers share a number of common features including well-defined face-face and face-edge interactions between the guest and host. The chromatographic behavior of these tweezers and related compounds on a chiral chromatography column has been investigated. Variable temperature studies suggest that the differences in binding for different enantiomers of a tweezer lie solely in enthalpy differences for the binding of each enantiomer to the CSP. Generalizing this observation, establishing a definitive model for enantiomer recognition, refining solid-state inclusion chemistry and using these tweezers for catalytic and supramolecular constructs are under study. Results will be reported in due course.

Acknowledgements: We are grateful to the University of Missouri Research Board for partial support of this work. We thank the National Science Foundation for partial support of the NMR (PCM-8115599) facility at the University of Missouri-Columbia and for partial funding for the purchase of a 500 MHz spectrometer (CHE-89-08304) and an X-ray diffractometer (CHE-90-11804).

References

1. Zimmerman, S.C. *Top. Curr. Chem.* **1993**, *165*, 71.
2. Harmata, M.; Barnes, C.L. *Tetrahedron Lett.* **1990**, *31*, 1825.
3. Data on **3**: mp > 300 ^{o}C; ^{1}H NMR ($CDCl_3$, 500 MHz) δ 7.70 (d, 2H, J = 8.3 Hz), 7.62 (d, 2H, J = 7.8 Hz), 7.54 (d, 2H, J = 8.4 Hz), 7.34, (t, 2H, J = 7 Hz), 7.28 (t, 2H, J = 7 Hz), 7.13 (d, 2H, J = 8 Hz), 6.75 (s, 2H), 5.37 (d, 2H, J = 6.4 Hz), 5.32 (d, 2H, J = 6.4 Hz), 3.61 (dd, 2H, J = 6.4, 16.2 Hz), 3.55 (dd, 2H, J = 6.4, 16.2 Hz), 3.02 (d 2H, J = 16.2 Hz), 2.81 (d, 2H, J = 16.2 Hz); ^{13}C NMR ($CDCl_3$, 125 MHz) δ 136.9, 134.9, 132.4, 131.8, 128.9, 128.4, 126.9, 126.3, 126.3, 126.1, 125.7, 125.2, 123.5, 122.5, 69.8, 69.3, 34.6, 32.7; HRMS exact mass calcd for $C_{34}H_{26}O_2$: 466.1933, found 466.1927.
4. Data for **3**/maleic anhydride: Formula: $C_{34}H_{26}O_2{\cdot}C_4H_2O_3$; MW = 564.63 Space Group: P21/c; a = 7.1332(10)Å, b = 20.343(2)Å, c = 18.834(3)Å; V = 697.6(7)Å^3; Z = 4; Dcalc = 1.393 Mg.m^{-3}; mu = 0.09 mm^{-1}; F(000)=1184; Radiation=MoKα (λ=0.70930Å); Temperature= -100(1)^{o}C; Final R=0.037 for 3174 reflections with I>2.5σ(I).
5. (a) Hunter, C.A. *Angew. Chem. Int. Ed. Engl.* **1993**, *32*, 1584. (b) Jones, G.B.; Chapman, B.J. *Synthesis* **1995**, 475.
6. Nishio, M.; Umezawa, Y.; Hirota, M.; Takeuchi, Y. *Tetrahedron* **1995**, *51*, 8665.
7. Data for **3**/p-chloronitrobenzene: Formula: $C_{34}H_{26}O_2{\cdot}C_6H_4ClNO_2$; MW = 624.13 Space Group: P21/n; a = 20.188(3)Å, b = 7.342(4)Å, c = 20.991(3)Å; V = 2972.8(17)Å^3; Z = 4; Dcalc = 1.395 Mg.m^{-3}; mu = 0.18 mm^{-1}; F(000)=1456; Radiation=MoKα (λ=0.70930Å); Temperature = 23(1)^{o}C; Final R=0.101 for 1998 reflections with I>2.5σ(I).
8. Data for **3**/1,3,5-trinitrobenzene: Formula: $C_{34}H_{26}O_2{\cdot}2C_6H_3N_3O_6{\cdot}0.5CHCl_3$; MW = 952.98 Space Group: P21/c; a = 13.326(3)Å, b = 17.834(2)Å, c = 18.314(4)Å; V = 4097.1(14)Å^3; Z = 4; Dcalc = 1.545 Mg.m^{-3}; mu = 0.20 mm^{-1}; F(000)=1964; Radiation=MoKα (λ=0.70930Å); Temperature = -100(1)^{o}C; Final R=0.054 for 4190 reflections with I>2.5σ(I).
9. Solution binding studies were conducted by titration of TNB with the host at concentrations below which the host self-associated. The change in the chemical shift in the ^{1}H NMR of the TNB as a function of host concentration was monitored and the data analyzed by a least squares fitting procedure.[10]
10. Wilcox, C.S.; Cowart, M.D. *Tetrahedron Lett.* **1986**, *27*, 5563.
11. We thank Mr. Sandy Pusalkar for obtaining this result.
12. Harmata, M.; Barnes, C.L. *J. Am. Chem. Soc.* **1990**, *112*, 5655.
13. Commercially available from Regis Technologies, Inc.
14. Pirkle, W.H.; Welch, C.J. *J. Chromatogr. A* **1994**, *683*, 347.

MOLECULAR TECTONICS : AN APPROACH TO ORGANIC NETWORKS

MIR WAIS HOSSEINI
Université Louis Pasteur
Laboratoire de Chimie de Coordination Organique
Institut Le Bel,
F-67000 Strasbourg, France

1. Introduction

Molecular solids are defined by the chemical nature of their molecular components and by their interactions with respect to each other in the crystalline phase. Whereas molecular chemistry deals with the design and the synthesis of the individual units, their assembly is governed by concepts of supramolecular chemistry [1].

The construction of large size molecules (10^{-6}-10^{-3} m scale) or hyper molecules with predicted and programmed structure may hardly be envisaged through stepwise classical synthesis using covalent linkages. However, the preparation of such higher-order materials may be attained through iterative process based on self-assembly [1] of individual tectons (from Greek TEKTON, builder) [2]. This particular area of research, called molecular tectonics [3], is currently under active investigation [4,5].

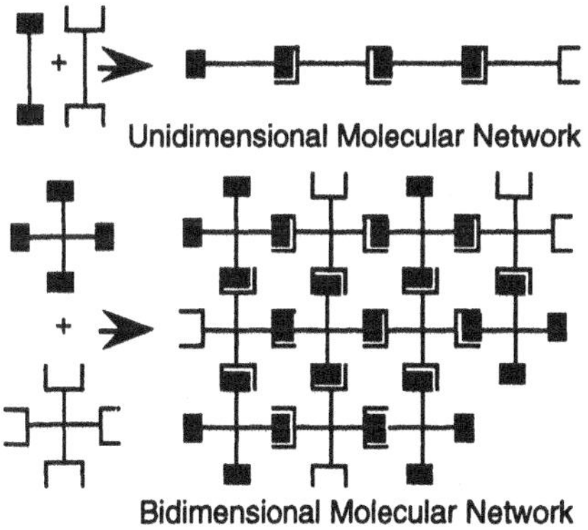

Fig. 1 Schematic representation of mono- and di-dimensional networks based on assembling of a self-complementary tectons

One of the most important aspect of molecular tectonics appears to be the fact that the assembling process is based on molecular recognition processes operating at the level of the complementary tectons. In other terms, the operational concepts in molecular tectonics are those defined for molecular recognition [6], as well as topological features allowing iteration.

The manufacture of molecuar networks may be achieved by an iterative assembling process of either a self-complementary tecton (one component system composed of a unique motif), or of several complementary modules (multi component system). In the later case the most viable approach would engage two complementary tectons (Figure 1).

In order to achieve an iterative assembling process, the building blocks or

A. W. Coleman (ed.), Molecular Recognition and Inclusion, 117–122.
© 1998 *Kluwer Academic Publishers. Printed in the Netherlands.*

tectons must fulfil both structural and energy criteria. In particular, the complementary tectons must recognise each other (molecular recognition) and furthermore should allow the repetition of the recognition pattern (iteration).

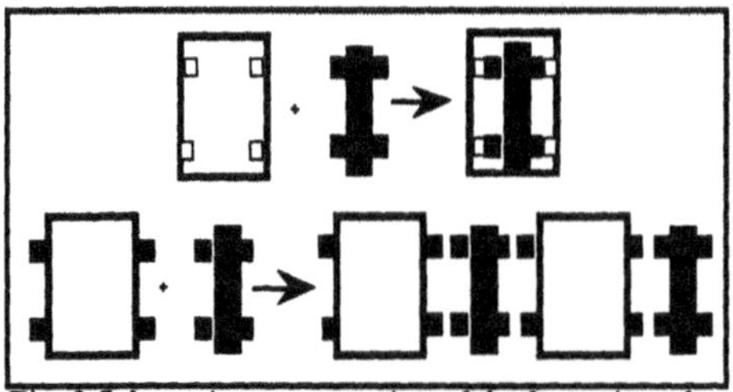

Fig. 2 Schematic representation of the formation of an endo-molecular complex (top) and of a one-dimensional network resulting from an iterative assembly of one exo- and one endo-tectons

These two requirements lead to molecular modules possessing connecting points or interaction sites located in a divergent fashion (Figure 2). Whereas an endo molecular receptor (endo-receptor), by conception, is designed to form only discrete molecular complexes with a selected exo-substrate, an exo-receptor may form molecular networks with appropriate exo-substrates.

2. Design of molecular networks based on weak interactions

The synthesis of solids based on iterative assembling of individual complementary tectons still remains a challenge. A strict control of the self-assembly of molecular modules in the solid state should lead to structurally strictly controlled molecular networks (one-, two- and three-dimensional solids).

Although in principle any type of weak interactions may be used to self-assemble complementary tectons, so far, we have focused our attention on hydrogen bonding and on van der Waals interactions. Only results obtained in the latter case will be presented here.

2.1. Linear molecular arrays using van der Waals interactions

The chemistry of inclusion complexes [7], *i.e.* the inclusion of a substrate within the cavity of a receptor molecule is an established area.

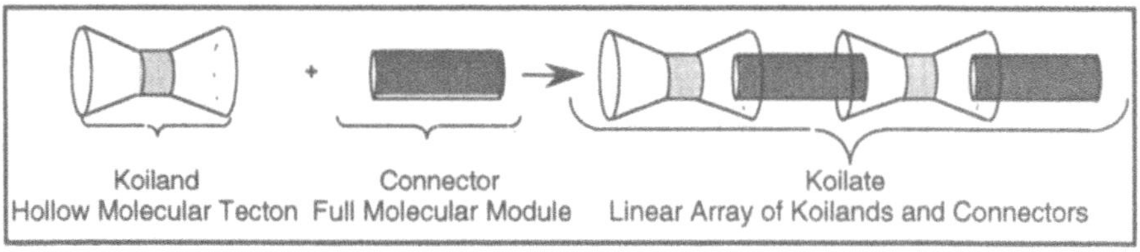

Fig. 3 Schematic representation of a linear array (koilate) resulting from interpenetration of a convex molecule (connector) into the cavity of a concave receptor (koiland)

One may extend the concept of inclusion in solution to the construction of new networks in the solid state. For example, a rigid and compact direceptor possessing two divergent cavities (linear *koiland* , from Greek koilos : hollow), and a linear connector, possessing two extremities capable each to be included within the cavities of the direceptor, may be assembled in the solid state, by an iterative process through non-covalent van der Waals interactions leading thus to a linear molecular array which may be called linear *koilate* [8,9] (Figure 3).

For the construction of linear koilates, the design of a koiland possessing two divergent cavities with an angle of 180° between them is crucial. For this purpose, calix[4]arene seems to be an interesting unit since, it possess both hydrophobic and hydrophilic sites[10]. The *p-tert*-butylcalix[4]arene **1** (Figure 4), in its favoured cone conformation below 20°C in chloroform [11-13] and in the solid state [14,15] offers a genuine hydrophobic pocket surrounded by four *p-tert*-butyl aryl moieties. This basket type cavity was proved to accommodate a variety of neutral substrates such as chloroform [13], toluene [14], anisole [15], xylene [16] and benzene [16]. Complexation of Cs^+ through metal-π-aromatic interactions has also been reported for **1** [17] in its cone conformation.

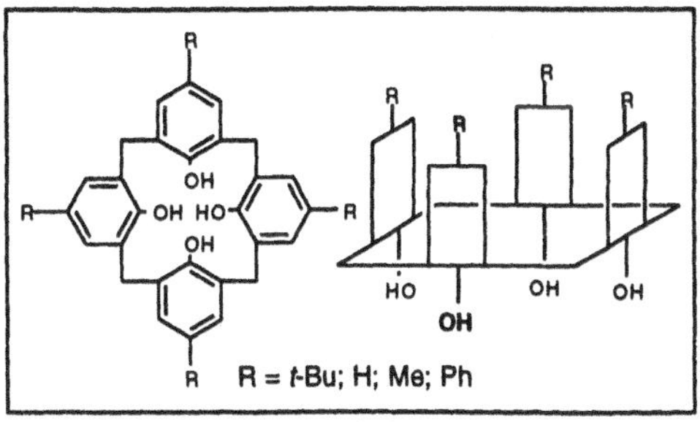

Fig. 4 calix[4]arenes (left) and a schematic representation of their cone conformation (right)

The fusion of two *p-tert*-butylcalix[4]arenes **1** in its cone conformation by two silicon atoms leads to a hollow brick (koiland) **2** [8,9] (Figure 5,6). Other similar compounds using titanium [18], niobium [19] and aluminium [20] have been reported.

As stated above, calixarenes seem particularly well suited for the design of the hollow modules since one may control both the entrance and the depth of their preorganised cavity by the nature of the substituant R at the *para* position, *i.e.* H, CH_3, $C(CH_3)_3$, Ph (Figure 4). The synthesis of different koilands (Figure 5) based on the double fusion of two *p*-methylcalix[4]arenes, and two *p*-phenylcalix[4]arenes with two silicon atoms was achieved (22). In order to control both the dimensionality and the directionality of koilates, one may use non-symmetrical koilands. Such hollow modules may be obtained by two different strategies. The differentiation of the two cavities may be achieved by using two different cavities (Figure 5). Mixed compounds based on the double fusion of two different calix units have been prepared [23].

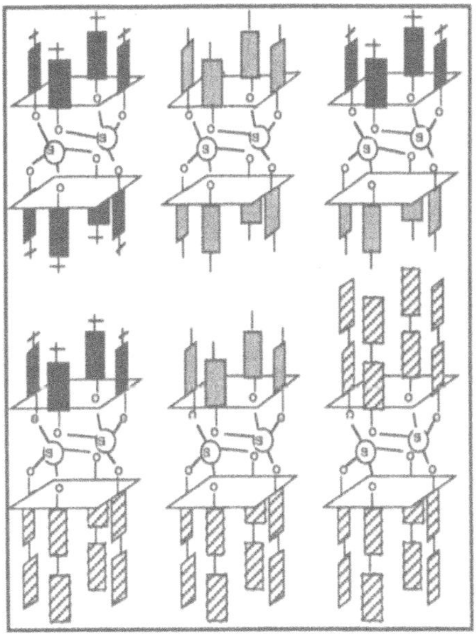

Fig. 5 *Schematic representation of koilands resulting from the fusion of two identical or different calix[4]arenes*

Another alternative to the formation of non-symmetrical koilates may be the use of two different metals. Thus, the heterobinuclear dimer **4** containing both a Si and a Ti atoms was prepared [9]. The non-symmetric nature of **4** is based on the induced difference between the two calix units by their coordination to two different metals with different elctronegativity. Indeed, one of the calix is triply coordinated to a Si atom whereas the other unit is triply coordinated to a Ti atom (Figure 5). In order to prepare **4** by a directed strategy, compound **3** was prepared. For **3**, due to possible free rotation around the Si-O bond, different rotamers may be expected. However, NMR studies revealed that, due to inter calix H-bonding between the hydroxy groups, the face to face (syn) rotamer was the preferred one (Figure 5). Treatment of the isolated **3** with $TiCl_4$ produced, in high yield, the heterobinuclear compound **4**.

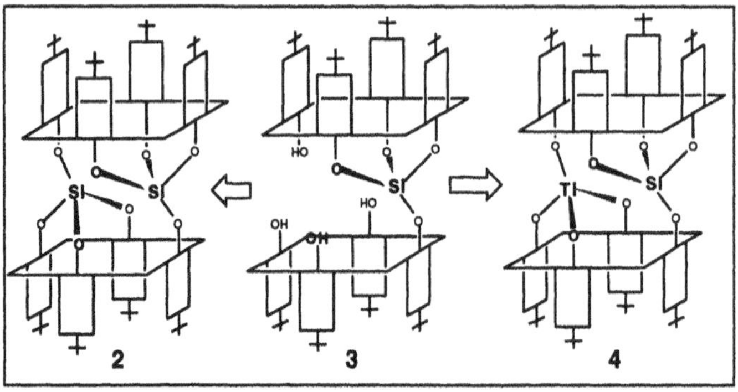

***Fig. 5** Schematic representation of koilands based on: the fusion of two p-tert-butylcalix[4]arene 1 by two Si atoms 2, the fusion of two tert-butylcalix[4]arene by one Si atom 3, or by both Si and Ti atoms 4.*

In the solid state, the structure of **2** was investigated by X-ray analysis. Compound **2** as shown in Figure 6 was indeed a centro-symmetric silicon aryloxy dimer consisting of 2 Si and 2 calixarenes. Both calixarene units were in cone conformation, thus, presenting two divergent cavities. The coordination geometry around both Si atoms was tetrahedral. Each of these two tetrahedrons was composed of 3 oxygen atoms belonging to one of the two calixarenes and the fourth one to the other calix unit.

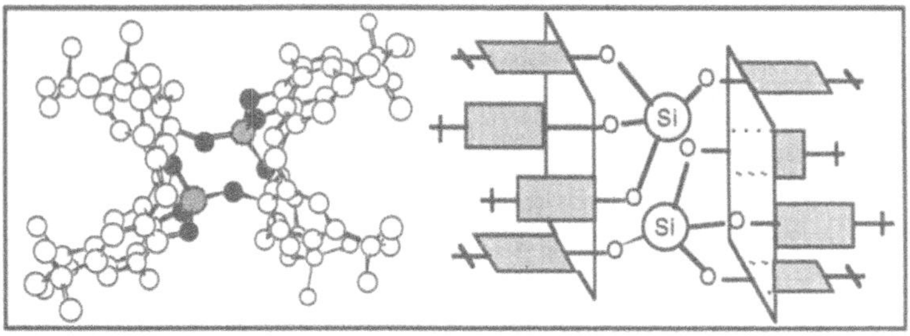

***Fig. 6** Schematic representation of 2 composed of two calixarenes fused by two silicon atoms yielding a double basket type receptor (right), and its crystal structure (left).*

As shown in the figure 6, in the solid state, the hollow ditopic receptor **2** indeed possesses two divergent cavities. In order to ascertain the possibility of double inclusion, binuclear complexes composed of one compound **2** and two molecules of anisol or *p*-xylene were prepared and their structures investigated by X-ray analysis (Figure 7). The crystal study showed that in both cases, each cavity accommodates, in a inclusive fashion, an anisol or a *p*-xylene molecule. Both substrates penetrate the cavity of **2** through their CH_3 group. It is worth noting that anisol, due to its reduced dimension and its bent shape, is not a well suited connector for bridging the hollow modules **2**, and thus leading to the linear koilate.

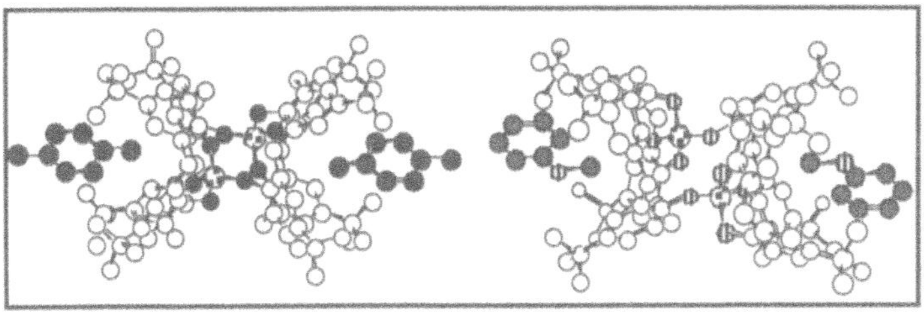

Fig. 7 X-ray structure of the binuclear inclusion complexes formed between 2 and p-xylene (left) and anisol (right)

However, for *p*-xylene, although due to the discrete nature of the complex, *i.e.* the absence of interconnection between the discrete bi-nuclear complexes, the linear koilate was not formed, nevertheless, the aligned positioning, with respect to each other, of binuclear complexes in the solid appeared to be rather interesting and encouraging.

Our aim was the preparation of linear arrays resulting from an iterative inclusion of both extremities of connectors within the cavities of koilands (Figure 8).

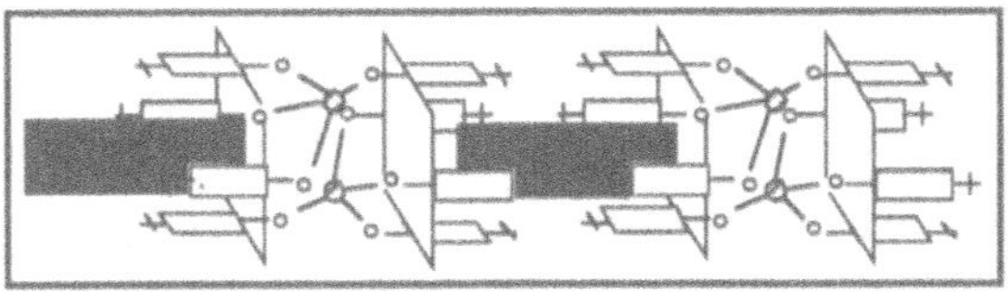

Fig. 8 Schematic representation of the formation of koilates resulting from an iterative assembling process between hollow molecular modules and connectors capable of penetrating the cavity of the ditopic receptors

Hexadiyne, a rod type molecule, possessing both the required linear geometry and a sufficient distance between its two terminal CH_3 groups (d_{C-C} = 6.65 Å) leads to the formation of the desired linear koilate (Figure 9) [21]. The latter crystallises from a $CHCl_3$/MeOH solution of koiland **2** and hexadiyne in large excess. The X-ray analysis revealed that in the crystal, in addition to **2** and hexadiyne present in 1/1 ratio, two $CHCl_3$ molecules were also present.

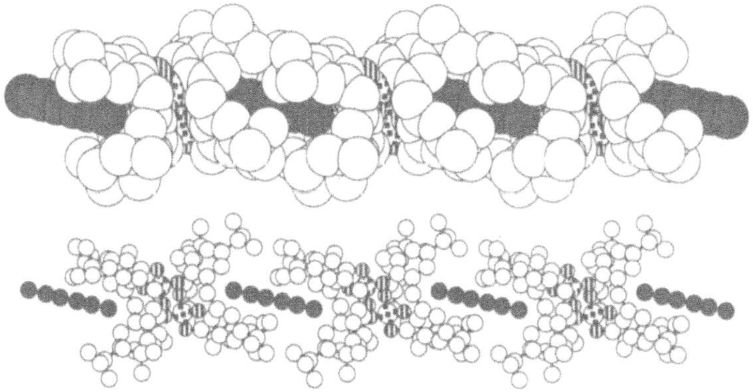

Figure 9: X-ray structure of the koilate formed between 2 and hexadiyne. space filling (top) and balls & sticks (bottom) representations.

The lattice is composed of linear arrays of koilates formed between koilands **2** and hexadiyne as connectors. Each connector bridges two consecutive koilands by penetrating their cavities through its terminal CH_3 groups. The methyl groups of hexadiyne are deeply inserted into the cavity of the koiland. Indeed, the shortest distance between the CH_3 group of the connector and the aromatic carbon atom of the koiland is 3.64 Å. It is worth noting that the hexadiyne molecules are extremely well encapsulated by two consecutive koilands **2** forming a sandwich, the shortest distance between the CH_3 groups of the *tert*-butyl moieties of **2** being 3.51 Å.

3. Conclusion

We have shown that our approach, based on van der Waals interactions, to assemble organic entities in the solid state with pre-programmed structures was viable. Further extension of these concepts to ionic solids using both directional hydrogen bonding and less directional electrostatic charge-charge interactions was also achieved [24]. Thus, the modular approach seems interesting for the confection of organic molecular networks. Although our approach has been so far restricted to structural features controlling the dimensionality of organic crystals, the next step, under investigation, will deal with functional aspects of organic crystals.

References

1. Lehn, J.-M. (**1995**) *Supramolecular Chemistry, Concepts and Perspectives,* VCH, Weinheim; Desiraju, G. R. (**1995**)*Angew. Chem. Int. Ed. Engl., 34,* 2311; see also Tsoucaris, G. *et al.*, Ed.,(**1996**) *Crystallography of Supramolecular Compounds,* Kluwer; Lindsey, J. S. (**1991**) *New. J. Chem., 15,* 153; Lawrence, D. S., Jiang, T., Levett, M. (**1995**) *Chem. Rev., 95,* 2229; Stoddart, J. F., Philip, D. (**1996**) *Angew. Chem. Int. Ed. Engl., 35,* 1155.
2. Simard, M., Su, D., Wuest, J.D. (**1991**) *J. Amer. Chem. Soc., 113,* 4696.
3. Mann, S. (**1993**) *Nature, 365,* 499.
4. Lehn, J.-M. (**1990**) *Angew. Chem. Int. Ed. Engl.,29,* 1304.
5. Whitesides, G. M., Mathias, J. P., Seto, C. T. (**1991**) *Science, 254,* 1312.
6. Lehn, J.-M. (**1973**) *Struc. Bonding., 16,* 1; *ibid,* (**1978**) *Pure Appl. Chem., 50,* 871.
7. Lehn, J.-M.(**1980**) *Leçon Inaugurale,* Collège de France, Paris.
8. Delaigue, X., Hosseini, M. W., De Cian, A., Fischer, J., Leize, E., Kieffer, S., Van Dorsselaer, A. (**1993**) *Tetrahedron Lett.,34,* 3285.
9. Delaigue, X., Hosseini, M. W., Leize, E., Kieffer, S., Van Dorsselaer, A. (**1993**) *Tetrahedron Lett., 34,* 7561; Delaigue, X., Hosseini, M. W., Graff, R., Kintzinger, J.-P., Raya, J.(**1994**) *Tetrahedron Lett., 35,* 1711.
10. Gutsche,C.D. (**1989**) *"Calixarenes"*, Monographs in Supramolecular Chemistry, Ed. J.F. Stoddart, R.S.C., London; J. Vicens and V. Böhmer, Eds. (**1991**) *Calixarenes. A Versatile Class of Macrocyclic Compounds,* Kluwer Academic Publishers, Dordrecht; Böhmer, V. (**1995**) *Angew. Chem. Int. Ed. Engl., 34,* 713.
11. Kämmerer, H., Happel, G., Caesar, F. (**1975**) *Makromol. Chem., 176,* 3317.
12. Gutsche, C. D., Bauer, L. J. (**1985**) *J. Amer. Chem. Soc., 107,* 6052.
13. Gutsche, C. D., Dhawan, B., No, K. H., Muthukrishnan, R. (**1981**) *J. Amer. Chem. Soc., 103* , 3782.
14. Andreetti, G. D., Ungaro, R., Pochini, A. (**1979**) *J. Chem. Soc., Chem. Commun.,* 1005.
15. Ungaro, R., Pochini, A., Andreetti, G. D., Domiano, P. (**1985**) *J. Chem. Soc., Perkin Trans. II,* 197.
16. Coruzzi, M. ; Andreetti, G. D.; Bocchi, V.; Pochini, A., Ungaro, R. (**1982**) *J. Chem. Soc. , Perkin Trans. II,* 1133.
17. Harrowfield, J. M., Ogden, M. I., Richmond, W. R., White, A. H. (**1991**) *J. Chem. Soc., Chem. Commun.,* 1159.
18. Olmstead, M. M., Sigel, G., Hope H.; Xu, X., Power, P. P. (**1985**) *J. Amer. Chem. Soc., 107,* 8087.
19. Corazza, F., Floriani, C., Chiesti-Villa, A., Guastini, C. (**1990**) *J. Chem. Soc., Chem. Commun.,* 1083.
20. Atwood, J. L., Bott, S. G., Jones, C., Raston, C. L. (**1992**) *J. Chem. Soc., Chem.Commun.,* 1349.
21. Hajek, F., Graf, E., Hosseini, M. W., Delaigue, X., De Cian, A., Fischer, J.(**1996**) *Tetrahedron Lett., 37,* 1401.
22. Hajek, F., Graf, E., Hosseini, M. W. (**1996**) *Tetrahedron Lett., 37,* 1409.
23 Hajek, F., Graf, E., Hosseini, M. W. , unpublished.
24 Hosseini, M. W., Ruppert, R., Schaeffer, P., De Cian, A. , Kyritsakas, N., Fischer J. (**1994**) *J. C. S. Chem. Comm.,* 2135; Brand, G., Hosseini, M. W., Ruppert, R., De Cian, A., Fischer, J., Kyritsakas, N. (**1995**) *New J. Chem., 19,* 9; Hosseini, M. W, Brand, G., Schaeffer, P., Ruppert, R., De Cian, A., Fischer, J. (**1996**) *Tetrahedron Lett., 37,* 1405.

NEW MACROCYCLIZATION REACTION BASED ON TRIS(2-AMINO-ETHYL)AMINE

JANUSZ JURCZAK,[a,b] PIOTR LIPKOWSKI,[a] DANIEL T. GRYKO,[a] JANUSZ LIPKOWSKI[c]

[a] *Institute of Organic Chemistry, Polish Academy of Sciences, 01-224 Warsaw, Poland*

[b] *Department of Chemistry, Warsaw Uniwersity, 02-093 Warsaw, Poland*

[c] *Institute of Physical Chemistry, Polish Academy of Sciences, 01-224 Warsaw, Poland*

Supramolecular chemistry deals with more or less rigid and organized molecules called molecular receptors. Owing to their noncovalent bonding, these species exhibit special abilities with respect to molecular receptors have to be properly organized in space. This specific organization can theoretically be achevied *via* step by step synthesis, using more or less elaborate synthetic procedures. The complexity of target molecules acting as molecular devices is, however, often too high to allow the chemist to build up the molecule stepwise. Moreover, the noncovalent interactions are of a fragile nature, and the classic synthetic strategies cannot be applied, as they use relatively hard tools to form the covalent bonds. The only way to obtain this type of desired molecules is to design the system, procure components and than to create suitable conditions to allow the system to preorganize [Stankiewicz *et al.*, 1992].

A. W. Coleman (ed.), Molecular Recognition and Inclusion, 123–128.
© 1998 *Kluwer Academic Publishers. Printed in the Netherlands.*

The self assembly approaches to the synthesis of various molecular receptors, such as diazacoronands [Jurczak *et al.*, 1993] and bicyclic [Pietraszkiewicz *et al.*,1984] or tricyclic [Jurczak *et al.*, 1992] cryptands, have been reported. For example, it has recently been found that dimethyl α,ω-dicarboxylates (*e.g.* **1**) react under ambient conditions with α,ω-diamino aliphatic ethers (*e.g.* **2**) in methanol as solvent, to afford the macrocyclic diamides of type **3** in good yields [Jurczak *et al.*, 1993] (Scheme 1).

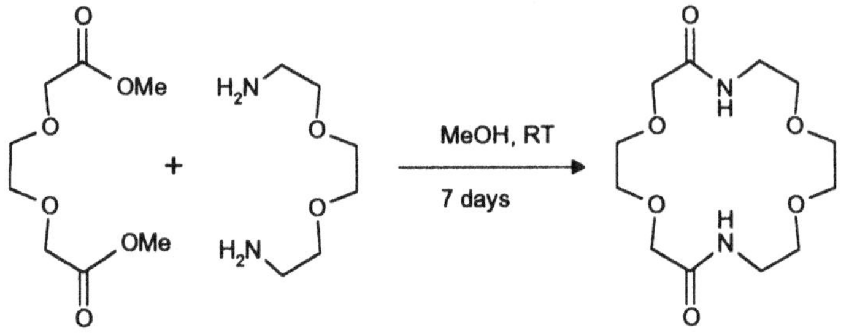

1 **2** **3**

Scheme 1

Recently, an interesting route to bicyclic cryptands of type **6** based on [2+3] Schiff-base condensation of tris(2-aminoethyl)amine **4** with dialdehyde **5** (Scheme 2) has been published [Jaźwiński *et al.*, 1987].

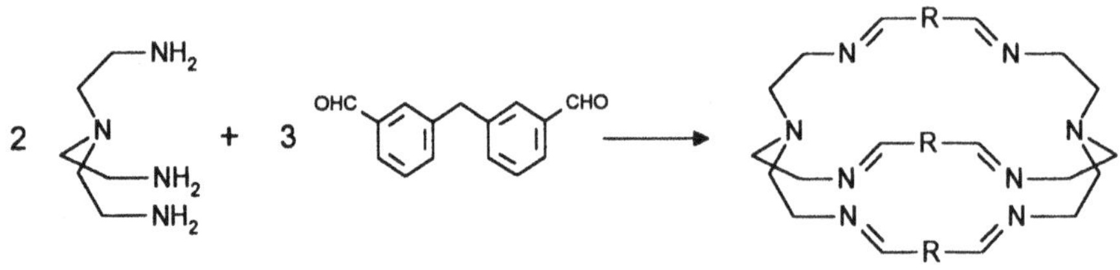

4 **5** **6**

Scheme 2

The above-presented results prompted us to apply tripodal amine **4** to preparation of various possible macrocyclic structures according to routes shown in Scheme 3.

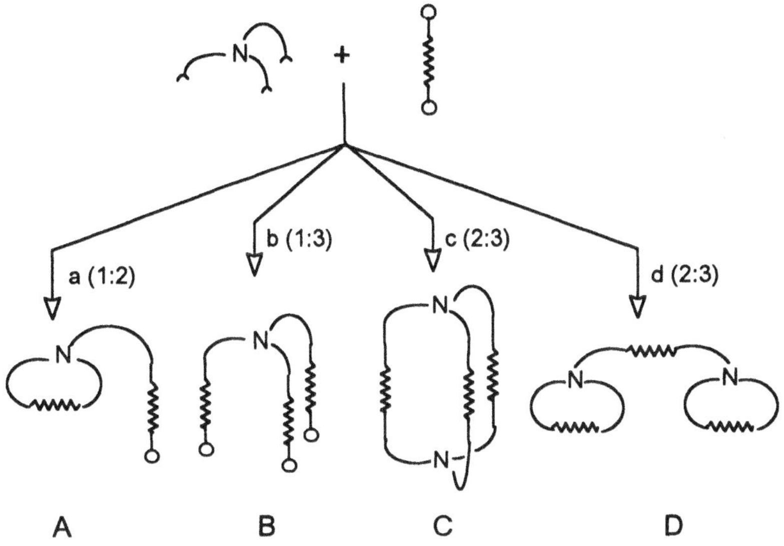

Scheme 3

The reaction of tripodal amine **4** with diester **7**, carried out in methanol under ambient conditions, afforded two products: armed monomacrocycle **8** (11% yield) *via* route a and bis-macrocycle **9** (4% yield) *via* route d (Scheme 4).

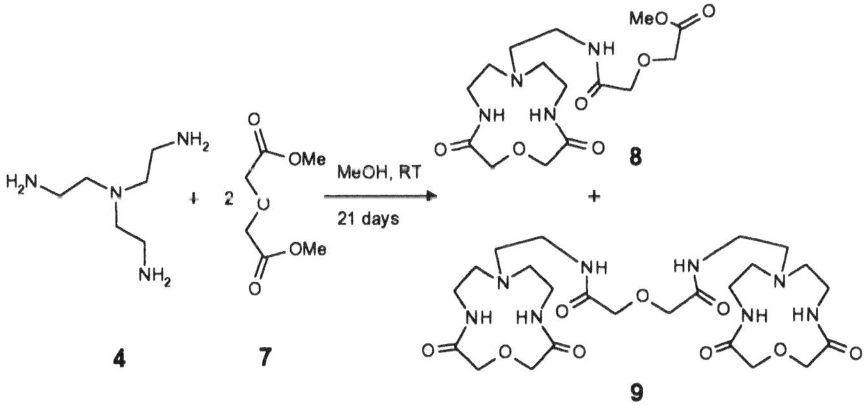

Scheme 4

The reaction of the same tripodal amine **4** with benzodiester **10**, carried out under analogous conditions, afforded only one product - crystalline bis-macrocycle **11** (34% yield) *via* route d (Scheme 5)

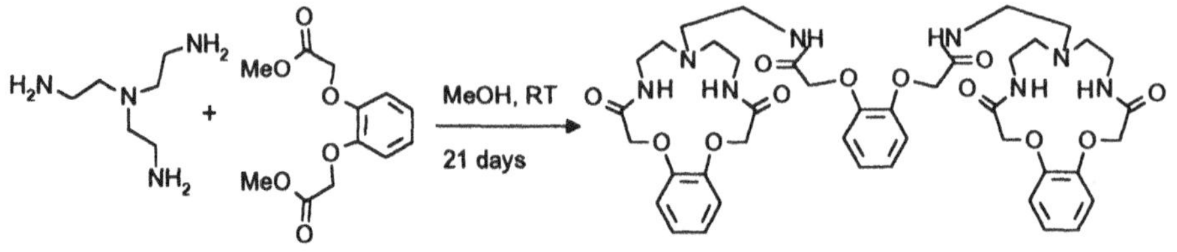

4 **10** **11**

Scheme 5

The compound **11** partly precipitated from the reaction mixture, and after a few attempts we managed to obtain the well-shaped crystals suitable for X-ray analysis. Figure 1 shows the selected projection of the molecule.

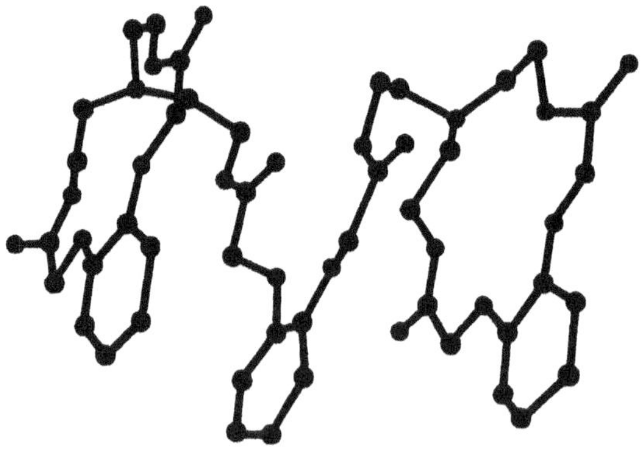

Figure 1

Finally, we carried out the reaction of tripodal amine **4** with tripodal ester **12**, expecting the formation of bicyclic cryptand **13** (Scheme 6). Instead of compound **13**, the crystalline tricyclic cryptand **14** was formed as a single product (50% yield).

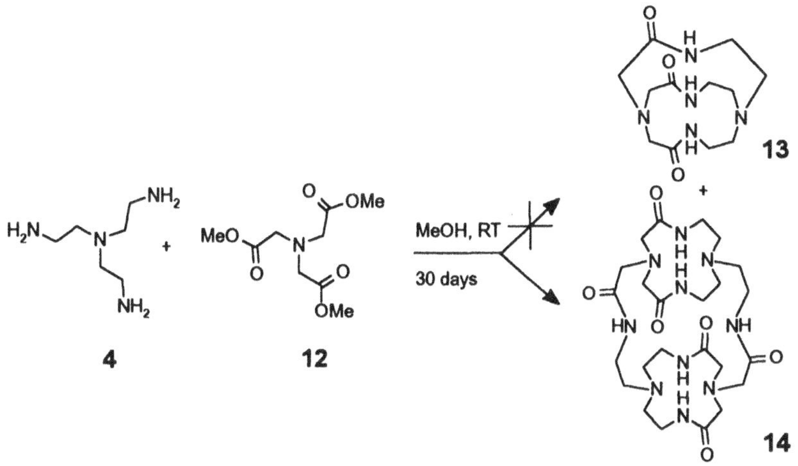

Scheme 6

The proposed structure of compound **14** the; based on spectral analysis (^{1}H and ^{13}C NMR and mass spectrometry). The ^{13}C NMR spectrum is very characteristic: it consists of four pairs of signals in a 1:2 ratio. The final proof of the structure was based on the X-ray analysis. Figure 2 shows the selected projection of the molecule.

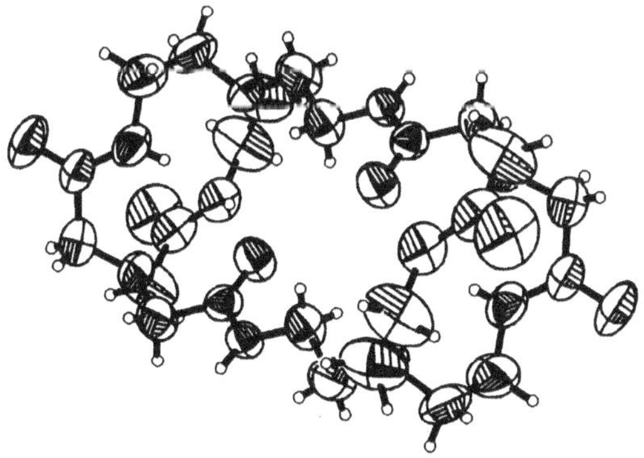

Figure 2

A similar product, appropriate tricyclic cryptand, was obtained in 39% yield from the reaction of tripodal amine **4** with another tripodal ester **15**.

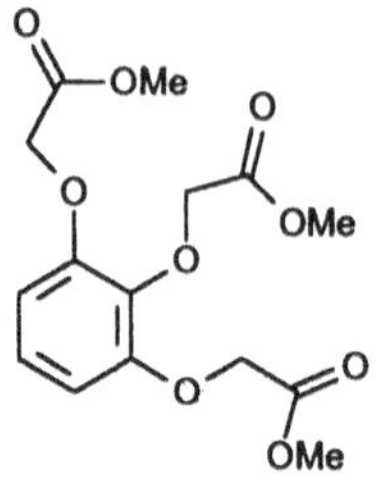

15

The present work deals with the syntheses of complex diazacoronands and tricyclic cryptands *via* the amidation reaction. The obtained compounds differ in the molecular architecture and size. The reactions studied proceed efficiently under ambient conditions, and their course is assumed to depend on the occurrence of self-assembly phenomena which are probably stimulated by a properly selected solvent.

References

1. Stankiewicz, T. and Jurczak, J. (1992) Self-assembly in Organic Synthesis, *Polish J. Chem* **66**, 1743-1779.
2. Jurczak, J., Ostaszewski, R., Sałański, P.and Stankiewicz, T. (1993) Synthesis of N,N'-Dimethyl Diazacoronands *via* Duble-quaternization Reaction, *Tetrahedron* **49**, 1471-1477.
3. Pietraszkiewicz, M., Sałański, P. and Jurczak, J. (1984) Synthesis of Novel Chiral [2.2.1] Cryptands Incorporating Sugars, *Tetrahedron* **40**, 2971-2973.
4. Jurczak, J. and Ostaszewski, R. (1992) A General High Presure Method for the Synthesis of Tricyclic Cryptands, *Polish J. Chem* **66**, 1733-1736.
5. Jurczak, J., Stankiewicz, T., Sałański, P., Kasprzyk, S. and Lipkowski, P. (1993) A New Method for the Synthesis of Diazacoronands *via* Double-amidation Reaction, *Tetrahedron* **49**, 1478-1488.
6. Jaźwiński, J., Lehn, J.M., Lilienbaum, D., Ziessel, R., Guilhem, J. and Pascard, C. (1987) Polyaza Macrobicyclic Cryptands: Synthesis, Crystal Structures of a Cyclophane Type Macrobicyclic Cryptand and of Its Dinuclear Copper (I) Cryptate, and Anion Binding Features, *J. Chem. Soc., Chem. Commun.* 1691-1694.

SIGNAL TRANSMISSION BY ARTIFICIAL RECEPTORS EMBEDDED IN BILAYER MEMBRANES

JUN-ICHI KIKUCHI
Institute for Fundamental Research of Organic Chemistry, Kyushu University
Hakozaki, Higashi-ku, Fukuoka 812-81, Japan

1. Introduction

Many metabolic processes in biological cells are triggered through recognition of signaling ligands by specific proteins, so-called receptors. The receptors are classified into two categories depending on their location in cells: intracellular receptors in cytoplasm and cell-surface receptors in biomembranes. Furthermore, there are at least three known classes of cell-surface receptors – channel-linked, G-protein-linked, and catalytic ones [1]. However, receptor functions remain to be clarified at the molecular level with emphasis on molecular recognition and the resulting responses. On these grounds, signal transduction based on molecular recognition by artificial receptors is one of the most attracting subjects in supamolecular chemistry [2]. While functional simulation of intracellular receptors in homogeneous solutions has been actively carried out with attention to various noncovalent interactions [3], signal transmission by artificial cell-surface receptors embedded in suplamolecular assemblies has been scarcely studied up to the present time, except for channel-linked receptors [4-6].

I would like to show here our recent approaches to develop artificial receptors being capable of performing functional simulation of signaling via G-protein-linked cell-surface receptors.

2. Molecular Design of Artificial Cell-surface Receptors

Receptor proteins placed in cell surfaces bind various signaling ligands, such as neurotransmitters, protein hormones, and growth factors, with high affinity and convert these extracellular events into intracellular signals that alter the behavior of the target cell. Accordingly, recognition of signaling ligands by receptors is a common and basic phenomenon in the cell signaling. As regards signaling via G-protein-linked receptors, response to the recognition is mediated by a G-protein and transmitted to an enzyme located in intracellular space. Thus, in order to simulate such signaling process the following requirements must be at least fulfilled for artificial receptors. (1) The receptor is capable of recognizing both a first signaling ligand and a signal transmitter. (2) Binding affinity of the receptor toward the signal transmitter changes depending on recognition of the signaling ligand. (3) The receptor effectively binds to bilayer membranes to form hybrid molecular assemblies. On these grounds, we have designed so-called steroid cyclophanes as artificial cell-surface receptors (Figure 1) [7-11].

A. W. Coleman (ed.), Molecular Recognition and Inclusion, 129–134.
© 1998 *Kluwer Academic Publishers. Printed in the Netherlands.*

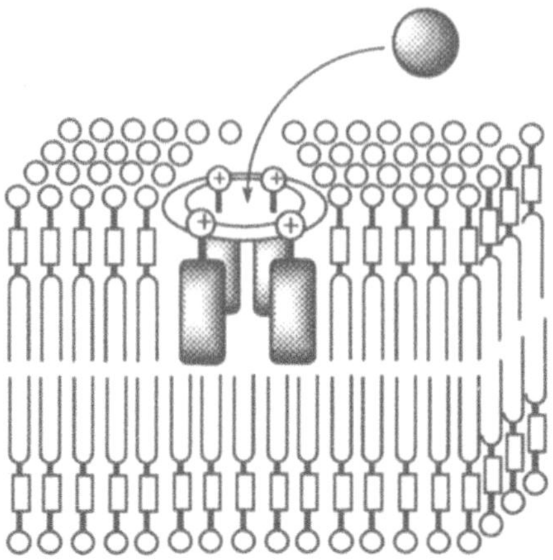

Figure 1. Schematic representation of molecular recognition by the steroid cyclophane embedded in a bilayer membrane.

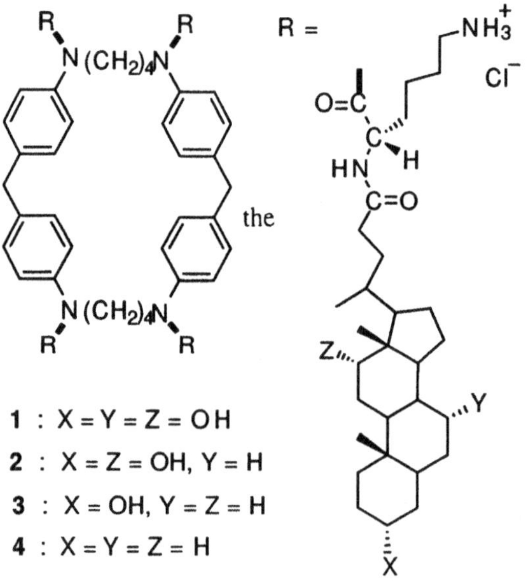

1 : X = Y = Z = OH
2 : X = Z = OH, Y = H
3 : X = OH, Y = Z = H
4 : X = Y = Z = H

Each of the steroid cyclophanes (**1**–**4**) is constituted in combination with three functional components; a 1,6,20,25-tetraaza[6.1.6.1]paracyclophane ring, four bile acid moieties, and four L-lysine residues connecting them. Bile acids are well-known natural compounds, and many studies have been carried out for characterization of them and their derivatives as hosts in various physical phases such as solid state, and aqueous and organic media [12-14]. In bilayer membranes, the bile acid moieties of the steroid cyclophane would act as hydrophobic anchors as well as functional components for molecular recognition. The CPK and computer-aided molecular model studies reveal that the steroid cyclophane is able to incorporate a hydrophobic guest molecule into its three-dimensionally extended inner cavity created by the four steroid moieties and the macrocyclic skeleton. We employed here hydrophobic aromatic guests and metal ions as signaling ligands and signal transmitters, respectively, for the steroid cyclophanes.

3. Molecular Recognition in Aqueous Solution

Prior to investigate chemical functions of the steroid cyclophanes as cell-surface receptors, we clarified their essential abilities for molecular recognition of various organic guests in aqueous media by means of ^{1}H NMR, fluorescence, and circular dichroism (CD) spectroscopies.

Upon complexation of 2-naphthyl phenyl ketone with the steroid cyclophane under conditions that allow the host to behave as a tetracationic species, **1** acted to induce marked upfield shifts of the proton signals of the guest, especially for 1-H, 4-H, 5-H, and 8-H on the naphthalene ring, while the signal splitting for aromatic protons of the host was much enhanced. Binding constant (K) for the host-guest complex and complexation-induced shifts (CIS), the shifts of NMR signals for the guest upon 100% complexation, were evaluated from the ^{1}H NMR titration curve by means of a numerical curve-fitting method: $K = 3.8 \times 10^4$ dm^3 mol^{-1} in D_2O–CD_3OD (8:2 v/v) at pD 5.0 and 313 K. Judging from the CIS values, the guest is incorporated into the macrocyclic cavity with the long axis of the naphthalene ring nearly parallel to the molecular axis of the steroid cyclophane (Figure 2). It is noteworthy that the phenyl ring of the guest molecule is presumably surrounded by four steroid moieties in order to minimize hydrophobic hydration in aqueous media. Thus, hydrogen-bonding between the carbonyl oxygen of the guest and the 12-OH on a steroid moiety of the host becomes to be possible in the complex as shown in Figure 2. Such hydrogen-bonding interaction in the host–guest complex is supported from the results that induced CD bands for the guest molecule can be observed upon complexation with **1** or **2** but not with **3** or **4**. The K value for the complex of the guest with **1** increased to be 1.2×10^6 dm^3 mol^{-1} in aqueous acetate buffer containing 1%(v/v) ethanol at pH 5.0 and 303 K, as evaluated by means of CD spectroscopy. In addition, studies on binding behavior of the steroid cyclophanes with various fluorescent guests revealed that the hosts effectively incorporate guests having a naphthalene moiety in a 1:1 stoichiometry through hydrophobic, electrostatic, and hydrogen-bonding interactions.

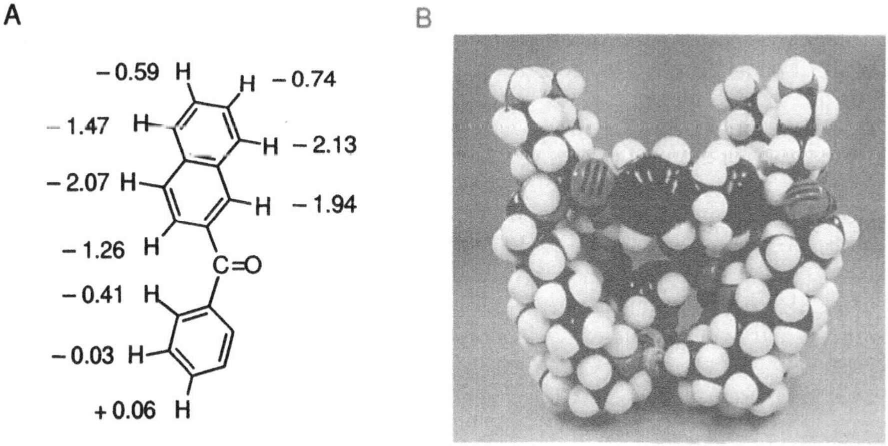

Figure 2. CIS values (ppm) for 2-naphthyl phenyl ketone bound to **1** (A) and a space-filling model of the host–guest complex (B).

4. Molecular Recognition in Bilayer Membranes

We have recently developed synthetic lipids bearing α-amino-acid residue(s), interposed between a polar head moiety and a hydrophobic double-chain segment through peptide bonds [15]. In general, such peptide lipids afford morphologically stable bilayer aggregates in aqueous media and are useful as molecular elements for simulation of biomembrane processes. When the steroid cyclophanes are embedded in the bilayer membranes formed with the peptide lipids, the hosts behave as artificial cell-surface receptors exhibiting marked discrimination toward guest molecules [7-10].

For example, 6-(*p*-toluidino)naphthalene-2-sulfonate (TNS) effectively bound to the hydrophobic binding site of **1** embedded in the bilayer vesicle formed with **5** through hydrophobic and electrostatic interactions. We evaluated stoichiometry and formation constant for the host–guest complexation by fluorescence spectroscopy under conditions that a composition of the steroid cyclophane relative to the peptide lipid was maintained constant in a molar ratio of 1:40. The continuous variation method applied to the present system clearly indicates that **1** and TNS forms a 1:1 complex in the bilayer vesicles as well as in aqueous solution. The K value was determined on the basis of Benesi-Hildebrand plot; 5.1 x 10^4 dm^3 mol^{-1}. The corresponding K value for the complex of TNS with **4** was 1.8 x 10^4 dm^3 mol^{-1}. Although the binding constants for the steroid cyclophanes embedded in the bilayer membrane were somewhat lower than those in aqueous solution, the superiority of **1** over **4** for molecular recognition of TNS was also retained in the bilayer membrane: the K values were 1.3 x 10^6 and 3.8 x 10^5 dm^3 mol^{-1} for **1** and **4**, respectively, in aqueous acetate buffer (10 mmol dm^{-3}) at pH 5.0 and 30.0 °C. The results imply that hydrogen-bonding effect originating from the hydroxyl groups placed in the steroid skeleton of **1** gives out a significant role in the guest recognition behavior.

1 is capable of providing a chiral binding site for TNS in bilayer membrane as well as in aqueous solution. The bilayer vesicle formed with **5** showed a CD band at 231 nm with $[\theta]$ of –2.3 x 10^3 deg cm^2 $dmol^{-1}$. No induced CD was observed upon addition of TNS to this vesicular solution. CD spectra for the hybrid bilayer assembly of **1** with **5** was nearly identical to a sum of the individual spectra, strongly suggesting that the steroid cyclophane was incorporated into the membrane without meaningful conformational change. In the present hybrid assembly, we observed induced CD bands due to TNS bound to **1**: λ_{max} ($[\theta]$) values for a guest upon 100% complexation with a host were 323 (+1.9 x 10^4) and 371 (–0.7 x 10^4). On the other hand, the induced CD band was not detected in the hybrid assembly composed of **4** and **5**.

1-Hydroxy-2-naphthaldehyde (HNA) was effectively recognized by the steroid cyclophane (**1**) embedded in the bilayer vesicle constituted with **6** to form an imine bond between the formyl group of the guest and the amino group of **1**, as confirmed by means of electronic absorption spectroscopy: the K value was 10^6 dm^3 mol^{-1} in aqueous HEPES buffer (100 mmol dm^{-3}) at pH 7.0 and 30.0 °C. The complexation is reversible and much enhanced upon addition of copper(II) ions due to formation of the corresponding metal complex.

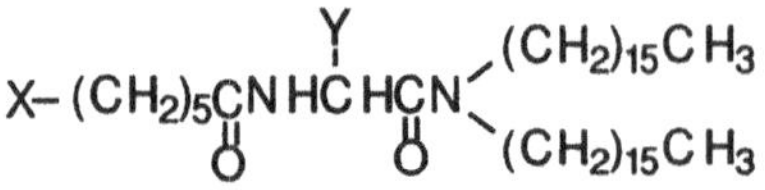

5 (X = Na^+ ^-O_3S– , Y = CH_3)
6 (X = $(CH_3)_3N^+$ Br^- , Y = H)

5. Signal Transmission by Artificial Receptors

NADH-dependent lactate dehydrogenase (LDH) catalyzes transformation between pyruvate and lactate and its activity is inhibited in the presence of metal ions [16]. We clarified that the inactivation of LDH by copper(II) ions is reversible process and that the enzyme binds to bilayer vesicles of cationic peptide lipids mainly through electrostatic interactions [17]. On these grounds, we constituted a supramolecular assembly as an artificial signaling system in combination with the steroid cyclophane (**1**), the cationic peptide lipid (**6**), and pig heart LDH, as schematically shown in Figure 3. The catalytic activity of LDH in the reduction of pyruvate was evaluated spectrophotometrically by following a consumption rate of NADH in the presence and absence of signaling ligands such as HNA and copper(II) ions.

Each bar chart in Figure 4 represents an extent of inactivation of LDH upon addition of copper(II) ions in the bilayer membrane. In the absence of **1**, we observed a 39% decrease in the LDH activity by copper(II) ions under the present conditions, and the LDH activity was not affected upon addition of HNA. On the other hand, LDH was

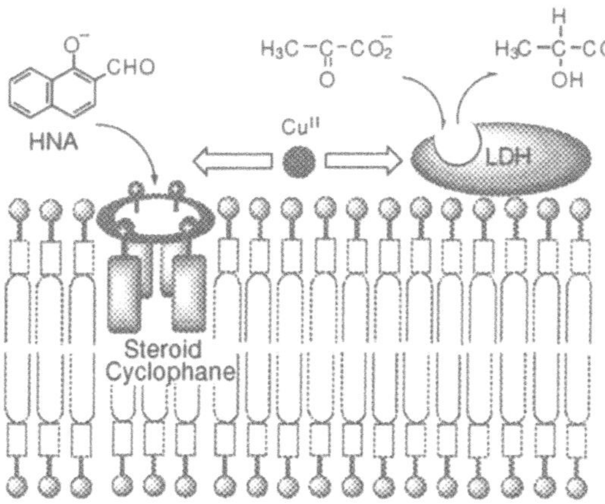

Figure 3. Schematic representation of signal transduction mediated by the steroid cyclophane as an artificial cell-surface receptor.

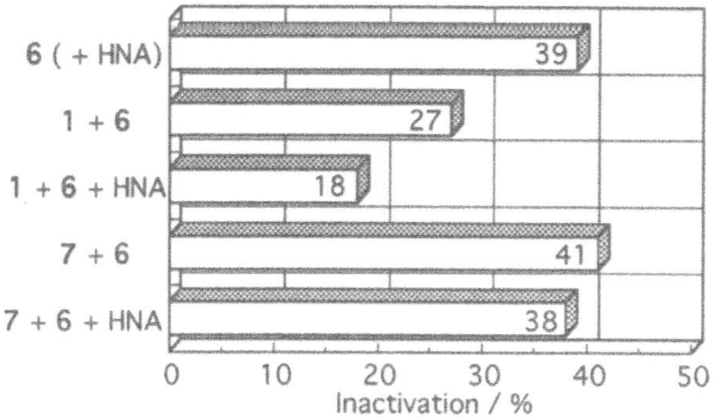

Figure 4. Inactivation of LDH by copper(II) ions in the presence of various molecular assemblies in HEPES buffer (100 mmol dm^{-3}) at pH 7.0 and 30.0 °C. Concentrations in mmol dm^{-3}: pyruvate, 3.0; NADH, 0.25; **6**, 0.8; $Cu(ClO_4)_2$, 0.01; HNA, 0.02; **1**, 0.02; **7**, 0.02. LDH, 170 μg dm^{-3}.

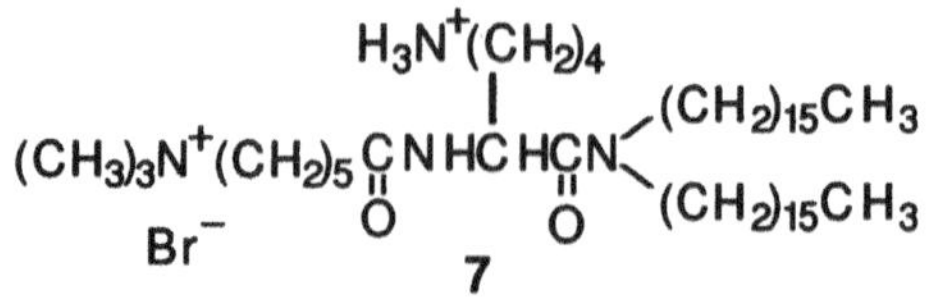

7

activated a 12% extent upon addition of **1**, reflecting that copper(II) ions bound to the enzyme were partly transferred to the receptor. It is noteworthy that the LDH activity recovered furthermore upon addition of HNA to the hybrid assembly. Such activation of the enzyme comes from that the metal binding affinity of the receptor was enhanced by recognition of the signaling ligand. When we replaced the steroid cyclophane with a peptide lipid having an L-lysine residue (**7**), we observed no recovery of the LDH activity under the comparable conditions.

In conclusion, we demonstrated here the first example of supramolecular assemblies in which an artificial cell-surface receptor transmits an external signal to an enzyme via a second messenger signal. Although we employed bilayer membrane without discrimination between the inner and the outer surfaces at present, I believe more intelligent supramolecular systems would make come true by further modification of structure and function of the artificial receptor.

6. References

1. Alberts, B., Bray, D., Lewis, J., Raff, M., Roberts, K., and Watson, J.D.: *Molecular Biology of the Cells*, Garland Publishing, New York, 1994, Chap. 15.
2. Lehn, J.-M.: *Supramolecular Chemistry*, VCH, Weinheim, 1995.
3. Lehn, J.-M. Ed.: *Comprehensive Supramolecular Chemistry*, Pergamon, Oxford, 1996, and references cited therein.
4. Kobuke, Y., Ueda, K., and Sokabe, M.: Artificial Non-Peptide Single Ion Channels, *J. Am. Chem. Soc.* **114** (1992), 7618-7622.
5. Åkerfeldt, K.S., Lear, J.D., Wasserman, Z.R., Chung, L.A., and DeGrado, W.F.: Synthetic Peptides as Models for Ion Channel Proteins, *Acc. Chem. Res.* **26** (1993), 191-197.
6. Ghardiri, M.R., Granja, J.R., and Buehler, L.K.: Artificial Transmembrane Ion Channels from Self-assembling Peptide Nanotubes, *Nature* **369** (1994), 301-304.
7. Kikuchi, J., Matsushima, C., Suehiro, K., Oda, R., and Murakami, Y.: Molecular Recognition by Steroid Cyclophane in Aqueous Solution and Synthetic Bilayer Membrane, *Chem. Lett.* (1991), 1807-1810.
8. Kikuchi, J., Matsushima, C., Tanaka, Y., Hie, K., Suehiro, K., Hayashida, O., and Murakami, Y.: Molecular Recognition by Macrocyclic Receptors Having Multiple Hydrophobic Branches in a Synthetic Bilayer Membrane, *J. Phys. Org. Chem.* **5** (1992), 633-643.
9. Kikuchi, J., Inada, M., Miura, H., Suehiro, K., Hayashida, O., and Murakami, Y.: Steroid Cyclophanes as Artificial Receptors Embedded in Synthetic Bilayer Membranes: Aggregation Behavior and Molecular Recognition, *Recl. Trav. Chim. Pays-Bas* **113** (1994), 216-221.
10. Kikuchi, J., Ogata, T., Inada, M., and Murakami, Y.: Circular Dichroism of an Aromatic Guest Induced by a Chiral Steroid Cyclophane in Aqueous Solution and Synthetic Bilayer Membrane, *Chem. Lett.* (1996), 771-772.
11. Kikuchi, J., Inada, M., Murakami, Y., Egami, K., and Suehiro, K.: Characterization of Three-dimensionally Extended Hydrophobic Cavities. Difference in Molecular Recognition Ability between Steroid and Octopus Cyclophanes, *J. Phys. Org. Chem.* in press.
12. Miyata, M., Shibakami, M., Chirachanchai, S., Takemoto, K., Kasai, N., and Miki, K.: Guest-responsive Structural Changes in Cholic Acid Intercalation Crystals, *Nature* **343** (1990), 446-447.
13. O'Connor, C.J. and Wallace, R.G.: Physico-chemical Behavior of Bile Salts, *Adv. Colloid Interface Sci.* **22** (1985), 1-111.
14. Davis, A.P.: Cholaphanes et al.; Steroids as Structural Components in Molecular Engineering, *Chem. Soc. Rev.* **22** (1993), 243-253.
15. Murakami, Y. and Kikuchi, J.: Supramolecular Assemblies Formed with Synthetic Peptide Lipids. Functional Models of Biomembranes and Enzymes, *Bioorg. Chem. Frontiers* **2** (1991), 73-113.
16. Steinbüchel, A. and Schlegel, H.G.: NAD-linked L(+)-Lactate Dehydrogenase from the Strict Aerobe *Alcaligennes eutrophus*, *Eur. J. Biochem.* **130** (1983), 321-328.
17. Kikuchi, J., Kamijyo, Y., Etoh, H., and Murakami, Y.: Catalytic Performance of a Supramolecular Bienzyme Complex Formed with Artificial Aminotransferase and Natural Lactate Dehydrogenase, *Chem. Lett.* (1996), 427-428.

INCLUSION COMPOUNDS: KINETICS AND SELECTIVITY

L. R. NASSIMBENI
Chemistry Department, University of Cape Town
Rondebosch 7700, South Africa

Many of the classical compounds which act as hosts in clathrate structures were discovered by chance [1,2]. However in the last twenty years considerable effort has gone into the synthesis of host molecules with specific properties, and Weber has recently reviewed the principles of directed host design [3,4]. He has shown that a successful host molecule should be bulky and rigid in order to provide suitable cavities which can accomodate a guest. In addition, it is helpful to have a high affinity functional group in the host molecule which can engage in specific host-guest interactions such as hydrogen bonding.

Non-covalent interactions such as hydrogen bonding are crucial components of "Supramolecular Synthons" which Desiraju describes in his concept of Crystal Engineering and Molecular recognition [5], and our laboratory has studied the physico-chemical properties of organic inclusion compounds in order to relate them to their structures.

A general scheme outlining the formation and decomposition of such compounds is shown in Fig 1. The solid host compound, in its non-porous, α-phase, is exposed to a guest, generally in the liquid or vapour phase, and recrystallises as the host-guest compound or β-phase. This may in turn decompose in a number of different ways:

1. It may lose the guest and revert to its original α-phase. Interestingly the host may recrystallise in different α phases, depending on the nature of the guest, and we have studied several examples of this induced polymorphism in the host structure [6].

2. It may lose only part of the guest and recrystallise as a new compound with a different host:guest ratio, labelled as the γ-phase.

3. It may lose all its guest but the host may retain its β structure, and thus behave like a zeolite, giving rise to an empty cage or β_0 phase. This situation is rare but has been shown to occur in the gossypol-dichloromethane clathrate [7].

A. W. Coleman (ed.), Molecular Recognition and Inclusion, 135–152.
© 1998 *Kluwer Academic Publishers. Printed in the Netherlands.*

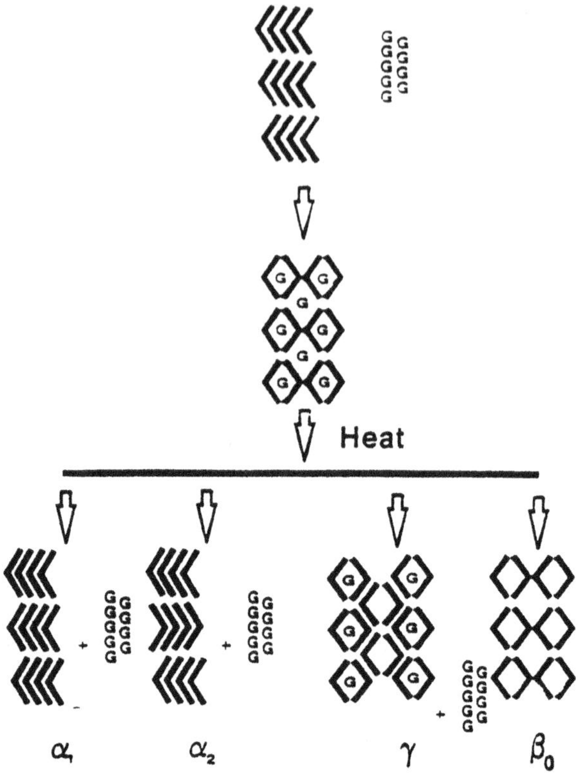

Figure 1. **Schematics for the formation and decomposition of an inclusion compound.**

Inclusion Compounds lend themselves to analysis by a number of complementary techniques, including structure determination by diffraction methods, thermal analysis, various spectroscopic techniques and molecular mechanics calculations. The main methods of analysis which we employed to monitor the structural changes which occur during their decomposition are depicted in Fig 2. This shows the weight loss of an inclusion compound when subjected to a controlled temperature programme, as measured by Thermal Gravimetry (TG). This technique is capable of high precision, often better than 1%, and allows us to obtain accurate Host:Guest ratios. This is important because Inclusion Compounds are often not stoichiometric, and the Host:Guest ratios thus derived are used to assign sensible site occupancy factors to the atoms of the guest molecules in the refinement of their crystal structures. The TG curves are complimented by the results of Differential Scanning Calorimetry (DSC), which, in principle, yield both the onset temperatures and the enthalpy changes associated with the guest release reaction.

Finally the structural changes accompanying the desolvation can be monitored by recording the X-ray powder diffraction patterns. This can be carried out continuously by using a modified Weissenberg goniometer which allows us to monitor the diffraction pattern from 20°C to 160°C on a cassette which moves in concert with the temperature programme [8].

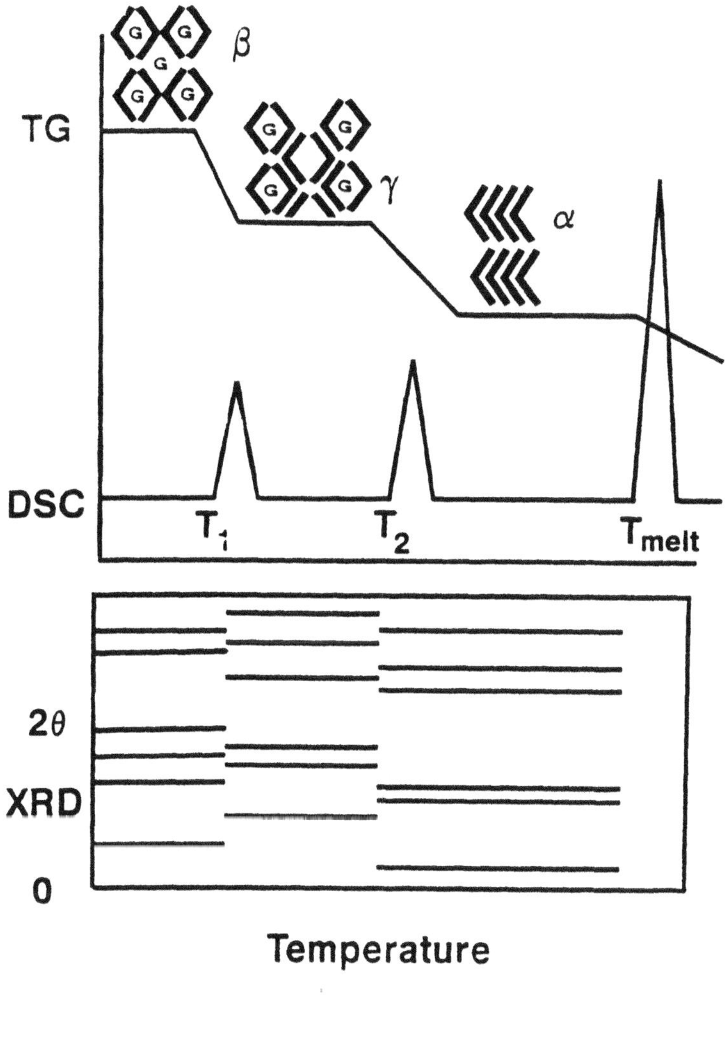

Figure 2. Combination of TG, DSC and XRD for monitoring the phase changes in the decomposition of an inclusion compound.

KINETICS OF ENCLATHRATION

The kinetics of inclusion between solid organic hosts and gaseous guests have received little attenton, because the measurements are experimentally difficult and the inclusion compounds formed are often unstable under ambient conditions. We have constructed an automated magnetic suspension balance which allows the measurement of mass gain with time under controlled conditions of temperature and pressure of the guest vapour [9].

However, this apparatus requires a minimum of 300 mg of host compound in order to yield satisfactory results, and such quantities of host are not always readily available. We have therefore constructed a Quartz Micro Balance which is highly sensitive and requires only a few micrograms of host compound which is thinly spread on the surface of a quartz crystal.

The method is based on the mass-frequency dependence of an oscillating crystal according to the formula

$$\Delta f = -k\, f_o^2\, \Delta m$$

where f_o is the frequency of the unloaded crystal, and k is a constant specific to each crystal. The frequency of the oscillation decreases with increasing mass of the crystal, and this is exploited to weigh very small quantities of the organic guest which deposits on the host covering the surface to form the inclusion compound:

$$H(s,\alpha) + nG(g) \rightarrow HG_n(s,\beta)$$

The gain in mass of the quartz crystal results in a decrease in oscillation frequency which can be monitored with time, thus enabling the kinetics to be measured.

The QMB apparatus is shown in Fig 3 and consists of a glass cell which can be evacuated and whose internal pressure can be measured with a transducer. There is an inlet for the guest vapour which is injected via a rubber septum. The cell is placed in a thermostat and the temperature is controlled to 0.1 K using a PID controller. The cell contains three probes. Each probe, shown in Fig 4, comprises a 10 MHz quartz crystal which has been mounted on a socket on a glass cone, with a crystal oscillator secured inside the cone. Each cone was sealed with a silicone sealant. An electronic circuit monitors the change in frequency of a given crystal and the signal is fed to a microcomputer which displays the change in frequency against time. In this way we were able to coat the three different crystals with different host compounds and analyse their kinetics with respect to a given guest simultaneously. Full details of the construction of the QMB have been published [10].

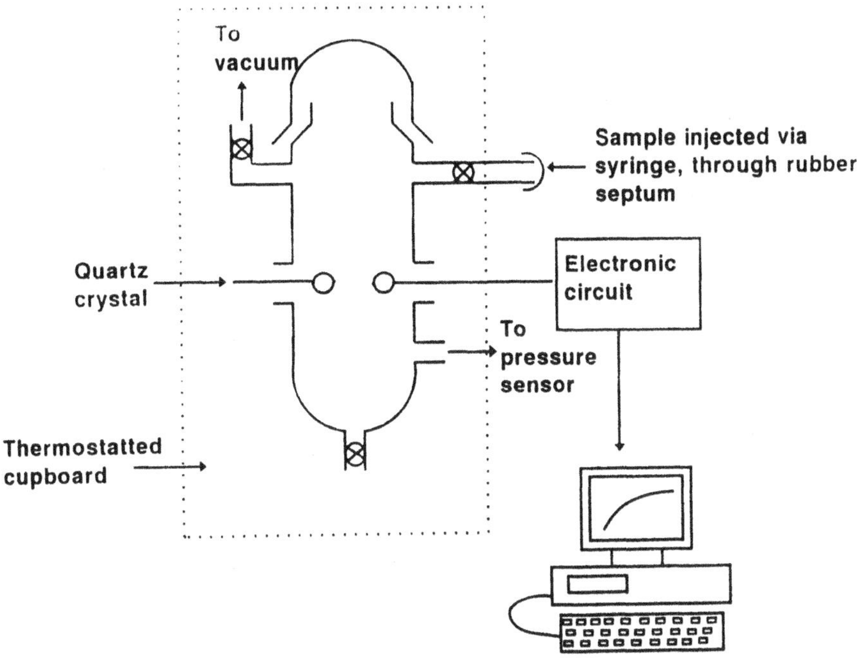

Figure 3. Principal features of the Quartz Microbalance (QMB) system.

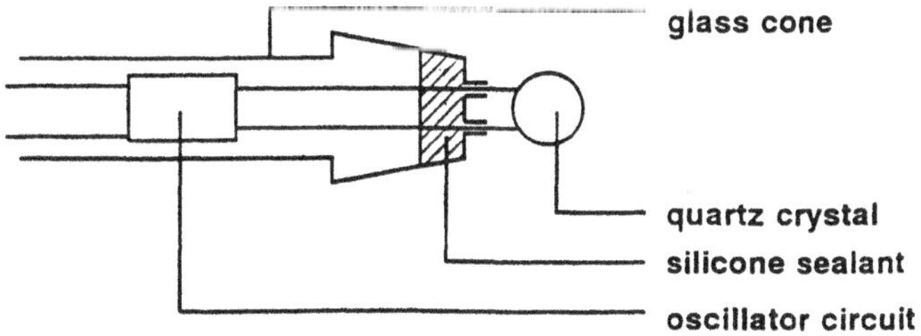

Figure 4. Schematic diagram of the quartz crystal probe.

The results for the reaction between the host H = 9,10-dihydroxy-9,10-diphenyl-9,10-dihydroanthracene (DDDA) with 1,3-dioxolane illustrate the use of the QMB.

The inclusion compound has a host:guest ratio of 1:1 and the structure is shown in Fig 5. The clathrate crystallises in the space group $P\bar{1}$ with Z = 1 and the host is located on a centre of inversion at Wyckoff position e. The dioxolane guest is statistically disordered and located at a centre of symmetry at Wyckoff position a. The structure is stabilised by (Host)–OH–O(Guest) hydrogen bonds [11].

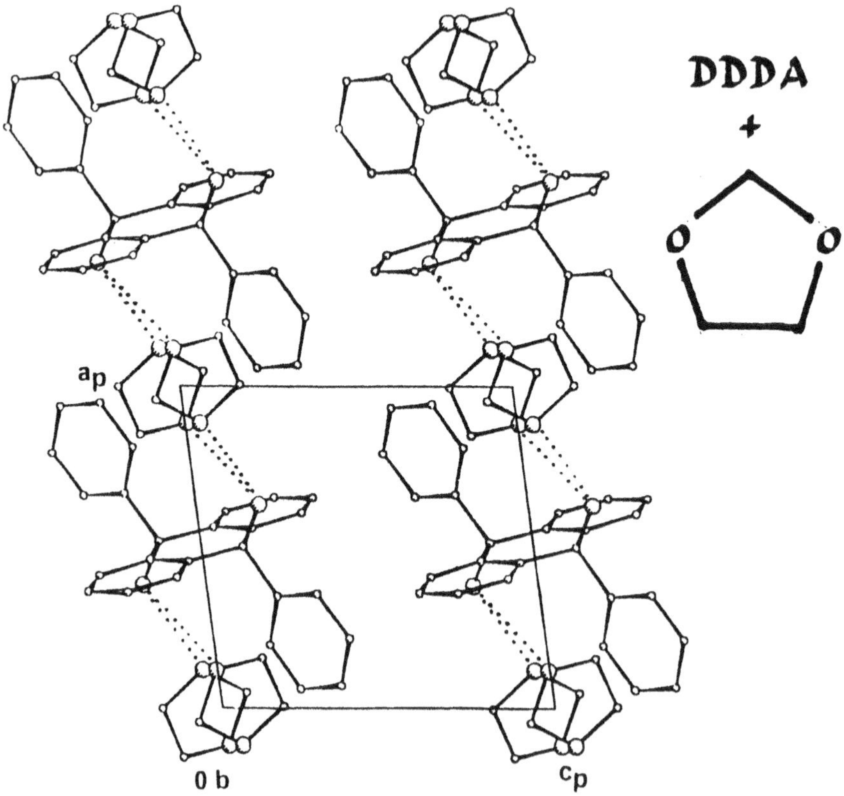

Figure 5. Packing of DDDA · dioxolane structure viewed along [101].

The curves showing the extent of reaction α versus time at 313 K for different values of the guest pressure are shown in Fig 6. They fit the Prout-Tompkins mechanism for sigmoidal curves:

$$\ln[\alpha/(1-\alpha)] = k_{obs}\, t$$

The velocity of the reaction increases with the pressure of the guest, an expected result. A more interesting, and unexpected result, is however obtained when we measure the kinetics of enclathration at a fixed pressure of the guest but at differing temperatures, Fig 7. Here we note that the reaction is *slower* at higher temperatures, and thus displays apparent anti-Arrhenius behaviour. This phenomenon has been noted before for the reaction of this host with acetone vapour [12].

A plot of k_{obs} versus pressure for various temperatures is shown in Fig 8. This shows that for a given temperature, there is a threshold pressure p_0 below which the reaction does not take place, and this has been confirmed experimentally. The reaction can then be explained in terms of the rate law

$$k_{obs} = k_f \frac{(P - Po)}{Po}$$

where k_f is the rate constant for the forward inclusion reaction. Using this treatment, the rate constants k_f do increase with increasing temperatures and a plot of ln k_f vs $^1/T$ yields a straight line with an activation energy of 140 kJ mol^{-1}.

We have also studied the kinetics of decomposition of this inclusion compound, DDDA · 1,3-dioxolane. This was carried out by analysing the weight loss versus time curves obtained under isothermal conditions over the temperature range 55°C to 80°C by Thermal Gravimetry. This method is too slow, however, when applied to the desolvation reaction at room temperature, since this requires several days to reach completion. We therefore followed it by recording the X-ray powder diffraction pattern and measuring the appearance and disappearance of selected peaks. It was gratifying to note that the resulting kinetic curves, measured by these two methods, could be successfully analysed by the same R2 mechanism for contracting area:

$$1-(1-\alpha)^{1/2} = kt$$

and the Arrhenius plot, shown in Fig 9, yields a good straight line over the complete temperature range, with an activation energy of 78 kJ mol^{-1}.

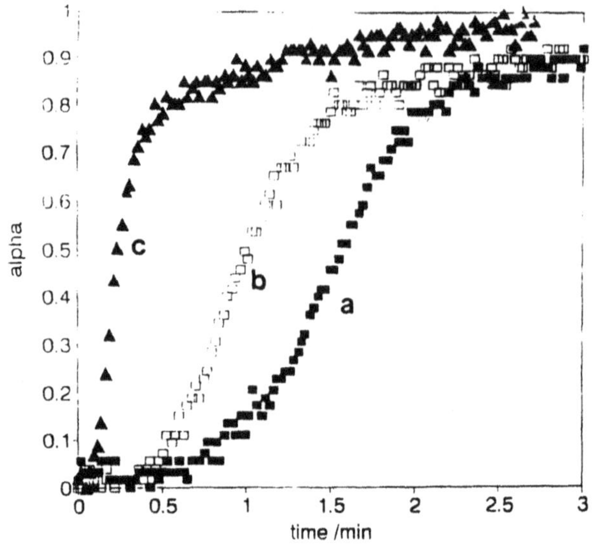

Figure 6. Isothermal kinetic curves (T = 313 K) for the reaction between DDDA and 1,3-dioxolane at various pressures.
(a) 82 mmHg (b) 97 mmHg (c) 120 mmHg

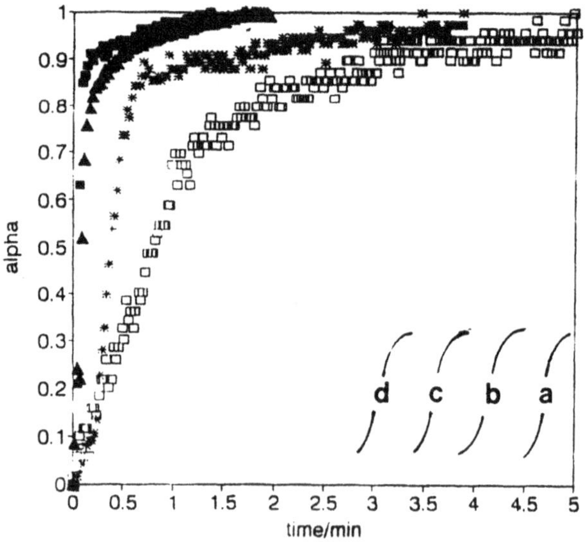

Figure 7. Temperature dependence of the kinetics at a fixed pressure (118 mmHg) of 1,3-dioxolane vapour.
(a) 318 K (b) 313 K (c) 308 K (d) 303 K

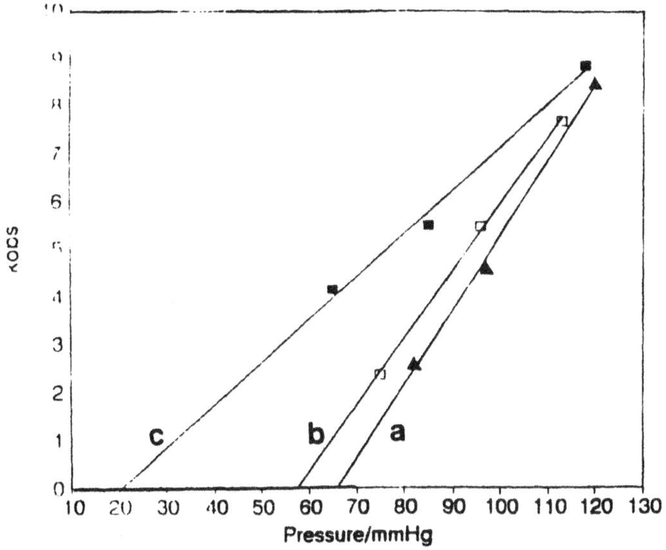

Figure 8. **Dependence of k_{obs} on the vapour pressure of 1,3-dioxolane at different temperatures.**
(a) 313 K (b) 308 K (c) 303 K

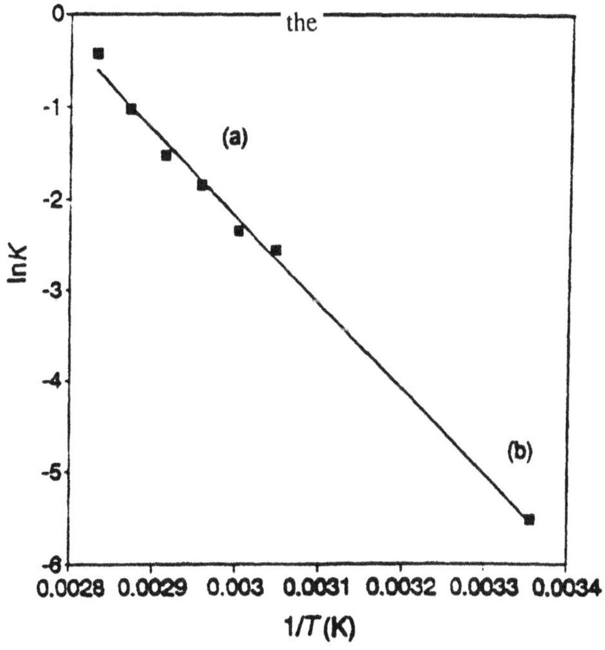

Figure 9. ln k vs T^{-1} for the DDDA-dioxolane desolvation.
(a) from TG (b) from XRD

We have analysed a number of desolvation reactions involving guests entrapped by bulky organic hosts [13,14], and have noted that when the guests are located in relatively open channels the activation energies of desolvation average ~ 60 kJ mol^{-1} while for guests which lie in cavities and are completely entrapped by the host, the avtication energies are higher, and average ~ 150 kJ mol^{-1}. We surmise that in the former reactions the guest molecules have a high mobility and can diffuse to the surface of the crystal with relative ease, while for the latter the host structure must be strongly disrupted in order for the guest to escape. In Fig 10 we have plotted a histogram of the frequency versus the magnitude of activation energies for decomposition reactions, including dehydrations and decompositions of oxides, hydrides, azides, carbonates and carboxylates [15], and compared these with the activation energies of desolvation for selected organic inclusion compounds. We note that the latter reactions have much lower activation energies, as expected from the weak interactions which stabilise the host-guest structures.

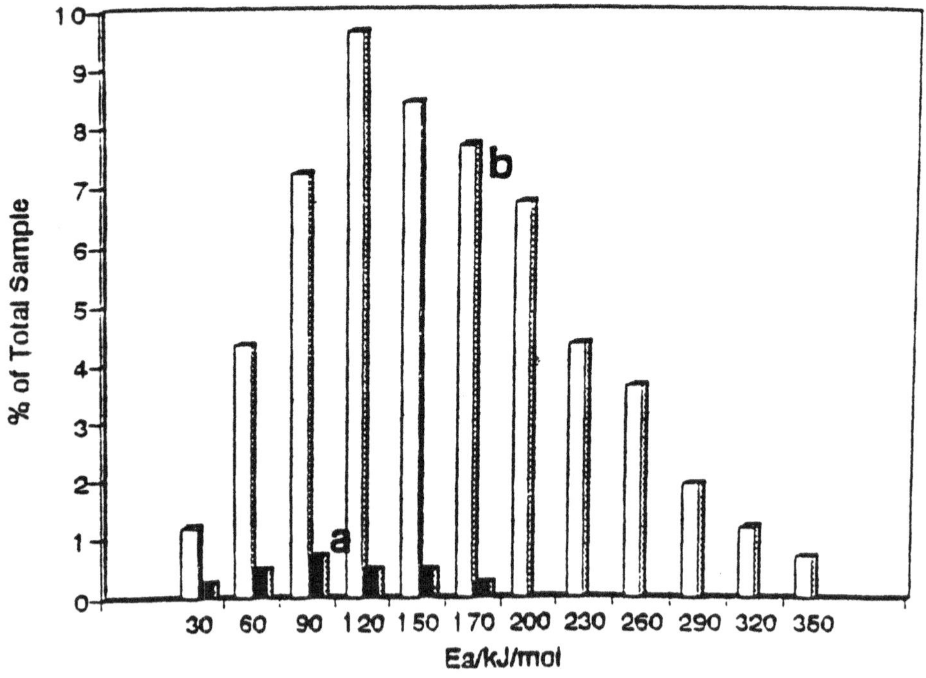

Figure 10. Histogram of activation energies for
(a) desolvation of inclusions compounds, and
(b) decomposition of compounds surveyed in [15].

SELECTIVITY OF ENCLATHRATION

The process of selective enclathration of a particular guest from a mixture, depends, ultimately, on the efficacy of the molecular recognition which occurs between host and guest. This is employed in the separation of close isomers by clathrate formation and the process is industrially important because it is simple, efficient and is not energy intensive.

We have studied the selectivity of enclathration by di-hydroxyphenylcyclohexane (DHPC) towards various groups of close isomers, including the phenylene diamines, benzene diols, picolines and lutidines.

The selectivity of a particular guest may be studied by setting up a number of solutions as shown in Fig 11. The quantity of the combined guest (say ortho- vs para-) should be at least ten times that of the host (moles G $\geq$ 10 moles H) and the mole fraction of one of the guests is varied systematically. If the guests are liquid, the host is dissolved in the liquid mixture, the host-guest compound is allowed to crystallise, filtered and analysed by a suitable technique such as GC or HPLC. If the guests are solid, a suitable common solvent which is *not* a competitor with the guests, must be employed. The experiment may also be carried out by shaking the host in a water suspension and this is a particularly attractive technique industrially, in that it avoids the tedious process of crystallisation on a large scale.

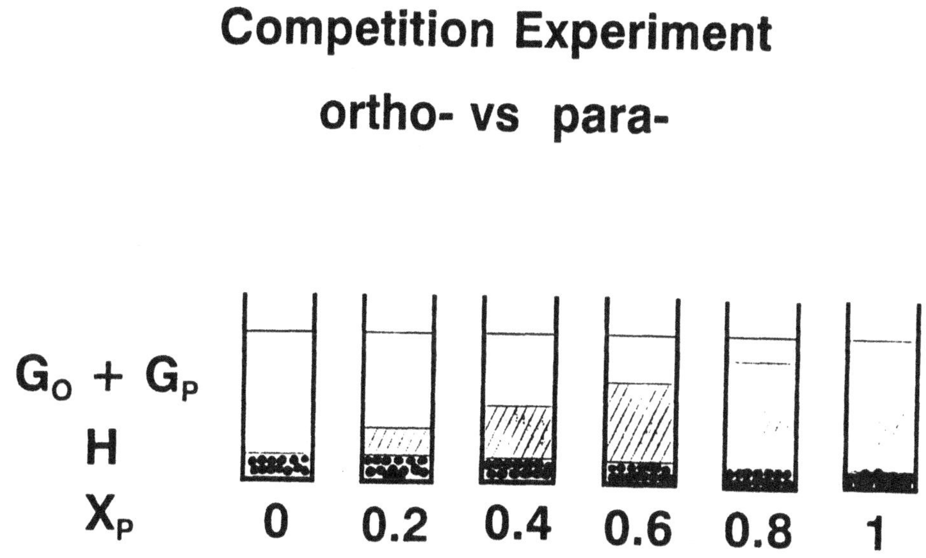

Figure 11. Scheme for competition experiment between close isomers.

The results of competition experiments for the benzene diols, catechol, resorcinol and hydroquinone with DHPC are shown in Fig 12. As can be seen for the two-guest competitions the *para-* isomer is strongly preferred over *ortho-* and *meta-*, but there is poor selectivity of the *ortho-* versus the *meta-* isomers. We extended the experiment to a simultaneous competition of all three isomers. The results are represented on the triangular graph of Fig 12, where the initial mixtures were taken at selected intervals on the circle and the final mixtures of the enclathrated guests are shown by the filled-in ellipse. The overall movement is clearly towards the *para-* isomer.

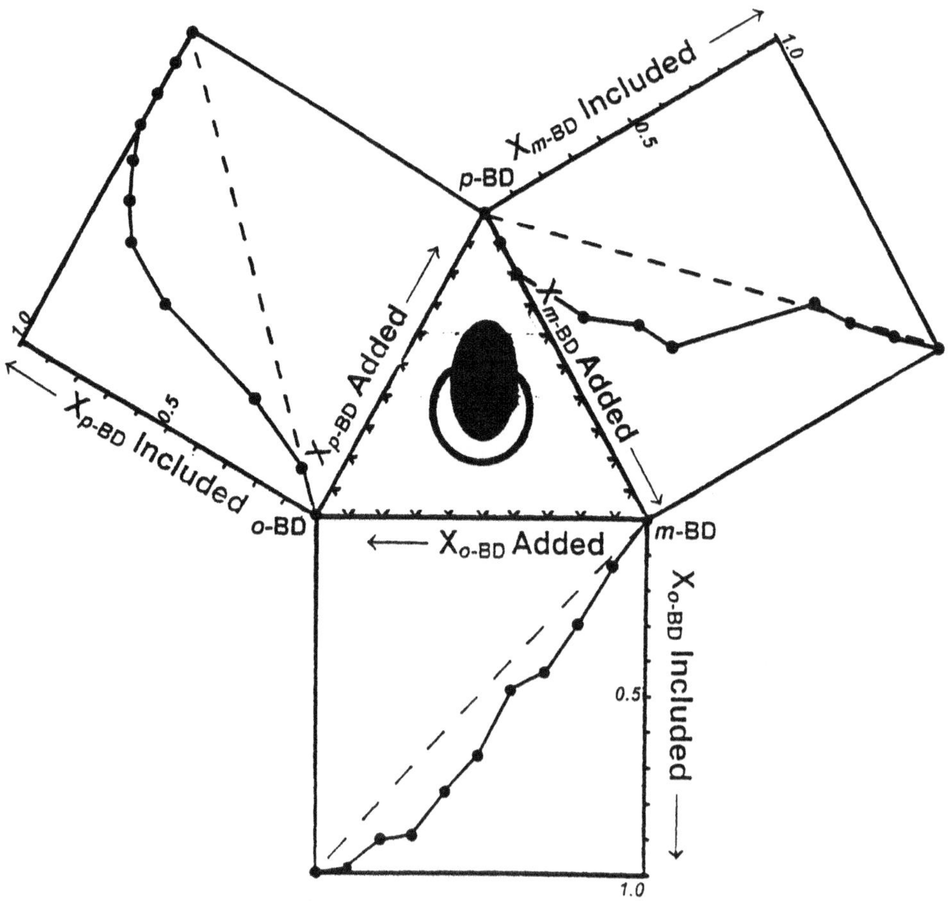

Figure 12. Selectivity curves for DHPC for the benzene diol isomers.

This can be explained in terms of the crystal structures. Fig 13 shows the hydrogen bonding scheme of the hydroquinone structure with DHPC. The compound crystallises in the space group C2/c, the guest lies on a diad, and the hydroxyl hydrogen is therefore disordered. The figure only shows one of the two configurations, and, on average, the guest is stabilised by

two (host)−OH···· O−(guest) and two (guest)−OH···· O−(host) hydrogen bonds.
(each O bearing H)

The O··· O distances vary from 2.64 to 2.73Å.

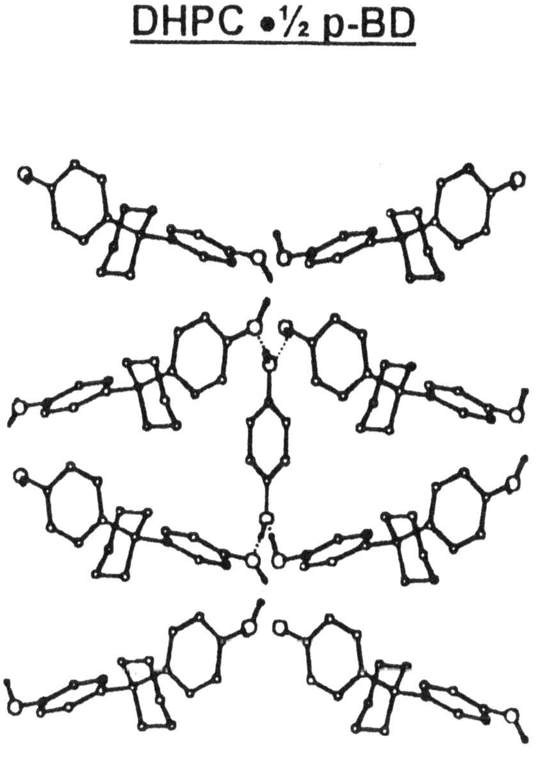

Figure 13. Hydrogen-bonding interactions in the DHPC · ½ hydroquinone structure.

The hydrogen-bonding scheme in the catechol is shown in Fig 14. This compound crystallises in the space group P$\bar{1}$ and both host and guest lie at general positions. Here the guest is stabilised by

two (guest)$-$OH $\cdots\cdot$ $\underset{\text{H}}{\text{O}}$$-$(host) and one (host)$-$OH $\cdots\cdot$ $\underset{\text{H}}{\text{O}}$$-$(guest) hydrogen bonds,

and the O$\cdots$O distances are somewhat longer, 2.70 to 2.90Å. The *para-* isomer is therefore stabilised by more and stronger hydrogen bonds than the *ortho-* isomer. This is not, however, a sufficient reason to state that the *para-* structure is unequivocally the more stable, and we have therefore calculated the lattice energy of both structures, using the technique of atom-atom potentials [16,17]. We summed the host-host, host-guest and guest-guest interactions occurring in both structures, incorporating the hydrogen bonding potential proposed by Vedani and Dunitz [18]. The lattice energies obtained were -264 kJ mol^{-1} for the *para-* and -250 kJ mol^{-1} for the *ortho-* structure. This is a convincing proof that the DHPC · ½ *para*-benzene diol structure is energetically preferred, and that the *para-* isomer will be favoured over the *ortho-* isomer in a competition experiment.

We failed to obtain single crystals of DHPC with resorcinol, but X-ray powder diffraction patterns showed this to be isomorphous to the catechol structure. We surmise that the hydrogen bonding and lattice energies of these two structures are similar and that there would be little preference for either isomer by the host. This was confirmed by the competition experiment.

DHPC • o-BD

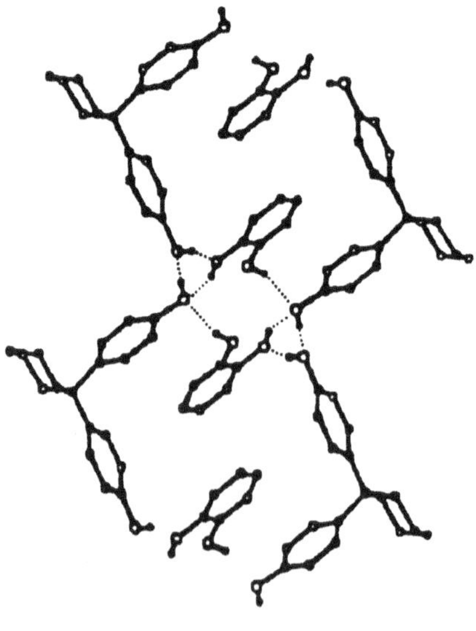

Figure 14. Hydrogen-bonding interactions in the DHPC · catechol structure.

SOLID-SOLID REACTIONS

We have successfully demonstrated the formation of a variety of host-guest compounds by direct interaction of the two components in solid-solid reactions. Thus we have shown that co-grinding sulfadimidine with several aromatic carboxylic acids produces 1:1 hydrogen bonded cocrystals which are identical to those obtained from reaction between the components in solution [19]. This technique has also been employed to prepare the inclusion compounds of alkylammonium bromides with phenol derivatives [20].

We have carried out the solid state reaction between DHPC and catechol and have demonstrated the formation of the 1:1 inclusion compound by measuring its X-ray powder diffraction pattern. The latter was compared with that simulated from the atomic positions derived from the single crystal structure, as shown in Fig 15 and there is an excellent match, both in peak position and relative intensities. The solid-solid reaction between DHPC and resorcinol yields the same powder diffraction pattern, showing these two inclusion compounds to be isomorphous.

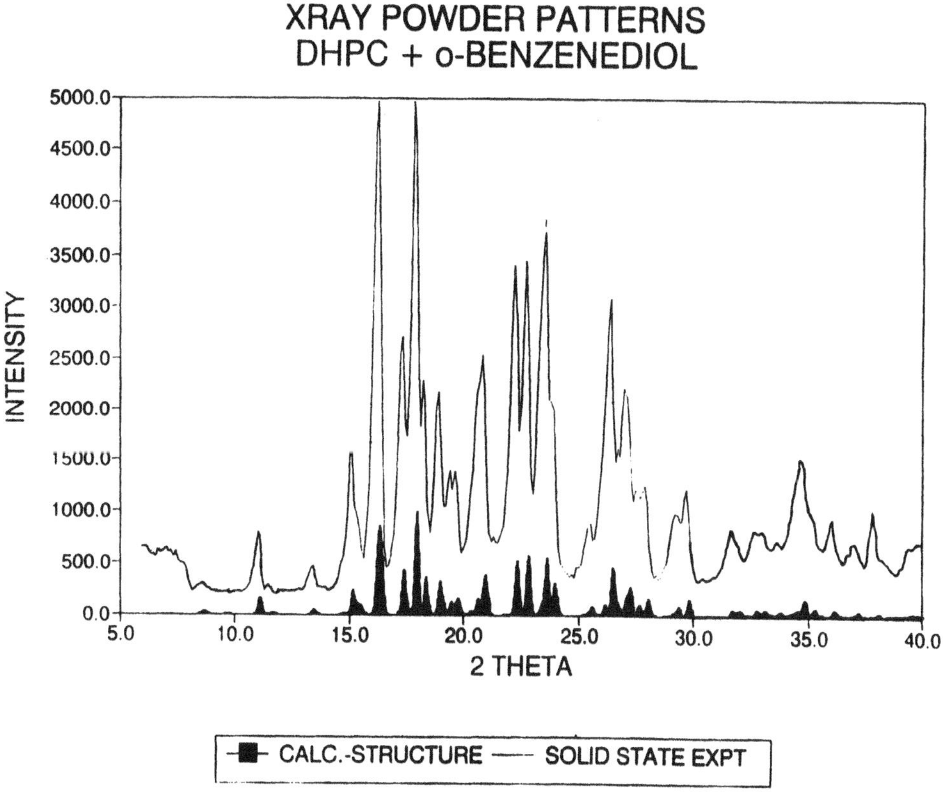

Figure 15. Powder diffraction patterns resulting from the DHPC–catechol solid-solid reaction.

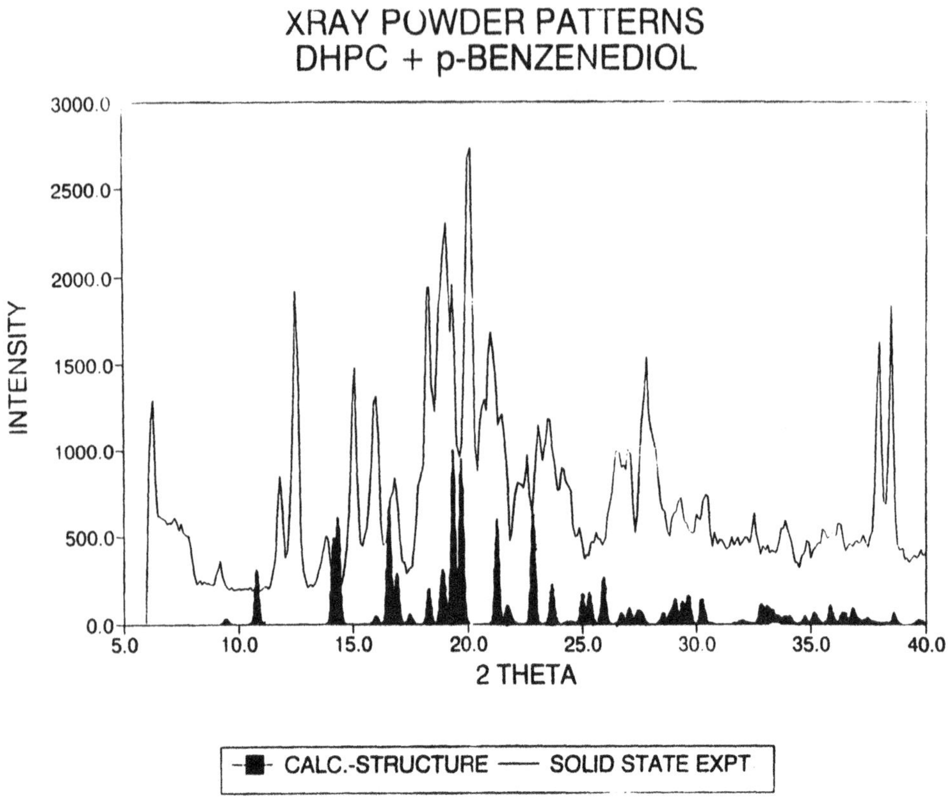

Figure 16. **Powder diffraction patterns resulting from the DHPC–hydroquinone solid-solid reaction.**

However, co-grinding DHPC with hydroquinone at room temperature yields a powder diffraction pattern which is simply the sum of two individual components, Fig 16 and is quite different from the simulated pattern derived from the DHPC-hydroquinone structure, showing that no reaction had taken place. We note that the melting point of hydroquinone is 170°C and is considerably higher than that of catechol at 105°C or resorcinol at 110°C. We have estimated the vapour pressure at 25°C for hydroquinone to be ~ 3×10^{-4} torr, and this is at least an order of magnitude lower than that of resorcinol, and two orders of magnitude lower than that of catechol. We suggest that the mechanism of the solid-solid reaction involves the guest in its vapour state, at least in the initiation stages of the reaction, and that the hydroquinone is insufficiently volatile for the reaction to occur. However we recognise that further co-grinding experiments under controlled conditions of temperature and particle size are needed to elucidate the mechanism of these solid-solid reactions.

References

1. Hagan, M. (1962) *Clathrate Inclusion Compounds*, Reinhold Publishing Corporation, New York.
2. Mandelcorn, L. (1964) *Non-stoichiometric Compounds*, Academic Press, New York.
3. Weber, E. (1987) *Molecular Inclusion and Molecular Recognition-Clathrates I*, Topics in Current Chemistry, Vol 140, Springer Verlag, Berlin.
4. Weber, E. (1991) *Inclusion Compounds*, Vol 4, Atwood, J.L., Davies, J.E.D. and MacNicol, D.D. Oxford University Press.
5. Desiraju, G.R. (1995) Supramolecular synthons in crystal engineering, *Angew. Chemie International Edition English*, **34**, 2311-2327.
6. Weber, E., Skobridis, K., Wierig, A., Nassimbeni, L.R., and Johnson, L. (1992) Complexation with Diol Host Compounds. Part 10, *J. Chem. Soc. Perkin 2*, 2123-2130.
7. Ibragimov, B.T., Talipov, S.A. and Aripov,T.F. (1994) Inclusion complexes of the natural product gossypol. Recognition by gossypol of halogeno methanes. Structure of the dichloromethane complex of gossypol and single crystal conservation after decomposition, *J. Incl. Phenom. Mol. Recog. Chem.*, **17**, 317-325.
8. Boeyens, J.C.A., Ferg, E., Levendis, D.C. and Schoning, F.R.L. (1991) X-ray diffraction analysis of the ammonium nitrate IV-III-II and IV-II phase changes under controlled humidity conditions, *S. Afr. J. Chem.*, **44**, 42-46.
9. Barbour, L.J., Achleitner, K. and Greene, J.R. (1992) A system for studying gas-solid reactions in controlled atmospheres, *Thermochimica Acta*, **205**, 171-177.
10. Coetzee, A., Nassimbeni, L.R. and Achleitner, K. (1996) A quartz microbalance for measuring the kinetics of guest uptake from the vapour, *Thermochimica Acta*, in press.
11. Caira, M.R., Coetzee, A., Nassimbeni, L.R. and Toda, F. (1996) Inclusion of dioxolane by a tricyclic diol host. Structure and kinetics, *J. Chem. Res.(S)*, 280-281.
12. Barbour, L.J., Caira, M.R. and Nassimbeni, L.R. (1993) Kinetics of inclusion, *J. Chem. Soc. Perkin 2*, 2321-2322.
13. Barbour, L.J., Caira, M.R., Coetzee, A. and Nassimbeni, L.R. (1995) Kinetics of desolvation of inclusion compounds of 9,10-derivatives of *trans*-9,10-dihydroxy-9,10-dihydroanthracene with benzene, *J. Chem. Soc. Perkin 2*, 1345-1349.
14. Caira, M.R., Coetzee, A., Nassimbeni, L.R., Weber, E. and Wierig, A. (1995) Complexation with diol host compounds, part 20. Kinetics of desolvation of inclusion compounds of 2,2′-bis(2,7-dichloro-9-hydroxy-9-fluorenyl)biphenyl with 1,4-dioxane and 1,3-dioxolane, *J. Chem. Soc. Perkin 2*, 281-284.
15. Galway, A.K. (1994) Magnitudes of Arrhenius parameters for decomposition reactions of solids, *Thermochim. Acta*, **242**, 259-264.

16. Pertsin, A.J. and Kitaigorodsky, A.I. (1987) *The atom-atom potential method*, Springer-Verlag.
17. Bourne, S.A., Johnson, L., Marais, C., Nassimbeni, L.R., Weber, E., Skobridis, K. and Toda, F. (1991) Complexation with hydroxy host compounds. Part 4. Structures and thermal stabilities of inclusion compounds with dioxane as guest, *J. Chem. Soc. Perkin 2*, 1701-1713.
18. Vedani, A. and Dunitz, J.D. (1985) Lone-pair directionality in hydrogen-bond potential functions for molecular mechanics calculations. The inhibition of human carbonic anhydrase II by sulfonomides, *J. Amer. Chem. Soc.*, **107**, 7653-7658.
19. Caira, M.R., Nassimbeni, L.R. and Wildervanck, A.F. (1995) Selective formation of hydrogen bonded cocrystals between a sulfonamide and aromatic carboxylic acids in the solid state, *J. Chem. Soc. Perkin 2*, 2213-2216.
20. Toda, F., Tanaka, K., Okada, T., Bourne, S.A. and Nassimbeni, L.R. (1994) Selective inclusion crystallisation of phenol derivatives with tetraalkylammonium salts, *Supramolecular Chem.*, **3**, 291-298.

KINETICS OF INTERCALATION IN LAMELLAR HOSTS USING TIME-RESOLVED X-RAY DIFFRACTION

DERMOT O'HARE, JOHN S.O. EVANS AND STEPHEN PRICE
Inorganic Chemistry Laboratory, South Parks Road,
Oxford, OX1 3QR, UK

1. Introduction

Some of the earliest and most extensive studies of the kinetics of intercalation reactions were first carried out using graphite crystals. In these early classic experiments, for example those by Hooley and co-workers [1,2], they investigated the gain in weight of a graphite flake under a bromine atmosphere. Until recently only two time-resolved diffraction studies on metal chalcogenide the intercalation have reported. [3-7] One of these pioneering studies by Riekel and Schollhorn measured time-resolved neutron diffraction data during the intercalation of 2H-TaS_2 with NH_3(g). However, the available levels of beam flux, whilst high by neutron standards, necessitated either very large sample sizes (~15g) or long acquisition times, making a thorough, quantitative study under varying reaction conditions of temperature, concentration and particle size unfeasible. Moreover, because of the large incoherent scattering from hydrogen atoms by neutrons, the samples studied had to be highly deuterated, an expensive and not always achievable prerequisite for other intercalation reactions. Even taking all these previous experiments into consideration we still have a fairly rudimentary understanding of the intimate mechanism of these reactions. The critical factors which control the rates of these reactions are still based on trial and error experiments. We still commonly use the schematic illustration shown in Figure 1 to describe these reaction which undoubtedly greatly oversimplifies the true picture.

In recent years, the technological and experimental advances in energy-dispersive powder diffraction (EDXRD) using synchrotron X-ray sources have provided new possibilities for the solid state kineticist.[8-11] The markedly higher intensity (several orders of magnitude) over conventional laboratory X-ray sources means that good quality spectra can be obtained from milligram samples on a time scale of seconds, making kinetic studies on fast intercalation reactions a realisable goal. We have recently

A. W. Coleman (ed.), Molecular Recognition and Inclusion, 153–158.
© 1998 *Kluwer Academic Publishers. Printed in the Netherlands.*

constructed and commissioned an environmental cell for measuring time-resolved energy dispersive X-ray diffraction data of intercalation reactions.[12-14]

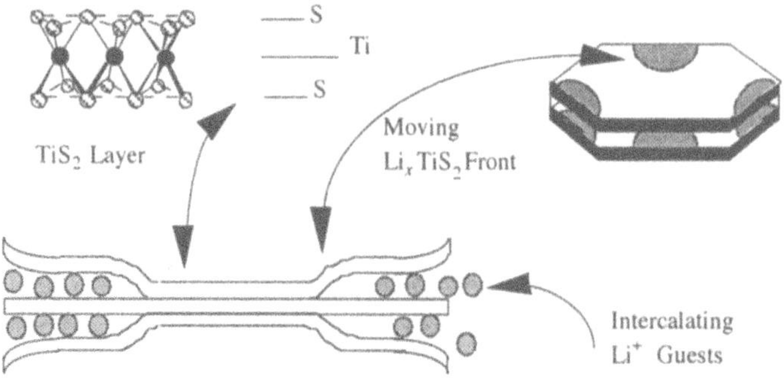

Figure 1. Schematic representation of the intercalation of a guest into a layered dichalcogenide host lattice.

In this paper we will review some of the significant observations we have made to date concerning our investigations into the kinetics of intercalation reaction using in situ, time-resolved, X-ray diffraction.

2. Kinetics of Intercalation Reactions

2.1 INTERCALATION OF CATIONS IN $MnPS_3$

The transition metal hexathiohypodiphosphates, MPS_3 (M = Mn, Fe, Ni, Cd) have been shown to intercalate a wide range of guest ions and molecules. In recent years they have been shown by Clement to exhibit a unique intercalation reaction. MPS_3 (M = Mn, Cd, Fe) react spontaneously with aqueous solutions of ionic salts, G^+X^- (where G can range from simple ionic species such as K^+ and NH_4^+ to species as large as, for example, cobaltocenium cations), to give compounds of general formula $Mn_{1-x}PS_3\{G\}_{2x}(H_2O)_y$.[15-17] In these compounds, charge balance has been maintained by the loss of one Mn^{2+} ion from the intralayer region for every two G^+ ions that are adopted in the interlayer region.

We have studied the kinetics of K^+, PyH^+ (Py = C_5H_5N) and NMe_4^+ intercalation into sub 90 μm crystals of $MnPS_3$ in water between room temperature and 60 °C using time-resolved, in-situ EDXRD.[18] In a typical experiment, solutions of the appropriate reagents were preheated to the reaction temperature, remotely injected into a 5 ml flask containing the stirred host lattice, and the X-ray diffraction pattern of the resulting suspension monitored as a function of time. Under these conditions the diffraction pattern could be recorded over a 20–4 Å d-spacing range in as little as 10 seconds, allowing the direct observation of the disappearance of the host lattice and growth of the product phase.

The integrated intensity of both the host and intercalate *001* reflections were then extracted from individual spectra, the time dependence for pyH^+ intercalation is shown in Figure 2. The kinetics of intercalation were analysed following the method of Hulbert, who showed that the general rate expression for Nucleation Growth reactions takes the form. [19]

$$\ln\frac{1}{1-\alpha} = (kt)^m \quad (1)$$

where α represents the extent of reaction at time t.

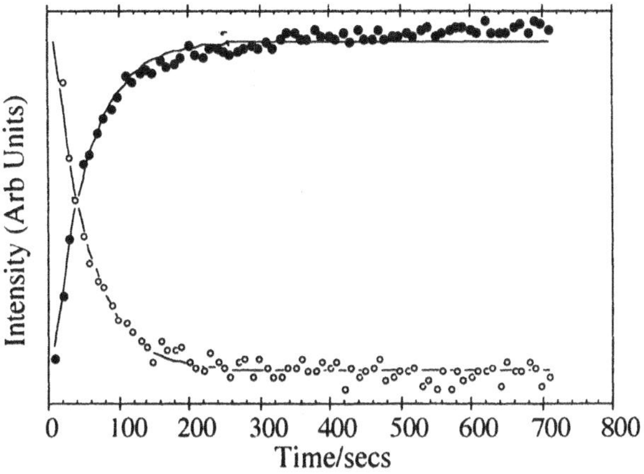

Figure 2. The time dependence of the intensities of the host *001* (○) and intercalate *001* (●) reflections for the intercalation of the pyH^+ by $MnPS_3$ under aqueous conditions at T=33°C.

TABLE 1. Kinetic rate parameters for the intercalation of various guest species into some lamellar hosts. Experimental rates have been fitted to a rate expression $\alpha = [1 - \exp(-kt)^m]$

Host	Guest	Temp/°C	m	k / $10^{-3}sec^{-1}$	$t_{1/2}$ / sec
$MnPS_3$	K^+	33	1.0	12.0	55
$MnPS_3$	pyH^+	33	1.0	20	43
$MnPS_3$	NMe_4^+	51	1.0	88	59
$MnPS_3$	NMe_4^+	60	1.0	20	35
ZrS_2	$Co(Cp)_2$	8	1.5	18.0	43
SnS_2	$Co(Cp)_2$	20	1.5	1.1	700
$SnSe_2$	$Co(Cp)_2$	50	1.5	0.6	1220
TaS_2	$Co(Cp)_2$	120	1.5	0.04	1700

Fits of such an expression to the experimental data and the rate parameters of the fits are given in Table 1. The exponents, *m*, corresponds to a reaction model involving a saturation of nucleation sites at the edge of the crystal so that the overall rate of intercalation is determined by the rate of diffusion of the guest molecules into the bulk of the host lattice.

A number of important conclusions can be drawn from this study. Firstly these intercalation reactions occur extremely rapidly, with half lives of the order of a minute or less. These exchange intercalation reactions appear to be some of the fastest known molecular intercalation reactions. Secondly the reactions occurred cleanly *via* a two-phase reactant to product route, and no unusual phenomena such as staging were observed. Finally the fact that these systems obey rate expressions similar to those found for the intercalation reactions of other host lattices, and the observation that the rate of formation of the product phase and disappearance of the host lattice obey identical rate laws seems to imply that these ion exchange reactions occur *via* a standard intercalation pathway, in which intercalation occurs simultaneously between each of the lattice layers.

2.2 KINETICS OF INTERCALATION OF COBALTOCENE INTO LAMELLAR MX_2 HOSTS

Redox-active organometallic compounds can be intercalated into the metal dichalcogenides by either direct reaction, ion-exchange, or by electrochemical routes.[20] This significant reaction was first reported in 1975 when Dines described the intercalation of the metallocenes cobaltocene {$Co(Cp)_2$; Cp = η-C_5H_5} and chromocene {$Cr(Cp)_2$} into a range of metal disulphides (MS_2: M = Ti, Zr, Nb, Ta and Sn).[21,22] Since then these materials have been extensively studied due to there fascinating structural and electronic properties. In general, the ability of an organometallic molecules to intercalate in a layered lattice has been based on empirical rules or by trial and error. Therefore we were excited at the possibility of being able the quantitatively determine the kinetics of these reactions and also being able investigate if intermediate novel phases are formed during the course of the reaction.

We have investigated the kinetics intercalation of $Co(Cp)_2$ into a range of other layered metal dichalcogenides using in-situ, time-resolved EDXRD.[23] We were particularly interested to determine if the variation in the rates of these intercalation reaction could correlated with the chemical properties of the hosts. Figure 2 summarises the extent of reaction $\alpha(t)$ vs. time (t) for the intercalation of $Co(Cp)_2$ into ZrS_2, SnS_2, $SnSe_2$ and 2H-TaS_2. In each case the reaction yields 1st stage intercalation compounds of chemical composition $MX_2\{Co(Cp)_2\}_{0.30\pm0.05}$. In all cases we observed the smooth growth of intensity of the reflections due to the intercalate phase phases with increased c-lattice constants of 5.35 ± 0.02 Å. No intermediate phases of higher order staging are observed in any of these reactions. What is immediately striking about the data shown in Figure 2 is the extreme variation in the rate of the reaction for the each of the host lattices. For all the reactions, we tried to keep the identical condition; each host was sieved to a particle size distribution between 90-250 μm and the same amount of $CoCp_2$ solution was injected into each sample. Unfortunately, we had to vary the temperature of the reactions in order to collect data on a reasonable timescale. A summary of the rate constants(k_{obs}/s^{-1}) and half-life ($t_{1/2}$) for these reaction are shown in Table 1.

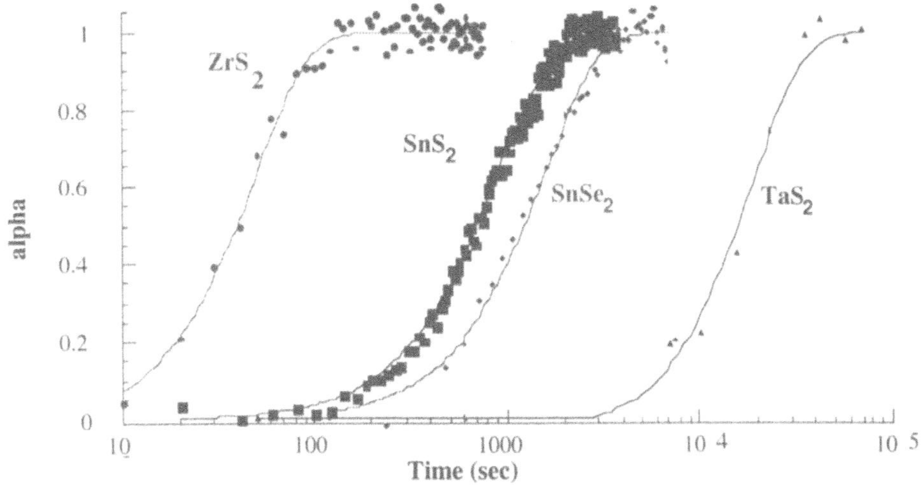

Figure 3. Extent of reaction (α) versus time for intercalation of $Co(Cp)_2$ into a range of metal dichalcogenides, ● ZrS_2, 8°C; ■ SnS_2, 20°C; ◆ $SnSe_2$, 50°C; ▲ $TaSe_2$, 120°C (—) are fits to the Avrami deceleratory nuclei-growth model, $\alpha = 1 - \exp(-k(t-t_0))^{1.5}$.

For SnS_2, ZrS_2 and $SnSe_2$ we have also investigated the rate of reaction in dimethoxyethane with respect to the initial $Co(Cp)_2$ concentration. In each case, varying concentrations of cobaltocene solution (enough volume to provide a 2:1 molar excess with respect to the host) were injected onto 200mg of the host lattice at a convenient temperature for kinetic analysis. The time dependence of the extent of reaction for the intercalation of cobaltocene into SnS_2 at 20°C for a range of initial $Co(Cp)_2$ concentrations could be fitted to the Avrami equation above and, for the complete range of initial $Co(Cp)_2$ concentrations studied (10-70 mg cm^{-3}), the calculated rate constants and reaction half-lives are invariant within experimental error. Therefore, we conclude that the nucleation and diffusion rate are independent of $Co(Cp)_2$ concentration over the range investigated. These experiments indicate that, at the concentrations studied, all the nucleation sites are saturated with potential guest molecules and that the guest molecules diffuse between the MX_2 layers at a rate independent of the concentration of the bulk solution. This result has also been obtained for similar experiments involving ZrS_2 and $SnSe_2$.

3. Conclusions

An apparatus has been developed to study the kinetics of intercalation into microcrystalline samples. The use of energy dispersive powder diffraction techniques enables reactions with half-lives of minutes to days to be examined. The design of the experiment allows a wide variety of solid-liquid reactions besides intercalation reactions to be studied with minimal turn-around time and experiments are currently underway to extend these opportunities to other classes of reactions.

The kinetics of ion-transfer intercalation appear to be solely controled by diffusion processes. This is in contast to the kinetics observed for the intercalation of $CoCp_2$ into a range of lamellar metal dichalcogenide hosts where the kinetics seems to be controlled by both nucleation and two-dimensional diffusion steps.

4. Acknowlegments

We would like to thank Leaverhulme Trust for financial support and EPSRC for access to the Synchrotron Radiation Source, Daresbury Laboratory. We also thank Dr S.M. Clarke, Daresbury Laboratory and the technical staff at the SRS for helping in the design and construction of the experimental cell.

5. References

1. Hooley, J. G. *Carbon* **1973**, *11*, 225.
2. Hooley, J. G. *Canad J Chem* **1962**, *40*, 749.
3. Riekel, C. *Prog. Solid State Chem.* **1980**, *13*, 89-117.
4. Riekel, C. Reznik, H. G., Schollhorn, R. *J. Solid State Chem.* **1980**, *34*, 253-262.
5. Paulus, W. Katzke, H., Schollhorn, R. *J. Solid State Chem.* **1992**, *96*, 162-168.
6. Riekel, C., Fischer, C. O. *J. Solid State Chem.* **1979**, *29*, 181-90.
7. Riekel, C., Reznik, H. G., Schollhorn, R. *J. Solid State Chem.* **1979**, *34*, 253-62.
8. Barnes, P. *Phase Transitions* **1992**, *39*, 1-2.
9. Hausermann, D., Barnes, P. *Phase Transitions* **1992**, *39*, 99-115.
10. Barnes, P., Clark, S. M., Hausermann, D., Henderson, E., Fentiman, C. H., Muhamad, M. N., Rashid, S. *Phase Transitions* **1992**, *39*, 117-128.
11. Sheridan, A. K., Anwar, J. *Chem Mater* **1996**, *8*, 1042-1051.
12. Clark, S. M., Irvin, P., Flaherty, J., Rathbone, T., Wong, H. V., Evans, J. S. O., Ohare, D. *Rev Sci Instrum* **1994**, *65*, 2210.
13. Clark, S. M., Evans, J. S. O., O'Hare, D., Nuttall, C. J., Wong, H. V. *J Chem Soc Chem Commun* **1994**, 809-810.
14. Clark, S. M., Nield, A., Rathbone, T., Flaherty, J., Tang, C. C., Evans, J. S. O., Francis, R. J., O'Hare, D. *Nucl Instrum Meth - Beam Interactions Materials & Atoms* **1995**, 97, 98-101.
15. Clement, R. *J Am Chem Soc.* **1981**, *103*, 6998.
16. Clement, R., Girerd, J. J., Morgenstern Badarau, I. *Inorg Chem* **1980**, *19*, 2852.
17. Clement, R. *J Chem. Soc Chem Commun.* **1980**, 647.
18. Evans, J. S. O., O'Hare, D. *Adv Mater* **1994**, *6*, 646-648.
19. Hulbert, S. F. *J Br Ceram Soc* **1969**, *6*, 11.
20. O'Hare, D. (1992) Inorganic Intercalation Compounds, D.W. Bruce and D. O'Hare (eds.), *Inorganic Materials*, J. Wiley & Sons: Chichester.
21. Dines, M. B., Gamble, F. R., Silbernagel, B. G. *Bull. Am. Phys. Soc.* **1975**, 289.
22. Dines, M. B. *Science* **1975**, *188*, 1210.
23. Price, S. J., Evans, J. S. O., Francis, R. J., O'Hare, D. *Adv Mater* **1996**, *8*., 2989-2993.

CONTROL OF PERMEATION OF IONS ACROSS VESICLES AND CHEMICAL DIFFERENTIATION OF THEIR BILAYER MEMBRANE

P. SCRIMIN,[a] F. FELLUGA,[a] G. GHIRLANDA,[b] PAOLO TECILLA,[b] U. TONELLATO,[b] AND A. VERONESE[b]
[a]University of Trieste, Department of Chemical Sciences, via L. Giorgieri, 1, 34127-Trieste, Italy; [b]University of Padova, Department of Organic Chemistry, via Marzolo, 1, 35131-Padova, Italy

1. Introduction

Vesicles are aggregates of amphiphilic molecules (lipids), typically bearing two paraffinic tails, which self-assemble in water under proper conditions when some form of external energy is provided (like swirling or sonication)[1]. The simplest morphology is that of a sphere made of a bilayer of lipids trapping in its interior a water pool. The bilayer (membrane) is organized in such a way (Figure 1) that the hydrophilic head-groups of the lipids are exposed to the aqueous solutions and the hydrophobic portions of the molecules, the hydrocarbon chains, are tightly packed providing an effective barrier against the permeation of polar species like hydrophilic ions (organic or inorganic).

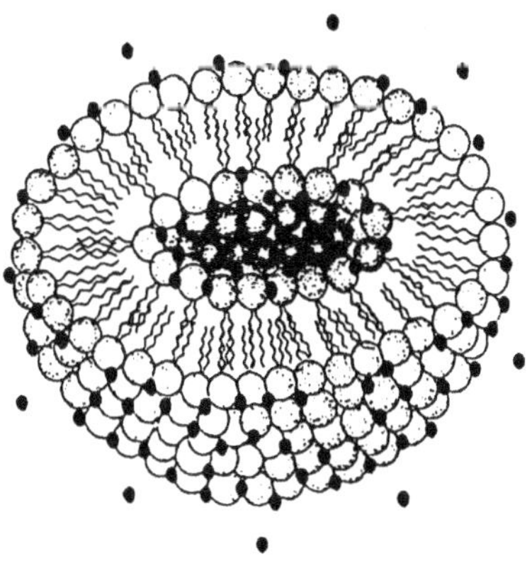

Fig. 1. Schematic representation of the cross-section of a unilamellar vesicle. Black dots represent counterions of the charged lipids of this model.

A. W. Coleman (ed.), Molecular Recognition and Inclusion, 159–166.
© 1998 *Kluwer Academic Publishers. Printed in the Netherlands.*

In our laboratory we have investigated these properties of vesicles, particularly as for the aspect of their permeability to metal cations, to differentiate the interior and the exterior of aggregates of custom-made amphiphiles and also to control permeation of ions. This short account reports a few examples of such studies.

2. Chemical differentiation of the bilayer

2.1 CATIONIC AGGREGATES

The synthetic amphiphilic molecule shown in Figure 2 presents an ester function which can be hydrolized releasing a *p*-nitrophenol-functionalized amphiphile in a process that can been easily followed spectroscopically at *ca.* 400 nm [2,3].

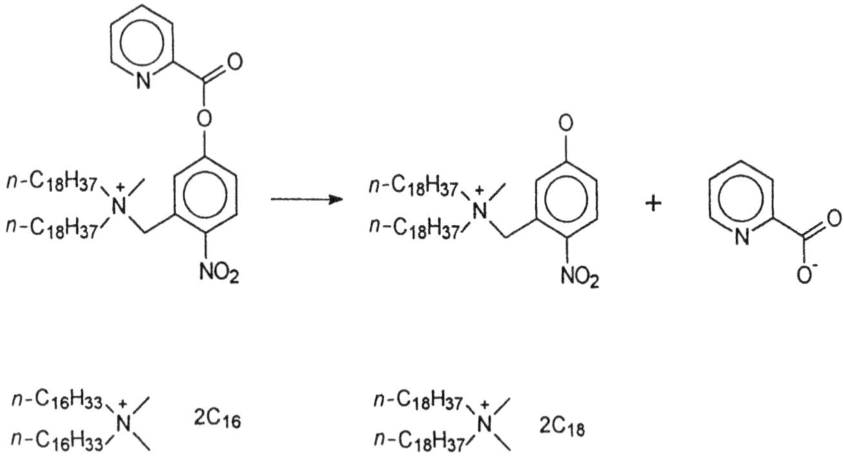

Fig.2. Picolinate-ester amphiphile used to study Cu(II) permeability acrsoss cationic vesicles.

The hydrolysis becomes much faster when transition metal ions (particularly Cu(II) ions) are added to the solution because of the well known lability of activated esters of picolinic acid in the presence of such ions even at neutral pH [4]. Vesicular aggregates of this ester-amphiphile may be viewed as "chemical sensors" of the presence of Cu(II) ions for their hydrolytic response to the presence of these ions. However a kinetic analysis of the hydrolytic reaction shows that when the metal ions are added after the creation of the vesicles, the amount of ester cleaved depends on the temperature. In particular, when the temperature is below the phase transition of the vesicles (*Tc*), partial cleavage (*ca* 65% of the total ester) occurs while when the temperature is above the *Tc*, all ester is cleaved (see Table 1).

TABLE 1. Percent of exo/endovesicular cleavage of picolinate-ester surfactant as a function of the temperature and *Tc* of the covesicular blend used.

Cosurfactant	T,°C	Tc	% exo	% endo
$2C_{16}$	10	23	69	31
$2C_{16}$	25	23	100	--
$2C_{18}$	25	35	62	38
$2C_{18}$	42	35	100	--

The temperature of the phase transition of a vesicle denotes the "melting" of the hydrocarbon chains from a highly packed *all-trans* conformation to a less ordered (fluid) situation when some *s-cis* conformations are present in the hydrocarbon chains. At the fluid state the bilayer may become more permeable and, as elegantly shown by Moss and his co-workers [5], the transbilayer movement of the single lipid (often called flip-flop) becomes much faster. The fact that only ca. 65% of the ester is cleaved when Cu(II) ions are added to the vesicles at a temperature below their *Tc* tells us that neither Cu(II) ions cross the bilayer nor the ester-amphiphile undergoes any sizeable flip-flop so that the ions interact only with the ester groups residing on the outer leaflet of the vesicles, cleaving them at an accelerated rate.

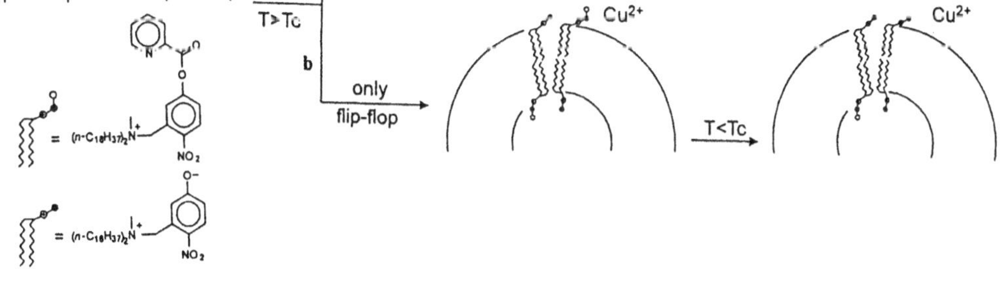

Fig. 3. Cu(II) ions do not cross cationic vesicles even above their phase transition temperature.

The rate of cleavage of the esters residing on the interior of the vesicles is not affected by the presence of the metal ions for the simple reason that they do not cross the bilayer. When the temperature is raised above *Tc* the experimental observation of a fast cleavage of all ester upon addition of Cu(II) may be explained either by assuming a substantial increase of permeability of the bilayer or by a fast flip-flop of the amphiphiles without permeation of the ions. In both cases the ester groups become exposed to the metal ions although for totally different reasons. We were able to show that what actually occurs above *Tc* is a fast flip-flop of the amphiphiles while permeation of Cu(II) ions does not occur at an appreciably higher rate. In fact when the temperature of a reaction started above *Tc* is lowered below the phase transition when half of the ester groups are cleaved we observe only partial cleavage of the remaining ester. This implies that flip-flop has slowed down upon cooling, and, most important, that Cu(II) permeation did not occur at high temperature. Partial cleavage means that the ions are still only outside as shown in Figure 3 (path b). The fact that the metal ions do not permeate across the vesicular membrane may by explained by considering that our aggregates, being cationic, provide not only a lipophilic barrier to cations but also an electrostatic barrier. In order to evaluate the relevance of electrostatic factors, we tried to answer the question: what happens if we switch from cationic to anionic aggregates?

2.1 ANIONIC AGGREGATES

In order to address this problem we studied, in collaboration with the group of Professor Moss, at Rutgers Univesity, the amphiphile shown in Figure 4.

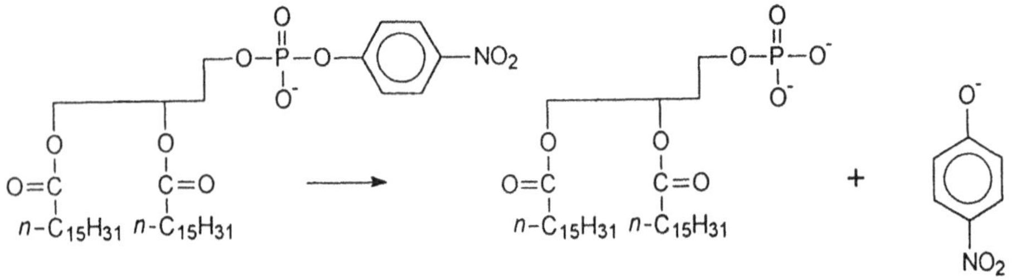

Fig. 4. Hydrolytic cleavage of vesicular aggregates of this anionic phosphate diester produces *p*-nitrophenol and an amphiphilic phosphate monoester.

This lipid is functionalized with an anionic phosphate diester which, upon hydrolysis, is converted in an amphiphilic ester monophosphate and releases *p*-nitrophenol[6]. The vesicles made of this functional lipid are hydrolytically stable at neutral pH and hardly affected by the addition of Cu(II) ions.

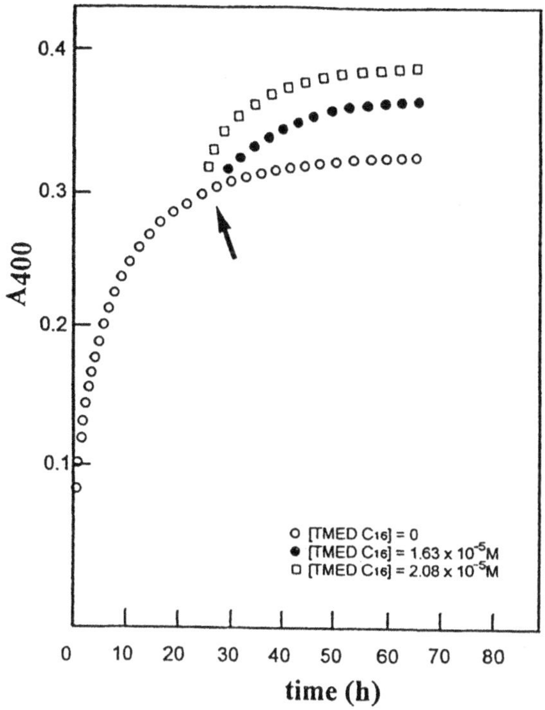

Fig. 5. Hydrolytic behavior of sonicated vesicles of the phosphate of Figure 4 (5.4×10^{-5} M) upon addition of Eu^{3+} (1.7×10^{-3} M) at 25°C and pH 7. The first kinetic process accounts for ca. 70% of the total phosphate. The remaining phosphate is cleaved by addition of N-*n*-hexadecyl N,N,N-trimethylethyl diamine (TMED C_{16}).

Moss discovered that addition of a lanthanide (Eu(III) for instance) results in the cleavage of outer phosphate groups leading to chemical differentiation of the bilayer. The remaining phosphate may be cleaved by raising the temperature above the phase transition or by adding N-hexadecyl-N,N,N-trimethylethylene diamine (see Figure 5)[7]. This lipophilic ligand may act as a carrier of the europium ions or, being a single-tail molecule, make the liposomal membrane more permeable to them. An experiment similar to that illustrated in Figure 3 showed that also in this case the metal ion does not permeate across the bilayer above *Tc*. The similarity of the results obtained in cationic and anionic aggregates indicates that the charge of the liposomes is not an important factor in controlling the permeation of the cations.

3. Control of permeability of liposomal membranes

The possibility to control the permeability of the membrane is quite challenging[8] and to realize this we need a more sophisticated system than a lipophilic diamine. For this purpose, biological membranes rely, among other systems, on membrane proteins[9] or amphipatic polypeptides which form channels or pores across the bilayer. Such peptides are present, for instance, in the venom of bees[10] or produced by toxic fungi. Peptaibols[11] are a class of peptides produced by fungi that alter the permeability of the membrane of liposomes. They adopt an helical conformation because they are particularly rich in α-amino isobutyric acid (Aib), an amino acid known for its ability to induce helical conformation even in very short peptides[12]. While longer peptaibols are long enough to span the entire length of the membrane in their helical conformation and are believed to form channels by clustering in such a way to expose the hydrophobic part of the helic to the lipids, shorter peptaibols require the presence of an hydrocarbon tail in order to show activity. For these systems the number of amino acids can be as small as ten (trichogin [13]) or even six (trichodecenin).

In collaboration with the group of Professor Toniolo at the University of Padova we realized a series of synthetic peptides representing fragments of increasing length of trichogin and connected them to tris (2-aminoethyl)amine (tren) *via* a benzoic acid spacer (Figure 6). Tren is a strong ligand for transition metal ions (with Zn(II) the affinity constant is *ca* 10^{15} M^{-1} for the 1:1 complex) and previous studies[14] had shown that a benzyl-functionalized tren·Zn(II) complex adopts a calix-like conformation with the three aromatic residues oriented on the same side with respect to the coordination plane of the metal ion. We thougth that these polypeptide templates could alter the permeability of a model membrane in analogy with trichogin and, most important, their effect could be modulated by the formation of the metal complex.

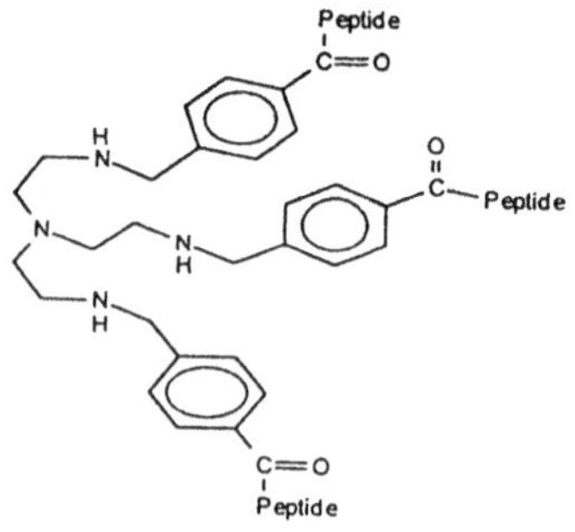

$T(10)_3$	**10**	Gly - Leu - Aib - Gly - Gly - Leu - Aib - Gly - Ile - Leu
$T(6)_3$	**6**	Gly - Leu - Aib - Gly - Ile - Leu
$T(5)_3$	**5**	Leu - Aib - Gly - Ile - Leu
$T(4)_3$	**4**	Aib - Gly - Ile - Leu
$T(3)_3$	**3**	Gly - Leu - Aib

Fig. 6. Polypeptide templates synthesized.

Our hypothesis proved correct[15] and, in fact, some of the templates devoid of the Zn(II) ion proved quite effective in promoting the release of carboxyfluorescein (CF) trapped inside unilamellar liposomes made of a blend of phosphatidylcholine and cholesterol (70/30). The effect was higher with the derivatives with longer peptide sequences (5-10 amino acids) and dropped considerably with the derivative with four amino acids for each peptide. Quite interesting, addition of a stoichiometric amount of Zn(II) ions to the templates resulted in the inhibition of the effect as shown in Table 2.

The control of permeability was reversible. For istance, the slow release of CF observed when $T(6)_3 \cdot Zn(II)$ is added, can be higly enhanced by addition of excess EDTA which competes quite effectively with the tren platform for the complexation of the metal ion. Analogously, inhibition of the effect can also be observed by progressive addition of zinc ions to the apopeptide

TABLE 2. Percentage of CF released by the different templates after 20' at 25°C. [Template]=$6x10^{-7}$ M; [lipid]=$4x10^{-5}$ M.

Template	%F(20') free	%F(20') Zn(II)
$T(10)_3$	83	18
$T(6)_3$	79	18
$T(5)_3$	65	15
$T(4)_3$	12	8
$T(3)_3$	7	5

The inhibition observed points to the formation of a 1:1 complex and we may speculate that the formation of the complex brings about a conformational change in the template from an extended conformation (Figure 7, top) to a globular one (Figure 7, bottom) with the three peptide chains aligned one with the other.

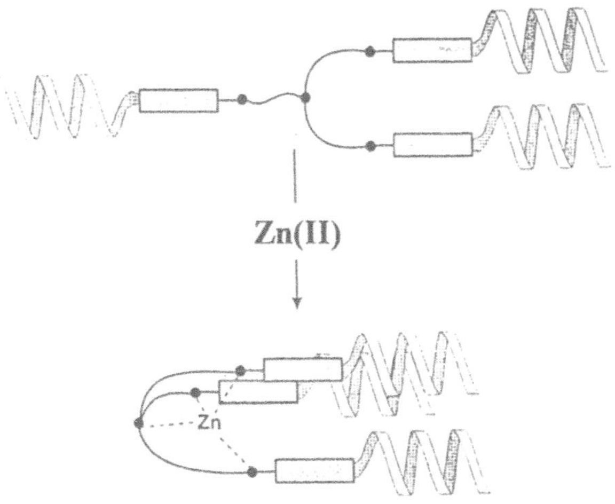

Fig. 7. Possible conformational changes induced by Zn(II) complexation in the $T(P10)_3$ template.

Such a conformational change results in a decrease of the length of the template. The extended conformation is long enough, according to molecular models, to span the whole bilayer while the folded conformation of the complex allows the spanning of only half the membrane. In the latter case, the formation of a channel requires the alignment of polypeptides residing on the outer and inner leaflet of the bilayer. This is a possible explanation of the diminished effect observed in the presence of Zn(II) ions.

Thus these polypeptides are quite effective in inducing the leakage of a dye trapped inside model liposomes and control of the permeability can be achieved through the modulation of the conformation of their templates. Although our molecules are potential antibiotics they are in fact not selective in their interaction with the membrane. A challenging goal would be that of targeting a specific membrane on the basis of its lipid composition or charge, for instance, leaving totally unaffected those of different structure.

Acknowledgments. We thank Professors Robert A. Moss and Claudio Toniolo and their groups for the fruitful collaboration that led to many exciting results here reported. Financial support by the Ministry of University, Scientific and Technological Research (MURST) is also acknowledged.

References

1. (a) J. H. Fendler, *Membrane Mimetic Chemistry*, Wiley, New York, 1982; (b) D. D. Lasic, *Liposomes*, Elsevier, Amsterdam, 1993; (c) P. Scrimin in *Supramolecular Control of Structure and Reactivity*, A. D. Hamilton (ed)., Wiley, New York, 1996, Chapter 3.
2. G. Ghirlanda, P. Scrimin, P. Tecilla, and U. Tonellato *J. Org. Chem.* **58** (1993), 3025.
3. For previous work on cleavable surfactants for chemical differentiation of endo- and exoliposomal membranes, see: (a) R. A. Moss, S. Bhattacharya, P. Scrimin, and S. Swarup *J. Am. Chem. Soc.* **109**(1987), 5740; (b) R. A. Moss, S. Bhattacharya, and S. Chatterjee *J. Am. Chem. Soc.* **111**(1989), 3680; R. A. Moss and Y. Okumura *J. Am. Chem. Soc.* **114**(1992), 1750.
4. H. T. Fife and T. Przystas *J. Am. Chem. Soc.* **107**(1985), 1041.
5. R. A. Moss *Pure Appl. Chem.* **66**(1994), 851 and references therein.
6. R. A. Moss and S. Swarup *J. Am. Chem. Soc.* **108**(1986), 5341.
7. R. A. Moss, B. D. Park, P. Scrimin, and G. Ghirlanda *J. Chem. Soc. Chem. Commun.* (1995), 1627.
8. G. W. Gokel and O. Murillo *Acc. Chem. Res.* **29**(1996), 425.
9. K. S. Åkerfeldt, J. D. Lear, Z. R. Wasserman, L. A. Chung, and W. F. DeGrado *Acc. Chem. Res.* 26(1993), 191.
10. P. Juvvadi, S. Vunnam, and R. B. Merriefield *J. Am. Chem. Soc.* **118**(1996), 8989.
11. E. Benedetti, A. Bavoso, B. Di Blasio, V. Pavone, C. Pedone, C. Toniolo, and G. M. Bonora *Proc. Natl. Acad. Sci. U.S.A.* **79**(1982), 7951.
12. A. Polese, F. Formaggio, M. Crisma, G. Valle, C. Toniolo, G. M. Bonora, Q. B. Broxterman, and J. Kamphuis *Chem. Eur. J.* **2**(1996), 1104.
13. C. Auvin-Guette, S. Rebuffat, Y. Prigent, and B. Bodo *J. Am. Chem. Soc.* **114**(1992), 2170.
14. P. Scrimin, P. Tecilla, U. Tonellato, G. Valle, and A. Veronese *J. Chem. Soc., Chem. Commun.* (1995), 1163.
15. P. Scrimin, P. Tecilla, U. Tonellato, V. Monaco, F. Formaggio, M. Crisma, and C. Toniolo *J. Am. Chem. Soc.* **118**(1996), 2505.

MOLECULAR RECOGNITION AND ARTIFICIAL ION CHANNEL WITH AMPHIPHILIC MACROCYCLES

Yasutaka TANAKA
"Field and Reaction", PRESTO, Japan Science and Technology Corporation (JST) and Department of Materials Science, Faculty of Engineering, Shizuoka University, Hamamatsu, Shizuoka 432, Japan

1. Abstract

Amphiphilic calix[6]arene (**1**) formed a ion channel in planar lipid bilayers. A stable conductance level was observed with well-defined transitions between open and closed states. The conductance was mostly one level, 8.1 picosiemens at symmetrical 0.5 M KCl. In addition to the channel activity observed in resorcinol-octadecanal cyclotetramer (**2**) in the bilayers, these results suggested that the cavity composed of aromatic rings and the consecutive inward surrounded by alkyl tails in **1** or **2** provided an ion conducting pore. The aromatic cage as a selectivity filter for voltage-gated K^+ channels seems to find a rational support.

2. Introduction

The structure elucidation of voltage-gated K^+ channels has gained a great advance by the application of site-directed mutagenesis and single-channel patch clamp studies: they suggest that the assembly of four subunits, each with six transmembrane segments, provides the central ion conduction pore. [1] The ion selectivity mechanism by these K^+ channels has attracted much attention. The mechanism whereby K^+ is discriminated from other cations has been understood that naked K^+ is generated easily in a relatively weak electric field and is molecular-sieved by the size effect. [2]

The recent progresses on identifying the ion selective filter of K^+ channels have focussed to suggest that the interactions between K^+ and the negative π-obverse of the aromatic residues of amino acids determine the K^+ preference. [3] Molecular mechanics calculations suggested that the aromatic residues, especially tyrosine rings in the GYG sequence, contribute the construction of the ion selective filter through their stabilizing interaction between cation and π-electron cloud. [4, 5] The calculation of interaction

A. W. Coleman (ed.), Molecular Recognition and Inclusion, 167–172.
© 1998 *Kluwer Academic Publishers. Printed in the Netherlands.*

free energy for cations with benzene ring by ab initio quantum mechanical methods and Monte Carlo simulations also substantiated the proposal above: the inward of a set of face-to-face benzene rings stabilizes K^+ best out of the monovalent cations (Li^+, Na^+, K^+, and Rb^+) in aqueous environment. [6]

Site-directed mutations aimed at the aromatic residues in the pore domain of the K^+ channel were successful at least in part to verify the hypothesis. [7, 8, 9] However, the lack of three-dimensional structural information of these proteins in the biological membrane still prevents us from elucidating what actually causes such substantial ion selectivity. In order to explain an ion-conducting mechanism in these proteins, some channel mimics with synthetic peptides and organic molecules have been reported. [10, 11] We previously showed a non-peptidic ion channel formed from an amphiphilic resorcinol cyclotetramer (**2**), where a stable and constant conductance (6.1 ± 0.8 pS) was measured. The ion selectivity for K^+ over Na^+ as well as for K^+ over Cl^- was also observed. [12] Here we wish to report the alternative approach to find a general artificial channel structure. A fully synthesized molecule, a structurally well-defined macrocyclic amphiphile (**1** in *Figure 1.*) with a symmetric aromatic cage, gave single ion channel currents upon incorporated into planar bilayer membrane.

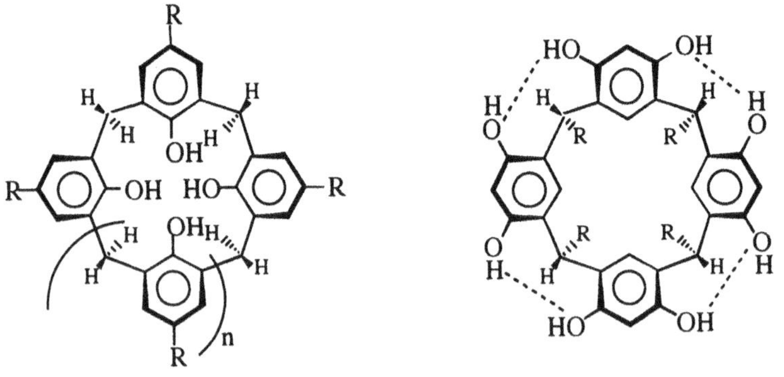

1: n=3, R=$CH_3(CH_2)_{14}CO$-

2: R = $CH_3(CH_2)_{16}$

Figure 1. Structure of the artificial ion channel molecule **1** and **2**.

3. Results and Discussion

Acylation of calix[6]arene with palmitoyl chloride afforded **1** in good yield. [13] The spectroscopic data including ^{1}H and ^{13}C-NMR, Mass, and IR spectra of the amphiphile synthesized were in full agreement with the proposed structures. The ^{1}H NMR spectra of **1** also suggest that the aromatic cage consisted of six phenol rings has a sixfold symmetry, each of aromatic rings being slightly inclined, in the light of the structure of other alkyl calixarenes. This configuration of **1** seems to fit in the membrane, aromatic rings being arranged at the surface of the membrane and alkyl chains being in parallel with lipid alkyl chains. In planar lecithin bilayers separating two 500mM KCl solutions, **1** exhibited typical characteristics of the single ion channel: a stable and constant conductance (8.1 pS) at various voltage with transitions between open and closed states

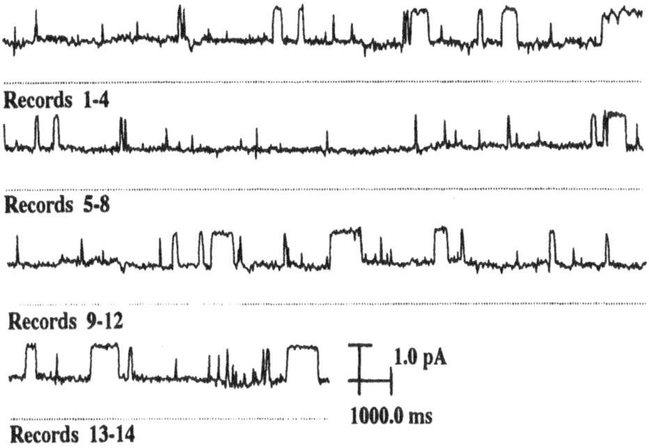

Figure 2. Typical records of **1** single-channel currents at -125.0 mV in symmetric solutions containing 500 mM KCl, 100 mM HEPES-Tris (pH 7.2).

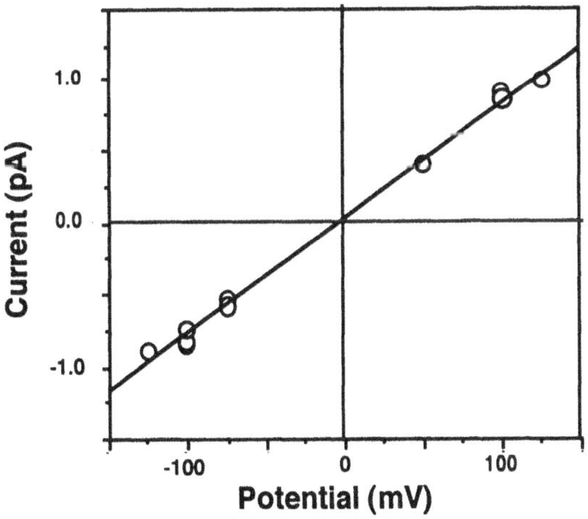

Figure 3. Single-channel current (*I*)-voltage(*V*) relations for **1** channel.in symmetric solutions containing 500 mM KCl, 100 mM HEPES-Tris (pH 7.2).

was observed (*Figure 2.*). The simultaneous operation of several single channels resulted in an integral multiple of the fundamental single channel currents. As shown in *Figure 3*, the symmetry in the channel current with respect to the voltage reversal across the membrane suggests a symmetric channel structure. Such a stable conductance was never observed in the absence of **1**, clearly indicating that neither leakage nor artifact currents had been taking place. In contrast to **1**, octadecanol, possessing almost same alkyl chain length as those in **1** and lacking the aromatic ring and the macrocyclic structure, failed to produce stable channel currents. It suggests that the macrocyclic structure composed of the aromatic rings is essential to forming a stable channel structure.

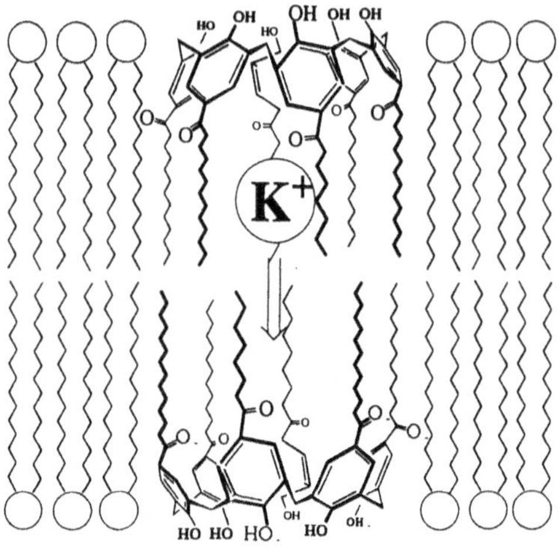

Figure 4. Hypothetical representation of **1** channel structure in the lipid bilayer. A lipid molecule is symbolized by a circle with two triangle waves: the circle and the triangle wave represent the polar (head) group and the alkyl chain in the lipid molecule, respectively. Potassium cation passing through the artificial ion channel and its flowing direction are also presented in the figure.

The conductance was only one type during more than 90 % of current records under variation of the molar ratio of **1** to lipid in the range of 1/200 ~ 1/3000. This is in marked contrast to the presence of several conductance levels observed for the molecular aggregate channels of oligoether ion-pair and synthetic peptides, [14, 15] where the different conductance levels were supposed to correspond to different aggregation numbers. These results may also signify that monomeric **1** penetrates nicely a single leaflet of the bilayers. The cavity formed by six aromatic rings, which is extended by the six alkyl chains in **1**, excludes the entry of lipid molecules and thus provides the channel pore for the passage of ions. Water molecules may also occupy the channel and coordinate to ions passing through the pore. Such a hypothesis that cations pass through the hydrophobic environment is not so radical because the sequences supposed to form the ion pore, the linker (H5 region) connecting S5 and S6 and S6 itself, in voltage-gated K^+ channels are highly conserved and moderately hydrophobic compared to what might be expected for the cation conducting pore. [4, 5, 16] On the assumption that the cavity formed by **1** recognizes the positioning of another molecule of **1** in the other half of the bilayer, a tail-to-tail dimer of **1**, as illustrated in *Figure 4.*, forms and thus gives rise to a symmetric transmembrane channel similar to the Gramicidin A channel. [17, 18, 19] In accordance with this dimer model, channel activities were not observed when **1** was added to only one side of the lipid membrane, probably because such a bulky amphiphile does not flip-flop easily.

Selectivity of ion permeation through membrane channels is determined primarily by the narrowest region of the channel pore termed the selectivity filter. The relation between the electric field strength of the selectivity filter and the dehydration energy of the permeating ion determines the extent of dehydration and the resulting ion with the corresponding hydration states is then discriminated by molecular sieving. [2] The narrowest ion conducting pore in **1** comprises six aromatic rings, while that of **2** does four aromatic rings. The size of this pore in **2** resembles to that of the pore of *p-tert*-butyl-calix[4]arenetetraamide, which permits K^+ (r=1.33Å) but not Cs^+ (r=1.69Å) to pass through, on the basis of molecular dynamics simulation [20]. The higher conductance (8.1 pS) observed in the **1** channel than that (6.1 pS) in **2** may reflect the difference of the pore size: **1** has the calix[6]arene skeleton (diameter = ca. 7.6Å), whereas **2** does calix[4]arene (diameter = ca. > 3Å).

This work presents the general condition for the design of channel forming molecules with a simple and stable synthetic molecule. This may serve as the simplest model for understanding the ion conduction in biological membrane and also give a useful guide for designing a more elaborate synthetic molecule or a system for molecular electronic devices .

3. References

1. MacKinnon, R.: Determination of the subunit stoichiometry of a voltage-activated potassium channel, *Nature* **350** (1991), 232-235.
2. Hille, B.: *Ionic Channels of Excitable Membranes* , Sinauer, Sunderland MA, 1992.
3. Heginbotham, L. and MacKinnon, R.: The aromatic binding site for tetraethylammonium ion on potassium channels, *Neuron* **8** (1992), 483-491.
4. Durell, S. R. and Guy, H. R.: Atomic scale structure and functional models of voltage-gated potassim channels, *Biophys. J.* **62** (1992), 238-250.
5. Bogusz, S., Boxer, A., and Busath, D. D.: An SS1-SS2 β-barrel structure for the voltage-activated potassinm channel, *Protein Eng.* **5** (1992), 285-293 .
6. Kumpf, R. A. and Dougherty, D. A.: A mechanism for ion selectivity in potassium channels. computational

studies of cation-π interactions, *Science* **261** (1993), 1708-1710.

7. MacKinnon, R. and Miller, C.: Mutant potassium channels with altered binding of charybdotoxin, a pore-blocking peptide inhibitor, *Science* **245** (1989), 1382-1385.
8. MacKinnon, R. and Yellen, G.: Mutations affecting TEA blockade and ion permeation in voltage-activated K^+ channels, *Science* **250** (1990), 276-279.
9. Yellen, G., Jurman, M. E., Abramson, T., and MacKinnon, R.: Mutations affecting internal TEA blockade identify the probable pore-forming region of a K^+ channel, *Science* **251** (1990), 939-942.
10. Nagaoka, Y., Iida, A., Kambara, T., Asami, K., and Fujita, T.: Role of Gln^7 in the ion-channel-forming properties of the peptaibol trichosporin-B-VIa, *Chem. Commun.* (1996), 1079-1080.
11. Deng, G., Merritt, M., Yamashita, K., Janout, V., Sadownik, A., and Regen, S. L.: Kinetic evidence for duplicity in ion transport, *J. Am. Chem. Soc.* **118** (1996), 3307-3308.
12. Tanaka, Y., Kobuke, Y., Sokabe, M.: A non-peptidic ion channel with K^+ selectivity, *Angew. Chem. Int. Ed. Engl.* **34** (1995), 693-694.
13. Shinkai, S., Nagasaki, T., Iwamoto, K., Ikeda, A., He, G.-X., Matsuda, T., and Iwamoto, M.: New syntheses and physical properties of *p*-alkylcalix[*n*]arene, *Bull Chem. Soc. Jpn.* **64** (1991), 381-386.
14. Kobuke, Y., Ueda, K., and Sokabe, M.: Artificial non-peptide single ion channels, *J. Am. Chem. Soc.* **114** (1992), 7618-7622.
15. Åkerfeldt, K. S., Lear, J. D., Wasserman, Z. R., Chung, L. A., and DeGrado, W. F.; Synthetic peptides as models for ion channel proteins, *Acc. Chem. Res.* **26** (1993), 191-197.
16. Lopez, G. A., Jan, Y. N., and Jan, L. Y.: Evidence that the S6 segment of the Shaker voltage-gated K^+ channel comprises part of the pore, *Nature* **367** (1994), 179-182.
17. Veatch, W. and Stryer, L.: The dimeric nature of the Gramicidin A transmembrane channel, *J. Mol. Biol.* **113** (1977), 89-102.
18. Arseniev, A. S., Barsukov, I. L., Bystrov, V. F., Lomize, A. L., and Ovchinnikov, Y. A.: ^{1}H-NMR study of gramicidin A transmembrane ion channel, *FEBS Lett.* **186** (1985), 168-174.
19. Ketchem, R. R., Hu, W., and Cross, T. A.: High-resolution conformation of Gramicidin A in a lipid bilayer by solid-state NMR, *Science* **261** (1993), 1457-1460 .
20. Guilbaud, P., Varnek, A., and Wipff, G.: Molecular dynamics study of *p-tert*-butylcalix[4]arenetetraamide and its complexes with neutral and cationic guests. Influence of solvation on structures and stabilities, *J. Am. Chem. Soc.* **115** (1993), 8298-8321.

CALIX[4]-*BIS*-CROWNS: FROM NUCLEAR WASTE TREATMENT TO MOLECULAR MACHINES

Z. ASFARI,[a] B. PULPOKA,[a] M. SAADIOUI,[a] S. WENGER,[a] M. NIERLICH,[b] P. THUÉRY,[b] N. REYNIER,[b] J.-F. DOZOL,[c] AND J. VICENS[a*]

(a) ECPM, Laboratoire de Chimie Analytique et Minérale, associé au CNRS, 1 rue Blaise Pascal F-67008 Strasbourg, France.

(b) CEA Saclay, SCM, Bât 125, F-91191 Gif-sur-Yvette, France.

(c) CEA Cadarache, F-13108 Saint Paul-lez-Durance, France.

Abstract. 1,3-Calix[4]-*bis*-crowns are ditopic receptors able to form 1:1 and 1:2 complexes with alkali and ammonium cations. X-ray structures of the Cs^+ complexes showed the 1,3-Calix[4]-*bis*-crown-6 to be highly preorganized prior to complexation. They showed very high Cs^+/Na^+ selectivity which allowed their used in the removal of cesium from radioactive wasted waters. Cation-tunneling has been evidenced through the cavity of the calixarene unit with 1,3-calix[4]-*bis*-crown-5. It is also presented the synthesis and complexation properties of 1,3-calix[4]-*bis*-aza-oxa-crowns and 1,3-Calix[4]-*bis*-cryptands.

Key Words. 1,3-Calix[4]-*bis*-crowns, 1,3-calix[4]-*bis*-aza-oxa-crowns, 1,3-*azobenzene*calix[4]crowns, Na^+/Cs^+ selectivity, thermodynamical parameters, mono- and bi-nuclear complexes, metal-tunneling, world globes.

1. Introduction

1,3-Calix[4]-*bis*-crowns correspond to calixcrowns[1-5] containing in their molecular frame one calix[4]arene unit and two polyether chains. The first member of the series was reported in 1990 as a by-product of the condensation of *p-tert*-butylcalix[4]arene on tetraethyelene glycol ditosylate to produce the *p-tert*-butylcalix[4]crown-5. [6]

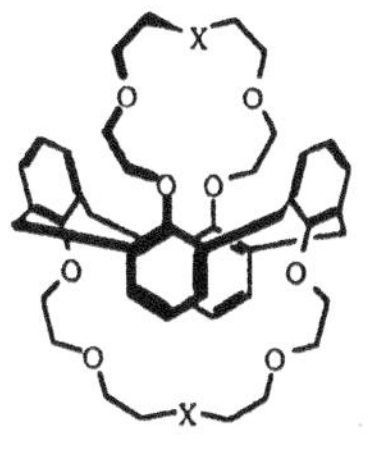

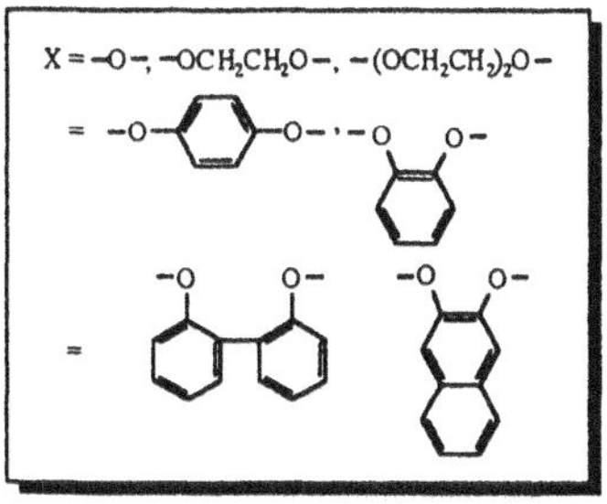

Chart I. 1,3-Calix[4]-*bis*-crowns.

A. W. Coleman (ed.), *Molecular Recognition and Inclusion*, 173–178.
© 1998 *Kluwer Academic Publishers. Printed in the Netherlands.*

We have prepared a series of calix[4]-*bis*-crowns (Chart I) by a one-pot procedure in ~50-80% yields by reacting calix[4]arene with various ditosylates in basic conditions.[7] The reaction in the presence of potassium carbonate mainly produced 1,3-calix[4]-*bis*-crowns in the *1,3-alternate* conformation.[7]

2. X-ray Structures

Their structures were confirmed by the determination of the crystalline structure of the 1,3-calix[4]-*bis*-crown-5[8] and 1,3-calix[4]-*bis*-crown-6.[9] 1,3-calix[4]-*bis*-crowns-n or **Bis-Cn**, have a globular arrangement in which the two polyether chains are equivalent on each side of the calix unit in the 1,3-alternate conformation. They are potential ditopic receptors and due to the presence of donor oxygen atoms mono- and binuclear complexes have been isolated with alkali metals. The crystal structures of mononuclear K**Bis-C6**(NO_3), Cs**Bis-C6**(NO_3) and binuclear Na_2**Bis-C6**$(NO_3)_2$.$2H_2O$, Cs_2**Bis-C6**$(NO_3)_2$, and Cs_2**Bis-C6**$(SCN)_2$ have been determined.[9] The following conclusions were drawn:

1. The ligands containing 5 O in the glycolic chains have a loop-size adapted to the complexation of potassium and rubidium. The ligands containing 6 O are convenient for complexation of cesium.

2. The complexation of cesium by **Bis-C6** occurred without changing the conformation of the glycolic chains showing the ligand to be highly preorganized for capturing this metal. Fig. 1 presents the very similar structures of **Bis-C6**, Cs**Bis-C6**(NO_3) and Cs_2**Bis-C6**$(NO_3)_2$.

3. There are π-metal interactions in the complexes of **Bis-C6** with potassium and cesium.

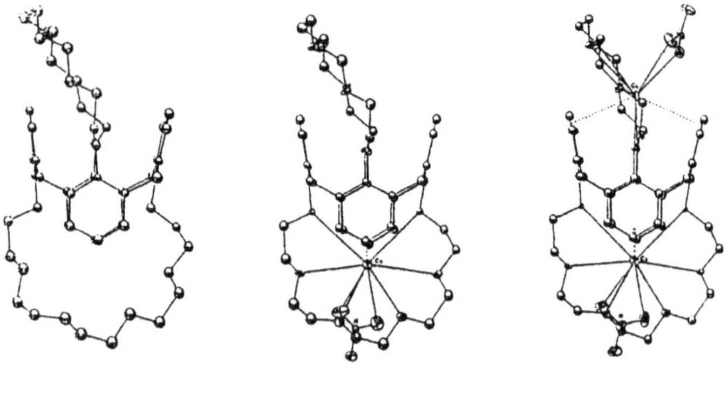

Figure 1. Molecular structures of **Bis-C6**, Cs**Bis-C6**(NO_3) and Cs_2**Bis-C6**$(NO_3)_2$.

3. Treatment of Nuclear Wastes

Bis-Cn have been used to remove radioactive cesium from nuclear wastes.[7,10] The treatment involves selective extraction of cesium from aqueous solutions that are 1 M in HNO_3 and 4 M in $NaNO_3$. Selective transport was carried out in supported liquid membranes (SLM's).[7] The procedure consisted in using *ortho*-NitroPhenylOctylEther as transport solvent with Cs+/Na+ selectivities ~ 50000.[7,10]

4. Photoresponsive 1,3-azocalix[4]crowns

To overcome difficulties in decomplexation with strong ligands we have designed a series of photosensitive 1,3-calix[4]crowns containing a photoisomerizable (E) ⇔ (Z) *azo* unit in the crown ether. The general scheme of synthesis is presented in figure 2.

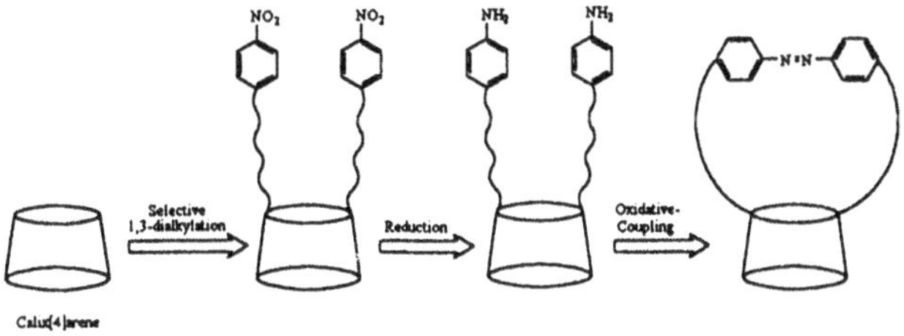

Figure 2 General synthesis of 1,3-*azo*calix[4]crowns.

(Z)-isomer of 1,3-*azobenzene*calix[4]-crown-6 in figure 3 was shown to improve the transport of cesium and rubidium through SLM's (30% to 60%) when compared to (E)-isomer.[11]

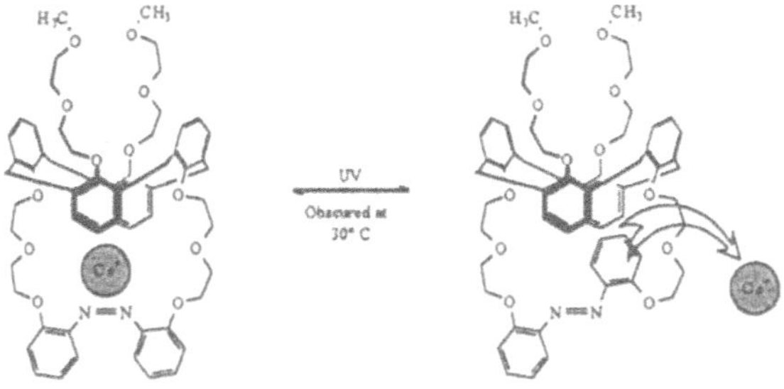

Figure 3. 1,3-*azobenzene*calix[4]-crown-6

5. Physico-Chemical Properties

The binding properties of **Bis-Cn** towards alkali cations were evaluated by extracting alkali picrates from water to dichloromethane and the determination of thermodynamical parameters (ΔGc, ΔHc, and TΔSc) of the complexation in methanol and acetonitrile.[12] The ligands with n = 6 showed a preference for the cesium cation in extraction and complexation with a selectivity $S = \beta(Cs^+)/\beta(Na^+) = 2 \times 10^3$ in the methanol for **Bis-C6**.[12] This selectivity was attributed to a good complementary between the radius of the cesium and the size of the polyether chain and to the impossibility of the chain to wrap efficiently the sodium cation.[12] The stabilization of Cs**Bis-C6** complex in the methanol is enthalpic: -ΔHc is strongly positive overcoming the negative value of TΔSc indicative of a strong desolvation of the cation and a small rearrangement of the glycolic chain during the complexation.[12]

6. Metal-Ligand Exchanges

In order to obtain more information on the behavior and structure of the alkali-complexes in solution we used NMR techniques. **Bis-C5** formed 1:1 and 1:2 complexes with alkali picrates CD_2Cl_2-DMF-*d7*.[13] 1:1 Complexes have been observed for K^+, Rb^+,

Cs^+ and NH_4^+ with association constants of log*Kass* ~ 2.9 - 4.6.[13] 1:2 Complexes were observed for Na^+ and NH_4^+ with log*Kass* = 2.0 and 2.7 respectively.[13] NMR Spectra of the 1:1 complexes with K^+, Rb^+, Cs^+ and NH_4^+ presented coalescence patterns. Temperature dependent NMR showed the existence of two temperatures of coalescence *Tc* corresponding to both intra- and intermolecular metal-ligand exchanges.[13] *Tc*,inter ranged 55 to 125° C while *Tc*,intra ranged 35 to 105° C. The intramolecular exchange was assumed because *Tc* was shown to be independent on the concentration of the sample.[13] The activation energies of the processes were ~ 16 kcal. mol^{-1}. Both exchange mechanisms are depicted in Fig. 4.

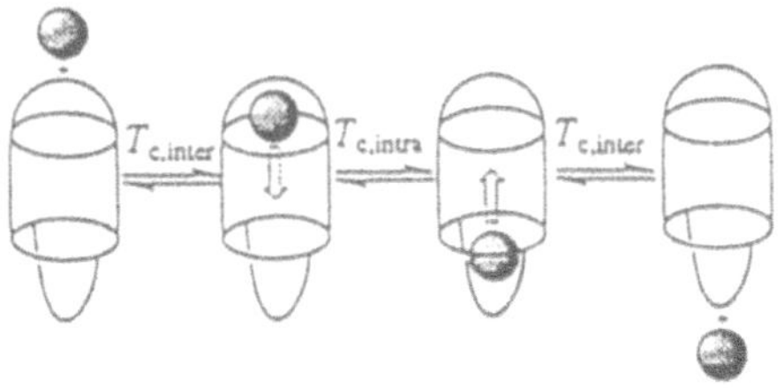

Figure 4. Metal-ligand exchange mechanisms of complexes of **Bis-C5**.

^1H- and ^{133}Cs-NMR techniques were used to investigate the mode of binding of **Bis-C6** towards cesium cation.[14] In 1:1 $CDCl_3$-CD_3OD, these ligands formed 1:1 and 1:2 complexes with cesium thiocyanate.[14] A study in function of the concentration of the metal showed a cation-ligand exchange between the mononuclear and binuclear species. The mode of complexation and the exchange process are shown in Fig.5.

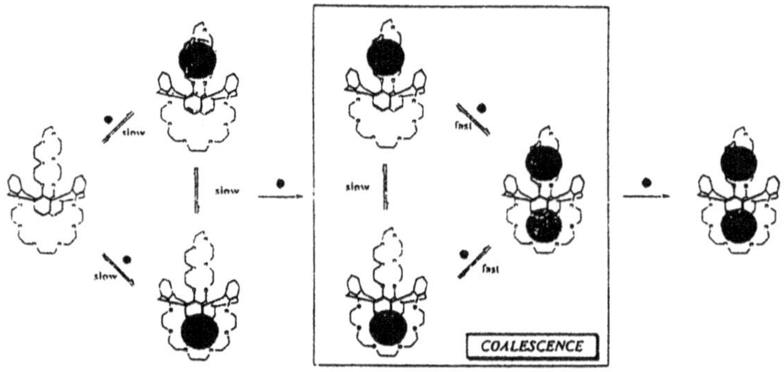

Figure 5. Mechanism of formation of mono- and binuclear species of **Bis-C6**.

7. Calix[4]-*bis*-aza-oxa-crowns

In a previous work we have shown that calixarenes capped with a diaza-oxa-crown are able to complex soft cations.[15,16] This lead us to synthesize receptors of 'calix[4]-*bis*-crown type' containing nitrogen atoms in N-Ts form. We named them calix[4]-*bis*-aza-oxa-crowns (Fig. 6) indicating the presence of N atoms in the glycolic part of the ligand.[17]

In a general manner the N-Ts ligands did not complex the alkali cations and quaternary ammonium salts.[17] Only was observed the 1:1 complex with NH_4^+ accompanied of cation-ligand intra- and intermolecular exchanges.[17]

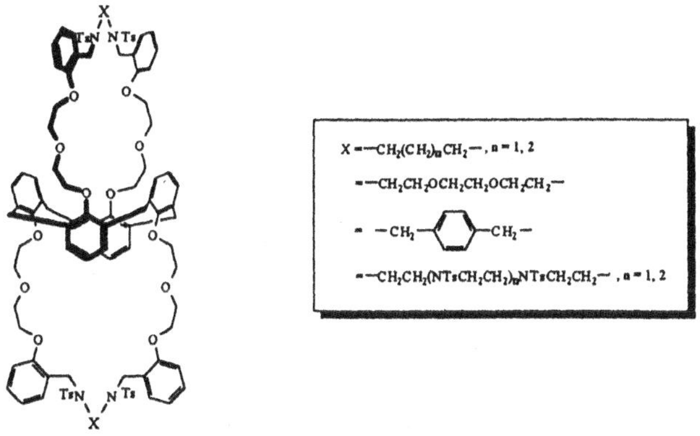

Figure 6. Calix[4]-*bis*-aza-oxa-crowns.

8. Mixed Receptors

Mixed receptors have been prepared consisting of ligands (Fig. 7) containing one polyether and one poly-aza-oxa-ether chains on each side of the calix unit.[18-20] These ditopic ligands present one 'soft' and one 'hard' ligating sites that can communicate through the calix unit.[18-20]

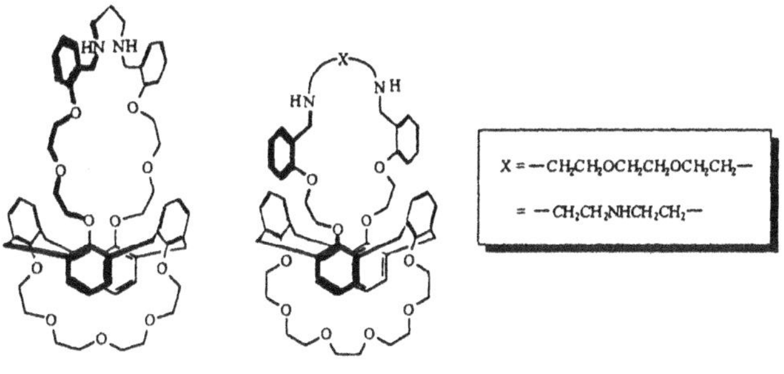

Figure 7. Mixed Receptors.

9. Calixcryptands

The obtaintion of ligands with reactive N-H groups lead to further cyclisation into 1,3-calix[4]cryptands[21] and 1,3-calix[4]-*bis*-cryptands[22] (see figure 8). These ligands combine in the calix unit one or two oxa-aza [18]-N_2O_4 subunits.

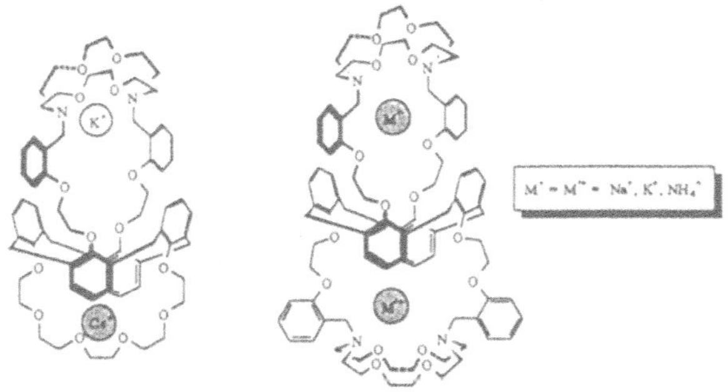

Figure 8 1:1:1 heterobinuclear complexes

They are able to form 1:1 and 1:2 complexes with alkali and ammonium cations. We also observed the formation of 1:1:1 heterobinuclear complexes (see figure 8). In the 1:1 complex of NH_4^+ and the 1,3-calix[4]-*bis*-cryptand of figure 8 the cation was observed to probably oscillate between the [18]-N_2O_4 subunits demonstrating the receptor to be an approach of synthetic *'nano-tubes'*.[22]

10. Molecular Machines

The globular shape of **Bis-Cn** suggested molecular architectures looking like a *world globe*.[23] The concept and the molecular achievement are presented in Fig. 9.

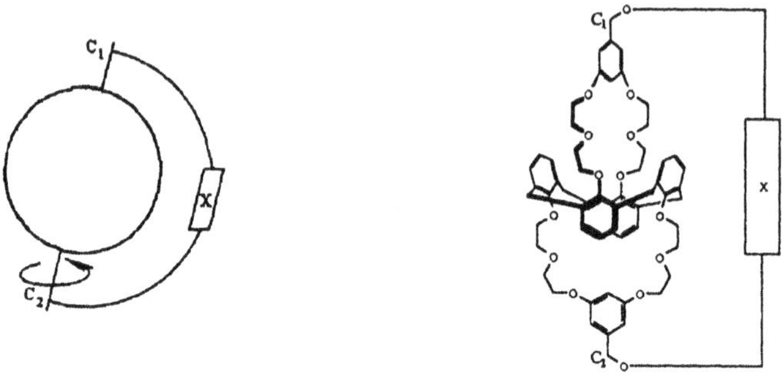

Figure 9. Molecular architectures looking like a *world globe*.

Work is in progress to stop the globe spinning by complexation of primary ammonium bearing aliphatic alkyl chains.

References.

1. Asfari, Z.; Wenger, S.; Vicens, J. *Pure & Appl. Chem.* **1995**, *67*, 1037.
2. Asfari, Z.; Wenger. S.; Vicens, J. *J. Incl Phenom. Mol. Recogn. Chem.* **1994**, *19*, 137.
3. Asfari, Z.; Wenger. S.; Vicens, J. *Supramol. Sci.* **1994**, *1*, 103.
4. Böhmer, V. *Angew. Chem. Int.Ed. Engl.* **1995**, *34*, 713.
5. Lhotak, P; Shinkai, S. *J. Synth. Org. Chem. Jpn* **1995**, *53*, 963.
6. Djistra, P. J.; Brunink, J. A.; Bugle, K.-E.; Reinhoudt, D. N. *J. Am. Chem. Soc.* **1990**, *112*, 1597.
7. Asfari, Z.; Bressot, C.; Vicens, J.; Hill, C.; Dozol, J.-F.; Rouquette, H.; Eymard, S.; Lamare, V.; Tournois, B. *Anal. Chem.* **1995**, *67*, 3133.
8. Abidi, R.; Asfari, Z.; Harrowfield, J. M.; Sobolev, A. N.; Vicens, J. *Aust. J. Chem.* **1996**, 49, 183
9. Thuéry, P.; Nierlich, M.; Bressot, C.; Lamare, V.; Dozol, J. F.; Asfari, Z.; Vicens, J. *J. Incl. Phenom. Mol. Recogn. Chem.* **1996**, 23, 305.
10. Hill, C.; Dozol, J. F.; Lamare, V.; Rouquette, H.; Eymard, S.; Tournois, B.; Vicens, J.; Asfari, Z.;, Bressot, C.; Ungaro, R.; Casnati, A. *J. Incl Phenom. Mol. Recogn. Chem.* **1994**, *19*, 399.
11. Saadioui, M.; Asfari, Z.; Vicens, J.; Reynier, N.; Dozol, J.F. *J. Incl. Phenom. Mol. Recogn. Chem.* submitted for publication
12. Souley, B.; Arnaud, F.; Asfari, Z. Vicens, J. *New J-Chem.* 1996, 20, 453.
13. Koh, K. N.; Araki, K.; Shinkai, S.; Asfari, Z. Vicens, J. *Tet. Lett.* **1995**, *36*, 6095.
14. Unpublished results from this laboratory.
15. Seangprasertkij, R.; Asfari, Z.; Arnaud, F.; Vicens, J. *J. Org. Chem.* **1994**, *59*, 1741.
16. Seangprasertkij, R.; Asfari, Z.; Vicens, J. *J. Incl Phenom. Mol. Recogn. Chem.* **1994**, *17*, 111.
17. Wenger, S.; Asfari, Z.; Vicens, J. *Tet. Lett.* **1994**, *35*, 8369.
18. Wenger, S.; Asfari, Z.; Vicens, Z. *J. Incl Phenom. Mol. Recogn. Chem.* **1995**, *20*, 293.
19. Wenger, S. *Nouvelle Thèse* **1996** Strasbourg, France
20. Pulpoka, B.; Asfari, Z.; Vicens, J. *J. Incl Phenom. Mol. Recogn. Chem* accepted for publication.
21. Pulpoka, B.; Asfari, Z.; Vicens, J. *Tet. Lett.* **1996**, 37, 6315.
22. Pulpoka, B.; Asfari, Z.; Vicens, J. *Tet. Lett.* submitted for publication.
23. Asfari, Z.; Naumann, C.; Kaufmann, G.; Vicens, J. *Tet. Lett.* **1996**, 37, 3325

TRIPODAL COORDINATION COMPLEXES AS SCAFFOLDS FOR MOLECULAR RECOGNITION AND CATALYSIS

J.W. CANARY*, C.S. ALLEN, J.M. CASTAGNETTO, C.-L. CHUANG, A.R. LAJMI, O. DOS SANTOS, X. XU
Department of Chemistry, New York University
New York, NY 10003-5180

The Scaffold: Tris((2-pyridyl)methyl)amine (TPA)

There is currently much interest in the utilization of coordination complexes as scaffolds for the design of metallohost-guest complexes and metalloenzyme mimics.[1] Anderegg et al. reported that the robust ligand tris(2-pyridylmethyl)amine (TPA, Figure 1) forms stable complexes with various divalent metal cations (e.g., $\log(K_{ZnTPA}) = 11$ and $\log(K_{CuTPA}) = 16$) in aqueous solution, and that $[M(TPA)OH_2]^{2+}$ ($M = Zn^{II}$, Cu^{II}) complexes dissociate at mildly alkaline pH to form the hydroxo species.[2] Karlin and coworkers have reported crystallographic, electrochemical, and solution thermodynamic studies of various Cu^{I} and Cu^{II} complexes of TPA and related compounds.[3-6] We[7] and others[8] have reported structural characterization of Zn^{II} complexes of TPA, observing that Zn^{II} complexes are usually 5-coordinate and nearly isomorphous with Cu^{II} complexes. We have reported 6- and 7-coordinate Cd^{II} complexes of TPA.[7] That TPA is geometrically suited for complexation of ions of small ionic radius was seen in the demonstration of the selectivity of the ligand for the complexation of Li^{+} ion over the other alkali metal cations.[9,10] It is notable that the conformation of the ligand in complexes with trigonal bipyramidal metal ions displays a C_3-symmetry, while that in octahedral metal ion complexes displays a distorted C_{σ}-symmetry.[7]

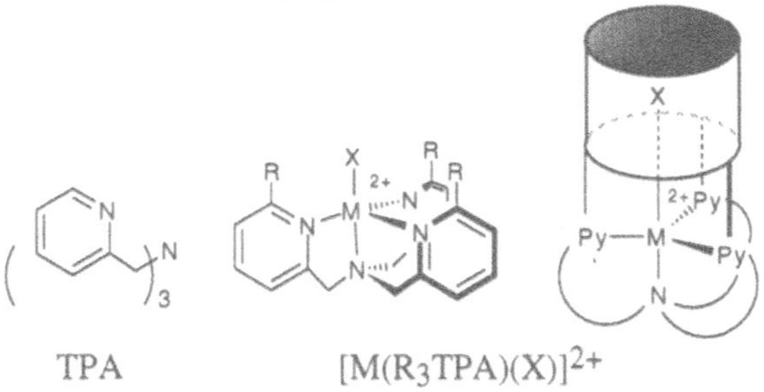

Figure 1. Elaboration of supramolecular coordination complexes from TPA.

A. W. Coleman (ed.), Molecular Recognition and Inclusion, 179–184.
© 1998 *Kluwer Academic Publishers. Printed in the Netherlands.*

Metal-Dependent Induction of Conformational Chirality

PROPELLER-LIKE TPA DERIVATIVES

The current interest in chiral C_3-symmetric coordination complexes[11-14] and the potential for enantioselective molecular recognition studies led us to find a way to introduce chirality into the TPA scaffold.[15] A method was sought that would take advantage of inherent conformational properties of certain TPA complexes. In the C_3-symmetric conformation of TPA (as seen in $[Zn(TPA)Cl]^+$), the tetradentate TPA encompasses the metal ion with a helical asymmetric twist producing a spatial arrangement of the three pyridyl groups, tilting with respect to the metal-amine axis, which resembles a propeller. The pyridyl rings can tilt in either a clockwise or counter-clockwise direction, affording conformational enantiomers that interconvert rapidly at room temperature in solution via rotation about the C_α–C_2 and C_α–N_{am} bonds.[7]

Attachment of a single alkyl substituent to one of the arms of the tripod results in the ligand adopting a single helical conformation in trigonal bipyramidal coordination complexes. In the absence of metal ion, a ligand such as α–MeBQPA, **1**, may adopt many conformations. Upon presentation of a metal ion, the ligand wraps around the metal, forming a propeller-like complex with a handedness dictated by the asymmetric carbon atom. The amplified chirality of the complex has been observed by several X-ray crystallographic and circular dichroism studies as well as by extensive molecular modeling studies. No helical chirality was observed in octahedral Cd^{II} complexes in which the ligand would be expected to adopt the C_σ-conformation.

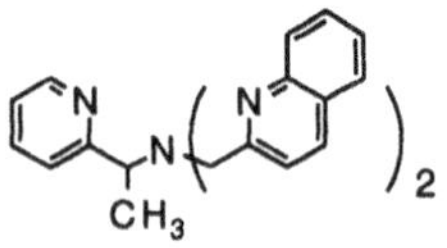

1 (α–MeBQPA)

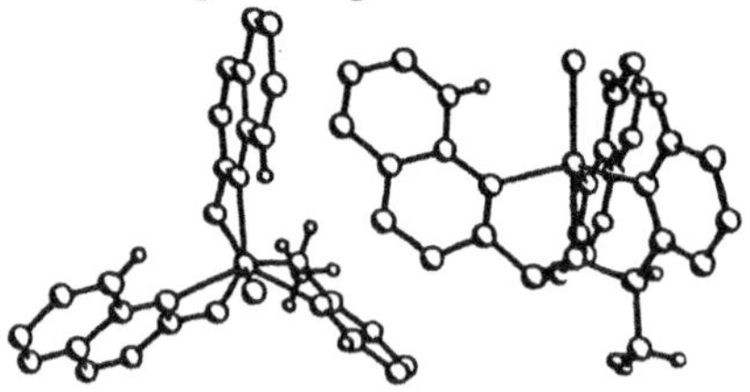

$[Zn(\alpha\text{–MeBQPA})Cl]^+$

Figure 2. Formula of **1** and crystallographic structure of $[Zn(\alpha\text{–MeBQPA})Cl](ClO_4)$.

A METAL ION-DIFFERENTIATING FLUORESCENT CHEMOSENSOR

The ligand α–MeBQPA **1** is an intrinsic fluorescent chemosensor for diamagnetic metal ions such as Zn^{II} and Cd^{II}, with increased relative fluorescence intensity for the complexes compared with the free ligand.[16] Diamagnetic Cu^{II} or Fe^{II} complexes show no increased fluorescence. Fluorescence titration experiments show that the binding of Zn^{II} and Cd^{II} to α–MeBQPA is quite strong ($K_a > 10^6\ M^{-1}$) in aqueous solution at neutral pH. Simultaneous evaluation of the fluorescence and circular dichroism properties of the complexes results in the interesting situation that the ligand can not only report that a metal ion is present, but can identify the metal. Detection of both fluorescence enhancement and anisotropic absorption would, for example, distinguish between Zn^{II}, Cd^{II}, Cu^{II}, and Fe^{II} (Figure 3).

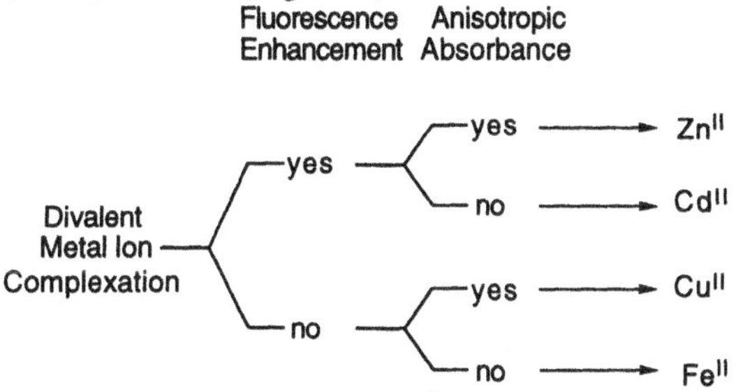

Figure 3. Differentiation between divalent metal ions by the ligand α–MeBQPA.

Ternary Complexes: Molecular Recognition and Catalysis

COMPLEXATION

A Redox-Switched Binding Site

The synthesis and characterization of the ligand, tris(6-phenyl-2-pyridylmethyl)amine (TPPA) and some of its copper complexes were described.[17,18] The complexes $[Cu^{I}(TPPA)]BPh_4$ (**2**) and $[Cu^{II}(TPPA)(CH_3CN)](ClO_4)_2$ (**3**) (Figure 4) were prepared and their X-ray crystal structures and redox potentials were determined.

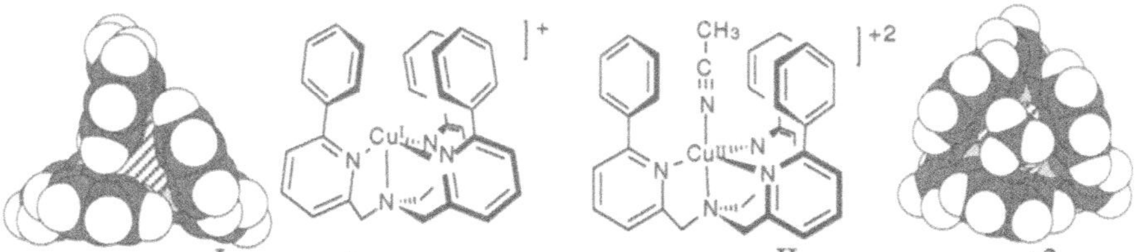

$[Cu^{I}(TPPA)]^{+}$ $[Cu^{II}(TPPA)(MeCN)]^{+2}$

2 3

Figure 4. Structures and space-filling models generated from X-ray coordinates of compounds.

Contacts were observed between the bound CH_3CN and the phenyl substituents in the Cu(II) complex **3** that were considerably shorter than the sum of the van der Waals radii; the observed distances were close to those previously reported for stacked aromatic rings engaged in favorable π-π interactions. Figure 4 shows that in the copper(I) complex **2**, two phenyl rings stack in a "T" configuration as observed in solid benzene and protein aromatic side chains. The space-filling models indicate that the copper(I) is completely encapsulated by the TPPA ligand. In **3**, the ligand contains a cavity in which a CH_3CN solvent molecule is bound. To create the cavity, the ligand undergoes a helical "twist." The average angle between the best axis of the 6-phenylpyridine "arms" and the Cu-N1 bond in **2** is 14.8°; in **3** it is 36.6°. These angles are achieved solely by rotation about the 2-pyridyl-CH_2 bond. Thus, the conformational change in the ligand that would be required to create the cavity for the CH_3CN in **3** resembles the motion of a flower undergoing anthesis.

Chiral Solvating Agent

Zinc perchlorate complexes of chiral ligands such as α–MeBQPA in acetone-d_6 have been shown to act as chiral solvating agents for sulfoxides.[15] The ligand is more efficient than the corresponding compound with all-pyridine arms, possibly due to the greater degree of twist or due to the larger shielding cones that would be expected for the quinoline rings. Baseline resolution of methyl vinyl sulfoxide and other sulfoxides was observed.

Amino Acid Binding

The $Cu(ClO_4)_2$ complex of the ligand CRTPA was observed to form complexes with amino acids with structure $^+H_3N–(CH_2)_n–CO_2^-$ (Figure 5).[19] Binding constants were observed on the order of 10^2-10^3 M^{-1} in 50% aqueous acetonitrile, with the best binding observed for n = 5.

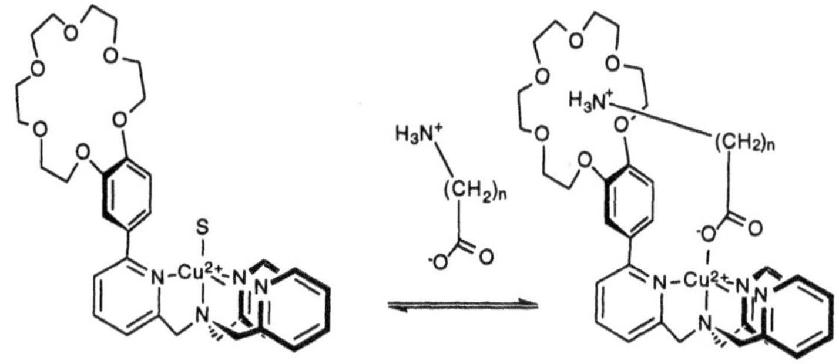

Figure 5. Complexation of $^+H_3N–(CH_2)_n–CO_2^-$ by $[Cu(CRTPA)](ClO_4)_2$.

Electrochemical Detection of Cation Binding

Cyclic voltammetry studies of copper complexes of the ligands CRTPA and **4** indicated that guest binding played a role in the determination of the copper redox couple (Figure 6).[20] Studies of the alkoxyphenyl ligands indicated that small but reproducible changes in the copper redox couple occurred upon presentation of a guest cation that would be expected to form a complex with the copper/ligand complex. In $[Cu(CRTPA)]^{+/2+}$, binding of a cation causes a positive shift in the redox potential of the copper(I/II) couple. In $[Cu(\mathbf{4})]^{+/2+}$, a similar result is observed but the magnitude is somewhat larger for NH_4^+. This may result from the closer proximity of the guest cation to the copper center and geometrical communication between the guest and the copper ion through the covalent structure of the scaffold. For many interesting guests, the electrochemical behavior becomes irreversible due to variation of the copper coordination number upon complexation.

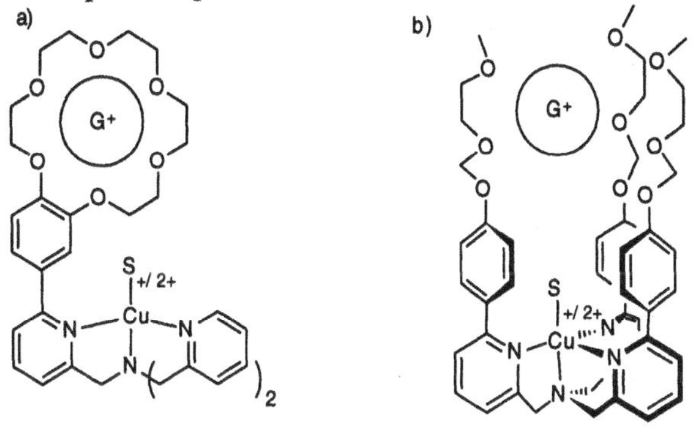

[Cu(CRTPA)(S)G]+2/3

[Cu(4)(S)G]+2/3

Figure 6. Complexes with N_4-binding sites for a redox-active metal ion such as copper and a second binding site for cations (G = guest, S = solvent).

CATALYSIS

Influence of Hydrophobic Surface Area on Zn-OH2 Acidity

The Zn–OH functional group is a critical component of the catalytic machinery employed by many hydrolytic enzymes. We are engaged in the study of Zn–OH basicity and nucleophilicity in a variety of model systems. In one recent study, the enthalpies of ionization (ΔH_a) of zinc-bound water in the complexes $[Zn(tren)OH_2]^+$ (tren = tris(2-aminoethyl)amine) and $[Zn(Me_6tren)OH_2]^{+2}$ (Me_6tren = tris(N,N-dimethyl-2-aminoethyl)amine) were determined by isothermal titration calorimetry and potentiometric methods in 1 M $NaClO_4$.[21] The latter complex encases the Zn-OH_2 by a ring of hydrophobic methyl substitutents. The $\Delta G°$ (14.6 kcal/mol) for $[Zn(tren)OH_2]^{2+}$ is mainly due to large positive ΔH (11.0 kcal/mol), while that for $[Zn(Me_6tren)OH_2]^{+2}$ (12.1 kcal/mol) is attributable equally to positive ΔH (6.1 kcal/mol) and negative ΔS at 298 K. The increase in acidity of $[Zn(Me_6tren)(OH_2)]^{2+}$ relative to $[Zn(tren)OH_2]^{2+}$ corresponds to a $\Delta\Delta G°$ of -2.5 kcal/mol. The $\Delta\Delta H$ is -4.9 kcal/mol, and is partially offset by a 2.4 kcal/mol contribution from $\Delta(T\Delta S)$. These values differ from ones previously reported,[22] particularly in that the thermodynamic basis of the increased acidity of the latter complex is not related to entropic effects, as was previously believed. The results of the study invalidate the original data that led to the view that solvent structure mediates the pK_a of zinc-bound water in zinc hydrolytic enzymes.

Phosphodiester Hydrolysis by Heterobinuclear Coordination Complexes

One enzyme mechanistic motif that is poorly understood yet extremely common involves the use of two metal ions in the hydrolysis of phosphodiesters and other substrates. For example, in the enzymatic mechanism for the 3'-5' exonuclease reaction of DNA polymerase I, a Zn^{+2} ion is proposed to facilitate the formation of an attacking hydroxide ion oriented towards the phosphorus by interactions with a Zn^{2+} ion. A Mg^{2+} ion is hypothesized to facilitate formation of the pentacovalent, dianionic intermediate and leaving group departure by electrostatic interactions. Similar two-metal ion mechanisms have been proposed for several nucleases and polymerases as well as for RNA splicing and ribozymes.

The ligand CRTPA was found to bind two different metal ions in aqueous solution.[19] Complexes of the ligand involving one zinc(II) ion and an alkaline earth ion (e.g., Ba^{2+}) catalyze the hydrolysis of bis(p-nitrophenyl)phosphate (BNPP) by at least 2000-fold over the background rate. Several lines of evidence suggest that the zinc ion delivers hydroxide as nucleophile while the second metal ion increases the binding of the substrate.

(1) Canary, J. W.; Gibb, B. *Prog. Inorg. Chem.* **1996**, *45*, 1.

(2) Anderegg, G.; Hubmann, E.; Podder, N. G.; Wenk, F. *Helv. Chim. Acta* **1977**, *60*, 123.

(3) Zubieta, J.; Karlin, K. D.; Hayes, J. C. In *Copper Coordination Chemistry: Biochemical and Inorganic Perspectives*, ; Karlin, K.D. and Zubieta, J., Ed.; Adenine Press: New York, 1983; pp 97.

(4) Wei, N.; Murthy, N. N.; Karlin, K. D. *Inorg. Chem.* **1994**, *33*, 6093.

(5) Wei, N.; Murthy, N. N.; Chen, Q.; Zubieta, J.; Karlin, K. D. *Inorg. Chem.* **1994**, *33*, 1953.

(6) Karlin, K. D.; Wein, N.; Jung, B.; Kaderli, S.; Niklaus, P.; Zuberbühler, A. D. *J. Am. Chem. Soc.* **1993**, *115*, 9506.

(7) Allen, C. S.; Chuang, C.-L.; Cornebise, M.; Canary, J. W. *Inorg. Chim. Acta* **1995**, *239*, 29.

(8) Murthy, N. N.; Karlin, K. D. *J. Chem. Soc., Chem. Commun.* **1993**, 1236.

(9) Chuang, C.-L.; Frid, M.; Canary, J. W. *Tetrahedron Lett.* **1995**, *36*, 2909.

(10) Brownstein, S. K.; Plouffe, P. Y.; Bensimon, C.; Tse, J. *Inorg. Chem.* **1994**, *33*, 354.

(11) Nugent, W. A.; Harlow, R. L. *J. Am. Chem. Soc.* **1994**, *116*, 6142.

(12) LeCloux, D. D.; Tokar, C. J.; Osawa, M.; Houser, R. P.; Keyes, M. C.; Tolman, W. B. *Organometallics* **1994**, *13*, 2855.

(13) Tor, Y.; Libman, J.; Shanzer, A.; Felder, C. E.; Lifson, S. *J. Am. Chem. Soc.* **1992**, *114*, 6653.

(14) Adolfsson, H.; Wärnmark, K.; Moberg, C. *J. Chem. Soc., Chem. Commun.* **1992**, 1054.

(15) Canary, J. W.; Allen, C. S.; Castagnetto, J. M.; Wang, Y. *J. Am. Chem. Soc.* **1995**, *117*, 8484.

(16) Castagnetto, J. M.; Canary, J. W., manuscript submitted for publication.

(17) Chuang, C.-L.; Lim, K.; Chen, Q.; Zubieta, J.; Canary, J. W. *Inorg. Chem.* **1995**, *34*, 2562.

(18) Chuang, C.-L.; Lim, K.; Canary, J. W. *Supramol. Chem.* **1995**, *5*, 39.

(19) Lajmi, A. R.; dos Santos, O.; Canary, J. W. manuscript submitted for publication.

(20) Chuang, C. L.; dos Santos, O.; Xu, X.; Canary, J. W., manuscript submitted for publication.

(21) Canary, J. W.; Xu, J.; Castagnetto, J. M.; Rentzeperis, D.; Marky, L. A. *J. Am. Chem. Soc.* **1995**, *117*, 11545.

(22) Coates, J. H.; Gentle, G. J.; Lincoln, S. F. *Nature* **1974**, *249*, 773.

(23) We gratefully acknowledge the National Institutes of Health (GM49170) for support of our research program.

PHOTOCHROMIC MOLECULAR RECOGNITION OF CYCLODEXTRINS BEARING SPIROPYRAN MOIETY FOR ORGANIC GUESTS

FUMIO HAMADA
Department of Materials Engineering and Applied Chemistry, Mining College, Akita University, Tegata, Akita 010, Japan

Abstract

The photochromic behaviors and guest binding properties of spiropyran modified cyclodextrins (**1**, **2** and **3**), have been investigated. Compound **1** and **3** are derivatives of β-cyclodextrin and **2** is obtained from γ-cyclodextrin. It is found that those derivatives show a different host-guest binding mechanism and photochromic response. In the system of **1**, the appended moiety acts as a hydrophobic cap to elevate the binding ability of **1**, whereas those appended moieties of **2** and **3** act as a probe to display the binding ability and mechanism together with acting as a hydrophobic cap. In the system of **3**, the spiropyran moiety is included into the cyclodextrin cavity whenever a larger or smaller guest is used with a different intensity of the binding ability.

1. Introduction

When we study a host-guest complexation properties of native cyclodextrins by spectroscopic method such as fluorescence, induced circular dichroism or absorption intensities changes, spectroscopically active guests should be used because cyclodextrins themselves are basically spectroscopically inert. Cyclodextrins, however, become spectroscopically active compounds by modification with chromophores, which results spectroscopically inert guests can be recognized by the spectral change of modified cyclodextrins. It is found that the chromophores act as a probe of the host-guest complexation behavior and together with as a cap to elevate a hydrophobic interaction with a guest or as a spacer to regulate cyclodextrin cavity size. On the other hand, when we use a photochromic residue such as spiropyran or azobenzene as a chromophore, it seems that photochromic changes of the chromophores probably can affect the affinity of the binding site for a guest molecule. A good example of such a modified cyclodextrin is azobenzene capped β-cyclodextrin reported by Ueno and co workers[1]. This host displays the different complexation properties of the *trans* and *cis* isomers. As an extension of this work, azobenzene capped g-cyclodextrin derivative was prepared by Hamada et al, in which markedly larger binding value for (-)-borneol than the values of cis form and γ-cyclodextrin itself were recognized[2]. Although, spiropyran is a well known photochromic unit, it has not been used as a photochromic part of cyclodextrins. The photochromism of spiropyran involves isomerization

A. W. Coleman (ed.), Molecular Recognition and Inclusion, 185–190.
© 1998 *Kluwer Academic Publishers. Printed in the Netherlands.*

between closed and opened forms(Scheme 1), in contrast to that of azobenzene in which *trans* and *cis* isomers. The closed form is hydrophobic, while the opened form, which is an ion pair form, is less hydrophobic, so spiropyran when linked with cyclodextrin, can be used to change the environment around the cyclodextrin cavity by photoirradiation.

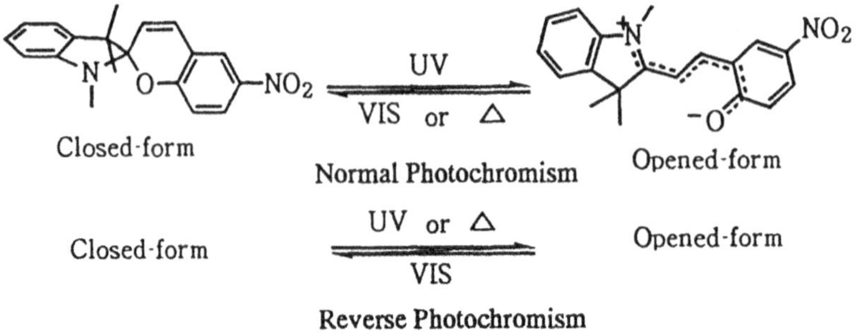

Scheme 1. Photochromism of spiropyran

To study the photochromic responses and guest binding properties of spiropyran modified cyclodextrins, three kinds of derivatives (**1**,**2** and **3**) have been prepared[3,4,5]. In these hosts, they exhibit a different host-guest binding property and photochromic response.

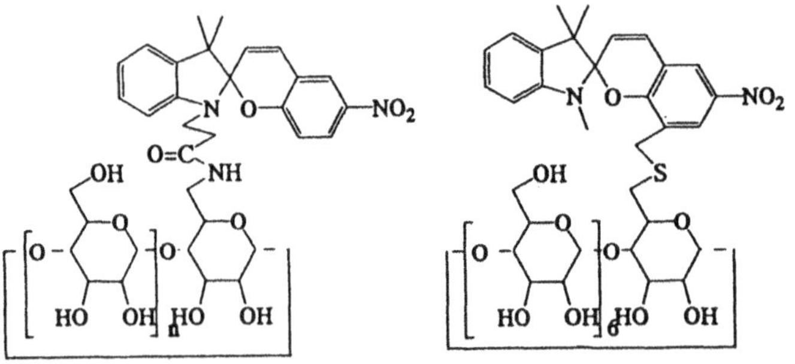

n=6; **1** n=7; **2**

3

2. Experiment

Compound **1** and **2** were prepared from primary hydroxyl side mono amino β- and γ-cyclodextrins, respectively, with 1'-(2-carboxyethyl)-3,3'-dimethyl-6-nitrospiro[2H-benzopyran-2,2'-indoline] in the presence of dicyclohexycarbodiimide. Compound **3** was prepared from 6-mercapto-b-cyclodextrin and 1,3',3'-trimethyl-6-nitro-8-chloromethylspiro[2H-1-benzopyran-2,2'-indoline]. The ultraviolet (UV) and induced circular dichroism (ICD) spectra were measured at 25°C with a Hitachi U-2000 or Shimazu UV-250 spectrophotometer and a JASCO J-700 spectropolarimeter, respectively. The ICD intensities were expressed as molar ellipticity([Θ]) (in degree·cm^2 $dmol^{-1}$). Photoirradiation was performed with 500W Xenon lamp (Ushio Electric Inc.UI-502Q) using cut-off filters, Corning 7-37 and 3-73 for isolating UV(320 nm to 380nm) and visible (425 nm to 500 nm) light, respectively.

3. Results and Discussion

3.1 Photochromism.

Figure 1 shows the absorption spectra of **1**. Because of the solubility of **1** in a pure water, we used 10 vol. -% ethylene glycol aqueous solution as the solvent for the spectroscopical measurements. When 10 vol. -% ethylene glycol aqueous solution of **1** was placed in the dark or irradiated by UV,**1** shows an absorption band at 515 nm, which is characteristic of the opened form, while after visible light irradiation, the absorption band disappeared associated with isomerizaton to the closed form.

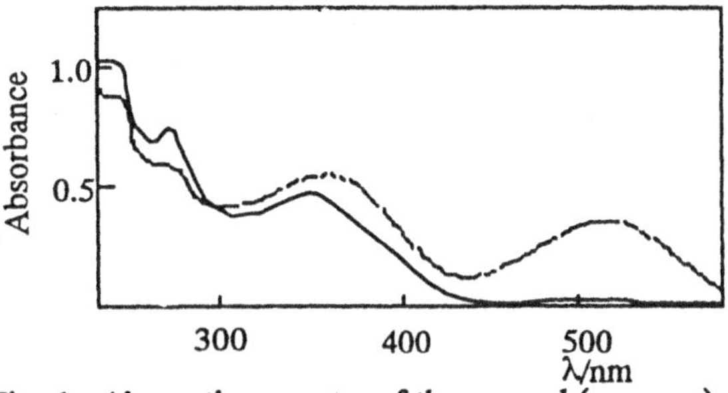

Fig. 1 Absorption spectra of the opened (— — –) and closed (———) forms of **1** in aqueous 10vol. -% ethylene glycol solution ($4.85 \cdot 10^{-5}$ mol dm^{-3}).

The conversion from the closed to opened form was achieved completely in keeping in the dark for 12 hr. This photochromism behavior, named reverse photochromism, is the same as observed for the spiropyran-modified enzymes in aqueous solution. It is found that the absorbance at 513 nm of the opened form of **2**, which is produced by keeping in the dark, is one third times that of **1** at the same concentration. It is suggested that the spiropyran residue of **2** is inside of the cyclodextrin cavity, which probably results to depress the isomerization of **2** with steric hindrance. On the other hand, the spiropyran of **1** is not inside of the cavity because β-cyclodextrin cavity is too small to form intramolecular complex with the spiropyran residue. Figure 2 shows the time-course of increase of the absorbance of **2** in the presence of (-)-borneol and without a guest at 513 nm. The increase of the absorbance of **2** obtained in keeping in the dark in the presence and absence of a guest molecule was examined.

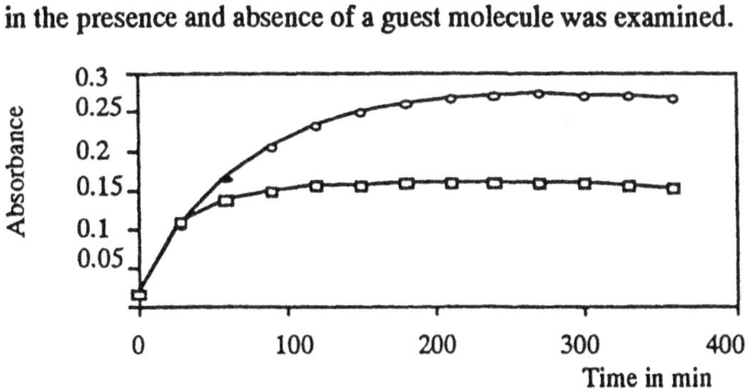

Fig. 2 Time-courses of increaseof the absorbance at 513 nm of **2** ($4.85 \cdot 10^{-5}$ mol dm^{-3}) in the absence (-□-) and in the presence of (-)-borneol ($2 \cdot 10^{-3}$ mol dm^{-3}) (-○-).

The absorbance increase is the highest when (-)-borneol was used as a guest. Higher values than without a guest were obtained with (-)-fenchone and (+)-fenchone and lower ones with cyclohexanone and cyclohexanol. The guest-induced variation of the absorbance at 513 nm was used to obtain the binding constant of **2** for a guest. The binding constant is given by Eq.(1).

$$\frac{1}{Ia - Ia_0} = \frac{1}{(b\text{-}a)\mathrm{Ch}} + \frac{1}{(b\text{-}a)\,\mathrm{Ch}K}\;\frac{1}{\mathrm{Cg}} \qquad (1)$$

Here, I=absorbance at 513 nm (Ia_0 for the host alone, Ia for the complex kept in the dark for 6h), Ch= total host concentration and Cg =total guest concentration, a and b are constants. When (-)-borneol was used as a guest, the binding constant was estimated to be 1400 $mol^{-1}\ dm^{3}$. When aqueous solution of **3** was placed in the dark or irradiated by UV,**3** shows an absorption band at 513 nm. Figure 3 shows the increase of the absorbance intensity at 513 nm of **3** in aqueous solution in the presence or absence of a guest. The conversion of **3** alone from the closed to opened form was achieved completely in keeping in the dark for 9h. On the other hand, it takes a longer time to complete the isomerization when (-)-borneol or adamantanol was used as a guest. This fact means that the photochromic response of **3** can be affected by the presence of a guest.

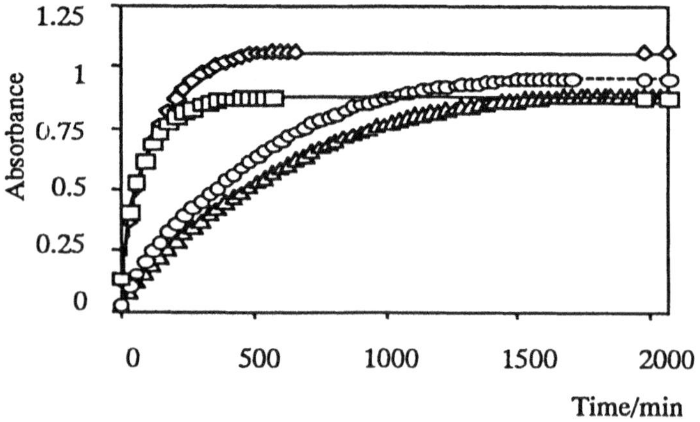

Fig. 3 Time courses of increase of the absorbance at 513 nm of **3** ($1{\cdot}10^{-4}$ mol dm^{-3}) in the absence of a guest (—□—)and in the presence of nerol($1{\cdot}10^{-3}$ mol dm^{-3})(—◇—), and, (-)-borneol ($1{\cdot}10^{-3}$ mol dm^{-3}) (---○---), and adamantanol ($1{\cdot}10^{-3}$ mol dm^{-3}) (—△—).

3.2 Induced circular dichroism.

Figure 4 shows the ICD spectra of the closed and opened forms of **1** in 10 vol. -% ethylene glycol aqueous solution. The spectra, which have a positive ICD band at 242, 365 nm and a negative one around 288 nm, are very similar to each other except for the intensity of the molar ellipticity. The large similarity of the ICD spectra of the closed and opened forms suggested that two possibilities that the method of placement of the solution of 1 in the dark overnight results in an equilibrium mixture of the opened and closed forms or the spectrum of the opened one is similar to that of the closed one.

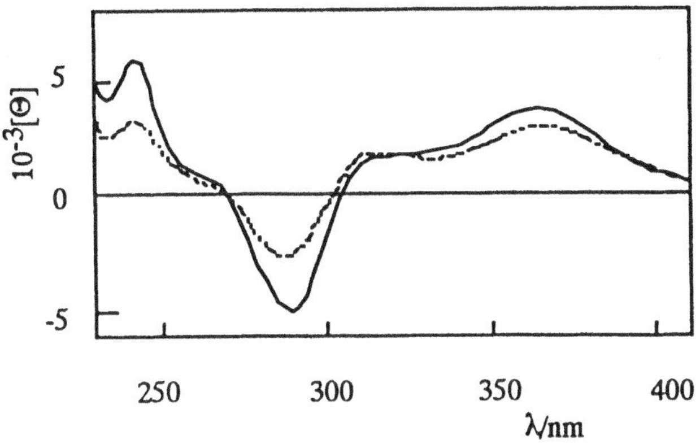

Fig.4 ICD spectra of the closed (———)and opened (— — -) forms of 1 in 10 vol.-% ethylene glycol aqueous solution ($4.85 \cdot 10^{-5}$ mol dm^{-3}).

The guest-induced enhancement of the ICD band in the 250-270 nm region have been observed for both forms of 1, which suggests that the photochromic moiety comes close the chiral cyclodextrin cavity as a hydrophobic cap when a host-guest complexation occurs. The binding constants of both forms of 1 were obtained by the analysis of the guest-induced ICD variations using Eq. 2.

$$K = \frac{\theta h - \theta x}{(\theta x - \theta c)\left[Cg - Ch\,\dfrac{\theta h - \theta x}{\theta h - \theta c}\right]} \qquad (2)$$

$$\frac{\theta h - \theta x}{Cg} = k\theta x - k\theta c \qquad (3)$$

Here, [Θ] is the molar ellipticity, Θx for the sample, Θh for the host alone, Θc for the complex, Ch is the total host concentration, and Cg is the total guest concentration. When the guest is in large excess, we used Eq. 3 which is obtained from Eq. 2.

Tab.1 Binding constants (K) of the closed and opened forms of 1 at25°c

Guest	K/(mol^{-1} dm^3)	
	closed-1	opened-1
cyclohexanol	388±9	454±16
(-)-borneol	12000	15000

The ICD spectra changes of the closed form of 3 alone or in the presence of a guest were examined. When adamantanol was used as a guest, the [Θ] value of the positive band around 238 nm decreases and the new negative band at 274 nm appears with large intensity together with increase of the negative bands at around 325 and 370 nm. On the other hand, the change of the [Θ]values in the presence of nerol is smaller than those of adamantanol. These results display that 3 can recognize adamantanol than nerol, because the ICD spectra change indicates the magnitude of a host-guest complexation. It is well known that a increase of ICD means the appended moiety is coming close to the chiral environment or included into the cyclodextrin cavity, while a decrease means the appended moiety is going far from the chiral cyclodextrin cavity. However, theICD spectra of 3 in the presence of a guest shows that a decrease of the negative band at 238nm and a increase of the negative band at 274 nm with large intensity. This phenomena suggests a movement of the appended moiety on accommodation of a guest in the cavity is not simple. It is expected that some part of the spiropyran moiety is excluded and the other part is included into the cavity with accommodation of a guest.

The guest-induced variation of the ICD intensities at 238 nm was used to obtain the binding constants of 3 for these guests. The binding constants are roughly in the order of guest size. It is interesting that the magnitude of the binding constants and the half-life time values of 3 for those guests are comparable.

Tab. 2. Biding constants and half-life of 3 at 25°C.

Guest	$K/mol^{-1}\ dm^3$	t1/2/min
no guest	———	70
nerol	2,300±230	80
(-)- borneol	12,600±170	340
adamantanol	18,800±400	400

4. Conclusion

The photochromic responses observed in the systems of 2 and 3 with a guest molecule can be used as a photochromic indicator to display a host-guest binding ability or property of the cyclodextrins.

References

[1] Ueno A, Yoshimura H, Saka R, and Osa T. (1979) Photocontrol of binding ability of capped cyclodextrin,*J. Am. Chem. Soc.*, **1 0 1**, 2779-2780.

[2] Hamada F, Fukushima M, Osa T, Ikeda H, Toda F, and Ueno A. (1993) Azobenzene-capped γ-cyclodextrin as a photo-switchable host*Makromol. Chem., Rapid Commun.* **1 4**, 287-291.

[3] Hamada F, Fukushima M, Osa T, and Ueno A. (1993) Photo-switchable hosts. Photochromism and inclusion complex formation of β-cyclodextrin bearing a spiro[2H-benzopyran-2,2'-indoline] moiety. *Makromol. Chem., Rapid Commun* **1 4**, 279-286.

[4] Hamada F, Suzuki I, Osa T, and Ueno A. (1994) Photochromic moleculr recognition of γ-cyclodextrin bearing a spiro[2H-benzopyran-2,2'-indoline] moiety. *Macromol. Raoid Commun.* **1 5**, 531-536.

[5] Hamada F, Hoshi K, Hioguchi Y, Murai K, Akagami Y, and Ueno A. Photochromic molecular recognition of β-cyclodextrin bearing spiropyran moiety for organic guests. submitted for *J. Chem. Soc., Perkin Trans 2.*

INTERACTIONS OF PORPHYRINS WITH CYCLODEXTRINS. PORPHYRINS AS PROBES FOR STUDYING INCLUSION PHENOMENA

Koji KANO,* Naoya TANAKA, and Hideo MINAMIZONO
Department of Molecular Science and Technology
Doshisha University
Tanabe, Kyoto 610-03, Japan

1. Introduction

Our research interest is mechanisms for intermolecular interactions in water. Especially, we have been interesting in hydrogen bonding, electrostatic, van der Waals, and dipole-dipole or ion-dipole interactions which play very important role for forming molecular complexes in water.

It has been known that anionic guest molecules can be included into a cyclodextrin (CDx) cavity while cationic ones such as anilinium, pyridinium, and quinoliniuim compounds hardly penetrate into the hydrophobic CDx cavity. However, no reasonable explanation for such a difference between anionic and cationic guests has been given so far. In the present study, the attempt has been done to interpret the nature of CDx in inclusion of ionic guest molecules.

There are four modes in penetration of a monosubstituted benzene into CDx cavity as shown in Figure 1. If R is an ionic substituent, the ionic group has to come

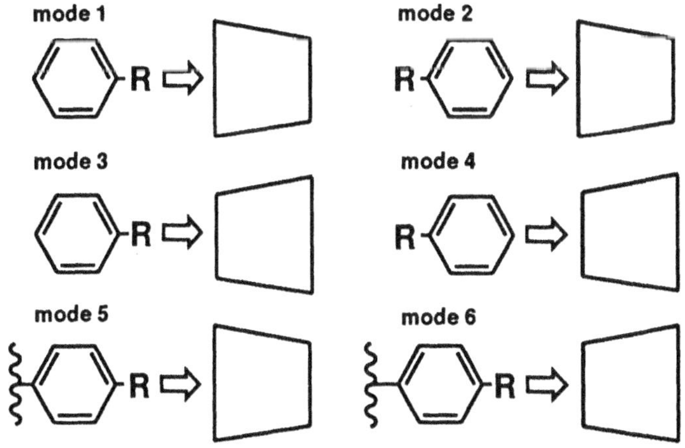

Figure 1. Modes of penetration of monosubstituted benzene into a CDx cavity.

A. W. Coleman (ed.), Molecular Recognition and Inclusion, 191–196.
© 1998 *Kluwer Academic Publishers. Printed in the Netherlands.*

through the hydrophobic CDx cavity in modes 1 and 3. Meanwhile, R does not need to penetrate into the host cavity in modes 2 and 4. We cannot distinguish mode 1 from mode 4 or mode 2 from mode 3 from the structure of inclusion complex which can be determined experimentally. If there is a wall at the one side of the guest molecule, the penetration modes are reduced to two types (modes 5 and 6). The porphyrin ring might be suitable for such a wall. In the present study, the tetraarylporphyrins were used as the guests whose aryl groups at the peripheries can act as the guests for CDx (see Table 1). The tetraarylporphyrins have another advantage. The pK_a value of the

TABLE 1. Tetraarylporphyrins used in the present study

Porphyrin	R
TSPP	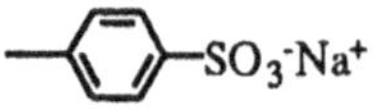
TAPP	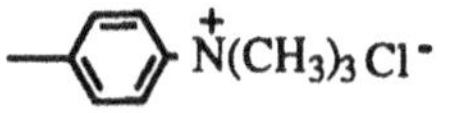
TMPyP	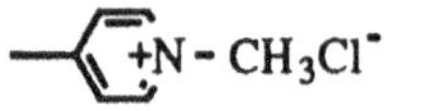
TPAP	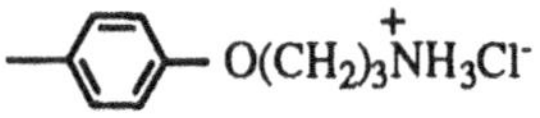
TGPP	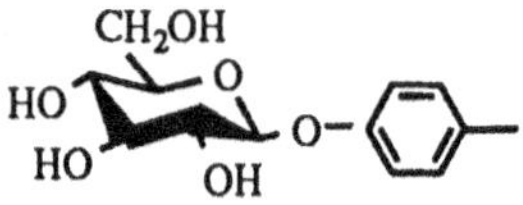
TMPP	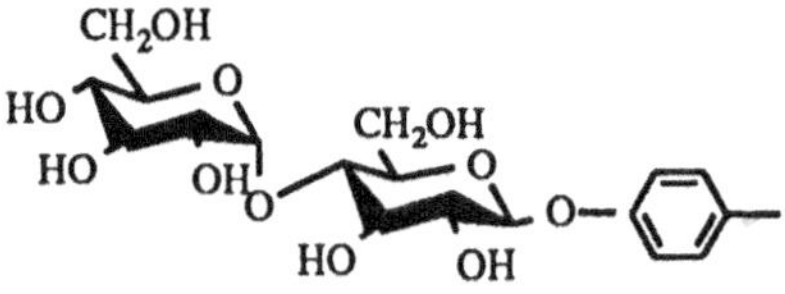

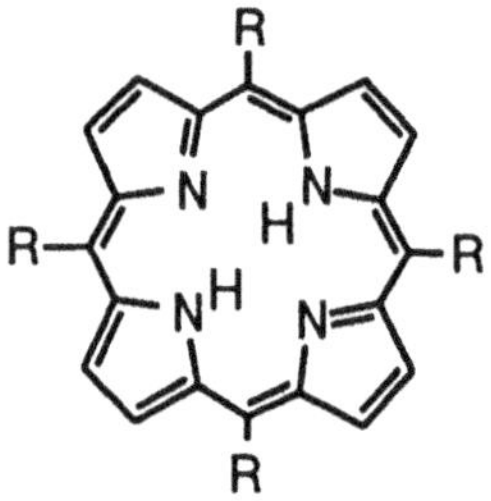

porphyrin free base seems to be used as a measure to estimate the extent of inclusion. If a deep penetration occurs, the pK_a value of the porphyrin should be lowered because of an extent dehydration from the center of the porphyrin upon inclusion.

As the host molecules, β-CDx, heptakis(2,6-di-*O*-methyl)-β-CDx (DMe-β-CDx), and heptakis(2,3,6-tri-*O*-methyl)-β-CDx (TMe-β-CDx) were used. The hydrophobic nature of the CDx cavity increases in the order of β-CDx, DMe-β-CDx, and TMe-β-CDx while the dipole moment of the host decreases in this order (*vide infra*). Therefore, we would be able to study the effects of dipole-dipole or ion-dipole interaction on inclusion of ionic guests using these CDxs. A part of this study has already been reported [1].

2. pK_a Values of Porphyrins

Table 2 shows the pK_a values of anionic TSPP in water in the absence and the presence of CDxs. It has been known that TSPP forms its dimer in water when the concentration of TSPP is high or the solution contains salt [2]. Since the pK_a values

TABLE 2. The pKa values of porphyrins (5 x 10-6 M) in the absence and the presence of CDxs (0.01 M)

Entry	Porphyrin	CDx	pK_a	-ΔpK_a
1	TSPP	none	5.4	0
2	TSPP	β-CDx	4.2	1.2
3	TSPP	DMe-β-CDx	2.2	3.2
4	TSPP	TMe-β-CDx	0.4	5.0
5	TAPP	none	3.0	0
6	TAPP	β-CDx	2.7	0.3
7	TAPP	DMe-β-CDx	2.2	0.8
8	TAPP	TMe-β-CDx	2.8	0.2
9	TMPyP	none	1.3	0
10	TMPyP	β-CDx	1.3	0
11	TMPyP	DMe-β-CDx	1.3	0
12	TMPyP	TMe-β-CDx	1.3	0
13	TPAP	none	5.2	0
14	TPAP	β-CDx	3.9	1.3
15	TPAP	DMe-β-CDx	2.4	2.8
16	TPAP	TMe-β-CDx	0.8	4.4
17	TGPP	none	4.6	0
18	TGPP	β-CDx	4.2	0.4
19	TGPP	DMe-β-CDx	2.7	1.9
20	TGPP	TMe-β-CDx	1.2	3.4
21	TMPP	none	4.4	0
22	TMPP	β-CDx	4.2	0.2
23	TMPP	DMe-β-CDx	2.7	1.7
24	TMPP	TMe-β-CDx	1.2	3.2
25	TPP	TMe-β-CDx	-0.2	--

were determined at low TSPP concentration in water without salt by means of absorption spectroscopy, TSPP did not form the dimer under the present conditions. The pK_a value of TSPP decreases moderately in the presence of β-CDx while the pK_a value drastically decreases upon addition of TMe-β-CDx. These results clearly indicate that the sulfonatophenyl groups of TSPP are included shallowly in the β-CDx cavity and are included deeply into the TMe-β-CDx cavity. It has been reported that TSPP forms the 1:2 complex of β-CDx where the sulfonatophenyl groups at the 5 and 15 positions are included in mode 5 [3].

The pK_a value of TAPP is scarcely affected by the addition of CDx. No change in pK_a was observed in the case of TMPyP. In these cases, the positive charges are located near the benzene rings or delocalized on the pyridinium rings. The results clearly indicate that the cationic aryl groups such as anilinium and pyridinium hardly penetrate into the hydrophobic CDx cavities. On the other hand, TPAP, whose positive charges exist far from the benzene rings, complexes with β-CDx and TMe-β-CDx as indicated by the results shown in Table 2.

TGPP and TMPP are nonionic porphyrins whose hydrophilic groups are glucopyranosyl and maltopyranosyl groups, respectively. On the basis of spectroscopic data, it is concluded that these nonionic porphyrins form self-aggregates in water under the present conditions. The pK_a values of TGPP and TMPP slightly decrease upon addition of β-CDx and markedly decrease upon addition of TMe-β-CDx. TPP (R = H) can be solubilized in water only by TMe-β-CDx and the inclusion complex shows a very low pK_a.

3. NMR Spectroscopy

The pK_a measurements exhibit that β-CDxs include anionic and nonionic benzene rings but hardly interact with cationic anilinium and pyridinium compounds. Especially, TMe-β-CDx forms very stable inclusion complexes with TSPP, TGPP, TMPP, and TPP. These results are well consistent with those of NMR measurements. Figure 2 shows the ^1H NMR spectra of TSPP in D_2O containing various amounts of β-CDx and TMe-β-CDx. In order to measure the ^1H NMR spectra, the 3 x 10^{-3} M solutions of TSPP were used. At such a high concentration, TSPP forms its face-to-face dimer in D_2O without CDx. The Ho protons of the sulfonatophenyl groups of the TSPP dimer appear at 7.83 ppm as a broad singlet which shifts to 8.51 ppm (doublet) upon addition of 8 x 10^{-3} M β-CDx. Such a large up-field shift is interpreted in terms of the dissociation of the TSPP dimer to the monomer upon complexation with β-CDx. The

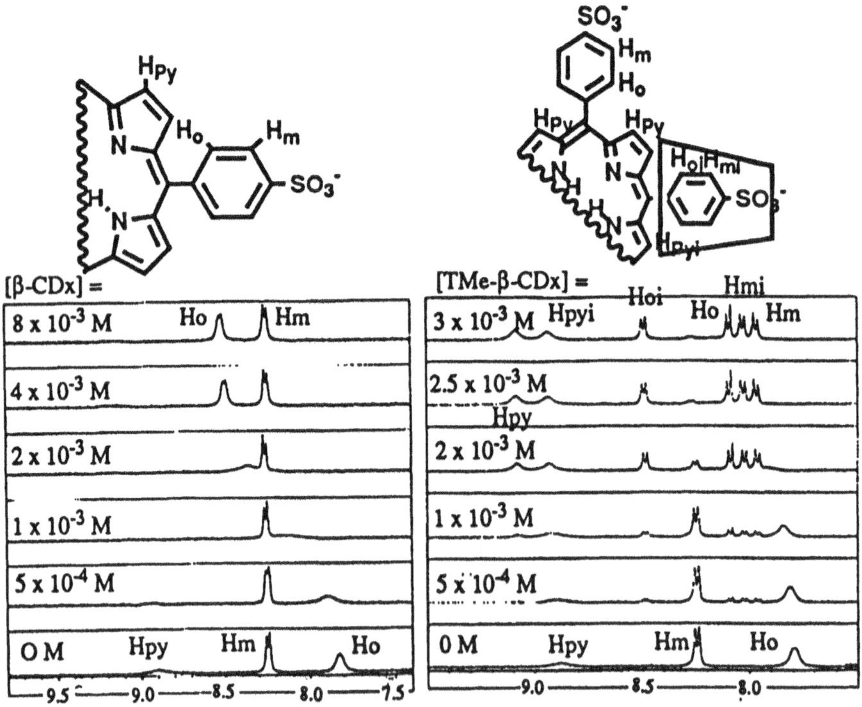

Figure 2. ^1H NMR spectra of TSPP (2 x 10^{-3} M) in D_2O in the absence and the presence of β-CDx (left) and TMe-β-CDx (right) at 25 °C (an external standard: TPS).

results shown in Figure 2 reveal that the sulfonatophenyl groups of TSPP are shallowly included in the β-CDx cavity and there is a relatively fast exchange between the complexed and free TSPP molecules. On the other hand, in the presence of TMe-β-CDx, the two-sets of the NMR signal were observed for each proton of TSPP as shown in Figure 2. Each signal was assigned from ROESY spectrum. Such an NMR spectral behavior clearly indicates that TSPP forms a very stable inclusion complex with TMe-β-CDx and the dissociation rate of the inclusion complex is significantly low. In addition, the NMR spectral data exhibit the formation of the complex of TMe-β-CDx having a high symmetry.

In the cases of cationic TAPP and TMPyP, however, the proton signals of these guests are not affected by the addition of both β-CDx and TMe-β-CDx, indicating that such cationic porphyrins do not interact with these CDxs. The results are in good agreement with those of the pK_a measurements. Then we prepared tri(*N*-methylpyridinium-4-yl)phenyl porphyrin (TriMPyP), 5,15-di(*N*-methylpyridinium-4-yl)diphenyl porphyrin (5,15-DiMPyP), and 5,10-di(*N*-methylpyridinium-4-yl)diphenyl porphyrin (5,10-DiMPyP) and measured the pK_a values and ^{1}H NMR spectra. The results of the pK_a measurements are shown in Table 3. The pK_a values of these pyridinium porphyrins remarkably decrease upon addition of TMe-β-CDx. The decrease in the pK_a values should be ascribed to the inclusion of the benzene ring(s) of

TABLE 3. The pKa values of pyridiniumporphyrins in water with and without CDx.[a]

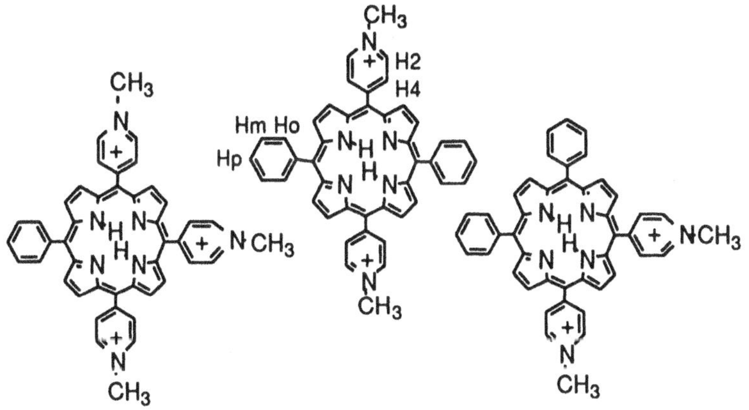

pK_a (H_2O)	1.9	2.8	2.9
pK_a (β-CDx)	1.2	2.3	2.5
pK_a (TMe-β-CDx)	0.8	0.8	1.1

[CDx] = 0.01 M, [Porphyrin] = 5 x 10^{-6} M

the porphyrin. The low pK_a values of 5,10- and 5,15-DiMPyPs suggest that the positive charges at the peripheries are delocalized on the porphyrin ring. The ^{1}H NMR of these pyridinium porphyrins were measured in D_2O in the absence and the presence of β-CDx and TMe-β-CDx. As mentioned above, no change was observed with

TMPyP. Slight down-field shifts were measured for the Ho, H2, and H3 protons of TriMPyP upon addition of these CDxs. The marked down-field shifts were observed for the Ho, H2, and H3 protons of 5,15-DiMPyP upon addition of β-CDx and TMe-β-CDx. Such large shifts of Ho and H2 protons upon addition of CDx are ascribed to the dissociation of the face-to-face dimer of 5,15-DiMPyP in D_2O upon complexation with CDx. In other words, 5,15-DiMPyP forms its face-to-face dimer in water at high porphyrin concentration, but TMPyP does not. In 1972, Pasternack and his co-workers reported that TMPyP does not form its dimer in water because of the electrostatic repulsion between the cationic porphyrin molecules [2]. Meanwhile, on the basis of the novel fluorescence behavior of TMPyP, Kano *et al.* assumed that TMPyP forms its dimer even at very low concentration [4]. The present results of the pK_a and ^1H NMR measurements clearly reveal that TMPyP does not form its dimer. It is concluded that the monomer model presented by Pasternack *et al.* is correct with TMPyP. Although TMPyP does not aggregate in water, the experimental data obtained in the present study exhibit that 5,15-DiMPyP as well as 5,10-DiMPyP forms its dimer in water in spite of the existence of the electrostatic repulsion. It is quite interesting that the van der Waals interaction between the cationic porphyrin rings is so strong in water as it overcomes the electrostatic repulsion.

The present study shows that anionic and nonionic phenyl groups can penetrate into the CDx cavity, but the cationic aryl groups such as anilinium and pyridinium groups do not. What is the difference between anionic and cationic guests in inclusion phenomena? One possible answer is participation of dipole-dipole interaction between the host and the guest. Then we calculated the dipole moments of CDxs. The dipole moments of β-CDx, DMe-β-CDx, and TMe-β-CDx were calculated to be 15.4, 6.5, and 0.71 D, respectively, by using a MOPAC software. The very small dipole moment of TMe-β-CDx denies the participation of the dipole-dipole interaction in stabilization of anionic guest in the TMe-β-CDx cavity.

It is quite reasonable to consider the electrostatic interaction between the nonionic host and the ionic guest to stabilize or instabilize the inclusion complex. Then we assume the microscopically polarized environments of the CDx cavities [1]. Since all carbon atoms of CDx are bound to the electronegative oxygen atoms which are located outside of the CDx cavity except for the oxygen atoms at the 1-positions. Therefore, the inside of the CDx cavity is composed of the positively polarized carbon and hydrogen atoms. The microscopically positive environment of the CDx cavity is suitable for penetration of anionic guest but is not preferable for staying cationic guest. Assuming the positively polarized microenvironment of the CDx cavity, all results obtained for the ionic guests can be explained reasonably.

4. References

1 Kano, K., Tanaka, N., Minamizono, H., and Kawakita, Y. (1996) *Chem. Lett.*, in press.

2 Pasternack, R. F., Huber, P. R., Boyd, P., Engasser, G., Francesconi, L., Gibbs, E., Fasella, P., Venturo, G. C., and Hinds, L. deC. (1972) *J. Am. Chem. Soc.*, **94**, 4511-4517.

3 Ribó, J. M., Farrera, J.-A., Valero, M. L., and Virgili, A. (1995) *Tetrahedron*, **51**, 3705-3712.

4 For example: Kano, K., Hayakawa, T., and Hashimoto, S. (1991) *Bull. Chem. Soc. Jpn.*, **64**, 778-784.

SYNTHESIS AND EVALUATION OF NEW IONOSELECTIVE MATERIALS

A. FAVRE-RÉGUILLON, B. DUNJIC, N. DUMONT, M. LEMAIRE*
Institut de Recherches sur la Catalyse, Laboratoire de Catalyse et Synthèse Organique, Université Claude Bernard, Lyon I, CPE
43 Bld du 11 novembre 1918, 69622 Villeurbanne, France

1. Introduction

Despite recent improvements in both sensitivity and selectivity of analytical methods, separation and preconcentration of traces of ions remain essential steps in many chemical analyses. Tailor-making specific ligands by attaching them to polymeric carriers is an excellent way of meeting the increasing demand for materials suitable for the separation of metal ions from complex sources.
Polymer-supported macrocyclic ligands have been reported in the literature, and reviews have been published on ionoselective polymers. [1,2] The polymer supports used were usually polystyrene, [2] polyacrylate. [3] the ne of these resins are commercially available and exhibit fair selectivity. However low extraction rates are frequently displayed because of the polymer support which is hydrophobic, preventing wetting of the material.
With a view to obtaining a material with both high ionoselectivity and high capacity we have synthetized materials with pendant macrocycles as complexing agents and hydrophilic backbones such as poly(oxyethylene), poly(alkoxyaluminate) or poly(urea).

2. Synthesis of poly(oxyethylene) based ionoselective material

The first monomer synthetized was functionalized 12-crown-4 (Fig. 1). The monomer was synthetized from allyl glycidyl ether **1** (AGE). The nucleophilic attack of triethyleneglycol on AGE led to only one isomer **2**. A two step intramolecular cyclisation of **2** allowed the formation of the (allyloxy)methyl-12-crown-4 **3** which was epoxidized with *mCPBA* to give monomer **4**. The overall yield for this 4 steps synthesis was 35 %.

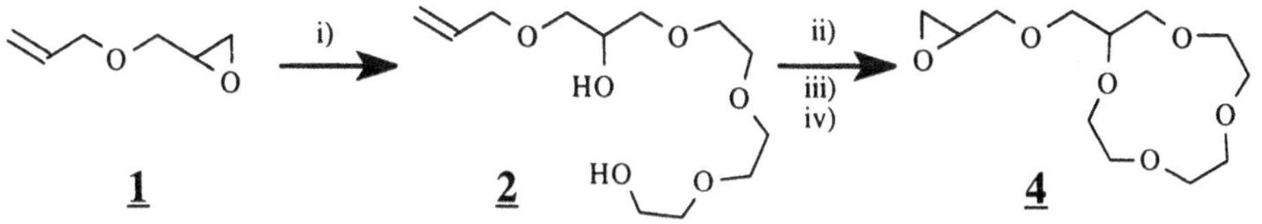

i) triethyleneglycol, KOH, 74% ; ii) NaOH, H_2O, pTosCl, 91% ; iii) NaH, THF, 56% ; iv) mCPBA, CH_2Cl_2, 97%

Figure 1. Synthesis of 2-(2,3-Epoxypropyloxy)methyl-12-Crown-4

A. W. Coleman (ed.), *Molecular Recognition and Inclusion*, 197–202.
© 1998 *Kluwer Academic Publishers. Printed in the Netherlands.*

KOH catalysed anionic polymerisation of **4** was carried out in bulk with quantitative conversion. The number-average molecular weight (M_n) of polymer **5** was 4500 g/mol. The polymer **5** is a viscous brown liquid, soluble in all organic solvents and partially in water. Terminal OH groups can be used in a cross-linking reaction. The cross-linking of **5** was carried out in bulk at room temperature, using diphenylmethane diisocyanate as a cross-linking agent, giving the polyurethane elastomer **6** (Fig. 2).

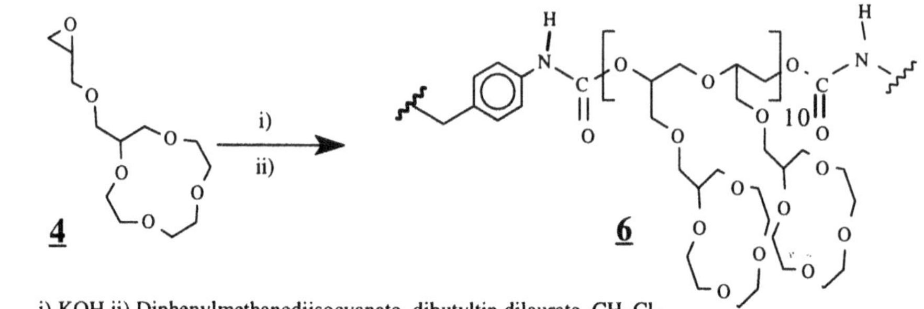

i) KOH ii) Diphenylmethanediisocyanate, dibutyltin dilaurate, CH_2Cl_2

Figure 2. Synthesis of Polyurethane Elastomer **6**

Glycidyl ether with more than 12 atoms size macrocrocyle showed poor reactivity under the conditions of anionic polymerisation, despite the variety of conditions tested. The inhibition of the polymerisation could be explained by the formation of a hindered ion-pair.

Material containing macrocycle as pendant functional group and alkoxy aluminate as matrix **7** could be readily obtained [4] by reaction of monomer **4** with an excess of Vanderberg catalyst (Fig. 3).

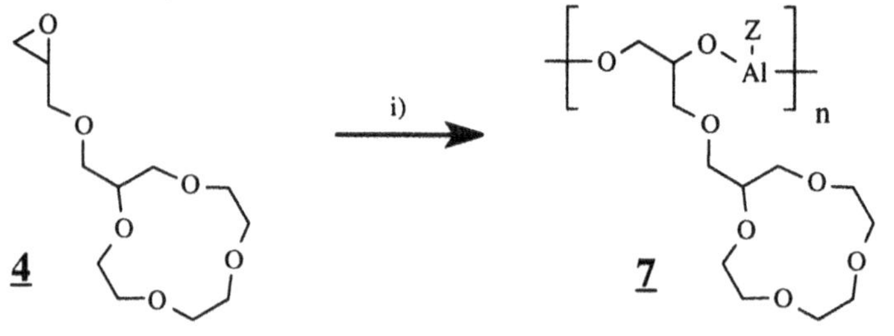

i) $AlEt_3$ / H_2O (1 / 0.6) Et_2O, hexane , "catalyst" / monomer = 1 / 1

Figure 3. Coordinate Cationic Polymerisation

Solid NMR of ^{13}C and ^{27}Al, elemental analysis and IR spectroscopy were used to determine the proposed structure.

3. Synthesis of ionoselective material with poly(urethane) as the main chain and macrocycle as the pendant groups.

Polyurethanes can be prepared by the catalysed polyaddition of diisocyanate with diol-functionalized crown ethers. Desired diol monomers were obtained by dihydroxylation of allyl groups. Thus, the syntheses of (allyloxy)methyl-crown with 12 to 24 atoms size macrocycle were studied. (allyloxy)methyl-18-crown-6 and 21-crown-7 **10** were synthetized by the cyclisation of the diol **8**, obtained by hydrolysis of AGE **1**, using the corresponding ditosylate (Fig. 4). In producing the (allyloxy)methyl-24-crown-8 **11** we used the diol **2** obtained by treatment of **1** with tetraethyleneglycol ditosylate and NaH in THF. Then, the functionalized 24-crown-8 could be obtained in two steps from AGE **1** with improved yield compared to the method previously reported [5].

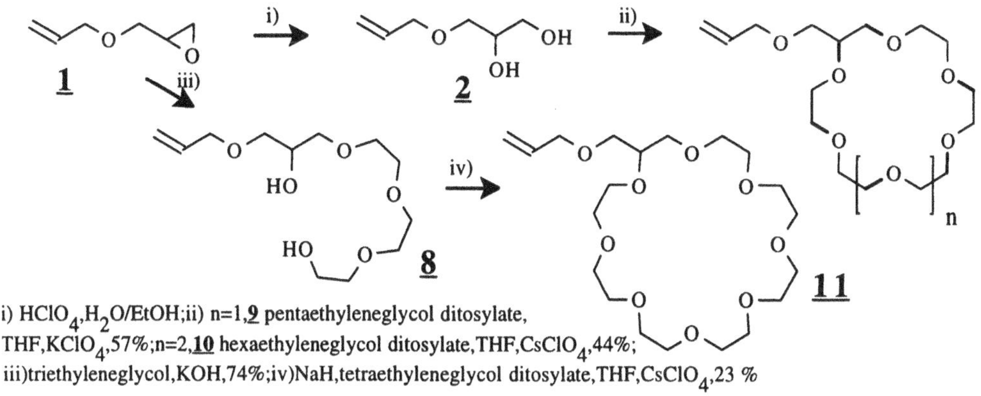

Figure 4. (Allyloxy)methyl Crown Ether from Allyl Glycidyl Ether

The dihydroxylation step was carried out using osmium tetraoxide catalyst. Polymers were prepared by catalysed polyaddition with diisocyanate [6] (Fig. 5). In order to determine the influence of the rigidity of the matrix on the accessibility of active sites, we synthetized two polyurethanes which were derived from the same macrocycle (15-crown-5). In the case of **13**, we used flexible hexamethylene diisocyanate while **14** was obtained by using rigid diphenylmethane diisocyanate. For this evaluation the cross-linking of polymers was achieved by using a 20% excess of isocyanate groups. The procedure described above is the most direct way of preparing immobilized pendant crown compounds on polymers. [6]

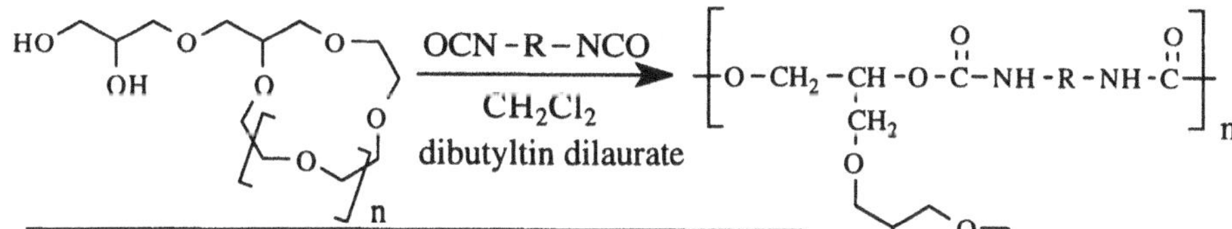

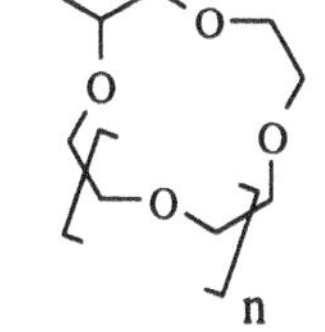

n	R	Macrocycle	Sample
1	$-(CH_2)_6-$	12-crown-4	**12**
2	$-(CH_2)_6-$	15-crown-5	**13**
2	$-p,C_6H_4-CH_2-p,C_6H_4-$	15-crown-5	**14**
3	$-(CH_2)_6-$	18-crown-6	**15**
4	$-(CH_2)_6-$	21-crown-7	**16**
5	$-(CH_2)_6-$	24-crown-8	**17**

Figure 5. Polyaddition of Diisocyanates with Diol-Functionalized Crown Ethers

4. Evaluation of ionoselective properties

The different polyurethanes were immersed in liquid nitrogen and then crushed. The cation binding properties of the polymers were determined with a competitive extraction of five alkali-metal cations. The distribution coefficient (D) is defined as the ratio of the concentrations of the cations on the polymer and in the solution after 8 hours of contact. The distribution coefficient (D), calculated from the concentration of each cation in solution before and after extraction, is given by Eq. 1:

$$D = \frac{[M^+]_i - [M^+]}{[M^+]} \times \frac{V}{m} \tag{1}$$

The concentration of the metal cations before and after adsorption was determined by atomic emission spectroscopy. The subscript i refers to initial conditions, V (mL) to volume of solution and m (g) to mass of resin. The results achivied for the resins are detailed in Table 1.

Table 1 : Distribution Coefficients for Polymers

Sample	Macrocycle	D (L/kg)				
		Li^+	Na^+	K^+	Rb^+	Cs^+
6	12-crown-4	0 *	**3.8** *	0.3 *	0.1 *	0 *
12	12-crown-4	0	**0.8**	0	0	0
13	15-crown-5	0	0.3	**1.4**	0.5	0.2
14	15-crown-5	0	0	**0.1**	0	0
15	18-crown-6	0.4	0.5	**5.0**	1.7	0
16	21-crown-7	0	0	1.0	**2.4**	**2.1**
17	24-crown-8	0	0	0.2	0.4	**0.6**

Conditions: $[Li^+]=[Na^+]=[K^+]=[Rb^+]=[Cs^+]=0.25$ M, $[HNO_3]=1.4$ N, V=2 ml, M= 1 mmol macrocycle
* $[Li^+]=[Na^+]=[K^+]=[Rb^+]=[Cs^+]=0.25$ M, V=2 ml, M= 0.45 mmol macrocycle
Margin of error: ± 0.2 L/kg

The polymers containing crown ether units with different cavities bound metals according to the "size-fit" concept (samples 15 to 17) with exceptions. In the presence of different alkali-metal cations, extraction of lithium by polymers 6 and 12 was not detectable, one could expect from macrocycle diameter. This can be explained by the higher degree of hydration of lithium and by the fact that unlike the other alkali cations, lithium forms a more stable six-membered chelate ring rather than five-membered chelate ring [7]. Polymer 6 strips off sodium with sodium/lithium and sodium/potassium selectivity greater than 18 and 7 respectively. This high loading capacity and selectivity for sodium can be explained by the formation of 2:1 complexes by cooperative action of two adjacent 12-crown-4 ether moieties. Conversely, the very small capacity of 12 is due to the impossibility of forming 2:1 complex as in the case of 6. Polymer 13, bearing 15-crown-5 ether bound potassium more than sodium. This is probably due to the presence of extracyclic oxygen. As illustrated by samples 13 and 14, greater capacity is reached by increasing active site accessibility and mobility. Polymer 13 with the flexible chain has a loading capacity ten times greater than 14 and also an increased selectivity towards potassium. Polymer **16** bearing 21-crown-7 possess good binding properties for cesium. When the size of the cavity is increased, e.g. polymer **17** bearing 24-crown-8, the capacity and selectivity drop.

5. Preparation of polyurethane-coated silica-gel

Polyurethane based crown ethers are elastomers. Improvement of physical properties and a much higher utilisation efficiency of the crown compound can be achieved with immobilized crown compounds bound chemically to the surface of a solid support. However, many steps are needed to synthetize silica-gel supported crown compounds [8] or polymeric crown ethers. [9]

In an attempt to overcome this disadvantages, monomers were reacted on the surface of a solid carrier to form a thin coating film of crown based resin. Such materials have crown ethers in the main chain and this limits their mobility and accessibility. Our approach allows simple preparation of supported flexible polymers with mobile crown ethers in pendant chains. Functionalized crown ether was reacted with hexamethylene diisocyanate on the surface of a solid carrier (silica gel) to form thin coating film of crown ether. The immobilization method is illustrated in figure 6.

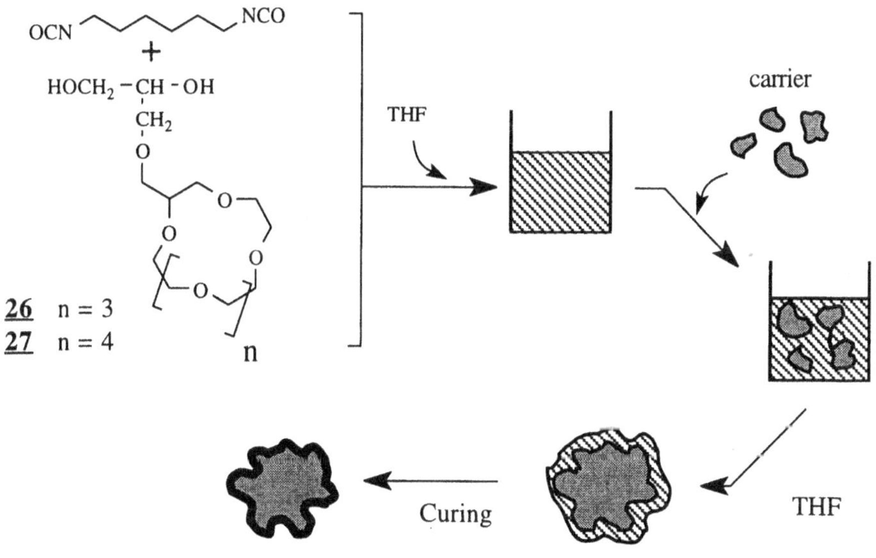

Figure 6. Preparation of Polyurethane-Coated Silica-Gel

The adhesion of the crown based resin is stable in protic and aprotic solvents. Furthermore, during the polyaddition and curing process, excess of isocyanate group could react with the silanol groups present on the surface of silica-gel. Such linkage could also contribute to improving the adhesion, even if such linkage is a hydrolytically labile one.

Two kinds of polyurethanes bearing 18-crown-6 and 21-crown-7 were immobilized. The amount of polymer present on silica-gel (i.e. 0.2 mmol/g) was confirmed by ATG analysis.

6. Evaluation of polyurethanes-coated silica-gel

Cation-complexing abilities for alkali-metal cations of samples **18** and **19** supported on silica gel are given in table 2. The results of the extraction show that the polymers supported on silica-gel kept their selectivity. Those silica gel supported polymers can be reused without loss of capacity and selectivity.

Table 2 : Distribution Coefficients for Polymers

Sample	Macrocycle	D (L/kg)				
		Li^+	Na^+	K^+	Rb^+	Cs^+
18 / SiO_2	18-crown-6	0	0	**4.2**	1.8	1.5
19 / SiO_2	21-crown-7	0	0	2.8	**5.5**	**5.1**

Conditions: $[Li^+]=[Na^+]=[K^+]=[Rb^+]=[Cs^+]=0.0025$ M, V=5 mL, M= 250 mg
Margin of error: ± 0.2 L/kg

The results of this study demonstrate that polyurethane obtained by catalysed polyaddition of diisocyanate with diol-functionalized crown ethers provides hydrophilic polymers for the selective extraction of ions. The ionoselective properties of these materials are based on the recognition of ions by highly selective synthetic organic ligands covalently attached to hydrophilic polymers In addition to exhibiting excellent selectivity, those polymers could be immobilized on silica gel. Those potential chromatographic materials are more easily prepared than poly(crown ether) modified silica [9] and could provide higher selectivity compared to crown ether modified one. [10]
These features provide a separation system with the selectivity required to capture extremely low levels of ions. These new materials could be useful in analytical purposes in several ways: concentrating analytes, removing unwanted matrix ions to purify solutions and allowing concentration even in solutions of moderate acidity.

References

1. Takagi M., Nakamura H. (1986) Analytical application of functionalized crown ether-metal complexes, *J. Coord. Chem.* **15**, 53-82
2. Kimura K., Shono T. (1992) Crown ethers and analogous compounds, in M. Hiraoka (ed.), Studies in Organic Chemistry, Vol. 45, Elsevier, Amsterdam, 198-264
3. Kimura K., Maeda T., Shono T. (1981) Cation-binding ability of tactic poly(crown ether)s, *Makromol. Chem.*, **182**, 1579-1586
4. Dunjic B., Favre-Reguillon A., Duclaux O., Lemaire M. (1994) New polyether-based ionoselective materials, *Adv. Mater.* , **6**, 484-486
5. Czech B., Czech A., Bartsch R. A. (1983) Synthesis of hydroxymethyl-21-crown-7 and hydroxymethyl-24-crown-8,*Tetrahedron Let.* , **24**, 1327-1328
6. Favre-Réguillon A., Dumont N., Dunjic B., Lemaire M. (1995) Synthesis and evaluation of new polyurethane based material for ion separation, *Tetrahedron Let.* , **36**, 6439-6442
7. Handock R.D. (1994) A molecular mechanics study of the selectivity of crown ethers for metal ions on the basis of their size,*J. Inclus. Phenom. Molecul. Recognit. Chem.* , 17, 63-80
8. Lauth M., Gramain P. (1985) Ion chromatographic separation on silica grafted with benzo 18-C-6 crown ether, *J. Liquid. Chromatogr.* , 8, 2403-2415
9. Nakajima M., Kimura K., Shono T. (1983) Ion chromatographic behavior of silica gels modified by poly and bis (crown ether)s of benzo-18-crown-6, *Bull. Chem. Soc. Jpn.* , 56, 3052-3056
10. Bradshaw J. S., Bruening R. L., Krakowiak K. E., Tarbet B. J., Bruening M. L., Izatt R. M., Christensen J. J. (1988) Preparation of silica gel-bound macrocycles and their cation-binding properties, *J. Chem. Soc., Chem. Commun.* , 812-814

CONCAVE REAGENTS AND CATALYSTS:

FROM LAMPS TO SELECTIVITY

ULRICH LÜNING
Institut für Organische Chemie, Christian-Albrechts-Universität zu Kiel
Olshausenstr. 40, D-24098 Kiel (Germany)
Fax: +49-(0)431-880-1558, e-mail: noc03@rz.uni-kiel.d400.de

The concave shielding of the active site of an enzyme is one of the causes of the high selectivity of enzymatic reactions. Therefore we have taken this geometry and accordingly have synthesized Concave Reagents [1, 2]. Like a light bulb in a lampshade, the functional group of a Concave Reagent (the tool) is located in the centre of a bimacrocyclic molecule.

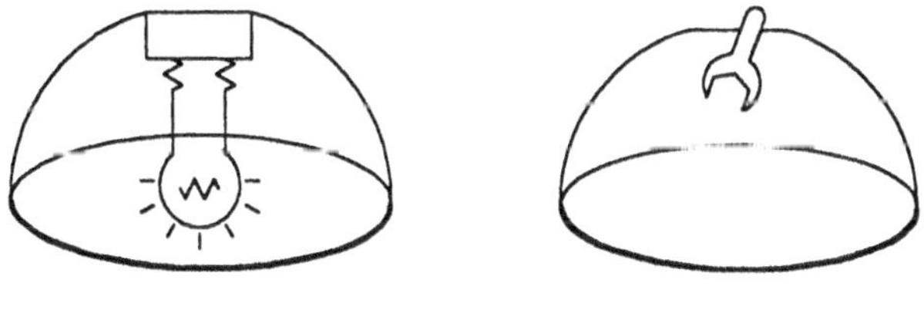

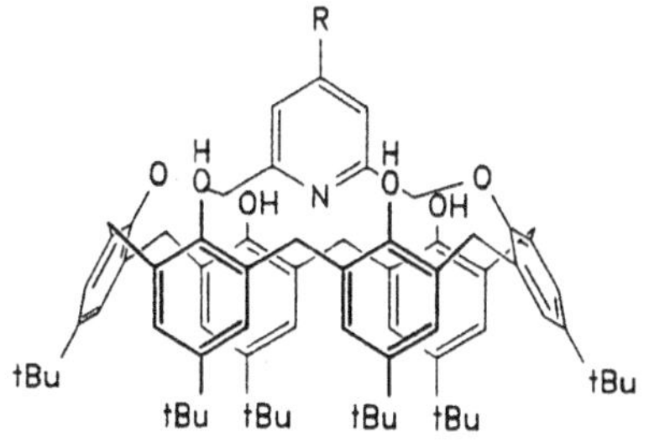

R = H, OMe

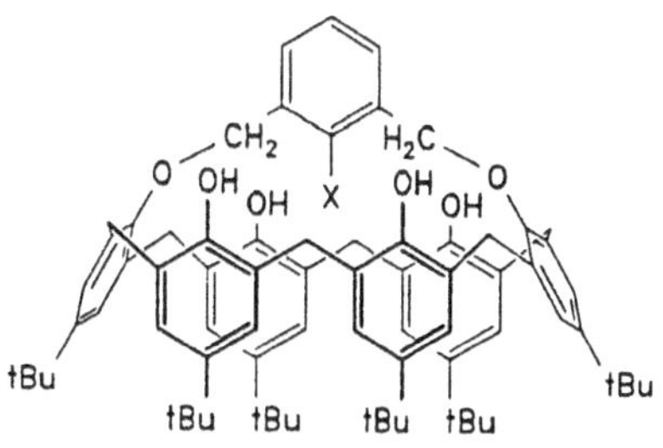

$X = NO_2, NH_2$

A. W. Coleman (ed.), Molecular Recognition and Inclusion, 203–208.
© 1998 Kluwer Academic Publishers. Printed in the Netherlands.

The functional group can be varied widely: concave acids, concave bases and concave metal ion complexes have been synthesized [2] and have been tested in model reactions. Latest developments include bimacrocyclic calix[6]arenes in which basic centers (pyridine [3] or aniline [4]) are incorporated into a bridge connecting opposite aryl rings of the calix[6]arene.

Base Catalyses: By formation of a hydrogen bond between a base center, e. g. a pyridine nitrogen atom and the OH group of an alcohol, the nucleophilicity of an alcohol can be enhanced and its addition to ketenes is catalyzed. The catalytic activity of concave bases depends on their basicity *and* on their size. Larger macrocycles are better catalysts than smaller ones in this acylation reaction [5]. In competition reactions, primary alcohols are acylated much faster than secondary alcohols [6]. The most remarkable selectivities were found for sugar derivatives as substrates. In a glucose and a chinovose derivative with only secondary and equatorial hydroxyl groups, concave pyridines were able to acylate only one OH group with good selectivities [7].

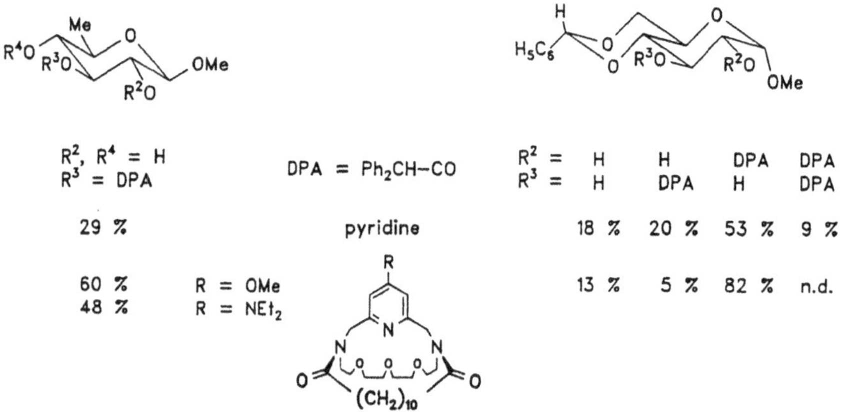

Metal Ion Complex Catalyses: Concave bases, especially concave 1,10-phenanthrolines, possess nitrogen atoms with lone pairs which are able to coordinate to a metal ion. Accordingly strong metal ion complexes could be obtained [8]. In two model reactions the influence of a concave ligand on a metal ion catalyzed reaction has been studied.

Transition metal nonafluorbutanesulfonates (manganese, iron, cobalt, nickel, copper) have been checked as catalysts in the Diels-Alder reaction between cyclopentadienes and pyrazol acrylates [9]. Surprisingly nickel and cobalt were the best catalysts. Without ligands, the *endo*-norbornenes were the main products. In Figure 1 the results for three different bimacrocycles and an open chain 1,10-phenanthroline are compared to the catalysis by metal ions alone. The use of a proper concave ligand changed the selectivity towards *exo*. A rational for this is given in Figure 1.

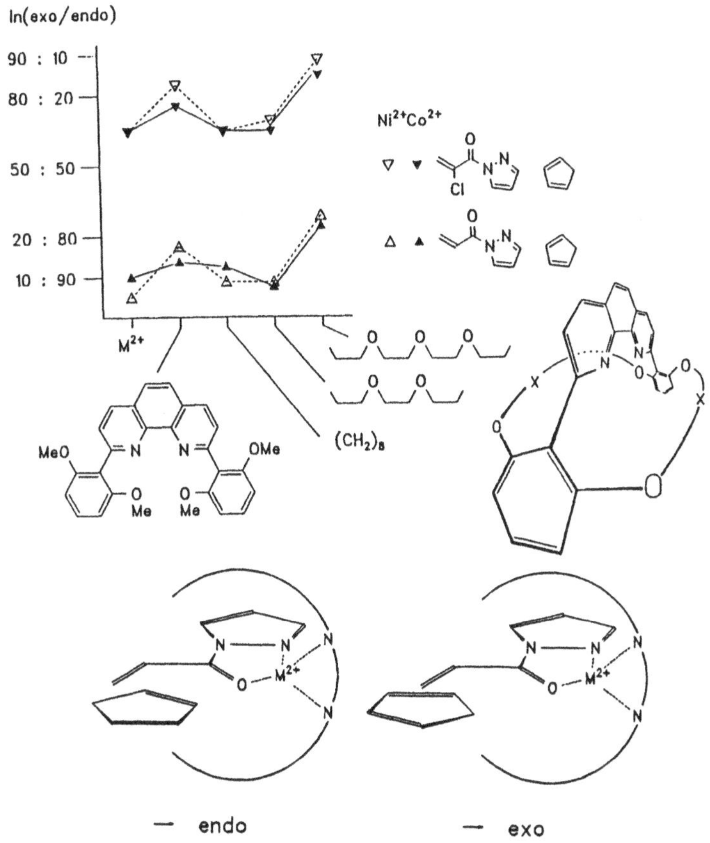

Figure 1. Results and transition states for the formation of *exo*- and *endo*-norbornenes by Diels-Alder reaction of cyclopentadiene with pyrazol acrylates. The acrylate is bound to the metal ion (nickel or cobalt) which is located in the cavity of the concave 1,10-phenanthroline. The cyclopentadiene has two possibilities to react with the dienophile. The approach with the C-2 and C-3 atoms first (leading to the *endo*-products) has a larger sterical hindrance than the approach with the C-5 atom first (leading to the *exo*-products).

In the second model reaction, copper(I) is the catalytically active metal ion which is able to catalyze the reaction between alkenes and diazo compounds to form cyclopropanes. This reaction was carried out with the alkenes shown in Figure 2 [10]. The use of sterically hindered 1,10-phenanthrolines leads to very good selectivities, especially to good *exo*/*endo*-selectivities in the case of cyclic alkenes. Transition state models are shown in Figure 3.

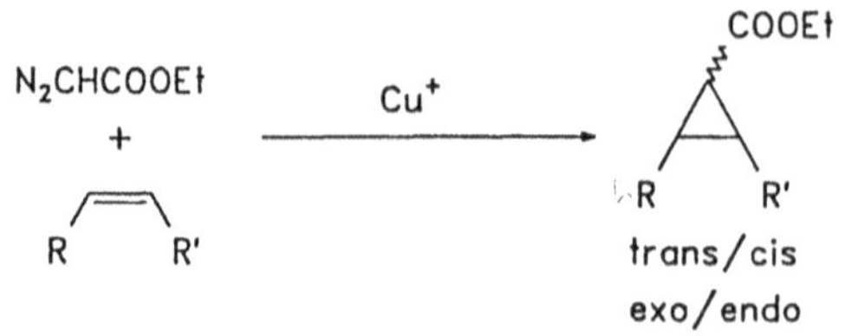

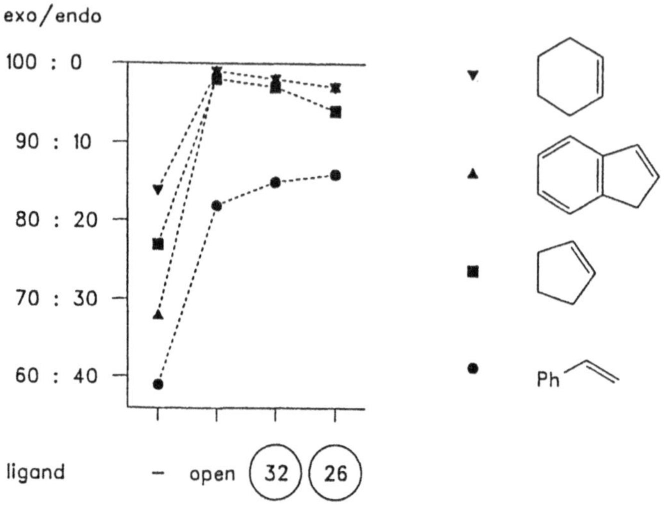

Figure 2. The *exo/endo*-selectivity of the reaction of cyclic alkenes with ethyl diazoacetate can be drastically increased if sterically shielded 2,9-disubstituted 1,10-phenanthrolines are used as catalysts. The circled numbers indicate the ring size of the concave 1,10-phenanthrolines with polyether chains, see Figure 1. 2,9-Bis[bis(2,6-methoxyethoxy)phenyl]-1,10-phenanthroline was used as open chain ligand.

Protonations: The transfer of a proton from an acid to an anion should be controllable by the nature of the reagent if the proton is directly transfered from the acid to the anion. Then regio- and stereoselective protonations can be expected.

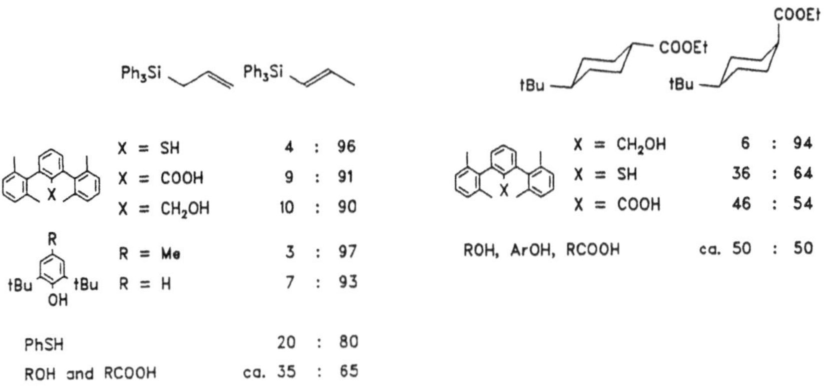

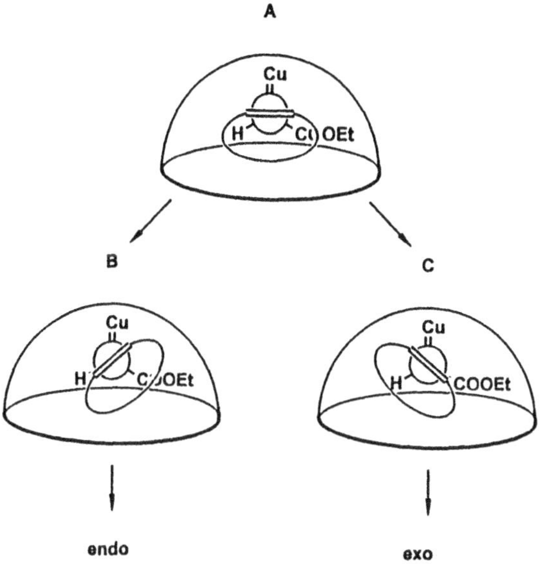

Figure 3. The intermediate of the reaction of a diazo compound with an alkene to form a cyclopropane is a copper bound carbenoid. If copper is complexed to a concave 1,10-phenanthroline the carbenoid is located in the cavity of the complex. The least hindered approach of a cyclic alkene to this carbenoid is shown in the Newman-projection **A**. But this orientation is unproductive because the double bonds of the alkene and the carbenoid are orthogonal to oneanother. A clockwise turn (**C**) will lead to the *exo*-product, a counterclockwise turn (**B**) to the *endo*-product. But in the latter case the interactions between the alkene and the ester substituent will increase explaining the *exo*-selectivity of the reaction.

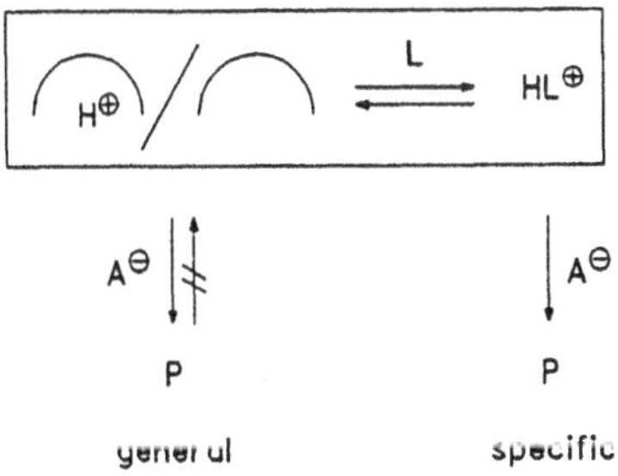

Figure 4. In a buffer system of a sterically shielded acid, an equilibrium between this acid and protonated solvent L (or co-solvent) always exists. A general protonation of an anion A⁻, a direct proton transfer from the employed acid (reagent control) always is in competition with a specific protonation by protonated solvent (or co-solvent) molecules (shuttle mechanism). The more acidic the acid (carboxylic acid instead of alcohol) and the more basic the solvent (or co-solvents) are, the more protonation will occur via the shuttle mechanism and the less reagent-control will be observed. Therefore an acid for a reagent-controlled general protonation should be as little acidic as possible. However, the conjugate anion must not be too basic to avoid a deprotonation of the products P which would lead to the thermodynamic equilibrium between the regio- or stereoisomers.

We have investigated the protonation of allylic systems (silyl stabilized allyl anions [11], nitronate anions [12]) and of cyclohexyl anions (ester enolates [13], nitronate anions [12]). Indeed reaction conditions could be found to observe reagent-controlled protonations. While a triphenylsilyl stabilized allyl anion can be protonated regioselectively by a variety of shielded acids like 2,6-di-*tert*-butyl-phenols or tetra-*o*-methyl substituted *m*-terphenyls carrying an acidic function in 2′-position (SH, COOH, CH_2OH), a reagent-controlled protonation of the anion of ethyl 4-*tert*-butylcyclohexane carboxylate was only possible with one of the *m*-terphenyls, the methanol derivative. It is the least acidic compound of the three *m*-terphenyls used. A closer look to general and specific protonations [14] may explain the results and argues that acids with the smallest acidity possible should be used for reagent controlled protonations (Figure 4).

Acknowledgements: This work has been carried out with the help of many *talented co-workers* whose names are listed in the references. The financial support of the *Deutsche Forschungsgemeinschaft* and the *Fonds der Chemischen Industrie* is gratefully acknowledged. H. Ross and M. Hagen thank the state of *Schleswig-Holstein* for stipends.

1 Lüning, U. (1987) Concave Reagents 1, *Liebigs Ann. Chem.* 949 - 955.

2 Lüning, U. (1995) *Top. Curr. Chem.* **175**, 57 - 99.

3 a) Ross, H., Lüning, U. (1995) Concave Reagents 16, *Angew. Chem.* **107**, 2723 - 2725, *Int. Ed. Engl.* **34**, 2555 - 2557. - b) Ross, H., Lüning, U. (1996) Concave Reagents 18, *Liebigs Ann.* 1367 - 1373.

4 Ross, H., Lüning, U. (1996) Concave Reagents 19, *Tetrahedron* **52**, 10879 - 10882.

5 Lüning, U., Baumstark, R., Schyja, W. (1991) Concave Reagents 9, *Liebigs Ann. Chem.* 999 - 1002.

6 Schyja, W., Petersen, S., Lüning, U. (1996) Concave Reagents 21, *Liebigs Ann. Chem.*, in press.

7 Lüning, U., Schyja, W (1995) in J. S. Siegel (ed.) *Supramolecular Stereochemistry*, Kluwer Academic Publisher, Dordrecht, pp. 223 - 226.

8 Lüning, U., Müller, M., Gelbert, M., Peters, K., von Schnering, H. G., Keller, M. (1994) Concave Reagents 15, *Chem. Ber.* **127**, 2297 - 2306.

9 Gelbert, M. (1995) *Ph. D. Thesis*, University of Freiburg.

10 Hagen, M., Lüning, U. (1996) *Chem. Ber.*, in press.

11 Lüning, U., Wangnick, C., Kümmerlin, M. (1994) *Chem. Ber.* **127**, 2431 - 2433.

12 Lüning, U., Müller, M. (1992) Concave Reagents 10, *Angew. Chem.*, **104**, 99 - 102, *Int. Ed. Engl.* **31**, 80 - 82.

13 Lüning, U., Baumgartner, H., Manthey, C., Meynhardt, B. (1996) Concave Reagents 20, *J. Org. Chem.*, in press.

14 The terms general and specific protonation are used in the same sense as general and specific acid catalysis. See: Maskill, H. (1993) *The Physical Basis of Organic Chemistry*, Oxford University press, New York. See also [11]).

ORDER-DISORDER PHENOMENA IN INCLUSION COMPOUNDS: A SOLID STATE NMR STUDY

S. TERNIEDEN, J. SCHMIDER AND K. MÜLLER
Institut für Physikalische Chemie, Universität Stuttgart
Pfaffenwaldring 55, D-70569 Stuttgart, Germany

Abstract

The molecular characteristics of two selected guest molecules in thiourea and urea inclusion compounds are examined using solid state NMR techniques. Particular emphasis is given to the guest properties in the vicinity of solid-solid phase transitions which are known to be accompanied by a distortion of the host lattice. The analysis of the NMR data reveals at such phase transitions also significant changes for the motional features of the guest molecules. It is concluded that with respect to the guest components they can be assigned to order-disorder transitions.

1. Introduction

Urea and thiourea are well known host components for inclusion compounds [1]. Numerous studies employing a variety of experimental techniques such as X-ray, calorimetry, IR, Raman or NMR spectroscopy gave a clear picture of such systems on a molecular level. It has been shown that the urea and thiourea host channels first of all are distinguished by a different cross-sectional area which is responsible also for the selectivity of the incorporated guest species. Since the urea channels are rather narrow, only n-alkanes and some derivatives can be used as guest components. On the other hand, in the larger thiourea channels bulkier guest species like cycloalkanes can be found. X-ray studies performed on such systems have shown that the room temperature phases are characterized by a hexagonal symmetry of the urea or thiourea channels. Below room temperature quite often solid-solid phase transitions are observed which are accompanied by a distortion of the hexagonal symmetry of the channels [2]. From NMR and IR studies it is known that at room temperature the guest molecules display a rather high mobility which is in line with a large degree of orientational or rotational disorder. So far, little is known about the behaviour of the guest molecules in the low temperature phases. In this contribution we present solid state NMR investigations of two selected urea and thiourea inclusion compounds. In particular, we focus on the molecular dynamics of the guest molecules, bromocyclohexane and n-hexadecane, in the low temperature phases and on the behaviour in the vicinity of the solid-solid phase transitions.

A. W. Coleman (ed.), Molecular Recognition and Inclusion, 209–214.
© 1998 *Kluwer Academic Publishers. Printed in the Netherlands.*

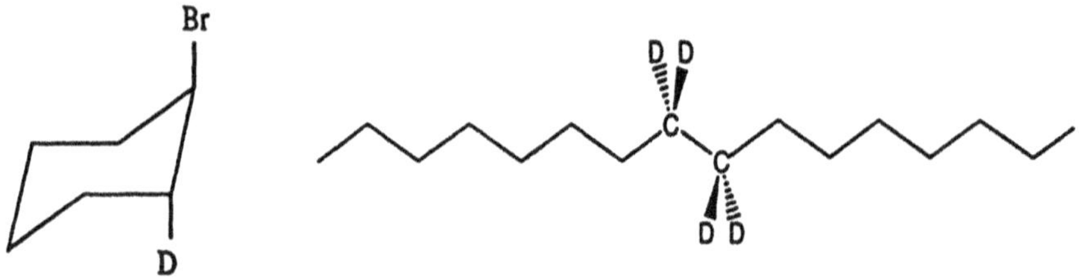

Figure 1. Selectively deuterated guest molecules.

2. Experimental

Selectively deuterated bromocyclohexane and n-hexadecane (see Fig. 1) were synthesized as described elsewhere [3]. Samples of inclusion compounds were prepared according to standard procedures. Calorimetric studies were carried out with a Netzsch DSC 200 differential scanning calorimeter. NMR measurements were performed on a basic Bruker CXP 300 spectrometer (equipped with a Tecmag control unit) operating at a frequency of 46.07 MHz for deuterium and 75.47 MHz for carbons, respectively. ^{2}H NMR spectra were obtained using a Bruker 5 mm probe employing the quadrupole echo sequence with $\pi/2$-pulses of 1.8 - 2.2 μs width and a time interval between the pulses of 20-30 μs. Spin-lattice relaxation times were measured by the inversion recovery method. For the ^{13}C MAS NMR investigations a Bruker 4 mm MAS probe was used. The experiments were carried out with the cross polarization technique followed by high power proton decoupling. The sample temperature during the variable temperature experiments was controlled with a Bruker BVT 1000 temperature control unit.

3. Results and discussion

3.1 BROMOCYCLOHEXANE IN THE THIOUREA INCLUSION COMPOUND

The bromocyclohexane/thiourea inclusion compound has been studied employing ^{13}C MAS NMR and dynamic ^{2}H NMR techniques. For the latter experiments a selectively deuterated bromocyclohexane (see Fig. 1) has been used. DSC investigations revealed for the bromocyclohexane/thiourea inclusion compound a solid-solid phase transition at T = 237 K. It is assumed that here, as for other thiourea inclusion compounds, the hexagonal host channels of the high temperature phase become distorted.

Fig. 2 shows a ^{13}C MAS spectrum recorded in the low temperature phase. Six strong carbon signals are visible which can be attributed to the axial conformer of bromocyclohexane. It is interesting that the carbons 2 and 6 as well as 3 and 5 become non-equivalent suggesting a different chemical environment of the two wings of the bromocyclohexane ring. In addition, three weak signals show up which stem from the equatorial conformer (relative amount: 2%). This result clearly shows that the unusual conformational equilibria, found in the high temperature phase for monosubstituted cyclohexanes in thiourea [4], are preserved in the low temperature phase. Former ^{13}C MAS investigations of the high temperature phase revealed an amount for the equatorial conformer of 5 %. Obviously, the conformational equilibria are almost unaffected by the

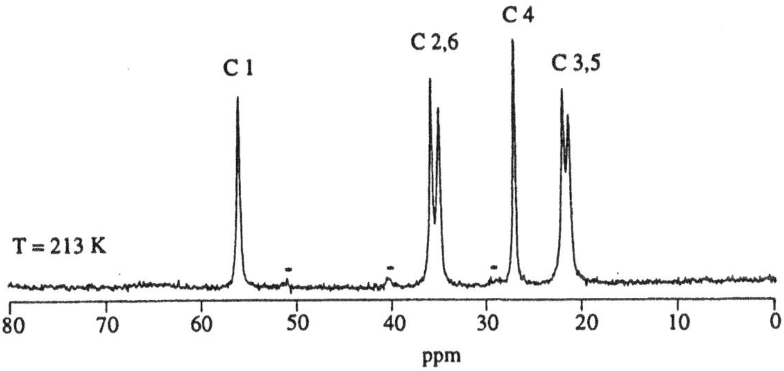

Figure 2. ^{13}C MAS spectrum of the bromocyclohexane/thiourea inclusion compound at 213 K and with a sample spinner frequency of 4.1 kHz. Asterisks indicate signals due to the equatorial conformer.

solid-solid phase transition.

Further information about the molecular dynamics can be obtained from the ^2H NMR studies of the selectively deuterated compound. A representative series of variable temperature ^2H NMR lineshapes (quadrupole echo sequence) from 107 K up to 333 K is given in Fig. 3. Inspection of this figure indicates the presence of almost rigid guest molecules at very low temperatures as can be taken from the Pake pattern at 107 K. Upon heating the spectra are smeared out. Eventually, close to the transition temperature a motionally narrowed, biaxial powder pattern is observed. This implies, together with the consideration of the experimental spin-spin (T_2) and spin-lattice (T_1) relaxation curves (not shown), that even in the low temperature phase the bromocyclohexane guests become rather mobile. In particular, the T_1-relaxation curve exhibits a pronounced minimum at 200 K which reflects the presence of a molecular process with a rate constant in the order of 10^9 s^{-1}. Preliminary simulations indicate that these motions are highly restricted overall motions, most probably reorientational motions of the guest molecules around their equilibrium positions [3a].

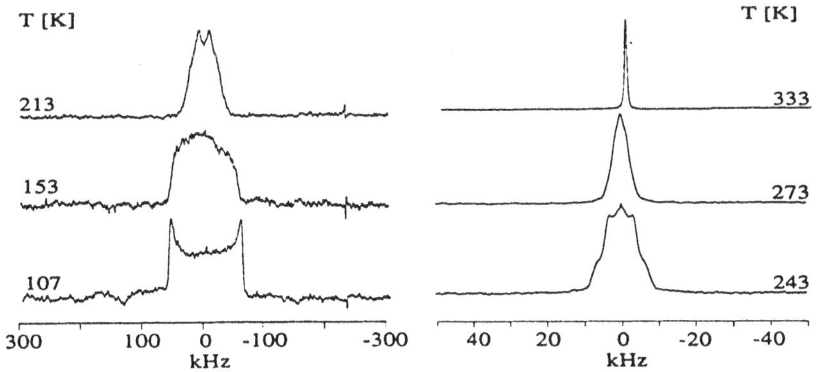

Figure 3. Variable temperature ^2H NMR spectra of the bromocyclohexane-d_1/thiourea inclusion compound.

Heating above the phase transition at 237 K causes a significant narrowing of the ^2H NMR spectra. At first, they show an axial symmetric pattern which directly reflects the hexagonal symmetry of the host channel. Upon further heating the lineshapes experience additional narrowing which results in an almost isotropic line at higher temperatures. These observations can be understood by the presence of a large spatial freedom for the guest molecules in the high temperature phase. Thus, the overall motions now are almost unrestricted giving rise to a large degree of orientational disorder as reflected by the narrow ^{2}H NMR signal. The main influence of the solid-solid phase transition on the guest molecules thus can be found in the unusual large increase in orientational disorder.

3.2 N-HEXADECANE IN THE UREA INCLUSION COMPOUND

A sample of the urea inclusion compound with n-hexadecane, selectively deuterated at carbons 8 and 9 (see Fig. 1), has been prepared and studied with dynamic ^{2}H NMR spectroscopy. Again, this compound is known to undergo a phase transition around 150 K [2a]. From our calorimetric studies the transition temperature was found to be 156 K.

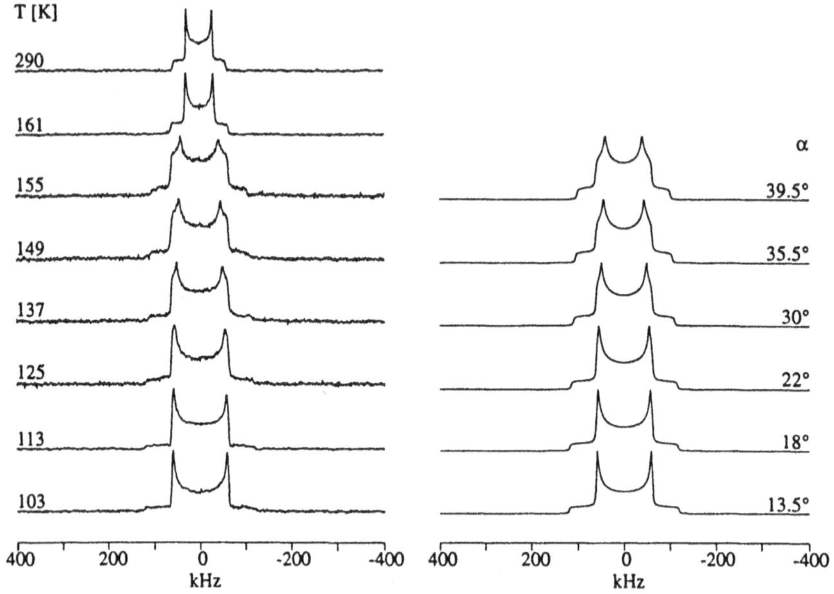

Figure 4. Experimental ^{2}H NMR spectra of n-hexadecane-d_4 in urea (left). Simulations of the low temperature spectra using a 2-site-jump model (fast exchange limit) and the jump angles α given in the figure (right).

Fig. 4 exhibits representative ^{2}H NMR spectra in a temperature range 100 K to 300 K. In the low temperature phase, again, the spectra start with an almost axial symmetric "rigid limit" powder pattern characteristic for completely immobile guest molecules. Upon heating within the low temperature phase ($100 \text{ K} < T < 156 \text{ K}$) the spectra exhibit an increasing biaxiality. Eventually, above the phase transition axial symmetric spectra are observed which now are reduced in spectral width by a factor of -1/2. This behaviour can be understood by assuming for the low temperature phase hexadecane chains in the all-trans configuration undergoing fast but very restricted rotational motions around their all-trans axes. The good agreement between the simulated spectra (see Fig. 4), obtained by using a simple two-site-jump process, and the experimental counterparts suggests that the model assumption is correct. Furthermore, a linear increase of the jump angle, i.e. reorientational angle, with temperature has been derived reaching an upper value of 39.5° close to the phase transition. As discussed before, the restricted molecular motions of the guest species in the low temperature phase is a direct consequence of the distorted symmetry of the local, molecular environment, i.e. the urea channels. The high temperature phase exhibits a hexagonal symmetry of the urea channels which allows the hexadecane chains to undergo unhindered (free) rotations around their long axes as well known from other studies of urea inclusion compounds [5]. A quantitative determination of the rate constants for the restricted and unrestricted overall motions is possible by the analysis of T_1-relaxation experiments. In Fig. 5 experimental, partially relaxed ^{2}H NMR spectra are shown together with their best fit counterparts.

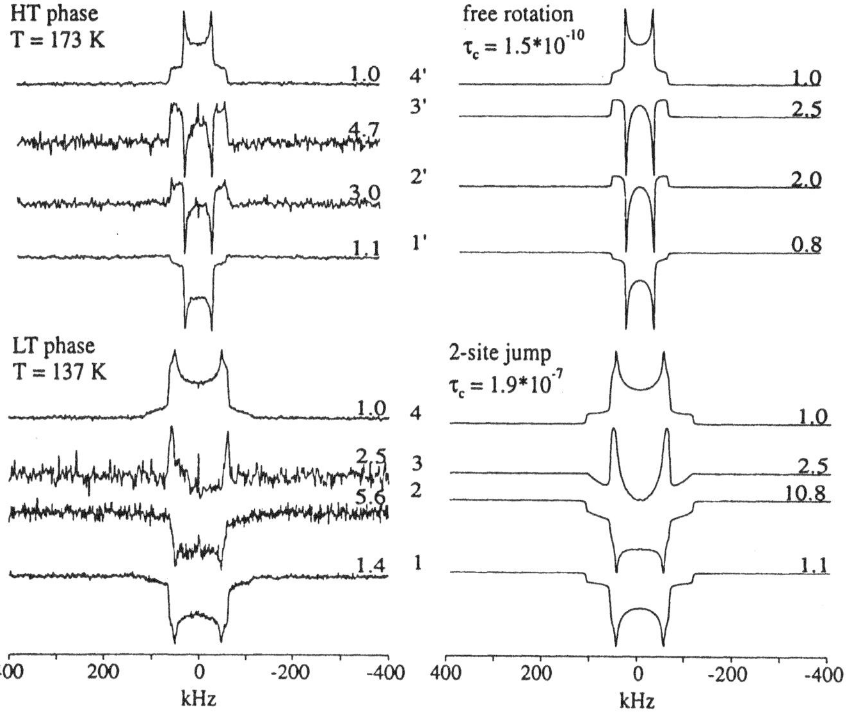

Figure 5. Experimental and simulated partially relaxed ^{2}H NMR spectra of n hexadecane-d_4 in urea for the high and low temperature phase. Pulse spacings are 10 ms (1), 120 ms (2), 200 ms (3), 2 s (4); 10 ms (1'), 60 ms (2'), 70 ms (3'), 300 ms (4'). The scaling factors of the spectra and the correlation times τ_c are given in the figure.

It is quite obvious that the T_1-anisotropy visible in such spectra is quite different for the low and high temperature phase. The general good reproduction of the experimental spectra by the model simulations is a further proof for the correctness of our model assumptions. From the simulations the activation energies have been derived which were found to be 6.2 kJ/mol for the restricted rotational motion (low temperature phase) and 12.5 kJ/mol for the unrestricted rotation (high temperature phase). Moreover, it should be mentioned that the rotational motion in the high temperature phase is about three orders of magnitude faster than the corresponding restricted rotational motion in the low temperature phase.

Similar experiments have been performed for a sample with hexadecane deuterated at the terminal methyl group. First of all, the spectra show a very fast methyl group rotation which reduces the spectral width by a factor of -1/3 and which dominates the T_1-relaxation in the low temperature phase. It is shown that the methyl group rotation and the restricted overall rotational motions, however, are not sufficient for a complete description of the experimental ^2H NMR spectra in the low temperature phase. Rather, the hexadecane chains possess a higher flexibility towards the chain ends which might be expressed by the presence of gauche conformers or restricted librational motions around various C-C bonds [5]. The high temperature phase again exhibits axial symmetric ^2H NMR spectra which, as discussed above, reflect the unrestricted rotational motion of the chains within the hexagonal urea channels. Here, the spin-lattice relaxation is determined by both the methyl group rotation and the rotational motions of the chains.

Summing up, the solid-solid phase transition for the hexadecane/urea inclusion compound can be ascribed to an order-disorder transition. Above this transition the hexadecane molecules, although they still adopt an almost all-trans configuration, are rotationally disordered.

4. Conclusion

Urea and thiourea inclusion compounds consist of almost rigid host channels in which suitable guest molecules can be incorporated. It has been shown that the guest species are highly mobile displaying a high degree of orientational or rotational disorder in the high temperature phase stable at room temperature. Solid-solid phase transitions for such inclusion compounds are accompanied by a distortion of the hexagonal symmetry of the host channels. As a direct consequence the guest molecule motions in the low temperature phase become highly restricted. However, they still are on a fast timescale being in the MHz- to GHz-region.

Acknowledgment

We thank Mrs. H. Seidel for the synthesis of the selectively deuterated guest molecules. Financial support by the Deutsche Forschungsgemeinschaft and the Fonds der Chemischen Industrie is gratefully acknowledged.

References

1. Atwood, J.L., Davies, J.E.D. and MacNicol, D.D. (Eds.) *Inclusion Compounds*, Vols 1-3 Academic Press, New York (1984); Vols 4 and 5 Oxford University Press, Oxford (1991).

2. a) Chatani, Y., Anraku, H. and Taki Y. (1978) Phase Transition and Structure Change of Urea Adducts with n-Paraffins and Paraffin-type Compounds, *Mol. Cryst. Liq. Cryst.* **48**, 219-231. b) Clement, R., Mazieres, C., Gourdji, M. and Guibe, L. (1977) Phase Changes and Molecular Motion in the Thiourea-Cyclohexane Inclusion Compound, *J. Chem. Phys.* **67**, 5381-5385.

3. a) Ternieden, S. and Müller, K., Solid State NMR Studies of the Bromocyclohexane/Thiourea Inclusion Compound, submitted, b) Schmider, J. and Müller, K. in preparation.

4. a) Müller, K. (1992) ^{13}C Magic Angle Spinning NMR Investigations of Thiourea Inclusion Compounds, *Magn. Res. Chem.* **30**, 228-234. b) Aliev, A.E. and Harris, K.D.M. (1993) Conformational Properties of Monosubstituted Cyclohexanes in Thiourea Inclusion Compounds and in Solution: Variable-Temperature One-Dimensional and Two-Dimensional ^{13}C NMR Investigations, *J. Am. Chem. Soc.* **115**, 6369-6377. c) Müller, K. (1995) Solid-State ^{13}C NMR Investigations of Monosubstituted Cyclohexanes in Thiourea Inclusion Compounds, *Magn. Res. Chem.* **33**, 113-123.

5. a) Greenfield M.S., Vold R.L. and Vold R.R. (1988) Deuterium Spin Relaxation and Guest Motion in the n-Nonadecane/Urea Clathrate, *Mol. Phys.* **66**, 269-298. b) Harris, K.D.M. and Jonsen, P. (1989) ^2H NMR Investigation of the Dynamic Behaviour of n-Hexadecane in its Urea Inclusion Compound *Chem. Phys. Lett.* **154**, 593-598.

ONE CLASS OF AZOCALIXARENE, DIFFERENT TYPES OF ASSEMBLIES IN SOLID STATE

N. EHLINGER and M. PERRIN
Université Claude Bernard LYON I, Laboratoire de Cristallographie - ESA 5078, 43, Boulevard du 11 Novembre 1918 69622 Villeurbanne Cedex

1. Introduction

The name of Azocalixarene is given for calixarene molecules having aromatic azo groups at the upper rim. The first azocalixarene described in the literature is the paratetrakis(4-nitrophenylazo)calix[4]arene[1]. Since this year, a lot of other azocalixarenes have been synthesized. These molecules according to their azo fonctionnalisation are chromogenic. They are most studied for selective extraction of ion.

Our objective is to acquire a better understanding of the conformation of azocalix[4]arenes in the solid state and to study their interaction with other organic substances. So we present here the first resolutions of structures of azocalixarenes by X-ray diffraction. To carry out this work, the first point was the preparation of five azocalix[4]arenes. Then we have tried to grow single crystals suitable for X-ray diffraction. Finally three structures have been resolved and studied.

2. Synthesis[2]

Five azocalix[4]arenes with R = H, 4-NO_2, 2,5-diCl, 3-CF_3, 4-SO_3H have been synthesised. The method applied was the diazo coupling of calix[4]arene with various aromatic diazonium ions (in the proportion 1:4) in the presence of pyridine at a temperature of 0°C (schema 1). The solvent usually used for such reaction is water. But, here the calix[4]arene is insoluble in water. So we used an organic solvent. The best yield, except for the sulfonated product, was obtained with tetrahydrofuran. At the end of the reaction four products are present: the mono, di, tri and tetrasubstitued azocalix[4]arenes. The tetrasubstitued product was isolated by washing and recrystallisation.

A special synthesis was applied for the preparation of the sulfonated product. First a sodium calixarenate was prepared by reaction of sodium methylate dissolved in methanol and calix[4]arene dissolved in acetone. After two hours, the product of this reaction was isolated by evaporation of the solvents. It was completely soluble in water, so the diazo coupling

A. W. Coleman (ed.), Molecular Recognition and Inclusion, 215–220.
© 1998 Kluwer Academic Publishers. Printed in the Netherlands.

took place in water and the tetrasubstitued azocalix[4]arene was isolated by acidification and filtration.

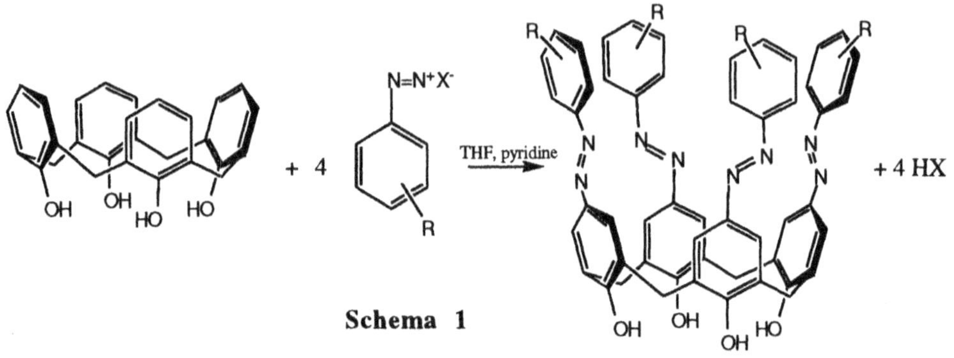

Schema 1

R	H	4-NO_2	2,5-diCl	3-CF_3
X	BF_4^-	BF_4^-	Cl^-	Cl^-
solvent of reaction	THF or acetone	THF	THF and water	THF and water
yield	41%	61%	29%	52%

All the products have been identified by elemental analysis, IR, ^{1}H NMR, mass spectroscopy. Their UV-visible spectra and their melting point (decomposition) have been measured.

R	H	4-NO_2	2,5-diCl	3-CF_3	4-SO_3H
melting point (°C)	309	340	307	325	310
λmax (nm) in DMF at 1.25 10^{-5} M	358	397	376	366	368

3. Crystallization and structure analyses

Single crystals were obtained by slow evaporation of a saturated solution in DMF for NC4H and 4-picoline for NC4N and NC4CL2. NC4F and NC4S afforded crystals too small for X-ray work.

The crystals structures were solved by direct methods (SHELX-76[3] for NC4H and MULTAN[4] for NC4N and NC4CL2) and refined (SHELX-76 for NC4H and NC4N and SHELXL-93[5] for NC4CL2).

4. Structure of NC4H, NC4N and NC4CL2 [6, 7]

The three azocalix[4]arenes adopt a cone conformation as it is usually the case for calix[4]arene with free OH groups. NC4N possesses a two fold axis. The two other macrocycles show no symmetry. Two phenylazo groups of NC4CL2 are turned so that the chlorine atoms in ortho position get closer. The intramolecular Cl...Cl distances are of 3.569(6) and 3.286(6) Å. The two aromatic nuclei, which correspond to the shorter Cl...Cl distances, are approximately coplanar.

R	H	4-NO_2	2,5-diCl
Compound	NC4H	NC4N	NC4CL2
Solvent	DMF	4-picoline	4-picoline
Complex	1:1	1:4	1:3
Crystal System	triclinic	monoclinic	monoclinic
Space Group	P1	C2/c	$P2_1/n$
Volume (Å^3)	2353	6961	7479
Z	2	4	4
Mv (g/cm^3)	1.290	1.330	1.240
R Factor	0.089	0.067	0.163

The cone conformation is generally characterized by a cycle of strong hydrogen bonds between the OH groups. This cycle is observed for the compounds NC4N and NC4CL2, although the intramolecular O...O distances for this last macrocycle are particular. For NC4H, this cycle is interrupt by a hydrogen bond between one hydroxyl group and the O atom of the guest molecule (intermolecular O...O distance of 2.70(1) Å).

The δ angles between the mean plane of the methylene bridges and the mean plane of the phenolic rings have been calculated. The cone conformation is approximately regular for the compounds NC4H and NC4CL2. But the shape of NC4N is comparable to a boat conformation. In fact, two opposite moieties are moved from each other, and the two others are brought closer.

compound	R	O...O (Å)	δ(°)	θ(°)
NC4H	H	2.693(7) 2.647(7)	128.5(2) 123.0(2)	9.0(3) 25.6(3)
		2.622(7) 2.885(7)	131.0(2) 114.4(2)	56.5(3) 4.9(3)
NC4N	4-NO_2	2.574(5) 2.606(4)	144.0(2) 116.9(2)	7.4(2) 5.0(2)
NC4CL2	2,5-diCl	2.497(9) 2.759(8)	126.5(2) 115.9(2)	4.3(2) 5.2(2)
		2.806(9) 2.535(7)	127.7(2) 129.0(2)	21.0(2) 33.0(2)

These cone conformations imply cavities. In the case of NC4H, this cavity is occupied by two phenylazo groups of two neighbouring macrocycles and partially by a guest molecule. This DMF molecule is held by a CH3 ... π interaction between one of its methyl group and one aromatic ring of a macrocycle. The cavities of NC4N and NC4CL2 are occupied by two 4-picoline molecules. usually when calix[4]arenes crystallise with aromatic compounds, only one guest molecule lies inside the cavity, the mean plane of the aromatic ring being perpendicular to the mean plane of the four methylene bridges. In the case of

NC4N and NC4CL2 the guest positions inside the cavity have unusual positions, approximately parallel to phenylazo groups.

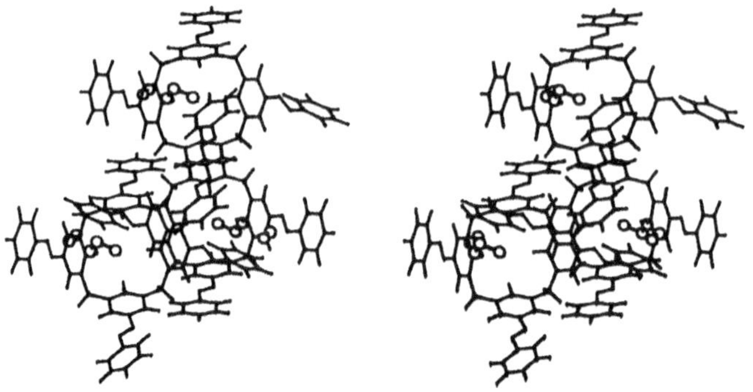

Figure 1. Interprenetration of NC4H molecules.

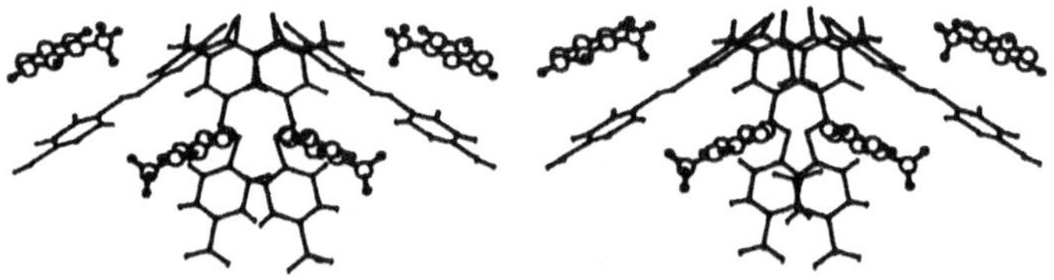

Figure 2. Conformation of NC4N and guest positions.

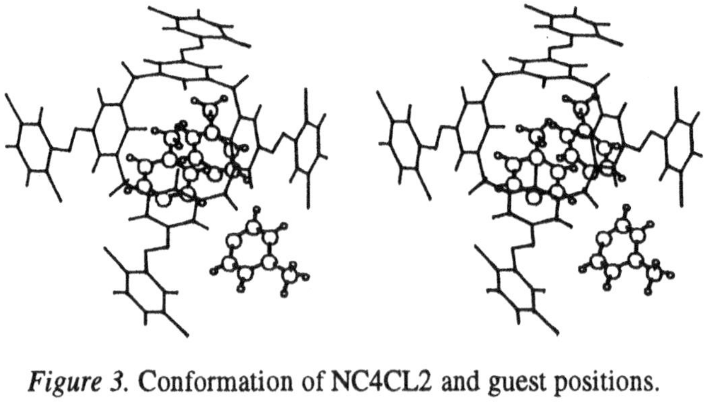

Figure 3. Conformation of NC4CL2 and guest positions.

Theoretically, the two phenyl groups linked by the azo group in the same moiety, which correspond to a conjugated molecule, should be coplanar. The θ angles between the two mean plane of phenyl group of each side of the azo group are listed . As expected, the moieties of NC4N are approximately planar. But, two moieties of NC4H and NC4CL2 deviates significatively from planarity. This can be explained for NC4H by the interpenetration of the macrocycles and for NC4CL2 by the existence of Cl...Cl van der Waals contacts.

The packing of the three azocalix[4]arenes are completely different.

The packing of NC4H can be described as an alternation of slices of azocalixarene molecules and slices of DMF molecules. These slices are held together both by CH3...π and OH...O interactions between host and guest.

NC4N crystallises with 4 picoline molecules. Two of them are located inside the cavity and the two others picoline molecules are outside. The mean planes of these four guest molecules are approximately parallel to one azobenzene group of the macrocycle and their methyl groups interact with the azo function. The NC4N molecules pile up with the aperture of the cones, half in one direction, half in the opposite one. The picoline molecules inside the cavity hold the macrocycles together by π...π interactions. The others picoline molecules hold the cones each to other both by π...π and CH3...N=N interactions.

NC4CL2 crystallises with 3 picoline molecules. Two of them are located in the cavity of the macrocycle with the methyl groups directing inward the cavity and are held together by π...π interaction. The guest molecule which is placed more deeply in the cavity is held by CH3...π interaction between its methyl group and the four phenolic nuclei. The packing can be described as the repetition of a group of two azocalixarenes which form a cage. Four guest molecules are inserted in this cage. The interhost space is also occupied by picoline molecules.

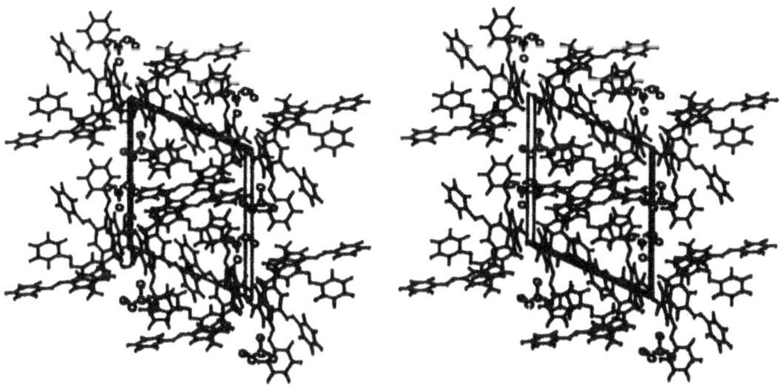

Figure 4. Packing of NC4H along [001].

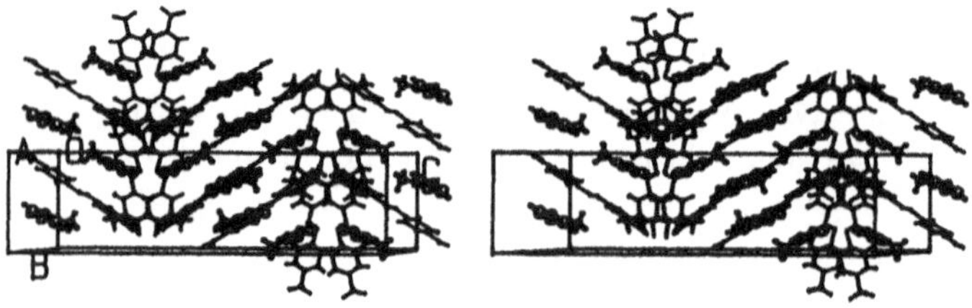

Figure 5. Packing of NC4N along [100].

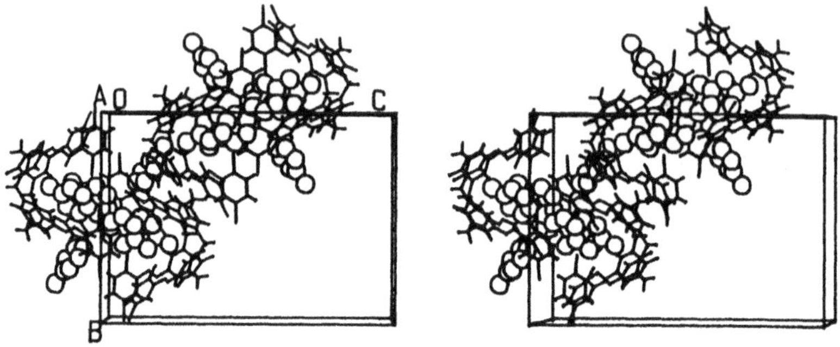

Figure 6. Packing of NC4CL2 along [100].

5. Conclusion

We have observed that the macrocycles adopt a cone conformation. These three azocalix[4]arene molecules are flexible, and their shape adapt oneself to the environment. Their cavities are large and deep enough to enclose two picoline molecules or two azobenzene residues.

References

1. SHINKAI, S., ARAKI, K., SHIBATA, J. and MANABE, O. (1989) *J. Chem. Soc. Perkin Trans.* 1, 195-196.
2. EHLINGER, N.(1995), *Thesis,* Université Claude Bernard LYON I, France.
3. SHELDRICK, G.M. (1976) *Shelx, Program for Crystal Structure determination,* Univ. of Cambridge, England.
4. MAIN, P., HULL, S.E., LEISINGER, L., GERMAIN, G., DECERCQ, J.P., WOOLFSON, M.M. (1980) *Multan 80.,* Univ. of York, England and Univ. of Louvain, Belgique.
5. SHELDRICK, G.M. ; *Shelxl* (1993) Univ.of Göttingen, Germany.
6. EHLINGER, N., LECOCQ, S., PERRIN, R., PERRIN, M. (1993) *Supramolecular Chemistry,* **2**, 77-82.
7. EHLINGER, N., PERRIN, M. (1995) *J. Inclusion Phenomena,* **22**, 33-40.

THE SIMPLE SYNTHESIS OF CHIRAL DIAZACORONANDS DERIVED FROM D-MANNITOL AND L-TARTARIC ACID

DANIEL T. GRYKO,[a] PIOTR PIĄTEK,[b] JANUSZ JURCZAK[a,b]

[a] *Institute of Organic Chemistry, Polish Academy of Sciences, 01-224 Warsaw, Kasprzaka 44/52, Poland*

[b] *Department of Chemistry, University of Warsaw, 02-093 Warsaw, Pasteura 1, Poland*

There is a continuous interest in the crown ether type of macrocyclic molecular receptors since their discovery by Pedersen.[1] The main feature of these compounds is their complexing behaviour. Introduction of nitrogen pivoting atoms into macrocyclic polyethers created a new class of host molecules. Diazacoronands and cryptands play a prominent role in molecular architecture; one may note that the incorporation of nitrogen pivoting atoms into molecules offers an enormous number of possible structures. Methods for the formation of diazacoronands have been extensively reviewed.[2-4] Among these methods the high-dilution technique,[5] the route based on the template effect,[6] and the high-pressure approach,[7] are frequently used as the most versatile procedures. At the beginning of the nineties, we found that α,ω-diamino aliphatic ethers react under ambient conditions with dimethyl

A. W. Coleman (ed.), Molecular Recognition and Inclusion, 221–226.
© 1998 *Kluwer Academic Publishers. Printed in the Netherlands.*

α,ω-dicarboxylates, in methanol as a solvent, to afford macrocyclic diamides in good yields[8-10] (Scheme 1). Isolated macrocyclic diamides can be readily transformed into macrocyclic diamines using $BH_3{*}Me_2S$ as a reducing agent.

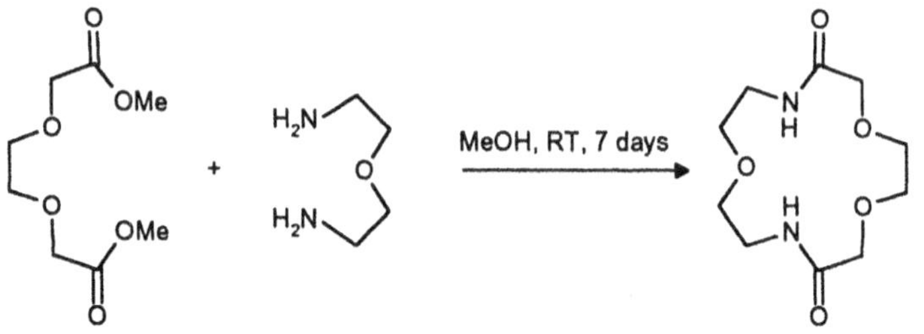

Scheme 1

The usefulness of our method prompted us to apply a similar approach to the synthesis of chiral diazacoronands derived from D-mannitol and L-tartaric acid. We would like now to report results of these studies.

Macrocyclic diamides derived from D-mannitol, precursors of interesting chiral diazacoronands, were selected as targets of the first part of our synthetic investigations. Bromoacetic acid was used as an alkylating agent in the first step of the presented synthesis. Treatment of the resulting crude diacid with methyl chloroformate in the presence of N-methylmorpholine provided diester **4**. The reaction of this ester with α,ω-diamines **2**, **5** and **6** afforded the respective diamides **7**, **8** and **9** in good yields (Scheme 2). Keeping in mind poor overall yields of macrocyclic diamides, it was decided to obtain the dimethyl ester **4** by another method. Therefore, conditions optimized in our laboratory for alkylation of D-glucose and D-mannose derivatives were applied.[11] The reaction of 1,2,5,6-di-O-isopropylidene-D-mannitol with *tert*-butyl bromoacetate afforded the corresponding di-*tert*-butyl ester

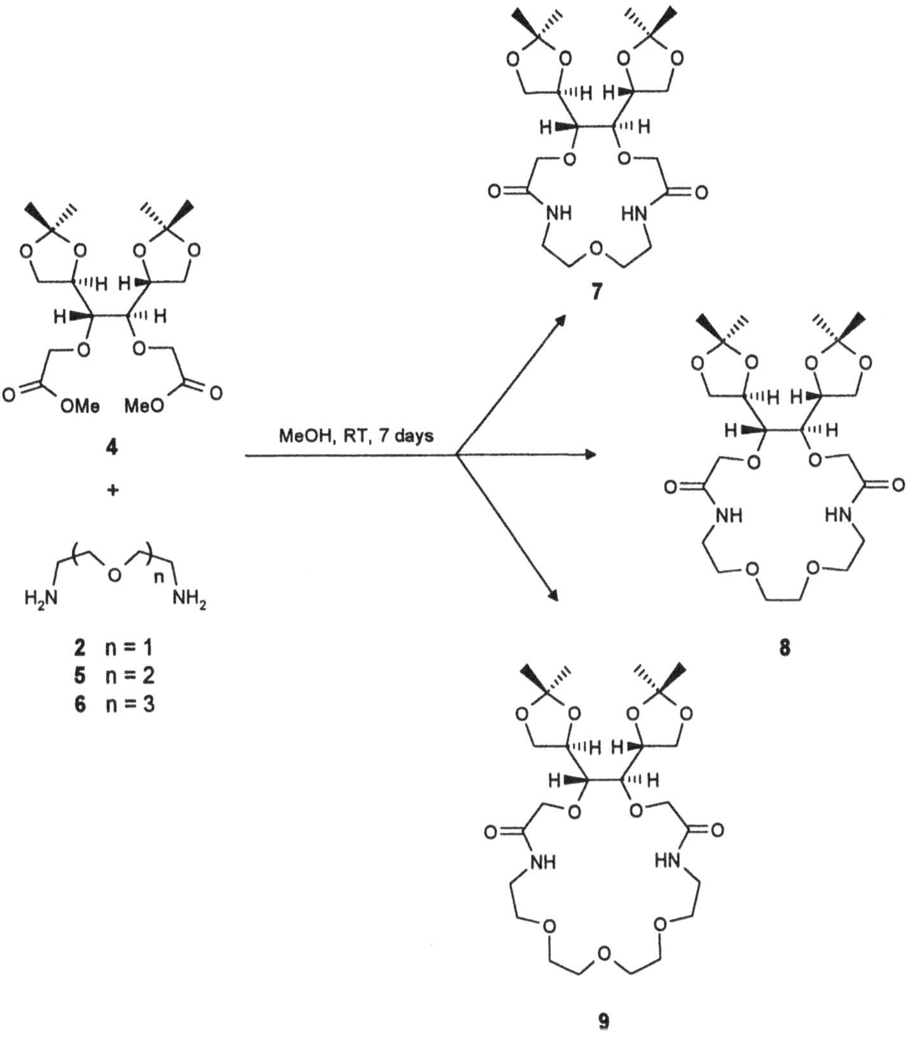

Scheme 2

in a yield of 80%. Unfortunately, all attempts to transesterification of this ester into dimethyl one, without destroying protecting groups, were unsuccessful. It is well known that treatment of di-*tert*-butyl ester with α,ω-diamines did not result in the formation of the respective diamides. In this situation it was resolved to utilise the high-pressure technique as an external cyclization enhancing factor. The reaction of di-*tert*-butyl ester with diamines **2** or **5** under 8 kbar pressure afforded the desired diamides **7** and **8** in a yield of 47% and 57%, respectively. On the basis of our earlier study, we believe that under high-pressure conditions a transesterification occurs. Keeping in mind these results it was resolved to try the macrocyclization under normal pressure, using typical basic transesterification catalysts. This reaction failed in the presence of NEt_3 or DABCO, whereas addition of DBU led to formation of thedesired products in a good yield. The reaction of di-*tert*-butyl ester and diamine **5** afforded diamide **8** in a yield of 42%.

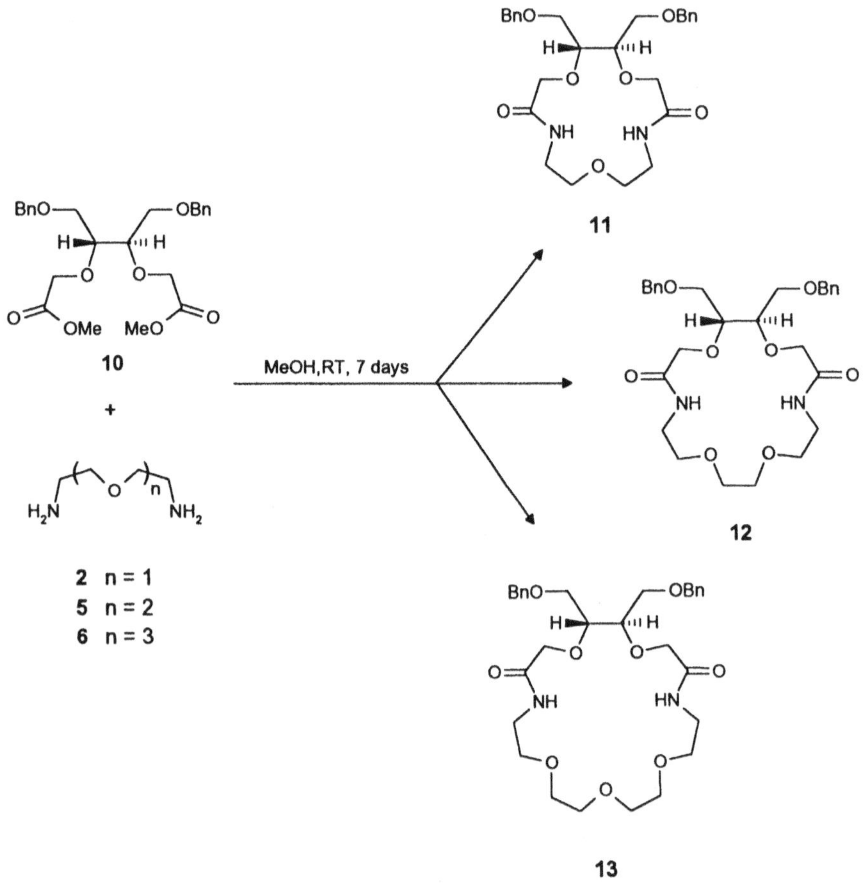

Scheme 3

In the second part of our synthetic studies, macrocyclic diamides derived from L-tartaric acid were selected as targets. The crucial substrate - chiral dimethyl α,ω-dicarboxylate **10** was obtained from L-tartaric acid in an analogous way as chiral α,ω-diester **4**. The reaction of α,ω-diester **10** with diamines **2**, **5** and **6**, carried out under similar conditions as in the previous case, afforded the respective diamides **11**, **12** and **13** (Scheme 3).

2,3-O-Benzylidene-L-threitol was transformed *via* di-*tert*-butyl ester **14** into diamides **15** and **16** using method which was very effective for D-mannitol derivatives. Thus, elongation with *tert*-butyl bromoacetate, followed by exposure of the resulting intermediate to amines **2** and **5** under high-pressure conditions, results in the formation of diamides **15** and **16** in very good yields (Scheme 4).

As in the previous case, we carried out the reaction of *tert*-butyl ester **10** with diamine **5** in the presence of DBU. We obtained diamide **16** in a yield of 53%.

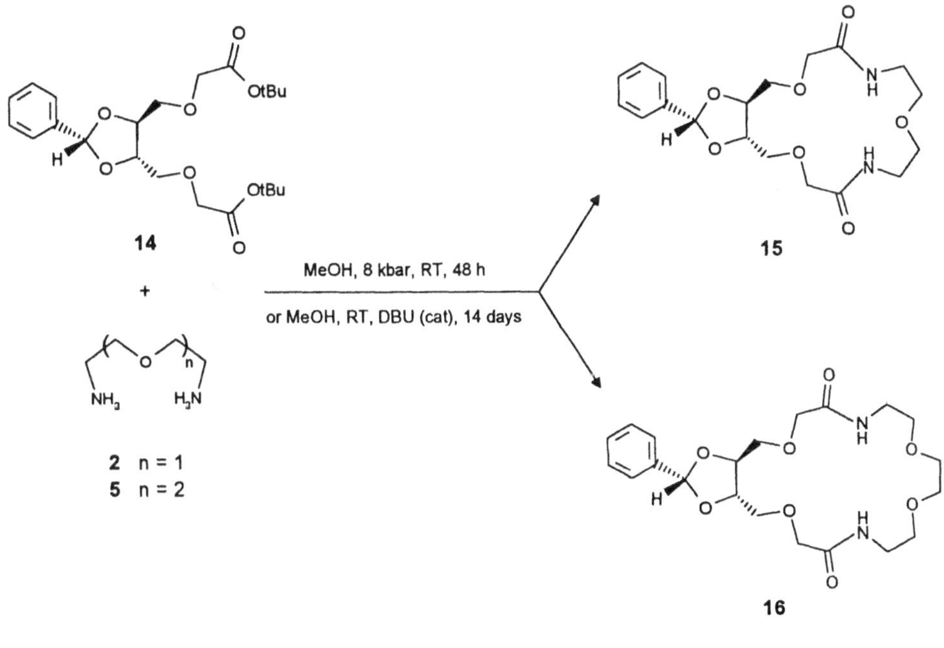

Scheme 4

References

1. Pedersen, C. J. (1967) Cyclic polyethers and their complexes with metal salts, *J. Am. Chem. Soc.* **89** 7017.

2. Gokel, G. W. and Korzeniowski, S. H. (1982) *Macrocyclic Polyether Synthesis*, Berlin, Springer.

3. Krakowiak, K. E., Bradshaw, J. S. and Zamecka-Krakowiak, D. J. (1989) Synthesis of aza-crown ethers, *Chem. Rev.* **89,** 929-972.

4. Jurczak, J. and Ostaszewski, R. (1992) High-pressure approach to the synthesis of diazacoronands and cryptands, *J. Coord. Chem.* **27,** 201-214.

5. Dietrich, B., Lehn, J.-M. and Sauvage, J.-P. (1969) Diaza-polyoxa-macrocycles et macrobicycles, *Tetrahedron Lett.* **10**, 2885-2888.

6. Kulstad, S. and Malmsten, L.A. (1979) Diaza-crown ethers. I. Alkali ion promoted formation of diaza-crown ethers and syntheses of some N,N'-disubstituted derivatives, *Acta Chim. Scand.* **B33,** 469-474.

7. Jurczak, J. and Pietraszkiewicz, M. (1985) High-pressure synthesis of cryptands and complexing behaviour of chiral cryptands, *Topics Curr. Chem.* **130,** 183-204.

8. Jurczak, J., Kasprzyk, S., Sałański, P. and Stankiewicz, T. (1991) A general method for the synthesis of diazacoronands, *J. Chem. Soc., Chem. Commun.* 956-957.

9. Jurczak, J., Kasprzyk, S., Sałański, P. and Stankiewicz, T. (1992) High pressure approach to the synthesis of macrocyclic diamides, *High Press. Res.* **11,** 139-143.

10 Jurczak, J., Stankiewicz, T., Sałański, P., Kasprzyk, S. and Lipkowski, P. (1993) A new method for the synthesis of diazacoronands via double-amidation reaction, *Tetrahedron* **49**, 1478-1488.

11 Pietraszkiewicz, M. and Jurczak, J. (1984) Synthesis of chiral diaza-crown ethers incorporating carbohydrate units, *Tetrahedron* **40**, 2967-2970.

BIOMIMETIC OXIDATION OF AROMATIC ALDEHYDES CATALYZED BY A BIS(COENZYME)-CYCLOPHANE

PATRIZIO MATTEI and FRANÇOIS DIEDERICH*

Laboratorium für Organische Chemie
ETH-Zentrum, Universitätstr. 16
CH-8092 Zürich, Switzerland

1. Introduction

As biological catalysts, enzymes moderate a large number of reactions with high efficiency but are limited to their natural substrates and physiological conditions. Chemical models of enzymes allow both an extension of the scope of the desired reactions and an enhanced understanding of how enzymes mediate these reactions. Pyruvate oxidase is a flavin- and thiamine diphosphate-dependent enzyme found in lactobacteria and catalyzes the reaction from pyruvate to acetyl phosphate [1]. The X-ray crystal structure of this enzyme has recently been determined [2].

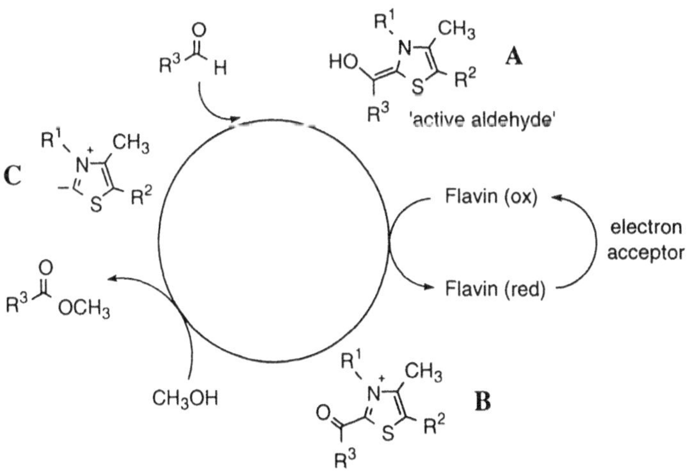

Scheme 1. Catalytic cycle for the oxidation of aldehydes to methyl esters.

* Author for correspondence

A. W. Coleman (ed.), Molecular Recognition and Inclusion, 227–232.

© 1998 *Kluwer Academic Publishers. Printed in the Netherlands.*

As outlined in *Scheme 1*, aldehydes may be elaborated in methanol to the corresponding methyl esters in a reaction sequence similar to that of pyruvate oxidase. These conversions have been catalyzed by simple thiazolium ions [3], thiazolium micelles [4], thiazolio-cyclodextrins [5] and a thiazolio-cyclophane [6] in the presence of various oxidizing agents such as nitrobenzene [3a-d], ferricyanide [5,6], and flavins [3e,4,6]. In basic methanolic solution, the aldehyde reacts with a thiazolium ylide to give an activated aldehyde moiety (**A**) which is subsequently oxidized to an acylthiazolium intermediate (**B**) with concomittant transfer of a hydride equivalent to a flavin. This intermediate rapidly reacts with the solvent to afford a methyl ester while regenerating the thiazolium ylide (**C**). The dihydroflavin is reoxidized by a remote electron acceptor such as dioxygen or a platinum anode, which completes the catalytic cycle.

This paper describes the synthesis and an evaluation of the catalytic performance of cyclophane **1**, which represents the first functional model for pyruvate oxidase combining a well-defined binding site with both flavin and thiazolium prosthetic groups. The tetra-oxa[6.1.6.1]paracyclophane moiety in **1** is an established receptor for benzene and naphthalene derivatives in polar solvents [7] and therefore should form complexes with aromatic aldehydes driven by *van der Waals* interactions and hydrophobic desolvation. As the two cofactors are covalently attached to the reactive site, this should lead to a high reaction rate and selectivity by maintaining intramolecularity in all key steps of the catalytic cycle.

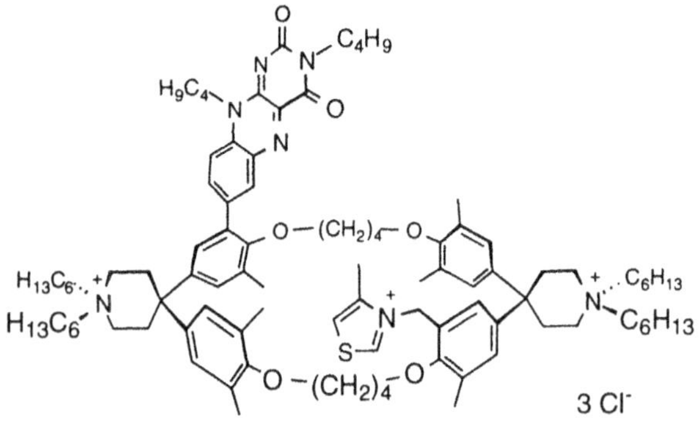

1

2. Synthesis

Receptor **1** was successfully prepared in gram quantities by an 18-step synthesis in a good overall yield of 9% [8]. The synthesis involved a *Suzuki*-coupling [9] between cyclophane boronic acid **2** and 7-bromoflavin **3**; after quaternization, the flavo-cyclophane **4** was obtained. The thiazolium moiety was introduced by chloromethylation of **4** and treatment of the resulting benzylic chloride with 4-methylthiazole (*Scheme* 2).

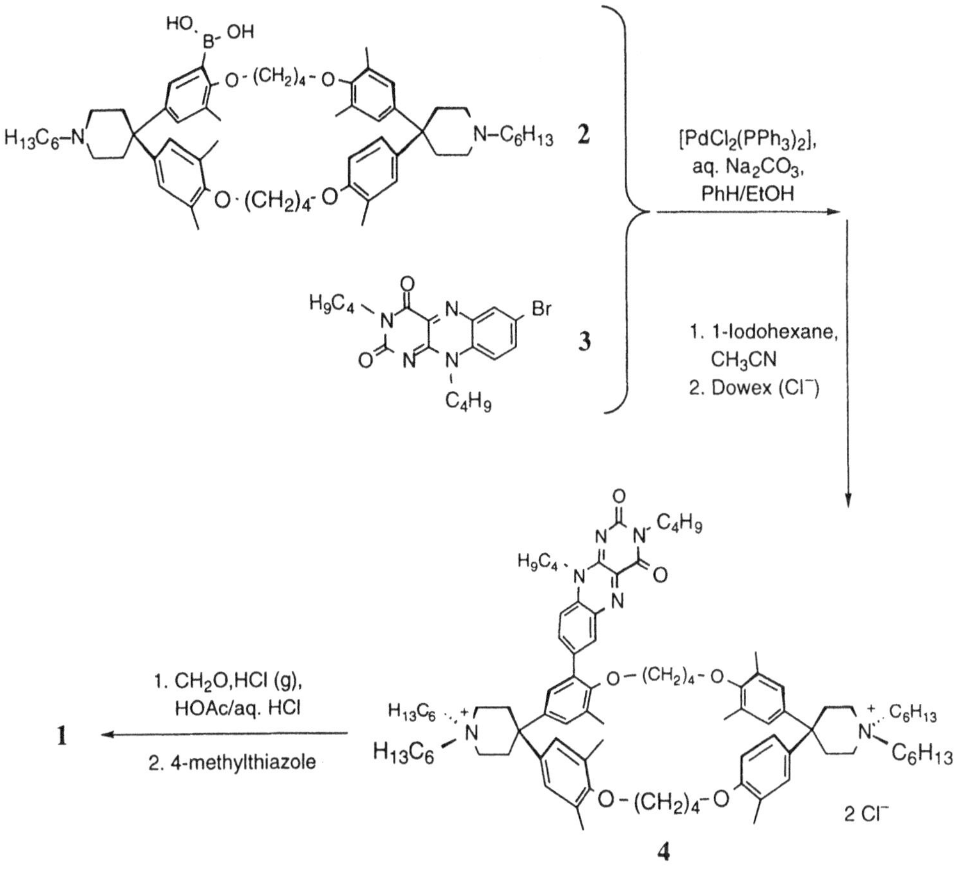

Scheme 2. Synthesis of the Flavo-Thiazolio-Cyclophane **1**.

3. Complexation Properties of the Flavo-Thiazolio-Cyclophane 1

The complexation of benzaldehyde and 2-naphthaldehyde with the flavo-thiazolio-cyclophane **1** was investigated by ^{1}H NMR binding titrations at 300 K. In D_2O/CD_3OD 3:2, the association constants K_a were measured to be 2900 and 160 L mol^{-1} for the comlpexes of 2-naphthaldehyde and benzaldehyde, respectively. In pure CD_3OD, K_a values of 26 and 2.2 L mol^{-1} for the complexes of 2-naphthaldehyde and benzaldehyde, respectively were obtained. These K_a values are in accord with those previously determined for thiazolio-cyclophane **5** [6]. The more strongly binding 2-naphthaldehyde was therefore chosen for kinetic studies to determine the maximum catalytic advantage that might be obtained by complexation.

4. Initial Rate of the Oxidation of 2-Naphthaldehyde to Methyl 2-Naphthoate

The oxidation of 2-naphthaldehyde to methyl 2-naphthoate was catalyzed by **1** in the presence of Et_3N. The reaction was initiated by adding Et_3N to a solution of **1** (0.5 mM) and 2-naphthaldehyde (2-50 mM) in MeOH under an argon atmosphere. The course of the reaction was followed by monitoring the decrease in absorbance of the flavin chromophore upon reduction to the colorless dihydro form. For comparison, this oxidation was also studied under similar conditions with **6** and {**7** + **8**}. The results are summarized in *Scheme 3*.

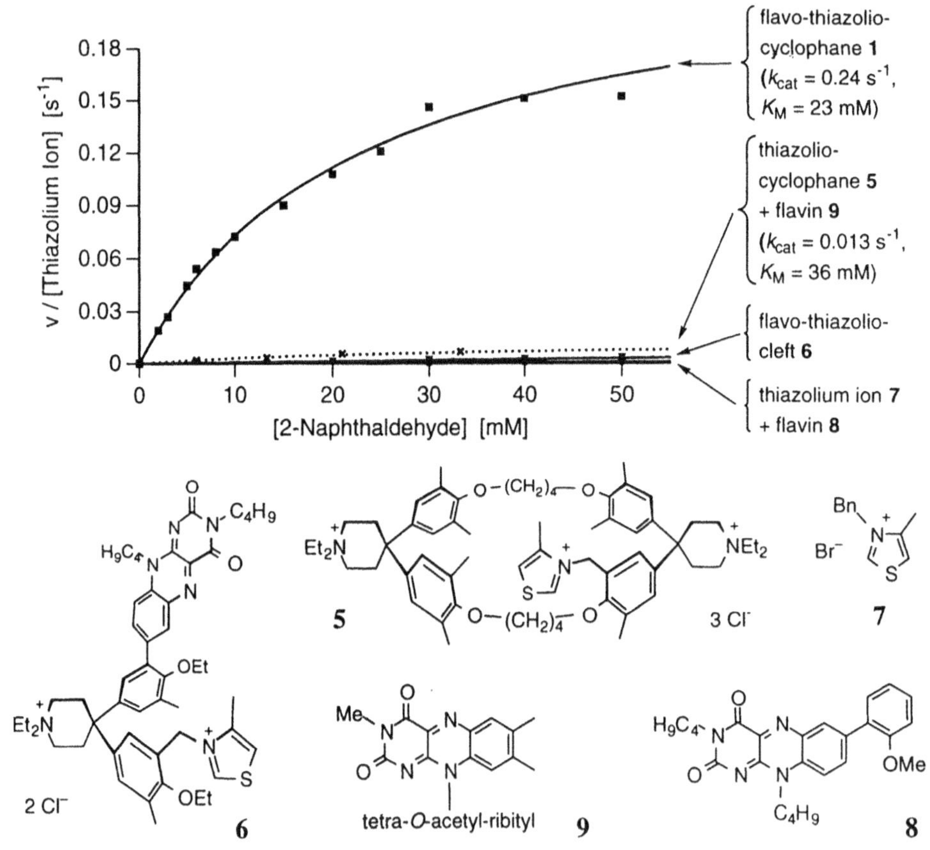

Scheme 3. Plot of relative initial rates *vs.* substrate concentrations in the oxidation of 2-naphthaldehyde to methyl 2-naphthoate catalyzed by different bis-coenzymes.

Saturation kinetics was observed with cyclophane **1**. The k_{cat} (0.24 s^{-1}) calculated represents approximately a 20-fold acceleration in comparison to the reaction with the bimolecular system {**5** + **9**} [6], and is one of the highest turnover numbers reported for artificial enzymes. In contrast, rates in the presence of **6** and {**7** + **8**} increased in a linear fashion with increasing substrate concentration. A comparison of the apparent bimolecular rate constant k_{cat}/K_M for **1** with the calculated second-order rate constants for the catalyst systems **6** and {**7** + **8**} lacking the macrocyclic binding site revealed that the rate acceleration is 170-fold between **1** and **6** and 670-fold between **1** and {**7** + **8**}. The large rate acceleration with **1** resulted from the intramolecularity of all key steps in the catalytic cycle. The formation of the methyl ester was confirmed by ^{1}H NMR spectroscopy, and the comparison with authentic samples of 2-naphthaldehyde and methyl 2-naphthoate showed that no by-product was formed.

5. One-Pot Synthesis of Aromatic Esters by Electrochemical Oxidation of Aldehydes Mediated by Bis(coenzyme)-Cyclophane 1

The use of the bis(coenzyme)-cyclophane **1** as a catalyst on a preparative scale requires the regeneration of the oxidized flavin (*Scheme 1*). Attempts to reoxidize the reduced flavin form by running the reaction under aerobic conditions (in analogy to pyruvate oxidase [1]) led largely to the oxidative destruction of the thiazolium catalyst, and no more than 1.1 equivalents of methyl ester were formed from one equivalent of catalyst. Previous work had shown that the dihydro form of flavin **9** can be oxidized electrochemically in the presence of the thiazolium ions **5** or **7** in basic methanolic solution [6]. The cyclic voltammogram of **1** shows the characteristics of both flavin and thiazolium, *i. e.*, a reversible reduction of the flavin ($E_{1/2}$ = -470 mV *vs.* Ag/AgCl) and an anodic peak at +230 mV indicating the oxidative destruction of the thiazolium moiety (*Fig. 1*).

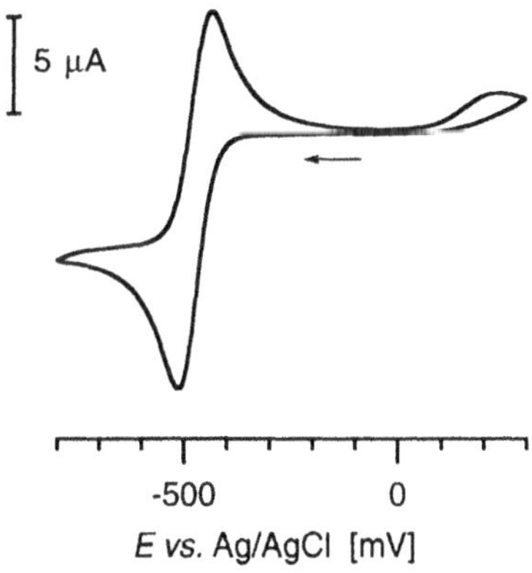

Figure 1. Cyclic voltammogram of **1** (1 mM) in a 50 mM solution of Et_4NBr in $MeOH/Et_3N$ 100:1 at glassy carbon working electrode. Scan rate 20 mV s^{-1}, 300 K.

$$\underset{(150\ \mathrm{mM})}{\text{R–CHO}} \xrightarrow[\substack{\mathrm{CH_3OH},\ 308\ \mathrm{K},\ \text{undivided cell,} \\ \mathrm{Pt}\ (-300\ \mathrm{mV}\ \textit{vs.}\ \mathrm{Ag/AgCl})}]{\substack{\text{Flavo-Thiazolio-Cyclophane } \mathbf{1}\ (5\ \mathrm{mM}), \\ \mathrm{Et_3N}\ (150\ \mathrm{mM}),\ \mathrm{Et_4NBr}\ (50\ \mathrm{mM})}} \text{R–COOCH}_3 \qquad (1)$$

Indeed, by applying the reaction conditions displayed in *Eq. 1*, the flavo-thiazolio-cyclophane **1** could be employed as a redox mediator for the one-pot electrochemical oxidation of aldehydes. The isolated yields of the corresponding carboxylic esters after a reaction time of 16 h were 78% and 48% for 2-naphthaldehyde and benzaldehyde, respectively. No ester was formed in the absence of the catalyst.

6. Conclusion

The flavo-thiazolio-cyclophane **1** is a functional model for the enzyme pyruvate oxidase. In initial-rate studies, **1** exhibits saturation kinetics and large rate acceleratons as a result of the covalent attachment of the two cofactors to a well-defined binding site. Aromatic esters can be prepared in a single step from the corresponding aldehydes by electrochemical oxidation mediated by **1**.

References

[1] a) B. Sedewitz, K. H. Schleifer, F. Götz, *J. Bacteriol.* **160** (1984), 273-278; b) F. Götz, B. Sedewitz in *Biochemistry and Physiology of Thiamin Diphosphate Enzymes*, eds. H. Bisswanger, J. Ullrich, VCH, Weinheim, 1991, pp. 286-293.

[2] a) Y. A. Muller, G. E. Schulz, *Science (Washington, D.C.)* **259** (1993), 965-967; b) Y. A. Muller, G. Schumacher, R. Rudolph, G. E. Schulz, *J. Mol. Biol.* **237** (1994), 315-335.

[3] a) J. Castells, H. Llitjós, M. Moreno-Mañas, *Tetrahedron Lett.* (1977), 205-206; b) J. Castells, F. Pujol. H. Llitjós, M. Moreno-Mañas, *Tetrahedron* **38** (1982), 337-346; c) H. Inoue, K. Higashiura, *J. Chem. Soc., Chem. Commun.* (1980), 549-550; d) H. Inoue, S. Tamura, *ibid.* (1985), 141-142; e) Y. Yano, Y. Tsukagoshi, *J. Chem. Res. (S)* (1984), 406-407.

[4] a) Y. Yano, Y. Hoshino, W. Tagaki, *Chem. Lett.* (1980), 749-752; b) S. Shinkai, T. Yamashita, Y. Kusano, O. Manabe, *Tetrahedron Lett.* **21** (1980), 2543-2546; c) S. Shinkai, T. Yamashita, Y. Kusano, O. Manabe, *J. Org. Chem.* **45** (1980), 4947-4952; d) S. Shinkai, T. Yamashita, Y. Kusano, O. Manabe, *J. Am. Chem. Soc.* **104** (1982), 563-568.

[5] D. Hilvert, R. Breslow, *Bioorg. Chem.* **12** (1984), 206-220.

[6] S.-W. Tam-Chang, L. Jimenez, F. Diederich, *Helv. Chim. Acta* **76** (1993), 2616-2639.

[7] F. Diederich, *Cyclophanes*, The Royal Society of Chemistry, Cambridge, 1991.

[8] P. Mattei, F. Diederich, *Angew. Chem.* **108** (1996), 1434-1437; *ibid. Int. Ed.* **35** (1996), 1341-1344.

[9] N. Miyaura, A. Suzuki, *Chem. Rev.* **95** (1995), 2457-2483.

SOME NEW CALIX[4]ARENE-BASED COMPLEXING AGENTS

J.-B. REGNOUF-de-VAINS
Laboratoire de Chimie Industrielle
UCBL Lyon I, La Doua, 69622, Villeurbanne - France.

With regards to the dense related literature, the calixarene familly plays an important role in supramolecular chemistry. Easily synthesized from phenol and formaldehyde, they are extensively studied as molecular receptors for neutral or ionic species, but also as synthetic platforms for more sophisticated ligands. In this sense, the pionnering work of Gutsche [1] has stimulated a variety of investigations on the structural modifications of calixarenes which, in their cone conformation, present an hydrophilic lower rim and a lipophilic upper rim.
Elaboration of the former with polar functional groups (*eg* acids, esters, amides,...) has produced spatially organized ligands tailor-made for the complexation of metallic cations.
Replacing such groups by the well known 2,2'-bipyridyl chelating unit or parent species should afford podands displaying specific complexing behavior towards transition metal and rare earth cations.

Coordination at the lower rim:

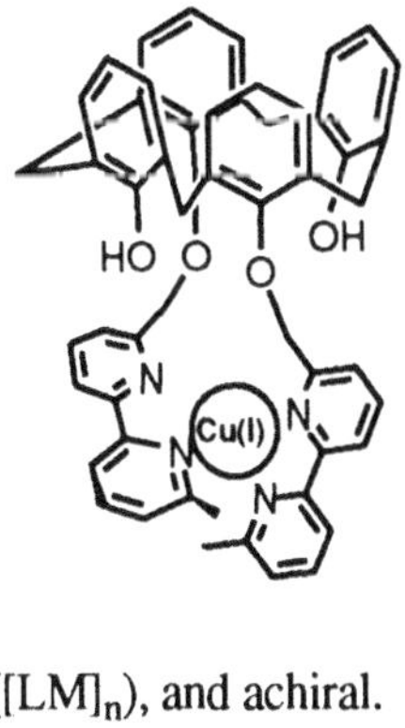

$[Cu(L)]^{+}$ or $[(Cu)_2L_2]^{++}$

In 1992, Beer [2] synthesized the first podand of this series, in which two 2,2'-bipyridines are alternatively disposed ($C_{2}v$ symetry) on the conic de-*ter*butylated calix[4]arene lower rim, in order to favour a tetrahedral complexation. The calixarene platform has to be considered in this case as a voluminous preorganizing substituent keeping the chelating subunits tightly close.
Upon reaction with a Cu(I) salt, the ligand afforded a stable complex which was supposed, based on structural analyses, to be mononuclear (LM) or dimetallic (L_2M_2) instead of polymeric ($[LM]_n$), and achiral.

Our aim in this domain was to develop a library of lipophilic ligands able to complex metallic cations in various kinds of organic solvents, notably alkanes, chlorinated alkanes and aromatics. We chosen for this purpose the more lipophilic *p-tert*-butylcalix[4]arene as platform.

A. W. Coleman (ed.), Molecular Recognition and Inclusion, 233–238.
© 1998 *Kluwer Academic Publishers. Printed in the Netherlands.*

Homo-armed species.

Dealing with the reactivity of the OH groups, and according to the results of Pappalardo *et al.*[3], we prepared by a direct base-strength driven regioselective O-alkylation process the full family of the O-[(2,2'-bipyridyl-6-yl)methyl]- substituted tetra-*p*-(*tert*-butyl)calix[4]arenes tetrols in their cone conformation and we studied their complexation behaviour toward the Cu(I) cation [4].

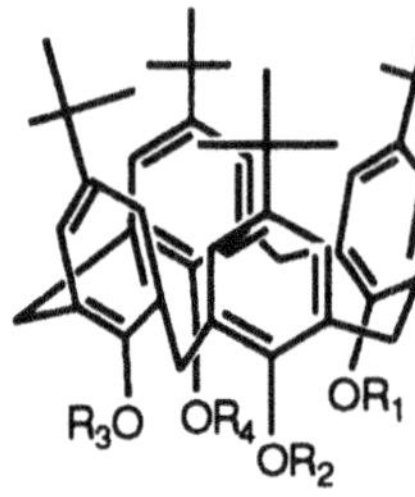

1: R_1 = bpy; R_2, R_3, R_4 = H
2: R_1, R_2 = bpy; R_3, R_4 = H
3: R_1, R_3 = bpy; R_2, R_4 = H
4: R_1, R_2, R_3 = bpy; R_4 = H
5: R_1, R_2, R_3, R_4 = bpy

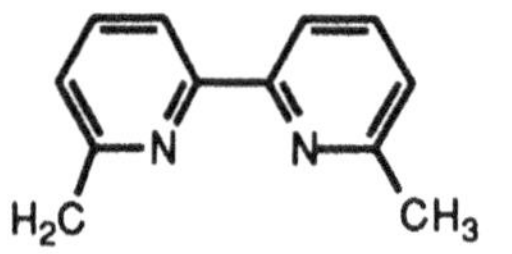

bpy

All the ligands, fully characterized, complexed Cu(I), but only **1**, **2** and **3** afforded TLC pure orange-red species, **6**, **7** and **8** respectively. Nevertheless, **7** and **8** displayed rather complicated ^{1}H-NMR spectra and FAB$^+$ or ES mass spectrometries were necessary to ascertain their structures. **6** presented as expected a L_2M stoichiometry while **7** and **8** where of the LM type. No polymeric structures were revealed in these cases, confirming the strong stabilizing effect of the calixarenic substituent on the complexation mode.

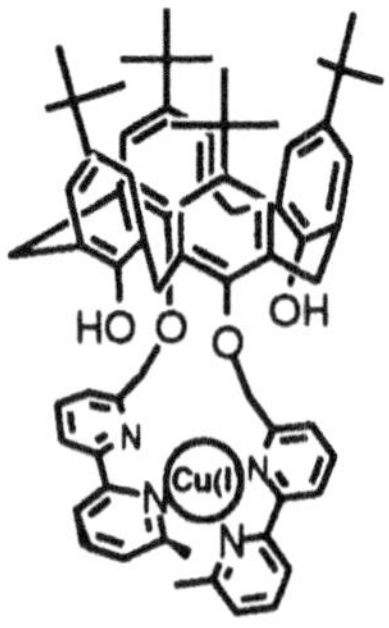

8

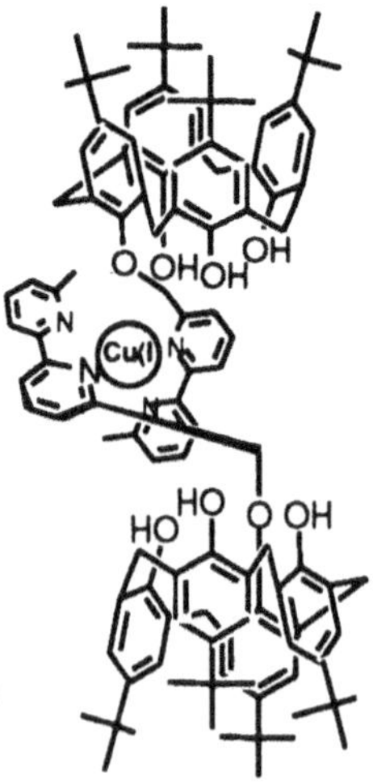

6

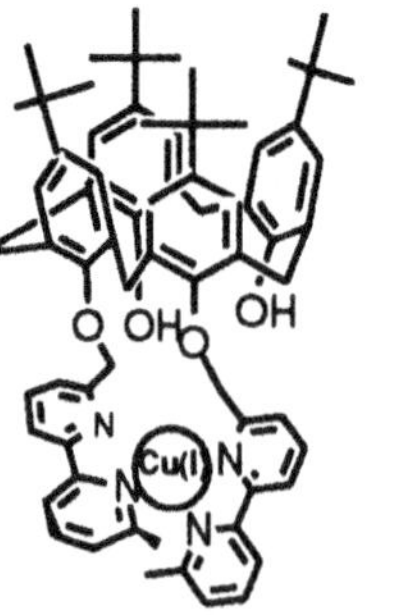

7

Replacing 2,2'-bipyridine by 2,2'-bithiazole afforded intrinsically fluorescent complexing podands which were studied as fluorescent probes [5] and as ligands for Cu(I). The quench of fluorescence by Cu(I) at concentrations of about 10^{-6} M was consistent with the stoichiometries L_2M and LM (0.5 eq. for the analogue of **1** and 1 eq. for the analogues of **2** and **3**). The analogue of **5** lost its fluorescence upon addition

of 2 eq. of Cu(I) salt, indicating that a LM_2 stoichiometry must be expected for this species and parent compounds. The corresponding complexes have been synthesized with the help of this technique and the suspected stoichiometries were verified [6].
This study has been developed with the introduction of other heterocyclic species, such as 2,9-dimethyl-1,10-phenantroline, 6,6'-dimethyl-2,2'-bipyrazine and 4-methyl-6,6'-dimethyl-2,2'-bipyridine.

Hetero-armed species: towards a versatile ligand.

All the ligands synthesized in the previous series, excepted the fully substituted species, present from one to three -OH groups which can be functionnalized selectively in a cascade mode with different heterocyclic or non-heterocyclic systems. Thus, the use of the same direct base-strength driven regioselective O-alkylation process allows us to design ligands with coordination properties adapted to our desired metallic species.

The ability to prepare calixarenes bearing different numbers and kinds of complexing heterocyclic subunits should lead to the synthesis of ligands able to complex efficiently both redox states of a metal during an oxydo-reduction process. To this end, we reported the synthesis and characterization of a C_{2v} symmetrically hetero-functionnalized *p*-(*tert*-butyl)calix[4]arene **9** bearing two opposed bipyridyl and two opposed picolyl substituants [7]. **9** was designed to allow a tetrahedral complexation mode for Cu(I) with only participation of the bipyridines, as well as a penta- or hexa-coordination for Cu(II) with participation of the bipyridines and one or two pyridine rings.

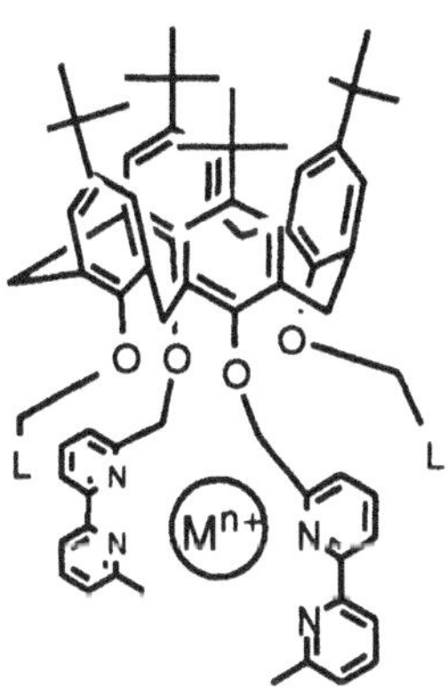

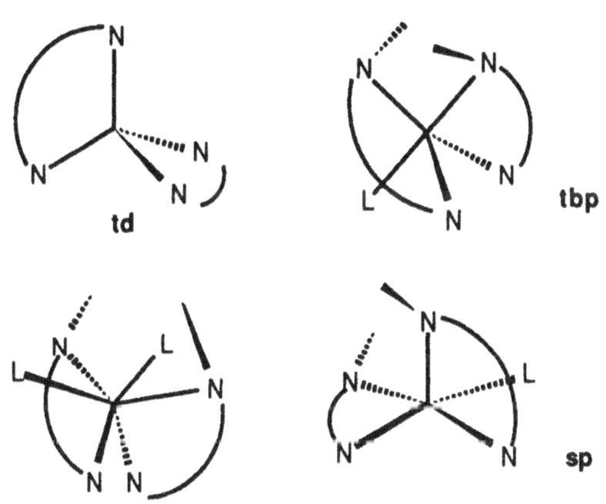

9 reacted as expected with Cu(I) salts to give the corresponding stable red-orange complexes **10** which were fully characterized by high resolution NMR techniques. It appeared clearly that the species containing the PF_6^- counter-anion was a racemate, the asymmetry being centered on the metal. Confirmation of the pro-helical structure of **10** was given by X-Ray crystal structure analysis.
In the presence of Cu(II) salts, **9** gave exothermically, in MeOH or MeCN, Cu(I) complexes of type **10**, indicating a strong reductive activity of the ligand in such

solvents. Nevertheless, some Cu(II) complex was formed in 1,2-dichloroethane, as demonstrated by ES-MS analysis of the reaction mixture.

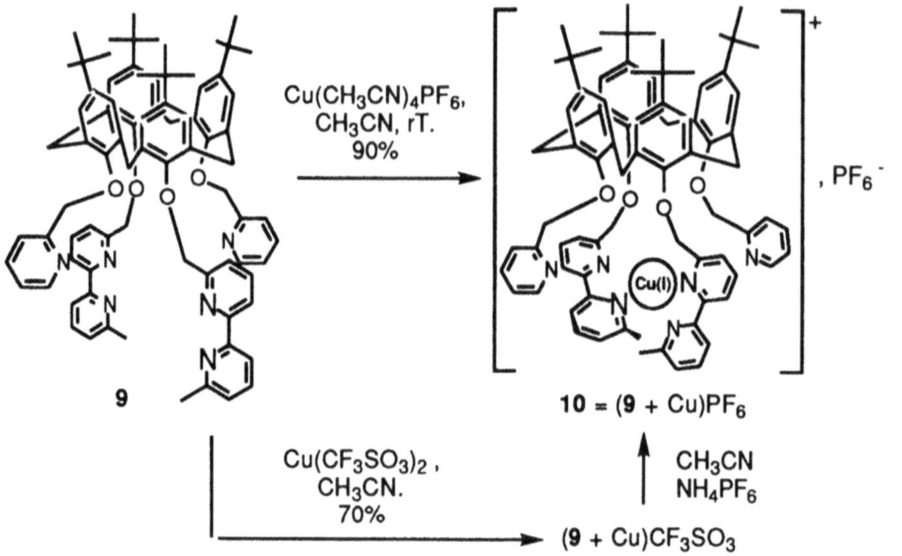

Replacing the bipyridyl units by 2,2'-bithiazole (podand **11**) allowed us to synthesize separatly both orange-red Cu(I) and light-green Cu(II) complexes, **12** and **13**. The formation of the former was not spontaneous and some dissociation occured at room temperature as demonstrated by ^{1}H-NMR in $CDCl_3$. According to the rather poor literature dedicated to bithiazole-containing complexes, the participation of sulfur atoms in the coordination of both cations is suspected but has to be improved, notably by NMR and X-Ray crystal analysis [8].

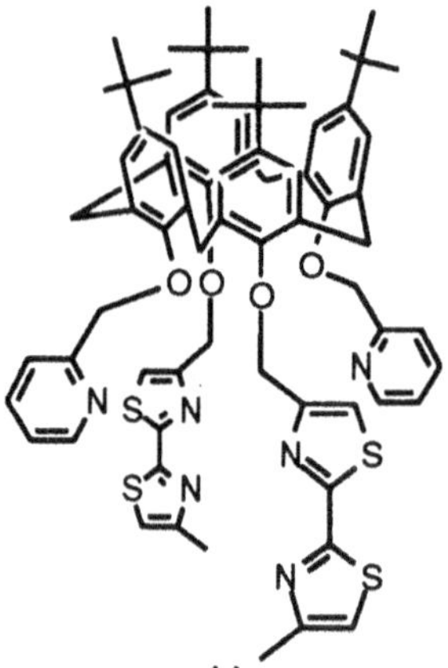

11

12 = (**11** + Cu)PF_6

13 = (**11** + Cu)CF_3SO_3

Nevertheless, the complexing behavior of **11** open an exciting synthetic and analytical field which could evolve towards the domain of homogeneous catalysis. In this sense, analogues of **10** incorporating 2,9-dimethyl-1,10-phenantroline, 6,6'-dimethyl-2,2'-

bipyrazine and 4-methyl-6,6'-dimethyl-2,2'bipyridine have been prepared and subjected to complexation studies.

Coordination at the upper rim.

Only a very little attention has been brought, until now, to the incorporation of heterocyclic chelates at the upper rim of calix[4]arene. Beer *et al.* [9] introduced notably bipyridyl subunits *via* amide bounds.The attachement of a pyridine moiety *via* an ethylenic linkage using a Grignard-like reaction *plus* dehydration process performed on a OH-protected formylcalix [4]arene has been recently described by Shinkai *et al.*.

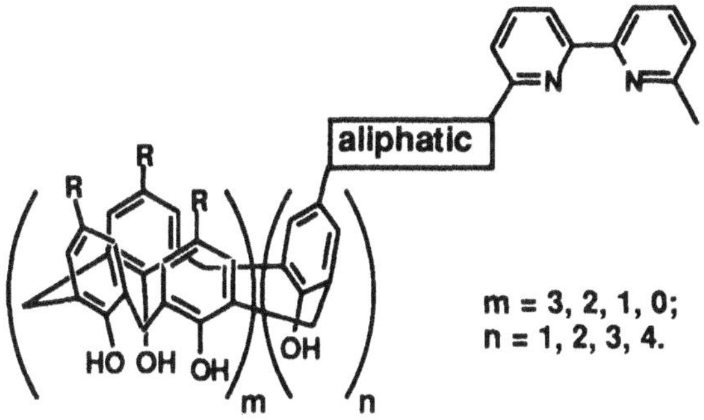

We thought that an easy way to introduce chelating properties *via* an aliphatic link at the upper rim of a calixarene platform was the Wittig reaction and we modelised our synthetic programm on the new *tris*(*p*-*tert*-butyl)calix[4]arene **14** [10].
We developed simultaneously two synthetic strategies based on this unprotected platform, one involving a calixarene phosphonium salt and various kinds of heterocyclic aldehydes, the other a calixarene aldehyde **15** and various kinds of heterocyclic phosphonium salts.

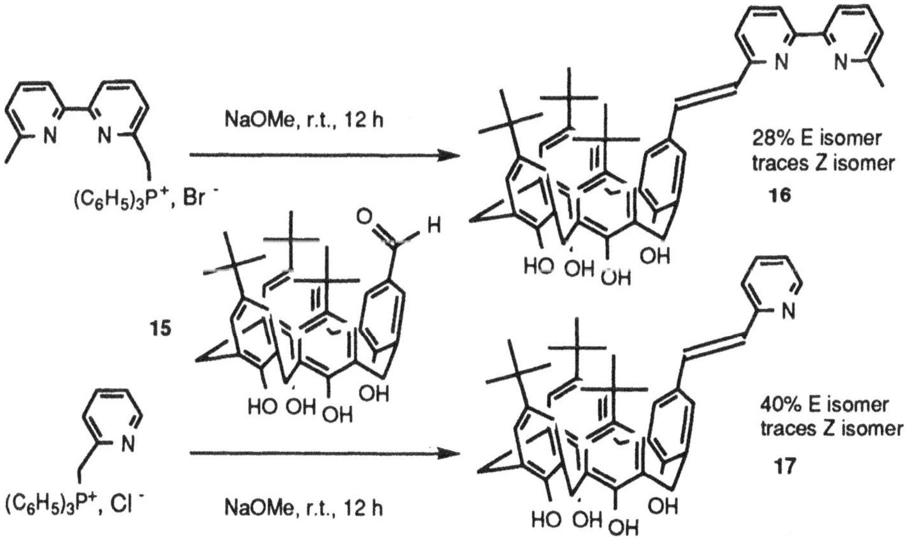

The latter was successful, affording the desired compounds [11]. Thus, reacting **15** with triphenylphosphonium salts of 6,6'-dimethyl-2,2'-bipyridine or α-picoline and MeONa in MeOH afforded in moderate yields the corresponding alkenes **16** and **17**. **16** reacted

fairly well with Cu(I) salt, affording an air stable red complex displaying, as expected from the results obtained with **1**, a L_2M stoichiometry (ES-MS: 1637.7; $[L_2Cu]^+$), but attempts to purify it by chromatography afforded decomposition products.
Catalytic reduction of the ethylenic linkage afforded readily the alkyl ligands which are actually under investigation.

In conclusion, the use of the preorganizing calix[4]arene platform seems to be a promising method for the controlled building of podand-like ligands which should present modulated complexing behavior [12]. Playing with the nature and the position of the chelating units could thus lead to tailor-made complexes displaying interesting properties, notably in the fields of cation transport and catalysis.

REFERENCES.

1. Gutsche, C. D. (1989) Calixarene, in J. F. Stoddart (ed), *Monographs in Supramolecular Chemistry Series*, Royal Society of Chemistry, Cambridge.
2. Beer, P. D.,Martin, J. P., Drew, M. G. B.: Calix[4]arene cryptand and new 1,3-bis-pyridyl, -bipyridyl and -alkylthioether calix[4]arenes designed to coordinate transition metal cations, *Tetrahedron* **45** (1992), 9917-9928.
3. Pappalardo, S., Giunta, L., Foti, M., Ferguson, G., Gallagher, J. F., Kaitner, B.: Functionalization of calix[4]arenes by alkylation with 2-(chloromethyl) pyridine hydrochloride, *J. Org. Chem.* **57** (1992), 2611-2624.
4. Regnouf-de-Vains, J. -B., Lamartine, R.: Synthesis and complexation properties of a new class of receptors based on a cone-configurated tetra-*p-tert*-butylcalix[4] arene and bipyridyl subunits, *Helv. Chim. Acta* **77** (1994),1817-1825.
5. Pellet-Rostaing, S., Regnouf-de-Vains, J. -B., Lamartine, R.: 2,2'-bithiazolyl-*p-tert*-butylcalix[4] arene podands. Synthesis and fluorescence properties, *Tetrahedron Lett.*, **37** (1995), 5889-5892.
6. Pellet-Rostaing, S., Regnouf-de-Vains, J. -B., Lamartine, R.,Meallier, P., Guittonneau, S., Synthesis and fluorescence properties of calixarene-bithiazolyl podands. Interaction with copper(I). ISMRI 9, Lyon, France, 7-12 Septembre 1996. Manuscript in preparation.
7. Regnouf-de-Vains, J. -B., Lamartine, R., Fenet, B., Bavoux, C., Thozet, A., Perrin, M.: Synthesis and reductive complexation properties of a cone configurated tetra-*p-tert*-butylcalix[4] arene incorporating bipyridyl and pyridine subunits. NMR characterization and crystal structure of its chiral copper(I) complex, *Helv. Chim. Acta* **78** (1995),1607-1619.
8. Pellet-Rostaing, S., Regnouf-de-Vains, J. -B., Lamartine, R., manuscript in preparation.
9. Beer, P. D.; Chen, Z.; Goulden, A. J.; Grieve, A.; Hesek, D.; Szemes, F.; Wear, T.: Anion recognition by novel ruthenium(II)bipyridyl calixarene receptor molecules, *J. Chem. Soc., Chem. Commun.* (1994), 1269-1271.
10. Berthalon, S., Regnouf-de-Vains, J. -B., Lamartine, R.: Selective Mono-deal-kylation of Tetra-*p-tert*-Butyl-Calix[4]arene at the Upper Rim, *Synthetic Commun.* **26** (1996), 3103-3108.
11. Regnouf-de-Vains, J. -B., Lamartine, R., Wittig Reaction on Calixarene Upper Rim. Access to Conjugated Bipyridyl and Pyridyl Podands, *Tetrahedron Lett.*, **37** (1996), 6311-6314 .
12. Some other recent contributions:
a) Grigg, R.; Holmes, J. M.; Jones, S. K.; Amilaprasadh Norbert, W. D.; *J. Chem. Soc., Chem. Commun.*, (1994), 185-187. b) Ulrich, G., Ziessel, R. *Tetrahedron Lett.*, **35** (1994), 6299-6302.

DEVELOPMENT OF RUTHENIUM PROBES DESIGNED TO BIND ENANTIO- AND STEREOSPECIFICALLY TO DNA

R. S. VAGG, K. A. VICKERY
School of Chemistry, Macquarie University, N.S.W. 2109, Australia.
and P. A. WILLIAMS
Faculty of Science and Technology, University of Western Sydney, Nepean, Kingswood, N.S.W. 2747, Australia.

1. Introduction

In the past decade, substitutionally-inert chiral ruthenium complexes have found valuable use as metalloprobes for the determination of causal factors governing small-molecule recognition by DNA. Considerable work has derived from the initial reports of enantioselectivity observed between the Δ- and Λ-$[Ru(phen)_3]^{2+}$ ions and DNA [1-3], where it was concluded that the right-handed (Δ) propeller form binds preferentially. The binding mode originally was depicted by one of the *phen* ligands stacking, or intercalating into DNA, with the remaining two making van der Waals contacts with the surface of the helical groove [2]. It was proposed that this enantiopreference arises from steric clashes that exist between the two external *phen* blades of the Λ-isomer and the ribo-phosphate backbone of the duplex. However, the structural basis for these effects remained uncertain and in order to investigate this, subsequent studies have led to the use of analogues of general form $[Ru(L)_2(L')]^{2+}$, where *L* is 1,10-phenanthroline (*phen*) or 2,2'-bipyridine (*bipy*) and *L'* is a planar bidentate with an enhanced aromatic surface area. The latter component serves as the intercalating chromophore and hence determines overall molecular binding affinity.

2. Design Rationalé

We have adapted a molecular design approach based on synthetic coordination chemistry, in which the probe molecules are tailored to enhance this enantioselectivity function. We reason that, in octahedral complexes, any resulting enantioselectivity would not depend on the remaining four coordination sites being occupied by two bidentates. Rather, selectivity could be better achieved by substituting the four sites with a tetradentate fragment which was not only incapable of intercalation activity, but which could offer selectivity through its design. In doing so, ambiguity over which bidentate moiety is responsible for intercalation is removed. Hence, the replacement of the two secondary *L* ligands in the tris-bidentate complexes by a bulky and inflexible

A. W. Coleman (ed.), Molecular Recognition and Inclusion, 239–244.
© 1998 *Kluwer Academic Publishers. Printed in the Netherlands.*

optically-active tetradentate, exemplified by *picchxnMe*$_2$ [5], has resulted in the synthesis of a series of novel cations of the type [Ru(tetradentate)(L')]$^{2+}$. Such complexes form the basis of the study reported on here.

2.1 STEREOSELECTIVITY

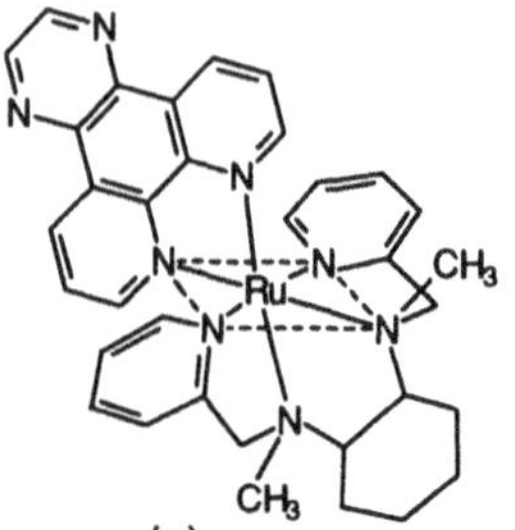

(I)

The substitution of two bidentates by one tetradentate in an octahedral sphere provides for two alternative forms of stereoisomerism unavailable to chelates which are solely bidentate-based. An example of the dissymmetric *cis*-α form, with C_2 symmetry, is represented in (I). The alternative asymmetric *cis*-β isomer (C_1 symmetry) which may result is shown in (II). We have synthesised a variety of complexes of Ru(II) of this general type, and have resolved the diastereoisomeric products into their pure α and β forms. Each would be expected to interact differently with DNA. For the α complex, the two pyridyl groups would make steric contact with the groove above and below the site of intercalation. A model illustrating this interaction is shown in Figure 1a.

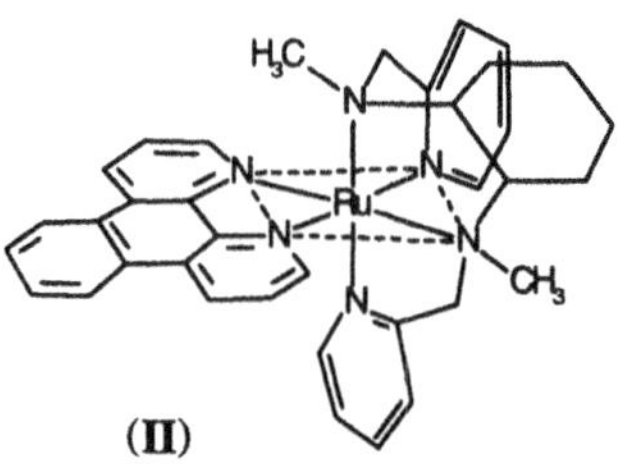

(II)

For the β complex, however, the *N*-methyl group in the fold of the tetradentate is directed above the intercalating bidentate. As a result, an unfavourable steric contact between this group and the duplex is seen (Fig. 1b). A comparison of the two models shows that at limiting van der Waals contact distances this complex would be excluded by approximately 0.7 Å further than the α complex from the centre of the intercalation site. Steric factors such as this

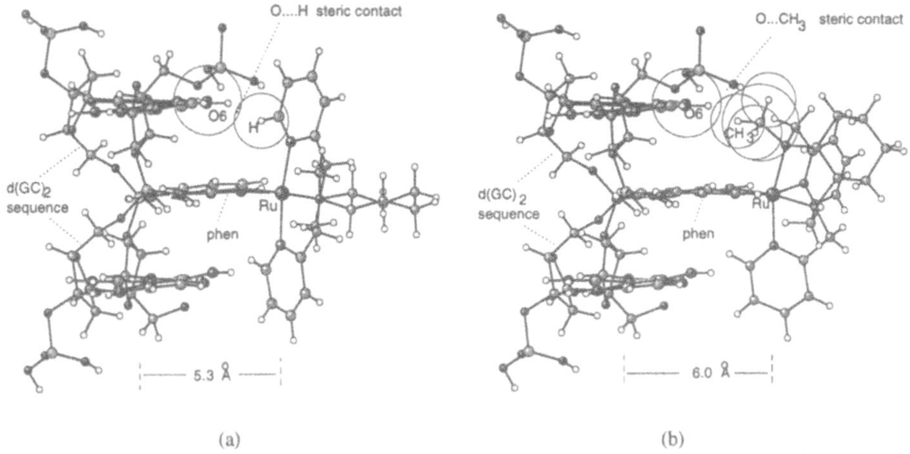

(a) (b)

Figure 1: Models of (a) Δ-α- and (b) Δ-β-[Ru(*RR*-picchxnMe$_2$)(phen)]$^{2+}$ ions intercalated into a [d(G_pC)]$_2$ site.

should contribute significantly to the expected discriminatory functions of these stereoisomers.

2.2 ENANTIOSELECTIVITY

The enantiomeric form of the metalloprobe molecule is able to be determined by the choice of the inherent chirality of the tetradentate precursor. The *picchxnMe*$_2$ ligand, for example, is based on optically active 1,2-diaminocyclohexane, with both N atoms being di-substituted to give rise to diastereoisomeric *R,R*- and *S,S*-tetradentate forms. When the *R,R*- hand is used as the synthetic precursor, Δ-*cis*-α chelates form exclusively with metal ions such as Co(III) [4] and Ni(II) [6], whereas both Δ-*cis*-α and Δ-*cis*-β isomers are isolated with Ru(II) [5, 6]. Use of the *S,S*- hand results in the corresponding Λ enantiomers. An example of an effective metalloprobe, the Λ-*cis*-α-[Ru(*SS*-picchxnMe$_2$)(dpq)]$^{2+}$ complex cation, is represented in (I). Here the *dpq* bidentate ligand [5] serves as the intercalating component.

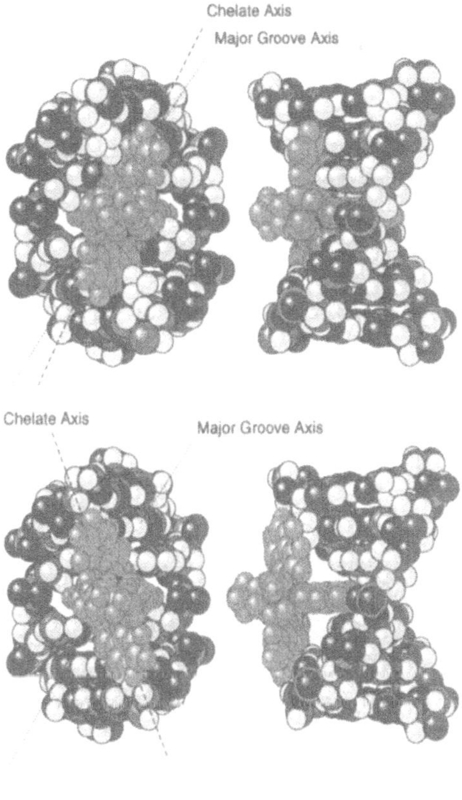

Figure 2. Space-filled molecular drawings, in orthographic projection, demonstrating the enantioselective intercalation of (top) Δ-*cis*-α- and (bottom) Λ-*cis*-α-[Ru(tetradentate)(bidentate)]$^{2+}$ ions into a DNA hexamer duplex fragment.

The enantioselective intercalation with an oligonucleotide of both Δ and Λ forms of a chelate of this type can be demonstrated by molecular modelling. In Figure 2 (top) it may be seen that the non-intercalated tetradentate component of a Δ-*cis*-α chelate would be disposed along the groove 'axes' of *B*-DNA. This behaviour contrasts with the Λ-*cis*-α isomer (bottom), where the chelate bulk lies opposed to the groove direction. Thus, the Δ hand provides a complementary fit for the intercalation site of the right-handed duplex, and the degree of enantioselectivity obtained thereby may be tuned by steric bulk incorporated into the design of the tetradentate.

2.3 INTERCALATING CHROMOPHORE DESIGN

In these compounds the bidentate *L'* operates as the intercalating component of the complex, with its surface area allowing DNA binding via stabilising electronic π-π and dipole interactions with the nucleobases. We have varied *L'* in order to optimise the area available for aromatic overlap and thereby to govern the overall binding abilities of these metalloprobes. For comparative purposes, *phen* has been included in the study to correspond with the simpler tris-bidentate analogues [1-3]. However, modelling experiments [5] indicate that the optimum area of overlap could best be achieved by extending the surface area of this ligand. Other bidentates which we have used, illustrated in Figure 3, include dipyrido[3,2-d:2'3-f]quinoxaline, (*dpq*) [5] and the fused-ring diimines 9,10-phenanthrenequinonediimine (*phdi*) and 2,3-naphthoquinonediimine (*npdi*).

Molecular models allow a comparison of the common areas of aromatic overlap between these bidentates and the nucleobases on intercalation. These are demonstrated by shading in Figure 3 for insertion of the probe molecules into a $[d(CpG)]_2$ duplex. In each of these diagrams the two molecular components are positioned at limiting van der Waals contact distances. The significant increase in intercalative binding affinity that likely would result from extension of the surface area of the bidentate from *phen* to *dpq* is implicit in these models. They suggest also that a further extension of this area to that of dipyrido[3,2-a:2'3-c]phenanzine (*dppz*) [5, 7] would not result in a significantly increased overlap area, and moreover that this could be achieved without the need for the tetradentate to make limiting steric contacts. This likely would result in a decrease in the degree of selectivity in the interaction. These models indicate that significant binding affinities might be expected also for metalloprobes based on the *npdi* and *phdi* bidentate chromophores.

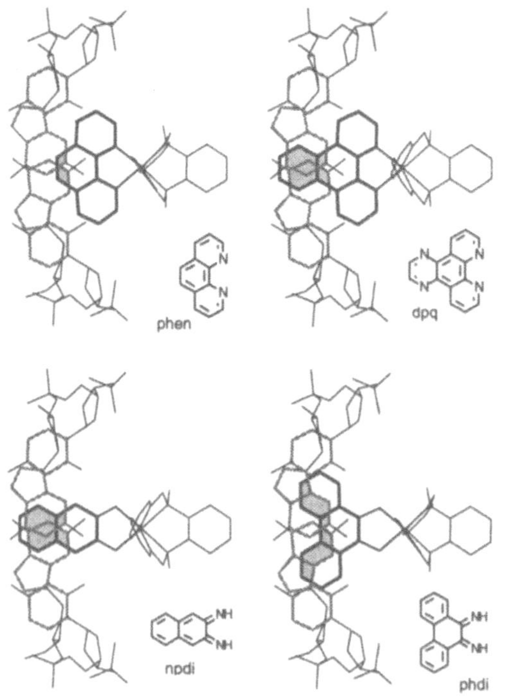

Figure 3. Models illustrating relative overlap of aromatic areas for Δ-*cis*-α-$[Ru(RR\text{-picchxnMe}_2)(\text{bidentate})]^{2+}$ ions intercalating in a $[d(GpC)]_2$ duplex sequence.

3. Experimental Studies

The chiral recognition functions between DNA (both natural calf-thymus and synthetic homopolymers and oligomers) and the four isomeric forms of each series, where *L'* is *phen*, *dpq* and *phdi*, have been investigated using ^{1}H NMR and electronic

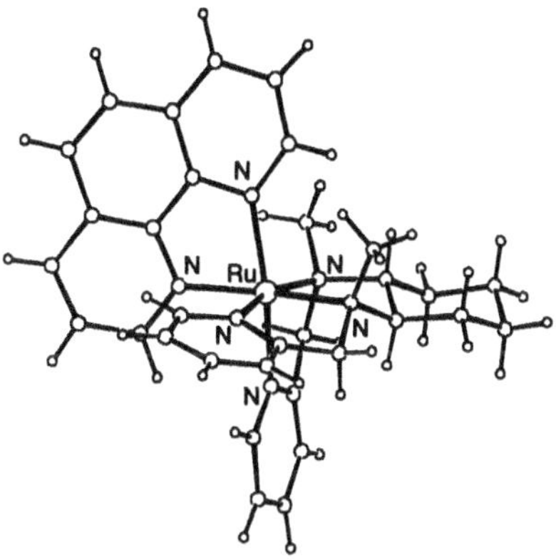

Figure 4. The crystal structure of the Δ-*cis*-β-[Ru(*RR*-picchxnMe$_2$)(phen)]$^{2+}$ cation.

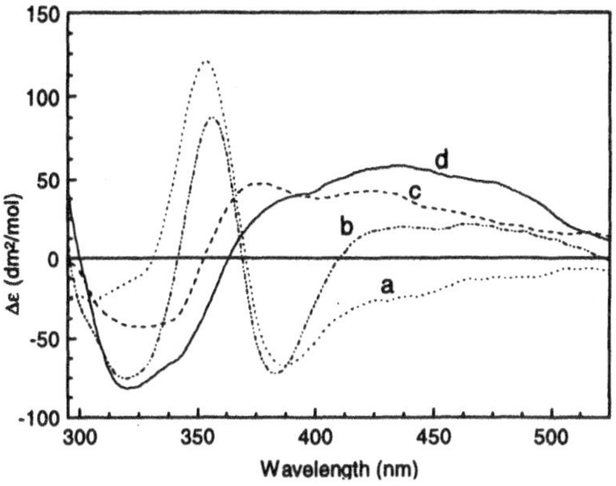

Figure 5. Circular dichroism spectra of [Ru(*RR*-picchxnMe$_2$)-(dpq)]$^{2+}$ cations. (a) Δ-*cis*-α; (b) Δ-*cis*-α + calf-thymus DNA; (c) Δ-*cis*-β; (d) Δ-*cis*-β + calf-thymus DNA.

and circular dichoism spectroscopic methods. These techniques provide structural information through their sensitivity to electronic interactions established between the metallointercalators and the host nucleic acid helices.

A good example of the molecular structures adopted by these metalloprobes is that of the Δ-*cis*-β cation derived from *phen*, determined by single-crystal x-ray analysis, which is shown in Figure 4. This provides proof of the Δ enantioselectivity adopted by the *RR*-picchxnMe$_2$ ligand in its coordination behaviour with Ru(II). The circular dichroism spectrum of its *dpq* analogue in the visible range is given in Figure 5, together with that of the Δ-*cis*-α isomer. The significant affects of the addition of calf-thymus DNA (10 base-pairs per probe molecule) on their charge-transfer-derived electronic transitions are illustrated here. Such affects also are mirrored by hypochromicity in the visible absorption spectra of the probes.

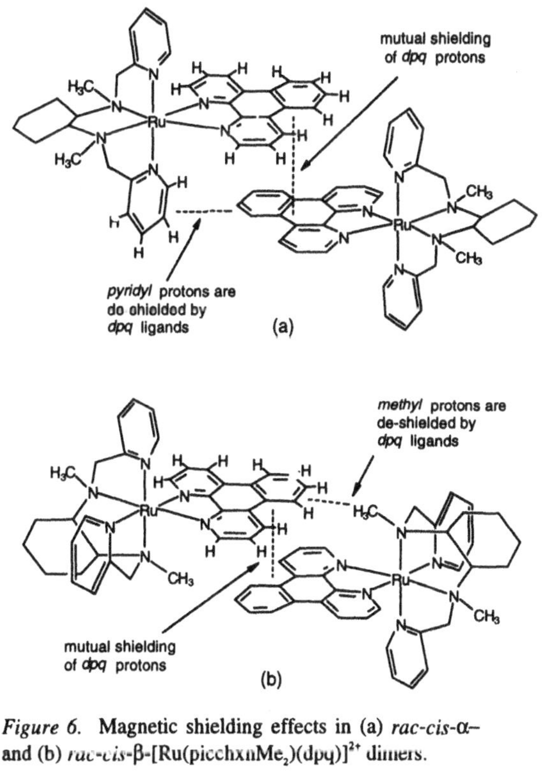

Figure 6. Magnetic shielding effects in (a) *rac-cis*-α– and (b) *rac-cis*-β-[Ru(picchxnMe$_2$)(dpq)]$^{2+}$ dimers.

In support of these studies an *in vitro* bacterial DNA system also has been utilised. Conventionally employed by regulatory bodies for the detection of mutagens, we have used the Ames *Salmonella*

Mutagenesis Test to detect and rank these intercalating complexes. Different frameshift-detecting *S. typhimurium* strains were used to investigate both the enantio- and stereoselectivity of these complexes for DNA [9]. For each of the series of complexes comparisons were able to be made between Δ *vs* Λ and of α *vs* β isomers. Consistent with our structural hypothesis, the Δ-*cis*-α isomers of each series demonstrate the highest activities in these test systems, indicating their higher intercalation affinities than for those of the respective Λ or β forms. Further, by comparing activities within the set of Δ-*cis*-α isomers a relative ranking of implied intercalation ability was obtained for the varying bidentates. The results mirror those obtained from the spectroscopic and modelling studies.

The propensity of these metalloprobes for intermolecular π-π hydrophobic interactions is demonstrated by their ^{1}H NMR behaviour in different solvents. Each of the various isomeric forms of the *dpq* and *phdi* chelates show evidence of self-association in aqueous solutions to form dimeric aggregates, similar to the simpler $[Ru(phen)_3]^{2+}$ cations [8]. This is evidenced by the concentration-dependence of proton chemical shifts, which are consistent with the dimeric structures illustrated in Figure 6 for the racemic *cis*-α- and *cis*-β-$[Ru(picchxnMe_2)(dpq)]^{2+}$ species. Up-field shifts in the aromatic protons of the bidentate fragments indicate their mutual magnetic shielding due to ring current effects, accompanied by de-shielding of the in-plane pyridyl and/or methyl protons. Similar effects are observed when the metalloprobes are allowed to interact with synthetic DNA fragments.

4. References

1. Yamagishi, A. (1984) Electric Dichroism evidence for stereospecific binding of optically active tris chelated complexs to DNA, *J. Phys. Chem.*, , **88,** 5709-5713.
2. Barton, J. (1984) Tris(phenanthroline)ruthenium(II): Stereoselectivity in binding to DNA, *J. Am. Chem. Soc.*, **106**, 2172-2716.
3. Hiort, C., Nordén, B. and Roger A. (1990) Enantiopreferential DNA binding of $[Ru^{II}(1,10\text{-phenanthroline})_3]^{2+}$ studied with linear and circular dichroism, *J. Am. Chem. Soc.*, **112**, 1971-1982.
4. Fenton, R.R., Stephens, F.S., Vagg, R.S. and Williams, P.A. (1993) Chiral Metal Complexes 34. Stereospecific *cis*-α coordination to cobalt(III) by the new tetradentate ligand *N,N'*-dimethyl-*N,N'*-di(2-picolyl)-diaminocyclohexane, *Inorg. Chim. Acta*, **182**, 67-75; Fenton, R.R., Stephens, F.S., Vagg, R.S. and Williams, P.A. (1995) Chiral Metal Complexes 44. Enantiomeric discrimination in ternary cobalt(III) complexes of *N,N'*-dimethyl-*N,N'*-di(2-picolyl)-1*S*,2*S*-diaminocyclohexane and α-amino acids; including the crystal structure of the *S*-prolinato complex, *Inorg. Chim. Acta*, **236**, 109-11.
5. Aldrich-Wright, J R., Greguric, I., Vagg, R.S., Vickery, K., and Williams, P.A. (1995) Development of DNA-immobilised chromatographic stationary phases for optical resolution and DNA-affinity comparison of metal complexes, *J. Chromatog. A*, **718**, 436-443.
6. Vagg, R.S., Vickery, K.A., and Williams, P.A., unpublished results.
7. Hartshorn, R.M. and Barton, J.K. (1992) Novel Dipyridophenazine Complexes of Ruthenium(II): Exploring Luminescent Reporters of DNA, *J. Am. Chem. Soc.*, **114**, 5919-5925.
8. Masuda, Y and Yamatera, H. (1984) Proton chemical shift study of the hydrophobic interaction between tris(1,10-phenanthroline)ruthenium(II) ions in aqueous solution, *Bull. Chem. Soc. Japan*, **57**, 58-62.
9. Vickery, K.A. (1995) Use of the Ames Test to provide structural information on DNA binding, *Abstracts of the 12th Conference of the Medicinal and Agricultural Chemical Division of the RACI and 14th Annual Conference of the Molecular Graphics and Modelling Society*, Cairns, Australia, A-27.

AFFINITY FOR BOTH 5-HT$_{1A}$- AND D$_1$-RECEPTORS AND ANXIOLYTIC ACTIVITY OF N-(ARYLPIPERAZINYLALKYL)-PHTHALIMIDES

S.A. ANDRONATI, T.A.VORONINA, V.M. SAVA,
G.M. MOLODAVKIN, S.YU. MAKAN, S.G. SOBOLEVA

A. V. Bogatsky Physico-Chemical Institute of the National Academy of Sciences of Ukraine
86 Lustdorfskaya doroga, 270080, Odessa, Ukraine
Institute of Pharmacology of Russian Academy of Medicinal Sciences
8 Baltiyskaya, 125315 Moscow, Russia

The aryl(hetaryl)piperazine derivatives form numerous group of anxiolytics of novel generation. They influence on the neuromediator processes of CNS through the binding with the serotonin and dopamine brain receptors. The analysis of buspiron (I) and of some its analogs influence on neuronal and behavioural reactions of animals allow to conclude that at least two components: serotoninomimetic and dopaminolytic are responsible for the demonstration of anxiolytic activity [1,2]. More over, analogs of buspiron: hepiron (II) and ipsapiron (III), which do not interact with dopaminergic brain systems, demonstrate more low anxiolytic activity in comparison with buspiron [2,3].

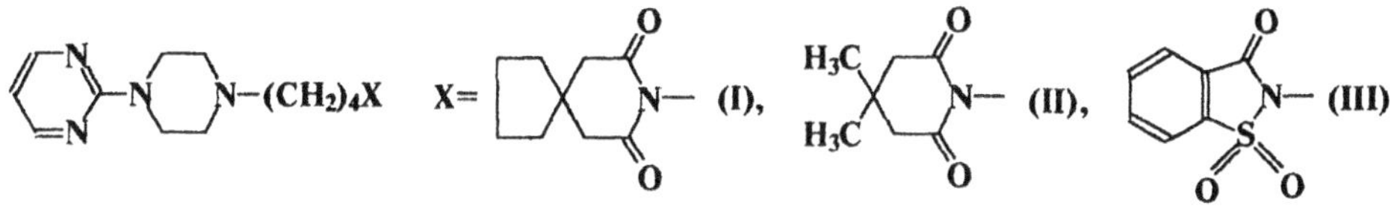

(I), (II), (III)

The plurality of subtypes and functional heterogeneity of the dopamine receptors

A. W. Coleman (ed.), Molecular Recognition and Inclusion, 245–249.
© 1998 *Kluwer Academic Publishers. Printed in the Netherlands.*

cause the necessity of concrete evaluation of relationships between dopaminergic and anxiolytic activity of buspiron and its analogs. Besides, study of the anxiolytics influence on the dopaminergic system is the important element of the direct search for anxioselective agents.

The purpose of the given work are the investigation of N-(arylpiperazinylalkyl)phthalimides affinity for 5-HT_{1A}- and for D_1-receptors as well as the evaluation their anxiolytic properties.

The series of compounds IV was synthesized for this purpose

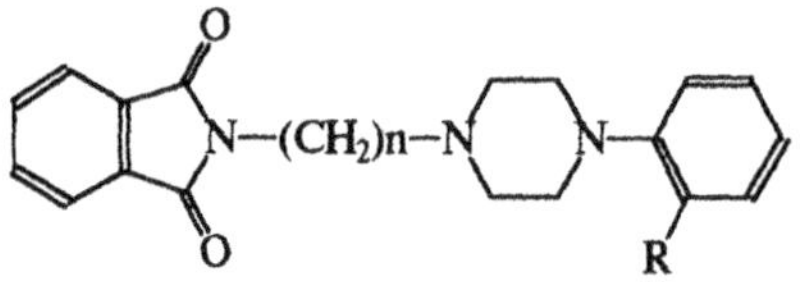

R = H, CH_3, Cl;
n = 1 - 6.

IV

Affinity for D_1 dopamine and 5-HT_{1A} serotonin receptors of compounds IV was determined by their ability to inhibite the specific binding [^{3}H]SCH-23390 [4] and [^{3}H]8-OH-DPAT [5] with the synaptic membranes of the rats brain (male rats 200±20g), respectively. Radioligands [^{3}H]SCH-23390 (70 Ci/mmol) and [^{3}H]8-OH-DPAT (73 Ci/mmol) were received from the Radium Institute (St. Petersburg, Russia).

The anxiolytic activity of substances was investigated in the conflict situation simulating the collision of drinking and defensive reflexes (the number of punished approaches) [6]. Substances were administrated intraperitoneally (10 mg/kg) 20 min before experiment.

The 5-HT_{1A}- and D_1-receptors recognition of substances IV are characterized by values of K_i = 5.2-14.0 nM and K_i = 95-1600 nM values, respectively (Table). Compounds 1 and 2 have the low affinity for both types of receptors (approximately on the same level). Increasing of the polymethylene chain length and inclusion of radicals (Cl or CH_3) in *o*-position of phenyl ring lead to increase of compounds affinity for D_1-receptors. Changes of compounds affinity for 5-HT_{1A}-receptors are more complicated. At the increase of the bridge length from n = 1 to n = 4 and inclusion of Cl atom in *o*-position of phenyl nuclea the sharp increase of affinity appears. At the subsequent increase of number of methylene chains to 5 or 6 some decrease of affinity of compounds IV for $5\text{-}HT_{1A}$-receptors has been observed.More higher affinity of compound 5 with Cl atom in o-position of the phenyl nucleus as compared with *o*-tolyl analog 6 attracts attention.

All compounds IV demonstrated the expressed anxiolytic properties (Table). The compounds with mono- and bis-methylene bridge (1 and 2) were less active as compared with buspiron, when compounds with tetra-, penta- and hexamethylene bridge (4-7) were similar to buspiron (3-4) or had higher activity (4-7). The data in Table testify that increase of polymethylene bridge and inclusion of substituent (Cl or CH_3) in *o*-position of phenyl ring lead to increase of anxiolytic activity of arylpiperazinylalkylphthalimides IV.

TABLE. The anxiolytic activity of N-(arylpiperazinylalkyl)phthalimides and their affinity for $5\text{-}HT_{1A}$- and for D_1-receptors.

Substances	R	*n*	The number of punished approaches	-log K_i	
				$5\text{-}HT_{1A}$	D_1
1	H	1	32±13	4.7	4.5
2	H	2	35±10	5.0	4.7
3	H	4	52±11	8.0	5.8
4	Cl	4	54±11	8.29	6.05
5	Cl	5	68±17	8.17	6.58
6	CH_3	5	65±20	7.89	6.44
7	Cl	6	67±15	7.86	7.03
Buspiron			53±17	7.83	
Control			11±5		

The cross-correlation of affinity for $5\text{-}HT_{1A}$- and for D_1-receptors and level of anxiolytic activity of compounds IV (Figure) testifies that affinity of compounds for D_1-receptors correlates with their anxiolytic activity in more degree than for $5\text{-}HT_{1A}$-receptors.

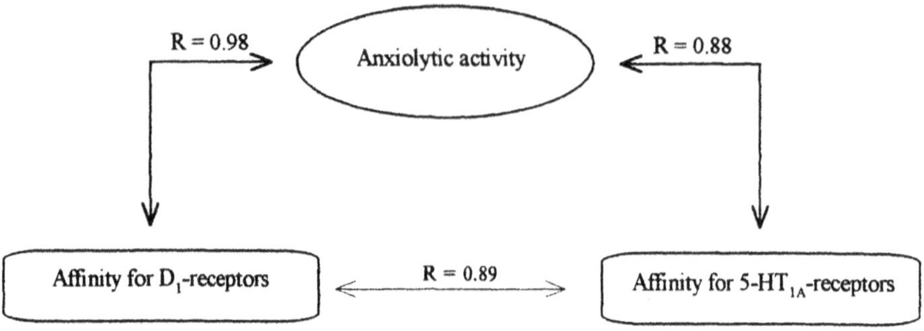

Figure. The correlation of anxiolytic activity of N-(arylpiperazinylalkyl)phthalimides and their affinity for D_1 and for 5-HT_{1A}-receptors.

The influence of the substituents on the affinity of N-(arylpiperazinylalkyl)phthalimides for D_1 receptors has been summarized within the Free-Wilson method [7]. For each compound of the series the affinity values A_i (used in the logarythmic scale) are expressed as the sum of the affinity contributions a_{ij} for the substituents R_i and $(CH_2)_n$ chain in the position j, referring to the overal average A_o:

$$A_i = A_0 + \Sigma\, a_{ij} \cdot X_{ij} \tag{1}$$

In equation (1), X_{ij} has value of 1 when the substituent R_i or $(CH_2)_n$ chain is present in the position j, otherwise its value is O.

$$\log Ki = -5.88 - 1.98 \cdot R_H - 2.42 \cdot R_{Cl} - 1.98 \cdot R_{CH_3} + 1.11 \cdot (CH_2)_6$$
$$+ 1.54 \cdot (CH_2)_5 + 2.42 \cdot (CH_2)_4 + 2.85 \cdot (CH_2)_2 + 3.28 \cdot (CH_2) \tag{2}$$
$$r = 0.972 \qquad s = 0.6$$

As it is seen from the equation (2), the change of substituents (H, Cl or CH_3) in phenyl ring influences more weakly on the affinity of compounds of this series than the change of the length of the polymethylene chain . Thus, the increase of chain length from n=1 to n=6 causes the increase of affinity nearly in 3 times.

Finally, these facts could be considered as evidence that D_1-receptors as well as 5-HT_{1A}- and D_2-receptors are involved in realization of anxiolytic effect of arylpiperazinylalkylphthalimides.

Acknowledgement - This work was supported by INTAS project № 94 - 839.

REFERENCES

1. New, J.S. (1990) The discovery and the development of Buspirone: A new approach to the treatment of Anxiety, *Medicinal Research Reviews* **10**, 286-326.
2. Abramets, I.I., Obraztsova, O.G., Samoylovich, J.M. and Kharin N.A. (1992) Serotonin- and dopaminergic mechanisms in action of the 1-pyrimidinylpiperazine derivatives, *Eksperimentalnaya i klinicheskaya farmakologiya* **55**, 8-11.
3. Dompert, W.U., Glaser, T. and Trafer, J. (1985) ^{3}H-TVX Q 7821: identification of 5-HT_1 binding sites as target for a novel putative anxiolytic., *Naunym-Shmidebergs Arch Pharmacol.* **328**, 467-470.
4. Kathy, M., O'Beyle and Waddington, J.L (1992) Agonist and antagonist interaction with D_1 dopamine receptors agonist - induced masking of D_1 receptors depends on intrinsic activity, *Neuropharmacology* **2**, 177-183.
5. Andronati, S.A., Varava, V.M., Soboleva, S.G., et al. (1992) 1-Phenyl- 4-[4-(N-naphthalimido)butyl]piperazin novel ligand of serotonin receptors, *Dokl. Akad. Nauk.* **327**, 341-344.
6. Voronina, T.A., Vikhliaev,Yu.I., Nerobkova L.N., et al.(1982) Characteristic of the pharmacologic properties of phenazepam, in A.V.Bogatsky (eds) *Phenazepam*, Naukova Dumka Publishers, Kyiv, p.p.146-147.
7. Free, S.M. and Wilson, J.M. (1964) A Mathematical contribution to structure-activity studies, *J. Med. Chem.* **7**, 395-399.

PHYSICO CHEMICAL STUDIES OF THE ADSORPTION PROCESS BETWEEN AMINATED SILICA WAFERS AND OLIGONUCLEOTIDES.

V.BALLADUR, B.MANDRAND, A.THERETZ
Unité Mixte CNRS/ bioMérieux.
Ecole Normale Supérieure de Lyon, 46, Allée d'Italie
69364 Lyon Cedex 07, FRANCE.
Tel: (33) 04-72-72-83-62
Fax: (33) 04-72-72-85-33
E-Mail: Veronique.Balladur@ens-bma.cnrs.fr

Abstract.

The adsorption of nucleic acid probes (NAP) onto aminated silica wafers has been investigated. This study aims to determine the driving forces of the binding reaction. Aminopropylsilane modified silica surfaces were characterized by contact angle measurements, XPS, ellipsometry and AFM. The adsorbed amount of oligonucleotides was estimated by radioactive counting. Adsorption kinetic of the process revealed a high affinity between NAP and aminated surfaces despite a low yield of adsorption. Varying adsorption conditions, namely oligonucleotides initial concentration, pH and ionic strength, we have demonstrated that two driving forces are controlling the adsorption process: electrostatic interactions and dehydration effect.

1. Introduction.

Nucleic acid probes (NAP) are extensively used for diagnosis, DNA purification and molecular recognition methods. Indeed, the understanding of adsorption process of these biological compounds onto surfaces is important to achieve efficiently their immobilization. Previous studies reported the adsorption of DNA onto mica [1], glass particles [2] or silica [3, 4]. On bare silica, it has been demonstrated that adsorption was favoured in a solvent of high chaotropic reagents concentration. In the present work, 20-mers oligonucleotides adsorption will be carried out on aminated modified silica wafers. The knowledge of this polyelectrolyte interactions will allow to perform, in the near futur, the covalent attachment study of NAP.

A. W. Coleman (ed.), Molecular Recognition and Inclusion, 251–254.

© 1998 *Kluwer Academic Publishers. Printed in the Netherlands.*

2. Materials and methods

Adsorption buffers were KH_2PO_4/ $NaHPO_4$ phosphate buffers. Using variable ratio of these coumpounds allowed us to set desired pH values. Ionic strength was controlled by adding NaCl in these solutions.

20-mers oligonucleotides were synthetized with an Applied Biosystems DNA/RNA Syntheziser, purified and characterized by reverse-phase HPLC (Kontron Instruments), and quantified by measuring UV adsorption at 260 nm.

Preparation of silica wafers. 64mm^2 silica wafers (Microsens, Neuchatel, CH) were silanized using aminopropylmonoethoxydimethylsilane (APMDS, $NH_2(CH_2)_3Si(CH_3)_2OEt$). Silanation process has been described elsewhere [5]. After 1 hour cleaning step in sulfochromic acid to remove organic contaminants, wafers were silanized (2% v/v in anhydrous toluene) for at least 4 hours at room temperature. Silica wafers were characterized using four kinds of analyses: AFM, XPS, ellipsometry and wettability studies.

Radiolabelled NAP adsorption and characterization of coated surfaces. A ^{32}P is introduced at the oligonucleotides 3'-end using the Terminal Transferase Kit from Boehringer. A mix of radiolabelled and non labelled NAP was diluted in adsorption buffer to give a standard concentration of 15 pmol/µl. A 40 µl drop of this solution was placed on the aminated wafer. Adsorption occured for 2 hours at 37°C. The adsorbed amount of NAP was estimated by counting the wafer radioactive emission.

3. Results and discussion.

3.1. CHARACTERIZATION OF SILICA WAFERS.

TABLE 1. Contact angle, roughness, ellipsometry and XPS measurements on silica wafers.

Surface	θmoy	Rrms (nm)	e (Å)	theo.%N	%N	%Si	%C	%O
Bare Silica	<10°	0,45	410	/	/	33	1,9	65
APMDS	64,5	0,8	420	0,6	0,6	31	5	63

Taking into account that theorical coverage of silanols on a silica wafer is estimated to be ca.~8×10^{-10} mol/cm^2 [6] and that theorical film thickness, *e*, after silanation process with APMDS is 10 Å [7], we were able to calculate theorical value of %N and %C after silanation and concluded that the silanation process was quite good and provide us smooth reactive surfaces. Furthermore, roughness values (Rrms) obtained by AFM are in good agreement with those obtained in previous publication [8].

3.2. ADSORPTION KINETICS OF NAP.

The adsorption reaction occured in adsorption buffer at pH=7 and 0,5 M ionic strength in standard reaction conditions. As shown on fig.1, the maximum adsorbed amount of NAP is obtained quickly. A 1 hour reaction time is enough to obtain the maximum oligonucleotide adsorption rate. This result revealed quite a good affinity of oligonucleotides towards surfaces under such conditions.

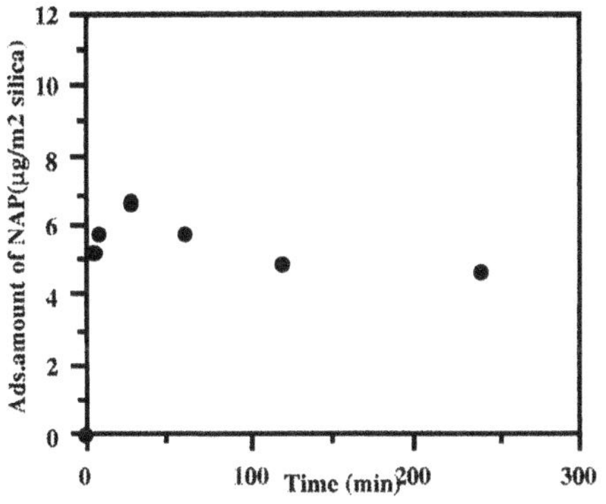

FIG.1: Adsorption kinetics of NAP (pH= 7, ionic strength 0,5 M).

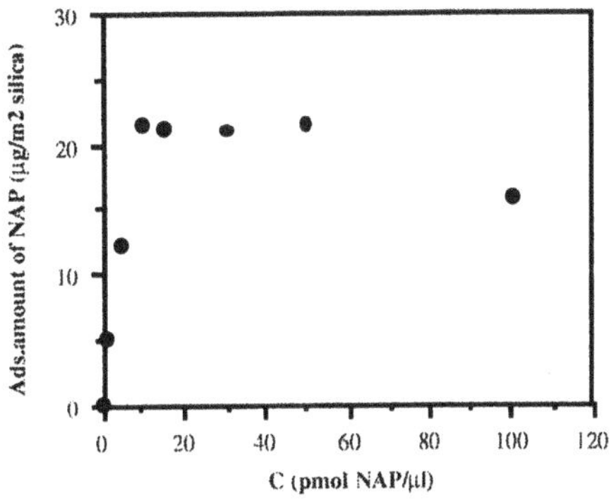

FIG.2: Adsorption isotherms of NAP (pH= 7, ionic strength 0,5M).

3.3. ADSORPTION ISOTHERMS OF NAP.

We have conducted the adsorption in standard conditions (pH=7, ionic strength 0,5 M) varying the initial bulk oligonucleotides concentration in the aim to define its effect on the adsorption process. The minimal value to reach the plateau of adsorption was found to be 20 pmol/µl. Despite the good affinity between NAP and aminated surfaces, we could notice that the adsorption yield is rather low.

3.4. EFFECT OF pH ON THE ADSORPTION OF NAP.

Varying the pH values of the bulk solution, we studied the charge effect on the adsorption process. The NAP are negatively charged polyelectrolytes in the pH range of the study. Aliphatic amino groups on the surface have a pK of approximetly 10. Results presented on fig.3 show that NAP adsorption is a charge dependent process. Decreasing pH value, surface charges are decreasing, so do electrostatic interactions. Moreover, the adsorbed amount obtained at pH 10 revealed

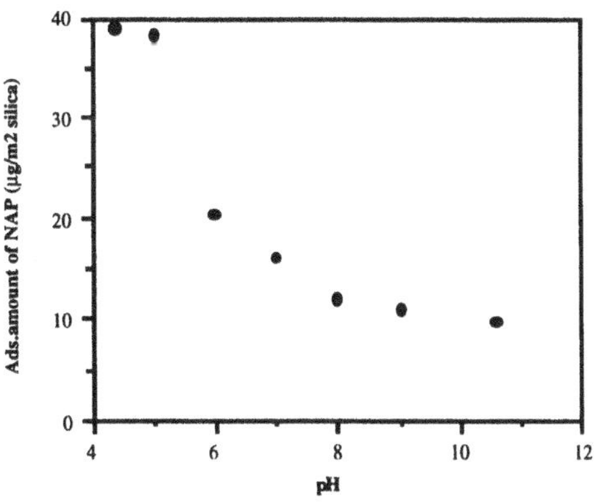

FIG.3: Effect of pH on the adsorbed amount of NAP (ionic strength 0,26M).

the existence of an other contribution in the adsorption process.

3.5. EFFECT OF IONIC STRENGTH ON THE ADSORPTION PROCESS OF NAP.

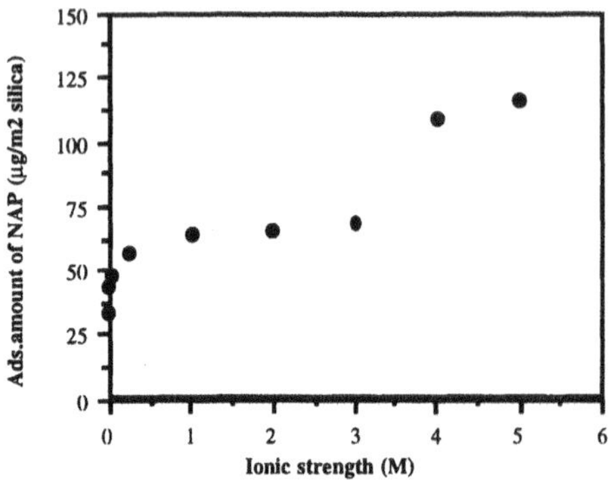

FIG.4: Effect of ionic strength on the adsorbed amount of NAP (pH=6).

Studying the effect of ionic strength (10^{e-3}M to 5M), we have demonstrate that the adsorption process is controlled by two phenomena. At a ionic strength below 3 M, NAP adsorption is driven by a charge effect. A plateau is obtained from 1 M which correspond of the all charges screening. On the other hand, at higher ionic strength, a dehydration effect [9] is controlling the adsorption. For entropic reasons, insoluble NAP sticked on the surface.

4. Conclusions.

This study has been achieved to define the interactions between oligonucleotides and aminated silica wafers. The results presented here have revealed that, at low salt concentration, the adsorption is conducted by electrostatic interaction forces. At high ionic strength, despite the charges screening, adsorption was shown to increase. Due to a high salt concentration, dehydrated NAP stick on the wafer. This approach to NAP immobilization has been achieved in the aim to provide covalent linkage conditions on modified silica wafers.

References

1- Benazilla, M., Manne, S., Laney, D.E., Lyubchenko, Y.L.and Hansma, H.G. (1995) Adsorption of DNA to mica, silylated mica, and minerals: characterization by atomic force microscopy, *Langmuir* **11**, 655-659.
2- Boom, R., Sol, C.J.A., Salimans, M.M.M., Jansens, C.L., Wertheim van Dillen, P.M.E.and van der Noordaa, J. (1989) Rapid and simple method for purification of nucleic acids, *J.Clinical Microbiology* **28**, 495-503.
3- Lorenz, G.M.and Wackernagel, W. (1987) Adsorption of DNA to sand and variable degradation rates of adsorbed DNA, *Applied and environmental microbiology* **53**, 2948-2952.
4- Melzak, K.A., Sherwood, C.S., Turner, R.F.B.and Haynes, C.A. (1996) Driving forces for DNA adsorption to silica in perchlorate solutions, *J.Colloid Interface* Sci **181**, 635-644.
5- Wasserman, S.R., Tao, Y.T.and Withesides, G.M. (1989) Structure and reactivity of alkylsiloxane monolayers formed by reaction of alkyltrichlorosilanes on silicon substrates, *Langmuir* **5**, 1074-1087.
6- Thompson, W.R. and Pemberton, J.E. (1994) Raman spectroscopy of covalently bonded alkylsilane layers on thin silica films immobilized on silver substrates, *Anal.Chem.* **66**, 3362-3370.
7- Porter, M.D., Bright, T.B., Allara, D.L.and Chidsey, C.E.D. (1987) Spontaneously organized molecular assemblies, *J.Am.Chem.Soc.* **109**, 3559-3568.
8- Bourdieu, L., Maaloum, M., Coulon, G. and Chatenay, D. (1992) Structure of free surface of thin organic films as revealed by atomic force microscopy, *Ann.Chim.Fr.* **17**, 229-239.
9- Damodaran, S. and Xu, S. (1996) The role of electrostatic forces in anomalous adsorption behavior of phostvitin at the air/water interface, *J.Colloid Interface Sci* **178**, 426-435.

HOW CAN X-RAY STRUCTURES BE HELPFUL FOR DESIGN OF IONOPHORES FOR ION-SELECTIVE MEMBRANE ELECTRODES?

J.F. BIERNAT and E. LUBOCH
Department of Chemical Technology
Technical University of Gdańsk, 80952 Gdańsk, Poland

1. Introduction

Theoretical aspects of function of ion-selective membrane electrodes (ISME) has been elaborated for years [1]. However, design of a particular structure of ionophore for ISME exhibiting desired properties presents a very difficult problem. It concerns, in particular, selectivity of the ionophore towards a particular ion in the presence of an interfering species. Definite relationship: ionophore structure *versus* selectivity properties of the ion-selective membrane electrode is still not known.

What kind of structural features of the ionophore (or its complex) are responsible for the doped membrane ability to recognize selectively the particular ion?

2. Stability Constants

Most commonly highly selective ionophores for ISME are searched in the midst of ligands forming complexes of highly differentiated stability constants. However, correlation of the complex stability constants with ISME selectivity coefficients is satisfactory only in few cases, e.g., potassium selective valinomycin electrode. The stability constants for sodium and potassium complexes in methanol are: $\log K_{Na}$ = 0.67 and $\log K_K$ = 4.90, resp. The $\log K^{pot}_{K,Na}$ = -3.5 to -4.5 [2]. Note, that valinomycin forms ionic complexes with potassium salts in which the complexed potassium cation has lipophilic exterior. By contrast, exterior of the free valinomycin is much more hydrophilic [3]. It helps to discriminate sodium by partition. Unfortunately, the X-ray structures badly reflect stability of the respective complexes.

3. Stoichiometry and Lipophilicity

For some ionophores the stability constants for sodium and potassium complexes are nearly equal, whereas they show relatively high selectivities in ISME's.

A. W. Coleman (ed.), Molecular Recognition and Inclusion, 255–258.
© 1998 *Kluwer Academic Publishers. Printed in the Netherlands.*

Benzo-15-crown-5 forms with sodium and potassium 1:1 (crown to cation stoichiometry) and a sandwich like 2:1 potassium complex. The 1:1 complexes are of comparable stabilities. The 2:1 potassium complex is more lipophilic than is the sodium complex. Benzo-15-crown-5 shows hundred times higher selectivity in ISME for potassium than for sodium. Highly lipophilic derivatives of this crown ether are up to five thousand times more selective for potassium [4]. Illustrative ISME selectivities and the respective stability constants [5] for compounds shown below are collected in the Table 1. These data show important role of stoichiometry (lipophilicity) of the charged complex cations on selectivity coefficient.

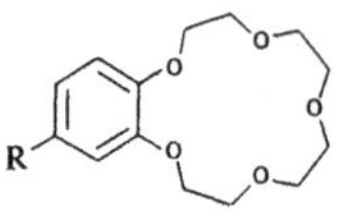

1 R = H
2 R = CH_3CO
3 R = $(CH_3)_3C$
4 R = $CH_3C(n\text{-}C_4H_9)_2$
5 R = $(CH_3)_2C(n\text{-}C_{12}H_{25})$

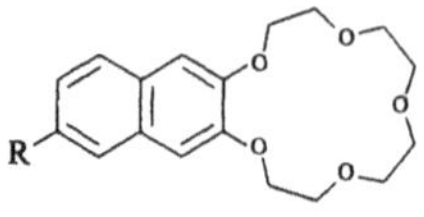

6 R = H
7 R = $(CH_3)_3C$
8 R = $CH_3C(n\text{-}C_4H_9)_2$
9 R = $(CH_3)_2C(n\text{-}C_{12}H_{25})$

TABLE 1. Stability constants for sodium and potassium complexes of benzo- and naphthocrown ethers, and the respective selectivity coefficients in ISME

Ligand	Na^+ complex	K^+ complex			
	log K_1	log K_1	log K_2	log β_2	$\log K^{pot}_{K,Na}$
1	2.78 (0.01)	2.73 (0.04)	3.40 (0.05)	6.13	**-1.9**
2	2.31 (0.01)	2.83 (0.03)	3.02 (0.06)	5.85	**-2.2**
3	2.78 (0.01)	2.53 (0.06)	3.66 (0.08)	6.19	**-2.7**
4	2.81 (0.01)	3.05 (0.02)	3.32 (0.03)	6.37	**-3.3**
5	2.86 (0.01)	3.06 (0.02)	3.17 (0.04)	6.23	**-3.4**
6	2.57 (0.01)	2.71 (0.04)	3.64 (0.05)	6.35	**-2.95**
7	2.83 (0.01)	2.75 (0.07)	3.76 (0.09)	6.51	**-3.30**
8	2.67 (0.01)	2.99 (0.03)	3.67 (0.04)	6.66	**-3.25**
9	2.63 (0.01)	2.73 (0.04)	3.98 (0.04)	6.71	**-3.50**
$K_1=[CrM^+]/[Cr][M^+]$		$K_2=[Cr_2M^+]/[Cr][CrM^+]$		$\beta_2=[Cr_2M^+]/[Cr]^2[M^+]$	

Note, that the sandwich like potassium complex of benzo-15-crown-5 is ionic [6]. The sodium complex possess a covalent bond between sodium and the respective anion [7]. The last species are uncharged and, if present on ISME interface, they do not contribute to the electrode potential. The structures and behaviour of naphtho-15-crown-5 complexes are similar [7, 8].

4. Stoichiometry and Molecular *Versus* Ionic Structure

In solution, benzo-12-crown-4 forms complexes with sodium and potassium cations of 1:1 and 2:1 stoichiometry. In acetonitrile the stability constants are $\log K_1 = 1.88$ and $\log K_2 = 2.62$ for sodium complexes, resp. For potassium complexes the stability constants are 1.76 and 1.08, respectively [9]. In a solid state both cations form 2:1 complexes. However, the potassium complex is molecular with the iodide anion tightly bonded to the complex sandwich cation, whereas the sodium iodide complex is ionic [10].

TABLE 2. Selectivity coefficients for ISME doped with derivatives of benzo- and naphtho-12-crown-4 [10]

	4-Alkylbenzo-12-crown-4	6-Alkylnaphtho-12-crown-4
Alkyl	$\log K^{pot}_{Na,K}$	$\log K^{pot}_{Na,K}$
$-C(CH_3)_3$	-0.95	-0.90
$-C(CH_3)_2C_5H_{11}$	-1.0	-0.95
$-C(CH_3)_2C_{12}H_{25}$	-0.92	-

Lipophilic derivatives of benzo- and naphtho-12-crown-4 are sodium selective (Table 2). The observed sodium selectivity could be explained assuming that the potassium complex shows higher tendency to exist (in membrane) as ionic pair, which does not contribute to the electrode potential.

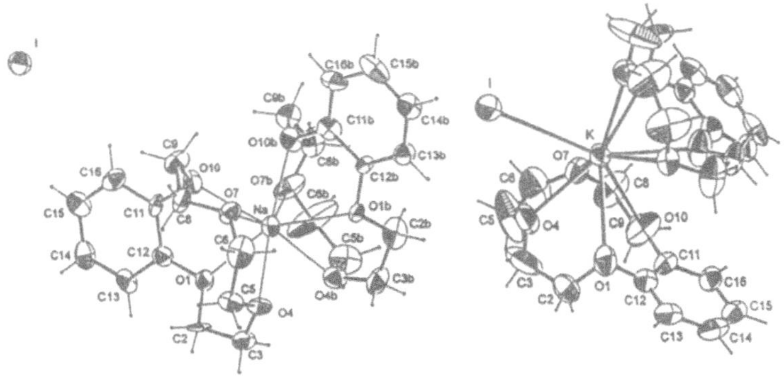

Figure 1. Structure of benzo-12-crown-4 complexes with NaI (ionic) and KI (molecular) [10]

Azocrown ethers present another example:

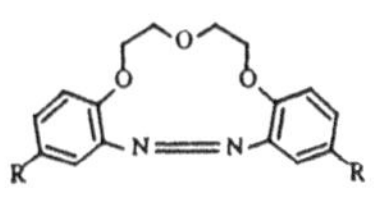

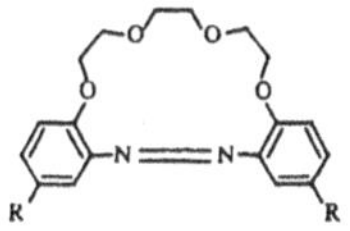

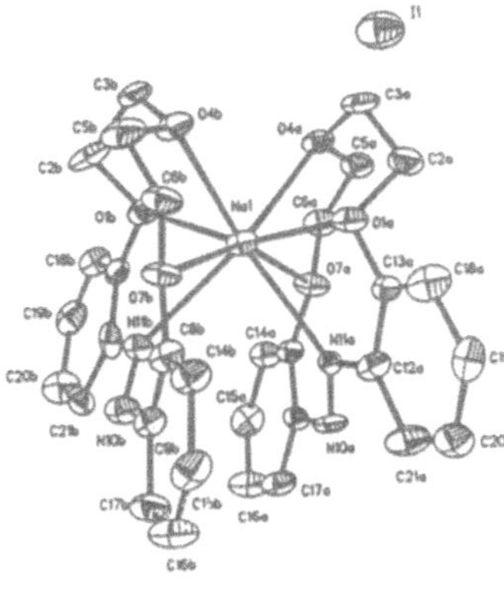

Figure 2. Structure of 13-membered azocrown ether complex with NaI [18]

The **13-membered azocrown** in acetonitrile forms complexes with NaI and KI, however the meaning values of stability constants are low [11]. ISME's based on this ionophore and on its lipophilic derivatives are highly sodium selective, $\log K^{pot}_{Na,K} \sim -2$. [12]. A solid NaI complex with the parent crown ether of 1:2 stoichiometry has been obtained in a facile way. The sandwich like complex is ionic [13]. Efforts to obtain the respective potassium iodide complex failed. It may indicate much lower stability constant for potassium than for sodium iodide complex of this ligand.

The **16-membered azocrown** in acetonitrile forms sodium and potassium complexes (1:1) of comparable stabilities [11]. The solid potassium iodide complex has 2:1 stoichiometry and is ionic. The sodium iodide complex has 1:1 stoichiometry and is molecular. ISME based on this ionophore and on its lipophilic derivatives are highly potassium selective. A typical $\log K^{pot}_{K,Na}$ value is -3.6 to -3.7 [14].

By structural features the 16-membered azocrown and its sodium and potassium complexes resemble benzo-15-crown-5 complexes; both ionophores possess similar properties in ISME's.

5. Conclusions

The observed correspondence between ionophore promoted selectivities of ISME's and the ionophore (complexes) structures in a solid state could be summarized as follows:

The selective response of ISME is high for a particular charged species if the species (1) is fully encapsulated, (2) possess a possibly high ratio of the ionophore to the ion in a complex of highly lipophilic exterior, (3) has low susceptibility to form ionic pairs, and (4) it forms complexes of highly differentiated stabilities.

Financial Support from KBN Grant No 3TO9A15508 is kindly acknowledged.

6. References

1. Morf, W.E. (1981) *The Principles of Ion-Selective Electrodes and of Membrane Transport*, Akadémiai Kiadó, Budapest.
2. Umezawa, K., and Umezawa, Y. (1983) *Selectivity coefficients for Ion-Selective Electrodes*, University of Tokyo Press, Tokyo.
3. Karle, I.L. (1996) Association of Helical Peptides and Ion Channels, in G. Tsoucaris, J.L. Atwood and J. Lipkowski (eds.), *Crystallography of Supramolecular Compounds*, NATO ASI Series, Series C:, Vol. 480, pp. 7-29, and *loc. cit.*
4. Luboch, E. Cygan A., and Biernat, J.F. (1990) The Synthesis of Some Aromatic Crown Ether Derivatives and Their Ion-Selective Electrode Properties, *Tetrahedron* **46**, 2461-2472; Wasilewski, J., and Biernat, J.F. (1991) Bis-Substituted Benzo-15-crown-5 Ethers as Ion Carriers in Potassium Ion-Selective Electrodes, *J. Incl. Phenomena* **10**, 109-118.
5. Cygan, A., Luboch, E., and Biernat, J.F. (1992) The Stability Constants of Some Benzo-15-crown-5 and Naphtho-15-crown-5 Derivatives, *J. Coord. Chem.*, **27**, 87-89.
6. Mallinson, P.R., and Truter, M.R. (1972) Crystal Structures of Complexes between Alkalimetal Salts and Cyclic Polyethers. Part V. The 1:2 Complex formed between Potassium Iodide and 2,3,5,6,9,11,12-Octahydro-1,4,7,10,13-benzopentaoxycyclopentadecin (Benzo-15-crown-5), *J. Chem. Soc., Perkin Trans., II* 1818; Mugnoli, A., Dauter, Z., Luboch, E., Cygan, A., and Biernat, J.F. (1986) Macrocyclic Polyfunctional Lewis Bases. XI. Structure of Trichlorocuprates of Crown Complexed Sodium, *J. Incl. Phenomena* **4**, 407-414.
7. Dreissig, W., Dauter, Z., Cygan, A., and Biernat, J.F. (1985) Macrocyclic Polyfunctional Lewis Bases. XI. Structure of Trichlorocuprates of Crown Complexed Sodium, *Inorg. Chim. Acta* **96**, 21-27, and *loc. cit.*
8. Yamauchi, M., Jyo, A., and Ishibashi, N. (1982) Potassium Ion-Selective Membrane Electrodes based on Naphtho-15-crown-5, *Anal. Chim. Acta* **136**, 399.
9. Strakhova, N.N., Solov'ev, V.P., and Raevskii, O.A. (1989) Comparison of Complexation of Benzo-15-crown-5 and Benzo-12-crown-4 with Metal and Ammonium Thiocyanates in Acetonitrile, *Koord. Khim (Russ)*. **15**, 483; *Chem. Abstr.* **111**, 13207 (1989).
10. Luboch, E., Dvorkin, A.A., and Simonov, Yu.A. (1995) 12-Membered Crown Ethers. Lipophilic Derivatives of Benzo- and Naphtho-12-crown-4 and Their Membrane Electrode Properties. Crystal Structures of Benzo-12-crown-4 NaI Complexes, *J. Incl. Phenom.* **20**, 335.
11. Shiga, M., Nakamura, H., Takagi, M., and Ueno, K. (1984) Synthesis of Azobenzo-crown Ethers and Their Complexation Behavior with Metal Ions, *Bull. Chem. Soc. Jpn.*, **57**, 412.
12. Luboch, E., Biernat, J.F., Muszalska, E., and Bilewicz, R. (1995) 13-Membered Crown Ethers with Azo or Azoxy Unit in the Macrocycle - Synthesis, Membrane Electrodes, Voltammetry and Langmuir Monolayers, *Supramolecular Chemistry*, **5**, 201-210.
13. Simonov, Yu.A., Luboch, E., Biernat, J.F., Bolotina, N.V., and Zavodnik, V.E. *J. Incl. Phenom.*, submitted.
14. Luboch, E., Biernat, J.F., and Simonov, Yu.A., to be published.

THE SYNTHESIS AND CONFORMATIONAL ANALYSES OF SOME DIBENZO[3n+2]CROWN-n ETHERS

Neval Bozkurtoğlu and Çakıl Erk*

Istanbul Technical University, Science Faculty, Chemistry Dpartment, Maslak, 80626, Istanbul, Turkey.

Abstract ; The some benzo[3n+2]crowns and dibenzo[3n+2]crowns were prepared from resorsin and polyglcol dichlorides in DMSO / Water and K_2CO_3. The underlying non-diamond lattice of cyclic oxyethylene structures were examined with molecular mechanics, MM+ methods.
Keywords; macrocyclics,dibenzo[3n+2]crown,stereochemistry,MM+ calculation.

1. Introduction

Despite the well known 1,2-dioxa-benzocrowns, [1-3] the 1,3-dioxa substituted aromatic macrocycles named dibenzo[3n+2]crown-n have been just studied by Weber [4], by Stoddart [5], by Gibson [6]. However, such compounds exhibited better solubilities and unusual cation binding properties compared to their 1,2-dioxabenzo analogous, [7,8a,b] In our work, some off the compounds were prepared with a reaction of resorsin with β,β'-dichlorodiethyl which afforded dibenzo[20]crown-6, and 1,9-dichloro-4,7-dioxononane afforded benzo[13]crown-4 and dibenzo[26]crown-8, preferably in DMSO and Na_2CO_3 or K_2CO_3..

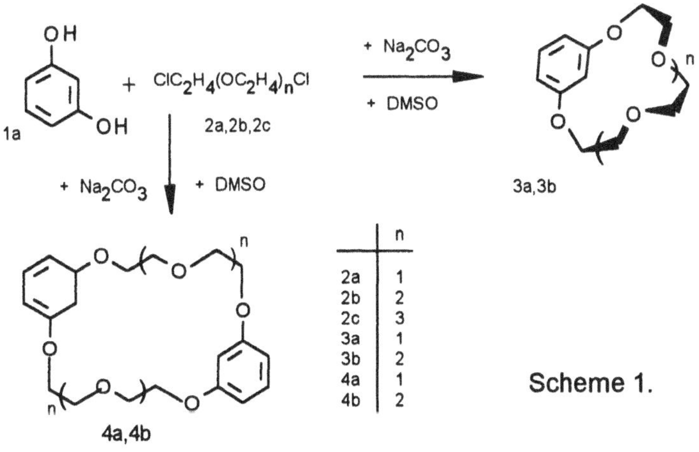

Scheme 1.

A. W. Coleman (ed.), Molecular Recognition and Inclusion, 259–262.
© 1998 *Kluwer Academic Publishers. Printed in the Netherlands.*

2. Experimental

The chemicals were from FLUKA. The structures were identified with the CI and EI Mass Spectra from FISONS, IR (KBr) with PERKIN ELMER, ^{1}H-and ^{13}C-NMR spectra (2D-COSY, HETCO) with 400 MHz BRUKER spectrometer. The conformation analysis were done with MM+ method in a commercial software MM+ in Hyper©Chem was used with Polak-Ribiere method for force field calculations.

2.1 Synthesis of benzo[3n+2]crown-n molecules.

The genaral example for of **dibenzo[20]crown-6;** Resorcinol (4.64 g, 0.042 mol), 2,2'-dichlorodiethyl ether (6.0 g, 0.021 mol) Na_2CO_3 (8.48 g, 0.042 mol) and DMSO (60 ml) were heated for 240 hr at 90-100 °C during strring. The mixture was acidified (100 ml, 2n) and extracted with $CHCl_3$ (2x100 ml). Extracts dried with Alumina (Basic) were evaporated and purified on a Alumina column (40 gr, 80x2 cm,). Yield 25 %, m.p 46-47 °C. For $C_{20}H_{24}O_6$ calc C % 66.59, H % 6.66; found C % 66.62, H % 6.60; IR (cm^{-1}) γ= 2953,2902,1625,1191,1140,1089, ^{1}H NMR 400 MHz ($CHCl_3$/TMS) δ = 3.89(m,8H,J;7.0 Hz) 4.15(m,8H,J;7.0 Hz) 6.5(m,6H,Ar) 713,(t,2H,Ar);^{13}C NMR 100 MHz ($CHCl_3$/TMS) δ = 68.69,70.93, Mass 360 (M^+), 180(M^+-180), 137(M^+–224+H^+), 92 (M^+–268).

3. Results and Discussions

3.1. The Stereochemistry of 1,3-Benzo Macrocyclics

The stereochemistry of the macrocycles is the most dominating factor for the molecular recognition . The X-ray studies have been conducted for preorganized macrocycle since the molecular geometry is valuable for the cation selectivity in solid complexes [1,2,9,10]. In solutions, the bond distances and bond angles are quite similar in complexes and in the free macrocycles. The changes in thorsional angles control the cavity size, therefore, crystal forces only minimally affect the molecular geometry which may be controlled by counter ions and water [11-13].

The MM+ analysis of dibenzo[14]crown-4 showed interesting results which explained the formation of smallest size of the macrocycle is practically too hard.. The isomers of dibenzo[14]crown-4 of different steric energies are represented in Figure 1A-D and Table 1, The *trans,trans* isomer with 22.14 kcal/mol total energy, Figure 1A is minimised to 10.69 kcal/mol of *±gauche, ±gauche* isomer, Figure 1D.

Table 1. The total energy of conformational isomers of 1,3-benzo[14]crown-4 from MM+

Isomers	Conformat. Sequence	Energy kcal/mol	1,3-Torsional angles
A	*trans, trans*	22.14	160.2, 160.1
B	*trans, eclipsed,*	18.82	-159.9, 160.0
C	*trans, gauche*	10.72	75.3 , 27.3
D	*gauche, gauche*	10.69	154.0, -38.8

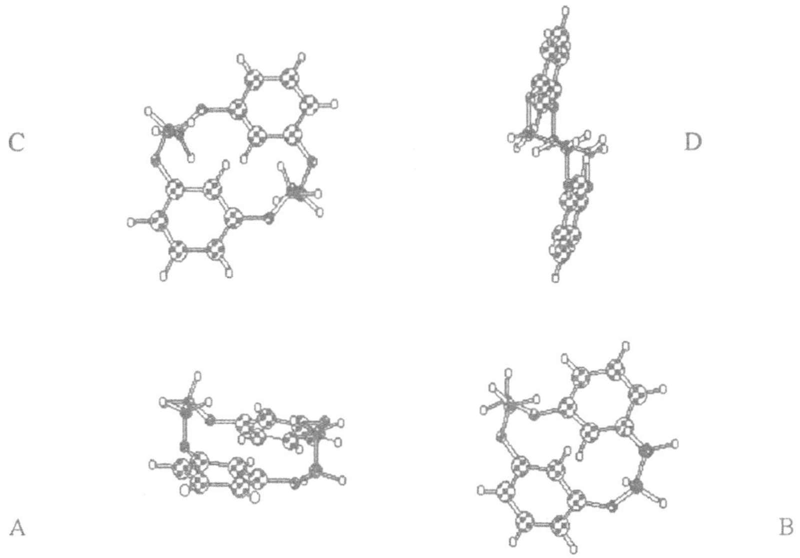

Figure 1. The different conformational isomers of dibenzo[14]crown-4 with MM+ .

The MM+ minimisation of dibenzo[20]crown-6 exhibited a maxima for the structure depending on the sequence of the oxyethylene units [12,13]. The *trans ,trans,trans,trans* isomer with 22.09 kcal/mol total energy, Figure 2A is minimised to 20.01 kcal/mol of *±gauche,trans,trans,trans* isomer, Figure 2B. It is, although, increased to 24.31 kcal/mol, see Figure 2C, with the *±gauche,trans, ±gauche, trans* isomer which is converted to 26.28 kcal/mol with a sequence of conformations, *±gauche, ±gauche, ±gauche, ±gauche* as depicted in Figure 2D, see also Table 2

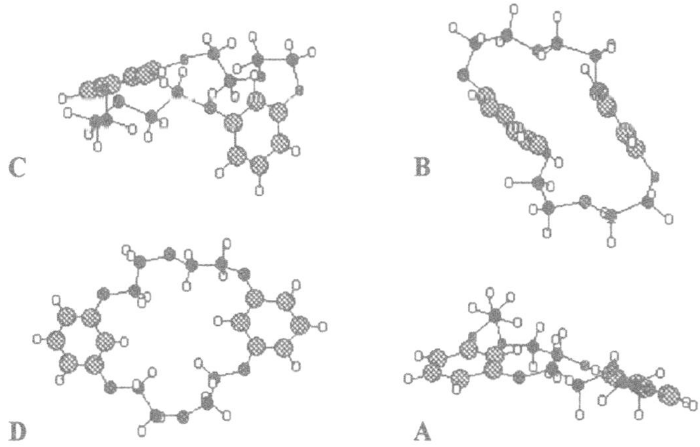

Figure 2. The structures conformers of dibenzo[20]crown-6 obtained with MM+.

Table 2. The total energy of conformational isomers dibenzo[20]crown-6 with MM+

Isomer	Comformation sequence	Energy (kcal/m)	1,4-torsional anglesequences
A	*trans,trans,trans,trans*	22.00	158.8, 178.3, 164.0,169.2
B	*gauche,trans,trans,trans*	20.01	161.8, 161.7, 165.6, 53.8
C	*gauche,gauche,trans,gauche*	24.31	179.0, -64.2, 175.8, -57.9
D	*gauche,gauche,gauche,gauche*	s28.28	178.3, 164.0, 169.2, 158.8

The some selectivity studies were done with the large macrocycle size of such seres compounds and they were unexpectedly found that they were more efficient to bind the cations in a polymer matrix [8a,b]. On the other hand the a comprehensive work regarding the molecular mechanics of Benzocrowns like benzo[15]crown-5, including the MOPAC calculations, is in progress to compare the structural facts between the two series of the compounds from the point ofereochemistry [8c].

Acknowledgement; Authors would like to express their thanks to Research Foundation of Istanbul Technical University for finanncial support. The MSc grant from TUBITAK is greatly acknowledged by N.B.

References

1. Y. Inoue and G. W. Gokel, *Cation Binding by Macrocycles*, Marcel Dekker, 1990, New York.
2. R. M. Izatt, K. Pawlak, J. S. Bradshaw and R. Bruening, *Chem. Rev.* 95,2529(1995)
3. a. A. Göçmen, M. Bulut and Ç. Erk, *Pure & Appl. Chem.* **65**(3)447(1993).
   b. A. Göçmen and Ç. Erk, *Fresenius J. Anal. Chem.* **347**(12)471(1993).
4. E. Weber, *Angew. Chem.*, Int. Ed.(Engl). **18**,219(1979).
5. a. J. F. Stoddart. *Pure & Appl. Chem.* **60**(4)467(1988).
   b. L. Alwood, H. Shahriari-Zavareh, J. F. Stoddart, and D. J. Williams, *J. Chem.Soc.,Chem. Commun.* 1058(1987).
6. J. G. Vinter, A. Davis and M. R. Saunders, *J. Comp. Aided Mol. Des.* **4**,132(1987).
7. P. Knops, N. Sendenhoff, H. -B. Mekelburger and F. Vögtle, *Top. Curr. Chem.* **161**,3(1991).
8. a. E. Luboch, A. Cygan, A. Biernat, S. Fonari, A. Yu. Simonov, and V. Belski, 4th Int. Summer School on Supramol: Chem., Poland, 1993.
   a. J. A. Biernat, Abstracts, 9th *Internat. Symp. on Mol. Recog. and Incusion,* ISMRI 9, Lyon, France, 1996.
   b. H. Nalbant and Ç. Erk, results to be published.
   c. N. Bozkurtoglu, H. Nalbant and Ç. Erk, Abstracts, 9th *Internat. Symp. on Mol. Recog. and Incusion,* ISMRI 9, Lyon, France, 1996.
9. a. F. R. Fronczek and R. D. Gondour, in *Cation Binding by Macrocycles,* Y. Inoue and G. W. Gokel (Eds.), Marcel Dekker, 1990, New York, pp 311.
   b. I. Goldberg, *The Chemistry of Functional Groups*, S. Patai, (Ed.), Wiley, New York, 1989
10. R. Holgenfeld and W. F. Sanger *Top. Curr. Chem.* **101**,1(1982).
11. R. Perrin, R. Decoret, G. Bertholet and R. La Martine, *Nouv. J. Chem.* 263(1983).
12. E. Pretsch, M. Badertscher, M. Welti, T. Maruzami, W. E. Morf, and W. Simon, *Pure & Appl. Chem.* **80(4)**,567(1988).
13. G. Wipf, P. Weiner, and P. Kollmann. *J. Am.Chem. Soc.* **104**,3251(1982).
14. Y. Delaviz and H. W. Gibson, *Macromolecules* **25**,4859(1992)

CALIX[4]RESORCINARENE DERIVATIVES AS IONOPHORES FOR CATIONS STUDIED IN POLYMERIC (PVC) MEMBRANE

ZBIGNIEW BRZÓZKA[b], EWA LISZEWSKA[b], MAREK PIETRASZKIEWICZ[a*], RAFAŁ GĄSIOROWSKI[a]
a) *Institute of Physical Chemistry, Polish Academy of Sciences 01224 Warsaw, Kasprzaka 44/52, Poland*
b) *Technical University of Warsaw, Department of Chemistry, Noakowskiego 3, 00664 Warsaw, Poland*

1. Introduction

Calixarenes constitute the class of molecular receptors capable of binding almost all conceivable chemical species; inorganic/organic cations and anions, neutral molecules, metal carbonyls. Rich chemistry of calixarenes has attracted considerable attention. Three monographs on calixarene chemistry have been published [1-3], and three reviews [4-6], one of them dedicated to metal complexes of calixarenes [6].
Some calixarene derivatives may display an interesting cation selectivities, and may be thus considered as selective ionophores for particular cations [6]. We have been interested in chemical modifications of calix[4]resorcinarenes and their inorganic ion selectivities. Thus we have prepared a range of lipophilic calix[4]resorcinarenes of Mannich base-type incorporating amino-, carboxamido-, sulfonamido-, and carboxhydrazido functions.

2. Experimental

The initial calix[4]resorcinarene was prepared from resorcinol and lauryl aldehyde [7]. The Mannich-type compounds were prepared according to the literature [8] using appropriate amino compounds: piperidine, N-methylpiperazine, *p*-nitroaniline, as well as *p*-toluenesulfonamide, isonicotinhydrazide, and isonicotinamide. The reactions were carried out in boiling ethanol in case of mp1 and mp2, and in the mixture of ethanol-benzene in case of mp3-6 in the presence of 10% aqueous NaOH as catalyst, with azeotropic removal of water. The products mp3-6 precipitated out from the solution after 2-5 hr, and after filtration were crystallized from acetonitrile. All new compounds (mp3-6) gave correct analytical data consistent with the proposed structures.
Electrochemical part: inorganic salts: nitrates of sodium, lithium, potassium, cesium, ammonium, calcium, magnesium; chlorides: of sodium, potassium, cesium. Acids: nitric and hydrochloric. Stock solutions: 10^{-1} M in redistilled water. Membrane composition: ionophore (1%wt), didecyl phthalate (65%wt), potassium tetra[3,5-bis(trifluoromethyl)phenyl]borate (50%mol), polyvinyl chloride (33%wt). The components were dissolved in freshly distilled THF, casted on 24 mm ring and allowed to stand overnight to evaporate the solvent. The membranes (7 mm) were placed in the

A. W. Coleman (ed.), Molecular Recognition and Inclusion, 263–266.
© 1998 *Kluwer Academic Publishers. Printed in the Netherlands.*

body of Phillips IS 561 electrode. Details concerning the instrumentation and SEM measurement can be found in ref [10].

Figure 1

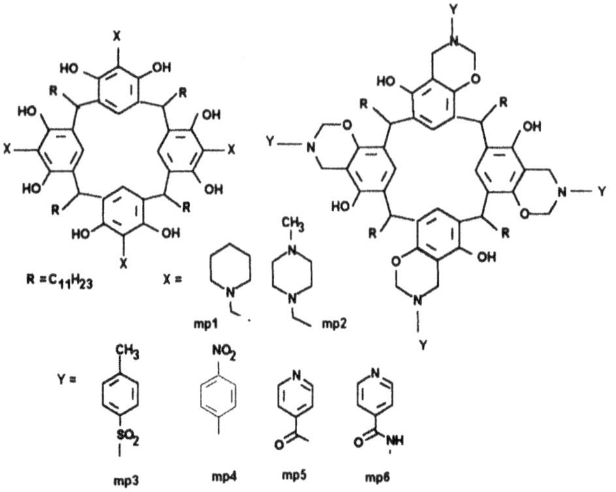

3. Results and Discussion

First two calix[4]resorcinarenes mp1 and mp2 were not particularly attractive as selective ionophores for alkali metal cations. The compounds mp3-6 with extended cavity displayed pronounced selectivities towards Cs^+ , probably due to cation-π interactions between cesium cation and aromatic walls of the receptor. In the case of aliphatic side-arms no cesium selectivity was observed. *Figure 2* shows the relative coefficients ($\log K_{Cs/M}$) of PVC membranes containing calix[4]resorcinarene derivatives

Figure 2

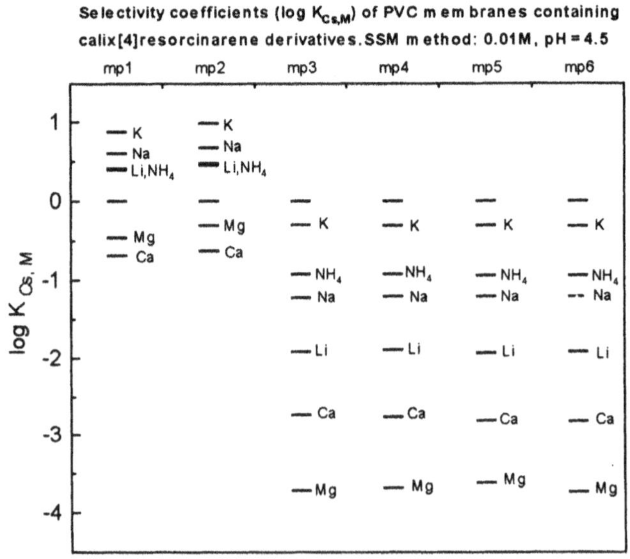

Figure 3 shows the response times of PVC membranes in the presence of variable concentrations of cesium nitrate, as well as pH dependencies of Cs^+ responses of PVC membranes for mp3 and mp4 in both cases. It is clear that the response time is short and SEM is stable in time. The SEM staibility is pronounced in the range of pH 4-9, which makes these ligands attractive for their possible applications as sensors for Cs^+.

Figure 3

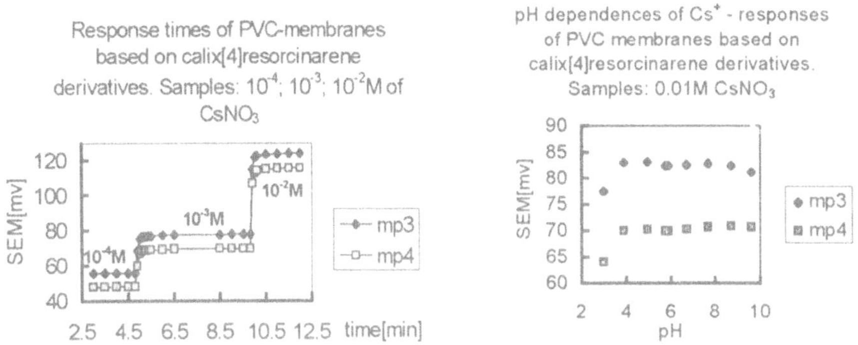

The *Figure 4* shows the cesium over sodium selectivity as a function of time, which reaches maximum after four days, and diminishes gradually for all ligands mp3-6. Also in the same Figure 4 there is the SEM dependence on the variable concentration of Cs^+. The SEM/log[Cs^+] is linear in the range of log Cs^+ -4.5 ÷ -2.5 for mp3 and mp4.

Figure 4

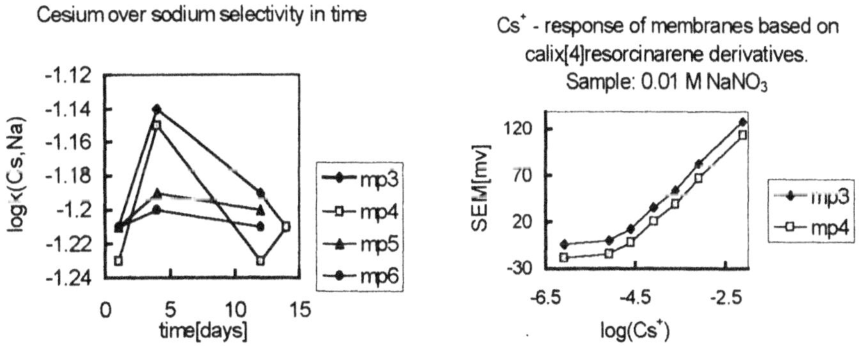

4. Conclusions

These results indicate that generally calix[4]resorcinarenes with extended cavity with aromatic substituents may be interesting from the point of view of selective complexation of Cs^+ cations. Further investigations with other calix[4]resorcinarene derivatives are in progress.

5. Acknowledgement

This work was supported by the Polish Committee for Scientific Research, Grant No 3 T09 A 025 011.

6. References

1. Gutsche, C. D. (1989) *Calixarenes, Monographs in Supramolecular Chemistry,* Royal Society of Chemistry, Cambridge.
2. Vicens, J. and Böhmer, V. (1991) *Calixarenes: a Versatile Class of Macrocyclic Compounds,* Kluwer Academic Publishers, Dordrecht.
3. Vicens, J., Asfari, Z., and Harrowfield, J. M. (1994) *Calixarenes 50th Anniversary: Commemorative Volume,* Kluwer Academic Publishers, Dordrecht.
4. Böhmer, V. (1995) Calixarenes, macrocycles with (almost) unlimited possibilities, *Angew. Chem. Int. Ed. Engl.,* **34,** 713-745.
5. Roundhill, D. M. (1995) Metal complexes of calixarenes, *Progr. Inorg. Chem.,* **43,** 533-592.
6. Timmerman, P., Verboom, W., and Reinhoudt, D. N. (1996) Resorcinarenes, *Tetrahedron,* **52,** 2663-2704.
7. Aoyama, Y., Tanaka, Y., Toi, H., and Ogoshi, H. (1988) Polar host-guest interaction. Binding of nonionic polar compounds with a resorcinol-aldehyde cyclooligomer as a lipophilic polar host, *J. Am. Chem. Soc.,* **110,** 634.
8. Matsushita, Y. and Matsui, T. (1993) Synthesis of aminomethylated calix[4]resorcinarenes, *Tetrahedron Lett.,* **34,** 7433-7436.
9. Pietraszkiewicz, M., Gąsiorowski, R., Brzózka, Z., and Wróblewski, W. (1992) Synthesis and selectivity towards divalent inorganic cations of macrocyclic dicarboxamides and disulfonamides bearing pendant electron donor groups, *J. Incl. Phenom. & Mol. Recogn. Chem.,* **14,** 237-245.
10. Brzózka, Z., Pietraszkiewicz, M., (1991) Mercury-ion-selective polymeric membrane electrodes based on substituted diaza crown ethers, *Electroanalysis,* **3,** 855-858.

Synthesis of Isoflavone Derivatives of Crown Ethers

Mustafa Bulut*, Birsen Yılmaz* and Çakıl Erk**
*) *Marmara University, Faculty of Science and Art, Chemistry Department, Kadiköy, Istanbul.* **) *Technical University of Istanbul, Faculty of Science, Chemistry Department, Maslak, Istanbul, TURKEY.*

Abstract. The crown ether derivatives of isoflavones have been prepared and characterized. 7,8-dihydroxyisoflavon prepared from pyrogallol and 4-methoxy phenylacetonitrile reacted with ditosylates or dichloride of polyglycols in K_2CO_3 / DMF / H_2O. The crown ether-isoflavon derivatives containing 3-6 oxygen member rings were prepared. The structure of the compounds were verified by elemental analysis, ^{1}H-NMR, ^{13}C-NMR, IR and Chemical Ionisation Mass Spectroscopy.
Key words; Macrocyclics, Isoflavones, Chromogenic compounds.

Received;

1. Introduction

Recently, the crown ether derivatives of some coumarins, coumestans and their structural analogues have been prepared [1,2] and structures as well as the ion binding roles have been recognized via potentiometry and fluorescence spectroscopy as reported from our laboratory [2,3].

We now report the synthesis of chromogenic crown ether derivatives of 7,8-dihydroxyisoflavone. The isoflavones have been recognized as antioxidants [4,5], fungicides [6] and anticontraceptives [7]. They were obtained by formylation of o-hydroxydesoxybenzoin prepared from pyrogallol and p-methxyphenylacetonitrile. The subsequent cyclo condensation is carried out with, hitherto, DMF, methanesulfonyl chloride and boron trifluoride-etherate as the formylating reagents. The identity of the substances has been ascertained by mp comparison and spectroscopy.

The macrocyclic derivatives were prepared from dichlorides or ditosylate of polyethylene glycol reacting with 7,8-dihydroxyisoflavone in the presence of K_2CO_3 in DMF / water (60/1) at 75-80°C for 36 hours, see Scheme 1.

A. W. Coleman (ed.), Molecular Recognition and Inclusion, 267–269.
© 1998 *Kluwer Academic Publishers. Printed in the Netherlands.*

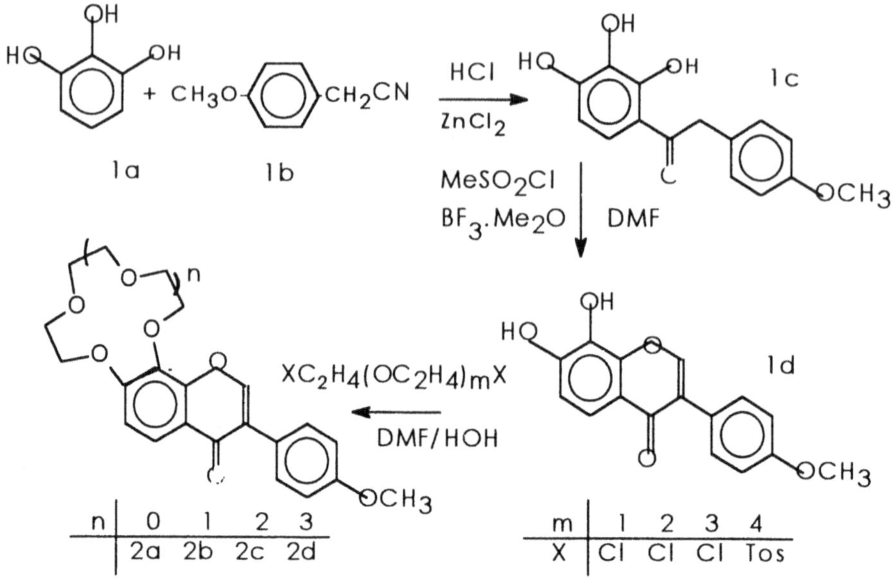

Scheme 1.

2. Experimental

All materials and solvents were of the analytical reagents grade. ^{1}H-NMR and ^{13}C-NMR spectra were recorded on VARIAN Gemini 200 MHz or BRUKER 400 MHz instruments in $CDCl_3$. IR spectra were recorded on a NICOLET 510P FTIR spectrometer. The purity of all the products have been checked using thin layer chromatography (Kieselgel 60, MERCK, aluminum foil) in a benzene /methanol (10/1) solvent system and iodine vapor or TLC with UV-light to detect the spots. Melting points are uncorrected.

1-(2,3,4-Trihydroxyphenyl)-2-(4-methoxypheyl)-ethanone **(1c)** : Into a mixture of pyrogallol, (1a) (17.7g, 0.14 mol) and anhydrous $ZnCl_2$ (17,5g ,0.11 mol) in diethylether (200 ml), p-methoxyphenylacetonitrile, (1b) (25g, 0,17 mol) was added drop wise. The mixture was cooled to 0 °C and dry HCl was passed through the stirred reaction flask for 4 hours. After 70 hours standing in refrigerator the ether phase was decanted and the residue was refluxed with water (½ l). After cooling, the product was filtered and crystallized from ethanol / water. Yield 7.0 g, % 18, mp:191-193°C. The authenticates have been checked by comparison of the mp and verified by spectroscopic methods for the structural elucidation.

7,8-dihydroxy-4'-methoxyisoflavone **(1d)** : **1c** (4.0 g, 14.6 mmol) in DMF was treated cautiously with $BF_3.Et_2O$ (7.5 g) a vigorous exothermic reaction was occurred. To this mixture a solution of $MeSO_2Cl$ (5.0 g) in DMF was added at 50°C and heated on a steam bath for 90 min. The cooled product poured with rapid stirring into water was solidified to give 7,8-dihydroxy-4′-methoxyisoflavone, (1d) which was purified on silicagel column. m.p 179-180, yield 0,8g (2,8 mmol), 19 %, (see Ref 4).

A typical procedure given below for [15]crown5 was used in general for the isoflavone derived macrocycle synthesis obtained in this work which is outlined in Scheme 1.

7,8-(15-crown-5)-4′-methoxyisoflavone (2c):: 7,8-dihydroxyisoflavone, (1d) (50m mol) and pentaethylene glycol ditosylate (50 mmol) and K_2CO_3 (100 mmol) was dissolved in water (5 ml) and DMF (75 ml) in a flask (250 ml). The mixture was heated for 36 hours at 75-80°C under nitrogen during stirring. The cooled mixture was diluted with water then extracted with chloroform. The evaporated extracts dried with magnesium sulfate finally purified by chromatography on silicagel. yielded **2c**, 0.33 g, 15 %, mp 86 °C (ethanol). Analyses for $C_{24}H_{26}O_8$ calculated C 65.16 %, H 5.88 %, found C 65.50 %, H 5.99 %; IR (cm^{-1}) γ = 3100,2947,1697,1514,1080,835 ; ^{1}H-NMR ($CHCl_3$/TMS) 400 MHz; ppm (δ) = 3.73(m,6H),3.77(m,2H),3.82(s,3H),3.94(m,2H), 3.99(m,2H),4.25(m,2H),4.31(m,2H),6.93(d,2H),7.00(d,1H),7.49(d,2H),7.97(s,1H),7.98 (d,2H); ^{13}C-NMR ($CHCl_3$/TMS) 100 MHz; ppm (δ) = 56.04,69.53,69.92,70.94, 71.10,71.21,71.72,72.04,74.59,111.92,111.05,120.03,122.68,125.31,125.51,131.31.13 7.30,152.14,153.5,157.15,160.95,177.42; Mass (m/z)442.2(M^+),310.2(M^+-3(C_2H_4O) x 44),178.2 ,132.2

Acknowledgement

The University of Marmara is greatly acknowledged by the author of M.B. due to the financial support for this work.

References

1 a. M. Bulut and Ç. Erk, *Syn. Commun.* **22**,1259(1992).
   b. A. Göçmen, M. Bulut and Ç. Erk, *Pure Appl. Chem.***65**,447(1993).
   c. M. Bulut and Ç. Erk, *Dyes and Pigments*, **30**,99(1996).
   d. A. Göçmen, M. Bulut and Ç, Erk, 1993, unpublished results
2. M. Bulut, B. Yılmaz, M. Yapıcı, and M. V. Kahraman, *J. Incl. Phenom.* 1996, in press.
3. S. Akyüz, A. Göçmen, M. Bulut and Ç. Erk, *Spectrosc. Lett.* **28**(4)603(1995).
4. R. J. Bass, *J. Chem. Soc. Chem. Commun.* **2**,78(1976)
5. P. Görgy, K. Murata and H. Ikehata, *Nature,* **870**,203(1964).
6. H. Fuami, M. Nakajima in M. Jacobsen, G. W. Crosby, *Naturally Occurring Insecticides*, Marcel Dekker, New York, 1971, p 71-97.
7. J. D. Biggers, in J. W. Fairbairn: *The Pharmacology of Plant Phenolics.* Academic Press, New York, 1969, p. 51-69.
8. H. J. Chandra, F. Zilliken and E. Breitmaier, *Angew. Chem.* Int. Ed. Engl. **20**,102(1981).

The Synthesis of Some Coumestan and Related Chromogenic Derivatives of Crown Ethers, Part II

M. Bulut * and Ç. Erk **

(*) Marmara University, Faculty of Science and Art, Chemistry Department, 81040, Kadıköy, Istanbul, Turkey. (**) Technical University of Istanbul, Faculty of Science, Chemistry Department, 80626, Maslak, Istanbul, Turkey.

Abstract ; The [18]crown-6 derivatives of 8,9-dihydroxy-6H-benzofuro[3,2-c][1]benzopyran-6-one, 1,3-dimethyl-8,9-dihydroxy-6H-benzofuro[3,2-c][1] benzo pyran-6-one, 6,7-dihydroxy-3,4-dihydro-2H-dibenzofuran-1-one, and 6,7-dihydroxy-3,3-dimethyl-3,4-dihydro-2H-dibenzofuran-1-one, were prepared from o-dihydroxy compounds, and polyethylene glycol ditosylates in the presence of K_2CO_3 / CH_3CN / H_2O. Some coumestan related 3-5 oxygen containing heterocyclic-crowns were also synthesized. The structures were identified by elemental analysis, ^{1}H-NMR, ^{13}C-NMR, IR spectra and CI mass spectra,

Key words ; Macrocyclics, chromogenic compounds, molecular recognition.

1. Introduction

The molecular recognition have been intensively studied for the synthetic structures as well as natural chemicals. Therefore, the ability of the synthesis of the novel structures of selective binding or inclusion property is the aim of the organic synthesis [1]

On the other hand the investigation of the molecular recognition process is of the prime importance of this field, therefore several analytical methods have been used [2].The intermolecular changes due to inclusion or molecular recognition of a macrocyclic may cause alterations at the electronic levels of chromogenic subunits available in the macrocyclic moiety [3].

Part I, reference 4.

A. W. Coleman (ed.), Molecular Recognition and Inclusion, 271–274.

© 1998 Kluwer Academic Publishers. Printed in the Netherlands.

Recently, we have synthesized some crown ether derivatives of 6H-benzofuro[3,2-c][1]benzopyran-6-one and some related *chromogenic* groups for the optical spectroscopy of the molecular recognition [4]. This work deals with synthesis of the different size of macrocyclics with the similar heterocyclic chromogenic moieties

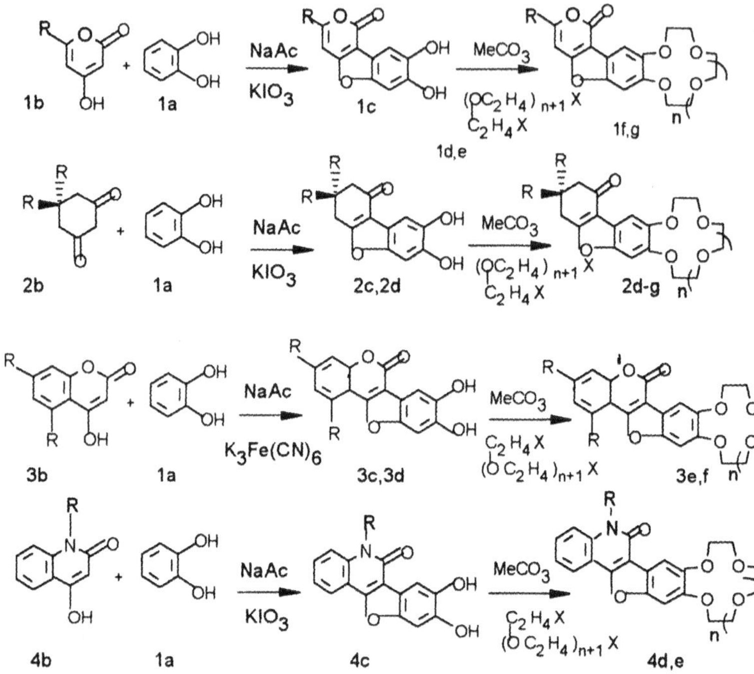

	n	R	x
1d	0	-	Cl
1e	2	-	Ts
1f	0	CH3	
1g	2	CH3	

	n	R
2d	0	H
2e	3	H
2f	0	CH3
2g	3	CH3

	n	R
3e	3	H
3f	3	CH3

	n	R
4d	0	CH3
4e	1	CH3

Scheme 1.

2. Experimental

The initially 4-hydroxycoumarins were prepared from phenols and malonic acid in $POCl_3$ / $ZnCl_2$. However, the reaction of coumarin with catechol in $K_3[Fe(CN)_6]$ / CH_3COONa / H_2O gave 8,9-dihydroxy-6H-dibenzofuro[3,2-c][1] benzopyran-6-one (3c) and 1,3-dimethyl-8,9-dihydroxy-6H-benzofuro-[3,2-c][1] benzopyran-6-one (3d) [3-5]. The other analogous (2c,2d) were obtained from cyclohexane-1,3-dione and 5,5-dimethylcyclohexane-1,3-dione and cathecol respevctively in the presence of KIO_3 / CH_3COONa / H_2O with the yields of 50-60 %.

The coumestan and coumestan related macrocycles were synthesized from o-dihyroxy compounds (1c,2c,2d,3c,3d,4c) mentioned above by reacting with ditosylates (or dichlorides) of polyethylene glycols and alkali carbonates dissolved in CH_3CN / H_2O (60:1) for 48-50 hours at 75-80 ^0C. The chromatography of solid row products on silica with chloroform gave desired pure macrocycles. The spectral data given at Table 1 are agree with the structures proposed in Scheme 1.

3. Results and Discussions

In connection with the series of our works for the preparation and the characterization of the chromogenic macrocyclics [8,9] we extended our work to synthesis the crown ether adjacent 8,9-dihydroxy-6H-benzofuro[3,2-c][1]benzopyran-6-one derivatives in this work. The optical spectra of similar compounds were effectively influenced from the cations in solutions [3,8]. Particularly effects on the fluorescence spectra of the cations are in accordance with the cation radii and the encapsulating size. [8-10].

However, it is also interesting to note that the [9]crown-3 derivatives were isolated mostly from the reaction of ethylene glycol dichlorides in acetonitrile solutions of high dilution conditions, see compounds, 1f, 2d, 2f and 4d. Accordingly the optical properties are still currently being studied in the presence of the cations due to low solubilities of some compounds obtained. The ^{1}H-NMR Spectra is also found to be interesting from the point of chemical shifts of the macrocycle chain.

Table 1. The structural and spectral data of chromogenic macrocycles.

No	Yield	m.p.	^{1}H NMR (ppm)	IR (cm^{-1})
1f	18	78	2.34(s,3H)3.59(t,4H)4.08(r,4H)6.40(s,1H), 6.90(s,1H)	3100,2940, 1690,1514, 1080,835
1g	32	146	2.33(s.3H)3.7(s,4H)3.90(m,8H)4.15(m,8H) 6.40(s,1H),6.99(s,1H)7.34(s,1H)	3100,2975, 1650,1210, 1150
2d	18	82	2.20(p,2H)2.53(t,2H)2.97(t,2H)3.90(m,4H)4.25(m,2H)4.36(m,2H)7.1(s,1H)7.6(s,1H)	3100,2850, 2990,1640, 1340,1150
2e	30	104	2.20(p,2H)2.53(t,2H),2.97(t,,2H)3.75(m,8H), 3.90(m,4H)4.25(m,4H)4.36(m,4H)7.10(s,1H), 7.50(s,H)	3100,2860, 2990,1635 1200-1300
2f	34	78	1.14(.s,6H)2.4(s,3H)2.8(s,3H)3.9(m,4H)4.3(m,2H)4.4(m,2H)7.1(s,1H)7.6(s,1H)	3100,2860, 1640,1280 1140
2g	36	oil	1.07(.s,6H)2.40(s,3H)2.80(s,3H)3.57(m,8H) 3.65(m,4H)3.9(m,2H)4.12(m,2H)7.30(s,1H)7.90(s,1H)	3100,2978, 2875,1675, 1200,1300,
3f	48	215	2.67(s,3H)2.30(s,3H),3.63.8.2(m,12H)3.97(m, 4H)4.17(m,4H),6.8(s,1H),6.9(s,1H)7.4(s,1H) 7.0 (s,1H)	3080,2960, 1750,1620, 1280,1076
4d	16	118	2.07(s,3H)3.77(m,4H)4.20(m,4H)6.9(m,4H) 7.32 (s,1H)7.82(s,1H)	3100,2880, 1685,1215, 1660,1160
4e	24	oil	2.07(s,3H)3.65(m,4H)3.8(m,4H)4.90(m,4H)6.8(m,4H)7.20,(s,1H)7.70(s,1H).	3100,2875, 1670,1483, 1300,1150

Acknowlwdgement; The financial support from Marmara University is kindly acknowledged by the author of M.B.

Referencss

1. R. M. Izatt, J. S. Bradshaw, S. A. Nielson, D. J. Lamb, J. J. Christensen and D. Sen *Currr. Contents* , 33,7(1993).
2. J. E. D. Davies and J. A. Ripmeester, *Comprehensive Supramolecular Chemistry*, Vol. 8, 1996,
3. H. G. Löhr and F. Vögtle, *Acc. Chem. Res.*. 18,65(1985).
4. M. Bulut, B. Yilmaz, M. Yapici and M V. Kahraman, *J. Incl. Phen.*, 1996, in press.
5. B. Harborne, *The Flavonoids-Advances in Research since 1980*, Chapman and Hall, London, 1988.
6 S. R. Hussain, J. Cillard and P. Cillard, *Phytochem.*, 26,2498(1987).
7. F. Ferraro, V. Martino, G. Borrajo and J. D. Coussio, *Phytochem.* 26,3092(1987).
8. M. Bulut and ç. Erk, *Syn. Communic.* 22,1259(1992).
9 A. Göçmen, M. Bulut and ç. Erk, *Pure and Applied Chemistry*, 65(3)447(1993).
10. ç. Erk, M. Bulut and A. Göçmen, 1995, unpublished results.
11. ç. Erk, M. Bulut and A. Göçmen, *Supramolecular Science*, 1966, submitted.

The Associaton Constants of Macrocyclic Ether-Cation Interactions in Dioxane / Water mixtures, Part II

Ümit Çakır*, Baki Çiçek* and Çakıl Erk**
(*) University of Balıkesir, Faculty of Art and Science, Department of Chemistry, Balıkesir, (**), Istanbul Technical University, Faculty of Science, Department of Chemistry, Maslak, 80626, Istanbul, Turkey.

Abstract : The association constants of Na^+ with the macrocycles of [12]crown-4, [15]crown-5, [18]crown-6, Diaza[18]crown-6 and PEG 600 were estimated in dioxane / water (50/50) using a Na^{2+} ion selective electrode with the relationship, $1/K_a\,[L_o]^{n+m-1} = (1\text{-}nP')^n\,(1\text{-}mP')^m / P'$ where, P' is the mole fraction of the complexed cation.
Key words : Macrocycles, cation association , potentiometry, ion selective electrode,

1. Introduction

The cation association of the macrocycles in polar solvents have been investigated [1,2]. However, the interactions of macrocycles with a cation is influenced from the solvent.

We have recently studied on Na^+ ion association constants with [18]crown-6 in various dioxane / water mixtures at different temperatures by potentiometry where the composition of the solvent could be arranged from aqueous to nonaqueous media to see the role of solvent on the macrocyle-cation interactions. We recently developed a formalism for the estimation of association constants as well as the measurement methods. [2-8]. The association constants, K_a were calculated using the Equations 1-4.

$$nA^+ + mL \rightleftharpoons A_n^+ L_m \qquad 1)$$

$$K_a = [A_n^+ L_m]/\{nA^+\}\,[mL] \qquad 2)$$

$$1/K_a\,[L_o] = (1\text{-}P)^2 / P \qquad 3)$$

$$1/K_a\,[L_o]^{n+m-1} = (1\text{-}nP')^n\,(1\text{-}mP')^m / P' \qquad 4)$$

This paper deals with the determination of association constants of Na^+ ion with the macrocycles of [12]crown-4, [15]crown-5, [18]crown-6, Diaza[18]crown-6 and PEG600 in dioxane / water (50/50) mixture with the potentiometry using a sodium selective electrode with the similar methods for the estimation and the calculation..

A. W. Coleman (ed.), Molecular Recognition and Inclusion, 275–278.
© 1998 *Kluwer Academic Publishers. Printed in the Netherlands.*

2. Experimental

The chemicals were from FLUKA and used without further purification. The associations constants were estimated using an Na^+ selective electrode [9-11]. Some experiments were also tried in the presence of K^+ ion to estimate the role of potassium ion on the Na^+ / K^+ selectivity with a sodium selective electrode [12]. Dioxane was dried on metallic sodium then mixed with deionized water 50:50 (v/v). The free cation and equivalent amounts of the cations and the macrocycle solutions were prepared in volumetric flasks and placed into the thermostated cell equipped with a combined electrode from ORION (model ROSS). The mV were measured (± 0.5 mV) with the ORION millivolt meter, model 920 [5,7-9]. K_a values were obtained form Equations 1-4 with the minimum correlation coefficient of 0.99, see Figures 1,2.

3. Results and Discussion.

The Tables 1,2 display the results. Especially, the most stable complexing of Na^+ ion with [18] crown-6 in dioxane/water (50/50). The results were furnished with the Figures 1,2 indicating the role of the method we used for the calculation of association constants, K_a. The results are fairly depended on the type of ion selective electrode as we tried but naturally depended on the structure of macrocycles in mixtures.

The higher stoichiometry of the interactions involved larger free energy contents. This is, also observed in our ^{13}C NMR dipole-dipole relaxation rate studies [2-4]. This is, however, thermodynamically explained in terms of enthalpy-entropy compensations originated from the mechanism of cation macrocycle interactions [1,2].

Table 1. The 1:1 Association Constants of Macrocycles with Na^+ in water / dioxane mixtures at 25 °C.

Crown Ether [a)]	*Salt* [b)]	*ln K_a*	*log K_a*	*ΔG^θ (kCal/M)*
[12]crown-4	$NaClO_4$	1.71	0.74	-1006.9
[15]crown-5	$NaClO_4$	3.81	1.65	-2247.9
[18]crown-6	$NaClO_4$	4.86	2.11	-2870.9
[15]crown-5	NaCl	3.43	1.49	-2024.1
[12]crown-4	NaCl (+ KCl) [c)]	2.35	1.02	-1384.7

a. With equivalent amount of salt and crown ether. c. in the presence of equivalent KCl.

Table 2. The 1:2 and 2:1 (n:m) Association Constants of some Macrocycles with Na^+ in water / dioxane [a)].

Crown Ether / anion [b)]	*ln K_a*	*log K_a*	*$\Delta G^{\theta c)}$ (1:2)*	*ln K_a*	*log K_a*	*$\Delta G^{\theta c)}$ (2:1)*
[12]crown-4 / ClO_4^-	6.36	2.82	-3837.5	5.77	2.50	-3406.4
[15]crown-5 / ClO_4^-	9.01	3.88	-5285.2	8.13	3.53	-4801.7
[18]crown-6 / ClO_4^-	10.25	4.45	-6052.3	9.81	4.26	-5790.5
[15]crown-5 / Cl^-	8.28	3.59	-4890.0	7.45	3.24	-4400.1

a. At 25 °C. b. With equivalent amount of crown ether. c. in kCal/M

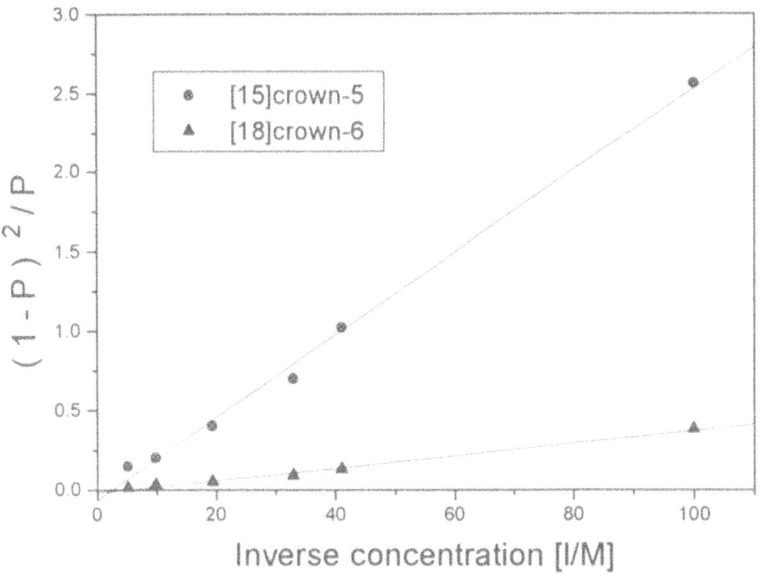

Figure 1. The plot of Na^+ for the estimation of 1:1 ratio of association onstants of macrocycles according to Equation 3.

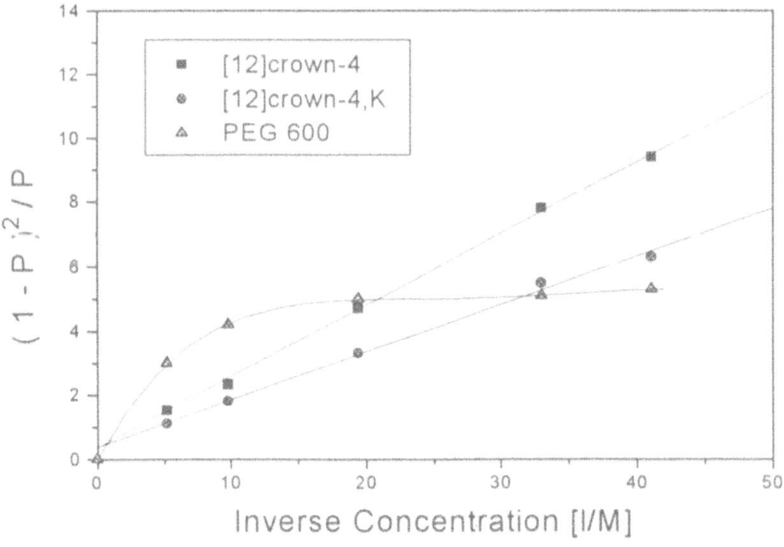

Figure 2. The plot of Na^+ for the estimation of 1:1 association constants of macro ligands according to Equation 3.

Accordingly, the results particularly displayed the selectivities so the role of macrocycle size fitting to cation radius. On the other hand, the mechanisms of cation macrocycle interactions are primarily solvent depended as reported by Izatt [2]. This is strongly supported in our studies reported with the ^{13}C-NMR, Potentiometry and recently using the fluorescence spectroscopy [3-8]. The Eigen-Winkler mechanism [13,14] concluded for NMR studies is the best to explaine the results in non-hydroxylic media regarding a *dissociative* reaction among the distinct species [3-5,14]. However, in polar solvents the *associative* exchange processes is likely to involve which is also related with conformational route of a macrocycle in solution [2].

The both of the mechanisms may be involved, therefore, the association constants in dioxane/water are rather high compared to results obtained in water as expected [1,6,14]. However, the dissociation process involved consecutive exchanges of the oxygen-metal bonds along with solvent-metal association [2]. The smaller size of macrocycles possessed low Log K_a values but more likely to form more stable 2:1 ratio of complexes, Table 2. We believed the mechanism is best explained with Equation 5 but what we measure experimentally is the Equation 6 as given below [3-7].

$$(A_{n+1}{}^{+}L_m)^+ \rightleftharpoons (A_n{}^{+}L_m) + A^+ \qquad 5)$$

$$A_n{}^{+}L_m \rightleftharpoons nA^+ + mL \qquad (m=1,2 \quad n=1,2) \qquad 6)$$

References

1. Y. Inoue, T. Hakushi and Y. Liu (1990) *Cation Binding by Macrocycles,*, Y. Inoue and G. W. Gokel (Eds), in Thermodynamics of Cation-Macrocycle Complexation, Enthalpy-Entropy Compensation. Marcel Dekker, New York,. pp 1-110.
2. R. M. Izatt, K. Pawlak and J. S. Bradshaw, *Chem. Rev.* **95,** 2629 (1995).s
3. Ç. Erk, *J. Mol. Liq.* **40,**1(1989).
4. Ç. Erk, *J. Phys. Chem.* **94,** 618 (1990).
5. Ç. Erk, *J. Mol. Liq.* **48,**17(1989).
6. Ü. Çakır and Ç. Erk, *Thermochim. Acta.* **178,** 67(1991), references cited.
7. A. Göçmen and Ç. Erk, *Fresenius J. Anal. Chem.* **347,**472(1993).
8. G. Yapar, G. Döner, A. Göçmen, S. Akman and Ç. Erk, Abstracts, *6th Internat. Seminar on Inclusion Comp.*, 1995, Istanbul, Turkey.
9. E. M. Eyring and S. Petrucci, (1990) *Cation Binding by Macrocycles*, Y. Inoue and G. W. Gokel (Eds), in Rates and Mechanisms of complexation Reactions of Cations with Crown Ethers and Related Macrocycles. Marcel Dekker, New York, pp 179-202.
10. G. Chaput and G. Jeminet, J. Juillard. Can. J. Chem. **53,** 2240(1975).
11. Y. Takeda and T. Kimura, *J. Incl. Phen.* **11,**159(1991).
12. G. W. Gokel, D. M. Goli, C. Minganti and L. Echegoyen, *J. Amer. Chem. Soc.* **105** 6786(1983).
13. M. Szwarc, *Pure and Appl. Chem.* **51,**49(1978).
14. C. Detellier and P. Laszlo, (1985) Topics in Molecular Interactions, W.J. Orville-Thomas, H. Ratajczak, C.N.R. Rao (Eds.) in Studies of Ion-Molecule Interactions by NMR Spectroscopy, Elsevier, Amsterdam, pp 291-336.

BINUCLEAR COPPER(II) COMPLEXES OF CYCLO-BIS INTERCALAND RECEPTORS. EFFECT OF THE LIGAND ON THE CRYSTAL STRUCTURE AND COMPLEXATION PROPERTIES

M. CESARIO [a] , J. GUILHEM [a] ,C. PASCARD [a] ,
M-P.TEULADE-FICHOU [b] , M. DHAENENS [b], J-P . VIGNERON [b],
J-M. LEHN [b]
a) Laboratoire de cristallochimie , Institut de Chimie des Substances Naturelles, CNRS, Gif sur Yvette, France.
b) Laboratoire de Chimie des Interactions Moléculaires, Collège de France Paris, France

the

1 Introduction.

There has been considerable interest in metal complexes of binucleating ligands as models for the coordination environment of the active sites in binucleating metalloenzymes. Numerous synthetic systems, like di-copper or di-cobalt complexes, have been demonstrated to function as catalysts for the transformation (oxidation, hydrolysis) of organic substrates other than the ligand itself. The catalytic efficiency and the selectivity depends critically on the distance between the metallic centers, the conformation of the spacer and the stability of the complex.

The cyclo-bisintercaland receptors (two large aromatic subunits linked by diethylene triamine chains) are a family of molecules topologically well adapted to the strong binding of metallic cations through N-coordination. Moreover the two aromatic systems define a flat walled lipophilic cavity of modulable size able to accomodate planar organic species.

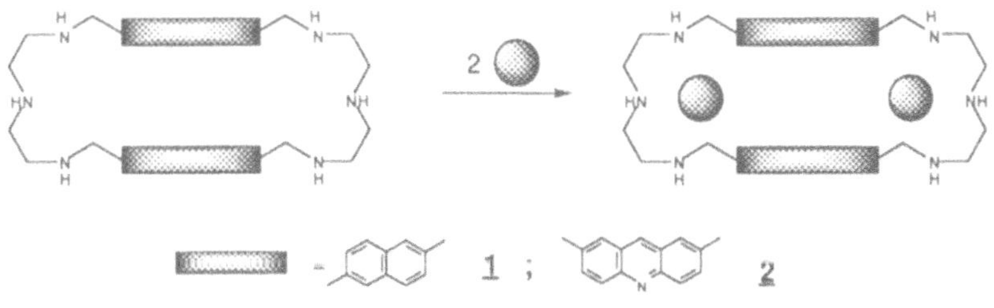

A. W. Coleman (ed.), Molecular Recognition and Inclusion, 279–282.
© 1998 *Kluwer Academic Publishers. Printed in the Netherlands.*

2 Crystal structures

The dicopper (II) complexes of cyclo-bisintercaland **1** and **2** have been synthesized and their crystal structure has been determined.

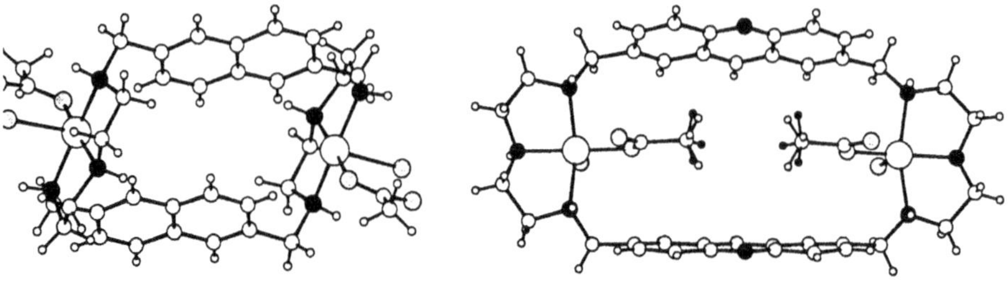

In *both cases* the two Cu(II) atoms are bound by the diethylenetriamine chains and each ion is further monocoordinated to the acetate ligand, the overall pentacoordination being achieved by a water molecule.
By *contrast* the spatial arrangement of the ligands bisnaphtalene or bisacridine around the two metallic centers is radically different.

Complex I

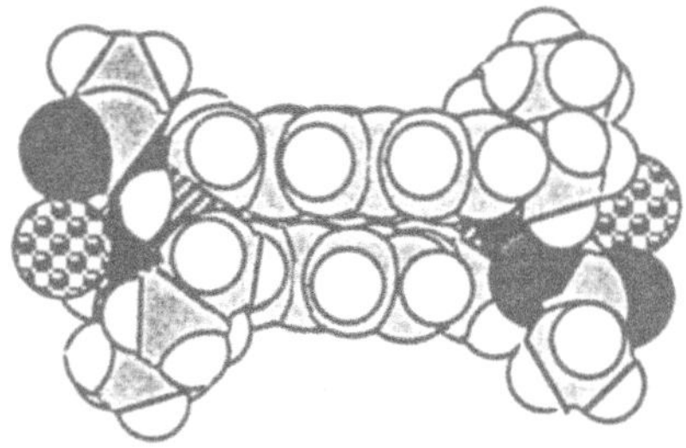

Complex II

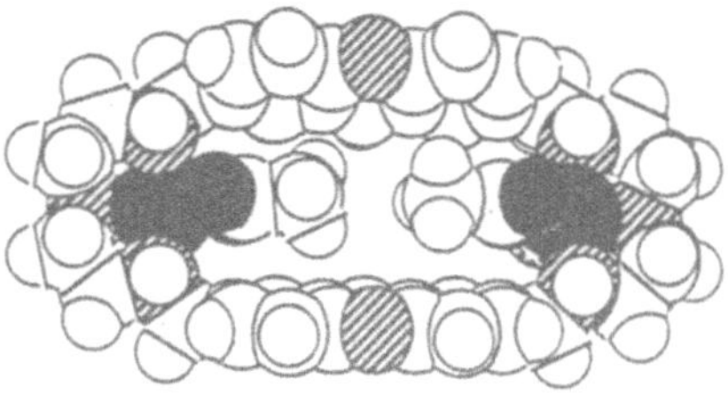

The smaller macrocycle **1** gives a **flattened** structure where the two naphtalene subunits are stacked in parallel planes [average d C(naphtalene)-C(naphtalene =3.8Å] There is no space available for the acetate groups.The whole coordination systems undergo a rotation of 90° around the Cu (II) ions, the two metal atoms being located on opposite sides of the macrocycle. The slipping of the parallel naphtalene subunits affords a better adaptability to the environment.

The larger bisacridine ligand **2** gives an **opened** structure with non parallel aromatic units [$\Phi = 32°$, d N(acridine)-N(acridine) = 7.43 Å] . The copper atoms are located on the same side of the macrocycle. A molecular cavity is thus defined in which the Cu (II) ions and the two acetate groups are included in a tail to tail arrangement, the distance between the two internal methyl groups being 4.0Å.

A second type of crystals has been isolated from the same sample for the dinuclear complex of **2**.The structure (**II bis**) has been found to be a polymeric assembly where dicopper macrocyclic units are coordinated by diacetate bridges.

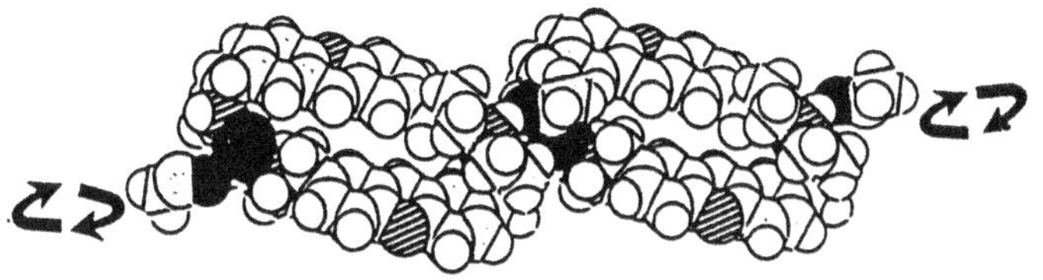

II bis

The macrocycle, under the polymeric form is **flattened** with two acridine subunits no more superimposed , as in structure **II** , but rotated of 180° one respect to another.The existence of these two different crystal packings highlights the flexibility of the macrocyclic ligand.

3 Complexation measurements.

The ability of compound **II** to accomodate anionic aromatic guests has been investigated by fluorescence and absorption spectroscopy in water/ methanol 1:1 (v/v). The gradual addition of small amounts of complex **II** to naphtalene 2,6-dicarboxylate (**NDC**$^{2-}$) results in the strong quenching of the fluorescence of the naphthalene ring, reaching a limiting value at 1 to 2 equivalents of the dicopper compound. Also, measurement of the absorbance of a 1/1 mixture of **II** and **NDC**$^{2-}$ shows a net hypochromism of the π–π* band of the acridine moieties of the macrocycle. Similar spectroscopic features are observed with the monoanionic guest naphthalene 2-sulfonate (**NMS**$^{1-}$) but with a lower affinity of the two species as a larger amount of compound **II** (10-15 eq) is necessary to quench completely the fluorescence of this substrate. Due to absorption of the bisacridine macrocycle at the excitation wavelength of the naphthalene (298nm), it was not possible to determine accurately the binding parameters (Kass, stoichiometries) from the fluorescence data. However these results clearly demonstrate the occurence of strong electrostatic and π/π interactions between compound **II** and naphthalenic anionic derivatives, which suggest the formation of ternary complexes. Further investigations are currently in progress to confirm this hypothesis.

Conclusion.

-The *whole non-covalent interactions* involved in the *binding* of *ligand*, *substrate* and also *medium* assume the *cohesion* of *all the pieces* of the *supramolecular architecture* but for a *same type* of [ligand, substrate and solvent] several *different schemes* could be devised or may occur.

- *Flexibility* of the *ligand* is necessary for a better structural adaption between macrocycle cavity and geometry of substrates

I

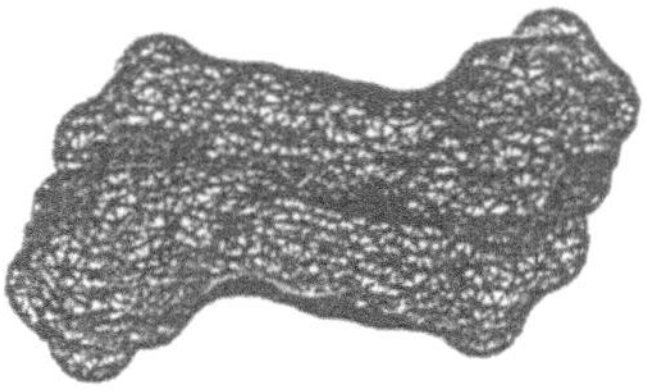

II

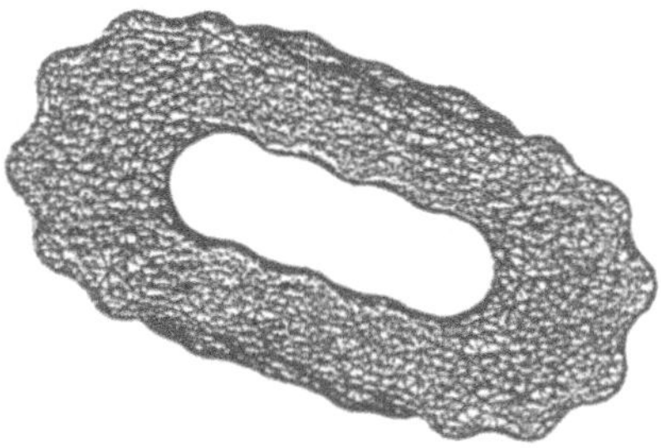

II bis

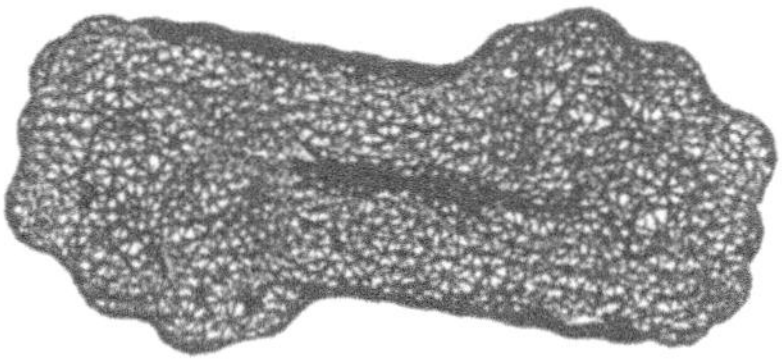

Macrocyclic envelops in structures **I**, **II** and **II bis**

-The dramatic difference between structures **I** and **II** emphasizes the strong influence of the size and of the ligand conformation on the arrangement in the solid state. The bisnaphtalene ligand can accomodate a pair of copper (II) cations but the two metallic centers bound in a close proximity are submitted to electrostatic and steric repulsions which reject them in an external position with subsequent flattening of the macrocycle. The bisacridine ligand of larger size but also sligthly different conformation (different position of the linkers on the aromatic ring) provides a more suitable distance between the two coordination sites and gives thus an open structure. This macrocycle seems more adapted for further catalytic investigations.

SPECIFIC INTERACTION OF β-CASOMORPHIN (HUMAN) WITH Cu(II) ION

E. CHRUSCINSKA,[a] G. MICERA,[b] D. SANNA,[b] and W. AMBROZIAK[a]
[a] *Institute of General Foqd Chemistry, Technical University of Lodz, 4/10 Stefanowski St., 90-924 Lodz, Poland.*
[b] *Department of Chemistry, University of Sassari, Via Vienna 2, 07100 Sassari, Italy.*

1. Introduction

Peptides with opioid activity have been described in partial enzymatic digests of proteins derived from foodstuffs and isolated from pepsin hydrolysates of wheat gluten, α- and β-casein [1]. Their appear quite resistant to the action of gastrointestinal enzymes. Therefore, once formed in the stomach, they could resist further degradation and elicit a physiological effect. The β-casomorphin sequence seems, to be of general occurrence in all the β-caseins so far investigated, although the structure of human β-casein slightly differs from that of its bovine counterpart. The differences occur within the β-casomorphin sequence for opiod activity. Thus, although the *N*-terminal tripeptide structure essential for opiod activity is conserved, human β-caseins are less potent than bovine β-casomorphins in the GPI assay [2]. β-Casomorphins show specific binding to μ-receptors probably due to their sequence Tyr-Pro-Phe. Apart from their possible role as neuro-transmitters and neuro-modulators, β-casomorphins may also influence neutral maturation, since pre-natal and post-natal administration of opiate receptor agonists can alter the development of various neutral pathways [3].

An earlier work on the opioid peptide β-casomorphin-5 isolated from casein and a number of its fragments containing Tyr and Pro residues showed a very specific coordination to Cu(II) ions [4]. It is also known that the stabilisation of complex species coordinated through the amide nitrogens of residues beyond the proline mojety or through the phenolic oxygens of Tyr side chains results in etheir large chelate rings or dimeric arrangements [5].

This study was undertaken in order to examine the Cu(II) coordination ability of the β-casomorphin-7 (human) sequence Tyr-Pro-Phe-Val-Glu-Pro-Ile. A comparison

A. W. Coleman (ed.), Molecular Recognition and Inclusion, 283–290.
© 1998 *Kluwer Academic Publishers. Printed in the Netherlands.*

with similar systems has been also made.

2. Experimental

The heptapeptide Tyr-Pro-Phe-Val-Glu-Pro-Ile was purchased from Bachem and used without purification. L-Tyrosylglycine was a Sigma product.

2.1. POTENTIOMETRIC STUDIES

Stability constants for complexes of H^+ and Cu^{2+} were calculated from titration curves obtained using total volumes of about 1.5 cm^3. Alkali was added from a 0.25 cm^3 micrometer syringe which had been calibrated by both weight titration and the titration of standardised materials. Experimental details were: peptide concentration 0.002 mol dm^{-3}, copper(II) concentration 0.001 mol dm^{-3}; ionic strength 0.1 (KNO_3); pH-metric titrations on a MOLSPIN pH-meter system using a micro combined glass/calomel electrode (Russell, CMAWL) calibrated in concentration using HNO_3; temperature 25°C. Calculations were made with the aid of the SUPERQUAD computer program [6] which allows for the refinement of total ligand concentrations. Titration data confirmed the purity of the peptide, in particular, the absence of acetate.

2.2. SPECTROSCOPIC MEASUREMENTS

Anisotropic X-band EPR spectra (9.15 GHz) of frozen solutions were recorded at 140 K, using a Varian E-9 spectrometer after addition of ethanediol to ensure good glass formation. Circular dichroism (CD) spectra were obtained with a Jobin-Yvon CD-6 spectropolarimeter. Concentrations used in the spectroscopic measurements were the same as those given for pH-metric titrations.

3. Results and Discussion

Over the measurable pH range the heptapeptide behaves as a tetraprotic acid H_4L. The ionizable protons belong to the carboxylic (pK = 3.56, terminal; pK = 4.68, chain), amino (pK = 7.34) and phenolic (pK = 9.73) groups (Table 1). The values are comparable to those measured previously on β-casomorphin-5 [4]. It is also worth to note that the acidity of the phenolic proton is comparable to that measured for *N*-terminal tyrosine residues [7].

Values calculated for the Cu(II) formation constants are also given in Table 1. Species distribution is shown in Figure 1. The $CuLH_2$ complex is a species where the peptide coordinates the metal ion only *via* carboxylate groups. Normally, these species have a very low concentration and are not easy to detect by conventional equilibrium studies (see ref. 8). However, a reliable indication of these species is given by EPR spectra where shoulders show up on the resonances of the aquaion even if precise spectral parameters cannot be extracted [8]. The deprotonation of the terminal ammonium group (pK = 4.97) makes possible the chelation of the metal through the terminal amino nitrogen and the CO oxygen of the neighbouring peptide linkage in CuLH. Usually, species with the (NH_2, CO) bonding mode are minor components of Cu(II)-oligopeptide systems because, as the pH is raised, the presence of the ionizable amide group adjacent to CO rapidly transforms the (NH_2, CO)-coordinated species into the much more stable (NH_2, N^-)-based complex [8]. The proline residue can act as a break-point to Cu^{2+} coordination when inserted into a oligopeptide chain, particularly in the second position [9-11]. In β-casomorphin this arrangement makes the (NH_2, CO) binding mode much more favoured than *e.g.* with Glygly and favours also the formation of a bis-complex CuL_2H_2 in the presence of excess ligand. The same behaviour was observed with glycylsarcosine where the methyl substitution excludes the amide nitrogen from coordination. The 1x and 2x(NH_2, CO) donor sets in CuLH and CuL_2H_2 complexes, respectively, result in EPR spectra which are in a fairly good agreement with those of the analogous species of Glysar and glycinamide [8].

A Cu_2L_2 dimeric species predominates in the pH region 7-9. Dimeric species are typical of oligopeptides containing the *N*-terminal tyrosine residue. For instance L-tyrosylglycine forms a dinuclear complex $Cu_2L_2H_{-2}$ where each metal ion is bound to the (COO^-, N^-, NH_2) donor set of a ligand and to the phenolate group of the other peptide molecule [12]. The arrangement is likely favoured by stacking interactions between the aromatic rings [13]. The magnetic interaction between the two copper(II) ions in this type of dimeric arrangement has been reported to be responsible for a decrease of the EPR signal intensity [14] and/or for the appearance of $\Delta m = 2$ resonances [5, 7, 15]. A distinct indication for the Cu(II)-Cu(II) interaction following phenolate bridging was obtained by our EPR analysis of the L-tyrosylglycine-Cu(II) system. In fact, the weak EPR spectrum which is concomitant to the formation of $Cu_2L_2H_{-2}$ exhibits parallel hyperfine features typical of a dimeric entity involving two I = 3/2 equivalent paramagnetic centers. Namely, in Cu(II) dimers, the hyperfine splitting is expected to consist of seven (four for a monomer) components with a coupling constant which is about one half of that expected for a monomer with the same coordination set. Actually, at least 5 lines are distinguished in the spectrum of $Cu_2L_2H_{-2}$ species of L-tyrosylglycine, with $A = 67 \times 10^{-4}$ cm^{-1} (Table 1). No zero-field splitting is measurable in the anisotropic spectrum. This indicates that the bridging phenolate residue, while is effective in pairing the copper(II) electrons and yielding S=0 (EPR-inactive) and S=1 (EPR-active) states, produces a low zero field splitting value in the S=1 state because of the rather large distance between the paramagnetic centres. Noticeably, the formation of Cu_2L_2 in the Cu(II)-β-casomorphin system yields S=1 EPR features analogous to those measured with Cu(II) L-tyrosylglycine (Table 1). This supports that the structure of the complex involves a couple of ligands, each of which bridges two copper ions by coordinating one *via* the (NH_2, CO) or (NH_2, 2 CO) set and another *via* the phenolate oxygen. On the whole, this results in a dimeric interaction analogous to that supposed for Cu(II)-L-Tyr-Gly. A support to the Cu(II)-phenolate interaction is provided by the CT band at *ca.* 420 nm observed in the CD spectra (in electronic spectra there is strong absorption in this region; however, the maximum is not resolved).

By pH 9 the fit of potentiometric data indicates a monomeric species CuL_2 in the presence of excess ligand. The species is also detected by spectral measurements. In particular, CD spectra support the bonding of Cu(II) to deprotonated amide through the N^--Cu(II) charge transfer transition and rule out phenolate-Cu(II) coordination. Therefore, deprotonation and coordination of the amide groups is supposed in this complex. It is probable that each of the two ligand molecules, with the undissociated phenolic group, binds the metal ion through the nitrogen of the amino end together with a deprotonated amide most likely belonging to the Val residue. A macrochelate ring may be assumed, as it occurs with other peptides bearing proline in the second position [7]. The CD band at *ca.* 540 nm and the EPR parameters may be indicative of a species where three nitrogen atoms are in the metal plane and the fourth is in the apical position of a hexa-coordinate geometry (see Scheme).

The stepwise deprotonation of the phenolic groups (pK values of 9.43 and 10.09) yields in sequence CuL_2H_{-1} and CuL_2H_{-2} in which the same metal coordination as in CuL_2 is retained, as substantiated by EPR and CD spectra. In particular, the absence of CD transitions around 400 nm excludes the phenolate-Cu(II) coordination, whereas the band at *ca.* 325 nm is supportive of the binding of the metal ion to the peptide nitrogen.

References

1. Schlimme, E., Meisel, H., and Frister, H. Bioactive sequences in milk proteins, in C. A. Barth and E. Schlimme (eds.), *Milk Proteins*, Steinkopff Verlag Darmsdadt, Springer-Verlag, New York, (1989), pp. 143-149.
2. Brantl, V. Novel opioid peptides derived from human β-casein: Human β-casomorphin, *Eur. J. Pharm.* **106** (1985), 213-214.
3. Handelman, G. E. Neuropeptide effects on brain development, *J. Physiol. Paris* **80** (1985) 268-274.
4. Formicka-Kozlowska, G., Pettit, L. D., Steel, I., Hartrodt, B., Neubert, K., Rekowski, P., and Kupryszewski, G. The coordination of copper(II) with β-casomorphin and its fragments, *J. Inorg. Biochem.* **22** (1984) 155-163.
5. Pettit, L. D., Steel, I., Kowalik, T., Kozlowski, H., and Bataille, M. Specific binding of the tyrosine residue in copper(II) complexes of Tyr-Pro-Gly-Tyr and Tyr-Gly-Pro-Tyr, *J. Chem. Soc., Dalton Trans.*, (1985) 1201-1205.
6. Gans, P., Sabatini, A., and Vacca, A. SUPERQUAD: An improved general program for computation of formation constants from potentiometric data. *J. Chem. Soc., Dalton Trans.* (1985) 1195-1200.
7. Kozlowski, H., Bezer, M., Pettit, L. D., Bataille, M., and Hecquet, B. Coordination abilities of tetrapeptides containing proline and tyrosine. A spectrophotometric and potentiometric study. *J. Inorg. Biochem.*, **18** (1983) 231-240.
8. Sóvágó, I., Sanna, D., Dessì, A., Várnagy, K., and Micera, G. EPR and potentiometric reinvestigation of copper(II) complexation with simple

oligopeptides and related compounds. *J. Inorg. Biochem.*, **63** (1996) 99-117.

9. Pettit, L. D., Steel, I., Formicka-Kozlowska, G., Tatarowski, T., and Bataille, M. The L-proline residue as a 'break-point' in metal-peptide systems. *J. Chem. Soc., Dalton Trans.*, (1985) 535-539.
10. Bezer, M., Pettit, L. D., Steel, I., Bataille, M., Djemil, S., and Kozlowski, H. Copper(II) interaction with tetrapeptides containing proline and phenylalanine. A potentiometric and spectrophotometric study. *J. Inorg. Biochem.*, **20** (1984) 13-21.
11. Livera, C., Pettit, L. D., Bataille, M., Krembel, J., Bal, W., and Kozlowski, H. Copper(II) complexes with some tetrapeptides containing the 'break-point' prolyl residue in the third position . *J. Chem. Soc., Dalton Trans.*, (1988) 1357-1360.
12. Hefford, R. J. W. and Pettit, L. D. Potentiometric and spectrophotometric study of the coordination compounds formed between copper(II) and dipeptides containing tyrosine. *J. Chem. Soc., Dalton Trans.*, (1981) 1331-1335.
13. Yamauchi, O., Tsujide, K., and Odani, A. Copper(II) complexes of tyrosine-containing dipeptides. Effects of side-chain groups on spectral and solution chemical properties and their structural implication. *J. Am. Chem. Soc.* **107** (1985) 659-666.
14. Radomska, B., Sóvágó, I., and Kiss, T. Tyrosinate and lysinate as bridging residues in copper(II) dipeptide complexes. *J. Chem. Soc., Dalton Trans.* (1990) 289-292.
15. Kiss, T. and Szücs, Z. Studies on transition-metal-peptide complexes. Part 12. Copper(II) complexes of dipeptides containing phenylalanine and tyrosine. *J. Chem. Soc., Dalton Trans.* (1986) 2443-2447.

Figure 1 Distribution curves of the complexes formed in the Cu(II)-β-casomorphin system at metal ion to ligand molar ratio 1:2, $c_{Cu(II)} = 0.001$ mol dm^{-3}.

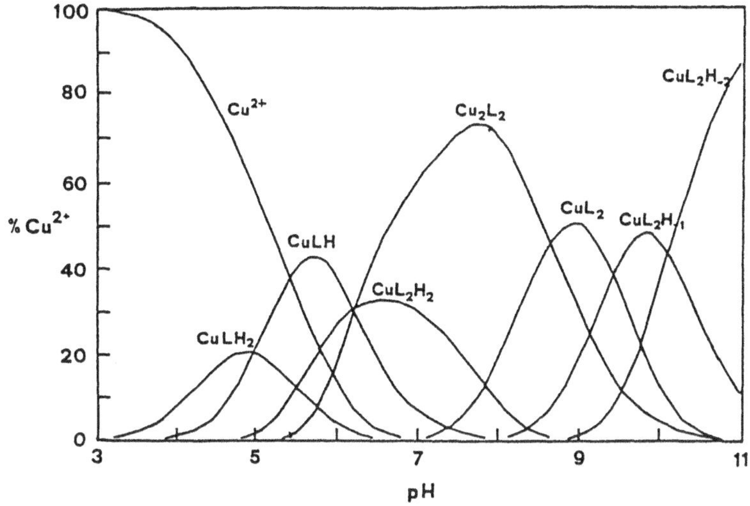

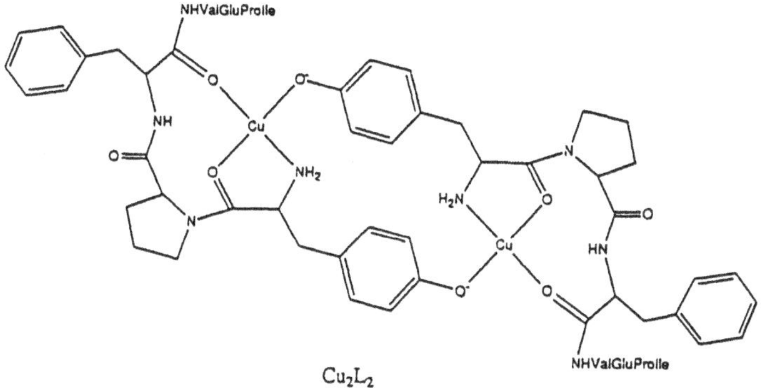

Cu_2L_2

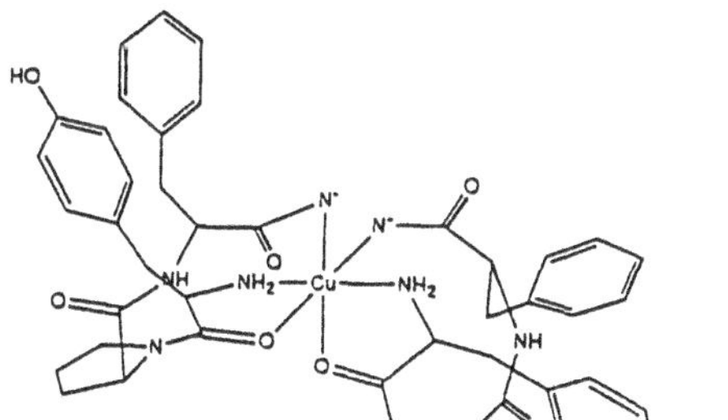

CuL_2

TABLE 1. Stability Constants (log β) and Spectroscopic Parameters.

Species	logβ	$g_{\parallel}$	$A_{\parallel}$ (Cu) (10^{-4} cm^{-1})	CD[a] λ (nm)	Donor Set
β-casomorphin					
H_4L	25.31(1)				
H_3L	21.75(1)				
H_2L	17.07(1)				
HL	9.73(1)				
$CuLH_2$	19.59(6)	b			
CuLH	14.62(3)	2.329	161		(NH_2, CO)
CuL_2H_2	28.66(3)	2.278	162		2x(NH_2, CO)
CuL_2	12.99(4)	2.195	187		(NH_2, N^-) (NH_2, CO) + CO, N^-
Cu_2L_2	20.06(6)	2.259	88	418(-0.04)	(O^-) (NH_2, 2 CO)
CuL_2H_{-1}	3.57(4)	2.195	187	324(+0.02) 540(-0.02)	(NH_2, N^-) (NH_2, CO) + CO, N^-
CuL_2H_{-2}	-6.53(4)	2.195	187	327(+0.07) 543(-0.07)	(NH_2, N^-) (NH_2, CO) + CO, N^-
L-Tyr-Gly					
CuL		2.244	184		(NH_2, N^-, COO^-)
$Cu_2L_2H_{-2}$		2.234	67		(O^-) (NH_2, N^-, COO^-)
$CuLH_{-2}$		2.243	151		(NH_2, N^-, COO^-)(OH^-)
$CuLH_{-3}$		2.217	203		(NH_2, N^-) (2 OH^-)

[a] Values of Δε (mol^{-1} dm^3 cm^{-1}) in parentheses measured at the maximum extent of formation of the species.

[b] Parameters were not measured because of the low concentration of the species.

THE STRUCTURE AND PROPERTIES OF THE NEW LIGAND FOR THE 5-HT_{1A} RECEPTORS

Yu.M.Chumakov,G.Bocelli, A.Cantoni, M.Gdaniec, V.M.Sava, S.G.Soboleva

Institute of Applied Physics of Academy of Science of Moldova, Academicheskaya 5 str., Kishinev Moldova A.V.Bogatsky Physico-Chemical Institute of Academy of Science of Ukraine, Lustdorsfcfaya doroga, 86, Odessa Ukraine, University of N.Copernic Poland, CSSD - CNR, Parma, Italy

The hetaryl(aryl) piperazines derivatives have been widely investigated in the role of potential psychothropic agents. Their neuroleptic and anxiolytic properties are caused by interaction with serotonine and dophamine receptors. It is necessary to determine the physicochemical factors that affect recognition and binding of these compounds to receptors. Ligand affinity to the receptor is found out by electron and X-ray ligand structure (Fig.1).

A. W. Coleman (ed.), Molecular Recognition and Inclusion, 291–295.
© 1998 *Kluwer Academic Publishers. Printed in the Netherlands.*

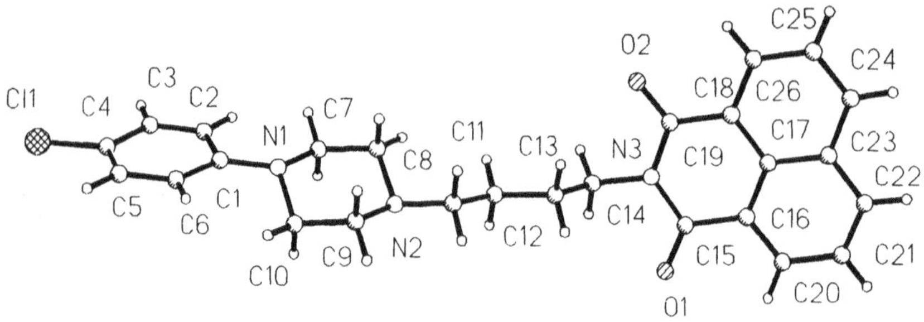

Figure 1. The X-ray structure of compound I

We have obtained the 1-(n-chlorphenyl-4-[4-N-naphtalimido) butyl] piperazine (I) ligand during the investigation of "structure-properties" relationship of potential serotonine and dophamine receptors ligands. The compound I was obtained by condensation of N-(ω-brombutyl)naphtalimide with N-(n-chlor)phenylpiperazine. Molecular recognition of the compound I to 5-HT_{1A} serotonine receptors was appreciated by us according to its ability to inhibit the ^3H-8-oxy-2-dipropylaminotetraline binding with the receptor. The received meanings of IC_{50} and K_i 46.312.2 and 38.712.9 HM correspondingly indicate to the high affinity of this substance to 5-HT_{1A} receptors.

In accordance with hypothetical ideas about 5-HT_{1A} receptors topography ligands of these receptors must content strongly basic nitrogen atom and aromatic nuclear at the distance about 5,6Å and they must have the stretched conformation [1,2]. In the investigated molecule the distance from N2 atom up to n-chlorphenyl nuclear centre is 5,67Å. The molecule has the linear (stretched) form as it was mentioned above. According to these characteristics compound I satisfies the structural and conformational requirements to 5HT_{1A} receptors ligands. It is considered that agonists and antagonists of 5-HT_{1A}-receptors differ conformationally. In the agonist molecules deviation of the basic nitrogen atom (N2 in our case) from the aromatic nuclear plane can be about 0.2Å, while the antagonist deflection reaches the value of 1,6Å.In the compound I the mentioned deviation is equal 0.69Å. That's why the behaviour of compound I as agonist 5-HT_{1A}-receptors is more likely. Aryl(hetaryl)piperazines are also able to interact with D_2-receptors. Ligands of these receptors must have claw like conformation. In such case it is reached the 4,5Å distance from N2 atom up to the lypophile fragment (naphthalimide nuclear in our case) centre. In order to reveal the peculiarities of the investigated molecule I and her protonated form we have made the quantum-chemical calculations of their electronic structure by CNDO/2 method.

For the aim of optimizing of protonated form geometry and conformational analysis of the molecule I there were used PCMODEL programs. The optimization of the protonated form geometry of compound I showed that its molecule has linear conformation. Conformation analysis of molecule I showed that during the rotation around the butylene chain bonds the N2-C17 distance became 6.53Å, while in the salt form of I this distance is equal 4.3Å. The donor properties of the molecule I are determined mainly by N1 atom's p_π-orbital and p_π-orbitals of the benzene ring. But donor-acceptor properties of the protonated form of the I are determinated by the atoms of the naphthalimide fragment. By CNDO method and conformational analysis it was shown that compound I may interact with D_2-receptor only in protonated form.

The interaction sites can be conveniently characterized with the aid of the molecular electrostatic potential (MEP). In compound I such positions are localized in vicinity of N1,N3 atoms, aromatic cycle and C7-C8,C22-C23 bonds, while in the protonated form - in vicinity of N3 atom, carbonyl oxygen atoms and N3-C14, N1-C1 bonds(Fig.2). Buspirone(II) and some of its structural analogues belong to the same class as compound I. In the central nervous system, the action of II as an anxiolytic is believed to proceed by way of binding to the 5-HT_{1A} subtype of the serotonin receptors. The MEP distribution analysis of II have shown that the protonated nitrogen atom, constituting the primery interaction site, belongs to the piperazine ring(N2). In the buspirone-receptor complex, the protonated nitogen is probably directed toward a hydrogen bond acceptor in the receptor protein chain. The second, third and fourth interaction sites are located in the vicinity of the cyclic imide moiety in front of the carbonyl oxygen atoms and in vicinity of the aromatic ring in front of the endocyclic nitrogen atom [2].

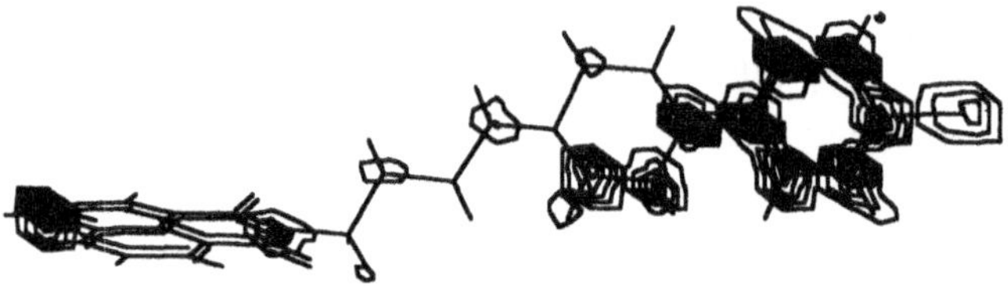

Figure 2. The MEP distribution for compound I

Therefore the extended rodlike form of molecule I, its donor-acceptor properties and the MEP distribution in the vicinity of the aromatic ring and naphtalimide fragment should help understand the recognition of ligands, via intermolecular hydrogen bonds, to 5-HT_{1A} receptor. The compound I may interact with D_2-receptor only in protonated form.

References:

1. Chidester, C.G., Lin , C.H., Lahti, R.A., Haadsma-Svensson, S.R., and Smith, M.W. (1993) Comparasion of 5-HT_{1A} and dopamine D_2 Pharmacophores. X-ray structures and affinities of conformationally constrained ligands, *J. Medicinal Chemistry* **36**, 1301-1315.

2. Chilmonczyk, Z., Les, A., Wozniakowska, A., Cybulski, J., Koziol, A.E., and Gdaniec, M. (1995) Buspirone analogues as ligands of the 5-HT_{1A} receptor. 1.The molecular structure of buspirone and its two analogues, *J.Medicinal Chemistry* **38**, 1701-1709.

Key words: piperazines , serotonine and dophamine receptors, psychothropic agents , electron and X-ray structure, structure-properties relationship, molecular recognition, quantum-chemical calculations, conformation analysis, CNDO method, molecular electrostatic potential.

PHASE TRANSITIONS OF CYCLOPHOSPHAZENE ADDUCTS DIRECTLY FOLLOWED BY SOLID STATE NMR

A. COMOTTI,[a] R. SIMONUTTI,[a] M.C. GALLAZZI,[b] and P. SOZZANI[a*]
a) Dip. Scienza dei Materiali, Università di Milano, Via Emanueli, 15, Milano;
b) Dip. di Chimica Ind., Politecnico di Milano, P[zza] L. da Vinci, 32, Milano.

1. Introduction

The Solid State NMR properties of guest and host species in inclusion compounds have been addressed by our research group in recent years. We have thus achieved a careful description of the dynamics of included chains of both high and low molecular weight.[1-4] We are now especially interested in matrices with tunable dimension of the space available for guest molecules.

Cyclotriphosphazenes are hosts in which even large side groups can be introduced. In particular we have studied tris(2,3-naphthalenedioxy)cyclophosphazene (TNP) inclusion compounds (IC) as compared with tris(o-phenylenedioxy)cyclotriphosphazene (TPP) ICs. TNP can show large channels (about 10 Å), with more than one molecule at the same cross section, "*Figure 1*".[5] Low boiling guests, such as benzene, THF and p-xylene, have been exploited for the formation of the inclusion compounds. Magic Angle Spinning (MAS) NMR can improve the knowledge of both the structures and the dynamics of the included molecules and the matrices.

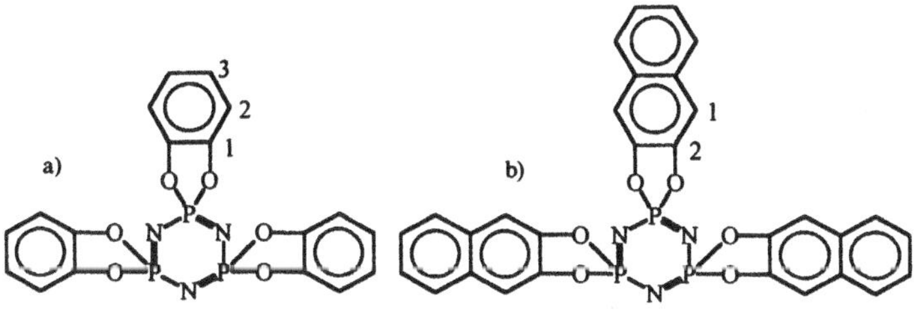

"Figure 1": a) (TPP); b) (TNP).

2. Experimental part

Tris(o-phenylenedioxy)cyclotriphosphazene (TPP) was prepared following the reported method.[6] Tris(2,3-naphthalenedioxy)cyclophosphazene was prepared as reported in the literature.[7] The adducts with THF, benzene and p-xylene were prepared by crystallization from hot solutions of the appropriate solvents. Pure cyclophosphazene was obtained by the THF adduct after washing with two portions of pentane.

High-resolution ^{13}C MAS NMR spectra were run on a MSL 300 Bruker instrument operating at a static field of 7.4 T. A Bruker magic angle spinning probe was used with 7 mm ZrO_2 rotors, spinning at a standard speed of 6000 Hz. The probe head was tuned to provide a

A. W. Coleman (ed.), Molecular Recognition and Inclusion, 297–300.
© 1998 *Kluwer Academic Publishers. Printed in the Netherlands.*

frequency of 300.13 MHz for ^{1}H (a 90° pulse length of 4 μs) and 75.47 MHz for ^{13}C. Cros Polarization MAS (CPMAS) experiments were performed with a contact time of 4 ms; 1000 2000 scans were performed on each sample.

3. Results and Discussion

3.1. TPP INCLUSION COMPOUNDS

^{13}C CP MAS NMR spectra of TPP and inclusion compounds with benzene, THF and p-xylene show, at room temperature, three signals for the host matrices, (Table 1) "*Figure 2*".

TABLE 1. Chemical shifts of the TPP adducts.

^{13}C δ (ppm)	HOST			GUEST		
	C1	C3	C2		INCL.	REF.
TPP	145.5	125.7	113.9			
TPP/benzene	145.4	125.6	114.0	benzene	127.4	128.5
TPP/THF	145.5	125.6	113.9	CH_2O	25.7	26.5
				CH_2	66.4	68.4
TPP/p-xilene	145.5	125.6	113.9	C	133.0	134.8
				CH	128.2	129.1
				CH_3	19.4	21.3

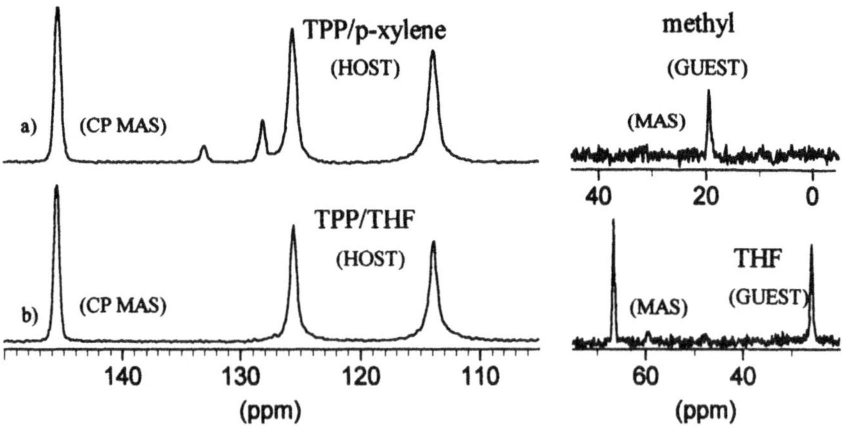

"Figure 2". ^{13}C MAS NMR spectra.

This result indicates that the high symmetry of the host molecules is maintained in the unit cell, this being compatible with the hexagonal unit cell. The spin-lattice relaxation times of the included molecules are short (a few seconds), indicating a relaxation mechanism due to the tumbling of the guest molecules. Accordingly, the number of guest signals follows the internal symmetry of the molecule, as in solutions.[5] Only a slight shift of 1-2 ppm was observed for the guest signals, compared to a 5% concentrated solution (Table 1). Even the pure matrix, once freed by the guest, shows the same multiplicity of signals, indicating no conformational changes and negligible NMR-active through-space interactions.

.2. TNP INCLUSION COMPOUNDS and POLYMORPHISM OF TNP MATRIX

THF and benzene form channel clathrates with hexagonal symmetry[8] as TPP/IC, generating in this case five signals for the host matrix *"Figure 3"*.
TNP/p-xylene clathrate presents a multiplicity of the matrix signals; three resolved peaks for each carbon nr. 1 and 2 (108-112 and 143-145 ppm) *"Figure 3"* are clearly observed. Signals marked by an asterisk correspond to the hexagonal phases. The spin-lattice relaxation times of p-xylene are longer than in TPP clathrate. A new crystal modification is therefore suggested, where guest molecules are arranged in a fixed position. A cage clathrate is consistent with this behavior.[9]

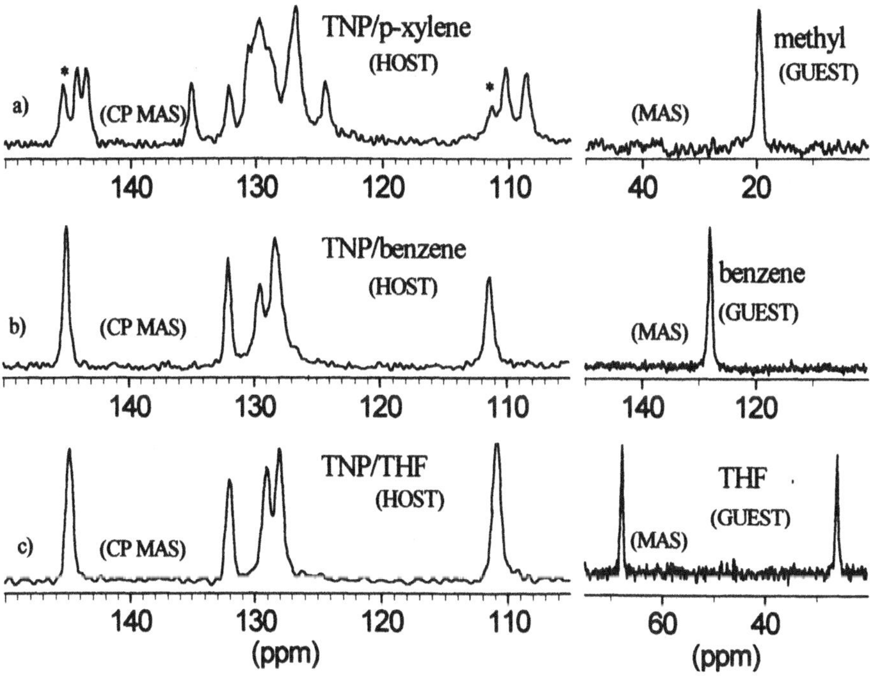

"Figure 3": ^{13}C MAS NMR spectra.

The multiplicity of signals of the pure TNP matrix is dramatically modified when the guest molecules are released. Two modifications of the matrix were obtained by different procedures: I) washing the reaction product with THF; II) washing the THF clathrate with pentane and vacuum drying, "*Figure 4*". Carbon nr.1 is particularly sensitive to the polymorphs. It presents signals centered at 107 ppm (two signals) and 117 ppm (two signals), respectively. Modification II was proved to be less stable, being relaxed to modification I after treatment at 400 K. Metastable states endowed with a certain degree of disorder were also observed during the transition.

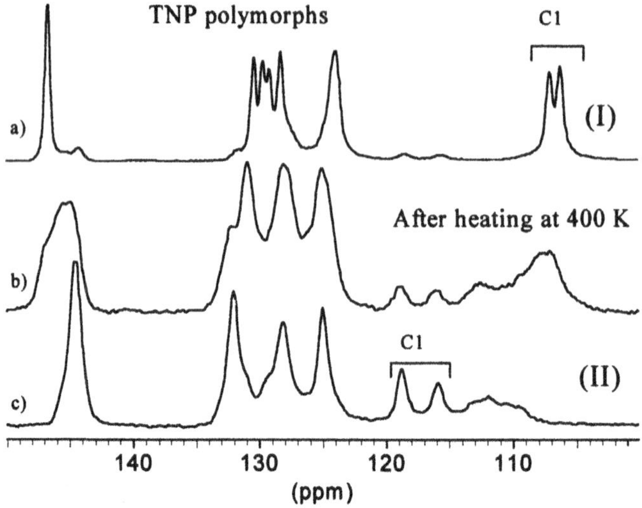

"Figure 4": ^{13}C CP MAS NMR spectra.

4. Conclusions

^{13}C MAS NMR allowed us to determine the solid state structure of a few inclusion compounds formed with TPP and TNP. The CS of each carbon atom in the TNP polymorphs and inclusion compounds covers, for some carbons, even 5 ppm and the splittings may be as wide as 2 ppm in the same spectrum. These marked effects are accounted for by conformational changes in the molecule, brought about by the supramolecular structure of the crystal.
Carbon nr. 1 seems to be the most sensitive; this carbon atom stands at a three bond distance from the phosphorus atom. The dihedral angle O-P-C(2)-C(1) when bent, changes the interaction of C(1) with P. Electronic delocalization is also to be taken into account. Thus the existence of a distortion of the five membered ring containing the oxygens is suggested. Bent conformations were described in the crystal structures proposed by Kobayashi.[10] Two modifications were recognized in the crystal structure of the pure matrix, and a third one is suggested by some spectra. In a few samples the modifications were not present in the pure forms, but the NMR spectra were valuable for the identification of the content of the single structures.

5. References

1. Sozzani, P., Bovey, F.A. and Schilling., F.C. (1989) *Macromolecules* **22**, 4225.
2. Sozzani, P., Schilling., F.C. and Amudsen, K.R. (1994) *Macromolecules* **27**, 6498.
3. Sozzani, P., Simonutti, R. and Comotti, A. (1996) *Mol.Cryst. Liq. Cryst.* **277**, 299-313.
4. Simonutti, R., Veeman, W.S., Runhau, F.C., Gallazzi, M.C. and Sozzani, P. (1996) *Macromolecules* **29**, 4958.
5. Allcock, H.R., Allen, R.W., Bissell, E.C., Smeltz, L.A. and Teeter, M. (1976) *J. Am. Chem. Soc.* **98**, 5120.
6. Allcock, H.R. (1964) *J. Am. Chem. Soc.* **86**, 2591.
7. Allcock, H.R. and Kugel, R.L. (1966) *Inorg. Chem.* **5**, 1016.
8. Allcock, H.R. and Stein, M.T. (1974) *J. Am. Chem. Soc.* **96**, 49.
9. Kubono, K., Asaka, N., Isoda, S. and Kobayashi, T. *Acta Cryst.* (1993) **C49**, 404.
10. Kubono, K., Asaka, N., Isoda, S., Kobayashi, T. and Taga, T. *Acta Cryst.* (1993) **C49**, 404.

MOLECULAR RECOGNITION IN SOLID INCLUSION COMPOUNDS OF NOVEL ROOF-SHAPED DIOL HOSTS

INGEBORG CSÖREGH
Department of Structural Chemistry, Arrhenius Laboratory, Stockholm University, S-106 91 Stockholm, Sweden, and
EDWIN WEBER
Institut für Organische Chemie, TU Bergakademie Freiberg, Leipziger Str. 29, D-09596 Freiberg/Sachsen, Germany.

1. Introduction

Crystallization is a good example of molecular recognition. The realized crystal structures are the result of a balance or compromise between intermolecular interactions of different strengths. In order to design materials with desired properties it is important to understand the laws governing the packing of organic molecules. Systematic X-ray analyses of cocrystals of synthetic host molecules provide excellent opportunities for investigation of non-covalent interactions between different groups in various environments.

Combination of the roof-shaped dihydroethanoanthracene skeleton with appended diarylmethanol clathratogenic groups yielded new crystalline inclusion hosts **1-4**.

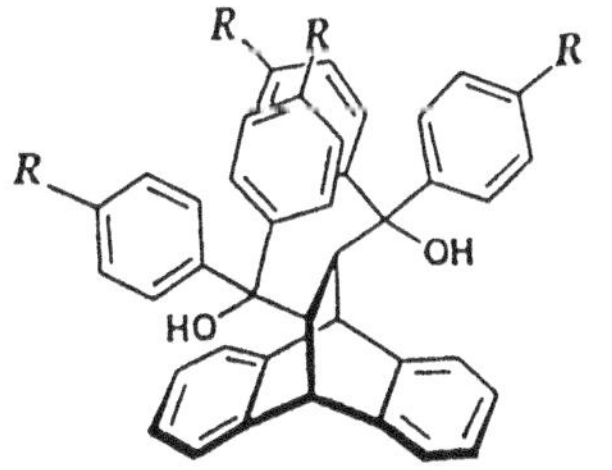

Host	*R*
1	H
2	Cl
3	CH_3
4	$C(CH_3)_3$

The clathrate formation ability of the new host family was tested with a broad variety of solvents including amines, alcohols, nitriles, nitro compounds and other aprotic dipolar molecules, heterocycles and various aromatic hydrocarbons (50 different species) [1]. 132 crystalline complexes of hosts **1-4** were isolated, with guests of varying polarity and H-bonding ability, indicating the present hosts to be substantial improvements upon the basic roof-shaped host type, previously studied by us [2,3].

X-ray studies of selected crystalline inclusion compounds of hosts **1-4** with different guests (Table 1) revealed not only crystalline architectures with a variety of symmetries

A. W. Coleman (ed.), Molecular Recognition and Inclusion, 301–304.
© 1998 *Kluwer Academic Publishers. Printed in the Netherlands.*

and host-guest stoichiometries, but also some unexpected intermolecular interaction modes and remarkable features, repeatedly appearing in different compounds.

TABLE 1. List of inclusion compounds studied by X-ray diffraction.

Host•Guest	*Stoichiometry*	*Space group*	*Z*	*R*	*for* N_{obs}
1•ethanol	1 : 2	*P2₁/c*	*4*	*0.042*	*3179*
1•nitroethane	1 : 1	*P2₁/c*	*4*	*0.059*	*3124*
1•benzene	2 : 1	*I-4*	*4*	*0.049*	*2066*
2•2-pentanol	1 : 1	*Fddd*	*16*	*0.053*	*1514*
2•1,4-dioxane	2 : 5	*P2₁/n*	*2*	*0.046*	*3877*
2•*o*-xylene	1 : 2	*Pc2₁n*	*4*	*0.041*	*2762*
3•methanol	2 : 1	*P-1*	*2*	*0.038*	*3366*
3•toluene	1 : 1	*Fddd*	*16*	*0.066*	*1239*
4•pyridine	2 : 3	*P-1*	*1*	*0.050*	*6234*

2. Molecular recognition by hosts 1-4

The present roof-shaped diol hosts exhibit two significantly different conformations in the studied solid inclusion compounds, most likely depending on the guest recognition modes. In one of these forms, called the 'active' conformation [Figure 1(a)], the two host OH groups form a rather strong intramolecular hydrogen bond as a part of the H-bond system, which involves also the guest functionalities. Host molecules **1**, **3** and **4** exhibit this conformation when including alcohols (**1**, **3**) or pyridine (**4**).

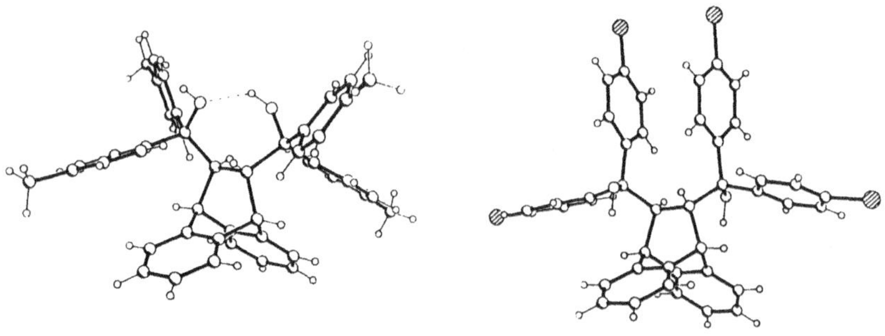

(a) (b)

Figure 1. Perspective illustration of the 'active' (a) and 'inactive' (b) conformation, shown by the present roof-shaped diol hosts.

In the other, more frequently occuring 'inactive' conformation [Figure 1(b)] the host OH groups are directed towards the nearest aryl rings of the anthracene moiety, thus possibly forming weak intramolecular O-H···π interactions.

Nevertheless, the chloro-containing host **2** proved to be an exception, exhibiting the latter 'inactive' conformation in all three of its studied inclusion compounds, independently of the proton donor/acceptor ability and polarity of the guest [4].

3. Molecular recognition in solid inclusions of the chloro-containing host 2

The solid inclusion compounds of host **2** have a very special structural feature in common with respect to the host matrix. This feature is that the host molecules are linked by simultaneous electrophile-nuclephile Cl···π interactions so as to form chains (Figure 2). The observed shorter Cl···π distances range between 3.33 and 3.75 Å [4].

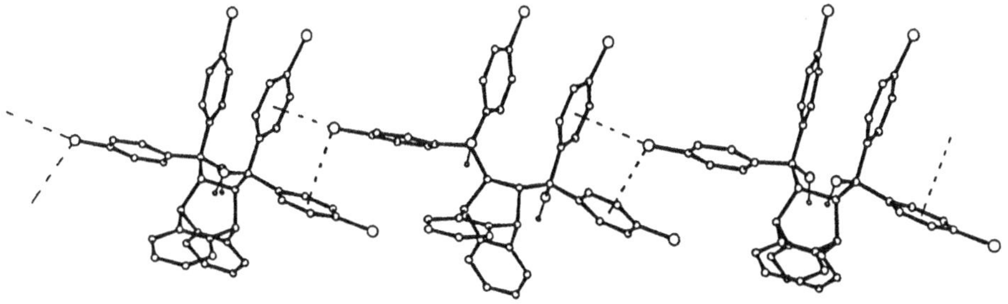

Figure 2. Characteristic Cl···π interaction pattern between the host **2** molecules.

The repeated appearance of the same host recognition pattern in three crystal structures with different symmetries, host-guest stoichiometries and packing relations suggests that these Cl···π contacts are structurally significant and important.

4. Recognition of polar guests with various proton donor/acceptor ability

Usually there is hydrogen bond interaction between the present roof-shaped diol hosts and guests exhibiting proton donor and/or acceptor abilities. The strength and the topology of the H-bond interactions as well as the size of the H-bonded host-guest associates vary, however, between the studied compounds, and the hosts may have either of the two conformations (*cf.* above).

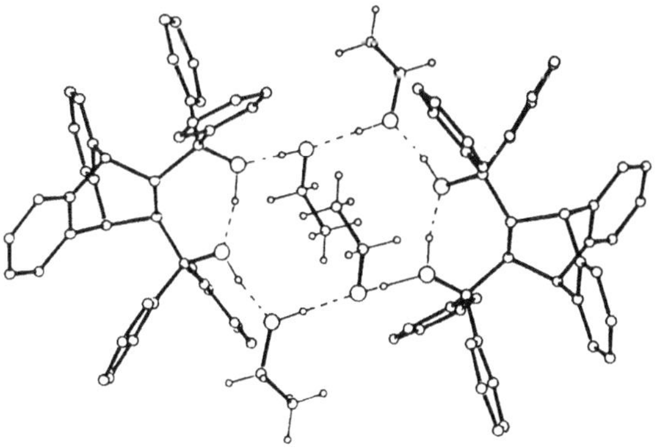

Figure 3. Closed ring of H-bonds involving two hosts and four ethanol guests in the **1**·ethanol (1:2) complex.

In the **1**·ethanol (1:2) complex, 2:4 host-guest associates are formed *via* closed rings of hydrogen bonds with full H-bond saturation (Figure 3), whereas in the

3·methanol (2:1) compound, 2:1 associates are created *via* host-host and host-guest H-bonds, without full H-bond saturation. Although each of the the compounds **2**·pentan-2-ol (1:1), **2**·1,4-dioxane (2:5) and **4**·pyridine (2:3) contains H-bonded 1:1 host-guest associates, they exhibit various guest recognition modes and packing relations. Surprisingly enough, host **2** functions only as a proton acceptor in its pentanol inclusion compound, but is a proton donor in the dioxane complex. Moreover, in the dioxane and pyridine inclusion crystals additional guests are trapped by lattice forces only. In the **1**·nitroethane (1:1) cocrystal, on the other hand, weak C-H···O interactions retain the guests in the host matrix.

In summary, the realized crystal structures seem to be a compromise between close packing on the one hand and hydrogen bonding on the other. Increasing size of the host molecules gives rise to increasing difficulties for the host-guest associates with full H-bond saturation to pack with enough density. As a consequence, increasing size of the host leads to a smaller number of hydrogen bond interactions and to decreasing size of the H-bonded supramolecular building blocks in the crystal structure.

5. Recognition of non-protic apolar guests

Hosts **1-4** in their 'inactive' conformation exhibit eminent clathrate formation ability for aromatic hydrocarbon guests, yielding lattice type inclusion compounds (*i.e.* true clathrates) with relatively high symmetries (*cf.* Table 1) and strictly stoichiometric host-guest ratios. Interestingly, the **3**·toluene (1:1) complex proved to be isostructural with **2**·pentan-2-ol (1:1), thus suggesting that the highly symmetrical host matrix (*Fddd*) is the result of shape recognition of the host molecules in both cases. The guest molecules, the apolar toluene as well as the H-bonded pentanol, are disordered and stabilize the crystals only by filling up the voids in the host framework. Hence, the isostructurality of these two in many respect different compounds well illustrates the crucial role that close packing plays in molecular crystals.

Acknowledgements

We are indebted to our coworkers, Mrs. Olga Helmle (formerly Gallardo) and Dr. Thomas Hens, with whom it has been a pleasure to cooperate. I. C. thanks the Swedish Natural Science Research Council (NFR) and E. W. the Deutsche Forschungsgemeinschaft and the Fonds der Chemischen Industrie for financial support.

References

1. E. Weber, T. Hens, O. Gallardo and I. Csöregh. (1996) *J. Chem. Soc., Perkin Trans.* 2,737-745.
2. E. Weber, S. Finge and I. Csöregh (1991) *J. Org. Chem.* **56**, 7281-7288.
3. I. Csöregh, O. Gallardo, E. Weber, S. Finge and C. Reutel (1992) *Tetrahedron Asymmetry* **3**, 1555-1562 and references therein.
4. I. Csöregh, E. Weber, T. Hens and M. Czugler (1996) *J. Chem. Soc., Perkin Trans. 2* (in press).

Success Rate in a Chiral Separation: Towards a Better Separation Machinery

MÁTYÁS CZUGLER, EDWIN WEBER and PETROS P. KORKAS
Institute für Organische Chemie, Technische Universität Bergakademie Freiberg, Leipziger Straße 29, D-09596 Freiberg/ Sa., Germany

1. Introduction

Optical resolution is a very fundamental object in chemistry. Although there are different techniques for the enantiomer separation of acids and bases the methods for optical resolution of non-acidic and non-basic compounds in a preparative scale are only rare. Making use of the crystalline inclusion phenomena is a promising new approach to fill this gap[1]. Host compounds having this particular feature of crystalline inclusion formation may be designed according to some basic principles such a rigid framework, attached functional groups and bulky substituents that make a host compund hard to crystallize without a proper spatial/functional complement thus

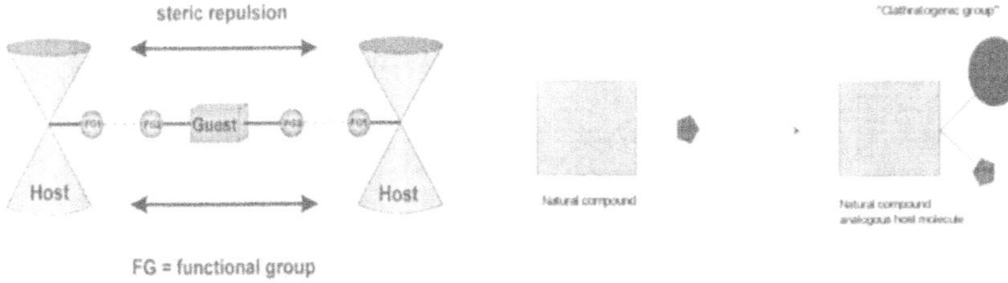

Fig.1. Design ideas of inclusion hosts and embedded chirality from natural pool.

forming a clathrate (Fig. 1) [2]. The clever installment of chirality gives rise to high enantioselectivity at clathrate formation. Ready availability of these blocks in optically resolved form arising from a natural chiral source makes them attractive [3].

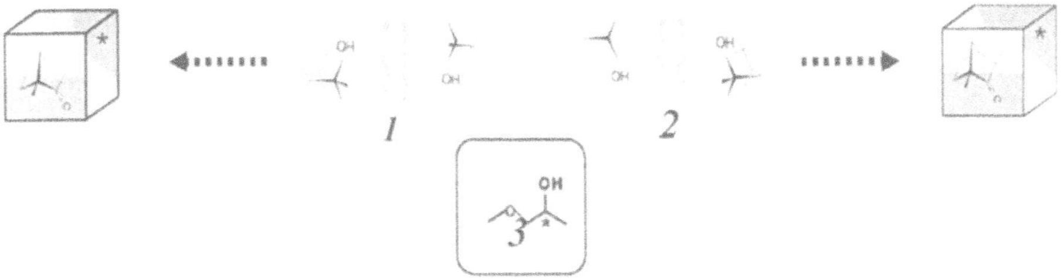

Fig. 2 Hosts **1** and **2** design concept and the studied guest 1-methoxy-2-propanol (**3**)

2. Results

A series of host molecules synthesized recently[4] that rely on a terpene chiral building block such as camphor and fenchone of which **1** and **2** are typical examples (Fig. 2). These hosts show distinct binding of chiral molecules in the crystalline state

A. W. Coleman (ed.), Molecular Recognition and Inclusion, 305–308.
© 1998 *Kluwer Academic Publishers. Printed in the Netherlands.*

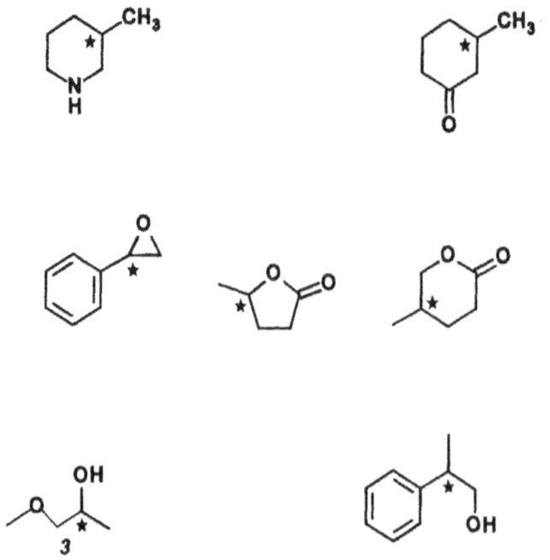

Fig. 3. Guest molecules enclathrated with perfect (100%e.e.) chiral recognition by **1**, **2** and another host.

with high stereoselectivity (Fig. 3). For instance, host **1** with borneol termini is capable of resolving 1-methoxy-2-propanol guest **3** in one single recrystallization step to yield a pure (*S*)-enantiomer. The finding that host compound **2** is unable to exert any chiral discrimination while **1** performs perfectly well towards racemate **3** points to the delicate balance governed by the displacement of two geminal methyl groups in the isomeric borneol (position 7) and fenchol (position 3) blocks. The reason for this peculiar behaviour became clear upon X-ray structure determination of the inclusion complexes **1**· **3** (1:1) and **2**· **3** (1:2). An X-ray structure model of the **1**·**3** (1:1) complex (Fig. 4) reveals that one ***S*** configured guest molecule is bound to the

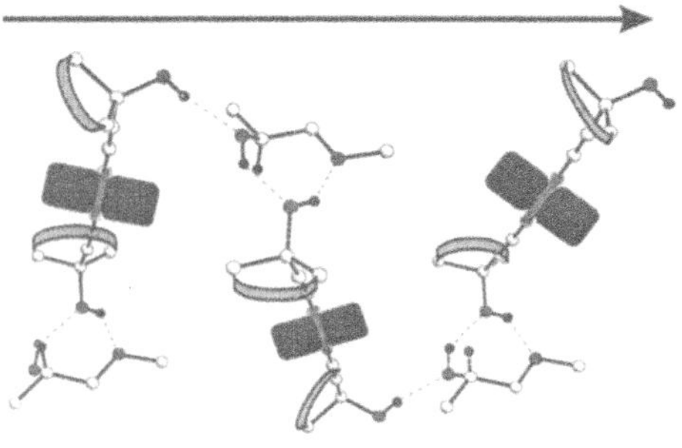

Fig. 4. Helical packing structure schematics of the **1**·**3**(*S*) 1:1 associate

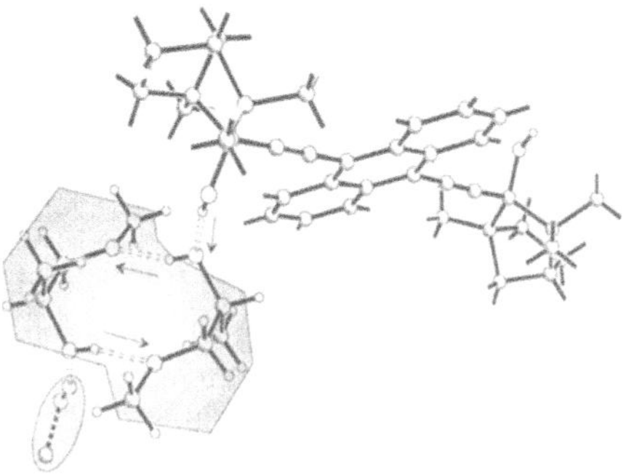

Fig. 5. Asymmetric unit / H-bonding model of the **2**·**3**(*R*,*S*) 1:2 associate

host while **2** yields to a 1:2 stoichiometry (Fig. 5) such that the two guest molecules of **3** in the **2** · **3** (1:2) crystal face each other through a *pseudo-symmetry center* thus forming a dimer of opposite (*R*,*S*) chirality of molecules **3**. These are engaged intimately in H-bonding with each other such that the alcohol function of **3** uses the opposite ether oxygen atom of the so formed dimer as acceptor thus loosing two binding sites for H-bonding in each guest molecule. There remains then only *one* further attachment for a guest (an inter-aggregate H-bond acceptor site at the alcoholic O atom of **3**) so that an obviously less effective *1D*-binding from one host molecule **2** takes place. Altogether this gives yield to two hydrogen bonds to the dimer from two -OH functions of two host molecules **2**. Apparently **2** is not able to break up the in solution conceivably prevalent *R,S*-dimers of **3**. In contrast to this the methoxy O atom of guest **3** accepts a hydrogen bond from the hydroxyl group of the host molecule **1** (Fig. 4) in the other complex. An anchoring effect of the hydrogen bonds from two hosts of **1** to a single molecule of **3** can be envisaged as affecting a *3D*-binding leading to the effective recognition of the chirality of the guest (cf. Fig. 4). This -OH group is also acceptor in an infinite helical chain of hydrogen bonds from the guest -OH function thus playing key role in establishing the crystal lattice being hydrogen bond donor and acceptor simultaneously.

To explain why these isomers must crystallize in different lattices from the same racemic solvent molecular graphics analyses were undertaken. Artificial Me-groups were generated for both hosts **1** and **2** in their crystal environments corresponding the alternative host molecule displacements. Two types of conflicts of close-contact kind arose in the case of **1**.

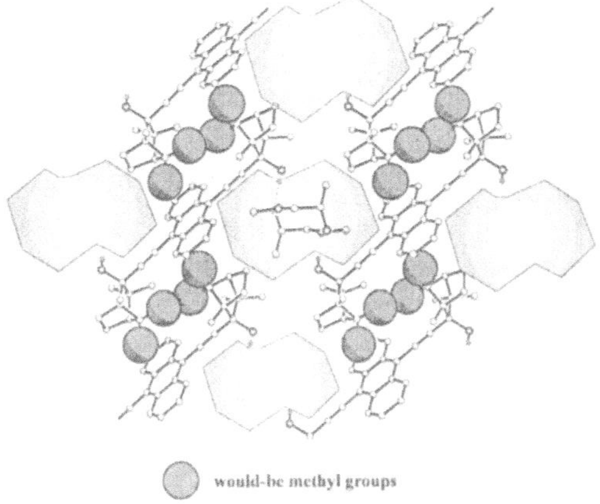

Fig. 6. Conflict sites (large spheres) of one kind in the packing of **2·3** 1:2 (*R,S*) associate

In the longitudinal molecular axis direction of molecule **1**, head-to-tail close approaches involving terminal Me-group region in both borneol groups symmetrically would occur. This would cause troubles in the *ac* diagonal translation directions. This is just the direction perpendicular to the base vectors of H-bonding spiraling around the crystallographic *b* axis in **1** (cf. Fig. 5). Another problem area is dislocated in a transversal molecular direction to the previous one. It would cause troubles just in the chiral center at position-1 in the borneol block, that has the Me-group on it, i.e. the "most chiral" part of the host is affected.The findings indicate that host **2** can not be packed the same way, i.e. building helical strips and with the dimensions as **1** does. It is clear from the like analysis for **2** (Fig. 6) that the borneol host **1** can not adopt itself to the same arrangement either like **2** does it in its crystal with the racemic pairs of **3**.

Apart from the intermolecular relations outlined above there are some conformational differences between the two associates in their crystals. The pseudo torsion angle relating the bridgehead C atom of the anthracene moiety through the C-O bond of the host to the anchored hydroxyl O atom of (S)-**3** in the **1·3** 1:1 complex is 71° while it is 44° for the **2·3** 1:2 associate. Mean planes of the carbon atoms and the ether oxygen atom (mean Δ: 0.019 Å) in the **1·3** 1:1 crystal make a dihedral angle of 85.4° to the anthracene least-squares plane indicating a nearly rectangular placement of an almost perfectly extended guest molecule. The same planes of these five atoms in the **2·3** 1:2 associate indicate a less straight placement of the backbone atoms (mean Δ: 0.20 Å and 0.19 Å, respectively). The dihedral angles between these nearly coplanar (dihedral angle: 6.0°) mean planes of *R,S*-guest molecules of the dimeric guest and that of the anthracene moiety of **2** are 30.2° and 27.8°. These all indicate that the success or failure in the molecular recognition is a concerted effect of inter-and intramolecular events manifested through a multitude of inter- and intramolecular conformation changes. It is apparent from the solid state structures of the inclusion complexes of **1** and **2** with 1-methoxy-2-propanol (**3**) why the latter host performs much worse in chiral recognition.

3. Conclusion

The structures also underline why and how minor isomeric differences in the chiral portions of host molecules cause such dramatic differences in their chiral separation behavior. One can also clearly see how the displacement of the two geminal Me-groups will lead to loss of the chiral separation power. The good performance of host **1** may be directly related to the dimensionality (*3D*) of the strongest (H-bonding) interactions. This finding is in accord with the observed *3D*-type fixing of another guest molecule by **1** reported recently [4], albeit in that case good performance is based on *3D*-type fixing due to one hydrogen bond and lateral (π–π stacking) type interactions. Thus the contrast to host **2** is obvious, since **2** is able to offer a *1D*-binding for a single guest molecule only. Changes in the relative displacements of host:guest molecules in the crystals reflect also differing deeds of the interactions between the given hosts **1** and **2** and that of the resolved (*S*)-**3** and those of the dimeric associated (*R,S*)-**3** molecules, respectively. Hence the mismatch in the chiral recognition for host **2** is explained [5].

Acknowledgement. The authors acknowledge financial support from the EEC COST to this project (Grant No. CIPA-CT93-01717(DG12-HSMU) and from the Fonds der Chemischen Industrie.

References

1. E. Weber, (1995), in *'Kirk-Othmer Encyclopedia of Chemical Technology'*, 4th ed., Vol. 14 (Ed.: J.I. Kroschwitz), Wiley, New York, pp. 122-154.
2. E. Weber, (1996), in *'Comprehensive Supramolecular Chemistry'*, 4th ed., Vol. 6 (Solid State Supramolecular Chemistry: Crystal Engineering), Eds.: R. Bishop, D.D. MacNicol, F. Toda, Elsevier Science, Oxford, Ch. 18, pp. 535-592.
3. E. Weber and C. Wimmer, (1993), Optical resolution by Crystalline Inclusion Formation Using New Lactic Acid Derived Hosts, *Chirality*, **5**, 315-319.
4. P.P. Korkas, E. Weber, M. Czugler, G. Náray-Szabó, New Terpenoid Hosts for Chrial Recognition: Crystal Structure and Molecular Modelling Study of an Inclusion Complex with (*S*)-(+)-Phenyloxirane, *J. Chem. Soc. Chem. Commun.*, **1995**, 2229-2230.
5. M. Czugler, P. P. Korkas, P. Bombicz, W. Seichter, E. Weber, (1996), Chiral Separation Machinery Using New Crystalline Inclusion Hosts. Match / Mismatch in the Enantiomer Recognition of (*R,S*)-1-Methoxy-2-propanol Effected by a Borneol / Fenchol Building Block Exchange in the Host Molecule *Chirality*, **8**, in the press.

NMR STUDY OF PER(3,6-ANHYDRO) α CYCLODEXTRIN AS A POTENTIAL AGENT FOR THE BIOLOGICAL DECONTAMINATION OF LEAD AS EVIDENCED BY NMR SPECTROSCOPY

J.C.DEBOUZY[1*], F.FAUVELLE[1], A.GADELLE[2], B.PERLY[3], C.BAUDIN[3]
1- CRSSA, U.Biophysique, 38702 La Tronche cedex -FRANCE.
2- CEN-Grenoble., SESAM, BP 85X, 38000-Grenoble.FRANCE.
3- CEN-Saclay, DRECAM, 91191 Gif/Yvette cedex. FRANCE.

1.Abstract.

The ability of per(3,6-anhydro) α cyclodextrin (A36CD) to capture lead from a preformed glutathion (GSH)-lead complex was investigated by NMR spectroscopy. This strongly depends on the nature and pH of the buffer used in the competition experiments. It was found that an almost complete removal of lead can be achieved at pH 5.5, especially when lead nitrate is used. The capture also strongly depends on the nature of the lead species as wall as of the counter ion present in the medium. These observations imply that decontamination of lead by this process will be optimal under acidic conditions, i.e. in the acidic tractus.

2.Introduction.

The levels of heavy metals present in the environment have generally decreased during the last decade. However, owing to the very high toxicity of the metal, even low levels of remaining (specially from waste incineration and lead containing batteries) is still a serious problem, as far as the public health is concerned[1]. Most of the various expressions of lead intoxications, i.e. chonic neuropathy, renal failure, hemolytic anemia..., have been ascribed to both a competition with calcium or activation of Ca-dependant proteins (*EF-Hand* motif[2] [3]) and to the reactivity towards sulfhydryl and amino groups of phospholipids or proteins[4]. Hence, the presence of lead reduces cellular glutathion (γ-L-glutamyl-L-cysteinyl-glycine, GSH) level[5], since it can directly interact with the thiol group of GSH[6]. This pseudo tripeptide was selected as a model of biological substrate in order to study the binding of lead and its removal using A36CD as capture agent. In a first step the removal of lead was shown using pure water as solvent. In a second step, the respective influence of counter ion (nitrate or acetate) of the pH and of the composition of the buffers was evidenced.

3.Materials and methods.

A36CD was synthetised according to Gadelle and coll[7]. GSH, lead acetate, lead nitrate (Sigma, France). were used as received. The 4mM solutions of GSH were prepared in 500 μL of pure water (H_2O/D_2O 90/10) or of sodium acetate or sodium nitrate buffers (70mM in H_2O/D_2O 90/10). pH were ajusted using NaOH or HCl.

A. W. Coleman (ed.), Molecular Recognition and Inclusion, 309–312.
© 1998 *Kluwer Academic Publishers. Printed in the Netherlands.*

All experiments were recorded at 293K. NMR spectra were recorded on a AM400 Bruker spectrometer using a presaturation of the water resonance and a spectral width of 10ppm.The chemical shifts were referenced by setting the water resonance at 4.75 ppm.

4.Results.

4.1.EXPERIMENTS IN THE WATER

Figure 1 shows that the resonance of the methylene $CH_2(5)$ next to the thiol group experiences shift (up to +40Hz) and broadening (up to 8Hz) upon additions of lead nitrate. The same effect is also observed for the methin proton CH(4), (not shown), while the other signals remained unaffected by the presence of lead. This observation fully supports that lead is bonded to the thiols groups of cystein and a fast exchange regime is observed. From the Job-plot analysis (data not shown) the 1:2 (Pb:GSH) stoichiometry was acertained as already reported[4]. This ratio will be used in the following experiments. The addition of A36CD at this system provided an almost complete restauration of the original spectrum (fig. 1) but requiered a double concentration of A36CD. The shifts of the signals 2', 3' and 4' of A36CD, support the complexation of lead. In the presence of equimolar concentrations of two potential complexants, most of the lead ions are indeed transeferred towards A36CD ; this implies that the affinity is by far higher for A36CD than for GSH.

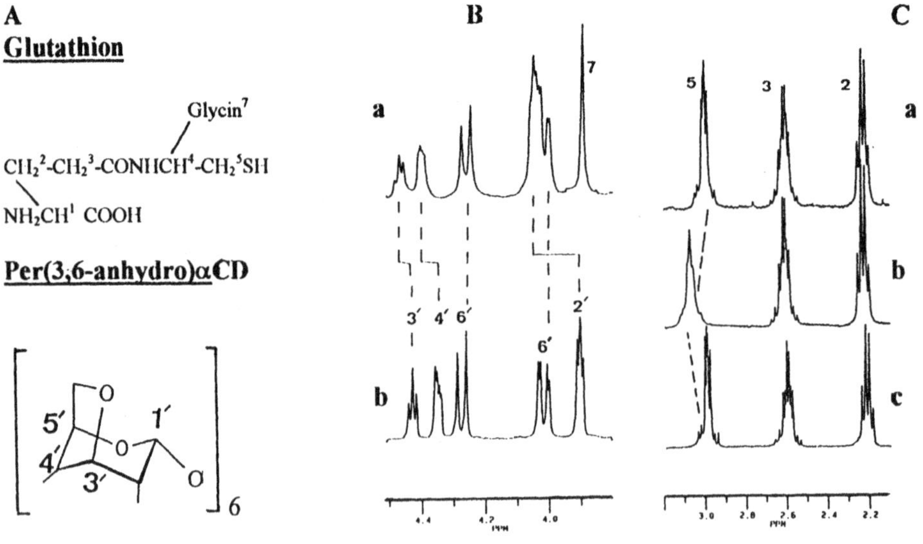

Figure 1 :.A: proton nomenclature, **B**: partial ^{1}H-NMR spectra (293K) of A36CD (2mM, pH 5.5) in the absence (a) and in presence of (b) $Pb(NO_3)_2$ and GSH (Pb/GSH :1/2). C: Partiel ^{1}H-NMR spectra(293K) of GSH (4mM, pH 5.5) in absence (a) and in the presence (b) of $Pb(NO_3)_2$. (2mM) or of equimolar $Pb(NO_3)_2$ and A36CD (2mM).

4.2.ROLE OF THE COUNTER IONS AND pH.

Two values for pH were selected as typical of a given ionizing state for lead : at pH 5.5, the Pb^{2+} form dominates [8] while the $Pb(OH)^+$ form is only present at pH 7.5. Lead salts were added to GSH in a first step, then A36CD was added to in order to observe lead removal (figure 2, see legend).

pH5.5 (two first columns): Some similarities are noted between the experiments using lead nitrate or acetate in the same experimental conditions : both broadening and shift of the $CH_2(5)$ line of GSH are observed when lead acetate is added ; however, the shift is limited to +20Hz while the broadening of the line predominates here (more than 100Hz). Moreover the recovery obtained in the presence of A36CD never excess 30% (even when more than 2/1 A36CD/Pb ratios are used), giving rise to a poorly resolved line of 30 Hz linewidth. As far as the A36CD spectrum is concerned, only a limited shift of $H_{2'}$ resonance is measured (less than 20 Hz). The former observations are in favour of either the formation of agregates, an exchange at an intermediate mode for the free/bound species. As a slight turbidity was observed in this

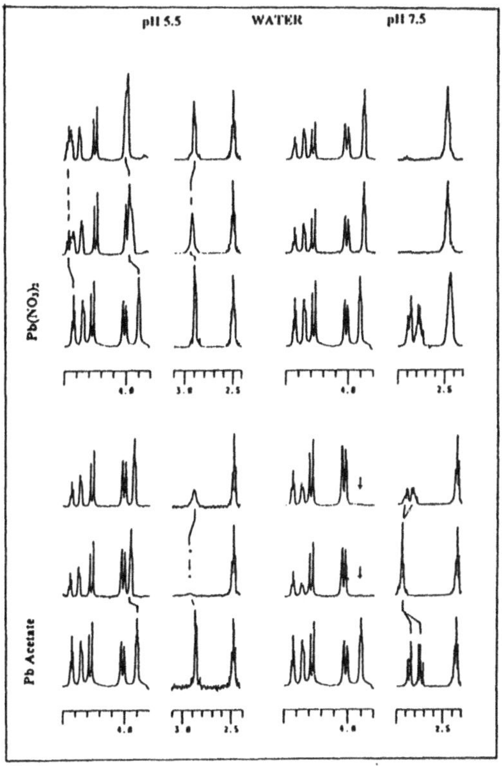

Figure 2 : ^{1}H-NMR ; the four parts of the figure are identically built as follows, and only differ by the pH and the salts of lead used, as respectivelly indexed in columns and lines : *left column* : spectra of pure A36CD, 2mM (bottom), and when added to GSH (4mM)/Pb(2mM) systems with A36CD/Pb ratio R=1/1 (middle) and R=2/1 (top). *right column* : aliphatic part of the spectrum of pure GSH, 4mM (bottom), after addition of Pb (Pb/GSH=1/2), (middle), and after addition of A36CD to the GSH:Pb system (A36CD/Pb, R=2)(top).

case (even at less than 4mM GSH) suggesting the formation of precipitates, the same experiment was also performed at 10mM GSH giving milky samples without yielding any increase in the extraction of lead by A36CD.

pH 7.5 (two last columns). The addition of lead nitrate induced a dramatic broadening of the $CH_2(5)$ line (and also of the CH(4) peak) that became undetectable while the GSH peak labeled (3) was unaffected, indicating a strong immobilization of the CH_2SH group. The addition of A36CD induced no significant spectral modification,and no shift was noted on the A36CD resonances by comparison with those of pure A36CD. Such a dramatic linebroadening of $CH_2(5)$ resonances was not observed in the presence of lead acetate : rather, a single shifted line was detected (average shift of +50Hz for the Pb/GSH molar ratio of 1/2 shown on the figure 2). Furthermore, the addition of equimolar concentrations of A36CD (A36CD/Pb, R=1) allowed the partial recovery (around 30% in intensity) of the native GSH lines. The loss of H_5 signal was attributed to the presence of unsoluble species, as suggested by the sligthly opalescent aspect of the NMR sample at the end of the experiment. Another striking point was the complete desappearence of the $H_{2'}$ resonance on the spectra of the A36CD/Pb/GSH systems, untill A36CD/Pb ratio exceeded 2/1. This strongly suggested the complexation of lead by A36CD at the $H_{2'}$ position.

5. Discussion and conclusion.

This study shows that the best removal of lead is observed in acidic medium and that, conver, no partial capture is obtained at pH this latter result is due to the likely formation of insoluble lead species in neutral medium.

The partial recovery of GSH obtained when human plasma was used fully supports these observations. This behaviour can be of major interest for digestive decontamination -i.e; gastric- since it is optimal for acidic conditions.

6.References

[1] Johanson K., (Ed.) Metals in the urban and rural areas-cycling and critical loads, *SNV, Rep.N°4382*, 10648 Stockholm, Sweden, 1994: 1-32.

[2] Dowd T.L., Rosen J.F., Gundberg C.M., Gupta R.K.,(1994) The deplacement of calcium from osteaoclacin at submicromolar concentrations of free lead, *Biochim.Biophys.Acta*, 1226(2), 131-137.

[3] Richardt G., Federolf G., Habermann E., Affinity of heavy metal ions to intracellular Ca2+-binding proteins, (1986) Biochem.Pharm., **35**(8), 1331-35.

[4]Nolan C.V., Shaikhy Z.A., Lead nephrotoxicity and associated desorders: biochemical mechanisms,(1992) *Toxicology*, **73**,127-146.

[5] Legare M.E., Barhoumi R., Burghart R.C., Tiffani-Castaglioni E., Low-level lead exposure in cultures astroglia : identification of cellular targets with vital fluorescent probes. (1993) *Neurotoxicology*, **14**(2-3),267-272.

[6] Christie N.T., Costa M., In vitro assesment of the toxicity of metal compounds.(1984) *Trace Elem. Res*, **6**, 139-145.

[7] Gadelle A., and Defaye J.,(1991) *Angew.Chem. Int. Ed. Engl.* **30**, 1.

[8] MinQL programm (2.21) written for PC(1991), *Environemental Research software*, Cincinati, USA.

THIOUREIDO β-CYCLODEXTRINS AS MOLECULAR CARRIERS FOR THE ANTICANCER DRUG TAXOTÈRE®

J. DEFAYE, C. ORTIZ MELLET, J.M. GARCÍA FERNÁNDEZ AND S. MACIEJEWSKI
CNRS and CEA, Département de Recherche Fondamentale sur la Matière Condensée, CEA Grenoble, F-38054 Grenoble, France

Abstract

Alkyl, glycosyl and glycopeptidyl deivatives of 6^{I}-deoxy-6^{I}-thioureido-cyclomaltoheptaose, which have been prepared in high yield starting from either 6^{I}-amino-6^{I}-deoxycyclomaltoheptaose or the related 6^{I}-isothiocyanate, exhibit considerably enhanced water solubilites compared to the native β-CD, e.g. the alkyl and glycosyl derivatives at aqueous concentrations of 64.2 and 57.3 mM raise the water solubilities of the hydrophobic anticancer drug Taxotère® from 0.004 up to 4.7 and 3.8 g L^{-1}, respectively.

1. Introduction

A variety of covalently modified cyclodextrins (CDs) has been synthetized over the last few years with the aim of improving their performances with regard of the bioavailability enhancement of drugs. When considering the more accessible β-cyclodextrin (β-CD) entity, a main feature to be improved is water solubility. A general strategy to face this problem consists in the binding of hydrophilic substituents onto the β-CD core through appropriate spacers. In addition, the use of such derivatives for the vectorized delivery of medicinally useful compounds would require incorporation of biological markers. With this problem in mind, we have now devised a more comprehensive approach based on the use of the thiourea functionality as highly hydrophilic spacer arm. The methodology relies on the high selectivity and efficiency of the condensation reaction of amines and isothiocyanates. Alkylthioureido, as well as more elaborated derivatives incorporating saccharide or glycopeptide substituents, have

A. W. Coleman (ed.), Molecular Recognition and Inclusion, 313–316.
© 1998 *Kluwer Academic Publishers. Printed in the Netherlands.*

been thus prepared following a relatively simple synthetic methodology and, remarkably, with no need for hydroxyl protection.

2. Results and Discussion

Isothiocyanation of the monoamine derivative **1** was effected with $CSCl_2$ following the procedure previously reported for per-6-deoxy-6-isothiocyanato-CDs.[1] The title compounds were synthetized by one of the two different routes given in Scheme 1. The condensation was efficiently achieved in pyridine at r. t., and the crude products showed a high degree of purity (>90%). Although route (a) is preferable for commercially available isothiocyanates (e.g. MeNCS), the use of the β-CD—NCS reagent **2** (route b) was advantageous for the reaction with unprotected amino sugars, the reverse procedure (a) leading to formation of cyclic thiocarbamates in this case.[2]

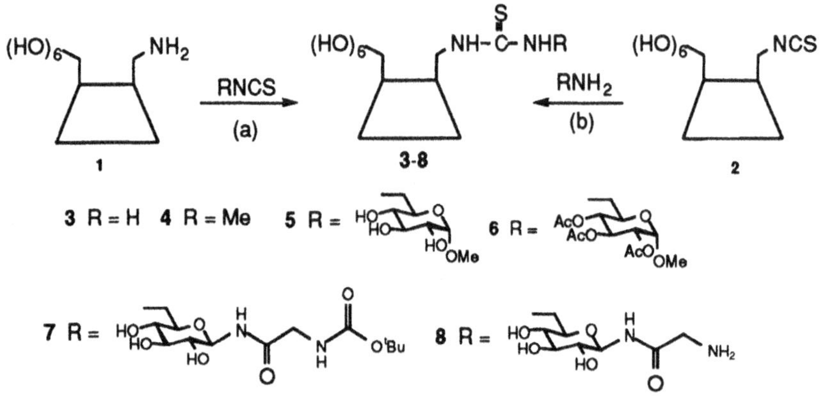

Scheme 1. Synthesis of thioureido β-CDs.

All the prepared thioureido—β-CDs, excepting **3**, exhibited a water solubility several times higher as compared to the native β-CD (Table 1). The observed solubility values also compare favourably with those reported for monosubstituted glycosyl—β-CDs or their *S*-linked counterparts.[3] Actually, they are similar to those for branched β-CDs persubstituted at the primary face, with the a priori advantage of the higher accessibility of the cyclodextrin cavity in monosubstituted derivatives. In agreement with that, the water solubility of the anticancer drug Taxotère® (~0.004 g L^{-1}) was raised up to 4.7 and 3.8 g L^{-1} in 64.2 and 57.3 mM solutions of the methylthioureido and α-methyl glucopyranoside derivatives **4** and **5**, respectively

From this set of results, it can be concluded that the increase in water solubility in **4-8** lies in the intrinsic hydrophilicity of the thiourea functionality, since similar values were measured for thioureido—β-CDs incorporating hydrophilic (e.g. **5** and **8**) or hydrophobic substituents (e.g. **4** and **6**). The simplicity of the synthetic methodology, the selectivity and high yield of the coupling reaction between amines and isothiocyanates, compatible with the presence of the free hydroxyl groups, and the stability of the thiourea functionality allows incorporation of relatively complex haptens suitable for biological recognition in a controlled and very convergent fashion. This possibility is illustrated by the preparation of the glycosylamino acid—β-CD conjugate **8**. Reaction of **2** with the selectively functionalized *N*-glycyl-β-D-glucopyranosylamine derivative precursor afforded the Boc-protected compound **7** (70% yield) which after deprotection with 90% aqueous TFA afforded **8** in almost quantitative yield.

TABLE 1. Water solubility at 25 °C (mmol L^{-1}) of thioureido—β-CD derivatives.

β-CD	**4**	**5**	**6**	**8**
15	640 (x 42.6)	518 (x 34.5)	373 (x 24.8)	>567 (x 38.0)

3. Experimental

3.1 GENERAL METHODS

^{13}C NMR spectra (D_2O, 50.3 MHz) were recorded at 323 K using a Bruker AC200 instrument. LC was carried out using a Perkin-Elmer 250 pump fitted to a LC-30 refractive index detector. A Nucleosil C-18, 5 μ column was used with 12:88 MeOH-H_2O as eluent.

3.2. 6^I-AMINO-6^I-DEOXYCYCLOMALTOHEPTAOSE (1)

To a solution of 6^I-azido-6^I-deoxycyclomaltoheptaose[4] (1.49 g, 3.69 mmol) in DMF (1 mL) was added 1,3-propanedithiol (62 mL, 6.15 mmol) and ethyldiisopropylamine (0.65 mL, 6.15 mmol), and the reaction mixture was stirred for 16 h at room temperature. Acetone (100 mL) was then added, the solid precipitate was collected, dissolved in water (100 mL), filtered through Celite and freeze-dried. Yield 1.34 g (92%). Compound **1** showed the spectroscopical properties already reported[5].

3.3. 6^I-DEOXY-6^I-ISOTHIOCYANATOCYCLOMALTOHEPTAOSE (2)

To a solution of **1** (1.13 g, 1 mmol) in water-acetone (3:2, 75 mL) was added $CaCO_3$ (0.9 g, 3.0 mmol) and $CSCl_2$ (0.15 mL, 1.5 mmol). The mixture was stirred overnight at room temperature, then concentrated to half volume, diluted with water (30 mL), demineralized with Amberlite MB-6113 mixed resine (H^+,OH^-; 15 mL, 15 min), filtered and freeze-dried. Yield 0.76 g (65%), $[\alpha]_D$ +112.1 (*c* 0.6, pyridine).

3.4. 6^I-DEOXY-6^I-THIOUREIDO-β-CD DERIVATIVES **3-7**

To a solution of **1** or **2** in pyridine was added the corresponding isothiocyanate or aminonucleophile, respectively. The solution was stirred for 48 h at room temperature, then concentrated, the residue was dissolved in water, washed with $CHCl_3$, and freeze-dried. The yield of the condensation reactions was >90%, as seen from LC chromatograms. Pure samples were obtained after LC purification. Data for: **3**, $[\alpha]_D$ +114.3 (*c* 0.8, H_2O); **4**, $[\alpha]_D$ +108.7 (*c* 1.0, H_2O); **5**, $[\alpha]_D$ +96.9 (*c* 0.7, H_2O); **6**, $[\alpha]_D$ +116.1 (*c* 0.6, H_2O); **7**, $[\alpha]_D$ +103.4 (*c* 0.9, H_2O).

3.4. GLUCOSYL AMINO ACID THIOUREIDO-β-CD **8**.

A solution of **7** (77 mg, 51 μmol) in TFA-H_2O (9:1, 1 mL) was stirred at room temperature for 1 h, then concentrated. Traces of acid were eliminated by coevaporation with water and further treatment with Amberlite IR-904 (OH^-, 2 mL) anion exchange resin. Yield 66 mg (91%), $[\alpha]_D$ +82.2 (*c* 1.8, H_2O).

4. Acknowledgement

This work was supported by the European Commission DG12 under the FAIR programme contract no. FAIR-0300.

5. References

1. García Fernández, J.M., Ortiz Mellet, C., Jiménez Blanco, J.L., Fuentes Mota, J., Gadelle, A., Coste-Sarguet, A. and Defaye, J. (1995) Isothiocyanates and cyclic thiocarbamates of α,α'-trehalose, sucrose, and cyclomaltooligosaccharides, *Carbohydr. Res.*, **268,** 57-71.
2. García Fernández, J.M., Ortiz Mellet, C. and Fuentes, J. (1993) Chiral 2-thioxotetrahydro-1,3-O,N-heterocycles from carbohydrates. 2. Stereocontrolled synthesis of oxazolidine pseudo-*C*-nucleosides and bicyclic oxazine-2-thiones, *J. Org. Chem.*, **58,** 5192-5197.
3. Lainé, V., Coste-Sarguet, A., Gadelle, A., Defaye, J., Perly, B. and Djedaïni-Pilard, F. (1995) Inclusion and solubilization properties of 6-*S*-glycosyl-6-thio derivatives of β-cyclodextrin, *J. Chem. Soc., Perkin Trans. 2,* 1479-1487.
4. Petter, R.C., Salek, C.T., Sikorski, G., Kumaravel, G., Lin, F.-T. (1990) Cooperative binding by aggregated mono-6-(alkylamino)-β-cyclodextrins, *J. Am. Chem.* Soc., **112,** 3860-3868.
5. Brown, S.E., Coates, J.H., Coghlan, D.R., Easton, C.J., van Eyk, S.J., Janowski, A.L., Lincoln, S.F., Luo, Y., May, B.L., Schiesser, D.S., Wang, P. and Williams, M.L. (1993) Synthesis and properties of 6^A-amino-6^A-deoxy-α- and β-cyclodextrin, *Aust. J. Chem.*, **46,** 953-958.

CONTINUITY AND DISCONTINUITY IN THE THERMODYNAMIC PROPERTIES OF SOLID β–CYCLODEXTRIN VERSUS HYDRATION. A COMPARATIVE STUDY

C. DE BRAUER, M. DIOT, P. GERMAIN, J.M. LETOFFE
Mineral Thermochemistry Laboratory, UPRESA CNRS 5079,
National Institute of Applied Sciences.69621 Villeurbanne cedex FRANCE.

ABSTRACT

The thermal behavior of solid β–cyclodextrin (βCD,nH_2O) has been studied at low temperature by adiabatic and differential scanning calorimetry.
Energy of bonding is equivalent for each water molecule in all the hydration range : $0<n<12$.
On the other hand, three types of different behavior are observed according to the number n of fixed water molecules : $0<n<7$, $7<n<10$ and $n>10$. These observations are discussed in terms of intra- and intermolecular interactions.

1. INTRODUCTION

β-cyclodextrin is well known for its ability to form many inclusion compounds. A significant number of scientific papers describe inclusion reaction data, generally in aqueous medium. The litterature is rather poor, excepted the studies of STEINER [1], concerning the fundamental aspect of the β-cyclodextrin molecule and its interactions with water in the solid state. From an energetic point of view, the mechanisms of inclusion cannot be explained without a complete thermodynamic study of pure hydrated β-cyclodextrin at various hydration rates.

2. MATERIALS AND METHODS

β-cyclodextrin was supplied by ROQUETTE under the reference βCD 16. The hydration rates are determined by thermogravimetric analysis using a TGA analyser Setaram Labsys.
All the data given in this work are obtained or calculated from direct calorimetric measurements. The enthalpies of dissolution are obtained with an isoperibolic calorimeter [2].
The thermal behavior of βCD is followed by two different technics : adiabatic calorimetry using a home built calorimeter, and DSC using a Mettler TA8000 apparatus.
A Mac Bain type thermobalance [2] is used to follow the hydration of βCD versus water vapor pressure.

A. W. Coleman (ed.), Molecular Recognition and Inclusion, 317–320.
© 1998 *Kluwer Academic Publishers. Printed in the Netherlands.*

3. RESULTS AND DISCUSSION

3.1. Apparent continuity of some properties of hydrated βCD

In a precedent work [2], we assume that some properties of solid β-cyclodextrin (βCD,nH_2O), have a regular evolution according to the number n of fixed water molecules.
From isoperibolic dissolution calorimetry (in water, at 298K), we observe that the enthalpy of dissolution of βCD,nH_2O varies linearly according to n for n = 0 to 12 water molecules. In this range of hydration : $\Delta_{diss}H$ βCD,nH_2O = -91.2 + 10.5 n (in kJ/mol anh.βCD). We deduce that from an enthalpic point of view, each water molecule seems to be equivalent. The energy of bonding is -10.5 kJ/mol H_2O.
From isothermal thermogravimetric analysis at 302K, we conclude that the average number of fixed hydration water molecules has a linear evolution from 10 to 12 in the range of water vapor pressure between 0.013 and 0.031 bar.

3.2. Discontinuities in the thermal behavior of solid βCD at low temperature versus its hydration

3.2.1. Adiabatic calorimetry results

HANABATA and Al. [3] have studied the thermal behavior of a sample of βCD,10.9H_2O in the temperature range between 13 and 300K. They pointed out a first order transition with a maximum at 227K.
We do the same experiment, using a sample of βCD,9.7H_2O. In this case, the evolution of the βCD heat capacity versus temperature shows a significantly different shape. No effect is observed at 227K, but a broad transition occurs from 250 to 300 K with a maximum around 273K. The two curves are plotted on figure 1. The observed differences cannot be explained in terms of cristallographic transitions.

Figure 1 : Comparison of the molar heat capacities of two differently hydrated βCD samples

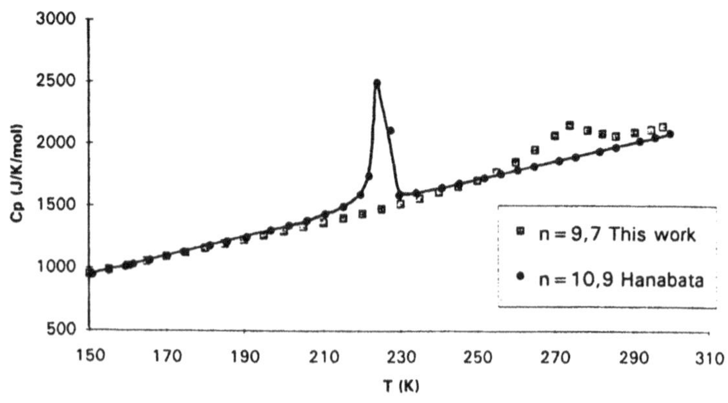

3.2.2. Differential scanning calorimetry results

DSC experiments have been carried out in the temperature range of 170-310K with a heating rate of 0.083 Ks^{-1}. Controlled dehydration of βCD samples have been done by heating at 50°C during various duration. Before each DSC run, a thermogravimetric analysis was performed from 298 up to 470K in order to evaluate the water content. Reproducibility of this determination is ± 0.2 wt %. DSC curves are shown in figure 2. It is obvious that the thermal behavior at low temperature is directly depending on the amount of water molecules contained in each sample of βCD.

Three hydration states must be considered :

- 0-$7H_2O$: No significant thermal event is detected ;
- $7H_2O$-$10H_2O$: Transition observed by our adiabatic measurements is also detected. Depending on the water content, the maximum of transition is observed around 277K with a low value of enthalpy of 2.9 ± 0.4 Jg^{-1}.
- $10H_2O$-$11H_2O$: As it is shown on figure 2, this transition fastly disappeared to be replace by a near first order transition previously found by HANABATA and Al. [3]. Maximum of transition is 224K with an enthalpic value of 5.1 ± 0.3 Jg^{-1}.

Figure 2 : DSC curves of βCD,nH_2O

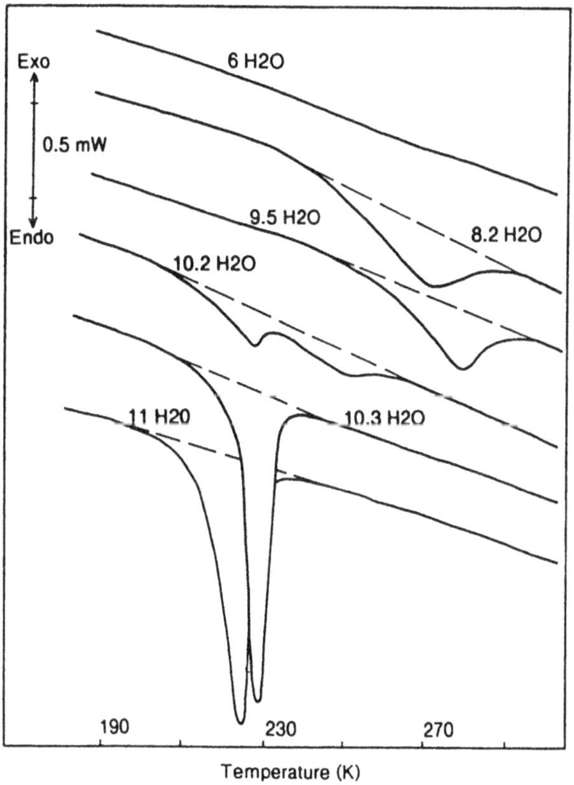

4. CONCLUSION

All the informations obtained by DSC and adiabatic calorimetry at low temperature give us major informations on the βCD behavior according to the number of hydration water molecules. These informations allowed us to confirm new hypothesis concerning intra- or intermolecular interactions.

For n<7, the water molecules are principally located inside the βCD host cavity. The considerable disorder in positions and orientations of water molecules and hydroxyl groups is confirmed by the lack of phase transition on the corresponding DSC curve (figure 2). We assume that there is no possibility of intermolecular arrangments.

For n>10, the shape and the enthalpy change of the transition around 225K can be interpreted as a minor structural change. Three, or more, external water molecules probably induce a long distance order between the macrocycles. This order is modified as it is pointed out by the phase transition observed at 225K.

In the intermediate hydration state, for 7<n<10, the number of external water molecules cannot stabilize a long distance order. These molecules are located around one macrocycle and distributed over a great number of energetic sites. So, we can consider that these two or three water molecules are almost "free". Taking this hypothesis, the broad transition observed around 273K can be attributed to the fusion of water in an environment comparable to a macroporous structure. Further measurements are needed to develop an accurate calculation based on these considerations.

The existence of, at least, two structuraly distinct βCD : isolated or chained macrocycles, must be taken into account to interprate its reactivity. The thermodynamic parameters of an inclusion reaction by βCD are obviously influenced by the initial hydration state of βCD molecules.

REFERENCES

[1] STEINER, T., SAENGER, W., Geometry of C-H...O Hydrogen bonds in Carbohydrate crystal structures. Analysis of neutron diffraction data, *J. Am. Chem. Soc.*, **114**, 10146-10154, (1992) and Ref.

[2] BILAL, M., DE BRAUER, C., CLAUDY, P., GERMAIN, P., LETOFFE, J.M., β-Cyclodextrin hydration : a calorimetric and gravimetric study, *Thermochim. Acta*, **249**, 63-73, (1995).

[3] HANABATA, H., MATSUO, T., SUGA, H., Calorimetric study of β-Cyclodextrin Undecahydrate, *Journal of Inclusion Phenomena*, **5**, 325-333, (1987).

ACKNOWLEDGEMENTS

The authors are grateful to M. SERPELLONI from Roquette Frères Company for his contribution and stimulating discussions. Thanks are due to Muryèle Champenois who greatly assisted in preparing the manuscript.

Phosphorylated Cavitands : Encapsulation of Hard Cationic Guests

PASCALE DELANGLE, JEAN-CHRISTOPHE MULATIER and
JEAN-PIERRE DUTASTA.
Stéréochimie et Interactions Moléculaires, Ecole Normale Supérieure de Lyon, UMR CNRS 117, 46 allée d'Italie, 69364 Lyon Cedex 07, France

1. Introduction

The cone-shaped cavity of the calixresorcinolarene based cavitands opened the route to promising new receptors. Recent studies showed that it was possible to introduce phosphorus donor groups at the upper rim of these molecules. According to their structure the inward *(i)* and outward *(o)* orientations of the phosphorus substituents define the six stereoisomers *iiii, iiio, iioo, ioio, iooo, oooo.*

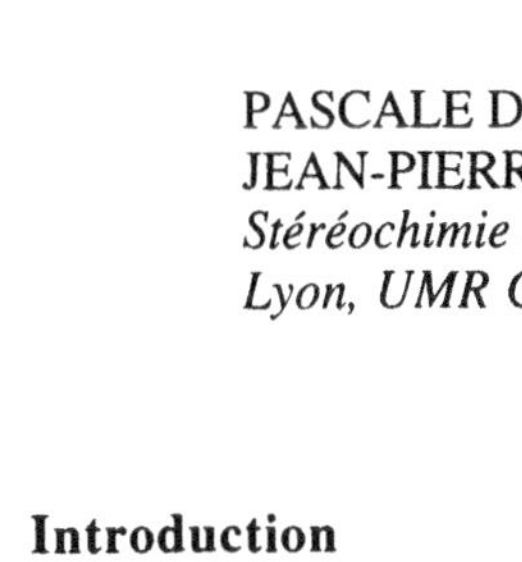

Phosphorylated cavitand
X = lp, O, S

A thiophosphorylated cavitand [X = S, R = $N(CH_2CH_3)_2$, R' = CH_3] has been obtained by reaction of the calixresorcinolarene with tris(diethylamino) phosphine and treatment with solid sulfur. Its X-ray structure showed that the four sulfur atoms are oriented toward the center of the macrocyclic cavity (*iiii* isomer).[1] The phosphonitocavitand synthesized by Puddephatt *et al* (X = lp, R = Ph, R' = CH_2CH_2Ph) exists mainly as the *iiii* isomer with all phenyl groups oriented outwards. It forms polymetallic complexes with transition metal cations and have been used for selective anion complexation.[2] With the phosphorylated host synthesized by Dalcanale *et al* (X = O, R = OC_7H_8, R' = CH_3) the six stereoisomers were characterized. The *iooo* isomer with one phosphoryl group inward is the major compound thus obtained.[3]

We report here our contribution to the design of new phosphorylated cavitands for the complexation of hard cationic guests. In order to favor the formation of the *iiii* isomer with the four P=O groups inwards, we synthesized the tetraphosphonatecalix[4]-resorcinarenes **I** and **II** which bear four phenyl groups bonded to the phosphorus atoms.

2. Synthesis

The tetraphosphonate-calix[4]resorcinarenes **I** and **II** were synthesized by reacting tetraalkyl resorcinolarene[4] with phenyl phosphonic dichloride (Figure 1).[5] Nmr analysis of the main product isolated reveals that it has C_{4v} symmetry and corresponds to the expected *iiii* isomer, with the four P=O groups pointing inwards ($\delta^{31}P$ = 9.15 ppm for **I** in $C_2D_2Cl_4$, $\delta^{31}P$ = 7.73 ppm for **II** in $CDCl_3$).

A. W. Coleman (ed.), Molecular Recognition and Inclusion, 321–324.
© 1998 *Kluwer Academic Publishers. Printed in the Netherlands.*

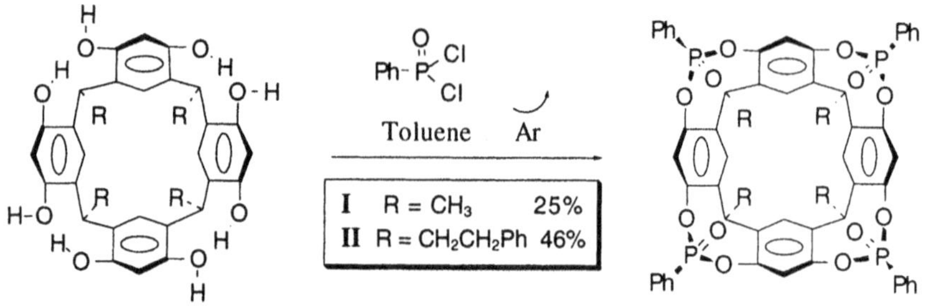

Figure 1. Synthesis of hosts **I** and **II**.

3. Complexation of alkali metal and ammonium cations

The most interesting feature of cavitands **I** and **II** is their potentiallity to form complexes with cationic guests. The complexation of various alkali metal and ammonium ions by these receptors was investigated by performing liquid liquid extraction experiments.[6] The association constants (K_a) and the free energies of complexation ($-\Delta G°$) were calculated for 1:1 complexes at 291 K for **I**, and 294 K for **II**, in $CHCl_3$ saturated with H_2O, and are reported in Table 1.

TABLE 1. Association Constants (K_a, L mol^{-1}) and free energies of complexation ($-\Delta G°$, kJ mol^{-1}) for the complexes of **I** and **II** with alkali metal and ammonium cations.

	Li^+	Na^+	K^+	Rb^+	Cs^+	NH_4^+	$MeNH_3^+$	$t\text{-}BuNH_3^+$
I) K_a	3,7 10^8	1,2 10^9	6,5 10^9	4,0 10^8	1,1 10^{10}	8,6 10^8	8,1 10^9	2,7 10^8
$-\Delta G°$	47,9	50,7	54,8	48,1	56,2	49,9	55,4	47,2
II) K_a	5,0 10^7	7,8 10^7	2,9 10^8	1,2 10^8	5,4 10^8	4,9 10^8	2,0 10^9	8,5 10^8
$-\Delta G°$	43,3	44,4	47,6	45,4	49,2	48,9	52,4	50,2

Binding constants with alkali metals reach a maximum with Cs^+ ion. The cesium cation fits perfectly in the cavity of **I** and **II** and we suggest a strong participation of π-cation interactions in this complex with encapsulation of the cation into the host cavity.[7] The binding constants are comparable to those reported for calix[4]arene-crowns where two of the benzene rings of the 1,3-alternate cone conformation interact with the charged guest.[8] The origin of the relatively low selectivity of **I** and **II** for the alkali metal cations and the drop in stability of the rubidium complex are not clear.

It is interesting to compare the binding constant of **II** (stereoisomer *iiii*) with $t\text{-}BuNH_3^+$ in chloroform (K_a = 8.5 10^8 L mol^{-1}) and that of the *iiio* isomer of the tetraphosphate cavitand previously described (X = O, R = OC_7H_8, R' = CH_3), complexing the c-hexylammonium cation (K_a = 1370 L mol^{-1}).[3] These results clearly demonstrate the higher efficiency of the *iiii* isomer with the four phosphoryl groups pointing inwards.

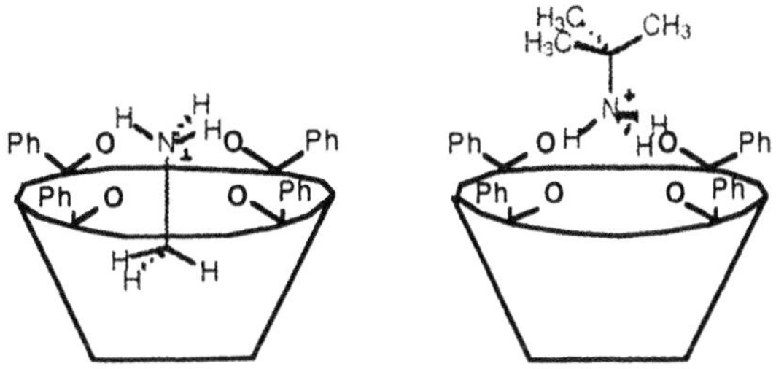

Figure 2.

Hosts **I** and **II** are remarkable efficient binders of ammonium cations. Table 2 gives the ^1H chemical shifts of the methyl groups of the $CH_3NH_3^+$ and t-$BuNH_3^+$ cations in their complexes with **I**. The larger shift observed with $CH_3NH_3^+$ indicates that the methyl group is exposed to the shielding effect of the aromatic rings of the cavity of **I** (Figures 2 and 3). The cation is thus encapsulated into the host cavity. This is not possible for the bulkier t-$BuNH_3^+$ guest (Figure 2). In both cases the H_3N^+ protons form tight hydrogen bonds with the phosphoryl groups of the host.

TABLE 2. ^1H chemical shifts (ppm) of the methyl and t-butyl resonances of the free and complexed guest with **I**, in 1,1,2,2-tetrachloroethane-d_2

Guest	$CH_3NH_3^+$	$(CH_3)_3CNH_3^+$
$\delta(CH_3)$ free	2,90	1,41
$\delta(CH_3)$ complexed	-0,91	1,14
$\Delta\delta$	-3,81	-0,27

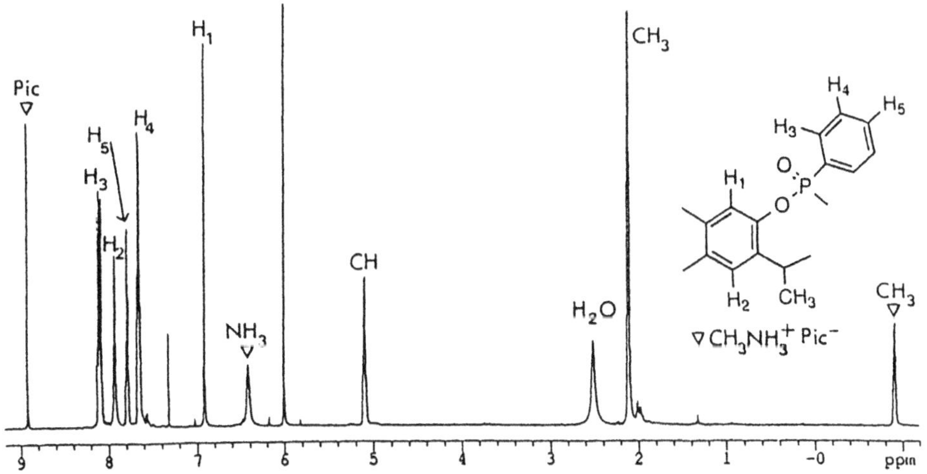

Figure 3. ^1H NMR spectrum of the complexe **I•CH$_3$NH$_3$$^+Pic^-$** in $C_2D_2Cl_4$.

4. Extraction of alkali, alkaline-earth and europium cations

The percent extractions of the alkali, alkaline-earth and europium picrates by **II** in chloroform, were measured from liquid liquid extraction experiments[9] and are defined as the ratio of the concentration of complex in the organic phase to the initial concentration of guest in the aqueous phase (10^{-3} mol L^{-1}). The results are depicted in figure 4.

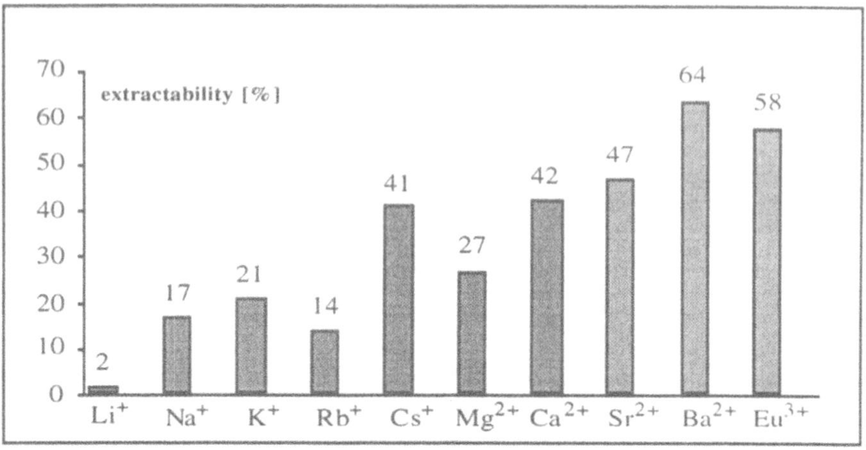

Figure 4. Extractability of alkali, alkaline earth and europium cations using host **II**.

5. Conclusion

Tetraphosphonate calix[4]resorcinarenes have proved to be a new and promising family of molecular receptors for the extraction of hard cationic guests. The phosphorylated upper rim of the host and the aromatic cavity act cooperatively, by means of ion-dipole and specific π-interactions for encapsulation of size adapted species.

References

[1] E. E. Nifant'ev, V. I. Maslennikova, L. K. Vasyanina, E. V. Panina, *Russ. J. Gen. Chem.*, **1994**, 64, 142. V. I. Maslennikova, E. V. Panina, A. R. Bekker, L. K. Vasyanina, E. E. Nifant'ev, *Phosphorus, Sulfur, and Silicon*, **1996**, 113, 219. E. E. Nifant'ev, V. I. Maslennikova, E. V. Panina, A. R. Bekker, L. K. Vasyanina, K. A. Lysenko, M. Y. Antipin, Y. T. Stuchkov, *Mendeleev Commun.*, **1995**, 131.

[2] W. Xu, J. J. Vittal, R. J. Puddephatt, *J. Am. Chem. Soc.*, **1993**, 115, 6456. W. Xu, J. P. Rourke, J. J. Vittal, R. J. Puddephatt, *J. Chem. Soc. Chem. Commun.*, **1993**, 145. W. Xu, J. J. Vittal, R. J. Puddephatt, *J. Am. Chem. Soc.*, **1995**, 117, 8362. W. Xu, J. P. Rourke, J. J. Vittal, R. J. Puddephatt, *Inorg. Chem.*, **1995**, 34, 323.

[3] T. Lipmann, E. Dalcanale, G. Mann, *Tetrahedron Lett.*, **1994**, 35, 1685. T. Lipmann, H. Wilde, E. Dalcanale, L. Mavilla, G. Mann, U. Heyer, S. Spera, *J. Org. Chem.*, **1995**, 60, 235.

[4] Högberg, A.G.S. *J. Org. Chem.* **1980**, *45*, 4498-4500; *J. Am. Chem. Soc.* **1980**, *102*, 6046-6050.

[5] P. Delangle, J.- P. Dutasta, *Tet. Lett.*, **1995**, 36, 9325.

[6] E. P. Kyba, R. C. Helgeson, K. Madan, G. W. Gokel, T. L. Tarnowski, S. S. Moore, D. J. Cram, *J. Am. Chem. Soc.*, **1977**, 99, 2564. S. S. Moore, T. L. Tarnowski, M. Newcomb, D. J. Cram, *J. Am. Chem. Soc.*, **1977**, 99, 6398. K. E. Koenig, G.M. Lein, P. Stuckler, T. Kaneda, D. J. Cram, *J. Am. Chem. Soc.*, **1979**, 101, 35.

[7] A. McCurdy, L. Jimenez, D. A. Stauffer, D. A. Dougherty, *J. Am. Chem. Soc.*, **1992**, 114, 10314. R. A. Kumpf, D. A. Dougherty, *Science*, **1993**, 261, 1708. K. S. Kim, J. Y. Lee, S. J. Lee, T. K. Ha, D. H. Kim, *J. Am. Chem. Soc.*, **1994**, 116, 7399. F. Inokuchi, Y. Miyahara, T. Inazu, S. Shinkai, *Angew. Chem. Int. Ed. Engl.*, **1995**, 34, 1364.

[8] A. Casnati, A. Pochini, R. Ungaro, F. Ungozzoli, F. Arnaud-Neu, S. Fanni, M. -J. Schwing, R. J. M. Egberink, F. de Jong, D. N. Reinhoudt, *J. Am. Chem. Soc.*, **1995**, 117, 2767. A. Casnati, A. Pochini, R. Ungaro, C. Bocchi, F. Ungozzoli, R. J. M. Egberink, H. Stuijk, R. Lugtenberg, F. de Jong, D. N. Reinhoudt, *Chem. Eur. J.*, **1996**, 2, 436.

[9] F. Arnaud-Neu, S. Cremin, D. Cunningham, S. J. Harris, P. McArdle, M. A. McKervey, M. McManus, M.- J. Schwing-Weill, K. Ziat, *J. Incl. Phenom.*, **1991**, 10, 329. F. Arnaud-Neu, M. -J. Schwing-Weill, K. Ziat, S. Cremin, S. J. Harris, M. A. McKervey, *New J. Chem.*,**1991**, 15, 33.

THE CATION COMPLEXATION PROPERTIES OF PER-3,6-ANHYDRO-α AND β-CYCLODEXTRINS STUDIED BY THIN LAYER CHROMATOGRAPHY AND ^{1}H NMR

F. FAUVELLE[1], A. GADELLE[2] , J.C. DEBOUZY[1] and B. PERLY[3]
1 CRSSA, Biophysique, BP 87, 38702 La Tronche cedex, FRANCE.
2 CENG, SCIB, Avenue des Martyrs, BP 85X, Grenoble, FRANCE.
3 CEN-Saclay, DRECAM, 91191 Gif sur Yvette cedex, FRANCE.

1. Introduction

Natural cyclodextrins and most of their derivatives can form inclusion complexes with hydrophobic molecules, and numerous applications are developping[1] as well in research as in pharmaceutical, food and cosmetic industries. However, the per-3,6-anhydro-cyclodextrins[2] have amphiphilic caracteristics and exhibit cation binding abilites.[3] Although several artificial hosts such as crown ethers and cryptands bind cations strongly and selectively, few molecules possess the bioavailability for decontaminating purpose in human pathologies. The chelating agent should be firstly not itself toxic and secondly selective for toxic ions. Such requirements could be fullfilled by the per-3,6-anhydro-cyclodextrins, which exhibit promising properties of cation affinities[3] and poor toxic effects.[4]

For this aim, a large set of cations -vital such as Ca^{2+}, Na^+, K^+ or poisonous heavy metals such as Hg^{2+}, Pb^{2+}, Sr^{2+}- have been tested with per-3,6-anhydro-α andβ-cyclodextrins (3,6αCD and 3,6βCD respectively) by Thin Layer Chromatography (TLC). Then a restricted set of ions -the most reactive on the NMR point of view- was selected for the complete NMR analysis.

2. Experimental section

3,6αCD and 3,6βCD were synthetised according to Gadelle et al.[2]

Polygram ionex-25 SA-Na 5×20cm plates (Macherey-Nagel, Germany) for TLC experiments were loaded with the cations to test by one hour bath in a 5% salt solution. The plates with 3,6CD spots were then immersed in water until the front was at 20cm (approximately 2hours), then dried. This cycle was repeated four times. 3,6CD was revealed by spraying a 10% sulfuric acid solution, followed by 15minutes incubation at 120°C. The distance **r** from the initial spot was then measured in millimeters.

The ^{1}H NMR spectra were recorded at 400.13 MHz on an AM-400 WB Bruker spectrometer. All the solutions were in pure D_2O. Chemical shifts were measured relative to HDO at 297K, 4.8ppm.

A. W. Coleman (ed.), Molecular Recognition and Inclusion, 325–328.
© 1998 *Kluwer Academic Publishers. Printed in the Netherlands.*

3. Results

3.1 THIN LAYER CHROMATOGRAPHY

The method used is based on 3,6CD cation binding properties: it implies a direct complexation between the cyclodextrin and the ionic support. Migration of 3,6CD on plates loaded with different cations can then be related to the strength of the complex, providing rather relative than absolute results. In this way this method can be conveniently used for a pre-testing study.

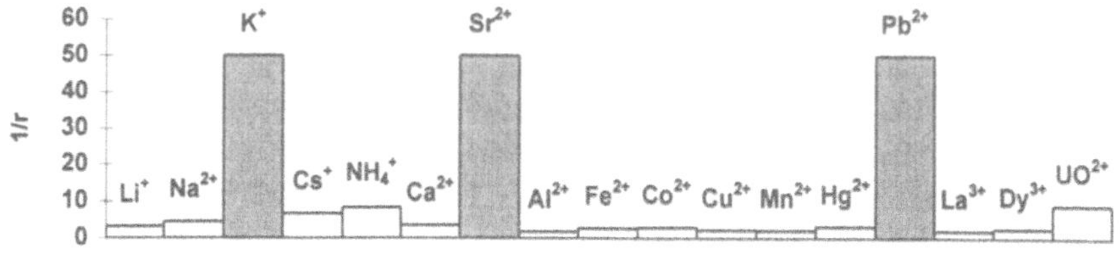

Figure 1: The inverse of the migration distance **r** of 3,6αCD on plates loaded with the cations to test.

As shown figure 1, mainly three ions: K^+, Sr^{2+} and Pb^{2+} strongly retained the 3,6αCD molecule, while all the other one tested did not or slightly hindered the 3,6αCD migration. Cs^+ and K^+ were the most efficient to retain the 3,6βCD molecule. A detailed investigation of these complexes was then undertaken by ^1H NMR

3.2 STOICHIOMETRIES AND ASSOCIATION CONSTANT DETERMINATION FROM NMR ANALYSIS

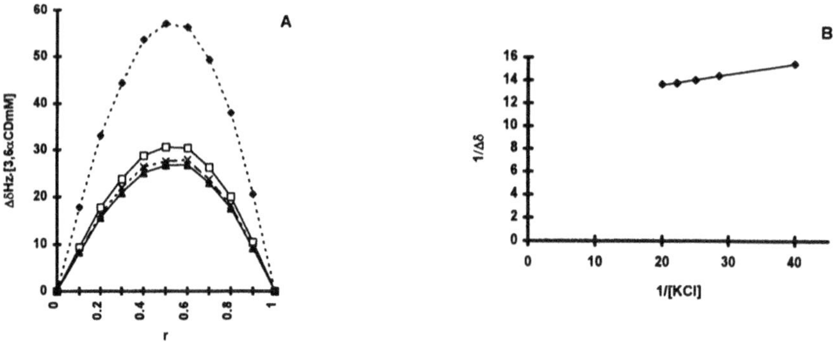

Figure 2: A/ The Job Plot built from chemical shift variations of protons H_2(◆), H_1(□), H_5(×) and H_3(▲) of 3,6αCD under Sr^{2+} addition; the total concentration was maintained at 5 mM for all experiments. B/ The Benesi-Hildebrand straight line obtained from H_2 variations in the K^+/3,6αCD complexation study; the 3,6CD concentration was maintained at 1 mM and the cation concentration varied from 25 to 50 mM.

The addition of Sr^{2+} and Pb^{2+} salts, and to a lesser extend of NH_4^+, K^+ and Cs^+ salts, to aqueous solution of 3,6αCD led to downfield chemical shift variations, mainly important shifts of H_2 on the ^1H NMR spectra. The other ions like Ca^{2+}, Cu^{2+}, Hg^{2+} and Na^+ for example did not affect at all the 3,6αCD resonances. Concerning the 3,6βCD molecule, only K^+ and Cs^+ induced slight variations of the same nature. Under these conditions, a dynamic process of binding can be suggested, and the Continous Variation Method[5] can be used for the stoichiometry determination. The Job plots[5] were then built for the five most reactive cations with 3,6αCD and the two with 3,6βCD. Highly symetric curves (fig 2A) were obtained for all elements tested with a maximum at $r = 0.5$, indicating 1:1 stoichiometries and that no side effects are present.

The apparent binding constant k was evaluated from NMR parameters (chemical shifts) using graphical procedure with the Benesi-Hildebrand[6] method for K^+, NH_4^+ and Cs^+, and with the computer assisted iterative method Complex[7] for Pb^{2+} and Sr^{2+} since the graphical procedure becomes unprecise with very weak slopes. The results are summerized in table 1:

Table 1: Apparent affinity constant k in M^{-1}.

Cations	3,6αCD	3,6βCD
K^+	130	100
Cs^+	80	110
NH_4^+	40	–
Pb^{2+}	2500	–
Sr^{2+}	400	–

4. Discussion

The H_2 chemical shift variation is the caracteristic shared by all the ^1H NMR spectra recorded on cation/3,6CD mixtures. Even in an isotonic salt solution of $NaNO_3$ 0.154M (data not shown), the Job plot[5] and binding constant of Pb^{2+} with 3,6αCD were strictly the same than in pure water although all the chemical shifts were slightly different (10Hz), may be due to a general solvent salt effect. This strongly supports the hypothesis that a real complexation process occurs in the bulk, and that the ring of hydroxyl groups OH_2 -grouped on one side of the 3,6CD- is the preferred site of cation fixation.

However most of the binding constant given table 1 are very weak. More interesting is the affinity of Pb^{2+} for 3,6αCD. The evolution of ^1H NMR spectra under lead addition is displayed figure 3. H_1 shows exactly the same chemical shift variations than H_2, with 0.2ppm downfield difference measured for a ratio r of 0.2 (fig.3 top). H_4, H_5, and H_3 resonances are also but much less shifted. On the contrary H_6 and $H_{6'}$ are not at all affected by the complexation. The proton coupling constant $^3J_{H1-H2}$ is slightly reduced (from 3Hz to 2.5Hz) but the others remain readily unchanged. The heteronuclear coupling constant $^1J_{13C-1H}$ measured on undecoupled ^{13}C NMR spectra (not shown) were also unchanged. Under these conditions, it can be suggested that the anhydroglucose units do not really undergo conformationnal constraints under complexation. On the other hand the modifications of the chemical shifts of H_1 and H_4 with Pb^{2+} binding could suggest that a better host-guest fit is obtained by the entire cyclic molecule deformation. Indeed the 3,6-anhydro-cyclodextrins are more flexible than the natural one, with $^1C_4$ glycosidic bond instead of $^4C_1$. The ether link is more rigid and far from the fixation site so that protons 6 are unaffected by the presence of Pb^{2+}.

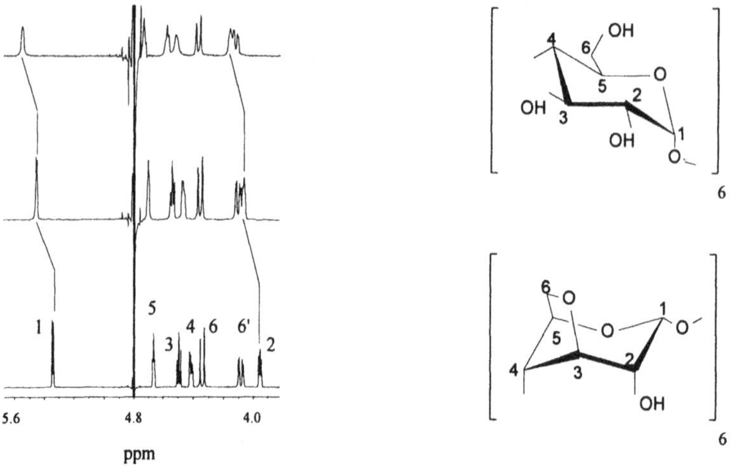

Figure 3: ^1H NMR spectrum of pure 3,6αCD, 5mM in D_2O (bottom); 3.5mM 3,6αCD and 1.5mM $Pb(NO_3)_2$ spectrum (r=0.7) (middle); 1mM 3,6αCD and 4mM $Pb(NO_3)_2$ (r=0.2) (top). Each spectrum has been enlarged differently for convenience. On the right the natural αCD (top) and 3,6αCD (bottom) are represented with the ^1H NMR nomemclature.

5. Conclusion

A step scale affinity of cations for 3,6αCD can be deduced from the NMR binding constant determination, in agreement with TLC results: $Pb^{2+}>>Sr^{2+}>K^+>Cs^+>NH_4^+$. The other ions tested, like Na^+ and Ca^{2+} did not induce any observable spectral modifications on the NMR timescale. The 3,6αCD molecule is then selective for Pb^{2+}, and a thermodynamic study has been undertaken to determine the driving forces in the complexation process. Nevertheless this selectivity could be interesting for biological purposes since the physiological ions do not complex -or complex very weakly- with this cyclodextrin. Conversely, 3,6βCD has poor cation binding properties: only K^+ and Cs^+ are complexed. Moreover the weakness of the binding constants and the absence of selectivity are not in favor of a biological use.

6. References

1. Saenger, W. (1980) Cyclodextrin inclusion compounds in research and industry, *Angew. Chem., Int. Ed. Engl.* **33**, 803-822.
2. Gadelle, A. and Defaye, J. (1991) Selective halogenation at primary positions of cyclomaltooligosaccharides and a synthesis of per-3,6-anhydro cyclomaltooligosaccharides, *Angew. Chem., Int. Ed. Engl.* **30**, 78-80.
3. Yamamura, H., Nagaoka, H., Kawal, M., Busugan, Y. (1994) Synthesis and alkali metal ion binding of poly(3,6-anhydro)-α-cyclodextrins, *Tetrahedron Lett.* **36**, 1093-1094.
4. Fauvelle, F. (1996) unpublished results.
5. Job, P. (1925) Etude spectrographique de la formation de complexes en solution et de leur stabilité, *C. R. Acad. Sci.* **9**, 113-134.
6. Hildebrand, J.H and Benesi, H.A. (1949) A spectrophotometric investigation of the interaction of iodine with aromatic hydrocarbons, *J. Am. Chem. Soc.*, **71**, 2703-2707.
7. Djedaini, F. and Perly B. (1991) in D. Duchêne (Ed), *New Trends in Cyclodextrins and Derivatives*, Editions de Santé, Paris, pp. 217-246.

CAESIUM-SELECTIVE IMPRINTED PHENOLIC RESINS

A. FAVRE-RÉGUILLON, B. DUNJIC, N. DUMONT, M. LEMAIRE*
Institut de Recherches sur la Catalyse, Laboratoire de Catalyse et Synthèse Organique, Université Claude Bernard, Lyon I, CPE
43 Bld du 11 novembre 1918, 69622 Villeurbanne, France

1. Introduction

The molecular imprinting technique is a promising way of preparing highly selective host polymers since it requires no multistep synthesis. This in contrast with supramolecular design. This direct polymerisation method, called "template polymerisation" has been reported for the synthesis of ion-selective polymeric reagents. [1] Nishide *et al.* [2] have prepared a resin by crosslinking poly(4-vinylpyridine) in a presence of a transition metal ion. The resin competitively adsorbed the metal ion which was used as the template. This promising technique has been known for 20 years [2], but only few applications dealing with imprinting resins have been described. [1] This could be explained because resins prepared in such a way suffer from fundamental problems. The capacities of the resins are low, extractions are limited to transition metal ion and the polymer-metal complex is formed in organic media whereas the extractions are performed in aqueous solutions. Presence of organic solvents during the cross-linking process could modify the cavity conformation. Thus, in aqueous phase, cavity and imprinted ions may not match.
In order to overcome these disadvantage, new metal ion imprinted resins have been prepared by surface template emulsion polymerisation in water. [3] Increased adsorptions have been found but selectivity in relation to non targeted metal ions remains low. [4] Furthermore this methodology can only be applied to multivalent ions.

2. Results and discussion

Interest in resorcinol based ion-exchange resins, for the removal of caesium from basic media, has increased. Phenolic resins, in particular resins synthesized by alkaline polycondensation of resorcinol with formaldehyde, showed high affinity for caesium. [5,6] To further probe the potential for template polymerisation, an investigation of the influence of the nature of the base (alkali hydroxide), which serves as the basic catalyst upon the alkali-metal sorption properties of phenolic resins, was undertaken.
Template polymerisation with resorcinol-formaldehyde resins has been studied, [7] but despite the use of radio tracers the effect on ionoselective properties couldn't be brought to the fore. Unlike phenol, high reactivity of resorcinol with formaldehyde didn't allow the formation of a definite product. Nevertheless, resorcinol-acetaldehyde

A. W. Coleman (ed.), Molecular Recognition and Inclusion, 329–332.
© 1998 *Kluwer Academic Publishers. Printed in the Netherlands.*

cyclotetramer, resorcinarene **1**, could be prepared in high yields via simple, one-step procedures without using templates or high dilution techniques. [8] Recently, the selective complexation of caesium by resorcinarene **1** has been described. [9] It is conceivable that the possibility of forming resorcinarene **1** in the resorcinol-acetaldehyde resin could make the characterisation of imprinting effects easier, although the presence of macrocyclic structures in phenolic resins remains conjectural. [10]

In this study, two resorcinol-acetaldehyde resins were synthesized using sodium and caesium hydroxide as the respective catalysts via methods which have been previously described. [5] These resins were then compared with resorcinarene **1**-acetaldehyde resins synthesized using sodium hydroxide under the same conditions.

After curing, the polymers were crushed, washed, and conditioned by subjecting them to two 1 N HCl/0.1 N NaOH cycles. The resins were finally converted to acid form and washed thoroughly with water until neutral. The adsorption equilibrium of the five alkali-metal ions into the different resins was examined by competitive extraction. [5] A known amount of resin was shaken for 5 hours with the solution of the five alkaline salts. The distribution coefficient was calculated as follows:

$$D = \frac{[M^+]_i - [M^+]}{[M^+]} \times \frac{V}{m} \quad (1)$$

The concentration of the metal cations before and after adsorption was determined by atomic emission spectroscopy. The subscript i refers to initial conditions, V (mL) to volume of solution and m (g) to mass of resin. The results achivied for the resins are detailed in figure 1.

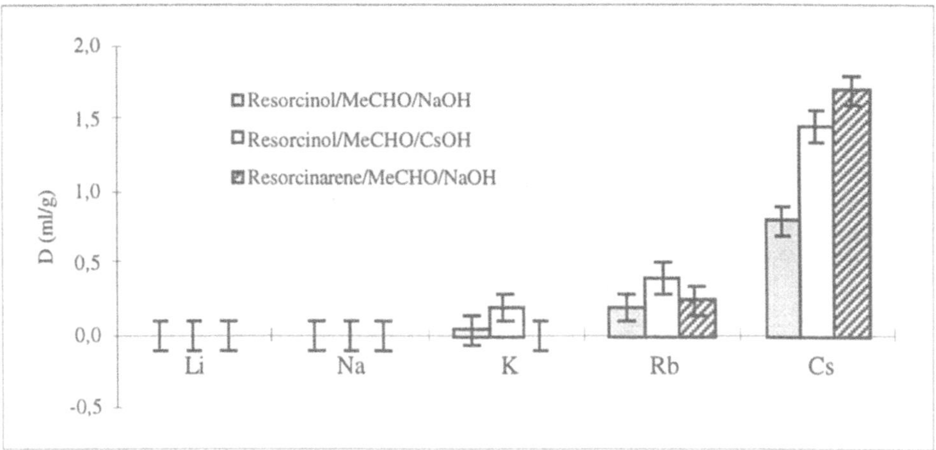

Figure 1. Distribution coefficients of the alkali-cations using differents resins
Conditions: $[LiNO_3]=[KNO_3]=[RbNO_3]=[CsNO_3]= 0.25$ M; $[NaNO_3]= 0.15$ M; $[NaOH]= 0.1$ M; V= 4 mL; m= 100 mg; Initial pH=13; Final pH=6

First, resorcinol based resins were considered. If a comparison is made between the two resins, caesium is most effectively adsorbed by resins synthetized using caesium hydroxide as base, with improved selectivity. This demonstrates the effect of caesium imprinting. Under these conditions, the selectivity of the two resins for caesium towards lithium and sodium was high. The introduction into the resin of a highly organised macrocycle which is selective towards caesium, resorcinarene 1, leads to an improvement in the selectivity for caesium. Nevertheless, the ionoselective properties were close to these of the imprinted resin. From this one can conclude that, when caesium was adopted as a guest, a metal ion-imprinted structure, similar to resorcinarene 1, was constructed in the bulk of the resorcinol-acetaldehyde resins. This was achieved with the aid of specific interactions of the resorcinol groups with the caesium ions. Such specific interactions are well documented in the literature. [11,12]

3. Solid-state NMR studies

The three resins were then compared by ^{13}C solid-state NMR spectroscopy. The aim of this study was to establish whether ^{13}C-CPMAS NMR [13] is capable of detecting changes in structure as a function of the nature of the base. ^{13}C solid-state NMR spectra of the three resins were compared to ^{13}C solution and solid-state NMR spectra of resorcinarene 1 (spectra a). Knowing these peak assignments, we were able to compare them with the spectra obtained from the resins which were prepared.

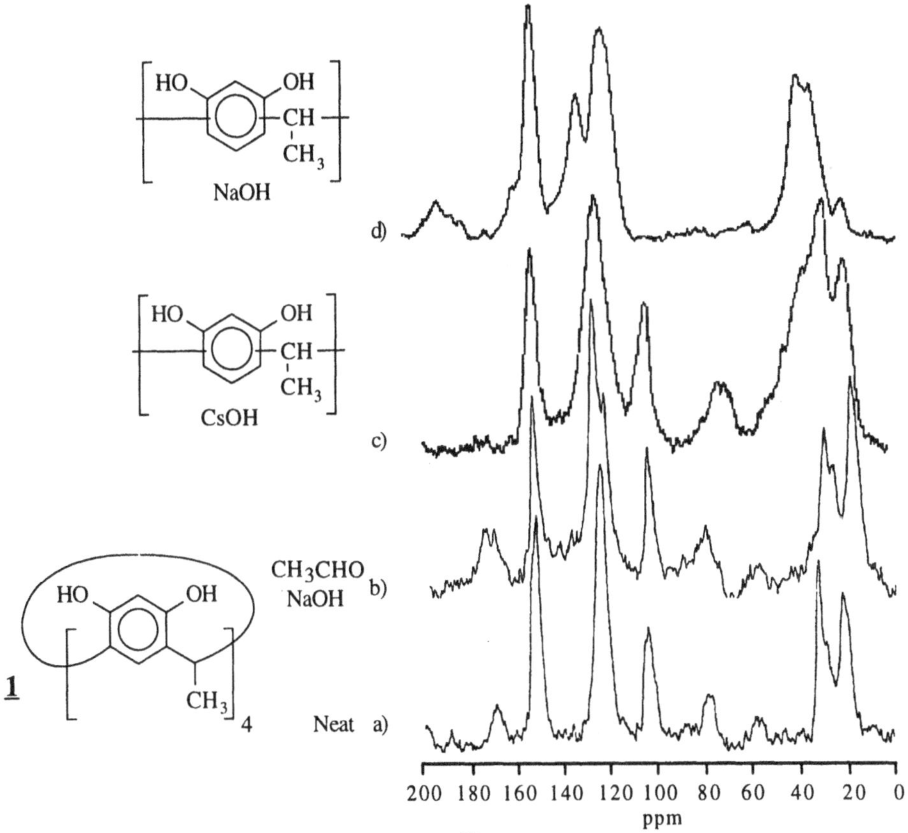

Figure 3. Comparison of the resin by ^{13}C-CPMAS solid state NMR spectra

Cross-linking of 1 with acetaldehyde in the presence of sodium hydroxide (spectra b)) didn't change the chemical shift of the carbon. The peak around 155 ppm was assigned to a carbon bearing OH group and the peak around 125 ppm was assigned to unsubstitued *meta* carbons. Intensity in the 105 ppm region was assigned to the remaining *ortho* phenol-ring carbons. Those peaks in the 35-20 ppm region can be assigned to the sp^3 carbon, CH and CH_3 respectively. These structural units results from resorcinarene and acetaldehyde. When caesium hydroxide was used for the cross-linking of resorcinol with acetaldehyde (spectra c) no significant difference was noticed. The chemical shift of the aromatic carbon didn't change but a slight alteration appears in the region of aliphatic carbon that could be attributed to higher degree of cross-linking. With the resin synthesized using sodium hydroxide, changes occur in the region of aromatic carbon. These differences occurred due to the modification of the type of substitution involved. Thus, when caesium hydroxide is used as catalyst for the synthesis of resorcinol-acetaldehyde resins, substitution occurs mainly at the same position as with resorcinarene 1, i.e. at the 4 and 6 positions, leading to better selectivity towards caesium.

4. Conclusion

The results presented here underline further the great potential of molecular imprinting in ionoselective material synthesis. In this report, novel metal ion-imprinted resins were prepared by alkaline polycondensation of resorcinol and acetaldehyde. Caesium imprinting resins manifested the effect of caesium ion-imprinting. The guest-induced modification of the structure was successfully demonstrated. Data from ^{13}C solid state NMR and extraction studies gave prominence to the presence of a structure similar to 1 in the imprinted resin. The imprinted resin is expected to be a useful material for the separation and concentration of metal ions.

References

1. Wulff, G. (1995) Molecular imprinting in cross-linked materials with the aid of molecular templates - a way towards artificial antibodies, *Angew. Chem., Int. Ed. Engl.* **34**, 1812-1832
2. Nishide, H., Deguchi, J., Tsuchida, E. (1976) Selective adsorption of metal ions on crosslinked poly(vinylpyridine) resin prepared with a metal ion as a template, *Chem. Lett.* 169-174
3. Yu, K. Y., Tsukagoshi, K., Maeda, M., Takagi, M. (1992) Metal ion-imprinted microspheres prepared by reorganization of the coordinating groups on the surface, *Anal. Sci.* **8**, 701-703
4. Tsukagoshi, K., Yu, K. Y., Maeda, M., Takagi, M. (1993) Metal ion-selective adsorbent prepared by surface-imprinting polymerisation, *Bull. Chem. Soc. Jpn* **66**, 114-120
5. Dumont, N., Favre-Réguillon, A., Dunjic, B., Lemaire, M. (1996) Extraction of cesium from an alkaline leaching solution of spent catalysts using an ion-exchange column, *Sep. Sci. Technol.* **31**, 1001-1004
6. Samanta, S. K., Ramaswamy, M., Misra, B. M. (1992) Studies on cesium uptake by phenolic resins, *Sep. Sci. Technol.* **27**, 255-267
7. Kaczvinsky, J. R., Fritz, J. R., Walker, D., Ebra, M. A. (1987) The effects of reaction conditions on porous chelating polymers designed for the decontamination of nuclear waste, *J. Radioanal. Nucl. Chem., Articles* **116**, 63-75
8. Timmerman, P., Verboom, W., Reinhoudt, D. N. (1996) Resorcinarenes, *Tetrahedron* **52**, 2663-2704
9. Gaubert, E., Barnier, H., Favre-Réguillon, A., Nicod, L., J., Lemaire, M. (submitted) Selective cesium removal from a sodium nitrate aqueous medium by nanofiltration-complexation, *Sep. Sci. Technol.*
10. Gutshe, C. D. (1987) Calixarenes and the art of molecular basketmaking, in R. M. Izatt, J. J. Christensen (eds.), Progress in Macrocyclic Chemistry, J. Wiley & Sons, Vol. 3, Chap. 3, pp. 95
11. Samanta, S. K., Misra, B. M. (1995) Ion exchange selectivity of a resorcinol-formaldehyde polycondensate resin for cesium in relation to other alkali metal ion, *Solv. Extr. Ion Exch.* **13**, 575-589
12. Ungaro, R., Casnati, A., Ugozzoli, F., Pochini, A., Dozol, J. F., Hill, C., Rouqueete, H. (1994) 1,3-dialkoxycalix[4]arenecrowns-6 in 1,3-alternate conformation: cesium-selective ligands that exploit cation arene interactions, *Angew. Chem. Int. Ed. Engl.* **33**, 1506-1509
13. Neiss, T. G., Vanderheiden, E. J. (1994) Solution and solid-state NMR analysis of phenolic resin cure kinetics, *Macromol. Symp.* **86**, 117-129

SUPRAMOLECULAR SYNTHESIS WITH CARBOXYL-SUBSTITUTED SECONDARY DIALKYLAMMONIUM SALTS AND MACROCYCLIC POLYETHERS

MATTHEW C. T. FYFE, J. FRASER STODDART
School of Chemistry, The University of Birmingham
Edgbaston, Birmingham B15 2TT, UK

ANDREW N. COLLINS
ZENECA Specialities,
Blackley, Manchester M9 3ZS, UK

DAVID J. WILLIAMS
Department of Chemistry, Imperial College
South Kensington, London SW7 2AY, UK

Self-assembly [1] — the spontaneous generation of ordered supramolecular aggregates or arrays using intermolecular bonds — has of late become the noncovalent craftsman's primary tool for the supramolecular synthesis of organised, functioning superstructures. A recent report [2] from our laboratories has detailed an elementary self-assembling system in which secondary dialkylammonium ions — such as the dibenzylammonium ion **1**-H$^+$ — thread through the cavities of macrocyclic polyethers to afford pseudorotaxanes of varying stoichiometries (Scheme 1). These complexes are assembled using primarily [N$^+$–H···O] hydrogen bonds, though occasional supplementary stabilisation may be provided by [C–H···O] hydrogen bonds in conjunction with aromatic edge-to-face and face-to-face interactions. The monotopic crown ether dibenzo[24]crown-8 (**DB24C8**) can thread only one cationic strand through its macrocyclic cavity, while its ditopic counterpart bis-*p*-phenylene[34]crown-10 (**BPP34C10**) — possessing two dis-associated arcs of oxygen atoms in order to interact with the ammonium sites — allows two secondary dialkylammonium ions to be inserted through its macrocyclic interior to form a double stranded [3]pseudorotaxane.

We considered that the introduction of carboxyl groups — already exploited thoroughly in supramolecular, noncovalent synthesis [3] — on to the secondary dialkylammonium component of these systems would provide excellent access into a field concerned with the *crystal engineering* [4] of pseudorotaxane aggregates and arrays. Displayed here are our initial results, which reveal the noncovalent synthesis of novel interwoven supermolecules and supramolecular polymers [5], produced by the association of a finite or multitudinous number of molecular building blocks, respectively.

The X-ray crystal structural analysis (Figure 1) of cocrystals of the **2**-H$^+$ cation with the crown ether **DB24C8** confirms the formation of a [2]pseudorotaxane — *i.e.* the cationic species is threaded through the macrocycle's central void in such a way that each of its termini are projected away from the macroring's centre. The 1:1 complex is stabilised by both [N$^+$–H···O] and [C–H···O] hydrogen bonds, accompanied by π-π stacking interactions between one of the macrocycle's π-electron rich catechol rings and the amm-

A. W. Coleman (ed.), Molecular Recognition and Inclusion, 333–336.
© 1998 *Kluwer Academic Publishers. Printed in the Netherlands.*

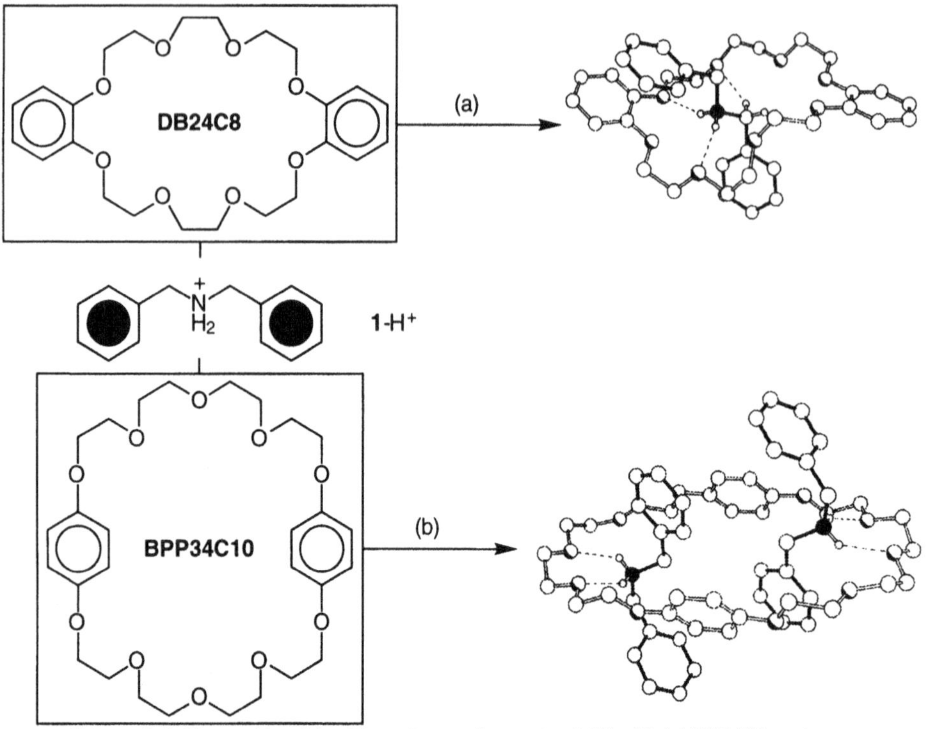

Scheme 1. Self-assembly of the dibenzylammonium cation **1**-H$^+$ with (a) **DB24C8**, and (b) **BPP34C10**, to afford a [2]pseudorotaxane and a [3]pseudorotaxane, respectively.

onium strand's π-electron deficient benzoic acid unit. However, at the next level of supramolecular hierarchy — *i.e. inter*pseudorotaxane interactions — an unforeseen situation has arisen. The desired and indeed expected formation of a dimeric pseudo-rotaxane — employing the carboxyl dimer supramolecular synthon — does not occur: instead, each carboxyl hydrogen atom is linked noncovalently to the oxygen atoms from one of the catechol units of a neighbouring pseudorotaxane *via* a bifurcated (three-centre) hydrogen bond. The ensuing supramolecular array (Figure 1) is reminiscent [6] of a hydrogen bonded, macromolecular daisy chain. Conversely, cocrystallisation of this cation with the ditopic crown ether **BPP34C10** affords the anticipated supramolecular product (Figure 2). This doubly encircled six-component organic supermolecule forms by the noncovalent dimerisation of a double stranded pseudorotaxane, since its benzoic acid subunits are oriented in the same direction by the macrocyclic polyether, leading to dimerisation using the carboxyl dimer supramolecular synthon. Interestingly, PF_6^- counterions are associated with the benzylic hydrogen atoms located at the supermolecule's extremities. It is possible that these counterions influence the outcome of the self-assembly process, by enforcing the codirectionality of the ammonium strands in the solid state.

We are of the opinion that two factors determine the differences in the hydrogen bonding motifs displayed by the carboxyl groups of the [**DB24C8**·**2**-H]$^+$ and the [**BPP34C10**·(**2**-H)$_2$]$^{2+}$ pseudorotaxanes presented here. Firstly, **DB24C8** has two sets

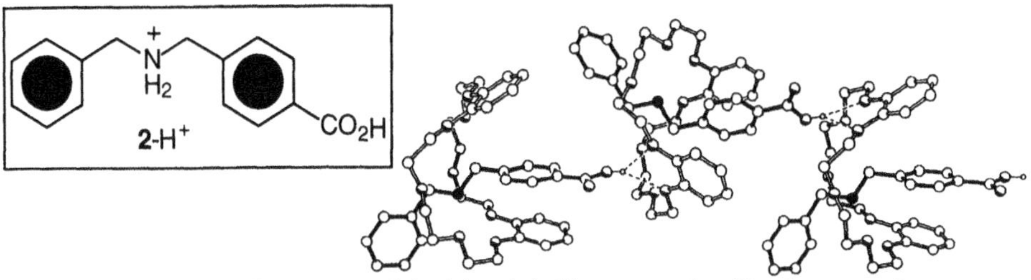

Figure 1. The structure of the daisy chain-like array produced by the noncovalent polymerisation of the [**DB24C8**·**2**-H]$^+$ pseudorotaxane in the solid state.

of preorganised catechol oxygen atoms which are rigidly juxtaposed in an orientation favourable for the formation of a three-centre hydrogen bond, whereas its ditopic relative **BPP34C10** has only hydroquinone oxygen atoms and $-OCH_2CH_2O-$ oxygen atoms, neither of which are preorganised in an orientation which can form bifurcated hydrogen bonds. Secondly, the π-electron rich catechol rings of **DB24C8** prefer to be involved in aryl-aryl stacking interactions with the π-deficient benzoic acid ring of the **2**-H$^+$ ammonium cation. Dimer formation — using the carboxyl dimer supramolecular synthon — would produce a four-component assemblage where catechol units from adjacent pseudorotaxanes would be in close proximity to one another, resulting in

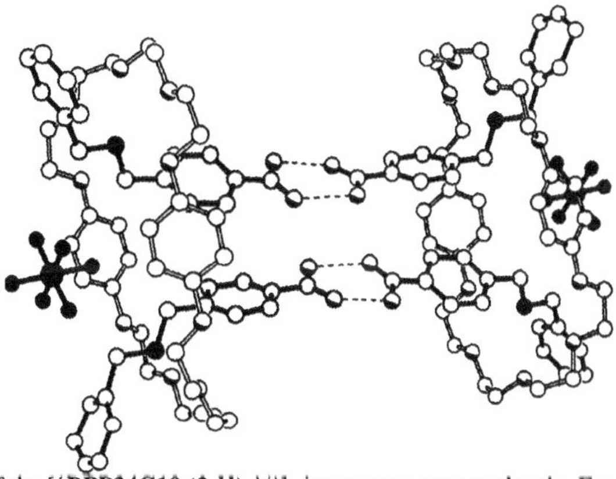

Figure 2. View of the [{**BPP34C10**·(**2**-H)$_2$}$^{2+}$]$_2$ interwoven supermolecule. For the sake of clarity, the only PF_6^- counterion displayed is the one associated with the benzylic methylene groups.

unfavourable nonbonding interactions. On the other hand, there are no π-π stacking interactions observed in the [**BPP34C10**·(**2**-H)$_2$]$^{2+}$ [3]pseudorotaxane.

The X-ray crystal structural analysis (Figure 3) of the 1:2 complex — formed between the **3**-H$^+$ ammonium ion and the ditopic crown ether **BPP34C10** — exhibits an extremely interesting superstructure, where the formation of a novel class of supramolecular polymer, *viz.* an interwoven supramolecular cross-linked polymer, is evident. Cross-linking of crinkled isophthalic acid tapes [7] is observed, since the macrocyclic polyether clutches two isophthalic acid units from separate **3**-H$^+$ cations in opposing directions, *i.e.* the tapes are interlinked *via* the [3]pseudorotaxane unit to form two-dimensional pseudopolyrotaxane [8] sheets.

To summarise, we have introduced a new paradigm for the rational solid state supramolecular synthesis — *i.e.* crystal engineering [4] — of interwoven systems [9] using

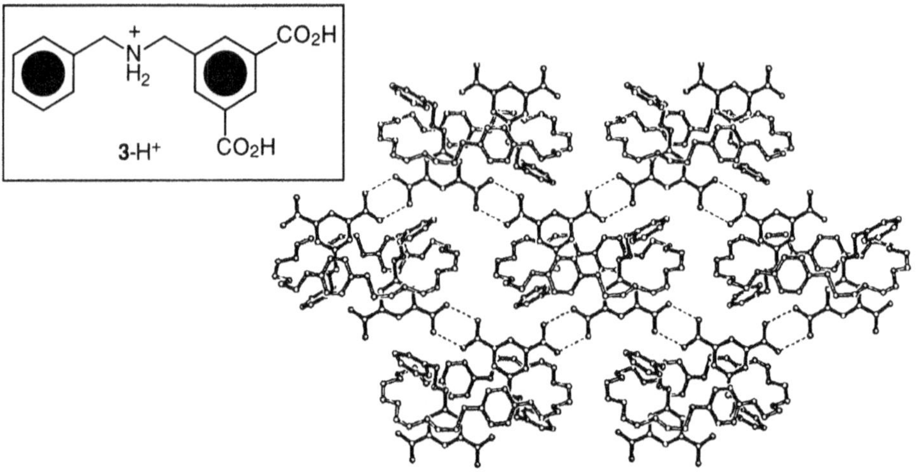

Figure 3. View of the interwoven supramolecular cross-linked polymer produced by the noncovalent polymerisation of the [**BPP34C10**·(**3**-H)$_2$]$^{2+}$ pseudorotaxane.

carboxyl-substituted secondary dialkylammonium ions and macrocyclic polyethers. The range of superstructures produced when the carboxyl-substituted ammonium cations co-crystallise with macrocyclic polyethers is extensive, traversing the whole superstructural kingdom, from an oligomolecular, six-component, doubly-encircled supermolecule to a macromolecular daisy-chain and an interwoven supramolecular cross-linked polymer. It is hoped that these systems will form organised supramolecular materials which might be incorporated into nanoscale devices; on the other hand, they could exhibit catalytic or zeolitic properties.

Further Reading

[1] Philp, D. and Stoddart, J.F. (1996) Self-assembly in natural and unnatural systems, *Angew. Chem., Int. Ed. Engl.* **35**, 1154–1196.

[2] Ashton, P.R., Chrystal, E.J.T., Glink, P.T., Menzer, S., Schiavo, C., Stoddart, J.F., Tasker, P.A., and Williams, D.J. (1996) Pseudorotaxanes formed between secondary dialkylammonium salts and crown ethers, *Chem. Eur. J.* **2**, 709–728.

[3] Kolotuchin, S.V., Fenlon, E.E., Wilson, S.R., Loweth, C.J., and Zimmerman, S.C. (1995) Self-assembly of 1,3,5-benzenetricarboxylic acids (trimesic acids) and several analogues in the solid state, *Angew. Chem., Int. Ed. Engl.* **34**, 2654–2657 and references cited therein.

[4] Desiraju, G.R. (1995) Supramolecular synthons in crystal engineering — a new organic synthesis, *Angew. Chem., Int. Ed. Engl.* **34**, 2311–2327.

[5] Lehn, J.-M. (1993) Supramolecular chemistry — molecular information and the design of new supramolecular materials, *Makromol. Chem., Macromol. Symp.* **69**, 1–17.

[6] Amabilino, D.B., Parsons, I.W., and Stoddart, J.F. (1994) Polyrotaxanes, *Trends Polym. Sci.* **2**, 146–152.

[7] Alcala, R. and Martinez-Carrera, S. (1972) The crystal structure of isophthalic acid, *Acta Cryst.* **B28**, 1671–1677.

[8] Raymo, F.M. and Stoddart, J.F. (1996) Polyrotaxanes and pseudopolyrotaxanes, *Trends Polym. Sci.* **4**, 208–211.

[9] Amabilino, D.B. and Stoddart, J.F. (1995) Interlocked and intertwined structures and superstructures, *Chem. Rev.* **95**, 2725–2828.

Cation Binding of Benzo Crown Ethers in acetonitrile using Fluorescence Spectroscopy, Part II *

Ayten Göçmen and Çakıl Erk
Istanbul Technical University, Faculty of Science, Department of Chemistry,Maslak, 80626, Istanbul.

Abstract ;The effects of Na^+, K^+, Li^+ and Mg^{2+} cations on the fluorescence spectra of benzo[12]Crown-4, benzo[15]Crown-5, and benzo[18]Crown-6 were investigated in dry acetonitrile. The alkali cation coused the complexation enhanced quenched fluorescence (CEQF) by increasing the fluorescence quenching rates of the cationic fluoroionophores. The 1:1 stoichiometry of ion association constants, $\ln K_a$ have been obtained. The results were ruled out by compatibility of cationic radii and macrocycle size. The order of $\ln K_a$ found for benzo[15]crown-5 was $Mg^+ > Na^+ > Li^+$ for the cations and the $Na^+ > Li^+$ was found for benzo[18]crown-6. The most of the cations have slight effects on the fluorescence spectra of benzo[12]crown-4 in acetonitrile.
Key words :Macrocyclic ethers,cation binding, fluorescence spectroscopy.

1. Introduction

The recent reports indicated the fluorescence spectroscopy could be an utility on molecular recognition of fluorogenic macrocycles [4]. The physical effects of cations on the fluorescence spectra have been has been reported by Morita [1] and by Sousa [2]. The chemical aspects were also reviewed by Vögtle [3].

Fluorescence spectroscopy, used in our laboratory for the coumarin derivatives of crown ether complexes with alkali and alkaline earth cations, has given good results for the molecular recognition in acetonitrile [5a-c]. Recently, we have examined the fluorescence spectra of some benzocrowns in the presence of alkali thiocyanates and reported the interesting cation binding roles on the spectra in agreement with ionic radii and macrocycle size relation [6b]

*) Part I, ref 6b.

A. W. Coleman (ed.), Molecular Recognition and Inclusion, 337–340.
© 1998 *Kluwer Academic Publishers. Printed in the Netherlands.*

In this work, the complexing effects of Na^+, K^+, Mg^{2+} and Li^+ with ClO_4^- anion on the fluorescence spectra of benzo[12]crown-4, benzo[15]crown-5 and benzo[18]crown-6 were investigated in dry acetonitrile. The complexation enhanced quenching effect (CEQF) of the cations on the fluorescence spectra of benzocrowns was mostly observed. The dynamics of the macrocycle is altered due to metal complex formation, therefore, the fluorescence is quenched in more rigid structure lowering the quantum yields [4].

2. Experimental

To estimate the association constants using Eqs 1,2 the fluorescence intensities of free and complexed crown ether [A_o]=[L_o] were measured. The $(1\text{-}P)^2$ / P values were plotted versus the inverse of initial concentrations, 1/[L_o] and the inverse of the slope gave the association constants. Fluorescence emission and excitation spectra were obtained with Perkin Elmer Luminescence spectrometer, model LS-50 in dry CH_3CN in the range of 10^{2} - 10^{6} M in 10 mm quartz cells.

3. Results and Discussion

The fluorophore macrocycles of the π-systems altered upon complexation involved relatively low energies [3,4]. However, the complexation enhanced quenching fluorescence (CEQF) is observed on the spectra of benzo-crowns. As a fluorescence probe the cation radius mainly governs the ion-dipole interactions depending on the average macrocycle size of the optimized energy of molecular geometry [1,2[.

Accordingly, the association constants were estimated from the free and altered fluorescence spectra of the solutions. The 1:1 association constants were calculated with Eqs. 1,2. P is the mole fractions of the complex and L_o is the initial concentrations [6a,b].

The association constants were found to be in the order of $Mg^{2+} > Na^+$ for benzo[15]crown-5, but the order for benzo[12]crown-4

$$nA^+ + mL \rightleftarrows A^+_n L_m \tag{1}$$

$$1/K_a [L_o] = (1\text{-}P)^2 / P \tag{2}$$

found was $Na^+ > Mg^{2+} > Li^+$ in acetonitrile. We were not able to estimate the potassium binding properly due to low solubility of $KClO_4$ in acetonitrile. However, ClO_4^- was also found to be more effective then SCN^- showing the role of large counter ion in complexing solutions, Table 1. However, such realtionships are more likely to be explained by the ion pairs and their aggregates theory of Szwarc [8].

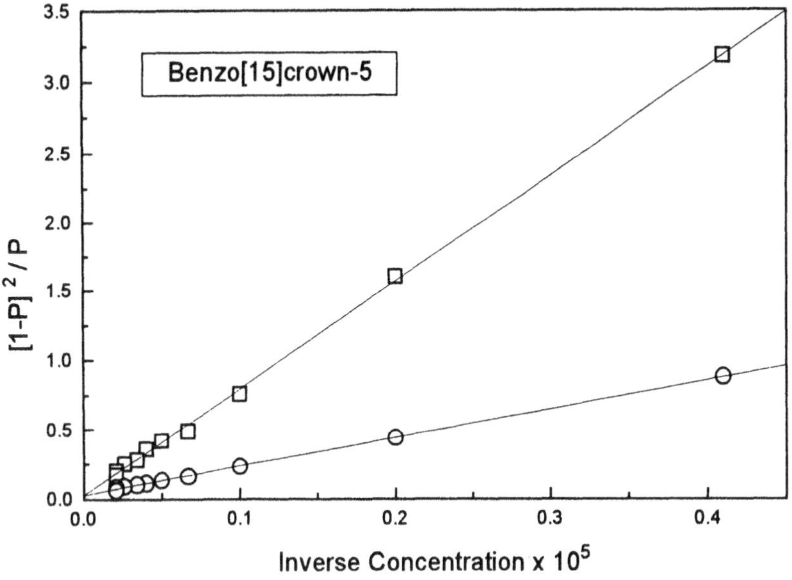

Figure 1. The plot of inverse concentration of Na^+ and Mg^{2+} in benzo[15]crown-5 versus $(1\text{-}P)^2 / P$ from emission spectra at $\lambda_{max} = 306$ nm (excitation $\lambda_{max} = 288$)

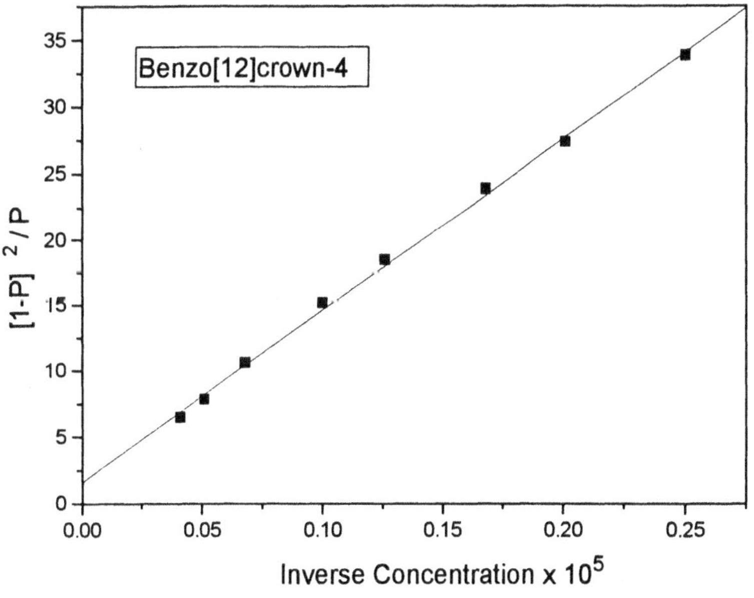

Figure 2. The plot of inverse concentration of Mg^{2+} in benzo[12]crown-4 in CH_3CN versus $(1\text{-}P)^2 / P$ from emission spectra at $\lambda_{max} = 306$ nm (excitation $\lambda_{max} = 288$ nm)s

Table1. The 1:1 association constants of the cations in acetonitrile at 25 °C.

Crowns	lnK_a [a)]	$-\Delta G$ [a)]	lnK [b)]	$-\Delta G$ [b)]	LnK [c)]	$-\Delta G$ [c)]
Benzo[12]Crown-4	2.85	3827.8	1.50	2045.4	0.505	740
Benzo[15]Crown-5	4.10	5577.5	4.68	6359.0		
Benzo[18]Crown-6	3.99	5413.7				

a) $NaClO_4$ complex, ΔG in J M^{-1}. b).$Mg(ClO_4)_2$, ΔG in J M^{-1}. c).$LiClO_4$, ΔG in J M^{-1}.

Acknowledgment

Appreciation is expressed by the authors to the Research Foundation of Istanbul Technical University who financially supported the research work.

References

1. Shizuka, H., Takada, K., Morita, T. : *J. Phys. Chem.* **54**,994(1980).
2. Sousa, L. R. and Larson, J. M. : *J. Am. Chem. Soc.* **99**,99(1977).
3. Weber, E. and Vogtle, F.: *Topics in Current Chemistry,* Vol 98, Springer Verlag, Berlin, 1981, pp 1.
4. Czarnik, A. W. : Acc. Chem. Res., **27**,302(1994).
5. a. Göçmen, A., Bulut, M. and Erk, Ç. : *Pure Appl. Chem.* **65** 447(1993).
   b. Akyüz, S., Göçmen, A., Bulut, M., and Erk, Ç. : *Spectrosc. Lett.* **28(4)**,603(1995).
6. a. Göçmen, A. and Erk, Ç., : *Fresenius J. Anal. Chem.* **347**,4 (1993).
   b. Göçmen, A. and Erk, Ç. : 1996, *J. Incl. Phen* , in press.
7. a. Göçmen, A., Bulut, M. and Erk, Ç. : *Supramol. Science*, 1996, submitted.
   b. Göçmen, A., Bulut, M. and Erk, Ç. : *Supramol. Chem.*, 1995, submitted,
8. Szwarc, M., in Ions and Ion Pairs in Organic Reactions, vol 1, J. Wiley, 1972,New York

CRYSTAL ENGINEERING WITH NOVEL AMINOBORATES. HYDROGEN-BONDED CYCLIC MOTIFS CONTAINING TETRAHEDRAL BORON AND NITROGEN

Z. GOLDSCHMIDT*, S. LEVINGER, I. BEN-ARIE, S. ALFI
Department of Chemistry, Bar-Ilan University, Ramat-Gan 52900, Israel
S. COHEN
Department of Inorganic and Analytical Chemistry, The Hebrew University of Jerusalem, Jerusalem 91905, Israel

1. Introduction

The design of molecular systems capable of self-organization to a well-defined supramolecular architecture is a subject of contemporary challenge [1]. Directed self-assembly of molecular units into supramolecular arrangements is often achieved via intermolecular hydrogen bonding, and this has been most widely encountered in crystal engineering [2].

The construction of supramolecular synthons in crystal engineering usually involves attaching covalently bonded hydrogen-bonding donor and acceptor groups to a *carbon atom backbone*. This, however, frequently limits the simultaneous binding of more than one good hydrogen donor group, such as OH or NH, to the same carbon atom, due to formation of labile compounds (e.g. aminohemiacetals).

We present here a simple approach to circumvent this problem by exchanging the backbone carbon atom with a *tetravalent boron atom*. This led to the design and synthesis of a new class of aminoboric acid derived supramolecular synthons **1** which are capable of enforcing complementary self-assembly into supramolecular structures directed by hydrogen bonding.

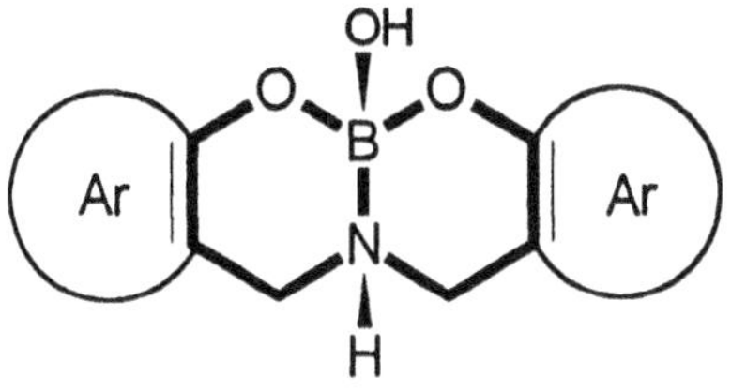

1. Ar = Substituted benzene, naphthalene

A. W. Coleman (ed.), Molecular Recognition and Inclusion, 341–344.
© 1998 *Kluwer Academic Publishers. Printed in the Netherlands.*

2. Synthesis

B-Hydroxy-[1,3,2]naphthoxazaborino[2,3-b][1,3,2]naphthoxazaborine (BNP) (**1**, Ar = naphthalene) was readily obtained in a one step synthesis following the previously reported Mannich reaction of β-naphthol with hexamethylenetetramine, in the presence of boric acid [3]. The reaction is general and could be applied to a variety of substituted electron rich phenols [4]. However, except for BNP none of the other hydrox-benzoxazaborines was structurally characterized. Furthermore, the stereochemical relationship between the two central 6-membered heterocyclic rings of BNP was also not determined [4]. The *cis*-fused configuration of BNP and related oxazaborines around the central bond, which is first reported here, turns out to be of crucial importance in the ability of these compounds to serve as supramolecular synthons in crystal engineering.

3. Crystal Engineering

When hydroxynaphthoxazaborine (BNP) is crystallized at room temperature from dioxane crystals of the 1:1 oxazaborine-dioxane complex (BNP-dox) are obtained. The analogous 1:1 oxazaborine-pyridine complex (BNP-py) was similarly obtained by crystallization from pyridine.

Single crystal X-ray analysis (Table 1) of both BNP-dox (Fig. 1) and BNP-py (Fig. 2) revealed a complimentary self-assembly of four components in a supramolecular network, directed by two hydrogen-bonded motifs. The central pattern, designated $\mathbf{R_2^2(8)}$ by graph set analysis [5], comprises a cyclic 8-membered ring formed by hydrogen bond dimerization of BNP. This is further bound to two dioxane (oxygen) or pyridine (nitrogen) acceptor molecules, via the remaining free BNP hydroxylic-hydrogen donors, in a finite two-atom **D** *motif.* Both crystal structures are specified by the first-order graph set $\mathbf{N_1 = DR_2^2(8)}$, and are thus *isographic.*

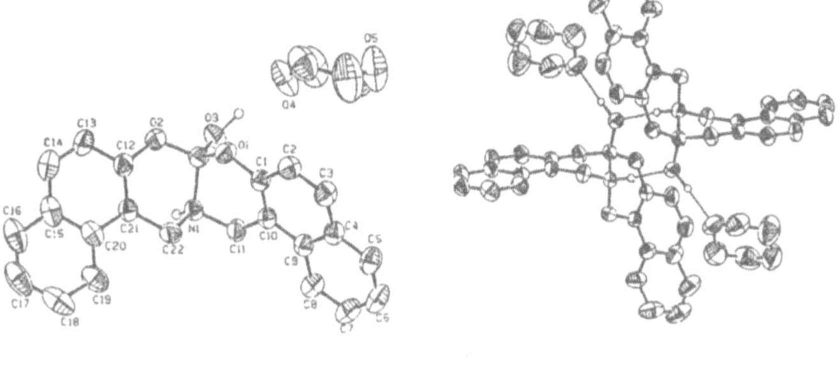

Fig. 1. X-ray crystal structure of BNP-dox (left) and hydrogen bonding pattern (right).

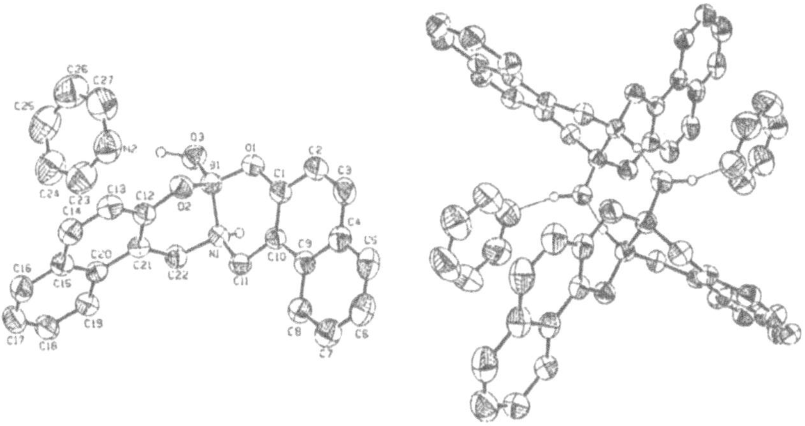

Fig. 2. X-ray crystal structure of BNP-py (left) and hydrogen bonding pattern (right).

In contrast, when BNP was dissolved in hot methanol and left to crystallize at room temperature, a 1:1 methoxyoxazaborine-methanol complex (BNPM-MeOH) was isolated (Fig. 3), in which the BNP hydroxy group has now been exchanged by a *methoxy* group. Furthermore, dimerization via a cyclic $\mathbf{R_2^2(8)}$ motif did not occur. Instead, self-organization of the two complementary species results in the formation of an infinite hydrogen-bonded chain, characterized by two descrete **D** motifs involving the methanolic OH group. One **D** motif of the methanol hydrogen donor and the boron methoxy oxygen acceptor, and the other of NH as donor and methanolic oxygen as acceptor. This difines a first-level graph set $\mathbf{N_1 = DD}$, and an overall hydrogen bonding second-level graph set arrangement $\mathbf{N_2} = \mathbf{C_2^2(6)}$.

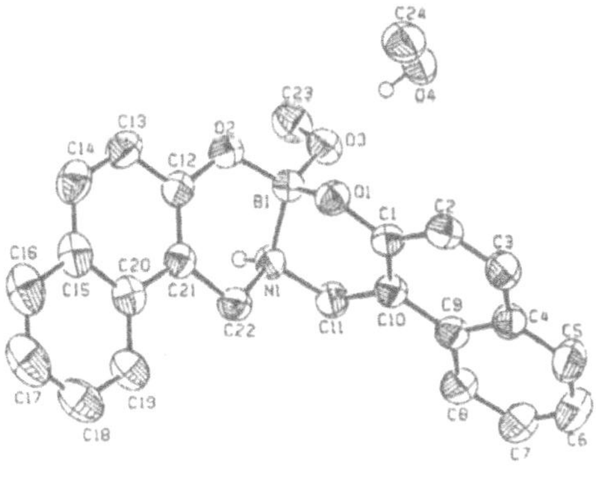

Fig. 3. X-ray crystal structure of BNPM-MeOH

Table 1. Crystal structure data for BNP-py, BNP-dox and BNPM-MeOH.

Crystal Structure Data	**BNP-py**	**BNP-dox**	**BNPM-MeOH**
Space group	$P2_1/n$	$P2_1/n$	$P2_1/c$
a, Å	24.566(4)	13.038(2)	17.190(5)
b, Å	7.263(4)	18.111(3)	10.333(2)
c, Å,	12.513(2)	10.314(4)	12.149(3)
β, deg.	101.07(2)	111.74(3)	106.41(3)
V, $Å^3$	2191.1(9)	2262.2(9)	2070.1(9)
Z	4	4	4
R	0.037	0.049	0.038
R_W	0.048	0.066	0.069
Selective Bond Distances (Å)			
Intramolecular			
B-N	1.614	1.603	1.610
Intermolecular H-bonding			
NH...O	1.790	1.828	1.762
OH...X	1.809 (X = N)	1.795 (X = O)	1.842 (X = O)

We conclude by noting that self-assembly of *B*-hydroxy-oxazaborines based on the new dimeric $\mathbf{R}_2^2(8)$ motif reported above, appears to be a general phenomenon. We have now evidence for complementary self-organization of hydroxyoxazaborines with molecules containing a variety of other acceptor groups, such as the carbonyl group. The design and properties of these new materials is currently under active investigation.

References and Notes

*E-mail: *goldz@gefen.cc.biu.ac.il*

[1] Vögtle, F. (1993) *Supramolecular Chemistry*, Wiley, Chichester; Lehn, J-.M. *Angew. Chem. Int. Ed. Engl.* **1990**, *29*, 1304.

[2] Desiraju, G.R. (1989) *Crystal Engineering*, Elsevier, Amsterdam; Desiraju, G.R *Angew. Chem. Int. Ed. Engl.* **1995**, *34*, 2311.

[3] Möhrle, H.; Züge, E. *Arch. Pharm. (Weinheim)* **1981**, *314*, 580.

[4] Duff, J.C.; Furness, V.I., *J. Chem. Soc.* **1951**, 1512; Zigeuner, G.; Yellinek, K. *Monatsh. Chem.* **1959**, *90*, 297.

[5] Bernstein, J.; Davis, R.E.; Chang N-.L. *Angew. Chem. Int. Ed. Engl.* **1995**, *34*, 1555; Etter, M.C. *Acc. Chem. Res.* **1990**, *23*, 120.

Novel Bis(phenoxyalkyl)sulfane Podands - Synthesis and Complex Formation with Thiophilic Metals Ions

B. HABERMANN[1], T. KRÜGER[1], H. STEPHAN[1], K. HOLLMANN[2], K. GLOE[1]
[1]*Institute of Inorganic Chemistry, Technical University Dresden, D-01062 Dresden, Germany (TK@cach05.chm.tu-dresden.de)*
[2]*Institute of Organic Chemistry, University Leipzig, D-04103 Leipzig, Germany*

Introduction

Preorganized host compounds are important in view of their specific binding possibilities for ionic and neutral guest molecules [1].
An interesting synthetic route to obtain novel sulfur containing macrocyclic and structure related open-chain compounds represents the reaction of allyl phenyl ethers with sulfur dichloride [2, 3]. Using this synthetic principle different thia crowns [4] and sulfane podands [5] were prepared and characterized by complexation studies.
In this paper the synthesis of modified bis(phenoxyalkyl)sulfane ligands and solvent extraction studies with Ag(I), Hg(II) and Pd(II) are described.

Synthesis

Allyl phenyl ethers react with sulfur dichloride at low temperature to bis(phenoxyalkyl)sulfane podands **1**. The resulting anti-MARKOVNIKOV products contain two ß-chloroethylthio groups. Owing to two asymmetric centres in the molecules mixtures of diastereomeres (meso-, d/l-forms) are formed. The reduction of **1** with $LiAlH_4$ in THF gives the dimethyl product **2** [5]. By substitution of chlorine atoms in **1** with different nucleophilic agents in the presence of Ag_2CO_3 the ligands **3** are obtained. Compound **1** reacts with NaOH/chloroform via elimination/dichlorocarben addition by phase transfer catalysis in dependence on reaction time to mono or dicyclopropyl sulfane derivatives **4** and **5**. The structure of the synthesized compounds are confirmed by mass, ^{13}C-n.m.r. and ^{1}H-n.m.r. spectroscopy. In case of **4** the molecular structure is determined by X-ray diffraction technique (Figure 1a).

A. W. Coleman (ed.), Molecular Recognition and Inclusion, 345–348.
© 1998 *Kluwer Academic Publishers. Printed in the Netherlands.*

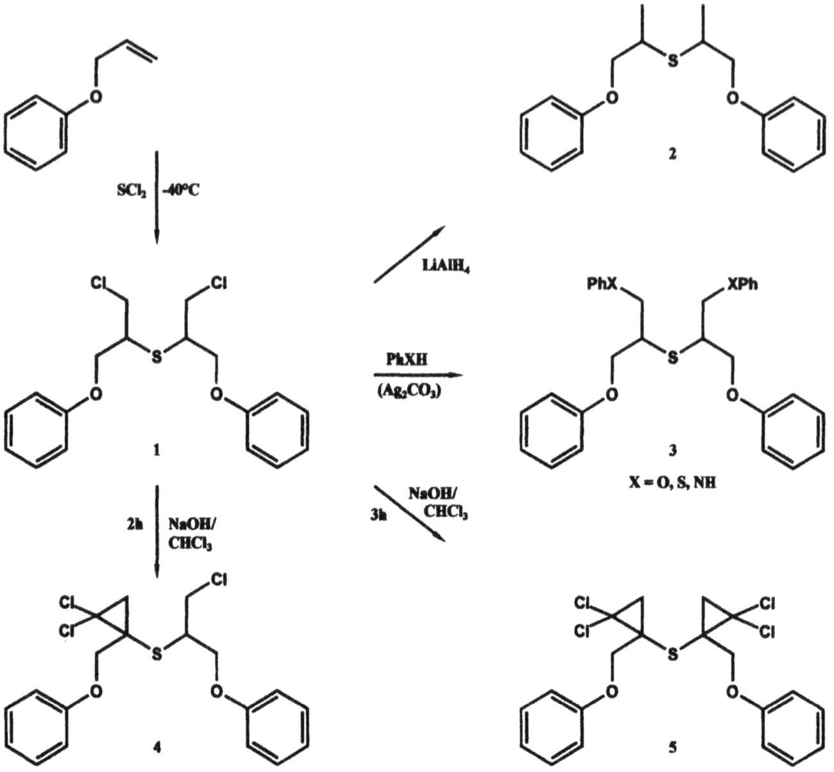

Molecular modeling calculations with the semiempirical MOPAC 6.0 program package can be used for describing the investigated ligand systems [6] (Fig. 1 b). The X-ray and the PM3 calculated structure of ligand **4** are in a good agreement. At present structure calculations of the novel ligands and their metal complexes are in progress.

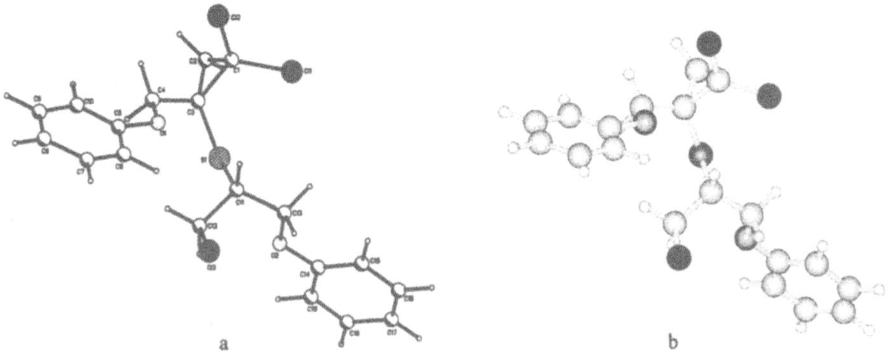

Figure 1: X-ray structure (a) and PM3 calculated structure (b) of ligand **4**

Solvent extraction studies

The synthesized podands **2** to **5** were characterized in view of their complexing properties toward Ag(I), Hg(II) and Pd(II) by liquid-liquid extraction investigations using the two-phase system metal salt-picric acid-H_2O/podand-chloroform.
The extraction studies were performed at 25 ± 1 °C in micro reaction vials by means of mechanical shaking. The phase ratio $V_{(org)} : V_{(W)}$ was 1:1 (0,5 cm^3 of each phase), the shaking time was 30 min. In this time the equilibrium is achieved in every case. The metal concentration in both phases was determined using the γ-radiation measurement of ^{110m}Ag, ^{203}Hg and ^{109}Pd (Medgenix Diagnostics) in a Na(Tl) scintillation counter (Cobra II, Canberra-Packard).

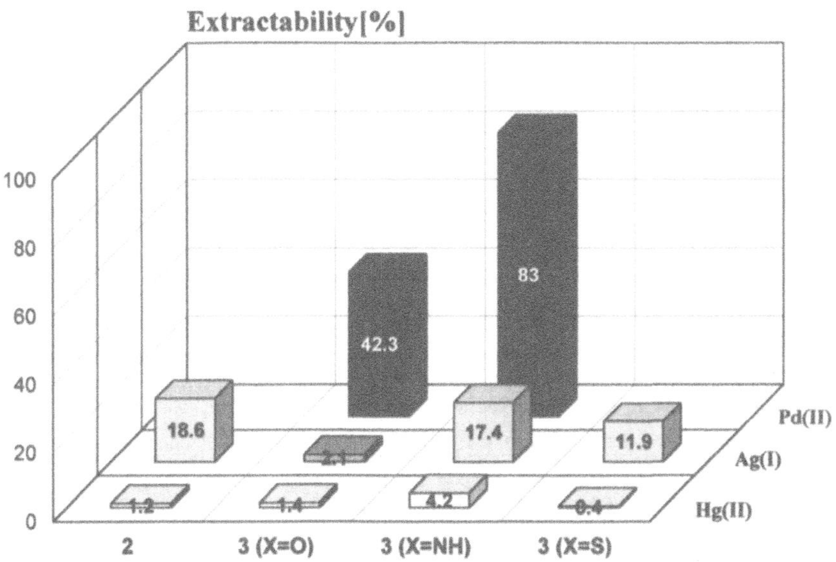

$c_M = 1\cdot10^{-4}$ M; $c_{HPic} = 5\cdot10^{-3}$ M; $c_L = 1\cdot10^{-3}$ M/$CHCl_3$

Figure 2: Solvent extraction of Ag(I), Hg(II) and Pd(II) with bis(phenoxyalkyl)sulfane podands

The investigated sulfur containing podands show a clear structure-extractability relationship. So the chlorine containing compounds **1**, **4** and **5** do not extract Ag(I), Hg(II) and Pd(II) under the chosen experimental conditions (extractability ≤ 0,1%). This fact is obviously caused by the decreasing basicity of the sulfur donor atom under the influence of the chlorine substituents and correlates with calculated charge density of the sulfur atom (**1**: -0.023, **4**: +0.049, **5**: +0.105; **2**: -0.047).
In contrast to this the extractability of the metal ions investigated is increased in case of compounds **2** and **3** (Figure 2). Especially the graduated extraction behaviour of the different substituted compounds **3** toward Ag(I) is remarkable. The highest extractability

is observed for the dianilid **3** (X = NH), whereas the analogous ether and thioether **3** (X = O; S) give significant lower values. The reasons for this observation should be related to structure differences of the extracted metal complexes. So the composition of the extracted Ag(I) complexes is changing from a metal to ligand ratio of 1:1 (**3**: X = O; **3**: X = S) to a ratio of 1:2 for **3** with X = NH. In case of **2** a preorganization of the ligand by a π-stacking effect was founding on the basis of molecular modeling calculations [5].

The novel synthesized podands give interesting possibilities for changing the selectivity during complexation reactions. Further investigations in view of stability and structure of the formed complexes should be lead to a better understanding of the observed behaviour.

Acknowledgement

We greatly thank the Deutsche Forschungsgemeinschaft and the Fonds der Chemischen Industrie for support of this work.

References

1. Lehn, J.-M. *Supramolecular Chemistry*, VCH Weinheim, 1995.
2. Hollmann, K.; Ecke, M. *Z. Chem.* (1989) *29*, 63.
3. Hollmann, K.; Mühlstädt, M.; Habermann, B.; Möder, M.; Kleinpeter, E. *J. prakt. Chem.* (1990) *332*, 503.
4. Gloe, K,: Mühl, P.; Beyer, M.; Mühlstädt, M.; Hoyer, E. *Solvent Extr. Ion. Exch.* (1986) *4*, 907.
5. Krüger, T.; Gloe, K.; Habermann, B.; Mühlstädt., M.; Hollmann, K. *Z. anorg. allgem. Chem.* in press.
6. The semiempirical calculations were performed with MOPAC 6.0 using the PM3 hamiltonian and the HyperChem™ 4.5 (Hypercube) program package. The molecules under study were completely geometry optimized at the SCF level until a gradient norm less 0.0001 $kcal \cdot mol^{-1} \cdot Å^{-1}$. The initial building of the molecular models were performed using the UNIVERSAL FORCE FIELD of the CERIUS^2 2.0© (BIOSYM/MSI) software package. All calculations were done with a DEC 3000 AXP/800 (200 MHz) workstation. see Krüger, T.; Gloe, K.; Stephan, H.; Habermann, B.; Hollmann, K.; Weber, E. *J. Mol. Model.* (1996) *2*, 386.

DESIGN OF COORDINATION ARRAYS AS POTENTIAL MOLECULAR MEMORY UNITS AND SWITCHES

GARRY S. HANAN, ULRICH S. SCHUBERT, DIRK VOLKMER, JEAN-MARIE LEHN*
Laboratoire de Chimie Supramoléculaire, Institut Le Bel, Université Louis Pasteur, 4 Rue Blaise Pascal, F-67000 Strasbourg, France

JÖRG HASSMANN, CATHERINE Y. HAHN, OLIVER WALDMANN, PAUL MÜLLER*
Physikalisches Institut III, Universität Erlangen-Nürnberg, Erwin-Rommel-Str. 1, D-91058 Erlangen, Germany

GERHARD BAUM, DIETER FENSKE
Institut für Anorganische Chemie, Universität Karlsruhe, Engesserstr. 1, D-76128 Karlsruhe, Germany

1. Introduction

The controlled arrangement of metal ions into specific arrays and patterns is one of the major goals in modern metallo-supramolecular chemistry [1]. Such programmed self-assembly of inorganic architectures from mixtures of organic ligands and metal ions could provide new materials for nano-technology such as molecular memory units. We investigated the synthesis, structure and properties of tetranuclear cobalt(II) complexes of the ligand *4,6-bis(6-(2,2'-bipyridyl))pyrimidine* and its derivatives, having a grid-type structure (*[2×2]-Grid*). Moreover, studies of the electronic, magnetic, and structural properties of these molecules on surfaces, as crystals, and in solution were performed.

2. Synthesis

The oligo-tridentate ligands **1** - **3** based on alterning pyridines and pyrimidines were synthesized by Stille type carbon-carbon bond forming reactions using organotin intermediates [2].

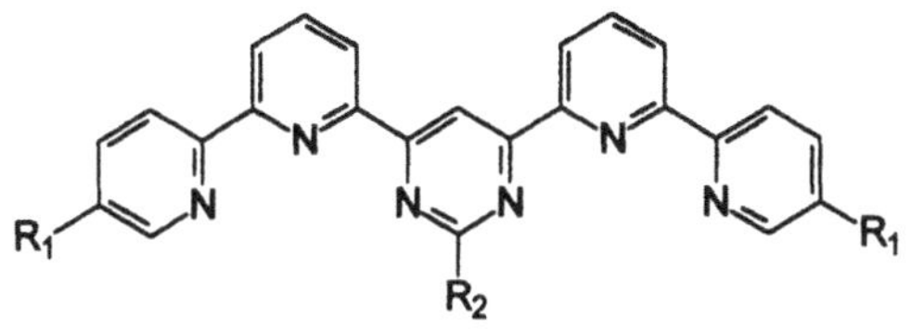

1: $R_1 = R_2 = H$

2: $R_1 = H, R_2 = CH_3$

3: $R_1 = CH_3, R_2 = H$

Variations were made on the 5-position of the terminal pyridine (R_1 = H or CH_3), and on the 2-position of the central pyrimidine ring (R_2 = H or CH_3).

A. W. Coleman (ed.), Molecular Recognition and Inclusion, 349–352.
© 1998 *Kluwer Academic Publishers. Printed in the Netherlands.*

3. Complexation

The reaction of equimolar quantities of the ligand *4,6-bis(6-(2,2'-bipyrid-yl))-pyrimidine* **1** (or the methyl analogs **2**, **3**) and Co(II)acetate in refluxing MeOH exclusively leads to the formation of tetranuclear complexes $[M_4(L)_4]^{8+}$, G_1 - G_3 (L = **1** - **3** respectively). The complexes were isolated as the PF_6 or SbF_6 salts.

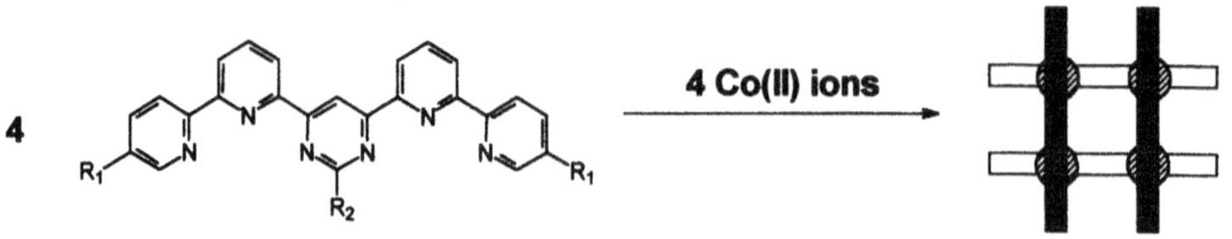

4. Structure

The grid type complex G_2 $[Co_4(\mathbf{2})_4](SbF_6)_8 \cdot 3\ H_2O \cdot 0.25\ MeOH \cdot 3.5\ CH_3CN$ crystallizes in the monoclinic crystal system, space group $P2_1/c$, a = 18.205(9), b = 20.566(7), c = 37.126(5) Å, β = 97.65(2)°. The asymmetric unit contains a single cation and eight SbF_6 anions (Figure 1) [3].

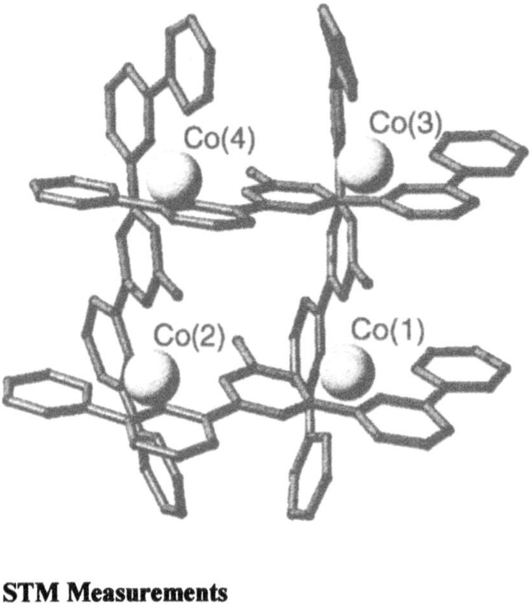

Figure 1: Structure of the $[Co_4(\mathbf{2})_4]^{8+}$ cation (wireframe model) showing the grid-like arrangement of the four ligands and the four metal ions.

5. STM Measurements

For any functionalization of the grids as electronic devices, it is neccessary to obtain ordered structures of the supermolecules on surfaces. One approach is the electrochemical preparation of thin films of the grids. The structure of the adsorbed layers depends strongly on the adsorption potential. In order to visualize the adsorption process, we have used the electrolytic in situ STM setup shown in Figure 2.

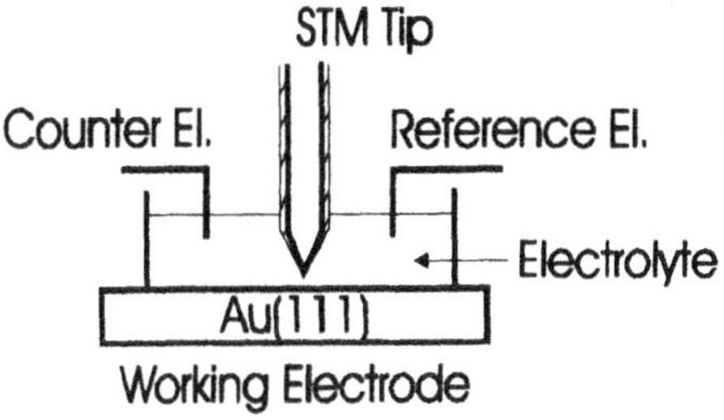

Figure 2: Electrochemical in situ STM setup: Counter and reference electrodes and the tunneling tip are immersed in a cell containing the electrolyte. The substrate serves as working electrode.

In Figure 3, an in situ STM image of the ordered structure of the [2x2]-Co(II) grids $\mathbf{G_2}$ on a Gold(111) suface is shown. Similar ordered structures were also observed ex situ on electrochemically prepared layers at characteristic adsorption potentials.

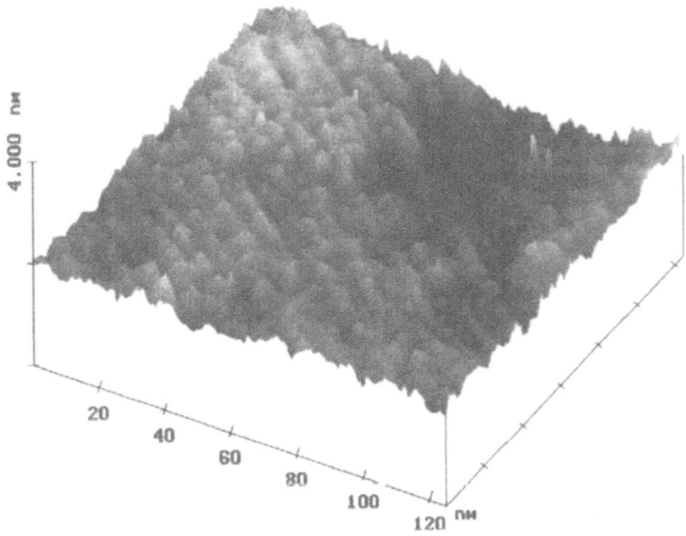

Figure 3: Electrolytic in situ STM image of a layer of [2x2]-Co(II) grids $\mathbf{G_2}$ deposited on a Au(111) surface.
Adsorptionpotential: -1 V vs Ag.

6. UV Photoelectron Spectroscopy (UPS)

For examination of the electronic structure of the different grids, we performed UPS experiments on the DESY synchrotron (the photon energy was 60 eV; 50 monolayers on a Gold (111) surface. The cross-sectional behaviour at various photon energies shows the position of the different contributions: Mainly metallic from approx. 0 to 8 eV, then organic with C-π-bonds at 16 eV and C-σ–bonds at approx. 18 eV) [4]. The spectra reveal isolating gaps of 2.5-3.5 eV. This semiconducting behavior demonstrates the possibilities for electronic storage/retrieval experiments.

7. Magnetic Measurements

SQUID studies on several crystals of the [2x2]-Co(II) grid $\mathbf{G_2}$ show a clear antiferromagnetic transition at a Néel temperature of approximately 10 K (see Figure 4). The intermolecular center to center distance is 2nm. The same behaviour and signal intensity was seen in a $2.85*10^{-3}$ m solution (intermolecular distance of 9 nm). A powdered sample of mononuclear $[Co(terpyridine)_2]^{2+}$ shows no transition down to a temperature of 2K, which suggests that grids form antiferromagnetic domains, which contain one grid molecule [5].

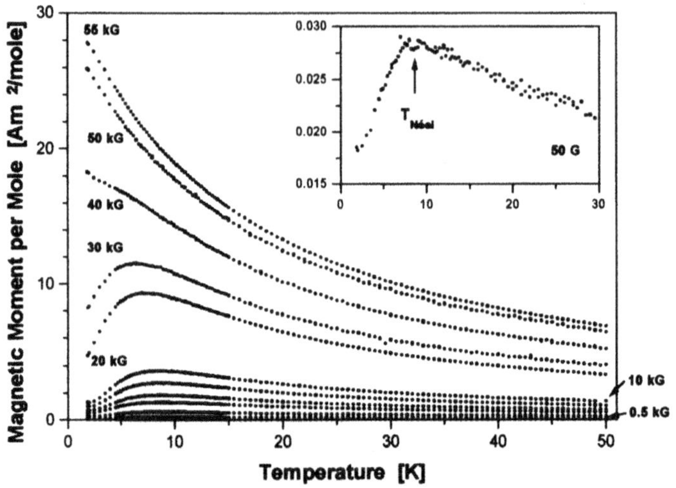

Figure 4: SQUID study of the magnetization of the [2x2]-Co(II) grid $\mathbf{G_2}$ at various external fields.

8. Conclusions

The supramolecular cobalt coordination arrays described here represent a new area in metallo-supramolecular science. The synthesis and self-assembly as well as the structural properties (x-ray diffraction) were studied in detail.

Ordered structures of electrochemically deposited [2x2]-Co(II)-grids $\mathbf{G_2}$ on gold surfaces were visualized with the STM. Furthermore, antiferromagnetic ordering of the unpaired Co(II)-spins of these grid systems was observed. The semiconducting properties of the grids demonstrate their potential utility as molecular device components in a futuristic information processing molecular electronics technology.

9. References

[1] Lehn, J.-M. (1995) *Supramolecular Chemistry: Concepts and Perspectives,* VCH, Weinheim.

[2] Hanan, G.S., Schubert, U.S., Volkmer, D., Riviere, E., Lehn, J.-M., Kritsakis, N., Fischer, J., *Can. J. Chem.*, in press.

[3] Hanan, G.S., Schubert, U.S., Volkmer, D., Lehn, J.-M., Baum, G., Fenske, D., to be published.

[4] Cooperation with Prof. Dr. R.L. Johnson, DESY, Universität Hamburg, Germany.

[5] Waldmann, O., Hassmann, J., Müller, P., Schubert, U.S., Hanan, G.S., Volkmer, D., Lehn, J.-M., to be published.

SYNTHESIS OF A FUNCTIONALIZED CHIRAL MOLECULAR TWEEZER

MICHAEL HARMATA* AND SRIRAM TYAGARAJAN
Department of Chemistry
University of Missouri-Columbia
Columbia, Missouri 65211 USA

Molecular tweezers are an emerging class of receptors characterized by two functional motifs or "pincers" separated by a rigid spacer to produce a cavity or molecular cleft into which a guest can fit. Seminal and extensive contributions in this area have come from the labs of Zimmermann who produced a series of molecular tweezers which are achiral.[1]

Our own work in this area has focused on the synthesis and study of molecular tweezers which are chiral, in the ultimate hope that recognition and/or functional properties of molecular tweezers of this type would be characterized by "enantioselectivity". In order to construct the appropriate systems, Kagan's ether **1** was chosen as a building block.[2] This molecule is not only chiral, but possesses an approximately 90° bend and is suitable for the construction of molecules with convergent functional groups.

1: Kagan's Ether

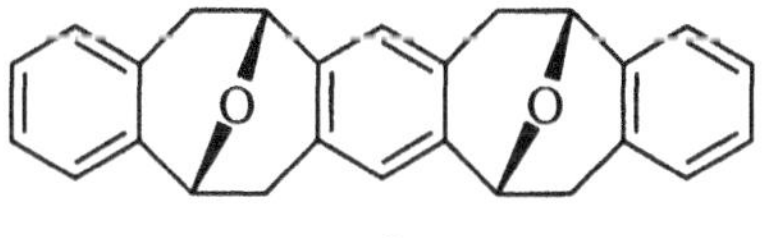

2

The first chiral (but racemic) molecular tweezer synthesized in our laboratories was **2**.[3] This molecule exhibited interesting interesting clathrate-forming properties in the solid state. However, its ability to bind guests in solution, as assessed by proton NMR, was virtually nil. In addition, the lack of functionality on **2** precluded its use as a building block for into other interesting receptors such as, for example, **3**. We thus set out to synthesize the functionalized molecular tweezer **4**.

A. W. Coleman (ed.), Molecular Recognition and Inclusion, 353–356.
© 1998 *Kluwer Academic Publishers. Printed in the Netherlands.*

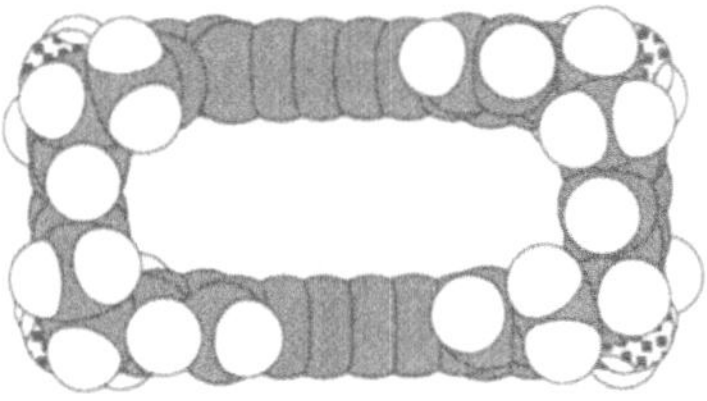

3

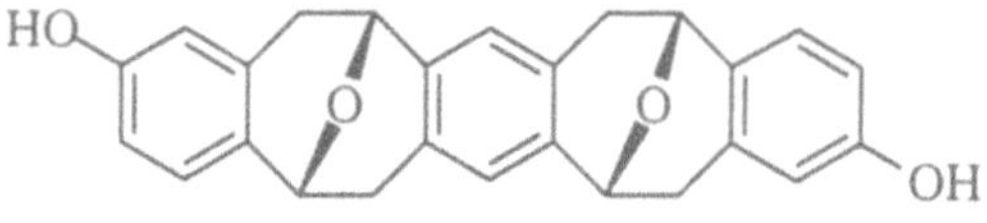

4

Our synthesis began with the readily available m-bromoanisole as illustrated in Scheme 1. Metalation and trapping with allyl bromide followed by ozonolysis gave m-methoxyphenylacetaldehyde **6** in 51% overall yield.

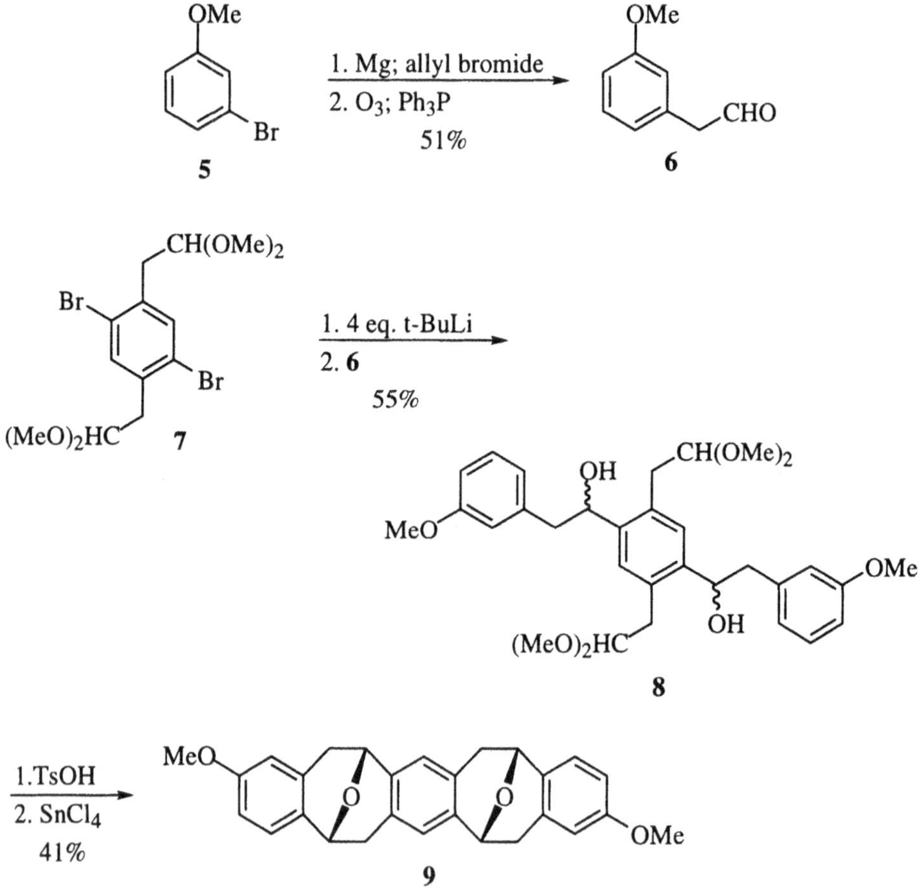

Scheme 1

Reaction of this aldehyde with the diorganolithium species derived from halogen-metal exchange on dibromide **7** gave diol **8** in 55% yield as a 1:1

mixture of easily separable meso and d,l isomers. The d,l isomer was treated with tosic acid to afford the corresponding cyclic acetal which, upon treatment with $SnCl_4$, afforded molecular tweezer **9** in 41% overall yield from **8**. Unfortunately, all attempts to remove the methyl groups from the phenolic oxygens in **9** were unsuccessful. Presumably, protocols involving either strong Lewis acids (e.g., BBr_3) or powerful, soft nucleophiles (e.g., thiophenoxide) resulted in ring-opening of the oxabicyclo[3.3.1] system and concomitant decomposition.

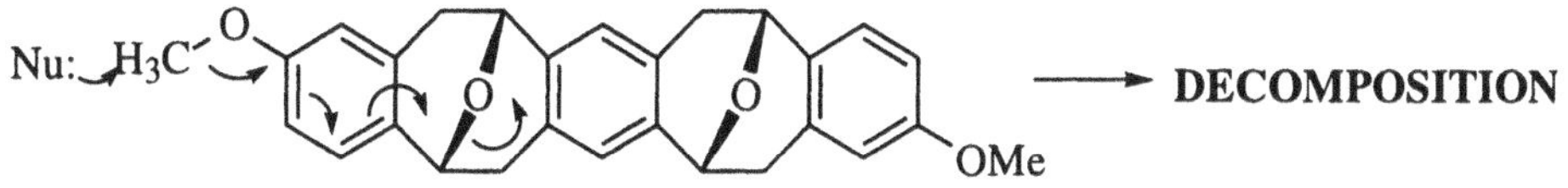

We thus sought a new protecting group and after some experimentation decided that the t-butyldimethylsilyl group was suitable to our needs.

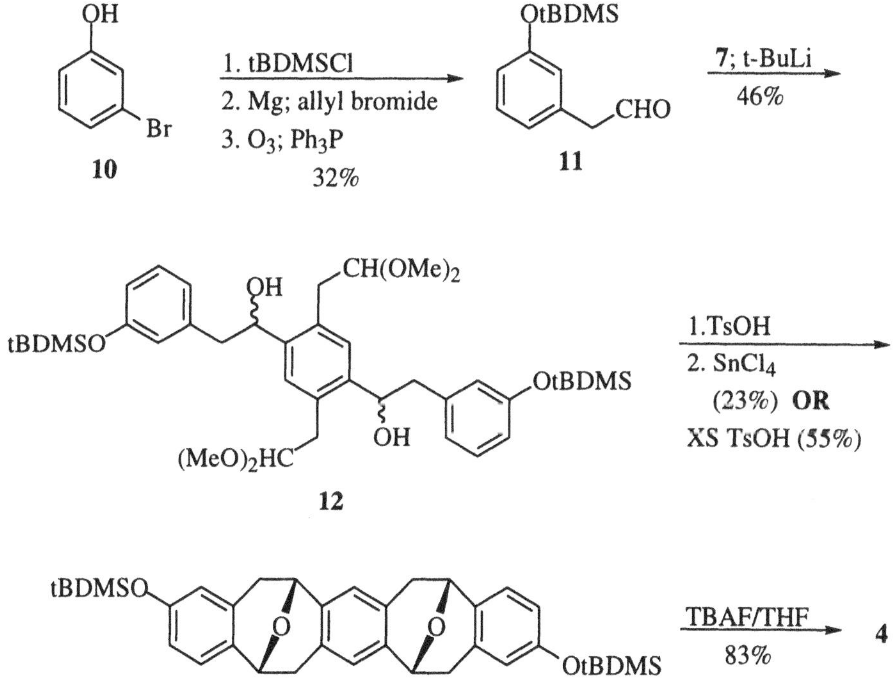

13

Scheme 2

Preparation of the appropriate substrate for tweezer synthesis began with m-bromophenol as shown in Scheme 2. Protection, allylation and ozonolysis gave the aldehyde **11** in 32% overall yield, the ozonolysis step being particularly problematic. Conversion to the diol took place as before in a yield of 46%. Treatment of the d,l diol with a catalytic amount of tosic acid followed by ring closure to the resulting acetal with tin tetrachloride gave **13** in a yield of only 23%. However, we found that reaction of **12** with excess tosic acid in dichloromethane resulted not in conversion to a cyclic acetal but to the target molecular tweezer **13** in a yield of 55%! The scope of this process has not been delineated, but it is clear that electron-rich aromatic rings are necessary for such a ring closure to take place. This first example of direct ring closure from a diol to a molecular tweezer represents an important simplification in the synthesis of compounds of this class and is being investigated further. Desilylation could be effected with tetrabutylammonium fluoride in THF or HF/CH_3CN to afford the molecular tweezer **4** in good yield. Unfortunately, **4** is a difficultly soluble compound which is apparently not crystalline and does not exhibit the clathrate-forming properties of its congener **1**. Nevertheless, it should be suitable for the synthesis of chiral molecular cavities such as **3** and studies along these lines are in progress.

Acknowledgements: We are grateful to the University of Missouri Research Board for partial support of this work. We thank the National Science Foundation for partial support of the NMR (PCM-8115599) facility at the University of Missouri-Columbia and for partial funding for the purchase of a 500 MHz spectrometer (CHE-89-08304) and an X-ray diffractometer (CHE-90-11804).

References

1. Zimmerman, S.C. *Top. Curr. Chem.* **1993**, *165*, 71.
2. Kagan, J., Chen, S.-Y.; Agdeppa, Jr., D.A.; Watson, W.H.; Zabel, V. *Tetrahedron Lett.* **1977**, *18*, 4469
3. Harmata, M.; Barnes, C.L. *Tetrahedron Lett.* **1990**, *31*, 1825.
4. Data on **4**: 1H NMR (CD_3CN) δ 6.84 (d, 2H, J = 8.4 Hz), 6.70 (s 2H), 6.47 (dd, 2H, J = 2.3, 8.3 Hz), 6.25 (d, 2H, J = 2.0 Hz), 5.06 (d, 2H, J = 6.0 Hz), 5.04 (d, 2H, J = 6.0 Hz), 3.25 (m, 4H), 2.56 (d, 2H, J = 16.4 Hz), w.46 (d, 2H, J = 16.4 Hz), 2.1 (br s, 2H); ^{13}C NMR (CD3CN) 156.2, 137.4, 133.7, 130.2, 129.8, 127.2, 126.3, 115.2, 114.1, 69.3, 69.1, 36.4; HRMS exact mass calcd for $C_{26}H_{22}O4$, 398.1518; found: 398.1517.

OPTIMAL POLYMER ARCHITECTURE FOR ADSORPTION AT THE SOLID–LIQUID INTERFACE: DENDRIMERS VERSUS LINEAR POLYMERS

ANDREW HOPKINSON
Unilever Research
Port Sunlight Laboratory, Bebington, Wirral, Merseyside L63 3JW, UK.

Abstract. The adsorption strengths of a dendrimer and a linear polymer having the same number of adsorbing groups were compared by Monte Carlo calculations. The dendrimer was calculated to adsorb more strongly than the linear polymer in the regime of weak attraction between the adsorbing groups and the surface.

1. Introduction

The adsorption coefficient of a polymer is determined by the balance, upon adsorption, between the number of attractive polymer–surface interactions and a loss in configurational entropy. This entropy loss should depend on polymer architecture, and dendrimers—polymers with a systematic branching structure [1]—represent an interesting extreme in terms of their architectural difference from linear polymers.

2. Calculations

The objective was to calculate the surface excess (Γ) for polymers with different architectures adsorbing onto flat surfaces using a lattice model and assuming no polymer–polymer interactions. Monte Carlo methodologies used recently by Mansfield [2] to determine the conformations of adsorbed dendrimers do not yield Γ. Instead the Monte Carlo methodology of Rosenbluth and Rosenbluth [3] was used. Many polymer conformations were each "grown" by a non-reversing random walk on a 3-dimensional lattice, one side bounded to represent a surface. Growth was started by placing an arbitrarily-selected "core" atom of the polymer onto the lattice some distance d from the surface. At each growth step a random choice was made between the *unhindered*, non-reversing directions. If a polymer conformation was completed successfully, then it was given a Boltzmann weight—the Rosenbluth weight, W_R—to correct for the bias introduced by choosing between only the unhindered directions. If growth stopped because there were no unhindered directions, then $W_R = 0$, otherwise W_R is given by

A. W. Coleman (ed.), Molecular Recognition and Inclusion, 357–360.
© 1998 *Kluwer Academic Publishers. Printed in the Netherlands.*

$$W_R = \prod_{i=2}^{l} \frac{z_i}{b_i}, \tag{1}$$

where z_i and b_i are the numbers of unhindered and non-reversing directions respectively at each step i, and l is the number of bonds in the polymer. The value of W_R averaged over many conformations, $\langle W_R \rangle$, is effectively a "concentration" or "occurrence probability" relative to a reference state of a non-reversing random walk.

Calculations were performed for the two polymer architectures shown in Figure 1 imposed upon a cubic lattice. These two architectures have the same number of bonds, and both have 24 endgroups. It is therefore interesting to compare them. $\langle W_R \rangle$ was calculated for different distances, d, between the core and the surface: the {100} plane of the cubic lattice. At each value of d, attempts to grow dendrimer and linear polymer conformations numbered 6,800,000 and 1,760,000 respectively. Calculations were performed on a Silicon Graphics "Power Challenge" computer.

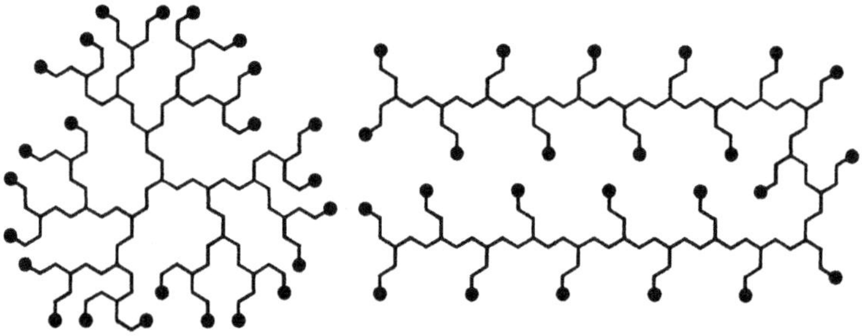

Figure 1. The dendrimer and linear polymer considered. The endgroups are shown by heavy dots.

3. Results and Analysis

For each value of d, $\langle W_R \rangle$ was calculated after dividing the conformations into groups according to the number of endgroup–surface contacts, n. The value of $\langle W_R \rangle$ for $n = 0$, $\langle W_{R,n=0} \rangle$, is lower for those short distances, d, at which the surface reduces the number of possible conformations, than for those longer distances at which the polymer cannot reach the surface: that is, in bulk solution denoted by $d = \infty$ below. Using just the values of $\langle W_{R,n=0} \rangle$, a surface excess (negative), Γ_0, normalised to $\langle W_{R,n=0} \rangle$ in the bulk, $\langle W_{R,n=0,d=\infty} \rangle$, can be defined by

$$\Gamma_0 = \frac{\Sigma_d(\langle W_{R,n=0,d} \rangle - \langle W_{R,n=0,d=\infty} \rangle)}{\langle W_{R,n=0,d=\infty} \rangle}, \tag{2}$$

where the summation is carried out over values of d for which the polymer can touch the surface. Note that n can only be zero at $d = \infty$. $\Gamma_0 = -6.04$ for the dendrimer and -7.77 for the linear polymer. In order to calculate the total surface excess, Γ, the conformations that make a finite number of surface contacts ($n > 0$) must be included.

In order to account for endgroup–surface attraction an *endgroup adsorption weight*, W_e, was introduced, defined as the concentration of isolated endgroups in the lattice plane next to the surface which would be in equilibrium with a concentration of unity in the bulk. Γ is given by $\Gamma = \Gamma_{n=0} + \Gamma_{n>0}$, where

$$\Gamma_{n>0} = \frac{\Sigma_{n>0,d}(\langle W_{R,n,d}\rangle . W_e^n)}{\langle W_{R,n=0,d=\infty}\rangle}. \tag{3}$$

$\Gamma_{n>0}$ represents the positive contribution to Γ made by those conformations which make at least one endgroup–surface contact. Figure 2 shows $\Gamma_{n>0}$ as a function of W_e. $\Gamma_{n>0}$ is greater for the dendrimer than for the linear polymer at $W_e < 9$. The relative difference is greatest at $W_e \approx 5$ at which $\Gamma_{n>0} = 240$ for the dendrimer and $\Gamma_{n>0} = 140$ for the linear polymer. Note that $\Gamma_{n>0} \approx -\Gamma_0$ ($\Gamma \approx 0$) at $W_e \approx 3$.

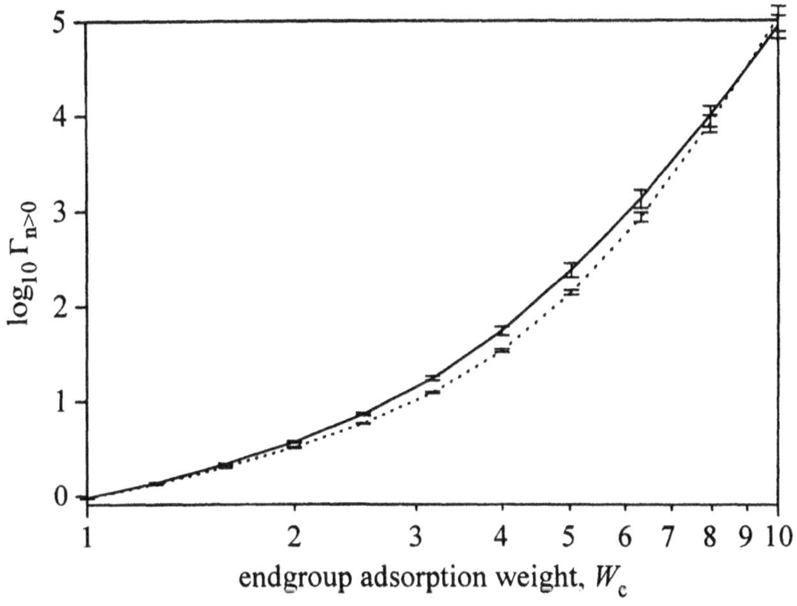

Figure 2. $\Gamma_{n>0}$ versus W_e for the 24-endgroup dendrimer (——) and for the 24-endgroup linear polymer (------). The error bars represent one standard deviation.

In order to understand these results, it is useful to analyse the contributions to $\Gamma_{n>0}$ made by conformations making different numbers of endgroup–surface contacts. Figure 3 shows this analysis. At low values of W_e, those conformations that make only one contact contribute most to adsorption. The decrease in Γ_n with increasing n attests to the entropic cost of forming increasing numbers of endgroup–surface contacts. At $W_e \approx 5$, those conformations that make between about 5 and 11 contacts contribute most to adsorption. Over this range, Γ_n is consistently higher for the dendrimer than for the linear polymer, which explains why the relative difference (ratio) in $\Gamma_{n>0}$ (and Γ) between the two architectures is greatest at $W_e \approx 5$. Figure 3 shows that at $W_e = 10$, those conformations that make even more contacts become important.

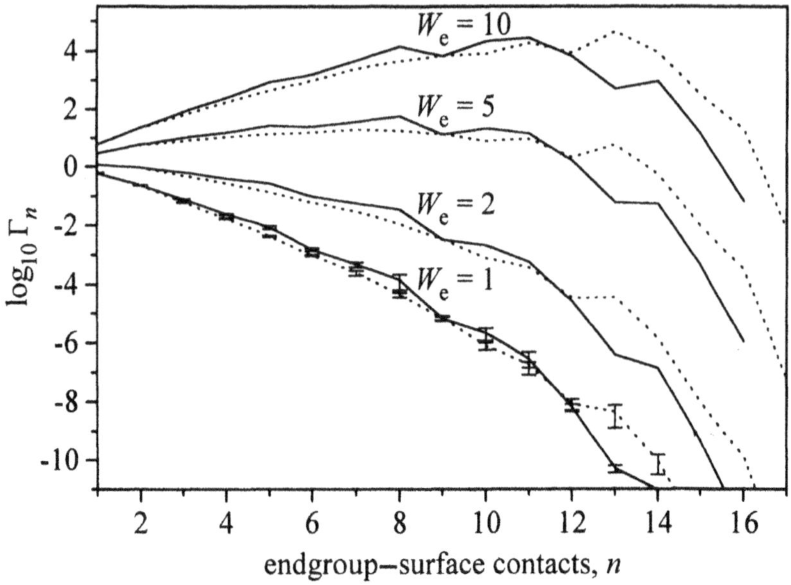

Figure 3. Γ_n versus n for the 24-endgroup dendrimer (——) and the 24-endgroup linear polymer (------), at the values of W_e indicated. The error bars represent one standard deviation.

The entropic cost of forming 12 or more contacts is greater for the dendrimer than for the linear polymer. At $W_e > 9$ the linear polymer adsorbs more strongly than the dendrimer (Figure 2), a situation that is expected in the limit of strong endgroup–surface attraction because all the endgroups of the linear polymer can reach the surface, whereas this may not be true, or at least be entropically costly, for the dendrimer.

4. Conclusions

The calculations show that the dendrimer adsorbs more strongly than the linear polymer, but only for endgroup–surface attractions, W_e, of less than 9. The relative difference passes through a maximum at $W_e \approx 5$. To put the meaning of W_e into perspective, consider isolated endgroups on a 2 Å lattice with $W_e = 9$. Then the surface excess (endgroups per unit area) is equal to the number of endgroups contained in a layer of bulk solution, in equilibrium, just 16 Å thick—that is very weak adsorption! The trends in the relative differences were understood by examining the positive contributions to the surface excess made by conformations that make different numbers of endgroup–surface contacts.

References

1. D. A. Tomalia, Advanced Materials, 1994, **6**(7/8), 529–539.
2. M. L. Mansfield, Polymer, 1996, **37**(17), 3835–3841.
3. M. N. Rosenbluth and A. W. Rosenbluth, J. Phys. Chem., 1955, **23**(2), 356–359.

ORGANIZATIONS OF TWO-DIMENSIONAL DNA-MIMETICS AT THE AIR-WATER INTERFACE

K. Ijiro*†, F. Nakamura†, M. Shimomura†

**PRESTO, Japan Science Technology Corporation (JST), Sapporo 060, Japan*

†Molecular Device Laboratory, Research Institute for Electronic Science, Hokkaido University, Sapporo 060, Japan

Double-helical DNA, which is one of natural products of "molecular-recognition directed self-assembly" [1], composed of one-dimensionally stacked base pairs, adenine-thymine and guanine-cytosine, formed by complementary multiple hydrogen bonds. Besides transferring molecular information as genetic signals, DNA is recently found out to transfer an electron through the stacked base-pairs [2,3]. A hydrophobic interface of a monolayer assembly formed on a water surface has been reported to be a sufficient environment for hydrogen bonding even though the chemical substances are surrounded by a large number of water molecules [4,5]. Here we report that nucleobase amphiphiles are organized to two-dimensional (2-D) molecular assemblies by the complementary base-pairing at the air-water interface.

Singly-alkylated nucleobase derivative, 1-octadecyl-cytosine (C_{18}Cyt) [6], was prepared by a nucleophilic substitution reaction of cytosine with octadecyl iodide in the presence of sodium hydride. As shown in Figure 1, the pressure-area (π-A) isotherm of C_{18}Cyt spread from chloroform solution onto a water surface was drastically changed when a small amount (5×10^{-4} M) of guanine nucleosides (guanosine and deoxyguanosine) was dissolved in the neutral water subphase, whereas other nucleosides show little effect on the isotherm (Fig.1) [7]. The monolayer on the guanosine subphase was transferred on the freshly deposited gold surface by the ordinary Langmuir-Blodgett technique under the constant surface-pressure condition (20 mN m^{-1}). Spectral splitting (1715 and 1666 cm^{-1}) of the C=O stretching band of the

A. W. Coleman (ed.), Molecular Recognition and Inclusion, 361–364.
© 1998 *Kluwer Academic Publishers. Printed in the Netherlands.*

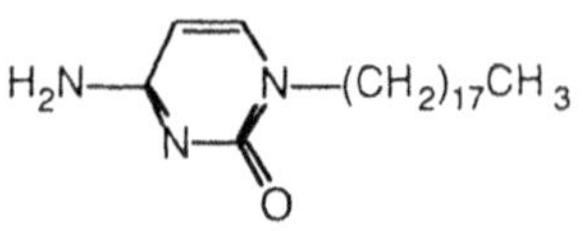

C18Cyt

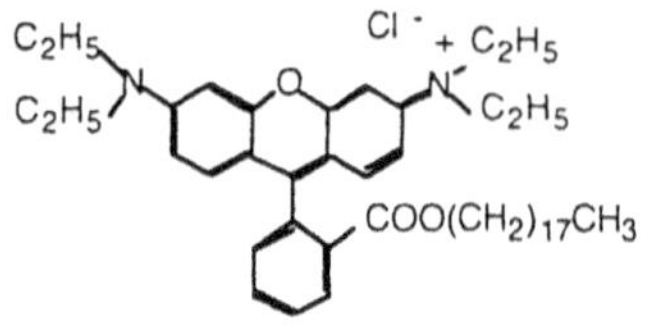

C18RB

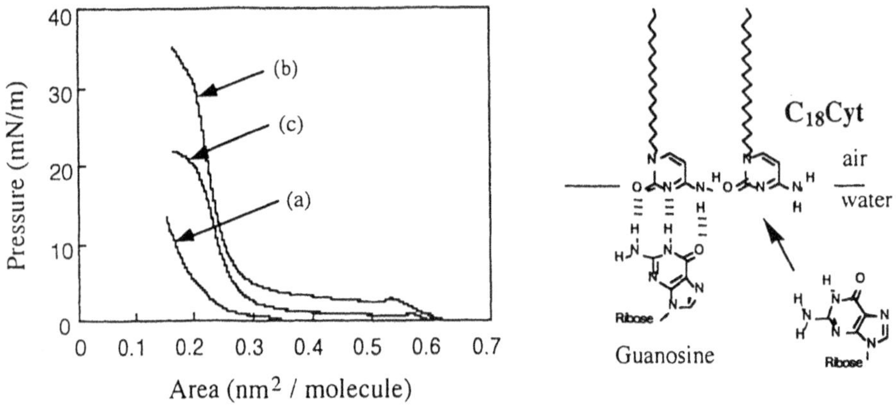

Figure 1. Pressure-area isotherm of C_{18}Cyt on (a) pure water, (b) guanosine, and (c) deoxyguanosine subphase at 20°C, and schematic illustration of Watson-Crick base-pairing at the air-water interface.

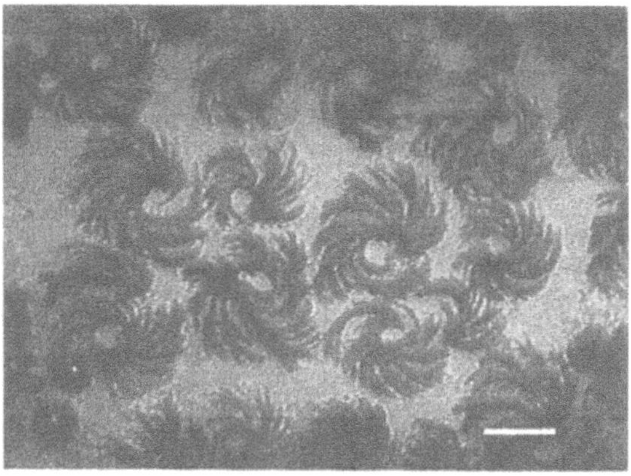

Figure 2. Fluorescence image of the C_{18}Cyt monolayer at 3 mN m^{-1} on guanosine subphase, scale bar: 25 μm.

guanine base (1694 cm^{-1} in crystalline guanosine) in the deposited monolayer monitored by FT-IR reflection absorption spectroscopy clearly indicates the complementary hydrogen bonding between guanosine and C_{18}Cyt, because a similar spectral splitting was found in an IR spectrum of the double-stranded DNA, poly (dC)•poly (dG) (1711, 1664 cm^{-1}). The Watson-Crick base-pair of C_{18}Cyt and the guanine base is assumed to be formed on the water subphase.

Fluorescence imaging of the water surface by using an epifluorescence microscope is an *in situ* observation method not only for the liquid-to-crystal phase transition of monolayers but also for the dynamic process of the molecular organization by specific molecular interactionas. A small amount (2 mole %) of octadecylrhodamine B (C_{18}RB), which is squeezed out from the crystalline phase to the liquid phase as an impurity [8], is added as a fluorescence probe for negative imaging of the molecular assemblies composed of C_{18}Cyt. Fishhook-shaped dark domains were observed on pure water surface (not shown here). Taking the result of the isotherm measurement into consideration, the dark domains on the pure water subphase are assumed to be three-dimensional crystallites of C_{18}Cyt. Fluorescence image was almost identical when the nucleosides, which did not affect the π-A isotherm, were added into the water subphase. While the monolayer was spread on the guanosine subphase and compressed to the plateau region of the π-A isotherm (see the Figure 1), spiral-shaped dark domains were generated in a bright red field of the fluorescence microscope (Figure 2). Similar shaped spiral domains were observed on the deoxyguanosine subphase, too. This is the first direct finding of the molecular organization by complementary base-pairing in the two-dimensional world. The chiral feature of the crystal domain is ascribable to the molecular chirality of the carbohydrate moiety of the guanine nucleoside.

Stacking of the base-pairs in double-stranded DNA shows hypo-chromism, diminution of absorbance, in the UV visible absorption spectra of the purine- and the pirimidine-chromophores. Absorption maximum of the guanine-cytosine pair in DNA, 256 nm in an aqueous poly (dG-dC) • poly (dG-dC) solution, shifted to 284 nm in the C_{18}Cyt-guanosine monolayer transferred on a quartz plate. The π-π interaction of the stacked base-pairs in the 2-D DNA-mimetics is stronger than in double-stranded DNA.

Formation of the chiral 2-D crystals based on the complementary base-pairing at the interface is one of the interesting examples of molecular organizations by specific intermolecular interaction. We would like to emphasize that the C_{18}Cyt molecule is endowed with organization ability at the interface by the guanine bases dissolved in the water subphase. In our case, the hydrogen bonding is simultaneously used as a tool of the molecular recognition and the structure formation. The 2-D spiral domain of the stacked base-pairs is expected to be an electron π-way, which is a suitable medium of long range electron- and/or energy transfer through the highly oriented π-electron array.

References

1. Lehn, J.M.: Supramolecular chemistry-scope and perspectives molecules, supramolecules, and molecular devices (novel lecture), *Angew.Chem.Int.Ed.Engl.* **27**(1988), 89-112.
2. Murphy, C., Arkin, M., Jenkins, Y., Ghatli, N., Bossmann, S., Turro, N., Barton, J.: Long-range photoinduced electron transfer through a DNA helix, *Science* **262**(1993), 1025-1029.
3. Meade, T., Kayyem, J.: Electron transfer through DNA: site-specific modification of duplex DNA with ruthenium donors and acceptors, *Angew. Chem. Int. Ed. Engl.* **34**(1995), 352-354.
4. Ahuja, R., Caruso, P., Möbius, D., Paulus, W., Ringsdorf, H., Wildburg, G.: Formation of molecular strands by hydrogen bonds at the gas-water interface: molecular recognition and quantitative hydrolysis of barbituric acid lipids, *Angew.Chem.,Int.Ed.Engl.* **32**(1993), 1033-1036.
5. Cha, X., Ariga, K., Onda, M., Kunitake, T.: Molecular recognition of aqueous dipeptides by noncovalently aligned oligoglycine units at the air/water interface, *J.Am.Chem.Soc.* **117**(1995), 11833-11838.
6. Tohda, K., Tange, M., Odashima, K., Umezawa, Y., Furuta, H., Sessler, J.: Liquid membrane electrode for guanosine nucleosides using a cytosine-pendant triamine host as the sensory element, *Anal.Chem.* **64**(1992), 960-964.
7. Shimomura, M., Nakamura, F., Ijiro, K., Taketsuna, H., Tanaka, M., Nakamura, H., Hasebe, K.: Chiral crystal formation of octadecylcytosine monolayer by complementary base-pairing, *Thin Solid Films*, in press.
8. Shimomura, M., Fujii, K., Shimamura, T., Oguchi, M., Shinohara, E., Nagata, Y., Matsubara, M., Koshiishi, K.: Effect of thermal treatment on crystal growth of the surface monolayer, *Thin Soli Films* **210/211**(1992), 98-100.

METAL-INDUCED "AGGREGATION-DEAGGREGATION" AND "COLOUR CHANGE" IN FULLERENE DERIVATIVES

Atsushi Ikeda and Seiji Shinkai
Department of Chemical Science & Technology, Faculty of Engineering, Kyushu University
Fukuoka 812-81, Japan

The exploitation of fullerene chemistry continues unabated. One problem facing the C_{60} chemist is, however, that C_{60} tends to aggregate in solution, [1] which is highly solvent-dependent and makes the C_{60} chemistry complicated and less reproducible. In compound **1** two C_{60} groups are introduced into the same side of a calix[4]arene platform. [2,3] Therefore, in polar solvents the aggregation of **1** is expected to occur more easily than that of **2** which has only one C_{60} group. Moreover, the deaggregation of **1** may be induced in the presence of Na^+ ion because the solubility of such metal complexes should be improved in polar solvents. [4]

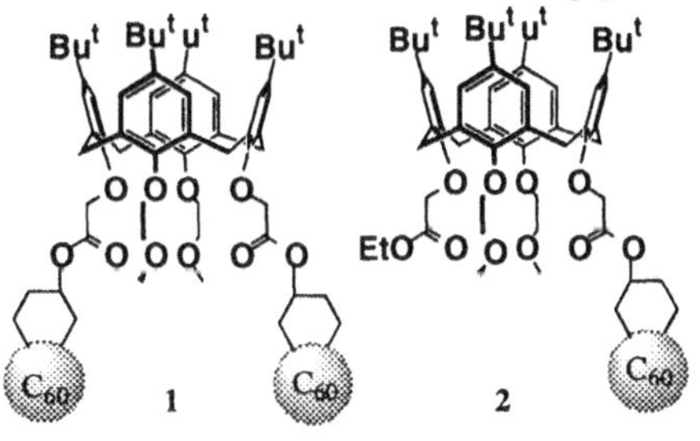

Firstly, we examined the solvent effect on the absorption spectra of **1** and **2**. As shown in Figure 1, the significant spectral difference among these two compounds was not observed in $CHCl_3$. In $CHCl_3$:MeCN = 1:1 v/v and $CHCl_3$:MeOH = 1:1 v/v which are poorer solvents for C_{60} than $CHCl_3$ itself, [5] the absorbance for **1** increased over wide wavelength region. This spectral change was visually detectable as a colour change from faint brown to deep brown. When $NaClO_4$ (10 M: 5.0 equivalents to **1** or **2**) was added to $CDCl_3$ solutions of **1** and **2**, the association with Na^+ could be detected by ^1H

A. W. Coleman (ed.), Molecular Recognition and Inclusion, 365-368.
© 1998 *Kluwer Academic Publishers. Printed in the Netherlands.*

NMR spectroscopy, but the solution colour (faint brown) was unaffected. In contrast, when $NaClO_4$ was added to these poor mixed solvent solutions, the deep brown colour disappeared in the $CHCl_3$-MeCN mixture but was affected to a smaller extent in the $CHCl_3$-MeOH mixture (Figure 1).

We measured the molecular-weight in solution by a light-scattering method. In $CHCl_3$ and the presence of $NaClO_4$ the formation of the aggregate particle was not observed for **1**, respectively. On the other hand, in the mixed solvents **1** aggregated into the particles with 240-390 nm diameter. Therefore, the behaviours observed in absorption spectroscopy and light-scattering commonly support the view that the "aggregation-deaggregation" equilibrium is responsible for the unique colour change.

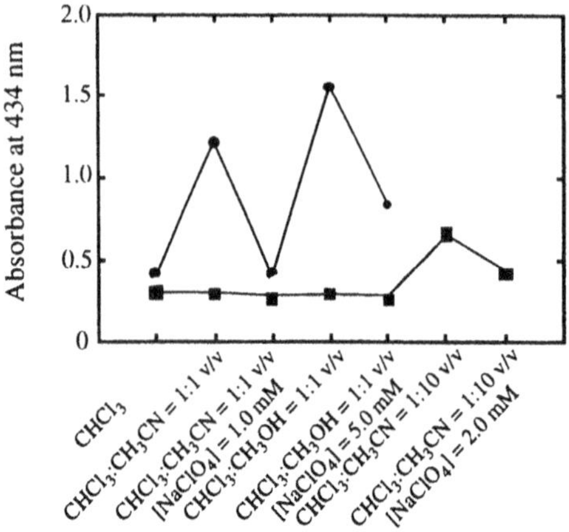

Figure 1. Solvent-induced and $NaClO_4$-induced spectral changes of **1** and **2** (●, **1**, 5.00×10^{-5} M at 25 °C; ■, **2**, 1.00×10^{-4} M at 25 °C).

Next, cation-π interactions have recently been attracting much attention as a new driving-force for the atom-molecule association. [6] Among them, the Ag^+-π interaction, which is facilitated by the exceptionally high affinity of Ag^+ with π-electrons, is the most visualizable and well-characterizable phenomenon. [7-9] Meanwhile, the surface of fullerene homologues is covered by such π-electrons. It is not so easy, however, to gain access to the fullerene surface with non-covalent interactions. [10] Here, it occurred to us that Ag^+ might be able to enjoy contact with the electron-rich fullerene surface. [11]

We mixed C_{60} and $AgCF_3SO_3$ in $CHCl_3$, but the absorption spectroscopic studies did not show any sign that they interact with each other.

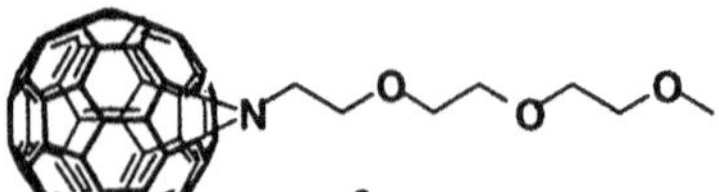

3

We thus used **3** [12] with an oligo(ethylene oxide) chain to retain Ag^+ near the C_{60} surface. When the metal salt (solid) was extracted with **3** in chloroform for 30 min under sonication, only $AgCF_3SO_3$ changed the absorption spectrum of **3**. [13] The spectral change could be visually followed as a colour change from deep orange to reddish orange.

The novel spectral change, particularly observable for Ag^+, can be rationalized in two different terms: *i. e.*, (i) Ag^+ is entrapped by the interaction with the oligo(ethylene oxide) chain (including that with the nitrogen atom) and the spectral change is due to the inductive effect through the N-C(C_{60}) bond or (ii) Ag^+ is adsorbed onto the C_{60} surface and the spectral change is induced by the direct interaction between Ag^+ and C_{60}. In the presence of CF_3COOD most signals for the methylene protons shifted to lower magnetic field, but the absorption spectrum was not affected (Table 1). These results mean that D^+ is entrapped only by the oligo(ethylene oxide) chain and does not interact with the C_{60} surface (as in Figure 2A). Furthermore, they provide evidence that the ND^+ scarcely affects the absorption spectrum of the C_{60} moiety through the inductive effect. On the other hand, Ag^+ affected only the δ for the NCH_2 methylene protons to a significant extent (Table 1). The result, together with the Ag^+-induced absorption spectral change, supports the view that Ag^+ is entrapped by the N and the C = C double bonds in the C_{60}

TABLE 1. ^{1}HNMR chemical shift (ppm) induced by M^+ addition in $CDCl_3$ at 25°C: [**1**] = 5.0 mM, 250MHz

	a	b	c	d	e	f	g
none	4.06	4.10	3.85	3.75	3.69	3.60	3.40
CF_3COOD	4.06	4.16	3.97	3.84	3.80	3.76	3.50
	(0.00)[a]	(+0.06)	(+0.12)	(+0.09)	(+0.11)	(+0.16)	(+0.10)
$AgCF_3SO_3$	4.36	4.11	3.87	3.79	3.74	3.65	3.39
	(+0.30)[a]	(+0.01)	(+0.02)	(+0.04)	(+0.05)	(+0.05)	(-0.01)

a) The values in parentheses are chemical shift changes. A plus sign (+) denotes a shift to lower magnetic field, whereas a minus sign (-) denotes a shift to higher magnetic field.

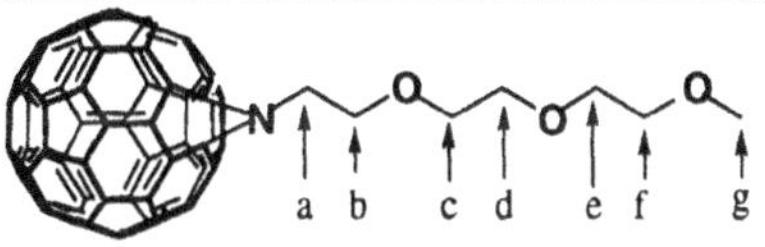

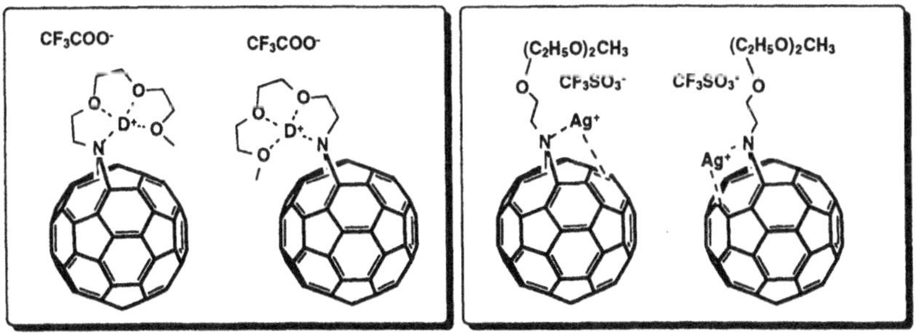

(A) (B)

Figure 2. Binding modes for (A) "hard" D^+ and (B) "soft" Ag^+. In (B) one must take two different binding sites into consideration for the interaction with different C=C double bonds. (A) may also require two different structures if one takes the direction of lone pair electrons in N into account.

moiety (as in Figure 2B). The foregoing results can be consistently explained as such that "hard" D^+ is favourably bound to "hard" oxygen bases whereas "soft" Ag^+ is favourably bound to "soft" π-electron bases.

In conclusion, the present paper has shown that the solvatochromic colour change thus observed is primarily ascribable to a change in the aggregation-deaggregation equilibrium induced by solvents or added $NaClO_4$ and that the colour change of **3** is induced by the driving-force that Ag^+, which *a priori* has the peculiar affinity for π-electrons, interacts with the π-electron-rich C_{60} surface with the aid of the 5,6-bridged nitrogen. We believe that the finding and its extended systems are of great importance to reversibly change the spectroscopic and redox properties of C_{60} and its derivatives by added metal cations.

1. Bezmelnitsin, V. N., Eletskii, A. V., and Stepanov, E. G. (1994) Cluster origin of fullerene solubility, *J. Phys. Chem.*, 6665- 6667.
2. An, Y.-Z., Anderson, J. L., and Rubin, Y. (1993) Synthesis of α-aminoacid derivatives of C_{60} from 1,9-(4-hydroxycyclohexano)-buckminsterfullerene, *J. Org. Chem.*, 4799-4801.
3. Ikeda, A. and Shinkai, S. (1996) Metal-induced aggregation-deaggregation equilibrium change in calix[4]arene-appended bisfullerenes, *Chem. Lett.*, in press.
4. Diederich, F., Jonas, U., Gramlich, V., Hermann, A., Ringsdorf, H., and Thilgen, C. (1993) Synthesis of a fullerene derivatives of benzo[18]crown-6 by *Diels-Alder* reaction: complexation ability, amphilic properties, and X-ray crystalstructure of a dimethoxy-1,9-(methano[1,2]benzonomethano)fullerene[60] benzene clathrate, *Helv. Chim. Acta*, 2445-2453.
5. Ruoff, R. S., Tse, D. S., Malhotra, R., and Lorents, D. C. (1993) Solubility of C_{60} in a variety of solvents, *J. Phys. Chem.*, 3379-3383.
6. Dougherty, D. A. (1996) Cation-π interactions in chemistry and biology: a view of benzene, Phe, Tyr, and Trp, *Science* , 163-168 and references cited therein.
7. Ikeda, A. and Shinkai, S. (1994) On the origin of high ionophoricity of 1,3-alternate calix[4]arenes: π-donor participation in complexation of cations and evidence for metal-tunneling through the calix[4]arene cavity, *J. Am. Chem. Soc.*, 3102-3110.
8. Ikeda, A., Tsuzuki, H., and Shinkai, S. (1994) NMR spectroscopic and X-ray crystallographic studies of calix[4]arene•Ag^+ Complexes. Infuence of bound Ag^+ on C_{2v}-C_{2v} interconversion in *cone*-calix[4]arenes, *J. Chem. Soc., Perkin Trans. 2*, 2073-2080.
9. Kang, H. C., Hanson, A. W., Eaton, B., and Boekelheide, V. (1985) [2_6](1,2,4,5)cyclophane (deltaphane) and related compounds. Simultaneous π-electron interaction among three benzene rings, *J. Am. Chem. Soc.*, 1979-1985.
10. Araki, K., Akao, K., Ikeda, A., Suzuki, T., and Shinkai, S. (1996) Molecular design of calixarene-based host molecules for inclusion of C_{60} in solution, *Tetrahedron Lett.*, 73-76 and references cited therein.
11. Ikeda, A., Fukuhara, C., and Shinkai, S. (1996) Fullerene (C_{60})-Ag^+ interactions which induce a solution color change, *Tetrahedron Lett.*, in press.
12. Hawker, C. J., Saville, P. M., and White, J. W. (1994) The synthesis and Characterization of a self-assembling amphiphilic fullerene, *J. Org. Soc.*, 3503-3505.
13. Metal salts tested herein are $LiClO_4$, $NaClO_4$, $AgCF_3SO_3$, $Ca(ClO_4)_2$, $Zn(ClO_4)_2$, and $Mg(ClO_4)_2$.

Allosteric Regulation in the Catalytic Activity of Cyclodextrin Dimer as an Artificial Hydrolase

H. IKEDA,* S. NISHIKAWA, A. UENO, and F. TODA†
Department of Bioengineering, Tokyo Institute of Technology, Nagatsuta-cho, Midori-ku, Yokohama 226, Japan and †Polytechnic College, 2-32-1, Ogawanisi, Kodaira 187, Japan

1. Introduction

Most enzymes have systems to control their activities in response to the concentration of substrates, products, or other effectors. Allosteric regulation is one of the sophisticate control systems, and allosteric enzymes or proteins are made up of several subunits. Each subunit is allowed to change its tertiary structure on substrate binding, thereby to affect the structure and the chemical activity of its neighbors. Although many artificial enzymes that had high activity or high selectivity were reported, there were not many artificial enzymes that had the systems to control their reactivities [1, 2]. Because the homotoropic allosteric enzyme consists of same structural sibunits and the binding sites of allosteric enzyme are located at the symmetrical position, a host molecule that has two same size binding site at the symmetrical position is expected to show allosteric effect, if substrate binding to a binding site of the host can change the binding ability of the other binding site. We already reported artificial enzyme (βCβH) that had a catalytic site between two β-cyclodextrins (β-CD) as an artificial hydrolase [3]. We investigate the substrate specificity for the acyl chain length and homotropic allosteric effect of βCβH.

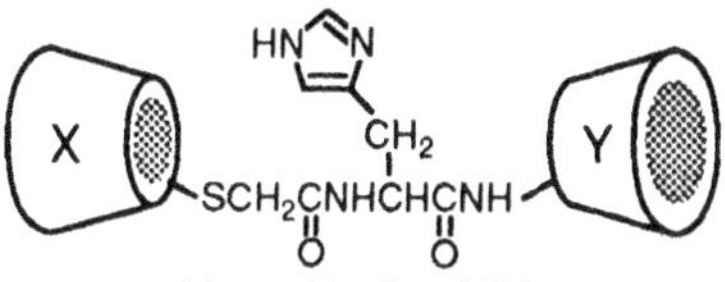

X = α, Y = β: αCβH,
X = β, Y = β: βCβH,
X = β, Y = α: βCαH

A. W. Coleman (ed.), Molecular Recognition and Inclusion, 369–372.
© 1998 *Kluwer Academic Publishers. Printed in the Netherlands.*

2. Materials and Methods

2.1. MATERIALS

Imidazole-appended cyclodextrin dimers were prepared by the condensation of 6-deoxy-6-(L-histidylamino)-cyclodextrin and 6-(carboxymethylthio)-6-deoxy-cyclodextrin with dicyclohexylcarbodiimide (DCC) and 1-hydroxybenzotriazole (HOBt). 6-Deoxy-6-(L-histidylamino)-cyclodextrin was obtained by the reaction between 6-amino-6-deoxy-cyclodextrin and N^{α}-*tert*-butyloxycarbonyl-N^{im}-tosyl-L-histidine in the presence of DCC and HOBt followed by deprotection. 6-(Carboxymethylthio)-6-deoxy-cyclodextrin was synthesized from 6-deoxy-6-iodo-cyclodextrin and methyl sodium sulfidoacetate followed by demethylation. These compounds were identified by elemental analysis, some kinds of NMR spectra including 2D NMR, and mass spectra.

2.2. METHODS

Ester cleavage reaction was followed by monitoring the appearance of nitrophenolate ion using a Simadzu UV-3100 spectrophotometer. The reaction was conduced in a quartz cell in the water-jacketed cell holder of the UV-3100. The temperature was maintained at 25 °C by a HAAKE F3 circulating water bath. The reaction was initiated by adding a stock solution of the ester in acetonitrile or DMSO to a buffer solution in the quartz cell. The rates used in the calculation of rate constants were averages of at least three determinations which agreed within 3 %.

the

3. Results and Discussion

3.1. CLEAVAGE REACTION OF *p*-NITROPHENYL ALKANOATES

The enzyme-like activities of three kinds of dimers were studied by measuring the rates for the cleavage reaction of some kinds of nitrophenyl alkanoates in a 16.1 % DMSO aqueous phosphate buffer solution of pH 7.8 at 25 °C under the condition of excess of substrate. The reaction proceeded by a Michaelis-Menten mechanism, showing saturation behavior with increasing substrate concentration. Table 1 shows the dependence of kinetic parameters on the acyl length of the esters. The transition state stabilization (K_{TS}) of the cleavage reaction

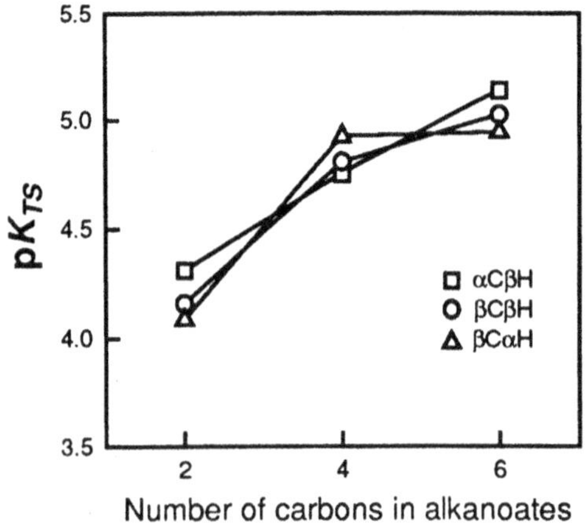

Figure 1. Dependence of pK_{TS} on acyl length of the substrates for the cleavage of p-nitropheyl alkanoates by the cyclodextrin dimers.

was evaluated using the Kurz approach as presented by O. S. Tee [4]. As the acyl chain length of the substrate was longer, the transition state was more stabilized by the dimers as shown in Figure 1. This results indicates that both phenyl part and alkyl part are included in the CD cavity.

TABLE 1. Kinetic parameters for the cleavage of *p*-nitrophenyl alkanoates by cyclodextrin homo- and hetero-dimers

catalysis	substrate	k_{cat} / 10^{-4} s^{-1}	K_m / 10^{-4} M	k_{cat}/K_m / $M^{-1}s^{-1}$	K_{TS} /10^{-5} M	k_{cat}/k_{un} / -
	C2	2.54	9.14	0.278	4.89	18.7
αCβH	C4	1.13	3.31	0.341	1.77	18.7
	C6	1.73	1.70	1.02	0.726	23.4
	C2	8.04	38.7	0.208	6.98	55.4
βCβH	C4	2.32	5.53	0.420	1.54	35.9
	C6	4.76	5.78	0.824	0.939	61.6
	C2	1.47	8.32	0.177	8.21	10.1
βCαH	C4	0.995	1.83	0.544	1.19	15.4
	C6	0.815	1.19	0.685	1.13	10.5

3.2. ALLOSTERY OF HOMO-DIMER

The presence of some cooperativity between two CD cavity of the CD dimer is suggested by the result of dependency on the acyl chain length of the substrates in the cleavage reaction, but it could not confirmed by the reaction with the alkanoate esters, because of low solubility of the ester in water. So, we synthesized a new ester (**2**) that can be more soluble in water, and the cleavage ability of three kinds of dimers were studied in high substrate concentration condition. The only homo dimer (**βCβH**) showed sigmoid behavior with increasing sub-

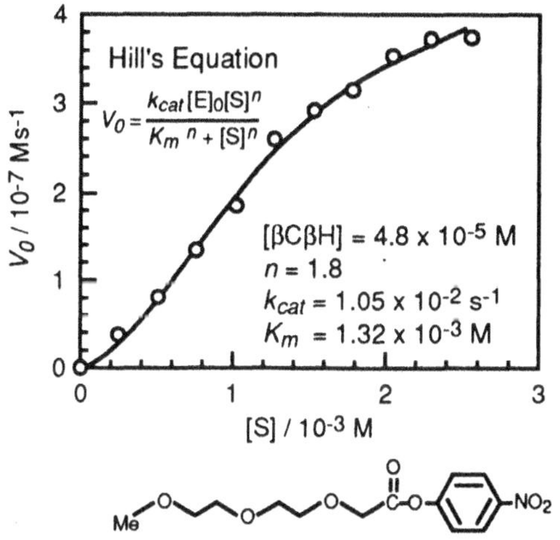

Substrate (**2**)

Figure 2. Dependence of the initial rate on the substrate concentration for the cleavage of the substrate (**2**) by βCβH in pH 7.4 phosphate buffer (0.1M) / acetonitrile (15 : 1) at 25 °C.

strate concentration as shown in Figure 2, whereas two kinds of hetero-dimers showed simple saturation behavior. This sigmoid behavior was fitted to the Hill equation and the Hill constant was 1.8. This means that high cooperativity exits between two β-CD cavity of the homo dimer but the hetero dimers have no allostery. The MWC model is one of models that can explain the allosteric interactions of natural allosteric proteins [5]. The most important assumption of the MWC model is symmetry of the structure of the protein and symmetrical structural change. So, our result, that the only homo-dimer showed allostery, might be supported by the MWC model.

4. Conclusion

Three kinds of CD dimers were synthesized as artificial hydrolases. As the acyl chain length of the substrates were longer, the transition state was more stabilized by the dimers. We constructed a new artificial allosteric system with 1.8 Hill constant by the use of CD homo-dimer and *p*-nitrophenyl methoxyethoxyethoxyacetate as the artificial enzyme and the substrate, respectively.

5. Acknowledgments

We are grateful to Nihon Shokuhin Kako (the _TD. for providing us with cyclodextrins.

6. References

1. Breslow, R.: Biomimetic Chemistry and Artificial Enzymes: Catalysis by Design, *Acc. Chem. Res.* **28**, (1995) 146-153.
2. Ikeda, H., Kojin, R., Yoon, C-j., Ikeda, T., Toda, F.: Artificial Hydrolase Using Modified Dimethyl-β-Cyclodextrin, *J. Incl. Phenom.*, 7 (1989) 117-124.
3. Akiike, T., Nagano, Y., Yamamoto, Y., Nakamura, A., Ikeda, H., Ueno, A., Toda, F.: Coupled Cyclodextrin Appending Imidazole as an Enzyme Model, *Chem. Lett.* (1994) 1089-1092 .
4. Tee, O. S.The Stabilization of Transition States by Cyclodextrin and other Catalysts, *Adv. Phys. Org. Chem.*, **29** (1994) 1-85.
5 Perutz, M.: *Mechanisms of Cooperativity and Allosteric Regulation in Proteins*, Cambridge University Press, Cambridge, 1991.

CONFORMATIONAL STUDIES ON ANTHRYL(ALKYLAMINO) - β - CYCLODEXTRIN COMPLEXES AND THEIR ABILITIES AS DNA INTERCALATORS

T. IKEDA, A. NAKAZATO, M. MORI, A. UENO, F. TODA*, and H-J. SCHNEIDER**

Department of Bioengineering, Tokyo Institute of Technology, 4259 Nagatsuta-cho, Midori-ku, Yokohama 226, Japan

**Tokyo Polytechnic College, 2-32-1 Ogawanishi-machi, Kodaira 187, Japan*

***FR Organische Chemie, Universität des Saarlandes, D 66041 Saarbrücken, Germany*

1. Abstract

Two kinds of anthryl(alkylamino)-β-cyclodextrins having different alkylamino chain length were synthesized as DNA intercalators. Guest-induced conformational changes of them were observed by measurements of ^{1}H-NMR spectra and circular dichroism spectra. Their different abilities as DNA intercalators were observed.

2. Introduction

Cyclodextrins (CDs) have been extensibly studied as host molecules for molecular recognition and encapsulations of drugs and drug delivery systems. But, CD itself has less specific affinities for biomolecules. For development of molecular recognition using modified CDs with specific and/or selective affinities for biomolecules and also for the interest of DNA intercalation and drug delivery systems, we have already reported anthryl(alkylamino)-β-CD complexes as DNA intercalators. [1] The binding ability of such complex to DNA was lower than that of known anthrylamines. [2] The lower DNA intercalative abilities of them might be caused by the effects of bulky CD moiety. To investigate of the effects of the conformations of anthryl(alkylamino)-β-CD complexes on the abilities of DNA intercalation, two kinds of anthryl(alkylamino)-β-CDs having different alkylamino chain length were synthesized.

A. W. Coleman (ed.), Molecular Recognition and Inclusion, 373–376.
© 1998 *Kluwer Academic Publishers. Printed in the Netherlands.*

3. Materials and Methods

Two kinds of anthryl(alkylamino)-β-CDs (1 and 2) were prepared from primary hydroxyl side mono tosylated β-CD as a starting material respectively. For **1**, to attach an anthracene moiety with alkylamino chains, ethylenediamine modified β-CD were synthesized from mono tosylated β-CD. To a solution of anthracene-9-carboxylic acid and DCC in dry DMF was added ethylenediamine modified β-CD at 0 °C. The resulting mixture was stirred for 5 hours at 0 °C and then 24 hours at room temperature. After reprecipitation with acetone, washed with chloroform and purification with ion exchange column chromatography (CM Sephadex C-25), **1** was obtained. For **2**, the same procedure was used except for the use of 1,4-diaminobutane instead of ethylenediamine. The purity of **1** and **2** were confirmed by TLC, ^{1}H-NMR and MALDI TOF MS.

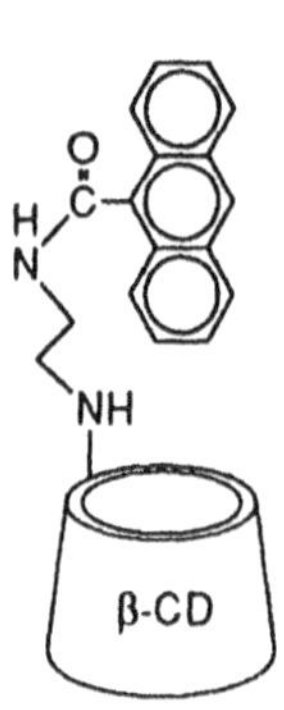

1

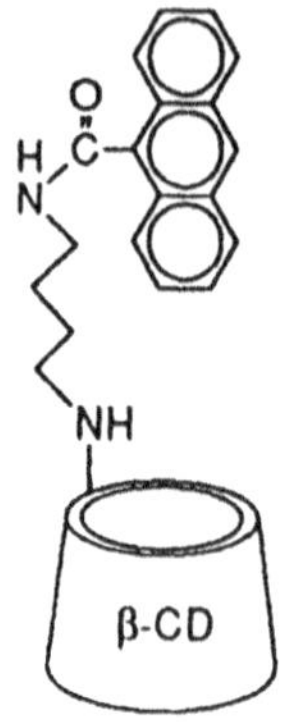

2

Calf thymus DNA was purchased from Pharmacia and purified by phenol extraction as a conventional method.

UV absorption, circular dichroism and fluorescence spectra in pH 7.5 Tris buffer were recorded on a SHIMADZU UV-3100 spectrophotometer, a JASCO-600 spectrophotometer and a HITACHI 850 spectro-fluorometer respectively using a 1 x 1 cm quartz cell at 25 °C.

^{1}H-NMR spectra were measured at 500 MHz and 400 MHz on VARIAN model VXR 5000 and UNITY*plus* 400 instruments respectively.

4. Results and Discussions

4.1. CONFORMATIONS OF ANTHRYL(ALKYLAMINO)-β-CDS AND THEIR COMPLEXES

As the aromatic ^{1}H-NMR signals of **1** and **2** are broadened and shifted compared with those of anthryl-alkylamines in D_2O which are similar to those of **1** and **2** after adding 1-adamantanol (**AOH**) as a guest molecule, the anthryl unit of **1** and **2** are included in their own CD cavity. (*Fig. 1*) This signifies the guest molecule (**AOH**) removes the anthryl unit out of the CD cavity. The different conformations of the anthryl unit of **1** and **2** are suggested from their own shifted signals without **AOH**. (*Fig. 1a and 1c*) Circular dichroism spectra of **1** alone gives a positive peak indicated that the long axis

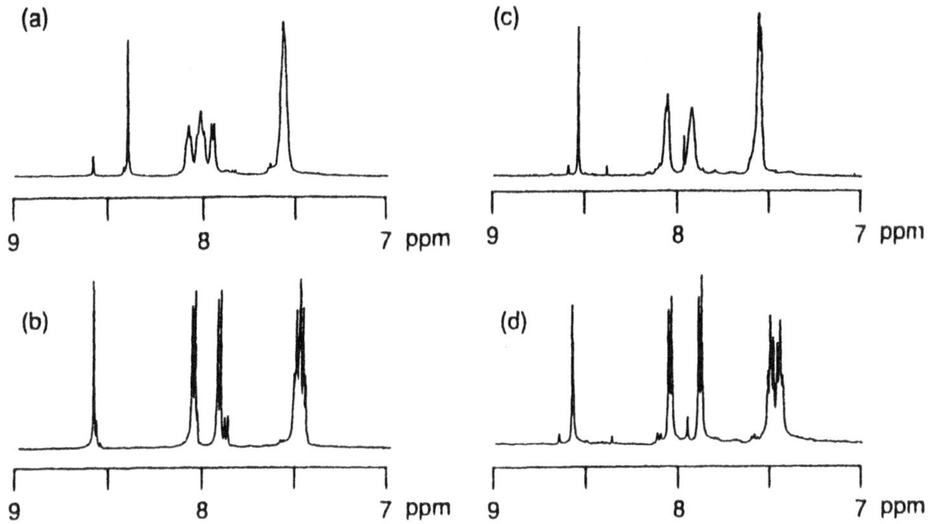

Figure 1 ^{1}H-NMR spectra of the aromatic region of (a) **1**, (b) **1** with AOH, (c) **2** and (d) **2** with AOH in D_2O

of the anthracene moiety of **1** is parallel to the axis of the CD cavity. After adding **AOH**, such a positive peak is changed to a negative one. This phenomena also suggested that anthracene moiety of **1** is moved to the out of the CD cavity and located perpendicular to the axis of the CD cavity after making complex with **AOH**. Circular dichroism spectra of **2** with **AOH** gives also a negative peak, however **2** alone shows no cotton effect. These data show the anthryl unit of **2** alone is in the equilibrium state between complexed and uncomplexed form by the CD cavity. These differences of conformations of **1** and **2** are caused by the differences of alkylamino chain length between **1** and **2**.

4.2. DNA INTERCALATION

UV absorption and fluorescence spectra of **1** are decreased with increasing concentration of calf thymus DNA only after adding **AOH**. The anthryl unit of **1** is bound to the CD cavity (intramolecular complex form) and an intercalation to DNA is inhibited. After making complex with **AOH**, the anthryl unit of **1** is moved to the outside of the CD cavity and **1** (with **AOH**) can intercalate to DNA.

UV absorption and fluorescence spectra of **2** are slightly decreased with increasing concentration of calf thymus DNA without adding **AOH**. After adding **AOH**, both spectra of **2** are more decreased. **2** itself can slightly intercalate to DNA, because the anthryl unit of **2** is located inside and outside of the CD cavity in the equilibrium state as mentioned before. After making complex with **AOH**, the anthryl unit of **2** is

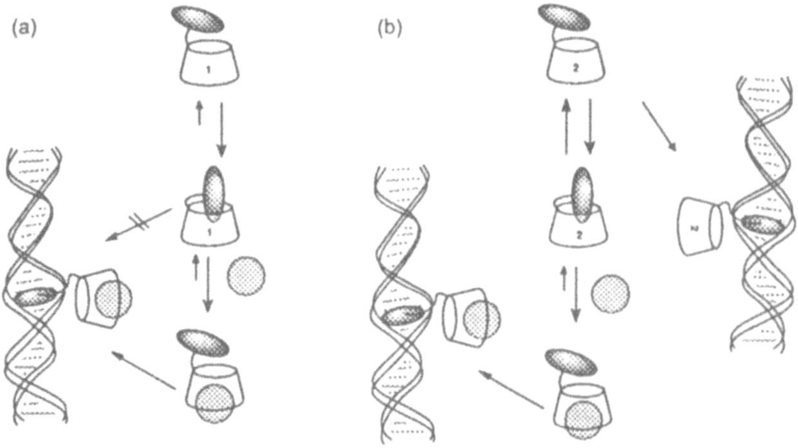

Figure 2

completely removed to the outside of the CD cavity and strongly intercalate to DNA. From the excitation spectra of **1** and **2**, the energy transfer from DNA to **1** and **2** are observed under the same conditions which DNA intercalation are occurred.

In conclusion, **1** can intercalate to DNA only after making complex with a guest molecule, that is, by a guest-induced conformational change, **1**-complex can intercalate to DNA. (*Fig. 2a*) On the other hand, **2** itself can slightly intercalate to DNA and the ability of intercalation to DNA is enhanced by making complex with a guest molecule. (*Fig. 2b*) These differences are caused by different chain length and different conformations of them.

More detailed studies using another guest molecules and the determination of the properties of DNA intercalation by ^{1}H-NMR measurements are now in progress.

5. Acknowledgments

We are grateful to Nihon Shokuhin Kako Co. LTD. for providing us with samples of CD's.

6. References

[1] Ikeda T., Yoshida K., and Schneider H-J., (1995) Anthryl(alkylamino)-cyclodextrin complexes as chemically switched DNA intercalators, *J. Am. Chem. Soc.*, **117**, 1453-1454.

[2] Kumar C.V. and Asuncion Emma H., (1992) Sequence dependent energy transfer from DNA to a simple aromatic chromophore, *J. Chem. Soc., Chem. Commun.*, 470-472.
Kumar C.V. and Asuncion Emma H., (1993) DNA binding studies and site selective fluorescence sensitization of an anthryl probe, *J. Am. Chem. Soc.*, **115** 8547-8553.

HOST-GUEST COMPLEXATION OF PHOSPHORUS CONTAINED CALIXARENES WITH AROMATIC MOLECULES IN *RP HPLC* CONDITIONS. THE STABILITY CONSTANTS DETERMINATION

O.I.KALCHENKO*, J.LIPKOWSKI, R.NOWAKOWSKI
Institute of Physical Chemistry Polish Academy of Sciences
Kasprzaka, 44/52, 01-224, Warsaw, Poland

V.I.KALCHENKO, M.A.VYSOTSKY, L.N.MARKOVSKY
Institute of Organic Chemistry, National Academy of Sci. of Ukraine
Murmanskaya, 5, 253660, Kiev-94, Ukraine

Abstract

Reversed-phase high-performance liquid chromatography has been applied to studies of the host-guest complexation of octakis(diethoxyphosphoryloxy)*tert*-butyl-calix[8]arene as well as tetraalkylcalix[4]resorcinarenoctols and their upper rim phosphoryl, sulfonyl, dialkylaminomethyl derivatives with some aromatic guests in mobile phase. It was shown that formation of the inclusion complexes results in changes of retention of aromatic guests and improves their separation. Stability constants of the complexes (30-2795 M^{-1}) were calculated from dependences of *1/k`* values of the aromatic guest *vs.* concentration of the macrocyclic host in the mobile phase.

Results and Discussion

Cavity shaped calix[n]arenes (CA) and calix[4]resorcinarenes (RA) made-up of phenol and resorcinol units linked *via* carbon atoms similar to cyclodextrins are promising compounds for design of host molecules able to bind organic and inorganic species [1]. In this work we investigated host-guest interaction of tetraalkylcalix[4]resorcinaren-octols **1-5**, some their upper rim nitrogen, phosphorus and sulfur derivatives **6-12** as well as

A. W. Coleman (ed.), Molecular Recognition and Inclusion, 377–380.
© 1998 *Kluwer Academic Publishers. Printed in the Netherlands.*

octakis(diethoxyphosphoryloxy)-*tert*-butyl-calix[8]arene **13** with various benzene derivatives (23 compounds possessing alkyl, halogene, hydroxy, aldehyde, carboxy groups in the aromatic ring) in conditions of reversed-phase high-performance liquid chromatography (LiChrosorb RP 18 or Separon SGX C 18, UV detection at 254 nm and acetonitrile-water 84:16, v/v as mobile phase). Stability constants of the host-guest inclusion complexes were determined.

Investigated all-*cis*-tetraalkylcalix[4]resorcinarenoctols **1-5** and tetrakis(diethylaminomethyl)calix[4]resorcinarenoctol **9** in solution exist in *crown* conformation with all resorcinol rings up-oriented (C_{4v} symmetry) and stabilized in this position by intramolecular hydrogen bonds at the upper rim of macrocycle. Tetra- and octasubstituion of hydroxyl protons in the calix[4]resorcinarenoctols by sulfonyl and phosphoryl groups (compounds **7,8,10-12**) [2] transforms the initial *crown* conformation into the *boat* one in which two opposite benzene rings are coplanar with the main macrocyclic plane formed by carbon atoms of the methine bridges and two others are perpendicular to the plane (C_{2V} symmetry). The *boat* conformation is stereochemically flexible. The rapid pseudorotation process consisting in site exchange of benzene rings with the coplanar and perpendicular orientation is observed in solution at the room temperature. The *crown* conformation is intermediate in this transformation. Octaphosphorylated calix[8]arene **13** is stereochemically flexible too. Fast rotation of the benzene rings through macrocyclic annulus is observed in solution at room temperature.

Like in the case of cyclodextrins the addition of RA **1-12** and CA **13** in the mobile phase decreases the retention and capacity factors of the solutes indicating the host-guest inclusion complexes formation in the acetonitrile-water solution. To determine the composition of the complexes formed the dependences $1/k'$ values of the solutes *vs*. the host concentration in the mobile phase were studied. For all investigated complexes a linear dependence $1/k'$ *vs*. host concentration in the range from 1×10^{-4} to 3×10^{-3} M is observed, indicating formation of the host-guest complexes in composition 1:1. From the obtained dependences $1/k'$ *vs*. the host additives concentration the stability constants of these complexes have been calculated by equation [3]

$$1/k' = 1/k'_0 + [\text{Host}]/K_D \times k'_0$$

where k'_0 is a capacity factor in the absence of host; [Host] is a

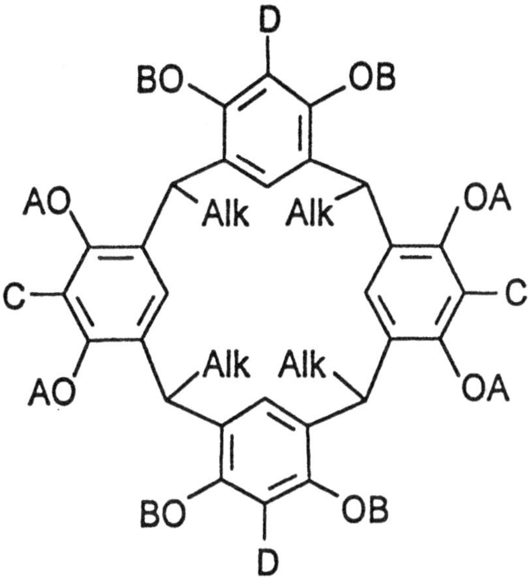

1-12

1	Alk = CH_3	A = B = C = D = H
2	Alk = C_3H_7	A = B = C = D = H
3	Alk = C_5H_7	A = B = C = D = H
4	Alk = C_7H_{15}	A = B = C = D = H
5	Alk = $C_{15}H_{31}$	A = B = C = D = H
6	Alk = CH_3	A = Ts, B = C = D = H
7	Alk = CH_3	A = p-$ClC_6H_4SO_2$, B = C = D = H
8	Alk = CH_3	A = Ts, B = C = H, D = $CH_2N(C_2H_5)_2$
9	Alk = CH_3	A = B = H, C = D = $CH_2N(C_6H_{13})_2$
10	Alk = C_3H_7	A = B = Ts, C = D = H
11	Alk = CH_3	A = $P(O)(OC_3H_7)_2$, B = C = H, D = $CH_2N(C_2H_5)_2$
12	Alk = CH_3	A = $P(O)(OC_3H_7)_2$, B = C = D = H

concentration of the host in mobile phase; K_D is a dissociation constant of the complex.

The stability constants of the inclusion complexes are changed in the range of 30 - 863 M^{-1} (RA **1-12**) and 608 - 2795 M^{-1} (CA **13**) depending on the host structure and nature of the aromatic guest molecules. Calix[4]resorcinarenes **4,5,7,9-12** additives in mobile phase in concentrations 1 x 10^{-3} - 3 x 10^{-3} M^{-1} essentially improve separation factors R_s and resolution of aromatic compounds with similar properties. The best values R_S were observed for separation of toluene and ethylbenzene in the presence of tetraheptylcalix[4]areneoctol **4**.

The results obtained indicate that calix[4]resorcinarenes, similarly to cyclodextrins, could be utilized in liquid chromatography analysis of aromatic compounds. The knowledge of the host-guest stability constants is also important for realization of allosteric effects in stereochemically flexible phosphorylated CA and RA. Effectively bonding aromatic guest molecules can change the macrocyclic skeleton conformation and organize the phosphoryl groups for cooperative complexation inorganic cationic species. This process may mimic action of enzymes and is important for design of their synthetic models possessing allosteric ionophore function;

Acknowledgements

This work is performed in accordance with scientific cooperation between Institute of Physical Chemistry Polish Academy of Sciences and Institute of Organic Chemistry National Academy of Sciences of Ukraine.

VIK, MAV, LNM thank INTAS (Grant No 94/1914) for partial financial support of the investigation. Authors thank to Prof.H.-J.Schneider for useful discussion.

References

1. Bohmer,V. (1995) Calixarenes. Macrocycles with (almost) unlimited possibilities. *Angew.Chem.* Int.Ed.Engl. **34**, 713; Schneider,H.-J., Schneider,U. (1994) Host-guest chemistry of resorcinarenes. *J.Incl.Phenom.* **19**, 67.

2. Kalchenko,V.I., Rudkevich,D.M., Shivanuk,A.N., Tsimbal,I.F., Pirozhenko,V.V., Markovsky,L.N., (1994) Phosphorylated octahydroxy[1_4]methacyclophanes. *Zh.Obsch.Khim.* **64**, 731.

3. Fujimura,K., Ueda,T., Kitagawa,M., Takayanagi., Ando,T. (1986) Reversed-phase retention behaviour of aromatic compounds involving cyclodextrin inclusion complex formation in the mobile phase. *Anal.Chem.* **58**, 2668.

SYNTHESIS OF CYCLODEXTRINS DERIVATIVES CARRYING BIO-RECOGNISABLE SACCHARIDE ANTENNAE

RIMA KASSAB, HELENE PARROT-LOPEZ*
Equipe Reconnaissance et Organisation Moléculaire, ESA 5078 CNRS
Université Claude Bernard-Lyon I, Bat. 305, 43 bd du 11 Novembre 1918, 69622 Villeurbanne Cedex, France.

Abstract

The use of coupling saccharide antenna onto cyclodextrins allows mobility for the biologically active galactose head group, and allows the recognition process by the lectin (KbCWL). This paper report a chemical synthesis of β-CD derivatives using spacer arms 3,4,5,6 and 9 carbon atoms. Preliminary results suggested that recognition is strongly dependent on the length of the spacer chain between the cyclodextrin and the sugar head group.

Introduction

Cyclodextrins (CDs) are interesting candidates for the transport of biologically active molecules.[1] They are a class of cyclic oligosaccharides having 6,7 or 8 glucopyranosyl units linked α, 1-4 (α–,β–, or γ-CD). The internal cavity of CD is slightly apolar. In this way, CDs are capable of including a wide range of organic molecules by interaction between host and guest molecules.[2] This type of molecular encapsulation may protect a great number of sensitive products. In particular, β-CD and its derivatives have been widely used for the solubilisation and non-specific transport of biologically active molecules.[3] In this work, the strategy consists of increasing the capacity and specifity of the transport and recognition properties by formation of a neo-glycoconjugate via an attached saccharide antenna.

In preceding work, we demonstrated the capacity of galactosyl-β-cyclodextrin derivatives to be recognized by a galactose specific cell wall lectin (KbCWL).[4] This preliminary work suggested that recognition is strongly dependent on the length of spacer chain between the cyclodextrin and the sugar head group and also on the number of sugars presented to the lectin.

This project considers the synthesis of carriers of biologically active molecules. In a first series, Gal head groups were connected to mono-6-amino-6-deoxy-β-cyclodextrin using varying spacer chain lengths (3,4,5,6 and 9carbon atoms). In the second part, a carrier having seven Gal head groups was constructed from the per-6-amino-6-deoxy-β-cyclodextrins.

A. W. Coleman (ed.), Molecular Recognition and Inclusion, 381–384.
© 1998 *Kluwer Academic Publishers. Printed in the Netherlands.*

Methodology of synthesis

The synthesis was carried out via standard peptide coupling of the β-D-Gal spacer molecule onto mono[5] and heptakis-6-amino-6-deoxy-β-cyclodextrin[6] (Figure1).

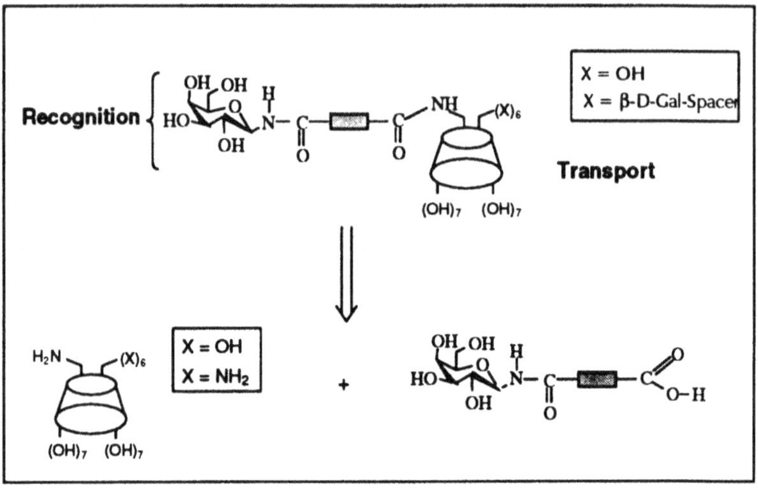

Figure 1. Synthesis of the transport systems.

The glycoconjugates were synthesised from the corresponding tetra-O-acetyl-α-D-bromo galactose in three steps. The synthetic route depends on the easy conversion of the saccharide isothiocyanate compounds into amides. The starting compounds are prepared in a phase transfer reaction of the corresponding halides with KSCN and tetrabutyl ammonium bromure ($Bu_4N^+Br^-$) in acetonitrile in the presence of molecular sieves (4·Å). The tetra-O-acetyl-β-D-galactosyl isothiocyanate (3.85mmol) is condensed with the mono-methyl-esters corresponding to malonic acid (C_3), succinic acid (C_4), glutaric acid (C_5), adipic acid (C_6) and nonanoic acid (C_9) (1.4 equiv) in anhydrous toluene (20ml) in the presence of triethylamine (0.1 equiv). This products were purified via flash chromatography (toluene/acetone: 8/2). The acetylated amido compounds were simultaneous de-acetylated and de-esterified with methanolic sodium hydroxide (1M) during 4h at room temperature (Figure 2).

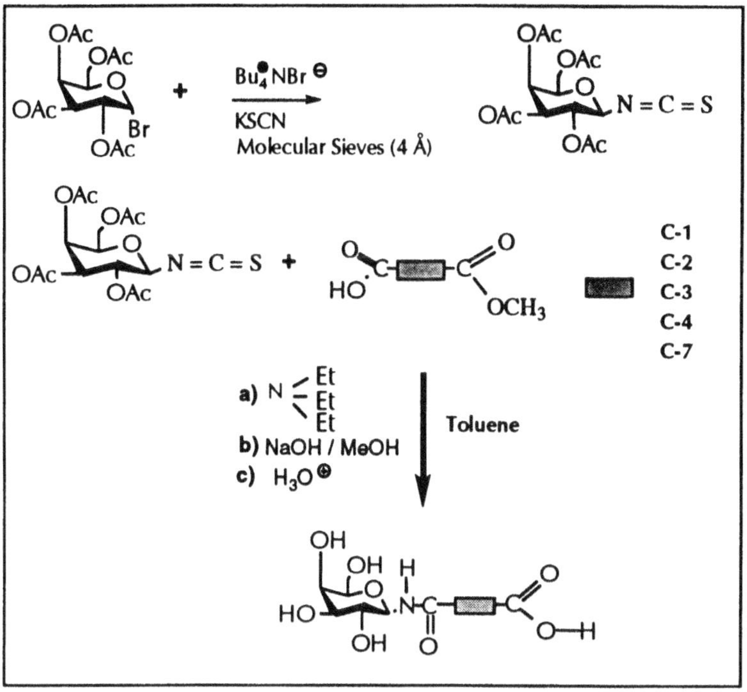

Figure 2. Synthesis of the glycoconjugates.

Results and discussions

The final products are obtained in 30-35% yields after purification on silica gel chromatography (n-butanol/EtOH/H_2O: 5/4/3) and characterised by ^1H NMR (300MHz and 400MHz)[7] and electrospray mass spectrometry. The H-1 signals of β-cyclodextrin were observed at 4.84 ppm (d, J=1.5Hz for the mono CD derivatives and non resolved signal for the per-CD-derivative) and the H-1 signals of the β-D-Gal residues at 4.68 ppm (J–7.5Hz). The broadening of the β-CD H-1 signals in the case of the per-substituted system, may arise from slow conformational rearrangement of the amido substituents as observed for polysubstituted amino-acid derivatives of β-CD by Stoddart.[6] In order to confirm complete substitution electrospray mass spectrosopy (negative mode) was carried out, this showed mass peaks corresponding to m/z= M-3H$^+$/3 at 1147. NOE difference experiments show interactions between the Gal H-1 proton and the adjacent amide proton. In the β-D-Gal-spacer-β-CD-derivatives no NOESY and ROESY effects are seen between the Gal-amide proton and the H-1 β-cyclodextrin protons. Thus saccharide antennae are directed out the cavity of the β-CD into the aqueous environment. Hence this system is available for the transport.

Preliminary results shows that in contrast to our expectations of obtaining a "clustering effect" per- substitution of β-CD by the β-D-Gal-C9 head group leads to only a 1.5 fold increase in recognition. This may on effect arise from cross linking of cells by the large (40Å diameter) molecule leading to cell floculation.

CONCLUSION

We have synthesised different types of cyclodextrin derivatives directed towards recognition by β-D-Gal specific cell wall lectins. Our results show that the chemical modification of β-CD by substitution of primary hydroxyl on the ring with galactose end arms induces their recognition by carbohydrate proteins. This is an "exo-recognition" process not involving the cyclodextrin cavity.

REFERENCES

1 Szejtli, J. (1988) *Cyclodextrin Technology*, Kluwer Academic Publishers, Dordrecht.

2 Wenz,G., (1994) Cyclodextrins as Building Blocks for Supramolecular Structures and Functional Units, *Angew. Chem. Int. Ed. Engl.* **33**, 803-822.

3 Duchêne, D. (1987) *Cyclodextrins and their Industrial Use*, Editions de Santé, Paris.

4 Attioui, F., Al-Omar, A., Leray, E., Parrot-Lopez, H., Finance, C., and Bonaly, R. (1994) Recognition ability and cytotoxicity of some oligosaccharidyl substituted β-cyclodextrins, *Biol Cell* **82**, 161-167.

5 Parrot-Lopez, H., Djedaini, F., Perly, B., and Coleman, A.W. (1990) Intramolecular Host-Guest Complexes of D- and L-mono-6-Phenylalanyl-Amino-6-Deoxy cyclomaltoheptaoses, *Tetrahedron Asym.* **1**, 367-370.

6 Ashton, P.R., Königer, R., and Stoddart, J.F.(1996) Amino Acid Derivatives of β-Cyclodextrin, *J. Org. Chem.* **61**, 903-908.

7 The new products were identified from its ^1H NMR spectrum (DMSO-d_6): **Mono-6-[6-(β-D-galactopyranosylamino)-1,6-dioxohexyl]amino-6-deoxy-β-cyclodextrin:** yield 52%; δ 2.41 (m, 4H, $2CH_2$ β to C=O); 2.51 (t, 4H, $2CH_2$-C=O); 2.74 (m, H Gluc β-CD, H-2 Gal); 2.90 (m, 7H, H-3 β-CD); 3.35 (m, 1H, H-3 Gal); 3.64 (m, 6H, OH-6 β-CD); 4.48 (d, 1H, H-1 Gal); 4.83 (d, 7H, H-1 β-CD); 5.7 (m, 16H, OH-2 and OH-3 β-CD and OH-2 et OH-3 Gal); 7.2 (m, 1H, NH-β-CD); 7.96 (d, 1H, J=10Hz, NH Gal). **Mono-6-[5-(β-D-galactopyranosylamine)-1,5-dioxopentyl]amino-6 deoxy-β-cyclodextrin:** Yield 54,5%, δ 2.09 (m, 2H, CH2 β to C=O); 2.51 (t, 4H, $2CH_2$-C=O); 3.4 (m, H Gluc β-CD and H-2 Gal); 3.61 (m, 7H, H-3 β-CD); 3.64 (m, 1H, H-3 Gal); 4.47 (m, 6H, OH-6 β-CD); 4.7 (d, 1H, H-1 Gal); 4.89 (d, 7H, H-1 β-CD); 5.76 (dd, 16H, OH-2, OH-3 β-CD and OH-2, OH-3 Gal); 7.2 (m, 1H, NH-β-CD); 7.6 (d, 1H, J=9.1Hz NH). **Mono-6-[4-(β-D-galactopyranosylamino)-1,4-dioxybutyl] amino-6-deoxy-β-cyclodextrin** : yield 46 %, δ 2.09 (t, 4H, $2CH_2$-C=O); 3.35-3.59 (m, H Gluc β-CD, H-2 Gal); 3.61 (m, 7H, H-3 β-CD); 3.65 (m, 1H, H-3 Gal); 4.52 (m, 6H, OH-6 β-CD); 4.67 (d, 1H, H-1 Gal); 4.84 (d, 7H, H-1 β-CD); 5.76 (m, 16H, OH-2, OH-3 β-CD and OH-2, OH-3 Gal); 7,58 (m, 1H, NH β-CD); 7.70 (d, 1H, J=9Hz, NHGal).
Mono-6-[3-(β-D-galactopyranosylamino)-1,3-dioxopropyl]amino-6-deoxy-β-cyclodextrin: yield (50%), δ 1.87 (s, 2H, CH_2-C=O); 3.06-3.35 (m, H Gluc β-CD and H-2 Gal); 3.64 (m, 7H, H-3 β-CD); 3.8 (m, 1H, H-3 Gal); 4.48 (m, 6H, OH-6 β-CD); 4;68 (d, 1H, H-1 Gal); 4.77 (d, 7H, H-1 β-CD); 5.71 (m, 16H, OH-2, OH-3 β-CD and OH-2, OH-3 Gal); 7.31 (m, 1H, NH β-CD); 7.58 (d, 1H, J=9Hz, NH Gal). **Per-6-[9-(β-D-galactopyranosylamino)-1,9-dioxononyl]amino-6-deoxy-β-cyclodextrin:** yield 35%, (D_2O) δ 1.3 (m, 42H, CH_2 spacer); 1.55 (m, 28H, CH_2 β to C=O); 2.3 (m, 28H, CH_2 α to C=O); 3.4-3.9 (brd, H Gluc β-CD and H Gal); 3.95 (d, 7H, J=1Hz, H-4 Gal); 4.87 (d, 7H, J=7.5Hz, H-1 Gal); 5.05 (brd, 7H, H-1 β-CD).

X-RAY AND ATOMIC FORCE MICROSCOPY STRUCTURES OF SHORT CHAIN AMPHIPHILIC CYCLODEXTRINS

I. NICOLIS[1], A.W. COLEMAN[2], M. SELKTI[1,] M. MUNOZ[2], A. KASSELOURI[1,**], S. ALEXANDRE[3], J.-M. VALLETON[3], P. CHARPIN[1], C. de RANGO[1].*

*[1]ERS128 CNRS, Centre Pharmaceutique, Université Paris-Sud, Châtenay-Malabry, F92296; [2]UPR412 CNRS, IBCP, 7, Passage du Vercors, Lyon F69367; [3]URA500 CNRS, Université de Rouen, F76821. Present Addresses: *Laboratoire de Physique, Faculté des Sciences Pharmaceutiques et Biologiques de Paris V, F75270; **Laboratoire de Chimie Analytique, Centre Pharmaceutique, Université Paris-Sud, Châtenay-Malabry, F92296.*

Cyclodextrins rendered amphiphilic by remarkably short substituents chains have been shown to form stable molecular layers[1,2]. The results deduced from compression isotherms of Langmuir films at the air-water interface led us to propose the formation of amphiphilic cyclodextrin multilayers and the existence of a highly rigid (solid) system. As the physical characteristics of the Langmuir films might be correlated with bulk solid-state structures, comparative crystallographic studies have been undertaken for α-, β- and γ-cyclodextrins selectively persubstituted at the primary face with very short hydrophobic groups.

The 6-bromo-6-deoxy-β-cyclodextrin (β-Br-CD) is the first amphiphilic cyclodextrin for which a single crystal analysis has been performed[1]. As for this compound, the 6-azido--6-deoxy-α-, β- and γ-cyclodextrins (N_3-CDs) crystallize in multilayer-type structures, with molecular areas similar to those determined from the compression isotherms. Comparison of the results from atomic force microscopy studies with Langmuir-Blodgett deposited layers on mica and the crystal structures, suggests that there may exist a close analogy between layer arrangement in the solid-state and in the monomolecular layers.

I - MONOMOLECULAR LAYERS

The compression isotherms of Langmuir films at the air-water interface of cyclodextrins persubstituted at primary face with bromine atoms[1] (Fig. 1) or azide groups [2] have shown that the molecular areas increase with the number of glucopyranose units (Table 1), while the stability decreases as: β->γ->α-(N_3/Br)-CD. In all cases, the formation of multilayer systems is evidenced.

A. W. Coleman (ed.), Molecular Recognition and Inclusion, 385–389.
© 1998 *Kluwer Academic Publishers. Printed in the Netherlands.*

	α-CD	β-CD	γ-CD
N_3-CD(Å2) Monomolecular layers	148	195	214
N_3-CD(Å2) Crystals	200	240	280
Br-CD(Å2) Monomolecular layers	180	220	260
Br CD(Å2) Crystals		250	

Table 1: Molecular areas deduced from compression isotherms and crystal structures

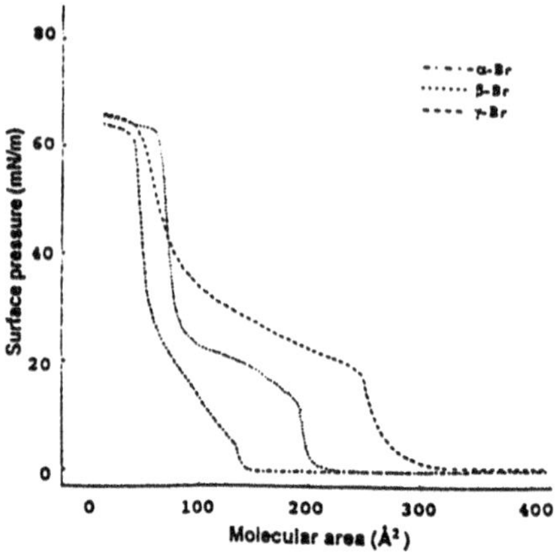

Figure 1: Compression isotherms of Langmuir films at the air-water interface for Br-CDs

II - Atomic Force Microscopy (AFM) Studies

AFM studies of Langmuir Blodgett deposited layers on mica give information on the structures of α-, β- and γ-N3 CD systems[3]. At low resolution, the images show rigid surfaces with apparent single step height of 8Å, indicating formation of multilayer systems. In the case of the β-N3 CD derivative, at high resolution a very high degree of organization at molecular level is observed with a regular structure of apparent periodicities 15Å.

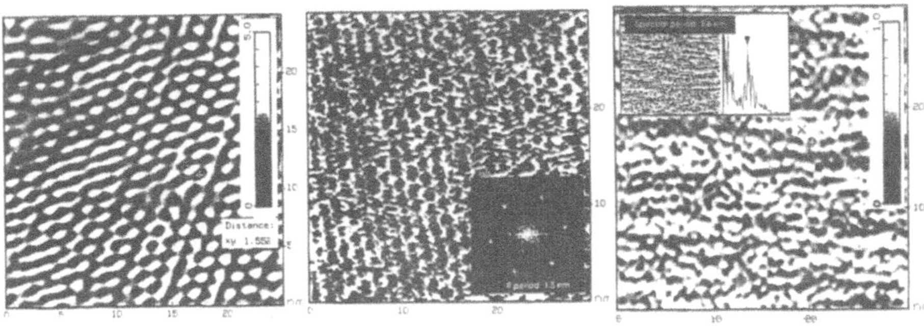

Figure 2: High resolution AFM images of α-, β- and γ-azido derivatives

III - First Single Crystal Analysis of Amphiphilic Cyclodextrins

Single crystals have been obtained for a set of derivatives[4] (Fig. 3) selected as a function of the nature of the substituent (azide groups, bromine atoms and dimethyl *t*-butyl silyl groups for **I**, **II** and **V**, respectively) and of the molecular symmetry (6, 7 and 8 for **III**, **I** and **IV**, respectively). Crystals of the β-N_3-CD derivative grown in solutions containing divalent cations salts ($ZnBr_2$, **VI**; $CoCl_2$, **VII**) have been obtained with significant differences in the unit cell parameters. The crystal structures of **I**, **II**, **III** and **IV** have been determined.

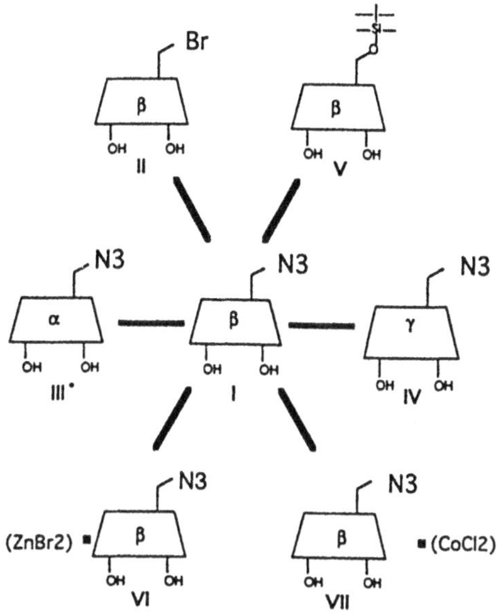

Figure 3: Comparative Crystallographic Studies.

The four derivatives crystallize in multilayer-type structures with different modes of "head-to-head" layer association; the molecular axes being perpendicular to the layers (Fig. 4). The structures present a common characteristic: the substituted primary faces play a predominant role in the formation and the stability of the multilayers through original association between dimers or monomers, involving van der Waals contacts between substituents.

- *Molecular Organization within the Layers*

The molecular areas are close to those deduced from compression isotherms (Table 1), although the crystal structures reveal different modes of molecular organization within the layers. The α-N_3-CD derivative (**III**) adopts a pseudo-centered square arrangement while **II** and **IV** form two different quasi-square patterns. In the case of the β-N_3-CD derivative (**I**), which shows a pseudo-centered rectangular arrangement, the crystal periodicities fit closely with the values deduced from high resolution AFM structures.

- Modes of "Head-to-Head" Layers Association

For all four derivatives, the layers have a height of 8Å, indicating that the multilayers systems in the bulk solid-state may be correlated with those observed from the AFM studies.

The relative settings of layers lead to four types of packings. Their common characteristic is the formation of alternate hydrophilic and hydrophobic interfaces. Two different packings of two dimer layers are observed, for β-(N_3 and Br)-CDs; they create infinite disjointed channels (**I**) or dimeric cages (**II**). In the case of the α- and γ-N_3-CD derivatives, the packing of four dimer layers (**III**) and four monomer (**IV**) layers gives rise to infinite channels and monomeric cages.

This suggests that the multilayer systems of N_3-CD derivatives might be porous (**III**), semi-permeable (**I**) or quasi-impermeable (**IV**), according to the molecular symmetry (6, 7 and 8, respectively).

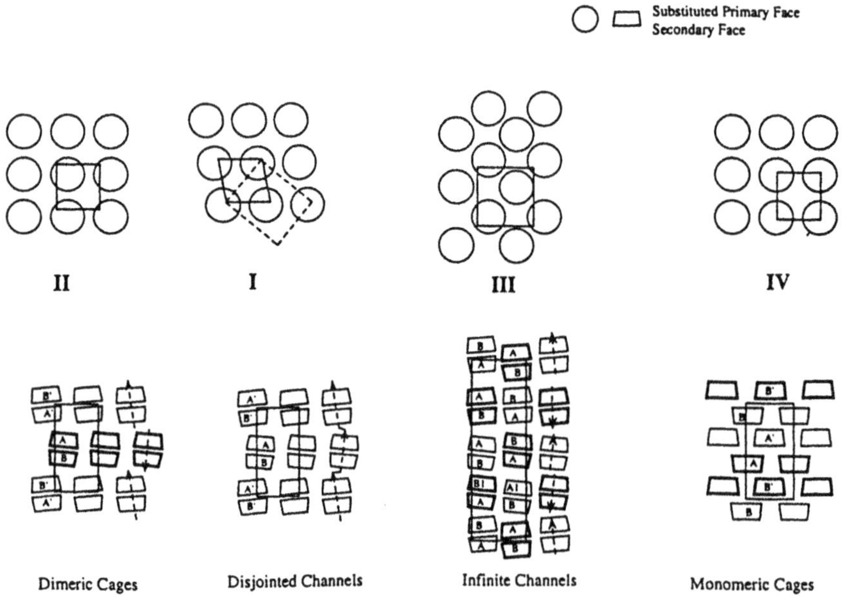

Figure 4: Modes of Assembly (top, within layers; bottom, stacking of layers); heavy lines denote translation of the dimers along the projection axis.

This comparative crystallographic study reveals that in the multilayer-type structures of short chain amphiphilic cyclodextrins, two different non covalent interactions are implied in distinct regions: the secondary face to face interactions via hydrogen-bonding which are essential in the "head-to-head" layer association as well as in the molecular arrangements within the layers and the primary face to face interactions via van der Waals contacts between substitutents which are fundamental in the stacking of the layers. For the β-CD derivatives, it has been shown that a few water and solvent molecules are dispersed in the intermolecular spaces at the level of the secondary faces but **none** are observed at the level of the **primary** faces.

REFERENCES

1 - Nicolis, I., Coleman, A. W., Charpin, P., Villain, F., Zhang, P., Ling, C.C. & Rango C. de (1993). *J.Amer.Chem.Soc.* **115**, *11596-11597*;

2 - Kasselouri, A., Munoz, M., Parrot-Lopez, H. & Coleman, A. W. (1993).*Polish J. Chem.* **67**, 1981-1985.

3 - Alexandre, S., Coleman, A. W., Kasselouri, A. & Valleton J. M. (1996).*Thin Solid Films* **284-285**, 765-768.

4 - Crystal unit cells. **I:** $P2_1$, Z=4, a=15.539(6), b=31.527(6) c=15.556(5) (Å), β=100.94(3)°; **II**: $P2_1$, Z=4, a=15.714(3), b=15.917(6à, c=30.102(4) (Å), β=90.55(2)°; **III:** $P4_32_12$, Z=16, a=b=19.959(3), c= 63.485(14) (Å), data collected with synchrotron radiation, Lure; **IV**: $P2_12_12$, Z=4, a=16.819(4), b= 17.154(11), c= 31.368(6) (Å); **V**: C2, Z=16, a=43.060(8), b=21.936(5), c= 43.466(5) (Å) β= 95.11(1)°; **VI**: $P2_1$, Z=4, a=15.595(11), b=30.409(17), c= 16.637(15) (Å), β= 97.40(6)°; **VII**: $P2_1$, Z=4, a=15.618(17), b=32.28(5), c= 15.637(15)(Å), β= 102.18(9)°.

STUDY OF INCLUSION OF COBALT(II) IN PER-6-O-(TER-BUTYL DIMETYLSILYL) β-CD USING PYRENE AS A FLUORESCENCE PROBE

A. KASSELOURI*, P. PROGNON*, A.W. COLEMAN+ and G. MAHUZIER*
*Lab. Chimie Analytique II, Faculté de Pharmacie, 5 rue J.B. Clement, 92290 Châtenay Malabry, FRANCE
+Institut de Biologie et Chimie des Proteines, 7 passage du Vercors, 693667 Lyon, FRANCE

1. Introduction

Cyclodextrins (CDs) are a family of cyclic oligosaccharides built up of six seven or eight glucopyranose units and known respectively as alpha, beta and gamma cyclodextrins (α-CD, β-CD, γ-CD). The most important property of these molecules is their capacity to include in their hydrophobic cavity a wide variety of organic molecules.[1-2]

The cyclodextrins may be modified by grafting to one of their hydrophilic faces (the primary or secondary hydroxyl groups) hydrophobic moieties. Thus, the obtained amphiphilic cyclodextrins may form micelles, vesicles, and mono-molecular layers and become of interest in view of biological applications (i.e.: vectorisation of drugs).[3-4]

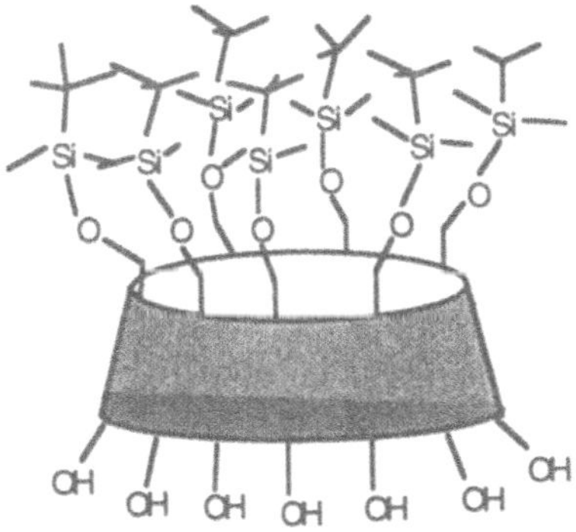

Figure 1: Schematic representation of β-CD(SiR)$_7$

We present here an example model of amphiphilic CD, synthetised from the natural β-CD by the grafting of a hydrophobic group at the primary face the per-6-O-(ter-butyl dimethylsilyl) β-CD (abbreviated: β-CD(SiR)$_7$, Fig. 1)

As cristallographic studies are difficult with such CD derivatives, the goal of this work is to explore if transition metals cations of biological interest may interact with an amphiphilic cyclodextrin. For this purpose, Co^{2+} and per-(ter-butyl dimethylsilyl) β-CD were selected as model, and pyrene is used as a fluorescent probe. Co^{2+} was chosen as a transition metal model of biological importance, due to its well Known quenching

A. W. Coleman (ed.), Molecular Recognition and Inclusion, 391–394.
© 1998 *Kluwer Academic Publishers. Printed in the Netherlands.*

properties towards arene compounds.[5] In order to insure the fiability of the results, two different spectroscopic approaches were used and the obtained results compared.

2. Experimental

Pyrene (P) and N,N-DiMethyl Formamide (DMF) were spectral grade reagents (Aldrich). Cobalt chloride hexahydrate was "pro analysis" reagent purchased from Merck. Pure water was obtained from an Alpha-Q ® Millipore apparatus.

Per-6-O-(ter-butyl dimethylsilyl) β-CD was synthetised from native β-CD. [6]

In all experiments pyrene concentration is 10^{-7}M; in order to allow β-CD$(SiR)_7$ solubility, all experiment were performed in a 90 H_2O/10 DMF mixture. All measurements were performed on a LS-50B luminometer (PERKIN ELMER) at 22±2°C.

2.1 PYRENE EMISSION SPECTRA

The excitation wavelength was λ_{exc}=335nm, the excitation and emission slits were set at 2.5 and 5nm respectively. Each measurement was repeated three times in three different samples (average of 9 values). The intensities of bands I and III of pyrene were mesured at 373.9nm and 383nm respectively.

2.2. POLARIZATION MEASUREMENTS

Excitation wavelength was at λ_{exc}=335nm, excitation and emission slits were set at 5 and 15nm respectively. Each measurement was repeated in three different samples (average of 3 values). Polarization values were automatically recorded by the LS-50B spectrometer as a function of the wavelength. [7] All the polarization values displayed below (part 4.2) are measured at the wavelength of emission of the I band (λ=373.9nm).

3. Methodology

Pyrene was chosen, first, due to the sensitivity of its five vibronic emission bands (I, II, III, IV, V) to the polarity of the micro-environment, especially the ratio I/III; second to its high quantum yield (Φ_F=0,7) which allows measurements at very low concentrations (10^{-7}M). [8, 9] It is demonstrated that the decrease of the I/III ratio in presence of CD is attributed to the penetration of the pyrene into CD cavity, traducing a less polar micro environment for the fluorescence probe. The idea here developed is that the change in emission signal of pyrene in presence of CD and Co^{2+} can contribute to a better understanding of the Co^{2+}/CD interaction.

4. Results and discussion

4.1 EVOLUTION OF THE VIBRATIONAL STRUCTURE OF PYRENE

The decrease of I/III ratio with the addition of cyclodextrin (Fig. 2, curve (1)) indicates the evolution to a less polar micro environment, and so demonstrates the inclusion of the pyrene into the cyclodextrin cavity.[10]

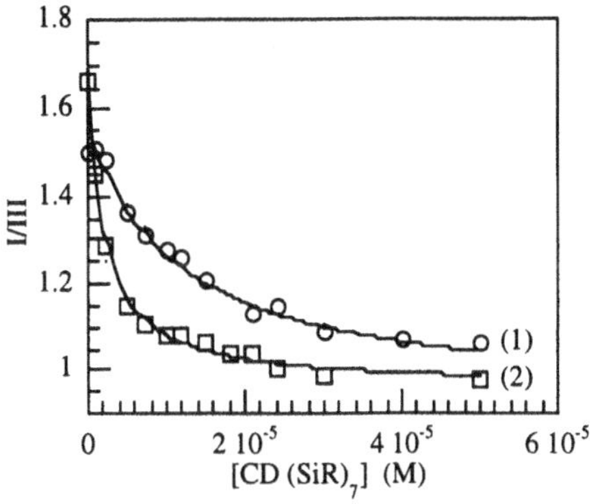

Figure 2 : Plot of I/III intensity ratio versus [CD], (1): water, (2): $[Co^{2+}]$=0.1M

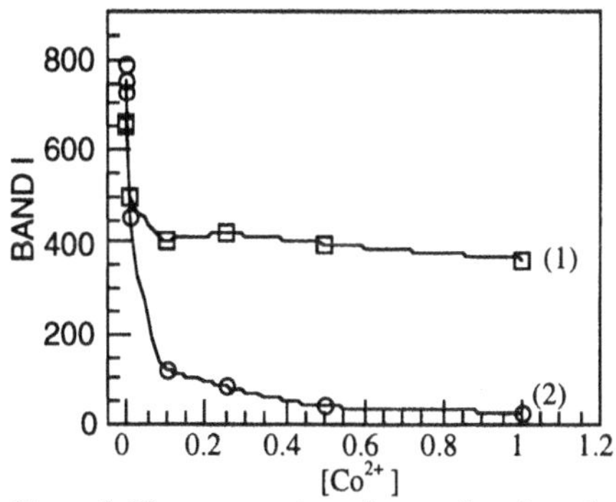

Figure 3: Fluorescence intensity as a function of $[Co^{2+}]$, (1): [CD]=3 10^{-5}M (2): [CD]=0

In the presence of Co^{2+} (0.1M) the I/III ratio decreases more rapidly (Fig. 2, curve (2)): the presence of the cation favorises the pyrene-CD association. At the end of the reaction the included pyrene has a I/III ratio different than in the absence of the cation, so guest micro environment is modified, indicating the presence of Co^{2+} in the vicinity of CD cavity.

Co^{2+} acts as a strong quencher towards the fluorophore: intensity of I band rapidly decreases with Co^{2+} concentration (Fig. 3). The phenomenon is probably due to well-known affinity of 3d metals toward aren the uctures. For $[Co^{2+}]$ >0.1M almost total extinction of the fluorescence intensity occurs (curve (2)), and the phenomenon is attributed to the formation of pyrene/nCo^{2+} non-emitting complexes.

When pyrene is included into CD cavity (Fig. 3, curve (1)), maintains about 60% of its initial fluorescence emission, obviously because it is protected, partially, from Co^{2+} cations. The loss in fluorescence intensity (40%) is due to Co^{2+} - pyrene interactions, probably as a consequence of the introduction of one Co^{2+} cation into CD cavity.

4.2 FLUORESCENCE POLARIZATION

Fluorescence polarization values increase with the inclusion, demonstrating [11] the decrease in the pyrene rotation rate induced by the inclusion (Fig. 4a).

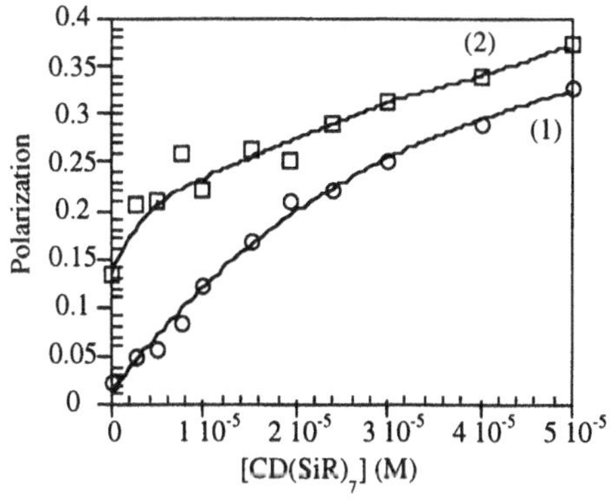

Figure 4a: Polarization as a function of [CD] in (1): pure water, (2): $[Co^{2+}]$=0.1M

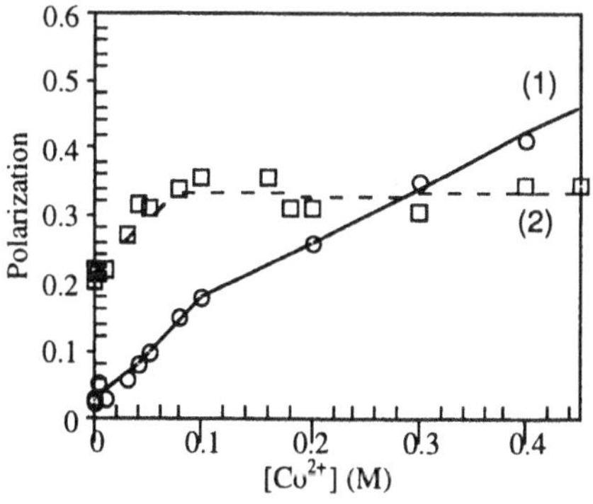

Figure 4b: Polarization as a function of $[Co^{2+}]$ in (1): [CD]=0, (2): [CD]=3 10^{-5}M

Polarization is higher in the presence of Co^{2+} (curve (2)), indicating that included pyrene rotation rate decreases upon Co^{2+} addition, which strengthen the assumption of Co^{2+} introduction in the cavity.

In Fig. 4b the evolution of polarization with $[Co^{2+}]$ shows that:

- polarization of free pyrene increases linearly with the $[Co^{2+}]$ (curve (1)). Two different regions can be distinguished according to their slopes (below and above $[Co^{2+}]$=0.1M), suggesting thus, two different stoichiometries in pyrene - Co^{2+} complex in solution.
- in the case of included pyrene (curve (2)), for $[Co^{2+}]$<0.1M, the polarization is higher than in absence of cyclodextrin, with a similar increasing slope: a similar association between pyrene - Co^{2+} is probable. Above 0.1M, the polarization remains constant, suggesting than no other type of pyrene - Co^{2+} association is possible. These results are in confirmity with the partial recuperation of fluorescence intensity with the CD addition, which suggests partial dissociation of the pyrene/nCo^{2+} complex upon inclusion.

5. Conclusion

We have confirmed the inclusion of pyrene into cyclodextrin cavity with: a) the evolution of I/III ratio, b) the increase in the polarization values.

We have demonstrated that in CD free solution, strong interactions between pyrene and Co^{2+} occur, resulting in fluorescence quenching of pyrene. These interactions are modulated upon inclusion. The polarity of the micro environment of included pyrene varies with the introduction of Co^{2+}, suggesting the presence of Co^{2+} into CD cavity, together with pyrene. This fact is confirmed by a fluorescence quenching study and fluorescence polarization measurements. Such metal - cyclodextrin interactions are of importance with respect to the use of amphiphilic cyclodextrins as drug carriers in biological media.

6. References

1. Szejtli, J.: *Cyclodextrin Technology*, Kluwer Academic Publishers, Dordrecht, 1988.
2. Duchêne D.: *Cyclodextrins and their Industrials Uses*, Editions de Santé, Paris, 1987.
3. Zhang P., Parrot-Lopez H., Tchoreloff P., Baszkin A., Ling C.C., de Rango C., Coleman A.W.: Self-Organizing Systems Based on Amphiphilic Cyclodextrin Diesters, *J. Phys. Org. Chem.*, **5**, (1992), 518.
4. Taneva S., Ariga K., Tagaki W., and Okahata Y.: *J. colloid and Interface Sci.*, **131**(2), (1988), 561.
5. Mortellaro M.A., Hartmann W.K., Nocera D.G.: *Angew. Chem. Int. Ed. Engl.*, **35**(17), (1996), 1945.
6. Zhang P., Ling C.C., Coleman A.W., Parrot-Lopez H., and Galons H.: Formation of Amphiphilic Cyclodextrins via Hydrophobic Esterification at the Secondary Hydroxyl Face, *Tetrahedron Lett.*, **32**(24), (1991), 2769.

7a. Pesce A.J., Rosen C.-G., Pasby T.L.: *Fluorescence Spectroscopy* Chapter 5, Marcel Dekker Inc., New York, 1971.

7b. LS 50B spectrometer, Perkin Elmer Ltd. *User's Manuel*, 1989.

8. Kalyanasundaram K., Thomas J.K., *J. Am. Chem. Soc.*, **99**(7), (1977), 2039.
9. Muñoz de la Peña A., Ndou T., Zung J.B., Warner I.M.: Stoichiometry and Formation Constants of Pyrene Inclusion Complexes with β– and γ-Cyclodextrin, *J. Phys. Chem.*, **95**, (1991), 3330.
10. Eddaoudi M., Coleman A.W., Prognon P., Lopez-Mahia P., Inclusion Capability of Some Alkylsilyl Cyclodextrins in Solution: A Steady State Fluorescence Study, *J.C.S. Perkin II*, (1996), 955.
11. Lakshman S., Mills R., Fang F., Patterson H., Cronan C., Use of Fluorescence Polarization to Probe the Structure and Aluminum Complexation of three Molecular Weight Fractions of a Soil Fulvic Acid, *Anal. Chim. Acta*, **321**, (1996), 113.

β-CYCLODEXTRIN COMPLEXES OF POLYMERS CONTAINING AROMATIC GROUPS

L. LECLERCQ[1], M. BRIA[2], M. MORCELLET[1], and B. MARTEL[1]
[1] : Laboratoire de Chimie Macromoléculaire Bâtiment C6 et [2] : Laboratoire de Chimie Organique Physique, Bâtiment C3, UA CNRS 351, Université des Sciences et Techniques de Lille 1, 59655 Villeneuve d'Ascq Cédex, France.

1. Introduction

Cyclodextrins (Cdx) are cyclic oligosaccharides consisting of six (α-CDx), seven (β-CDx), or eight (γ-CDx) D-(+)-glucopyranose units linked by α-(1,4) bonds. The presence of a central cavity with hydrophobic properties makes these molecules able to form inclusion complexes with organic or inorganic substrates. This fact opened the way to applications in many different fields, such as pharmacology, food science, analytical chemistry, and chemical synthesis and catalysis [1-6]. Cyclodextrins form inclusion complexes with a wide variety of low molecules weight compounds [3]. However, the guest molecules have been limited to small molecules, which can fit in the cavity of cyclodextrins. Inclusion complex formation has been studied by a variety of methods. Among these methods, ^{1}H NMR spectroscopy, UV-visible spectroscopy, and the CD method are considered to be the most popular. But only a few literature about the interactions between β-cyclodextrin and polymers in aqueous solution were reported. It particularly concerns Harada's studies who found that cyclodextrins form a complex with polymers, such as poly(ethylene glycol) [7-8], poly(propylene glycol) [9-10], poly(methyl vinyl ether) [11], or polyisobutylene [12], in a crystalline state. Only one study in aqueous solution with a polyamine was realized [13]. In our laboratory, the covalent and non covalent cyclodextrin-polyvinylamine system was studied [14]. The investigations with polymers in aqueous medium are more complicated to carry out, because problems may arise from limited solubility and high viscosity. In the present study, we have used Poly (4-sodium styrenesulfonate) NaPSS which is a polymer that doesn't take any compact structure in solution. The extended chain allows the aromatic groups to be easily accessible to the CD moieties. The interactions between β-cyclodextrin and NaPSS were investigated by measuring the changes of circular dichroïsm spectra. Then the probable structure of the inclusion complex was studied by considering ^{1}H NMR data. Furthermore, the viscosimetry changes induced by the formation of the inclusion complex were also investigated.

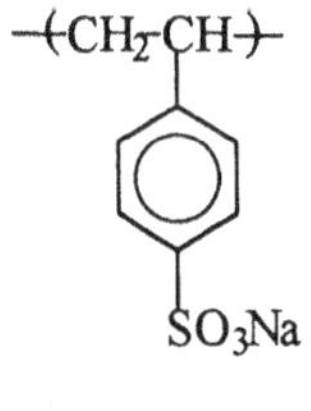

2. Results and Discussion

The internal cavity of the macrocycle provides an assymetric field which perturbs the electronic state of the included guest and induces optical activity [15-16]. The induced CD (circular dichroïsm) spectra have been investigated to elucidate the structure of host-guest interaction in the cyclodextrin complexes, especially in the complexes with optically inactive guests [17-18]. According to the coupled oscillator theory [19], the induced optical activity is interpreted in terms of dipole-dipole interaction between cyclodextrin and guest chromophore. Induced circular dichroïsm has been frequently used to prove or invalidate the inclusion of molecules in

A. W. Coleman (ed.), Molecular Recognition and Inclusion, 395–398.
© 1998 *Kluwer Academic Publishers. Printed in the Netherlands.*

cyclodextrin [18, 20, 21-22], to obtain the orientation of an aromatic molecule included in cyclodextrins [17-18, 21, 23-28], or to determine the direction of a transition moment of a chromophore [29-31]. In the presence of β-cyclodextrin, an induced CD signal for NaPSS appears, as seen in Figure 1. This CD signal was not seen when D-glucose, which is the constituent unit making up β-cyclodextrin, was added in place of β-cyclodextrin. Therefore, the induced CD spectrum can be attributed to the induced optical activity of NaPSS by formation of an inclusion complex with β-cyclodextrin. The CD spectrum shows a negative band at 232 nm and a positive band at 254 nm. The latter transition corresponds to the aromatic chromophore whereas the former one may be attributed to the sulfonate group on the basis of the UV spectrum.

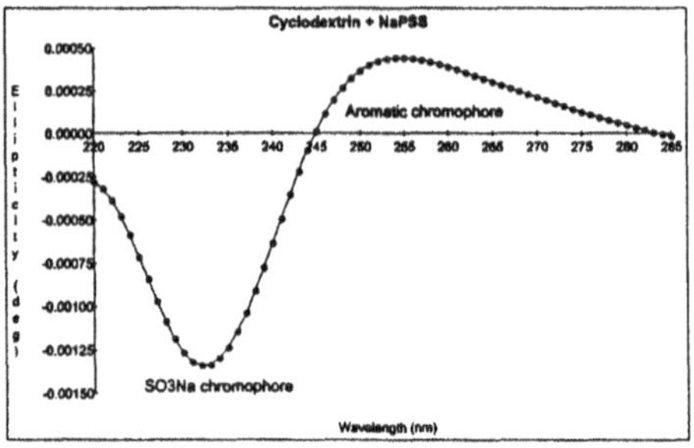

Fig. 1. Induced circular spectrum of NaPSS + β-cyclodextrin.

The ^{1}H NMR spectrum of free β-cyclodextrin in D_2O is shown in Figure 2 ❶. The spectrum of free β-cyclodextrin was initially assigned by Demarco and Thakker [32] and was confirmed in the present work. The free β-cyclodextrin resonance positions at 25°C relative to solvent line (D_2O, 4,79 ppm) are H_1 at 5.047 ppm (doublet), H_2 at 3.626 ppm (doublet of doublets), H_3 at 3.944 ppm (triplet), H_4 at 3.561 ppm (triplet), and H_5 at 3.829 ppm (doublet of triplets), and both H_6 protons nearly overlap at 3.858 ppm. In D_2O solution only resonances from the nonchanging hydrogens attached to the carbons are detected. The resonances for the active H_2, H_3, and H_6 hydroxyl protons were not observed. The effects of NaPSS on the spectrum of β-cyclodextrin were investigated by holding the concentration of β-cyclodextrin constant at 10^{-2} M, and changing the molar ratios of β-cyclodextrin to NaPSS from 0 to 1.65. Parts ❷-❺ of Figure 2 show the effect of NaPSS on the ^{1}H NMR spectrum of β-cyclodextrin. As expected, the lower field triplet assigned to the H_3 protons resonance is progressively shifted to higher field with increasing concentration of NaPSS in the solution. An upfield shift is also observed for the H_5 protons resonance which originally superimposed on the H_6 protons signals. No significant chemical shift or line broadening is observed for H_2, H_4, and H_6. Table 1 shows the different $\Delta\delta$ values. The strong molar ratio dependance of the $\Delta\delta$ values for H_3 and H_5 is the direct evidence for the inclusion complex formation between the cavity of β-cyclodextrin and the phenyl group of NaPSS, since only when the phenyl group on the NaPSS includes into the β-cyclodextrin cavity can the strong anisotropic shielding of the aromatic ring become accessible to the H_3 and H_5 protons. The $\Delta\delta$ value for the H_3 protons which form a ring near the larger opening of the cavity is related to the stability of the inclusion complex [33], while the $\Delta\delta$ value for the H_5 protons which form a ring near the smaller opening of the cavity can be taken as an indicator for the penetration depth of the aromatic group of the NaPSS. The $\Delta\delta$ values for the H_3 and H_5 protons observed in the presence of NaPSS suggest that NaPSS forms a stable inclusion complex with β-cyclodextrin, and the phenyl group of NaPSS penetrates into the cavity deeply enough to let the sulfonate group (hydrophilic group) to be on the exterior (partly at least) of the cavity (hydrophobic) of β-cyclodextrin.

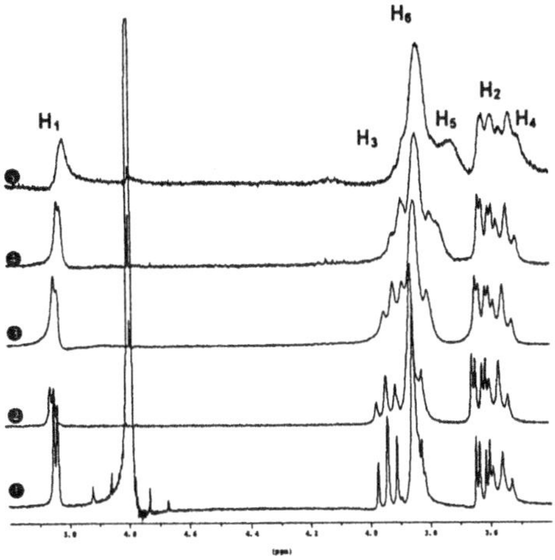

Fig. 2. The upfield region (3.4-5.2 ppm) of the ^1H NMR spectra of 1.0 X 10^{-2} M β-cyclodextrin and increasing concentrations of NaPSS in D_2O. ❶ 0 (β-cyclodextrin alone), ❷ 6.06 X 10^{-3} M, ❸ 1.75 X 10^{-2} M, ❹ 2.86 X 10^{-2} M, ❺ 4.0 X 10^{-2} M NaPSS.

Proton	Δδ (ppm)			
	❷	❸	❹	❺
H_1	0.000	0,004	0,014	0.032
H_2	0.000	0,000	0,012	0,046
H_3	0.005	0,023	0,052	0.101
H_4	0.000	0,003	0,016	0.025
H_5	0.009	0,021	0,049	0,097
H_6	0.002	0,009	0,012	0,015

Table 1. Δδ (δ (β-cyclodextrin alone) - δ (β-cyclodextrin + NaPSS)) values.

Viscosimetry studies were realised in order to describe the conformational changes induced in the polymer NaPSS by the addition of β-cyclodextrin. Experiments were also carried out with D-glucose, which is the constituent unit making up β-cyclodextrin. The inclusion of aromatic groups of NaPSS by β-cyclodextrin is expected to induce an increase in the hydrodynamic size of the polymer, and therefore an increase of the reduced viscosity. Figure 3 shows that the reduced viscosity (η/c) of NaPSS decreases in presence of β-cyclodextrin and D-glucose.

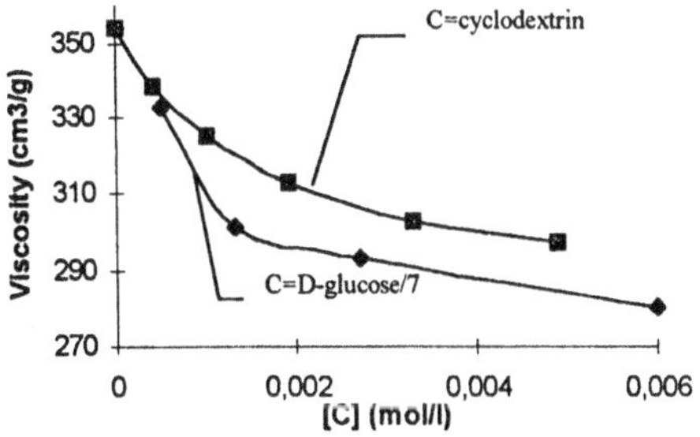

Fig. 3. Reduced viscosity measurements of NaPSS as a function of the concentration of D-glucose/7 and β-cyclodextrin. Polymer concentration : 0.02 g/dL.

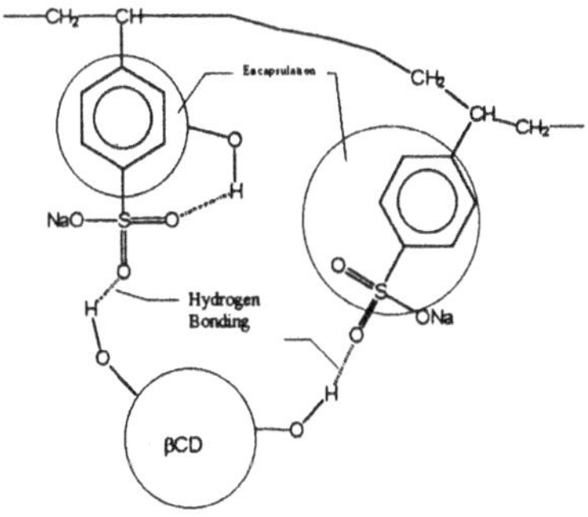

Fig. 4. Nature of NaPSS-β-cyclodextrin interactions (inclusion and hydrogen bonding).

As D-glucose cannot enclose molecules, the only explanation is the formation of a network of intramolecular hydrogen bonding between the sulfonate groups of NaPSS and the hydroxyl groups of D-glucose or β-cyclodextrin. Intramolecular hydrogen bonding induces a decrease in the hydrodynamic size of the polymer, and therefore a decrease in the reduced viscosity.

3. Conclusion

The experimental results show that the inclusion of the aromatic groups of NaPSS in the cavity of β-cyclodextrin is accompanied by the formation of a network of intramolecular hydrogen bonding between the sulfonate groups of NaPSS and the hydroxyl groups of β-cyclodextrin (Figure 4). Judging from investigation by the CPK model, the structure for the inclusion complex is : the whole benzene ring of NaPSS is enclosed in the cavity of β-cyclodextrin, the sufonate group being partly on the exterior of the cavity to form hydrogen bonding.

References

1. M. Bender and M. Komiyama : *Cyclodextrin Chemistry, Springer-Verlag, New York* (1978).
2. W. Saenger : *Angew. Chem., Int. Ed. Engl.* **19**, 344 (1980).
3. J. Szejtli : *Cyclodextrins and Their Inclusion Complexes, Akademiai Kiado, Budapest* (1982).
4. J. Szjetli : *Cyclodextrin Technology, Kluwer Academic Publishers, Dordrecht, Holland* (1988).
5. O. Beken, E. V. Uijtendaal, J. H. Beijnen, A. Bult, and W. J. M. Underberg : *Drug Development and Industrial Pharmacy* **17 (11)**, 1503 (1991).
6. G. Wenz : *Angew. Chem. Int. Ed. Engl.* **33**, 803 (1994).
7. A. Harada and M. Kamachi : *Macromolecules* **23**, 2821 (1990).
8. A. Harada, J. Li, and M. Kamachi : *Macromolecules* **26**, 5698 (1993).
9. A. Harada and M. Kamachi : *J. Chem. Soc., Chem. Commun.* 1321 (1990).
10. A. Harada, M. Okada, J. Li, and M. Kamachi : *Macromolecules* **28**, 8406 (1995).
11. A. Harada, J. Li, and M. Kamachi : *Chem. Lett.* 237 (1993).
12. A. Harada, J. Li, S. Suzuki, and M. Kamachi : *Macromolecules* **26**, 5267 (1993).
13. G. Wenz and B. Keller : *Angew. Chem., Int. Ed. Engl.* **31**, 197 (1992).
14. B. Martel, Y. Leckchiri, A. Pollet, and M. Morcellet : *Eur. Polym. J.* **31**, 1083 (1995).
15. K. Sensse and F. Cramer : *Chem. Ber.* **102**, 509 (1969).
16. M. Otagiri, K. Ikeda, K. Uekama, O. Ito, and M. Hatano : *Chem. Lett.* 679 (1974).
17. H. Shimizu, A. Kaito, and M. Hatano : *Bull. Chem. Soc. Jpn.* **52**, 2678 (1979).
18. K. Harata and H. Uedaira : *Bull. Chem. Soc. Jpn.* **48**, 375 (1975).
19. I. Tinoco : *Adv. Chem. Phys.* **4**, 113 (1962).
20. L. Flamigni : *J. Phys. Chem.* **97**, 9566 (1993).
21. H. Shimizu, A. Kaito, and M. Hatano : *Bull. Chem. Soc. Jpn.* **54**, 513 (1981).
22. G. Wulff and S. Kubik : *Makromol. Chem.* **193**, 1071 (1992).
23. H. Shimizu, A. Kaito, and M. Hatano : *J. Am. Chem. Soc.* **104**, 7059 (1982).
24. N. Kobayashi, R. Saito, H. Hino, Y. Hino, A. Ueno, and T. Osa: *J. Chem.Soc.,Perkin Trans. 2* 1031 (1983).
25. A. Ueno, F. Moriwaki, T. Osa, F. Hamada, and K. Murai : *J. Am. Chem. Soc.* **110**, 4323 (1988).
26. A. Ueno, I. Suzuki, and T. Osa : *J. Am. Chem. Soc.* **111**, 6391 (1989).
27. N. Kobayashi : *J. Chem. Soc., Chem. Commun.* 918 (1988).
28. M. Komiyama and Y. Takeshige : *J. Org. Chem.* **54**, 4936 (1989).
29. H. Yamaguchi, M. Higashi, and M. Oda : *Spectrochim. Acta* **44A**, 547 (1988).
30. N. Kobayashi, S. Minato, and T. Osa : *Makromol. Chem.* **184**, 2123 (1983).
31. M. Kodaka : *J. Phys. Chem.* **95**, 2110 (1991).
32. P.V. Demarco and A.L. Thakker : *Chem. Commun.* 2 (1970).
33. D. J. Wood, F. E. Hruska, and W. Saenger : *J. Am. Chem. Soc.* **99**, 1735 (1977).

INCLUSION OF NEUTRAL GUESTS IN A SELF-ASSEMBLING SUPERSTRUCTURE

SANG BOK LEE and JONG-IN HONG*
Department of Chemistry, Seoul National University, Seoul 151-742, Korea

Abstract: Self-assembled heterodimeric molecular capsule based on the charged hydrogen bonding interaction between two different CTV monomers (**1** and **2**) was developed. The formation of the molecular capsule was confirmed by inclusion phenomena of complementary neutral guests in DMSO-d_6. The encapsulated TMS was released by prolonged heating or pH adjustment. Molecular modeling and intermolecular NOEs suggested the three-dimensional structure of the termolecular inclusion complex.

Supramolecular structures such as enzyme complexes, ribosomes, protein filaments, viruses, and membranes are not made as single, giant, covalently linked molecules; instead they are formed by the noncovalent assembly of preformed macromolecular subunits.[1] Some outstanding examples of biological self-assembling molecular structures are the rodlike and fil the ntous viruses, in which the helical aggregation of the protein-coat subunits forms a cylindrical container for the virus's nucleic acid.[2] Recently, noncovalent self-assembly process used in the biological organization has been increasingly exploited in the design of synthetic receptors.[3] Herein, we describe a molecular assembly technique for constructing three-dimensional molecular capsule[4] based on the charged hydrogen bonding interaction (electrostatic interaction) between two different cyclotriveratrylene (CTV)[5] derivatives.

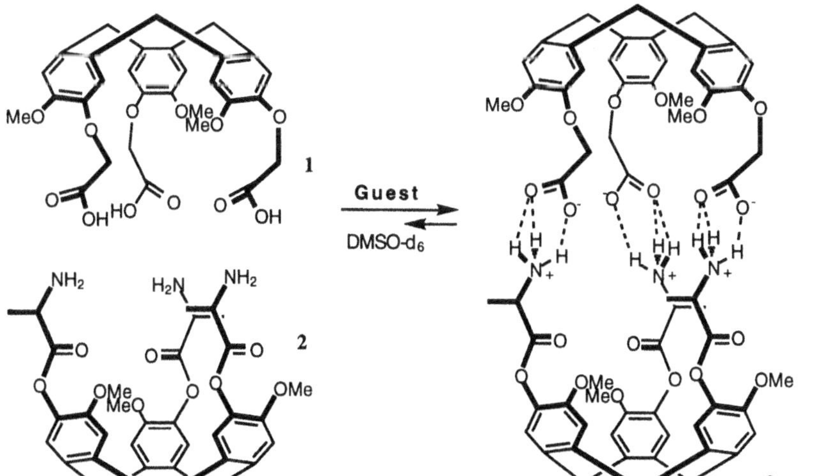

3

(The guest is included in the cavity)

A. W. Coleman (ed.), Molecular Recognition and Inclusion, 399–402.
© 1998 *Kluwer Academic Publishers. Printed in the Netherlands.*

Since bowl-shaped C_3-symmetric CTV monomers[6] can have functional groups on the upper rim that would recognize each other, CTV heterodimer assembled noncovalently from two different CTV monomers should result in a molecular capsule which can encapsulate suitable molecules of complementary size and shape.
As shown in the Figure, we designed a self-assembled molecular capsule based on the charged hydrogen bonding interactions between carboxylates on one CTV monomer (**1**) and ammoniums on the complementary CTV monomer (**2**).[7]

Each CTV trisacid (**1**) and trisamine (**2**) is soluble in acetone. However, the mixture of **1** and **2** is not soluble in almost all the organic solvent except DMSO. The NMR spectrum of 1:1 mixture of **1** and **2** at less than 1 mM concentration in DMSO-d_6 shows the simple combination of each unaffected spectrum of **1** and **2**, indicating formation of the potential dimeric structure.[8] In fact, the formation of a self-assembled molecular capsule was confirmed by inclusion phenomena of neutral guests at less than 1mM concentration in DMSO-d_6. Exchange of all guests in and out of the cavity is very slow at rt on the NMR time scale. The inclusion phenomena were elucidated by upfield shifts of included guest signal (see Table 1) because of its location in the shielding region of the aromatic cavity. To **1** dissolved in acetone was added 1% TMS in acetone and then **2** in acetone. Then solid immediately formed out of acetone. After removing acetone in *vacuo*, the NMR spectrum was taken in deuterated DMSO. In comparison with 1:0.4 and 1:0.6 mixture of **1** and **2**, 1:1 mixture is capable of capturing more TMS inside the cavity of the capsule, indicating the capsule formation by self-assembly.

Computer-generated structure of the vacant CTV heterodimeric assembly is compared with that of the inclusion complex of CTV heterodimer and TMS. Upon inclusion of TMS, the originally distorted unsymmetrical heterodimeric structure is reorganized around TMS to make maximum van der Waals contacts with the encapsulated TMS and form a spherical capsule.[9]

Since the hydrogen bonding interaction between carboxylate and ammonium is expected to be very strong, the rigidity of the CTV heterodimeric assembly was tested by release of the encapsulated TMS upon heating at 60 °C in the sealed nmr tube. With prolonged heating, the amount of released TMS increases. However, even after 2 weeks, less than 50% of TMS is released showing the rigidity of the capsule.[10]

Addition of trifluoroacetic acid to CTV heterodimer•TMS complex regenerated **1**, **2**, and free TMS. This demonstrates that formation of the complex is not completely irreversible and rather it can be switched by pH adjustment.[4c]

In order to elucidate three-dimensional structure of the termolecular complex, we performed intermolecular NOE study of CTV heterodimer and TMS complex. Irradiation of the methyl protons of the encapsulated TMS caused sizable increases in signal strengh in 5 different kinds of protons of CTV heterodimer (Ar*H* of **1**, 5%; Ar*H* of **2**, 8%; Ar-OCH_2-COOH of **1**, 10%; Ar-OCH_3 of **1** and **2**, 15%; alanine CH_3 of **2**, 16%). In fact, strong NOEs between TMS methyl protons and methoxy protons and methyl protons in alanine side chain indicates that both methoxy groups and methyl side chains in CTV heterodimer are pointing toward the cavity and therefore makes a good hydrophobic environment for encapsulating the neutral guest.

Table 1 shows relative stabilities (K_{rel}) of the termolecular complexes between CTV heterodimer and guests. The inclusion constant (K_{inc}) was determined by integrating the bound guest signal in the NMR spectra of 1:1 mixture of **1**, **2**, and the same amount of guest in DMSO-d_6. The inclusion constant (K_{inc}) is roughly correlated with guest volume except bromoform, indicating that van der Waals interaction is operative in the guest inclusion. However, since the molecular modeling indicates that the cavity can accomodate all the guest used,[11] the difference in the relative stabilities seems to depend not only on the van der Waals contacts with the concave inner surface, but also presumably on the solvation effect.[12]

Table 1. Inclusion of guest molecules in the cavity of the CTV heterodimer

Guest	K_{inc} (K_{rel})[a]	Δδ (ppm)[b]	Guest vol.(Å^3)[c]
tetramethylsilane	40 (200)	0.07	107
t-butyl chloride	6 (30)	0.21	96
1,1,1-trichloroethane	5 (25)	0.44	89
chloroform	3 (15)	3.20	69
bromoform	0.2 (1)	1.95	93

a.Inclusion constant (K_{inc}, M^{-1}) at 293 K. K_{inc} = [**1**•G•**2**]/[**1**•**2**][G], where **1**•**2** is CTV heterodimer and **1**•G•**2** is **3**•guest termolecular complex. b.Upfield shift of the included guest signal. c. The guest volume was calculated according to ref.13.

In conclusion, we developed a self-assembled heterodimeric molecular capsule based on the charged hydrogen bonding interaction between two different CTV monomers (**1** and **2**). The formation of the molecular capsule was confirmed by almost irreversible inclusion phenomena of complementary neutral organic molecules in DMSO-d_6, as evidenced by NMR spectroscopy. Release of the encapsulated TMS can be controlled by prolonged heating or pH adjustment. Computational molecular modeling and intermolecular NOEs suggested the three-dimensional structure of the termolecular inclusion complex. Since the charged hydı the n bonding interaction is so strong even in DMSO which is a strong hydrogen bonding acceptor and strongly precludes the formation of hydrogen bond based complex, we plan to attach water-soluble group to each CTV to achieve reversible encapsulation of suitable guest molecules of complementary shape and size in hydroxylic solvents such as methanol or water-containing solvent.

Acknowledgement. We gratefully acknowledge support of this work by the Basic Science Research Institute Program (BSRI-95-3416).

REFERENCES AND NOTES

1. For an excellent introduction on self assembling biological systems, see: (a) Lehninger, A. L. (1976) *Biochemistry*, Worth Publishers, Inc., New York, Chapter 36. (b) Alberts, B., Bray, D., Lewis, J., Raff, M., Roberts, K., and Watson, J. D.(1989) *Molecular Biology of the Cell*, Garland Publishing, New York.
2. (a) Namba, K.and Stubbs, G. (1986) Structure of Tobacco Mosaic Virus at 3.6 Å Resolution: Implications for Assembly, *Science*, 231, 1401-1406. (b) Klug, A. (1983) From Macromolecules to Biological Assemblies (Nobel Lecture), *Angew. Chem. Int. Ed. Engl.* **22**, 565-582.
3. For recent reviews, see: (a) Lehn, J.-M. (1995) *Supramolecular Chemistry*, VCH, Weinheim. (b) Lindsey, J. S. (1991) Self-Assembly in Synthetic Routes to Molecular Devices Biological Principles and Chemical Perspectives: A Review, *New J. Chem.* **15**, 153-180. (c) Lawrence, D. S., Jiang, T., and Levett, M. (1995) Self-Assembling Supramolecular Complexes, *Chem. Rev.* **95**, 2229-2260. (d) Philp, D. and Stoddart, J. F. (1996) Self-Assembly in Natural and Unnatural Systems, *Angew. Chem. Int. Ed. Engl.* **35**, 1154-1196.
4. (a) Grotzfeld, R. M., Branda, N., and Rebek, J., Jr. (1996) Reversible Encapsulation of Disc-Shaped Guests by a Synthetic, Self-Assembled Host, *Science*, **271**, 487-489. (b) Meissner, R. S., Rebek, J., Jr., and de Mendoza, J. (1995) Autoencapsulation Through Intermolecular Forces: A Synthetic Self-Assembling Spherical Complex, *Science*, **270**, 1485-1488. (c) Valdes, C., Spitz, U. P., Kubik, S. W., and Rebek, J., Jr. (1995) Pseudo-Spherical Host Molecules: Synthesis, Dimerization, and Nucleation Effects, *Angew. Chem. Int. Ed. Engl.* **34**, 1885-1887. (d) Valdes, C., Spitz, U. P., Toledo, L. M., Kubik, S. W., and Rebek, J., Jr. (1995) Synthesis and Self-Assembly of Pseudo-Spherical Homo- and Heterodimeric Capsules, *J. Am. Chem. Soc.* **117**, 12733 12745. (e) Branda, N., Grotzfeld, R. M., Valdes, C., and Rebek, J., Jr. (1995) Control of Self-Assembly and Reversible Encapsulation of Xenon in a Self-Assembling Dimer by Acid-Base

Chemistry, *J. Am. Chem. Soc.* **117**, 85-88. (f) Branda, N., Wyler, R., and Rebek, J., Jr. (1994) Encapsulation of Methane and Other Small Molecules in a Self-Assembling Superstructure, *Science*, **263**, 1267-1268. (g) Fujita, M., Ogura, D., Miyazawa, M., Oka, H., Yamaguchi, K., and Ogura, K. (1995) Self-Assembly of ten molecules into nanometre-sized organic host frameworks, *Nature*, **378**, 469-471. (h) Koh, K., Araki, K., and Shinkai, S. (1994) Self-Assembled Molecular Capsule Based on the Hydrogen-Bonding Interaction between Two Different Calix[4]arenes, *Tetrahedron Lett.* **35**, 8255-8258.
5. (a) Collet, A. (1987) Cyclotriveratrylenes and Cryptophanes, *Tetrahedron*, **43**, 5725-5759. (b) Collet, A., Dutasta, J.-P., Lozach, B., and Canceill, J. (1993) Cyclotriveratrylenes and Cryptophanes, in E. Weber (ed.), *Topics in Current Chemistry*, Springer-Verlag, Berlin, Vol. 165, pp. 103-129.
6. Spectral data for **1**: mp 167 ~ 170 °C (dec.); ^{1}H NMR (500 MHz, DMSO-d_6) δ 6.99 (s, 3 H, Ar*H*), 6.97 (s, 3 H, Ar*H*), 4.66 (d, *J* = 13.5 Hz, 3 H, Ar-$CH_{in}H_{out}$-Ar), 4.61 (s, 6 H, Ar-OCH_2-), 3.72 (s, 9 H, Ar-OCH_3), 3.46 (d, *J* = 13.5 Hz, 3 H, Ar-$CH_{in}H_{out}$-Ar); ^{13}C NMR (125.7 MHz, DMSO-d_6) δ 170.3, 147.4, 145.6, 132.7, 131.5, 115.0, 113.8, 65.2, 55.7, 35.0.; IR (KBr) 3340, 2960, 2912, 1734, 1603, 1507, 1328, 1257, 1139, 1083, 998, 934, 889 cm^{-1}.; MS (FAB) m/z 582 (M^+). **2**: mp 155 ~ 160 °C (dec.); ^{1}H NMR (500 MHz, DMSO-d_6) δ 6.83 (s, 3 H, Ar*H*), 6.81 (s, 3 H, Ar*H*), 4.57 (d, *J* = 13.5 Hz, 3 H, Ar-$CH_{in}H_{out}$-Ar), 3.88 (q, *J* = 7.5 Hz, 3 H, *C*HCH_3), 3.71 (s, 9 H, Ar-OCH_3), 3.35 (d, *J* = 13.5 Hz, 3 H, Ar-$CH_{in}H_{out}$-Ar), 1.25 (d, *J* = 7.0 Hz, 9 H, CHCH_3); ^{13}C NMR (50.3 MHz,$CDCl_3$) δ 174.9, 149.5, 138.3, 137.8, 131.3, 123.6, 114.0, 56.2, 50.1, 36.4, 20.6.; MS (FAB) m/z 622 (M^++1).
7. Since **1** and **2** exist in enantiomer pairs, there are two ways of self-assembly. The first one in Figure 1 is derived from the same CTV's. The second way of assembly is derived from two different enantiomeric CTV's.
8. NMR spectrum of 1:1 mixture of **1** and **2** at 7 mM concentration in DMSO-d_6 indicates formation of the oligomeric structure, indicating that formation of the dimeric or oligomeric structure from 1:1 mixture of **1** and **2** is concentration-dependent.
9. The energy-minimized structures were obtained with DISCOVER 95.0 of MSI on a Silicon Graphics INDY workstation.
10. In comparison with release rate, inclusion rate of outide TMS into the cavity of the preformed capsule upon heating also turned out to be very slow.
11. The cavity volume of the TMS-occupied CTV heterodimer (calculated from the average radius of the pseudo-spherical cavity, considering the van der Waal's half-thickness of the aromatic rings (1.77 Å)), which is the largest among those examined, is ~150 Å^3 which is ~40% larger than TMS.
12. Data of the solvation energy for guest dissolved in DMSO-d_6 are not available.
13.Bondi, A.(1964) van der Waals Volumes and Radii, *J. Phys. Chem.* 68, 441-451.

RECOGNITION AND TRANSPORT OF NUCLEOSIDE MONOPHOSPHATES WITH SYNTHETIC RECEPTORS

SANG BOK LEE, YONG-GYU JUNG, WOON-SEOK YEO, and JONG-IN HONG*
Department of Chemistry, Seoul National University, Seoul 151-742, Korea

Nucleotide analogues which have been known to have potential antiviral activity *in vitro* cannot penetrate across lipophilic cell membranes due to their highly charged and hydrophilic nature.[1] Therefore, there has been increasing effort for developing phosphate-binding receptors.[2] However, there are a few artificial carrier systems known capable of transporting phosphate-bearing nucleotides through organic liquid membrane.[3] Simple lipophilic ammonium ions bearing long hydrocarbon chains allow phosphate anion extraction into an organic phase and render liquid membranes permeable to the anion by an anion exchange process.[3a-c, f, g] However, structurally similar, lipophilic phosphonium ions have seldom been adopted as anion carriers.[4] In this paper, we report extraction and transport of organic anions, especially 5'-AMP (AMP) and 5'-GMP (GMP) by lipophilic phosphonium based carriers.[5]

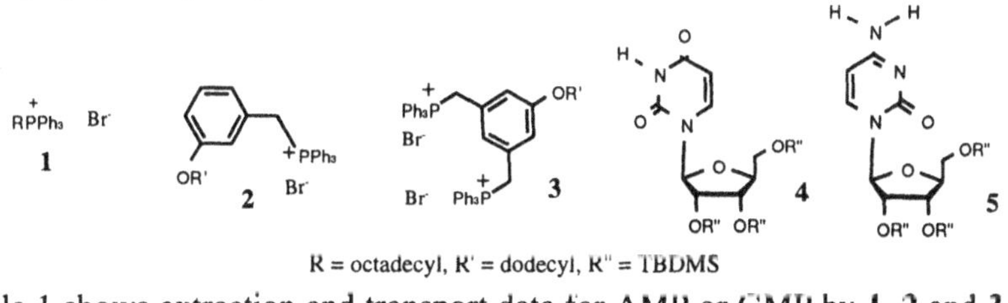

R = octadecyl, R' = dodecyl, R" = TBDMS

Table 1 shows extraction and transport data for AMP or GMP by **1**, **2** and **3** with co-carrier **4** or **5**. There exist qualitative correlations between the extraction and the transport data. As shown in Table 1, more hydrophilic GMP is extracted and transported slightly less efficiently than AMP. AMP (or GMP) is dissociated mostly into monoanion (AMP^- or GMP^-) at pH 5, while only one third of AMP (or GMP) exists as AMP^- or GMP^- at pH 7. A concentration-dependent extraction study provided valuable information on the mode of interaction between the monophosphonium carrier **2** (or diphosphonium carrier **3**) and nucleotide monophosphates. When aqueous solutions containing varying amounts of AMP (or GMP) were shaken with a chloroform solution of **2** (1.0×10^{-5} M), the maximum extractability of AMP (or GMP) was 97% (or 92%) at pH 5 and 65% (or 55%) at pH 7 respectively. This result supports that **2** undergoes formation of a 1:1 carrier **2**-nucleotide monophosphate complex (**6**) at pH 5 and a 2:1 carrier **2**-nucleotide monophosphate complex (**7**) at pH 7 in the organic phase. (Fig. 1) Similar concentration-dependent extraction studies in the case of the diphosphonium carrier **3** supports the formation of a 1:2 carrier **3**-nucleotide monophosphate complex

A. W. Coleman (ed.), Molecular Recognition and Inclusion, 403–406.
© 1998 *Kluwer Academic Publishers. Printed in the Netherlands.*

(**8**) at pH 5 and a 1:1 carrier **3**-nucleotide monophosphate complex (**9**) at pH 7 in the organic phase.[7] (Fig. 1) Since formation of the bimolecular complex (**6**) between **2** and monobasic form of AMP (or GMP) is entropically more favorable than that of the termolecular complex (**7**) between two **2**'s and dibasic form of AMP (or GMP), **2** transports either 5'-AMP or 5'-GMP more efficiently at pH 5 than at pH 7. However, the fact that the nucleotide transport at pH 5 by **3** doesn't differ from that at pH 7 shows that though the result is hard to rationalize, the termolecular complex **8** formed at pH 5 is energetically similar to the bimolecular complex **9** formed at pH 7 in the transport process. **3** extracts and transports more efficiently than **2** presumably because of higher effective molarity of the phosphonium ion in **3**. Especially, the transport rate for AMP or GMP by **3** at pH 7 is better than that by **2** at the same pH for the entropic reason: formation of the bimolecular complex **9** is entropically more favorable than that of the termolecular complex **7**. When comparison is made with transport rates obtained by **1**, **2**, and **3** at the same pH, **1** shows better extraction ability and transport rate for either AMP or GMP than **2** or **3**.

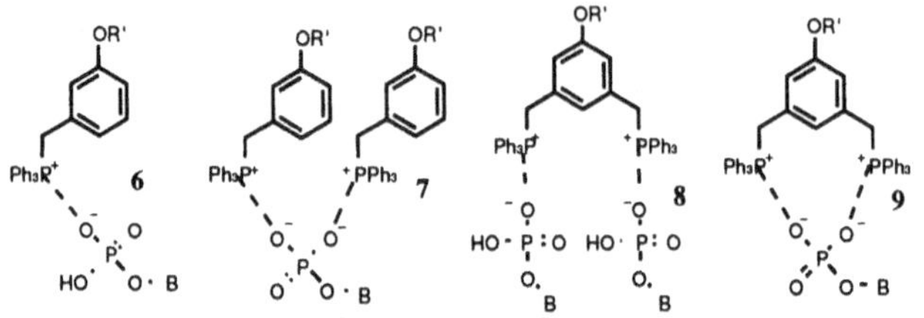

B = Adenosine or Guanosine

Figure 1. Proposed structures for the transport complexes determined by the concentration-dependent extraction study.

TABLE 1. Extraction and Transport of 5'-AMP or 5'-GMP by **1, 2** and **3** with co-carrier **4** or **5**.[a]

Carrier	pH	Extraction (%)[b]		Transport rate (10^{-10}mol/h•cm^2)[c]	
		5'-AMP[d]	5'-GMP[e]	5'-AMP[d]	5'-GMP[e]
2 + **4** (or **5**)[f]	5.0	15 (11)	14 (7)	1.2 (0.80)	1.2 (0.72)
	7.0	6 (5)	7 (4)	0.6 (0.40)	0.6 (0.40)
3 + **4** (or **5**)[f]	5.0	17 (13)	17 (10)	1.3 (1.0)	1.4 (0.96)
	7.0	20 (12)	15 (8)	1.7 (1.2)	1.4 (0.92)
1 + **4** (or **5**)[f]	5.0	42 (30)	35 (22)	3.7 (3.0)	5.6 (2.0)
	7.0	25 (19)	17 (12)	1.5 (1.2)	2.8 (1.4)

a. Values in parentheses are those in the absence of co-carrier. b. **1**, **2**, or **3** 0.010 M/$CHCl_3$ 5 ml, 5'-AMP (5'-GMP) 1.0 x 10^{-4} M/H_2O, pH 5.0 and 7.0. c. Source phase: 5'-AMP (5'-GMP) 1.0 x 10^{-1} M/H_2O (the specified pH values were adjusted by addition of a small amount of HCl or NaOH solution). Receiving phase: NaBr 2.5 x 10^{-2} M/H_2O 4 ml (pH 5.0 and 7.0). Organic phase: **1**, **2**, or **3** 2.5 x 10^{-3} M/$CHCl_3$ 8 ml. d. Co-carrier **4** was used for transport of 5'-AMP. e. Co-carrier **5** was used for transport of 5'-GMP. f. Concentration of the co-carrier (**4** or **5**) is the same with the carrier's (**1**, **2**, or **3**).

Our system is expected to mediate more efficient through-membrane transport of AMP or GMP with the aid of base-pairing in either inter- or intramolecular fashion.[3e, h, i, 8] In order to improve through-membrane transport of nucleotide monophosphates by a phosphonium cation, an organic-soluble nucleobase was used as a co-carrier.[8] Phosphonium cations (**1**, **2**, and **3**) and lipophilic nucleosides (**4** and **5**) function as a cooperative binary carrier system [8, 9] in which the nucleoside undergoes novel hydrogen bonding with the nucleobase portion of AMP or GMP. (Fig. 2) In fact, either at pH 5 or pH 7, in the presence of lipophilic uridine (**4**) or cytidine (**5**), moderate rate enhancements were observed. **4** or **5** was used as a co-carrier in the case of AMP or

GMP transport respectively. In order to make more effective and selective carriers for AMP or GMP compared to the joint co-carrier system, it would require constructing phosphonium-nucleobase conjugates, in which an adenine or guanine recognition unit is appended directly onto the phosphate-binding phosphonium center. In fact, the thymine(or cytosine)- bearing phosphonium carrier **12** (or **15**) was found to be much more effective for AMP (or GMP) transport either at pH 5 or pH 7. As shown in Table 2, receptor **12** displayed a higher transport rate for AMP (by a factor of 39 at pH 5 and

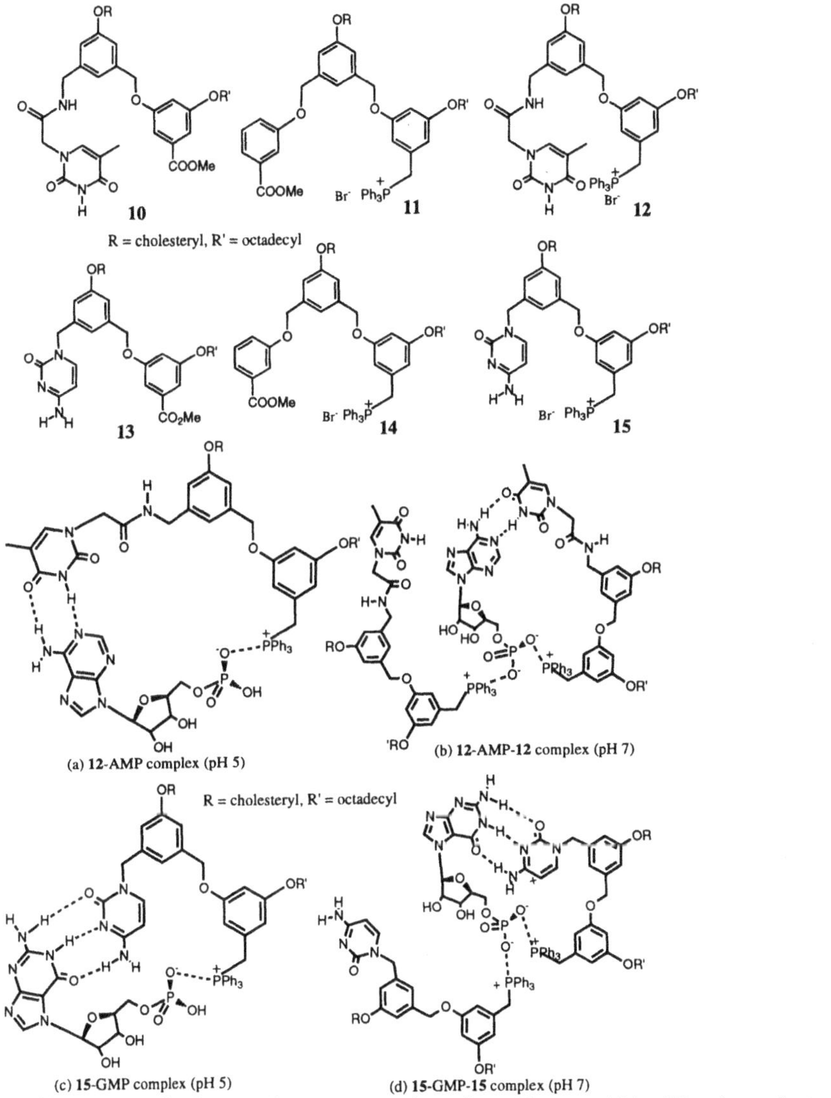

Figure 2. Possible structures for the neutral transport complexes formed between **12** (or **15**) and monobasic (or dibasic) nucleobase (AMP or GMP).

157 at pH 7) than the thymine-free phosphonium salt **11**. In the case GMP transport, the cytosine-substituted phosphonium receptor **15** showed similar rate enhancement (by a factor of 66 at pH 5 and 103 at pH 7) in comparison with the cytosine-free

phosphonium salt **14**. However, a phosphonium-free nucleobase **10** or **13** didn't transport AMP or GMP at all indicating importance of the phosphate group solubilization in $CHCl_3$ for transport of nucleotide monophosphates. Concentration-dependent extaction study supports possible structures for the neutral supramolecular transport complexes as shown in Fig. 2.[10]

TABLE 2. Extraction and transport of AMP and GMP by Synthetic Carriers **10** ~ **15**.

		Extraction (%)[a]		Transport rate (10^{-9} mol/h·cm^2)[b]	
Carrier	Guest	pH 5.0	pH 7.0	pH 5.0	pH 7.0
10		c	c	c	c
11	5'-AMP	7.4	6.3	0.14	0.03
12		9	9	5.4	4.7
13		6	2	c	c
14	5'-GMP	13	8	0.08	0.04
15		33	33	5.3	4.1

a. [Carrier] = 1x 10^{-2} M in $CHCl_3$, [Guest] = 1 x 10^{-4} M in deionized water. b. Source phase: 5'-AMP (or 5'-GMP) 0.05 M in 4 ml of H_2O. Receiving phase: NaBr 0.025 M in 4 ml of H_2O. Organic phase: Carrier 0.001 M in 8 ml of $CHCl_3$. c. Not detected.

In summary, we showed that the transport of a normally organic-insoluble AMP or GMP can be effected by using appropriately designed lipophilic nucleoside-phosphonium joint co-carrier system or nucleobase-substituted phosphonium carriers.

Acknowledgements: We thank Ministry of Education (BSRI-95-3416) and OCRC for financial support.

REFERENCES AND NOTES

1. (a) Robins, R. K.; *Chemical Engineering News*, Jan. 27, 1986, pp 28-40. (b) Harden, M. R., Ed.; *Approaches to Antiviral Agents*; VCH Publishers: Florida, U.S.A., 1985. (c) Martin, J. C., Ed.; *Nucleotide Analogues as Antiviral Agents*; American Chemical Society: Washington, D.C., 1989.

2. (a) Rudkevich, D. M.; Brzozka, Z.; Palys, M.; Visser, H. C.; Verboom, W.; Reinhoudt, D. N. *Angew. Chem. Int. Ed. Engl.* **1994**, *33*, 467. (b) Hosseini, M. W.; Blacker, A. J.; Lehn, J.-M. *J. Am. Chem. Soc.* **1990**, *112*, 3896. (c) Jubian, V.; Dixon, R. P.; Hamilton, A. D. *J. Am. Chem. Soc.* **1992**, *114*, 1120. (d) Kneeland, D. M.; Ariga, K.; Lynch, V. M.; Huang, C.; Anslyn, E. *J. Am. Chem. Soc.* **1993**, *115*, 10042. (e) Tabushi, I.; Imuta, J.-i.; Seko, N.; Kobuke, Y. *J. Am. Chem. Soc.* **1978**, *100*, 6287. (f) Galan, A.; de Mendoza, J.; Toiron, C.; Bruix, M.; Deslongchamps, G.; Rebek, J., Jr. *J. Am. Chem. Soc.* **1991**, *113*, 9424. (g) Deslongchamps, G.; Rebek, J., Jr.; Galan, A.; de Mendoza, *J. Angew. Chem. Int. Ed. Engl.* **1994**, *33*, 467. (h) Iverson, B. L.; Shreder, K.; Kral, V.; Sessler, J. L. *J. Am. Chem. Soc.* **1993**, *115*, 11022. (i) Schmidtchen, F. P. *Tetrahedron Lett.* **1989**, *30*, 4493. (j) Nishizawa, S.; Buhlmann, P; Iwao, M.; Umezawa, Y. *Tetrahedron Lett*, **1995**, *36*, 6483. (k) Galan, A.; Pueyo, E.; Salmeron, A.; de Mendoza, J. *Tetrahedron Lett*, **1991**, *32*, 1827.

3 (a) Tabushi, I.; Kobuke, Y.; Imuta, J.-i. *J. Am. Chem. Soc.* **1980**, *102*, 1744. (b) Tabushi, I.; Kobuke, Y.; Imuta, J.-i. *J. Am. Chem. Soc.* **1981**, *103*, 6152. (c) Maruyama, K.; Tsukube, H.; Araki, T. *J. Am. Chem. Soc.* **1982**, *104*, 3197. (d) Furuta, H.; Cyr. M. J.; Sessler, J. L. *J. Am. Chem. Soc.* **1991**, *113*, 6677. (e) Kral, V.; Sessler, J. L.; Furuta, H. *J. Am. Chem. Soc.* **1992**, *114*, 8704. (f) Diederich, F.; Li, T. *J. Org. Chem.* **1992**, *57*, 3449. (g) Li, T.; Krasne, S. J.; Persson, B.; Kaback, H. R.; Diederich, F. *J. Org. Chem.* **1993**, *58*, 380. (h) Sessler, J. L.; Furuta, H.; Kral, V. *Supramol. Chem.* **1993**, *1*, 209, and references cited therein. (i) Kral, V.; Sessler, J. L. *Tetrahedron*, **1995**, *51*, 539. (j) Andreu, C.; Galan, A.; Kobiro, K.; de Mendoza, J.; Park, T. K.; Rebek, J.; Salmeron, A.; Usman, N. *J. Am. Chem. Soc.* **1994**, *116*, 5501. (k) Yasuhisa, K.; Hatekayama, H.; Seshimo, H.; Ogoshi, H. *Supramol. Chem.* **1994**, *3*, 267.

4. Ion-selective electrodes based on the phosphonium salt: (a) Hadjiioannou, T. P.; Gritzapis, P. C. *Anal. Chim. Acta.* **1981**, *126*, 51. (b) Geissler, M.; Kunze, R. *Anal. Chim. Acta.* **1986**, *189*, 245.

5. All new compounds gave satisfactory ^{1}H NMR, ^{13}C NMR, and HRMS (FAB) analyses.

6. (a) Behr, J.-P.; Lehn, J.-M. *J. Am. Chem. Soc.* **1973**, *95*, 6108. (b) Tsukube, H. *J. Chem. Soc. Perkin Trans.* I, **1983**, 29.

7. When aqueous solutions containing varying amounts of AMP (or GMP) were shaken with a chloroform solution of **3** (1.0 x 10^{-5} M), the maximum extractability of AMP (or GMP) was 194% (or 188%) at pH 5 and 112% (or 95%) at pH 7 respectively.

8. (a) Furuta, H.; Morishima, T.; Kral, V.; Sessler, J. L. *Supramol. Chem.* **1993**, *3*, 5. (b) Riggs, J. A.; Hossler, K. A.; Smith, B. D.; Karpa, M. J.; Griffin, G.; Duggan, P. J. *Tetrahedron Lett*, **1996**, *37*, 6303.

9. Reetz, M. T.; Huff, J.; Rudolph, J.; Tollner, K; Deege, A. *J. Am. Chem. Soc.* **1994**, *116*, 11588.

10. Under the same condition as in ref. 7, the maximum extractability of AMP (or GMP) by **12** (or **15**) was 105% (or 119%) at pH 5 and 54% (or 49%) at pH 7 respectively.

STRUCTURE AND DYNAMICS OF GUEST MOLECULES IN CYCLOPHOSPHAZENE INCLUSION COMPOUNDS

A. LIEBELT, J. SCHUHMACHER AND K. MÜLLER
Institut für Physikalische Chemie, Universität Stuttgart
Pfaffenwaldring 55, D-70569 Stuttgart, Germany

Abstract. The structure and dynamics of the guest molecules in cyclophosphazene inclusion compounds are studied by ^{2}H and ^{13}C NMR spectroscopy. Despite the rather strong molecular constraints imposed by the host lattice, the guest species are highly mobile, comprising both overall reorientational and internal conformational motions. It is found that the conformational equilibria of monosubstituted cyclohexanes bearing polar substituents are significantly altered by the incorporation into the cyclophosphazene host channels.

1. Introduction

Cyclophosphazene derivatives possess a high potential to form inclusion compounds with a variety of organic guest molecules [1]. X-ray investigations have shown that generally channel structures are built up by the cyclophosphazene host components which are suitable for guest molecules of quite different chemical structure [2]. Although, these systems are known for years, so far little is

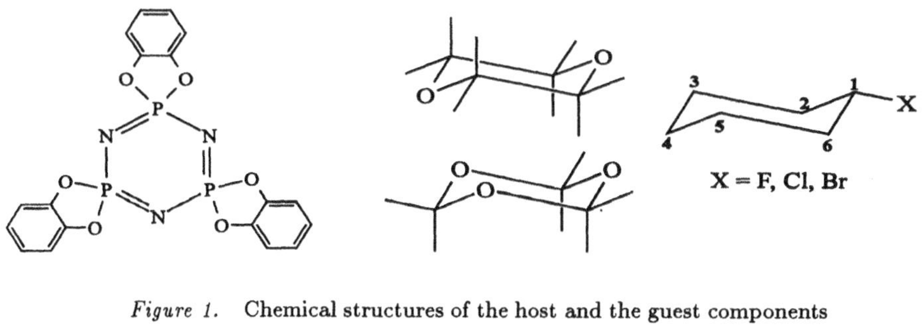

Figure 1. Chemical structures of the host and the guest components

known about the specific molecular properties of the guest components [3]. Here, we report on a solid state NMR study on cyclophosphazene inclusion compounds

A. W. Coleman (ed.), Molecular Recognition and Inclusion, 407–410.
© 1998 *Kluwer Academic Publishers. Printed in the Netherlands.*

where particular emphasis is given to the motional and ordering characteristics of the guest species. Explicitely, inclusion compounds with monosubstituted cyclohexanes, dioxane and trioxane (see Fig.1) have been subject to the ^{13}C MAS and ^{2}H NMR experiments (lineshape studies, relaxation and 2D exchange studies) presented here.

2. Experimental

Inclusion compounds of tris-(o-phenylenedioxy)-cyclotriphosphazene (CPZ) with the various guest components were prepared by standard procedures [1]. NMR measurements were performed on a Bruker CXP 300 spectrometer operating at a frequency of 46.07 MHz for deuterium and 75.47 MHz for carbons. The ^{2}H NMR spectra were obtained using a 5 mm homebuilt probe employing the quadrupole echo sequence with $\frac{\pi}{2}$-pulses of 2.7 - 2.9 μs width and a time interval between the pulses of 20 μs. For the ^{13}C MAS NMR investigations a Bruker 4 mm MAS probe was used. The experiments were carried out with the cross polarization technique followed by high power proton decoupling. Simulations have been performed using a FORTRAN simulation package developed to simulate the various types of dynamic NMR experiments.

3. Results and discussion

Dynamic ^{2}H NMR investigations have been carried out on deuterated 1,4-dioxane and 1,3,5-trioxane in cyclophosphazene inclusion compounds. In the following representative results will be reported for the sample with 1,4-dioxane-d_8. It should be noted that the other guest molecules behave quite similar. Figure 2 shows ^{2}H NMR spectra of 1,4-dioxane-d_8 in a temperature range between 193 K and 373 K. First of all, the ^{2}H NMR spectrum detected at 193 K exhibits a quadrupole splitting of 50 kHz which is far from the value of about 125 kHz, characteristic for an immobile species. This small value indicates the presence of fast overall dynamic processes of the guest molecules within the cyclophosphazene channels which dominate the relaxation properties between 20 and 150 K [4]. The additional spectral changes observed in this figure can be traced back to ring interconversions of the dioxane molecules. This model assumption is supported by the good agreement between the experimental spectra and their theoretical counterparts which were simulated on the basis of this molecular

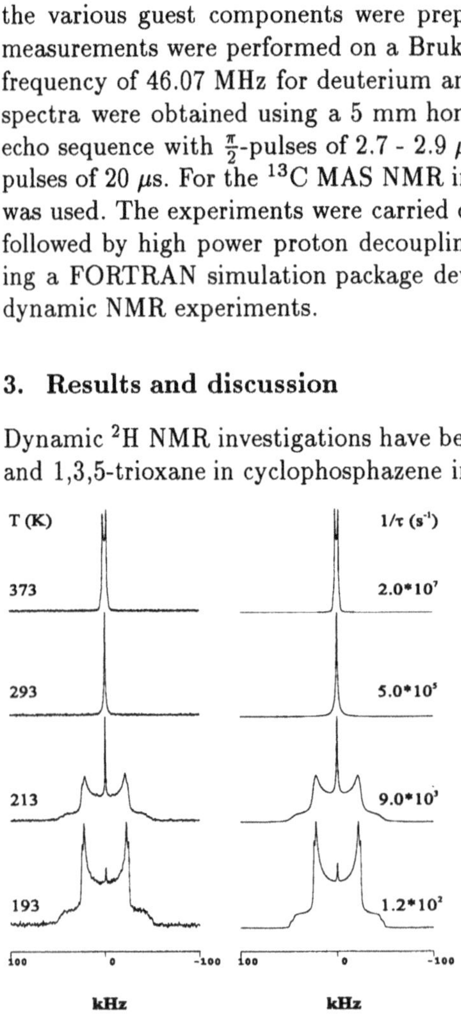

Figure 2. Experimental and simulated ^{2}H NMR spectra of 1,4-dioxane in CPZ

process. A further independent proof is given by the 2D exchange NMR spectrum in Figure 3 whose characteristic pattern also can be reproduced on the basis of a ring interconversion process. From the lineshape simulations the corresponding kinetic parameters have been derived in terms of rate constants and activation energy.

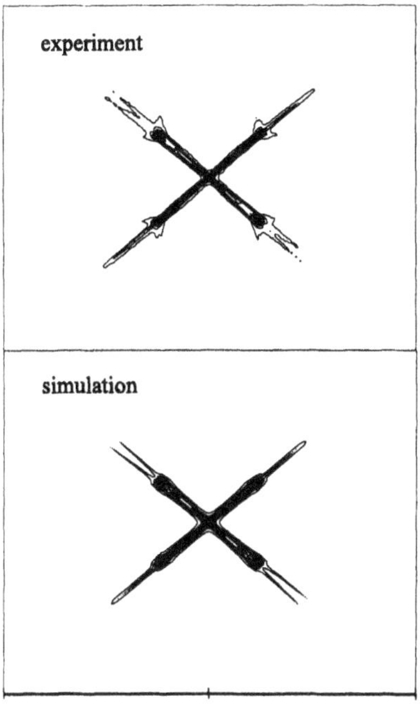

Figure 3. Experimental and simulated 2D exchange spectra of 1,4-dioxane in CPZ (contour plot)

The obtained value of 36 kJ/mol for the activation energy is very close to that reported earlier from solution NMR studies. From this, it can be concluded that the host matrix has only a minor influence on such an intramolecular process. In fact, that spectrum reflects different motionally averaged quadrupole coupling tensors of the non-equivalent axial and equatorial deuterons which strongly depends on the actual symmetry of guest molecule, i.e. orientation of the motional symmetry axis [5]. Likewise, the hexagonal channel symmetry is reflected by the axial symmetric ^{2}H NMR lineshapes which are encountered throughout the whole temperature range covered in this study.

In a further study we have focused on the conformational characteristics of monosubstituted cyclohexanes in cyclophosphazene inclusion compounds. Here, ^{13}C MAS NMR experiments have been carried out in order to evaluate the dynamics and the conformational order of cyclohexanes with fluoro-, chloro- or bromosubstituents.

Again, the ^{13}C NMR lineshapes in Fig.4 exhibit characteristic changes between 230 K and 280 K which can be attributed to ring interconversions of the guest molecules. The more interesting result refers to conformational equilibria. Here, a stabilization of the axial conformational state is detected (see Tab.1). Surprisingly, this effect is less pronounced than for the thiourea inclusion compounds [6] although the cyclophosphazene channel diameter is narrower. It thus is very likely that rather polar interactions between the various constituents of the inclusion compound than sterical reasons are responsible for this unusual molecular property.

TABLE 1. Population of the axial conformers of monosubstituted cyclohexanes in different inclusion compounds

	CPZ	thiourea
$C_6H_{11}F$	0.47	0.50
$C_6H_{11}Cl$	0.35	0.90
$C_6H_{11}Br$	0.59	0.95

4. Conclusions

The motional and ordering characteristics of guest molecules in cyclophosphazene inclusion compounds have been determined employing solid state NMR techniques.

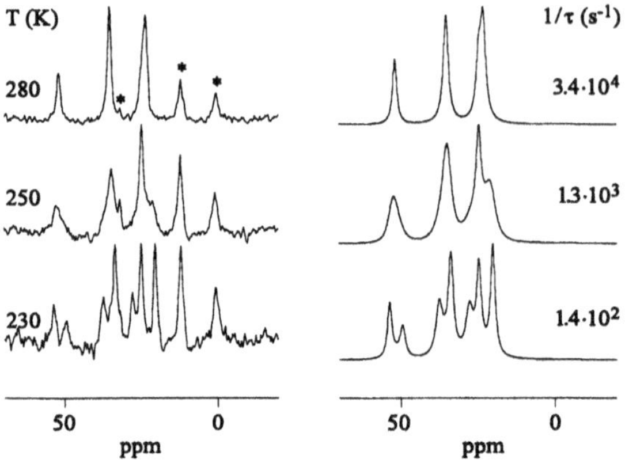

Figure 4. Experimental and simulated ^{13}C MAS NMR spectra of bromocyclohexane in CPZ (spinning sidebands are indicated with an asterisk)

The experimental results give evidence for the existence of both conformational and overall reorientational motions. Generally, the former process is dominant in the high temperature range between 200 and 300 K whereas the latter determines the relaxation behaviour between 20 and 150 K. Finally, for the conformational behaviour of monosubstituted cyclohexanes an unusual stabilization of the axial conformational state is reported.

Acknowledgements

We thank Mrs. D. Müller and Mrs. H. Seidel for the synthesis of the compounds used in the present study. Financial support by the Deutsche Forschungsgemeinschaft and the Fonds der Chemischen Industrie is gratefully acknowledged.

References

1. Atwood, J.L., Davies, J.E.D. and MacNicol, D.D. (Eds.), Inclusion Compounds, Vols. 1–3 Academic Press, New York (1984),Vols. 4–5 Oxford University Press, Oxford (1991)
2. Allcock, H.R., Levin, M.L. and Whittle, R.R. (1986), Tris(o-phenylenedioxy)cyclotriphosphazene: The Clathration-Induced Monoclinic to Hexagonal Solid-State Transition, *Inorg.Chem.* **25**, 41–47
3. Meirovitch, E., Rananavare, S.B. and Freed, J.H. (1987), Comments on the Interpretation of Dynamic Deuterium NMR Spectra from Solid Inclusion Compounds, *J.Phys.Chem.* **91**, 5014–5020
4. Liebelt, A., Müller, K., to be published
5. Poupko, R., Furman, E., Müller, K. and Luz, Z. (1991), Reinvestigation of the Thiourea Inclusion Compound by Deuterium NMR Spectroscopy, *J.Phys.Chem.* **95**, 407–413
6. Müller K. (1992), ^{13}C Magic Angle Spinning NMR Investigations of Thiourea Inclusion Compounds, *Magnetic Resonance in Chemistry* **30**, 228–234; Aliev, A.E. and Harris, K.D.M. (1993), Conformational Properties of Monosubstituted Cyclohexanes inTheir Thiourea Inclusion Compounds and in Solution: Variable-Temperature One Dimensional and Two-Dimensional ^{13}C NMR Investigations, *J.Am.Chem.Soc.* **115**, 6369–6377; Müller K. (1995), Solid-State ^{13}C NMR Investigations of Monosubstituted Cyclohexanes in Thiourea Inclusion Compounds, *Magnetic Resonance in Chemistry* **33**, 113–123

TOWARDS ROTAXANE-BASED METAL-ION SENSORS

OWEN A. MATTHEWS, J. FRASER STODDART
School of Chemistry, The University of Birmingham, Edgbaston, Birmingham B15 2TT, UK

NIGEL D. TINKER
British Nuclear Fuels plc, Springfields Works, Salwick, Preston PR4 OJX, UK

The self-assembly and dynamic behaviour of rotaxanes [1] is a fascinating outcome of supramolecular chemistry, which has enabled the genesis of simple molecular switches in the form of molecular shuttles [2,3]. We are now directing our attention towards the design and synthesis of metal-ion sensors using rotaxane-based compounds. The design logic for such sensors is based upon recent studies in our laboratory [4] which indicate that a solution containing equimolar quantities of the π-electron rich 'threads' **1** and **2**, and a tetracationic cyclophane **3**·$4PF_6$ is capable of acting as a K^+ ion sensor.

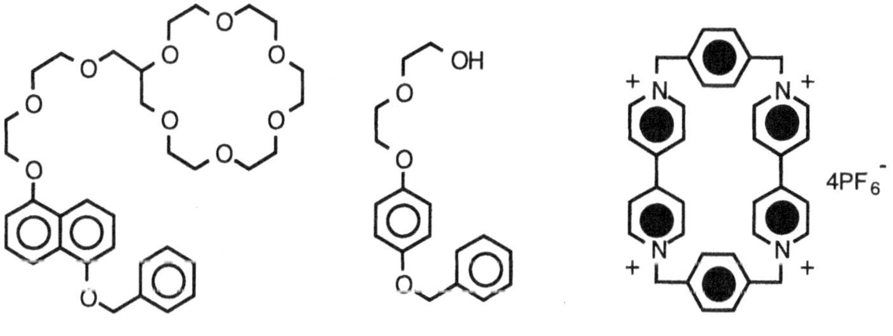

1 **2** **3.$4PF_6$**

In the absence of K^+ ions, a mixture of **1**, **2** and **3**·$4PF_6$ yields a purple-coloured solution (in MeCN at 298 K), indicating that cyclophane $\mathbf{3}^{4+}$ preferentially resides upon thread **1** yielding a characteristic charge transfer band at 518 nm in the visible region. Upon the addition of an excess of K^+ ions, the solution becomes orange with a charge transfer band centred at 475 nm (in MeCN at 298 K). This observation indicates that, in the presence of K^+ ions, **3** resides preferentially upon the hydroquinone-containing thread **2**. Such a change is probably the result of electrostatic repulsions between the tetracationic cyclophane and the K^+ ion bound within the cavity of the 18-crown-6 moeity, effectively destabilising the pseudorotaxane **1**·**3**·$4PF_6$. The encouraging results obtained for the prototypical pseudorotaxane-based metal-ion sensor prompted us to

A. W. Coleman (ed.), Molecular Recognition and Inclusion, 411–414.
© 1998 *Kluwer Academic Publishers. Printed in the Netherlands.*

design a rotaxane incorporating the two different 'stations' into the same thread. It was envisaged that such a molecular compound would be easier to handle since it would involve intramolecular rather than intermolecular shuttling.

The rotaxane **4**·$4PF_6$ was the compound which was targetted. It contains an aza-18-crown-6 metal-ion binding unit, located proximal to a 1,5-dioxynaphthalene 'station' but distal to a 1,4-dioxybenzene 'station', in the same thread. The terminal stoppers are the triisopropylsilyl group and the aza-18-crown-6 unit - the latter functions as a slippage stopper [5], allowing the tetracationic cyclophane **3** to be 'slipped' on to the thread by heating a mixture of thread and cyclophane at 323 K in MeCN for several days. The resulting rotaxane is stable since the cyclophane is 'thermodynamically trapped' on the thread.

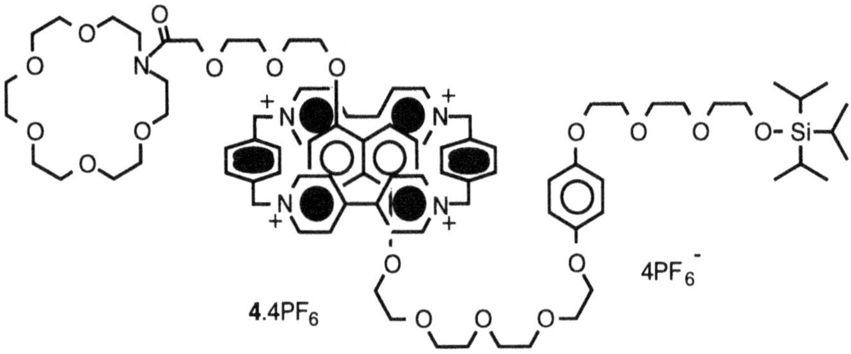

4.$4PF_6$

The synthesis of **4**·$4PF_6$ was carried out in moderate yield using a convergent strategy (Scheme 1). Readily available [6] diols **5** and **7** were monoprotected as their THP ether **6** and triisopropylsilyl ether **8**, respectively. The latter was tosylated to afford **9** which was then used to alkylate **6** using NaH in THF. Subsequent deprotection of the product, using PPTS in ethanol, yielded the monosilylated alcohol **10**. Compound **10** was then reacted with chloroacetylaza-18-crown-6 using TMEDA, NaH and THF to afford the crown-containing thread **11**. Heating equimolar amounts of **11** and **3** together at 323 K in MeCN for 21 days yielded, after chromatography, the required rotaxane **4**·$4PF_6$ as a purple solid in 56% yield.

Characterisation of **4**.$4PF_6$ was carried out using FAB mass spectrometry and UV/VIS and ^{1}H NMR spectroscopy. The ^{1}H NMR spectrum of **4**·$4PF_6$ is extremely broad at 303 K (in CD_3CN), suggesting that a variety of exchange processes are occuring at this temperature. Upon cooling the sample down to 243 K, the spectrum sharpens and several characteristic resonances become evident. For example, the multiplet at δ 2.22 was assigned to the H-4 and H-8 naphthalene protons bound 'inside' the tetracationic cyclophane. The large upfield shift of this proton signal (δ 5.59) is a result of the strong sheilding effect of the tetracationic cyclophane. The 'inside' hydroquinone protons were observed as a broad singlet at δ 3.28. Comparison of the integrals of the aformentioned signals (δ 2.22 and 3.28) indicates that the proportion of cyclophane residing upon the naphthalene 'station' is 64% with 36% on the hydroquinone 'station'. This observation is confirmed by the UV-VIS spectrum (in

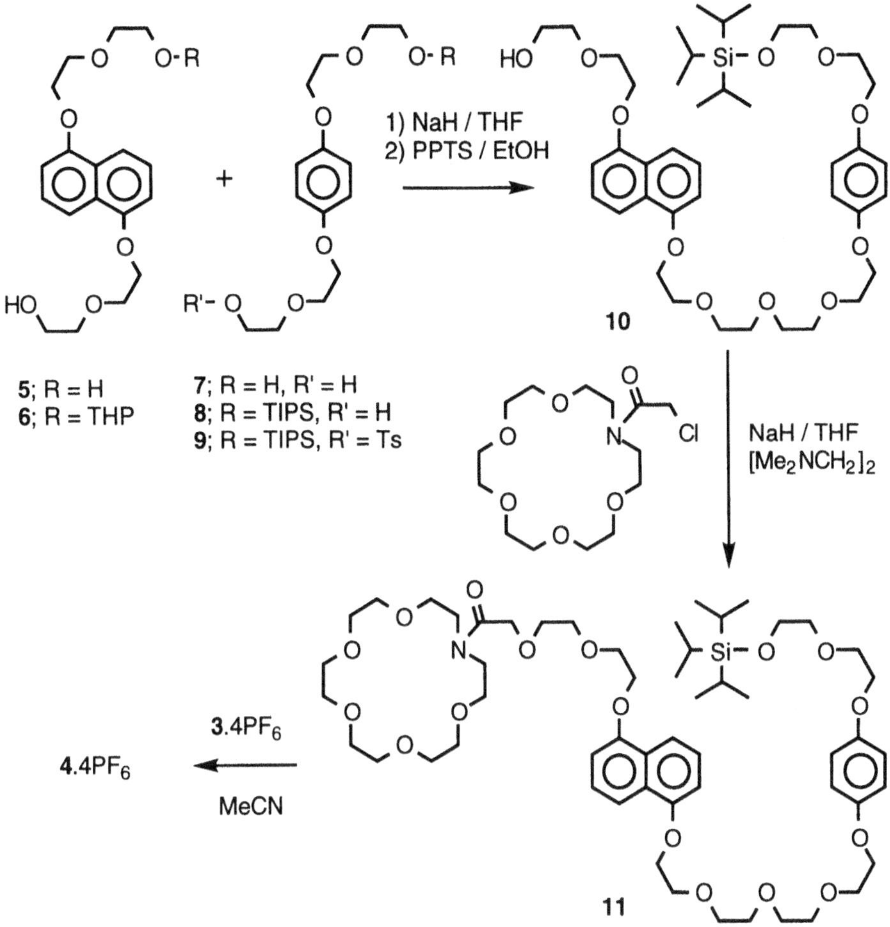

Scheme 1. The synthesis of the rotaxane **4**·$4PF_6$.

MeCN at 298 K). Under these conditions, a one-station [2]rotaxane in which the tetracationic cyclophane resides fully on the naphthalene 'station' displays a charge transfer band at 529 nm and the charge transfer band for a [2]rotaxane containing a hydroquinone 'station' and the tetracationic cyclophane is centred on 470 nm. The charge transfer band for **4**·$4PF_6$ is located at 520 nm, indicating that the tetracationic cyclophane is located mainly (85%) upon the naphthalene station. The FAB-MS of rotaxane **4**·$4PF_6$ reveals the expected fragmentation pattern with signals (*m/z*) at 2020 (M - PF_6), 1874 (M - $2PF_6$) and 1729 (M - $3PF_6$).

Upon the addition of an excess of KPF_6 to a solution of **4**·$4PF_6$ (in MeCN at 298 K), no change in colour is observed and the UV-VIS spectrum remains unchanged. The ^1H NMR spectrum of **4**·$4PF_6$ (243 K, CD_3CN) in the presence of an excess of KPF_6 changes significantly. Large changes are not unexpected upon the complexation of K^+ ions within the aza-18-crown-6 unit. However, there is no significant change in the occupancy of the tetracationic cyclophane upon the two different 'stations'. The FAB-MS of a mixture of **4**·$4PF_6$ and excess of KPF_6 clearly shows that the crown ether containing rotaxane complexes K^+ ions, displaying a peak (*m/z*) at 2187 which corresponds to the **4**·$4PF_6$.K^+ complex.

The experimental evidence which we have obtained indicates that rotaxane **4**·$4PF_6$ is capable of complexing K^+ ions but that the bound ions do not force the tetracationic cyclophane to shuttle off the naphthalene station on to the hydroquinone 'station'. The most probable explanation for this behaviour is that the cyclophane-naphthalene association is significantly stronger in rotaxane **4**·$4PF_6$ than it is in the pseudorotaxane formed between **1** and **3**·$4PF_6$. This enhanced stability is probably caused by two factors. Firstly, the extra polyether oxygen atoms which are present in the backbone of **4**·$4PF_6$ may increase the number of hydrogen bonds formed between the cyclophane and the thread. Such stabilisation has also been observed in closely related systems in which the number of oxygen atoms effects significantly the stability of pseudorotaxanes. The second reason, which may explain the unexpected behaviour of **4**·$4PF_6$, is that the flexible polyether chain linking the two 'stations' may allow the hydroquinone ring to interact with the cyclophane by 'outside' interactions. Both of these factors can be addressed effectively by modifying the chain linking the two π-electron rich 'stations'.

Financial support from British Nuclear Fuels plc is gratefully acknowledged. We thank Mr Peter Ashton and Mr Malcolm Tolley for their assistance in obtaining mass and NMR spectra, respectively.

References

[1] Amabalino, D. B., Stoddart, J. F. (1996) Interlocked and intertwined structures and superstructures, *Chem. Rev.* **95**, 2725-2828.

[2] Anelli, P. L., Spencer, N., Stoddart, J. F. (1991) A molecular shuttle, *J. Am. Chem. Soc.* **113**, 5131-5133.

[3] Bissell, R. A., Córdova, E., Kaifer, A. E., Stoddart, J. F. (1994) A chemically and electrochemically switchable molecular shuttle, *Nature (London)* **369**, 133-137.

[4] Asakawa, M., Iqbal, S., Stoddart, J. F., Tinker, N. D. (1996) A prototypical optically responsive molecular device, *Angew. Chem. Int. Ed. Engl.* **35**, 976-978.

[5] Asakawa, M., Ashton, P. R., Iqbal, S., Stoddart, J. F., Tinker, N. D., White, A. J. P., Williams, D. J. (1996) The self-assembly of some novel [2]rotaxanes and their alkali metal cation complexes *Chem. Commun.* 483-486.

[6] Amabalino, D. B., Anelli, P. -L., Ashton, P. R., Brown, G. R., Córdova, E., Godinez, L. A., Hayes, W., Kaifer, A. E., Philp, D., Slawin, A. M. Z., Spencer, N., Stoddart, J. F. S., Tolley, M. S., Williams, D. J. (1995) Molecular meccano. 3. Constitutional and translational isomerism in [2]catenanes and [*n*]pseudorotaxanes, *J. Am. Chem. Soc.* **117**, 11142-11170.

MICROCALORIMETRIC STUDIES OF LIGAND-INDUCED VANCOMYCIN DIMERISATION AND MOLECULAR RECOGNITION

D. MCPHAIL AND A. COOPER
Department of Chemistry, Glasgow University, Glasgow G12 8QQ, Scotland, U.K

The vancomycin group of antibiotics are an important method of control against Gram-positive bacterial infections and are effectively the last line in defence against methicillin-resistant *Staphylococcus aureus* (MRSA), the so-called 'superbug'. This family of antibiotics exerts its bactericidal effect by complexing with D-alanine-D-alanine precursor units on the surface of Gram-positive bacterial cells, preventing their incorporation into the major structural polymer of the cell wall, peptidoglycan. We have used microcalorimetry to study this recognition process directly by measuring the association/dimerisation constants and enthalpy changes of vancomycin with various cell wall analogue peptides under different conditions. Vancomycin has been shown to combine with various peptides, but most strongly with N-diac-lys-D-ala-D-ala, a ligand closely resembling its natural substrate. Dimerisation of antibiotic in the presence of this ligand is increased typically 10-fold (K_{dim} = 5050M^{-1}, ΔH = -39.3kJmol^{-1}) from dimerisation in the absence of ligand (K_{dim} = 475M^{-1}, ΔH = -29.2kJmol^{-1}). It is clear that vancomycin dimerisation is enhanced in the presence of ligand and this in turn suggests that dimerisation may play an important role in the biological function of these antibiotics.

1. Introduction

The incidence of Staphylococcal resistance to a wide range of antibiotics has resulted in an increase in the use of vancomycin to treat such infections, since vancomycin resistance is not yet as widespread. In view of this, it is important that we understand the way in which this important antibiotic functions as a simple recognition model, so that novel antibiotics against which resistance has not yet developed can be produced, keeping us one step ahead of the increase in bacterial resistance.

The binding of vancomycin to various cell wall analogue peptides has been carried out over a temperature and pH range using microcalorimetry as a direct method for determining the thermodynamics of the recognition process.

Previous microcalorimetric binding studies using vancomycin and N-acetyl-D-ala-D-ala showed that the apparent association constants and enthalpies depend on antibiotic

A. W. Coleman (ed.), Molecular Recognition and Inclusion, 415–418.
© 1998 *Kluwer Academic Publishers. Printed in the Netherlands.*

concentration. At higher concentrations, vancomycin is thought to form dimers or higher oligomers, which have been shown to enhance their associated binding affinity (Cooper and McAuley-Hecht, 1993). Using microcalorimetry, the energetics of this dimerisation process in the presence and absence of cell wall analogues over a range of temperature and pH conditions have been studied and it has been suggested that dimerisation may play an important functional role *in vivo* during interactions at the cell wall surface (Mackay *et al.*, 1994).

2. Materials and Methods

2.1. MATERIALS

Vancomycin, N-acetyl-D-alanine, N-acetyl-D-ala-D-ala and N_α, N_ε - diacetyl-Lys-D-ala-D-ala (LAA) were obtained from Sigma Chemical Co. and used without further purification. Buffers used were: 0.1M citrate at pH 3.0, 0.1M acetate at pH 5.0, 0.1M phosphate at pH 7.0, 0.1M TRIS at pH 8.0 and 0.1M phosphate at pH 11.0. Peptide and antibiotic concentrations were determined by weight and confirmed in the case of vancomycin by UV spectroscopy of diluted samples using the molar extinction coefficient E_{280nm} = 6690 at pH 7.0 (Nieto and Perkins[a], 1971).

2.2. MICROCALORIMETRY

Calorimetric data for the binding of vancomycin to cell wall analogues were obtained over a range of temperatures and pH using a Microcal Omega Isothermal Titration Calorimeter (ITC). A typical experiment involved sequential injections (20×5μl at 3 minute intervals) of ligand (approx. 3.5mM N-ac-D-ala-D-ala, 7mM LAA and 50mM N-ac-D-ala) solution into the stirred calorimeter cell containing vancomycin (0.07-0.3mM) solution. This gave rise to a series of exothermic heat pulses which, after correction for appropriate control experiments under identical conditions, were analysed using Microcal Origin software to give K_a and ΔH°. Calorimetric data for the dilution of vancomycin, both in the presence and absence of appropriate ligand were similarly obtained over a temperature and pH range. A typical dilution experiment involved injections (normally 12×20μl at 3 minute intervals) of concentrated antibiotic solution (2-3mM) into the stirred calorimeter cell initially containing buffer alone. In this case, a series of endothermic heat pulses were obtained which, after correction for appropriate control experiments under identical conditions, were analysed using Microcal Origin and specially-written software in terms of a simple dimer dissociation model to give K_{dim} and ΔH_{dim}. In experiments involving ligand, the free peptide concentration was kept the same in both cell and injection syringe solutions, typically 90-100mM for N-ac-D-Ala and 1-5mM for LAA.

3. Results and Discussion

The injection of ligand into the calorimeter cell containing vancomycin is an exothermic process consistent with complex formation. A sequence of injections, examples of which are shown in Figure 1, gives a series of exothermic heat pulses which, after integration and correction for appropriate control experiments, gives the absolute heat produced per injection. In such an injection series, successive injections become progressively less exothermic as antibiotic saturation is reached. The resulting thermal profile can be analysed to give association constant (K_a) and enthalpy of association ($\Delta H°$). Vancomycin, as expected, binds most strongly to the peptide sequence LAA i.e. K_a of $4.9 \times 10^5 M^{-1}$ and $\Delta H°$ of $-56.9 kJmol^{-1}$ compared with the dipeptide N-ac-D-ala with a K_a of $295 M^{-1}$ and $\Delta H°$ of $-36.6 kJmol^{-1}$, since LAA most closely resembles the antibiotic's natural substrate. With an increase in temperature, the stability of the complex is reduced, indicated by a decrease in association constants. Although molecular recognition occurs over a wide pH range, the optimum appears to be around neutral pH, since it has been shown that for complex formation to occur, the vancomycin carboxyl groups must be ionised and the phenolic groups unionised (Nieto & Perkins, 1971) and these conditions are met at pH 7.0.

Dilution of vancomycin solutions by injection of small aliquots into the calorimeter cell containing a larger volume of buffer is an endothermic process consistent with dissociation of vancomycin dimers or higher oligomers. In this case, a sequence of injections, examples of which are shown in Figure 2, gives a series of endothermic heat pulses, which after integration and correction for control mixing experiments, gives the absolute heat uptake per injection. Successive injections become less exothermic as the antibiotic concentration builds up in the calorimeter cell. The resulting thermal dilution profile is consistent with a simple monomer-dimer dissociation model and can be analysed by non-linear regression techniques to give estimates of the dimerisation constant (K_{dim}) and enthalpy of dimerisation ($\Delta H°_{dim}$). In the absence of ligand the aggregation of vancomycin molecules in solution is fairly weak e.g. at 25°C, pH 7.0, vancomycin has a K_{dim} of $475 M^{-1}$ and $\Delta H°_{dim}$ of $-29.2 kJmol^{-1}$. In the presence of ligand this dimerisation becomes both tighter and more exothermic. For example, in the presence of LAA the dimerisation constant is increased to $5050 M^{-1}$ with $\Delta H°_{dim}$ of $-39.3 kJmol^{-1}$, with significantly more heat required to dissociate the complex. An increase in temperature reduces the stability of the dimer complex, shown by a reduction in dimerisation constants and dimerisation falls significantly above pH 8.0. pK_a values of 8.7 and 9.6 have been assigned to groups on vancomycin involving phenolic dissociation (Popieniek and Pratt, 1991), therefore it seems possible that ionisation of these phenolic groups may be responsible for the reduction in dimerisation. These results and other microcalorimetric work (Cooper and McAuley-Hecht, 1993) show that cooperativity exists between ligand binding and dimerisation at high antibiotic concentrations *in vitro* and it is possible that dimerisation may play an important functional role *in vivo*. A dimer may be targeted more effectively than a monomer, if precursor units are concentrated locally on the bacterial cell wall (Waltho and Williams, 1989). It is unclear if such cooperativity exists *in vivo*, since much lower antibiotic concentrations are to be found in serum. However, it is possible that there is a chelate-like enhancement of binding affinity.

1.

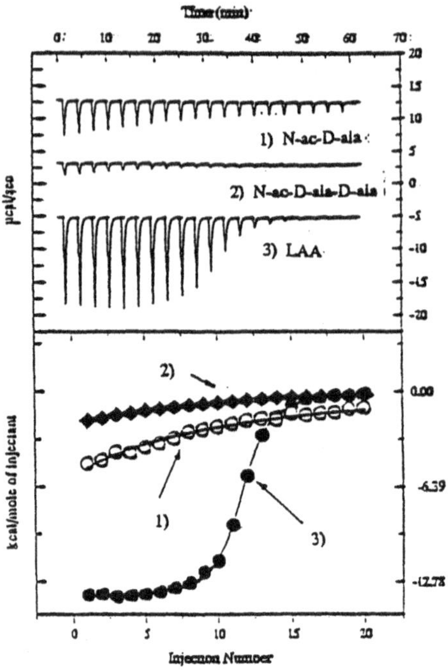

2.

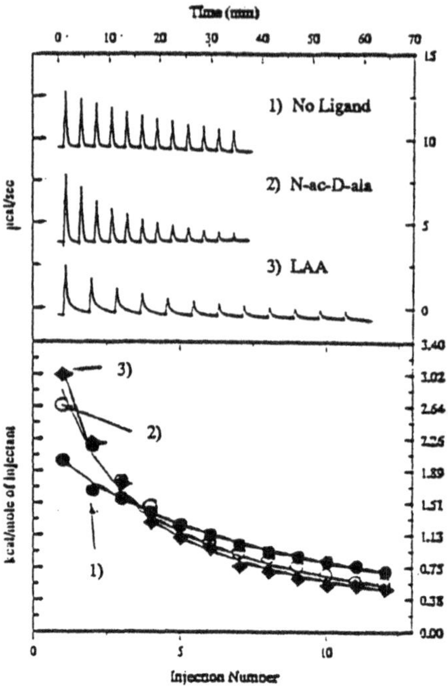

Figure 1 Vancomycin binding to various cell wall analogue peptides, using 20×5µl injections of ligand into vancomycin. Figure 2 Vancomycin dimerisation in the presence and absence of cell wall analogue peptides, using 12×20µl injections of vancomycin into the cell. All titrations were carried out at 25°C, pH 7.0.

4. References

Cooper, A and McAuley-Hecht, K. (1993) Microcalorimetry and the molecular recognition of peptides and proteins, *Philosophical Transactions of the Royal Society of London A*, **345**, 23-35.

Mackay, J.P., Gerhard, U., Beauregard, D.A., Westwell, M.S., Searle, M.S and Williams, D.H. (1994) Glycopeptide antibiotic activity and the possible role of dimerization; a model for biological signalling, *Journal of the American Chemical Society* **116**, 4581-4590.

Nieto, M and Perkins, H.R.[a] (1971) Physicochemical properties of vancomycin and iodovancomycin and their complexes with diacetyl-L-lysyl-D-alanyl-D-alanine, *Journal of Biochemistry* **123**, 773-789.

Nieto, M and Perkins, H.R. (1971) Modifications of the acyl-D-alanyl-D-alanine terminus affecting complex-formation with vancomycin, *Journal of Biochemistry* **123**, 789-803.

Popieniek, P.H and Pratt, R.F. (1991) Kinetics and mechanism of binding of specific peptides to vancomycin and other glycopeptide antibiotics, *Journal of the American Chemical Society* **113**, 2264-2270.

Waltho, J.P and Williams, D.H. (1989) Aspects of molecular recognition: solvent exclusion and dimerization of the antibiotic ristocetin when bound to a model bacterial cell-wall precursor, *Journal of the American Chemical Society* **111**, 2475-2480.

Self-Assembled Hydrogen Bonded Dimers of Calix[4]arenes

O. MOGCK[a)], V. BÖHMER[a)], M. PONS[b)], E.F. PAULUS[c)], W. VOGT[a)]
a) Institut für Organische Chemie, Universität Mainz, Becher Weg 34 SB1, D-55099 Mainz, Germany
b) Departament de Química Orgànica, Universitat de Barcelona, E-08028 Barcelona, Spain
c) Hoechst AG, D-65926 Frankfurt/Main, Germany

1. Introduction

Self-assembly of suitably functionalized molecules to well defined supramolecular structures has become a topic of current interest. Various molecules have been designed which are able to mimic in part the recognition and self-organisation processes which are found as a general principle in nature.[1]

The organisation of N,N'-dialkyl or diaryl urea derivatives in the crystal lattice via hydrogen bonds between the NH- and C=O-functions[2] has inspired the attachment of urea functions to a suitable calixarene skeleton in order to obtain molecules which are able to associate via such hydrogen bonds[3].

2. Self-Assembly of Urea-Calix[4]arenes

A series of calix[4]arenes **2**, substituted by various urea residues at the wider rim, could be easily prepared by reacting the readily available *p*-aminocalix[4]arenes **1** with isocyanates according to scheme 1.[4]

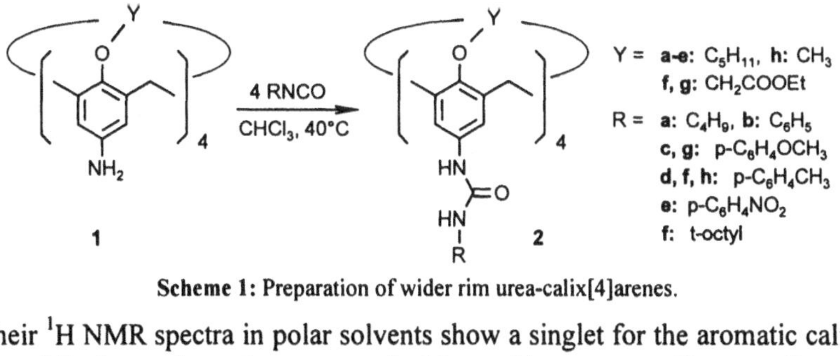

Scheme 1: Preparation of wider rim urea-calix[4]arenes.

Their ^{1}H NMR spectra in polar solvents show a singlet for the aromatic calixarene protons, while in apolar solvents two doublets with *meta*-coupling are found. This suggests that in polar solvents the monomeric calixarenes **2** with C_{4v}-symmetry are present while in apolar solvents hydrogen bonded dimers **2·2** with S_8-symmetry are formed by self-assembly (Fig. 1).

A. W. Coleman (ed.), Molecular Recognition and Inclusion, 419–422.
© 1998 *Kluwer Academic Publishers. Printed in the Netherlands.*

A first example for this reversible self-organisation phenomenon of urea-calix[4]arenes was described by Rebek et al., who already suggested the inclusion of guest molecules.[5]

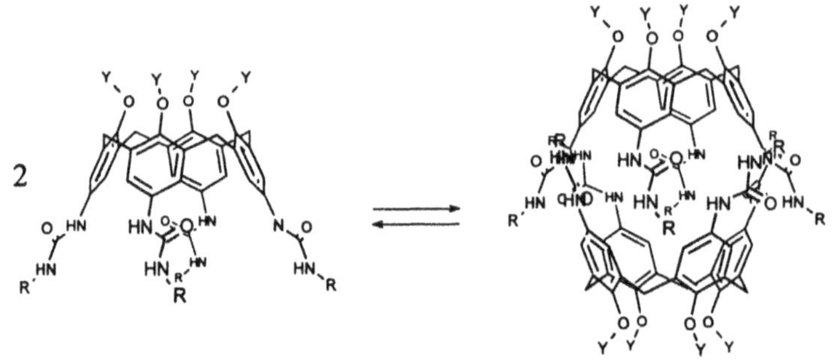

Figure 1: Reversible self-assembly of 2 to dimers 2·2.

2.1. FORMATION OF HETERODIMERS

Recently we could show by ^1H NMR spectroscopic methods that in solutions containing two different urea-calix[4]arenes heterodimers are present in addition to the two homodimers. This can be considered as an independent proof for the self-organisation to dimers in solution.[6] While each homodimer shows one pair of doublets for the diastereotopic Ar-CH_2-Ar protons the mixture shows additionally two pairs of doublets for the heterodimer, one pair for each calixarene part (Fig. 2).

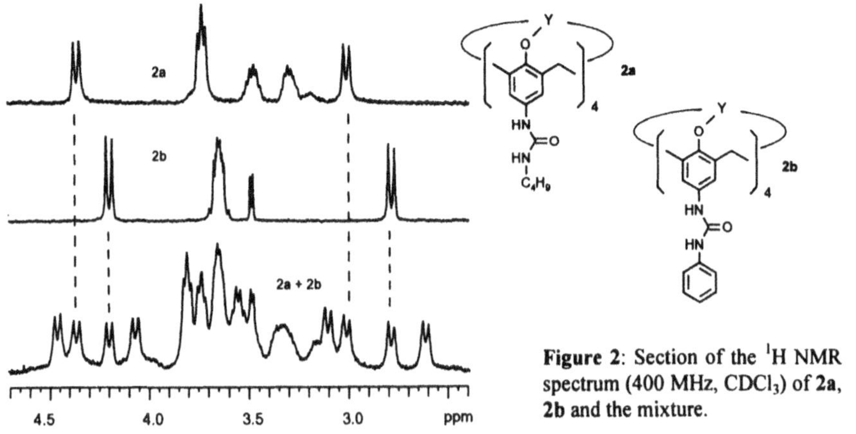

Figure 2: Section of the ^1H NMR spectrum (400 MHz, $CDCl_3$) of 2a, 2b and the mixture.

2.2. TEMPLATE EFFECT

The dimerisation is strongly favoured by the presence of suitable guest molecules which can be included into the cavity of the dimer. In solvents of wrong size or shape (e.g. CD_2Cl_2) just indefinite associates are formed as indicated by broad signals in the ^1H NMR spectrum, while the addition of small amounts of a suitable guest molecule such as e.g. benzene or $CHCl_3$ leads immediately to the formation of well defined dimers (Fig. 3).

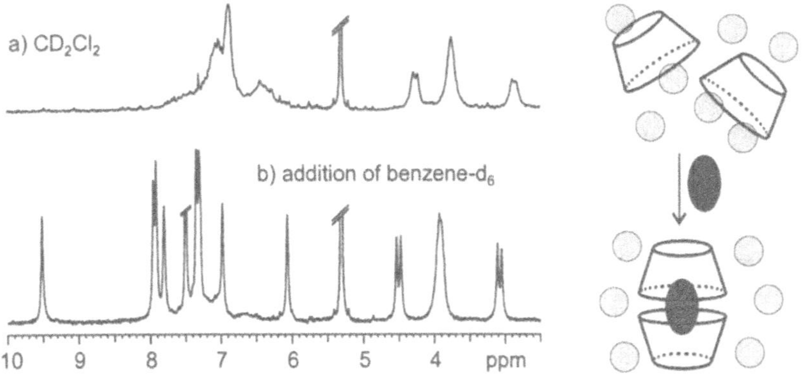

Figure 3: Section of the ^{1}H NMR spectrum (200 MHz) of **2d** in CD_2Cl_2 (a), and after the addition of ~5% benzene-d_6 (b).

2.3. SINGLE CRYSTAL X-RAY ANALYSIS

The shape of the dimer and the geometry of the cavity was so far just deduced from computational calculations.

We were able to confirm the dimerisation in the solid state by single crystal X-ray analysis.[7] Fig. 4 shows the dimer which is composed of two crystallographically independent calix[4]arenes. Both molecules lie on a fourfold axis turned by 43° with respect to each other. The inner cavity contains completely disordered benzene molecules (1.06 benzene per dimer) and further benzene molecules are found in voids of the crystal lattice. The dimers are held together by two hydrogen bonds per phenolic unit (altogether 16 hydrogen bonds) which are quite different in strength as indicated by different N···O distances (3.09/3.15 Å and 2.85/2.85 Å).

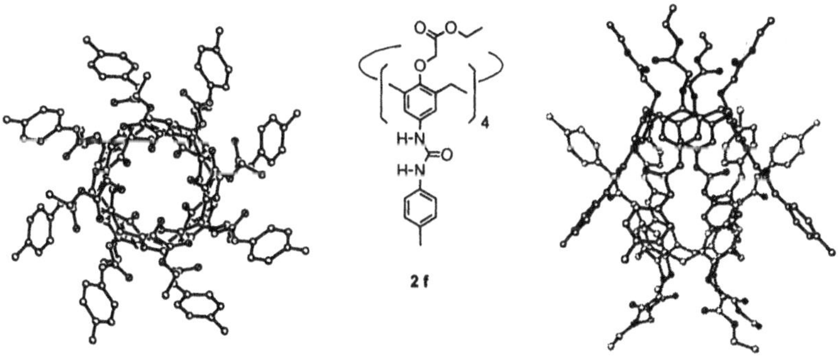

Figure 4: Single crystal X-Ray analysis. Dimer shown as top view (left) and side view (right). Disordered benzene molecules are omitted for clarity.

2.4. INCLUSION OF GUEST MOLECULES

The guest molecules included in the inner cavity can be directly observed in solution by ^{1}H NMR spectroscopy. The addition of undeuterated benzene to a solution of **2** in

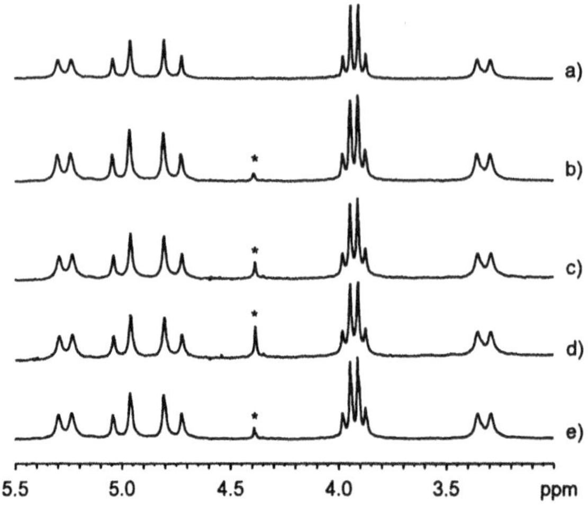

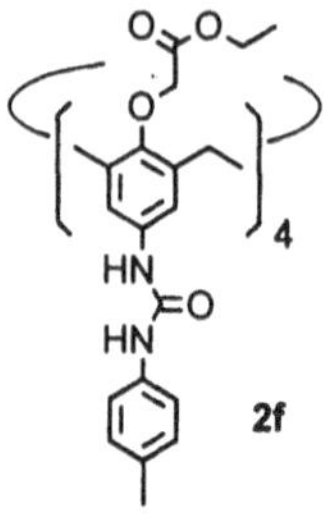

2f

Figure 5: a) Section of the ^{1}H NMR spectrum (200 MHz) of **2f** in benzene-d_6, b) - d) addition of increasing amounts of benzene-H_6, and e) addition of benzene-d_6 to d) (* = included benzene).

benzene-d_6 leads to an additional signal around 4.4 ppm arising from the *included* benzene molecules (Fig. 5). Exchange between free and trapped benzene could be observed in 2D-NMR spectra. This exchange is associated to the reversible dissociation of the dimer, a phenomenon which could be followed in EXSY spectra.[8]

The interiour volume of the cavity formed by dimerisation in solution is an important factor, for instance, if such dimers should be used as molecular reaction vessels, as shown by Cram et al. for carcerands.[9] From X-ray data a volume of roughly 200 $Å^3$ was calculated. This is in principle large enough to include two benzene molecules (V= 73.5 $Å^3$) but integration of ^{1}H NMR spectra recorded in undeuterated solvents (e.g. $CHCl_3$, benzene-H_6), which was performed for several calixarenes **2**, shows in all cases the formation of a 1:1 complex between the host and the guest. Obviously in solution not more than one molecule of the size of benzene or chloroform is included in the cavity of the dimer.

Acknowledgement. O. M., V. B. and W. V. acknowledge financial support from the Deutsche Forschungsgemeinschaft, and M. P. from DGICYT (Spain) (PB94-924).

REFERENCES

[1] D. S. Lawrence, T. Jiang and M. Levett *Chem. Rev.* **1995**, *95*, 2229-2260;
J. S. Lindsey *New J. Chem.* **1991**, *15*, 153-180.
[2] M. Etter, Z. Urbanczyk-Lipkowska, M. Zia-Ebrahimi and T.W. Panunto *J. Am. Chem. Soc.* **1990**, *112*, 8415-8426.
[3] B. C. Hamann, K. D. Shimizu and J. Rebek Jr. *Angew. Chem.* **1996**, *108*, 1425-1427.
[4] R.A. Jakobi, V. Böhmer, C. Grüttner, D. Kraft and W. Vogt *New. J. Chem.* **1996**, *20*, 493-501.
[5] K. D. Shimizu and J. Rebek Jr. *Proc. Nat. Sci. Acad.* **1995**, *92*, 12403-12407.
[6] O. Mogck, V. Böhmer and W. Vogt *Tetrahedron* **1996**, *52*, 8489-8496.
[7] O. Mogck, E. F. Paulus, V. Böhmer, I. Thondorf and W. Vogt *Chem. Commun.* **1996** *in press.*
[8] O. Mogck, M. Pons, V. Böhmer and W. Vogt *to be published.*
[9] D. J. Cram, M. E. Tanner and R. Thomas Angew. Chem. **1991**, *103*, 1048-1051.

CRYSTAL GROWTH OF MACROCYCLES IN GEL

Laure MOTTA VIOLA, Noëlle EHLINGER, Laurent GROSVALET

Université Claude Bernard LYON I
Laboratoire de Cristallographie - ESA 5078
43, Boulevard du 11 Novembre 1918 - 69622 Villeurbanne Cedex

Whereas X-ray diffraction needs high quality crystals for structural research, the size of macrocycles studied continually increases[1,2]. Usual methods of growing organic crystals (among them evaporation of a dilute solution at room temperature or cooling a boiling saturated solution) are no more adapted.

Gel growth techniques are an old process which have been revised and optimised for biological macromolecules. But proteins required hydrogels[3,4] like silica gels or aragose gels and calixarenes studied in our laboratory are not soluble in water.

Organic crystals growth in gels with organic solvents is unrecognized but Sephadex LH-20 seems to have proved it efficiency in some measure[5,6].

Preparation of Sephadex gel

A non aqueous gel is obtained immediately by adding a suitable quantity of Sephadex LH-20 to a mixture of an organic solvent with alcohol (methanol). The proportions used are :

0.45 g of Sephadex LH-20 for: -0.75 ml of organic solvent
-0.25 ml of methanol

No special conditions of temperature are required.
The polymerisation reaction proceeds with the formation of spherical particles.

Operational sequence for crystallisation with sephadex gel

The technique used is the liquid-liquid diffusion. Two solvents must be chosen: a good one (solvent A) and a bad one (solvent B); both must be miscible together.

1. Prepare a saturated solution of the product to crystallise in a 3:1 mixture of solvent A and methanol.
2. Add 0.45 g of Sephadex LH-20 for 1 ml of saturated solution. The gel sets immediately.
3. Add carefully solvent B on the top of the gel.

A. W. Coleman (ed.), Molecular Recognition and Inclusion, 423–426.
© 1998 *Kluwer Academic Publishers. Printed in the Netherlands.*

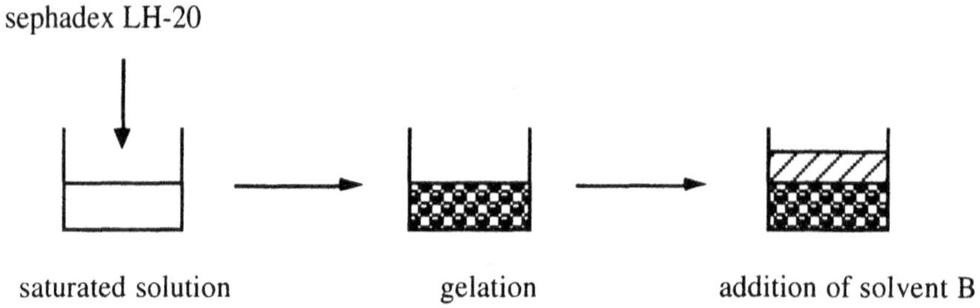

Some advantages of gel method for liquid-liquid diffusion technique

Reduction of excessive nucleation.
No convection current is present.
Mass transfer occurs only by diffusion.
Absence of sedimentation.
The crystallisation takes place in the interstices between the gel particles.
Crystals can grow free from strain exerted by wall holder and others crystals.

Comparison of crystals obtained

1) by slow evaporation of a solution at room temperature.
2) by liquid-liquid diffusion in Sephadex gel.

p-tert-butyldihomo-oxacalix[4]arene (CALO)

1. CALO crystals obtained by evaporation of a pyridine solution (x45).

2. CALO crystals obtained by gel method (x45)
(solvent A: pyridine, solvent B: methanol).

In the two cases, same shape and same size dimensions are observed for the crystals. But by the evaporation method, crystals are conglomerated, where as by the gel method, crystals are more dispersed. So, it is easier to pick up crystals for X-ray studies.

<u>hexa-*p*-*tert*-butyl-37,38,39,40,41,42- hexa[((-)-camphor-10-sulfonyl)-oxy]-calix[6]arene.</u>

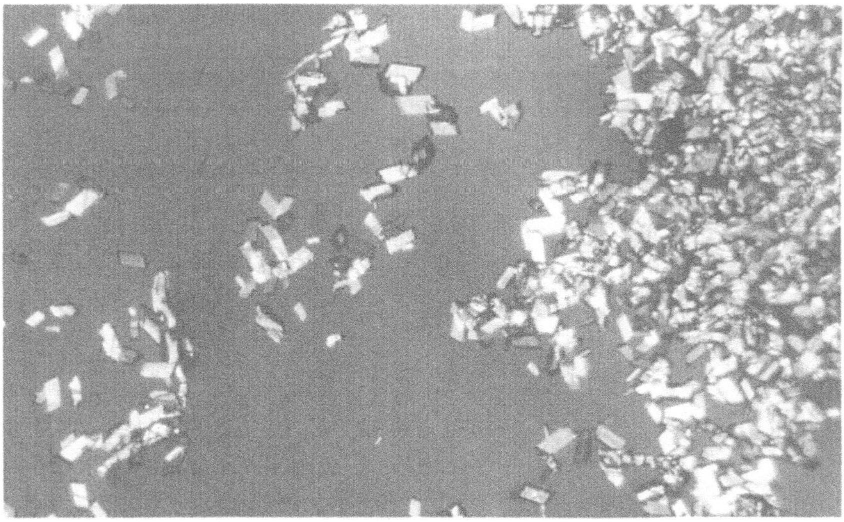

3. PTBU6-6CS crystals obtained by evaporation of a chloroform solution (x45).

4. PTBU6-6CS crystals obtained by gel method (x45).
(solvent A: chloroform, solvent B: ethanol).

Larger crystals are obtained by gel technic than by evaporation method. As for the first example, each crystal are isolated in the gel medium.

1 - EHLINGER, N.; LECOCQ, S.; PERRIN, R.; PERRIN, M. (1993) *Supra. Chem.*, 2, 77.
2 - MOTTA, L.; REGNOUF DE VAINS, J.B.; BAVOUX, C.; PERRIN, M. (1995) *J. Chem. Cryst.*, 7, 401.
3 - ROBERT, M.V.; LEFAUCHEUX, F.(1991) *J. Cryst. Growth*, 90, 350.
4 - PROVOST, K.; ROBERT , M.C.(1991) *J. Cryst. Growth*, 110, 258.
5 - DESIRAJU, G.R.; CURTIN, D.Y.; PAUL, I.C.(1977) *J. Am. Chem. Soc.*, 99,6148.
6 - ETTER, M.C.; JAHN, D.A.; DONAHUE, B.S.; JOHNSON, R.B.; OJALA, C.(1986) *J. Cryst. Growth*, 76, 645.

CHIRAL CALIXARENES FUNCTIONALIZED WITH CAMPHORSULFONYL GROUPS. SYNTHESIS, STRUCTURE AND INCLUSION PROPERTIES

Laure MOTTA VIOLA*, Jean-Bernard REGNOUF DE VAINS**, Claude BAVOUX*, Monique PERRIN*

**Laboratoire de Cristallographie et de **Chimie Industrielle - ESA 5078*
Université Claude Bernard Lyon I
43 boulevard du 11 Novembre 1918 - F - 69622 Villeurbanne cedex

1. Introduction

The interest for calixarenes is now well established in the field of supramolecular chemistry and in the synthesis of host compounds for ions and neutral molecules. The development of chiral calixarenes is an important step for the evolution of these macrocycles if we consider a possible enantioselective recognition or discrimination. The chirality in calixarenes can be obtained by several approaches [1, 2, 3].

2. Synthesis of the calixarenes

The synthesis of our calixarenes consists of treating p-tert-butylcalix[n]arene with (+) or (-) camphor-10-sulfonyl chloride in toluene and triethylamine.

In this way, we have obtained six differents macrocycles:

- (+) and (-) disubstituted calix[4]arene (PTBU4-2CS)
- (+) and (-) monosubstituted calix[6]arene (PTBU6-1CS)
- (+) and (-) hexasubstituted calix[6]arene (PTBU6-6CS)

3. Crystallization

We tried to obtain single crystals from saturated solutions in various solvents in order to determine their crystal structures by X-ray diffraction. Suitable crystals were obtained for the (+) disubstituted calix[4)arene and for its (1:2) complex with toluene.The two structures was solved by MULTAN[4] and refined using SHELX.[5, 6]

4. Crystal and Molecular Structure of (+) PTBU4-2CS (empty form)[7]

Crystal data are given in table 1.

The asymetric unit contains two independant molecules, A and B. Figure 1 shows the numbering scheme of atoms and rings for both molecules.

These two molecules are in cone conformation. The calixarene cavities have a pseudo C2v symmetry with the two opposite phenolic rings bearing the camphorsulfonyl groups almost parallel to one another (Figure 2).

A. W. Coleman (ed.), Molecular Recognition and Inclusion, 427–430.
© 1998 *Kluwer Academic Publishers. Printed in the Netherlands.*

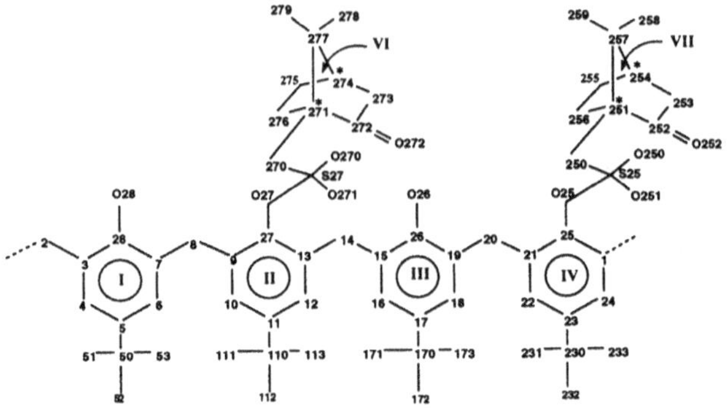

FIGURE 1. Numbering scheme

TABLE 1. Crystal data

$S_2C_{64}H_{84}O_{10}$		
P1	D = 1.172 g/cm^3	Z = 2
a = 14.671(4) Å	b = 17.530(3) Å	c = 13.085(9) Å
α = 103.69(3)°	β = 111.37(4)°	γ = 84.60(2)°
R = 0.084	S = 1.096	

The intramolecular O...O distances between two neighbouring oxygens are:

	Molecule A	Molecule B
O(25)...O(26)	3.077(4)	3.226(5)
O(26)...O(27)	3.089(6)	3.239(6)
O(27)...O(28)	3.134(5)	3.288(5)
O(28)...O(25)	2.968(6)	3.124(6)

These values are all larger than those found usually in calix[4]arenes. The norborane skelton of (+) camphorsulfonyl group is similar to ones found in literature. Its conformation is very similar for the two groups in molecule A and for the one connected on ring II in molecule B; on the contrary the group connected on ring IV in the molecule B is very twisted with regard to the orientation observed for the others: so the two independant molecules don't exhibit the same geometry (table 2).

The shortest intermolecular contacts concern oxygens of camphorsulfonyl groups; this fact confirms that they are oriented exo to the calixarene. This is due to the repulsion between polar oxygens in molecule.

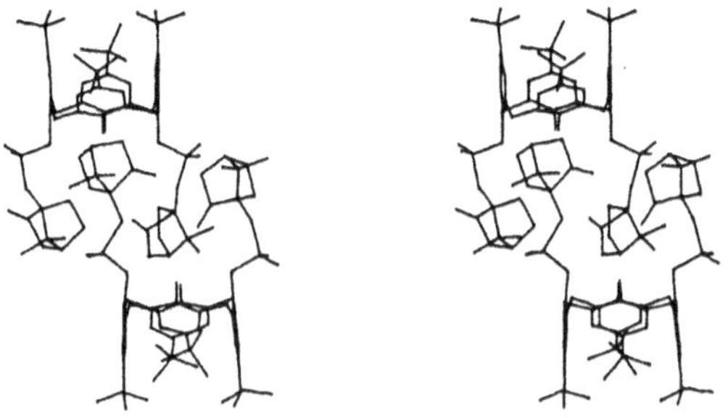

FIGURE 2. Stereoview of the molecules.

TABLE 2. Dihedral angles

	Molecule A	Molecule B
planes I - IV	6(1)°	1(4)°
planes I-III	107.3(1)°	107.9(1)°

Angles between phenolic units (I→IV) and the plane defined by the four methylene bridges (V).

Molecule A	Molecule B
146.5(2)°	146.8(2)°
94.2(1)°	90.5(1)°
140.8(1)°	141.0(1)°
91.5(1)°	90.8(1)°

Angles between the phenolic rings bearing camphorsulfonyl groups and the plane defined by the bridgehead (ex: C(251)-C(257)-C(254) or corresponding ones in other groups).

	Molecule A	Molecule B
II - VI	20.1(7)°	28.7(4)°
IV - VII	24.4(6)°	74.4(6)°

5. Crystal and Molecular Structure of the complex PTBU4-2CS-Toluene (1:2)[8]

Crystal data are given in table 3.

The unit cell contains two independent calixarene molecules in cone conformation and four toluene molecules. The atomic numbering scheme is shown in figure 1. Like in the empty form we find the same geometry with a pseudo C2v symmetry. The distances C(25)...C(27) and C(23)...C(11), near 5.50 Å, are indicative of the mean distance between the two parallel planes and thus of the width of the cavity. That's why the toluene occupies the interhost space and not the cavities (Figure 3). The relative orientations of the camphor groups with the phenolic unit to which they are connected are rather similar for molecule B but different in molecule A. The two calixarene molecules don't have the same geometry. It is also different from the orientation found in the uncomplexed structure: so the toluene molecule modifies the plane orientation (Table 4).

The intramolecular O...O distances are:

	Molecule A	Molecule B
O(25)...O(26)	3.228(6)	3.123(6)
O(26)...O(27)	3.032(6)	3.071(6)
O(27)...O(28)	3.217(6)	3.217(6)
O(28)...O(25)	3.004(8)	3.009(8)

TABLE 3. Crystal data

$S_2C_{64}H_{84}O_{10}$, $2C_7H_8$		
P1	D = 1.127 g/cm^3	Z = 2
a = 16.426(3) Å	b = 18.553(3) Å	c = 13.661(2) Å
α = 94.78(2)°	β = 110.76(2)°	γ = 72.83(2)°
R = 0.100	S = 1.142	

The shortest intermolecular contacts correspond to van der Waals contacts and concern particularly the oxygens of camphorsulfonyl groups.The shortest value is O(252)B...H(275)A=2.151(5)Å. This oxygen belongs to the sulfonyl group which is disoriented with respect with the conformation observed for the others. This short distance may correspond to a steric hindrance. No specific interactions have been found with the toluene.

TABLE 4. Dihedral angles

	Molecule A	Molecule B
planes II-IV	4(3)°	6(5)°
planes I-III	112.7(3)°	107.6(3)°

Angles between phenolic units (I → IV) and the plane defined by the four methylene bridges (V).

	Molecule A	Molecule B
	146.8(3)°	144.0(3)°
	92.7(3)°	93.2(2)°
	145.9(3)°	143.6(3)°
	91.41(2)°	92.8(2)°

Angles between the phenolic rings bearing camphorsulfonyl groups and the plane defined by the bridgehead (ex: C(251)-C(257)-C(254) or corresponding ones in other groups).

	Molecule A	Molecule B
II-VI	58.7(1)°	76(1)°
IV-VII	32(1)°	95.5(6)°

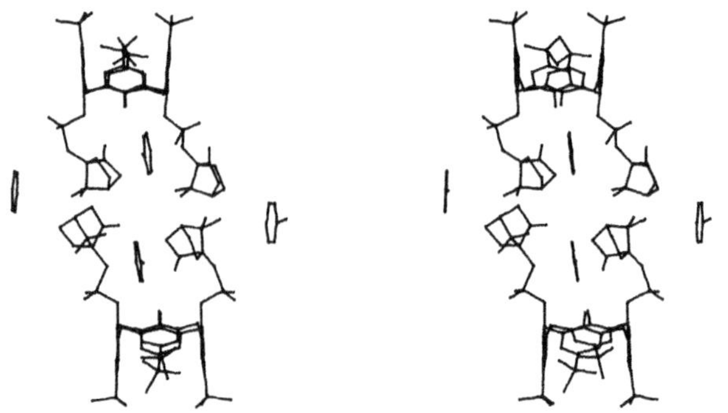

FIGURE 3. Stereoview of the molecules.

1 - MUTHUKRISHANAN, R.; GUTSCHE, C.D.(1979) *J.Org.Chem*, 44, 3962.
2 - BÖHMER, V.; MARSCHOLLEK, F.; ZETTA, L.(1987) *J.Org.Chem.*, 52, 3200.
3 - PAPPALARDO, S.; CACCAMESE, S.; GIUNTA, L.(1991) *Tet.Lett.*, 52, 7747.
4 - MAIN, P., HULL, S.E., LEISINGER, L., GERMAIN, G., DECERCQ, J.P., WOOLFSON, M.M. (1980) *Multan 80. A system of computer program for automatic solution of crystal studies from X-Ray diffraction data*, Univ. of York, England and Univ. of Louvain, Belgique.
5 - SHELDRICK, G.M. (1976) *Shelx, Program for Crystal Structure determination*, Univ. of Cambridge, England.
6 - SHELDRICK, G.M. ; *Shelxl* (1993) Univ.of Göttingen, Germany.
7 - BAVOUX, C.; PERRIN, M.; CHINO, P.(1994) *Supramol. Chem*, 4, 63.
8 - MOTTA, L.; REGNOUF DE VAINS, J.B.; BAVOUX, C.; PERRIN, M.(1995) *J.Chem.Cryst.*, 7, 401.

SYNTHESIS OF WATER SOLUBLE RESORCINARENES APPLICATION IN NANOFILTRATION-COMPLEXATION

L. NICOD[1], E. GAUBERT[1,2], H. BARNIER[2] and M. LEMAIRE[1]*
1 : Institut de Recherches sur la Catalyse, Laboratoire de Catalyse et Synthèse Organique, Université Claude Bernard, Lyon I, CPE, 43 Bld du 11 novembre 1918, 69622 Villeurbanne, France
2 : CEN de Cadarache DESD/SEP/LPTE 13108 Saint Paul lez Durance

Introduction

Considerable effort has been devoted to the development of methods for the long-term isolation of nuclear wastes from the environment. The most promising approach which is currently available involves incorporating radioactive materials into a glass matrix and placing this vitrified waste in a depository deep underground. The enormous expense of incorporating radioactive materials into glass, however, makes it desirable to perform a preliminary separation and preconcentration of key radionucleides, thereby reducing the volume of waste to be vitrified. Some of the isolated radionucleides may even find beneficial uses, like in the irradiation of medical products (caesium-137) or the construction of thermoelectric generators (strontium-90).
Located between ultrafiltration and reverse osmosis, nanofiltration membranes "cut-off" are in the range of 300 to 1000 kg.kmol-1 molecular weight [1]. To increase ionic separation, nanofiltration can be associated with a selective complexation step. In this way, retention of the largest element is improved because the ligand does not pass through the membrane. Water soluble ligands with low molecular weights (about 500 kg.kmol-1) can be used in this nanofiltration complexation association. Whereas, microfiltration or ultrafiltration complexation systems need much larger ligands, which induce viscosity. [2]
From the waste management point of view, the most attractive feature of this technique is its ability to partition radioactive effluents into two separate components, a small volume of concentrated waste containing most of the radioactivity and a relatively large volume of decontaminated effluent suitable for direct discharge. The achievement of higher volume reduction and decontamination factors are twin objectives, and efforts are underway in the development of new water soluble complexants which, it is hoped, will result in the realization of these aims.

Results and discussion

Nanofiltration is carried out with the FILMTEC NF 70 (2514) spiral wound membrane. (Figure 1). By totally recycling the permeate, the feed remains at constant composition during the experiments.

A. W. Coleman (ed.), Molecular Recognition and Inclusion, 431–434.
© 1998 *Kluwer Academic Publishers. Printed in the Netherlands.*

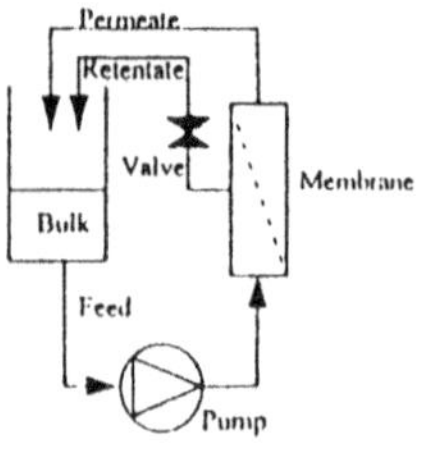

Retention (Ri) is calculated as following :
Ri = ((Cir-Cip) / Cir) . 100
with : - Cir : i solute concentration in the retentate
- Cip : i solute concentration in the permeate

Figure 1 : Nanofiltration apparatus

To concentrate caesium and strontium traces in a sodium nitrate medium, performances of the FILMTEC NF 70 membrane are insufficient. In order to increase membrane selectivity, the association of nanofiltration with complexation was investigated.
Phenolic resins, in particular resins synthesized by alkaline polycondensation of resorcinol with formaldehyde, showed high affinity for caesium. [3,4] Calix[4]resorcinarene (**R' = CH_3**) [5] is soluble in alkali phase [6] and it has a selectivity for caesium. [7]. Resorcinarenes, which have been studied by several researchers, can lead to the formation of interesting host guest complexes in solution. [8] The easy syntheses and functionalisations of the bowl-shaped octols could provide potential sources of molecular vessels for the preparation of polyfunctional host systems. The solubility of the cavitands should be manipulable by the introduction, at positions remote from the cavity, of a number of groups which will render the substances soluble in aqueous phase. [6,9]. The selective extraction of caesium may occur because of its dimensional matching to the cavity of resorcinarene [8], cooperative caesium-arene interactions [10,11], and selectivity of phenolic groups for caesium [12]. Resorcinarenes (**A**) are prepared via a simple one step procedure by acid-catalysed condensation. In this procedure, resorcinol (or one of its derivatives) was heated, along with an aldehyde, at reflux in a mixture of ethanol and concentrated HCl for several hours. For each aldehyde different optimal reaction conditions were apparent. [5,8] Caesium and strontium complexation was then studied using various resorcinarene ligands (table 1).

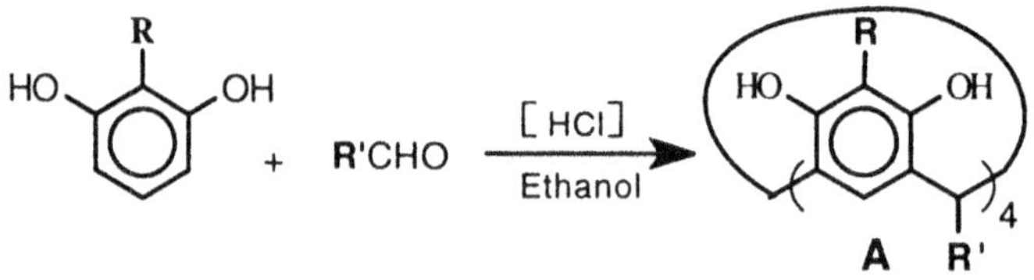

LIGANDS	R	R' m : meta ; p : para	Yield %
1	H	CH_3	87
2	H	C_3H_7	85
3	H	C_7H_{15}	77
4	H	$C_6H_4(OCH_3)p$	88
5	H	$C_6H_4(OH)p$	85
6	H	$C_6H_3(OH)_m(OCH_3)p$	87
7	OH	CH_3	60
8	CH_3	CH_3	75

Table 1 : Synthesis of resorcinarene derivatives

Different derivatives of resorcinarene were prepared and then used as ligands of caesium and strontium in nanofiltration-complexation ; the results achieved are compared in figure 2.

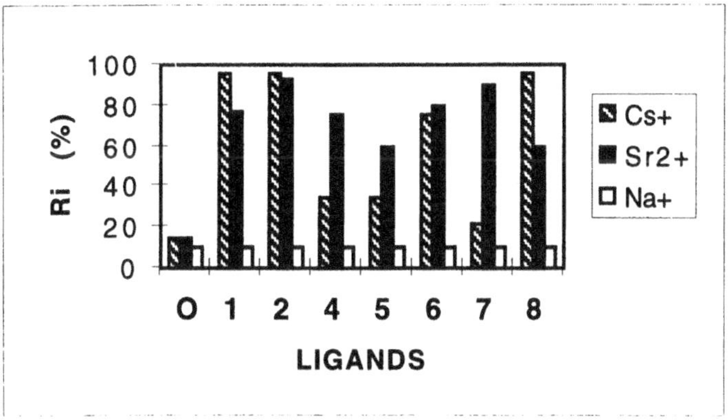

Figure 2 : Ligand effect on caesium and strontium retention
Conditions : $[NaNO_3] = 40\ g\ L^{-1}$; $[Cs+] = [Sr2+] = 0{,}075\ mmol\ L^{-1}$;
[resorcinarene] / [Cs+] or [Sr2+] = 50; $\Delta P = 1{,}5$ MPa ; T = 293 K
O = membrane retention

Without ligand, performances of the FILMTEC NF 70 membrane are insufficient. These resorcinarenes were prepared by the condensation of resorcinol and alkanals such as ethanal **1**, butyraldehyde **2**, octanal **3**. In nanofiltration-complexation, resorcinarene **1**, **2** exhibited the same selectivity of complexation with caesium ($R_{1,2}$ Cs+ = 95 %), but the solubility of cyclotetramers decreased with the length of the alkyl chain. Cyclotetramer **3** was insoluble in aqueous phase.

Resorcinarenes with aryl side chains were also prepared, and subsequently tested. **4**, **5** displayed higher levels of complexation with strontium than with caesium ($R_{4,5}$ Cs+ = 35 % for the both, R_4 Sr2+ = 75 %, R_5 Sr2+ = 60 %). The cyclotetramer **6** showed the same retention for caesium and strontium (R_6 Cs+, Sr2+ = 80 %). Ligand **4**, without the hydroxyl function, was less soluble than ligands **5** and **6**, but presented approximately the same caesium and strontium retentions as ligand **5**. The presence of a functional benzyl group did not improve caesium retention. Caesium retention is more favoured at position 2 of the resorcinol (**R** group) than at the methylidene bridges (**R'** group).

In order to increase the solubility of inclusion compounds, the acid-catalysed condensation of pyrogallol with 1,1-diethoxyethane in ethanol was realized. Cyclotetramer **7** was isolated and tested. It did not bind to caesium but complexed to strontium, selectivity due to the presence of the "5-membered chelate ring". [13]

Finally, tetramethyl cavitand **8** was synthesized and tested. It displayed good selectivity for caesium (R_8 Cs+= 95 %). The methyl group in position 2 of the resorcinol did not decrease the solubility of the corresponding resorcinarene. It also complexed to caesium, contrary to the result achieved using ligand **7**.

At the concentrations studied , the ligands used had no influence on sodium retention (R Na+ = 10).

Conclusion

The acid-catalysed condensations of resorcinols and aldehydes by the four fold oligomeric cyclisation reaction gave various conformationally stable octols in high yields. In association with nanofiltration these ligands improved the sodium / caesium separation by the FILMTEC NF 70 membrane. In an alkaline medium, resorcinarenes **1**, **2**, **8** are caesium selective ligands. At the present time, new resorcinarenes are being synthesized in which the resorcinol is substituted with a hydrophile group, at position 2, in order to improve the solubility. Within the context of the nanofiltration-complexation association, resorcinarenes will be further investigated because these ligands can be easily extracted from the effluent in acidic medium : the resorcinarenes precipitate and can then be reused after filtration.

References

1 Rautenbach, R., Groschil, A., (1990) Separation potential of nanofiltration membranes *Desalination*, **77**, 73-84

2 Lemaire, M., Foos, J., Guy, A., Gaubert, E., Bardot, C., Chomel, R., Radecky, J.J. Maurel, A., Barnier, H., (1995) *F. Patent*, 95 03137

3 Samanta, S.K., Ramaswamy, M., Misra, B.M., (1992), Studies on caesium uptake by phenolic resins, *Sep. Sci. Technol.*, , **27**, 255-267

4 Dumont, N., Favre-Réguillon, A., Dunjic, B., Lemaire, M., (1996) Extraction of caesium from an alkaline leaching solution of spent catalysts using an ion-exchange column, *Sep. Sci. Technol.*, **31**(7), 1001-1010

5 Sverker Högberg, A. G., (1980) Two stereoisomeric macrocyclic resorcinol-acetaldehyde condensation products, *J. Org. Chem..*, **45**, 4498-4500

6 Schneider, H. J.,Güttes, D., Schneider, U. (1988) Host-guest complexes with water-soluble macrocyclic polyphenolates including induced fit and simple elements of a proton pump, *J. Am. Chem. Soc*,, **110**, 6449-6454

7 Gaubert, E., Barnier, H., Nicod, L., Favre-Reguillon A., Foos, J., Guy, A., Bardot, C., Lemaire, M., (1996) Selective caesium removal from a sodium nitrate aqueous medium by nanofiltration-complextion, *Sep. Sci, Technol.*, submitted

8 Timmerman, P., Verboom, W., Reinhoudt, D.N., (1996) Resorcinarenes, *Tetrahedron*, **52**, 2663-2704

9 Lippmann, T., Wilde, H., Pink, M., Schäfer, A., Hesse, M., Mann, G. (1993) Host-guest complexes between calix[4]arenes derived from resorcinol and alkylammonium ions *Angew. Chem. Int. Ed. Engl.*, **32**, 1195-1197

10 Ungaro, R., Casnati, A., Ugozzoli, F., Pochini, A., Dozol, J. F., Hill, C., Rouquette, H., (1994) , 1,3 dialkoxycalix[4]arenecrowns-6 in 1,3-alternate conformation : caesium-selective ligands that exploit cation-arene interactions, *Angew. Chem. Int. Ed. Engl.*, **33**, 1506-1509

11 Harrowfield, J. M., Ogden, M. I., Richmond, W. R., White, A. H., (1991) Calixarene-cupped caesium : a coordination conundrum, *J. Chem. Soc., Chem. Commun*, 1159-1161

12 Samanta, S.K., Misra, B. M. (1995) Ion exchange selectivity of a resorcinol-formaldehyde polycondensate resin for caesium sin relation to other alkali metal ions, *Solv. Extract. Ion. Exchang*, **13**, 575

13 Samanta, S.K., Ramaswamy, M., Misra, B.M., (1992) Synthesis and caracterization of some phenolic resins for the selective removal of radiostrontium, *Radiochimi. Acta.*, **57**, 201-205

CRYSTAL ENGINEERING IN SOLID-STATE METAL SALT COMPLEXES OF CYCLODEXTRINS

I. NICOLIS[1], M. EDDOUADI[2], A.W. COLEMAN[2], M. SELKTI[1,], F. VILLAIN[1], C. de RANGO[1].*

*[1]ERS128 CNRS, Centre Pharmaceutique, Université Paris-Sud, Châtenay-Malabry, F92296; [2]UPR412 CNRS, IBCP, 7, Passage du Vercors, Lyon F69367. Present Address: *Laboratoire de Physique, Faculté des Sciences Pharmaceutiques et Biologiques de Paris V, F75270.*

We have developed extensive investigations on the nature of the solid-state cyclodextrin-salt interaction and its effect on the molecular self-organization. The first examples of complexes were provided by β-cyclodextrin with metal chlorides[1,2] (calcium and hydrated magnesium). Then, systematic crystallization experiments have been undertaken to determine the conditions for solid-state metal complexation, given the solubilities of α- β- and γ- cyclodextrins as a function of the salt concentration.

For calcium chloride, under certain limits crystals containing pure cyclodextrins with only slight variations in the hydration levels are obtained. Beyond these limits, complexes crystallize with distinct calcium/cyclodextrin stoichiometries in the different solubility phases. Crystal structures of a variety of complexes reveal different modes of molecular self-organization.

I - RELATION BETWEEN SOLUBILITY AND SOLID-STATE COMPLEXATION

For γ-cyclodextrin, the solubility curve shows three parts, as it was observed for other metal salts[3]: the solubility slightly decreases at low salt activities, then rapidly increases for medium activities before reaching a "plateau" for high activities. In the case of α-cyclodextrin, the solubility increases in a non linear way, again reaching a plateau at high salt activities.

Complexation takes place after the medium activities phases of solubility, for both α- and γ-cyclodextrins. For salt concentrations less than 1.5M, all crystal forms are isomorphous with those of the related hydrated cyclodextrin structures.

In the case of γ-cyclodextrin, two different complexes crystallize in the "plateau" area, with metal/cyclodextrin stoichiometries of (1/1) and (2/1), respectively at concentrations ranging between 1.75 to 2M and at a concentration of 2.25M. For α-cyclodextrin, at concentrations ranging between 1.75 and 2M, a (1/1) complex is obtained. At very higher concentrations, two different complexes, with a stoichiometry of (3/1), crystallize with different hydration levels.

A. W. Coleman (ed.), Molecular Recognition and Inclusion, 435–438.
© 1998 *Kluwer Academic Publishers. Printed in the Netherlands.*

II - CRYSTALLOGRAPHIC STUDIES OF CYCLODEXTRIN/$CaCl_2$ SYSTEMS

Single crystal analyses with a variety of $CaCl_2$/cyclodextrin complexes are in progress. Now, the structures of six novel organic/inorganic systems have been determined. In all systems, calcium complexation takes place at the exterior of the cavity, the cyclodextrins acting as uni- bi- or tridentate ligands. Cyclodextrin-cation and/or anion interactions are fundamental in the formation of two types of self-assembling systems.

- A modulation by the calcium ion of the basal herring-bone chain-type structure of the hydrated cyclodextrin was realized in the case of β-cyclodextrin[1]. A slighter modulation of the chain-like structure is encountered with complexes of γ-cyclodextrin, with (1/1) and (2/1) stoichiometries[4], which present different packings of the chains.

- Novel molecular composites showing distinct organic and inorganic substructures are created in the case of α-cyclodextrin. Totally different large molecular building blocks and inorganic frameworks are generated, according to the stoichiometry of the complexes (1/1 or 3/1), which all contain infinite columns filled by the majority of the mineral elements.

A - Modulation by the calcium cation of the basal structure of the hydrated cyclodextrin

The presence of 2 $CaCl_2$ per β-cyclodextrin inside the crystal lattice lead to the formation of a novel herringbone (HB) chain-like pattern of β-cyclodextrins, arranged in two "antiparallel" chains[1]. The ions serve as intermolecular bridges between parallel (*anions*) or antiparallel (*cations*) HB chains. These HB chains are tightly packed than those previously reported: the HB chain is elongated by a factor of 1.5, whereas the basic molecular area per chain is 30% smaller.

The structures of the two complexes of γ-cyclodextrin reveal different arrangements of parallel HB-chains. Several common characteristics are remarkable:

- none of the ions participate directly to the packing of the HB chains; direct interactions between cations and cyclodextrins are uniquely implied in the intra-HB chain structures while all the anions are positioned on partially occupied sites and not involved in the structural organization;
- the stacking mode of monomers within the HB chains is quasi invariant (Fig. 1); the chains are separated by an infinite interstitial space running parallel to the **b**-axis and filled by water molecules and anions;
- the second-sphere interactions displayed by the cations are fundamental in the self-assembly of the HB chains. In the (2/1) complex, the environment of one cation (black circle) is identical to that of the cation present in the (1/1) complex. Two water molecules coordinated to this cation and situated in the instersticial space, link the chains along the **c**-axis.

The major difference arises from the packing of the chains along the **a**-axis. This may be related to the presence of the second cation. While, in the (1/1) complex, these chains are intermolecularly hydrogen-bonded and related by translation along the **a**-axis, in the (2/1) complex they are not directly connected and are bridged by a water molecule coordinated to the second cation (grey circle). As a result, the chains are alternately stacked with a vertical shift of b/2 (6.5Å).

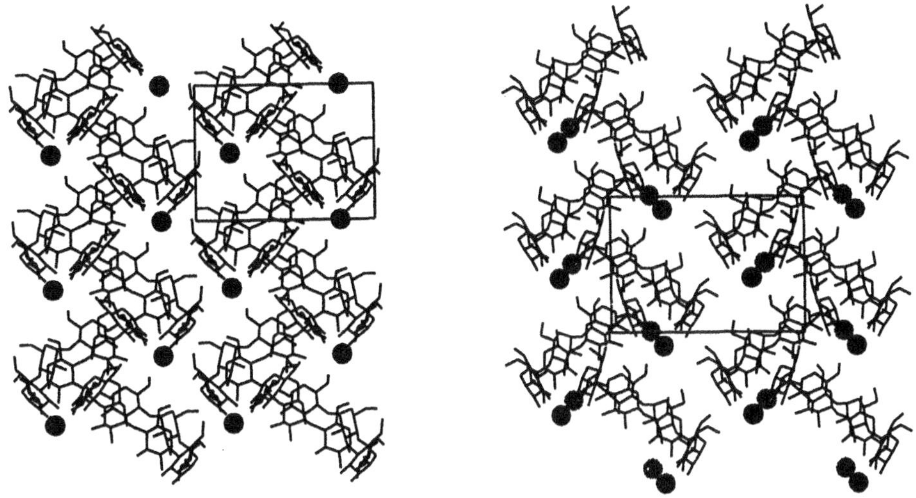

Figure 1: Comparison of the HB chains running along the vertical **b**-axis (**c**-axis horizontal; cations, black and grey circles). Left: (1/1) γ-CD; Right: (2/1) γ-CD.

B - New organic/inorganic assemblies of α-Cyclodextrin

First-sphere coordination of calcium ions by α-cyclodextrin gives rise to new organic-inorganic structural models based on self-assembly of different molecular building blocks. Two types of systems exhibiting large organized mineral areas are encountered.

- The layer-type structures of α-cyclodextrin monomers

The basic molecular motif of the two (3/1) complexes is a layer (height of 10Å) of α-cyclodextrins, as monomer units, with molecular axis perpendicular to the layers . Though the molecular association in layers is invariant, two modes of mineral component organization are observed[5]. In both systems, there are no short contacts between cyclodextrins from successive layers; structural motifs of mineral components span the intermolecular gap to join the layers by building up framework substructures.

- The multilayer-type structure of α-cyclodextrin trimers

The basic molecular motif of the (1/1) complex is a layer (height of 14Å) of α-cyclodextrins, as trimer units, with molecular axis parallel to the layers[6].

An original large molecular building block (198 non-H atoms) arises from the association of three symmetry-related α-cyclodextrin monomers, with molecular axes perpendicular to a 3-fold axis. The self-assembly of the trimers, which interact with anions and are intermolecularly hydrogen-bonded, gives rise to layers. The stacking of three symmetry-related layers creates infinite columns (repetition unit 42Å) filled by mineral components, which run along the 3-fold axis and are perpendicular to the layers. The anions, positioned on the 3-fold axis, are essential in the trimer association and in the organization in trimer layers as well as in the creation of inorganic junction zones, interconnecting the columns and joining the layers.

Work is ongoing to design solid-state complexes of cyclodextrin with specific physical properties, by extending the crystal engineering strategies exemplified with calcium chloride to a large range of metal salts.

REFERENCES

1 - I. Nicolis, A. W. Coleman, P. Charpin & C. de Rango *Acta B* **331**, 122-130 (1996).

2 - I. Nicolis, A. W. Coleman, P. Charpin & C. de Rango *Angew. Chem. Int. Ed. Engl.* **34, No 21**, 2381-2383 (1995).

3 - M. Eddaoudi, P.C. Junk & A. W. Coleman *J. Incl. Phenom. and Mol. Rec.*, in press.

4 - Crystal Data. (1/1) γ-CD complex: $P2_1$, Z=2, a=18.023(5), b=12.916(9), c=18.335(6) (Å), β=106.09(2)°, R=0.122 for 4582 unique reflexions with $I>4\sigma(I)$; (2/1) γ-CD complex: C2, Z=4, a=37.742(2), b=13.020(6), c=19.281(9) (Å), β=100.41(5)°, R=0.121 for 3683 unique reflexions with $I>4\sigma(I)$.

5 - I. Nicolis, A.W. Coleman, P. Charpin & C. de Rango (submitted paper).

6 - Crystal Data. R3, Z=9, a=b=18.876(4), c=41.82(2) (Å), R=0.070 for 4090 unique reflexions with $I>4\sigma(I)$.

Glycolipid Hydrolase Models.
D, L-Stereorecognition of Amino Acids

Yasukazu Ohkatsu and Miho Ozawa
Department of Applied Chemistry, Faculty of Engineering, Kogakuin University,
1-24-2 Nishishinjuku, Shinjuku-ku, Tokyo, 163-91 JAPAN

Introduction

A hydrolase, α-chymotrypsin, is studied relatively well on the structure and action mechanism and, for a long time, has been a target substance for enzyme models. Many researchers of hydrolase models have devoted their effort to the stereorecognition of substrates, especially such as D and L-phenylalanine esters. R. Ueoka et al. repored[1,2] the perfect D, L-recognition like an enzyme, for example, using a tripeptide enzyme model, Z-phe-his-leu, in a mixed vesicular system consisting of dipentadecyldimethylammonium bromide and cetyldimethylammonium bromide.

The authors, on the other hand, proposed glycolipids as new hydrolase model, which form vesicles in water by themselves, and reported[3] that $Man(Lau)_2$-His-COOH could distinguish esters of D, L-phenylalanines with p-nitro-phenol in a considerably high recognition ratio (about 100) in the presence of Man Lau. Recently, they also reported[4], in the study on hydrolyses of amides of amino acid with p-nitro-aniline, that an enzyme model system, consisting of $Man(Lau)_2$-His-COOMe and Man Lau, could recognize D, L-alanine amides perfectly, on the contrary that it did a little recognition of D, L-phenylalanine amides.

In this study, a few lipids were examined, on hydrolyses of amino acid amides, as reaction field in the place of Man Lau. D, L-Stereorecognition ability of an enzyme model was found remarkably to depend on the kinds of reaction fields. The recognition mechanism will be disscussed.

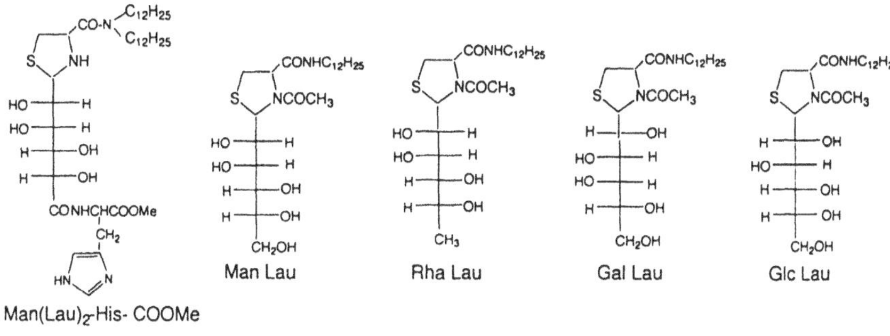

Figure 1. Enzyme model and lipids as reaction fields

A. W. Coleman (ed.), Molecular Recognition and Inclusion, 439–442.
© 1998 *Kluwer Academic Publishers. Printed in the Netherlands.*

Experiment

An enzyme model, Man(Lau)$_2$-His-COOMe, and lipids, Man Lan, Rha Lau, Gal Lau, and Glc Lau, as reaction field were synthesized as described elsewhere. Commercial phosphatidylcholine and cholesterol were used as reaction field as well. Substrates such as Z-phenylalanine, -valine, -methionine and -alanine, and Boc-alanine p-nitro-aniline amides were prepared by conventional methods.

Hydrolyses were carried out using Man(Lau)$_2$-His-COOMe at constant concentration, regardless of the presence of a lipid. The reaction was pursued by measuring the amount of p-nitro-aniline liberated by hydrolysis by means of a UV spectrometer, and was analyzed by overall rate constant k_2.

Results and Discussion

As described in a previous paper,[4] stereorecognition of D, L-phenyl-alanine, -valine, -methionine, and -alanine amides was studied using an enzyme model system [Man(Lau)$_2$-His-COOMe and Man Lau]. The model system recognized D, L-configuration of an amino acid, which has a more hydrophilic side group, in a higher recognition ratio, and distinguished the stereochemistry of D, L-alanine amides perfectly. Other lipids, such as Glc Lau, Gal Lau and Rha Lau, were then examined as reaction field, in hydrolyses of D, L-alanine amides (Fig. 2). In the Figure, results were shown at lipid concentrations of 0, 25, 50, 75, and 100% for most lipids, but it should be understood that both D- and L-alanine amides were not hydrolysed at the concentration, where points were not plotted.

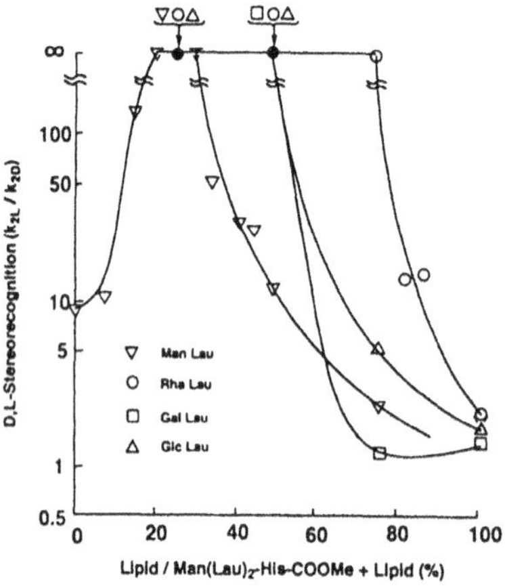

Figure 2. Stereorecognition in hydrolyses of alanine amides using an enzyme model consisting of Man(Lau)$_2$-His-COOMe and a lipid at 25℃.

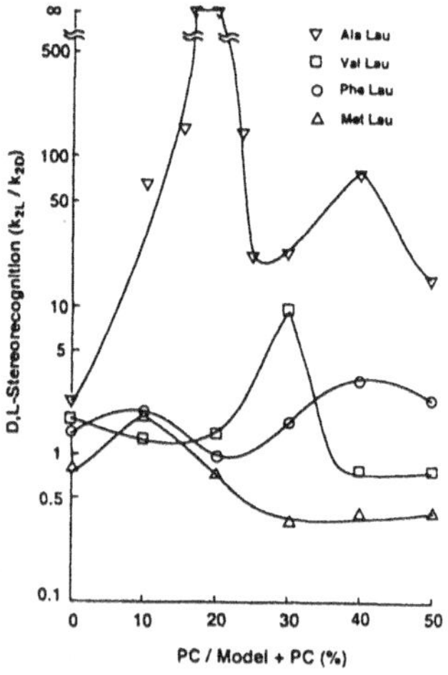

Figure 3. Stereorecognition in hydrolyses of amino acid amides using an enzyme model Man(Lau)$_2$-His-COOMe and phosphatidylcholine at 25℃.

It was found that the perfect recognition was observed in the presence of all lipids examined, and was much affected by kinds of lipids. This fact shows that D, L-recognition is ultimately attributable to the enzyme model, His-Man(Lau)$_2$-COOMe, while perfect D, L-recognition needs some assistance of a lipid reaction field. The effect of a lipid, however, does not seem to be concerned with the configuration of its sugar moiety. When comparing Man Lau with Rha Lau, the abilities of lipids to enhance D, L-recognition of the enzyme model were considerably different, although the sugar moieties of both lipids have the Fsame configuration. The different abilities, thus, may be ascribed to the groups on ω-position: a methylol group for Man Lau and a methyl one for Rha Lau. This fact suggests that the degree of hydrophobic nature around the hydrophilic region of a mixed vesicle of enzyme model system subtly controls the D, L-recognition abiliity of the enzyme model. It is an interesting fact that a slightly more hydrophobic rhamnose-type lipid provides perfect D, L-recognition over a wider range of the concentration (25-75%) than that for a mannose-type lipid (20-30%).

In order to elucidate the condition of hydrophilic region of vesicles, another lipid, phosphatidylcholine, was used as reaction field. Shown in Fig. 3 is the stereorecognition for four kinds of amino acids. In these cases, too, the perfect stereorecognition was obseved only for D, L-alanine amides, though the range of concentration of phosphatidylcholine enhancing the recognition ability of a model was narrower than that of a lipid mentioned above. The other amino acids were recognized only with slight recognition ratios.

Furthermore, phosphatidylcholine was examined as reaction field in the presence of cholesterol. Cholesterol is known to function in two ways, in a biobody, from the standpoint of flexibility of a biomembrane. That is, a rather rigid membrane is made soft,

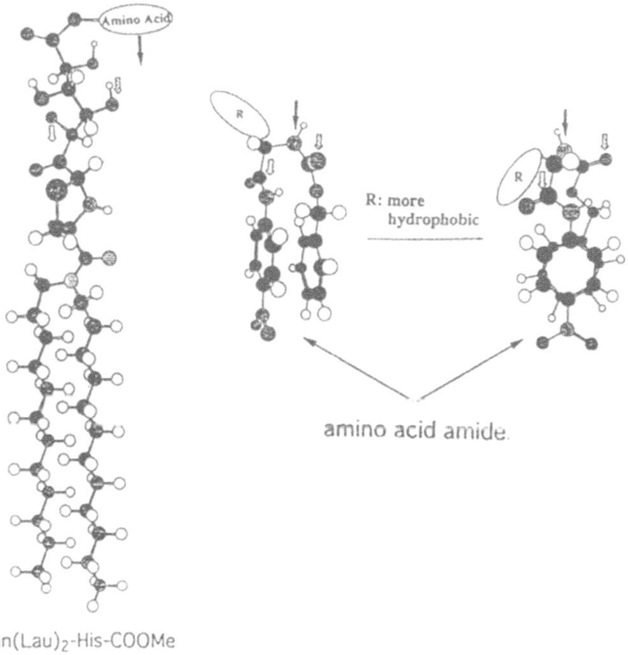

Figure 4. Postulated interaction of Man(Lau)$_2$-His-COOMe with an amino acid amide.

while a flexible membrane is rigid, resulting in the stabilization of the menbrane. Generally speaking, the addition of cholesterol, in this study, decreased D, L-stereorecog nition, probably because it made a vesicle more rigid as reflected by the decrease in hydrolysing velocity. For example, the D, L-recognition of alanine amides was decreased to the recognition ratio (k_2L/k_2D) of 9.0 (phosphatidylcholine: 20% added) in the presence of 5% of cholesterol, although the perfect recognition was observed at the same concentration of phosphatidylcholine without choresterol. Surprisingly it was found, however, that D, L-phenylalanine amides having a hydrophobic side group were perfectly distinguished when only cholesterol was used as reaction field at the concentration of 5%. This is the first discovery that only one enzyme model can perform D, L-stereorecognition of two kinds of amino acids, phenylalanine and alanine, by controlling the reaction field.

When the above-mentioned facts are discussed from a different point of view, it can be estimated that both hydrophilic nature and flexibility of a reaction field control the uptake of a substrate and the conformation of a substrate on being uptaken, to increase the stereorecognition ability of an enzyme model.

The authors have reported[5,6] that the 2-oxy anion derived from the dissociation of sugar moiety of an enzyme model attacks the carbonyl carbon of an ester bond of an amino acid with p-nitro-phenol, resulting in cleavage of the bond. When such an active oxy anion acts on an amino acid amide, a ball-and-stick molecular model exhibits that the high degree of hydrophobic nature of a side group of amino acid makes the group dive into the inside of a vesicle deeply (Fig. 4). Corrsequently, the methine hydrogen of amino acid changes its position by rotating the single bond between asymmetric carbon and carbonyl carbon, to widen the space around the carbonyl carbon of amino acid amide. Thus, the oxy anion can be easy to approach the carbonyl carbon and the hydrolysing velocity is increased, but the D, L-recognition ability of the enzyme model is losen. It is also understood by means of a ball-and-stick molecular model that the protection of amino group of a substrate by Z-group is necessary for D, L-stereo-recognition, since a phenyl (of Z) - phenyl (of p-nitro-phenyl) ineraction maintains a preferable conformation of the substarate (see, Fig. 4). This was confirmed using Boc-protected alanine amides in place of Z-protected ones. When hydrolysing the Boc-alanine amides by Man(Lau)$_2$-His-COOMe and Man Lau, they were hydrolysed only at the concentration of Man(Lau)$_2$-His-COOMe above 75mol%, and the hydrolysing rate was considerably low and the stereorecognition ratio was low as well, compared with those of Z-alanine amides.

In conclusion, the methine hydrogen of amino acid sterically controls the approach of the enzyme model and determines the stereorecognition ratio, being concerned with the chemical and physical nature of reaction field and Z-protection group of an amino acid.

References

1) Ueoka, R., Moss, R.A., Swarup, S., Matsumoto, Y., Strauss, G., and Murakami, Y.(1985) Extraordinary Micellar Enantioselectivity Coupled to Altered Aggregate Structure, *J. Am. Chem. Soc.*, **107**, 2185.

2) Ueoka, R., Matsumoto, Y., Yoshino, T., Watanabe, N., Omura, K., and Murakami, Y. (1986) Remarkable Steric Control for the Enabtioselective Cleavage of Amino Acid Esters, *Chem. Lett.*, 1743.

3) Kawaguchi, K., Isobe, K., Ohkatsu, Y., and Kusano, T. (1992) Synthesis and Catalytic Actions of Self-Vesicle and Micelle-Forming Enzyme Models of Hydrolase, *J. Jpn. Oil Chem. Soc.*, **41**, 458.

4) Ohkatsu,Y., Ozawa, M., Numata, Y., and Nakamura, N. (1996) Glycolipid Enzyme Models X. Catalysis of Glycolipids with Amino Acid Residue,*J. Jpn. Oil Chem. Soc.*, **45**, 29.

5) Ohkatsu, Y. and Nakamura, N. (1996) Glycolipid Enzyme Models IX. Effects of the Configuration of Sugar Moieties, *J. Jpn. Oil Chem. Soc.*, **45**, 21.

6) Ohkatsu, Y., Kubokoya, J., and Horibe, M. (1996) Glycolipid Enzyme Models XI. The Search for Active Sites of Glycolipids as Hydrolase Model, *J. Jpn. Oil Chem. Soc.*, in contribution.

ANTHRACENE CROWN ETHERS:
SYNTHESIS AND COMPLEXATION OF SELECTED CATIONS

RYSZARD OSTASZEWSKI

Institute of Organic Chemistry, Polish Academy of Sciences,

01-224 Warsaw, Poland

Complexing properties of several type of crown ethers are already well known [1]. Among them, anthracene crown ethers were shown to behave as sensitive fluorescent probes for metal cations. In most cases the syntheses of these compounds consist of several steps and desired compounds are obtained in small amount due to difficulties of synthesis. For example, in three step synthesis bis-(9,10-anthracenediyl)coronands were obtained in the yield of 20-30% [2]. Crown ethers incorporating the 9,9'-(ethane-1,2-diyl)bis(anthracene) groups were prepared in the yield of 7% [3]. Unfortunately, only poor selectivity towards single cations was observed for both types of fluoroionophores. Interesting results were obtained upon incorporation of nitrogen atom into the crown ether molecule. Azacrown ethers, based on anthracene cyclam, were prepared and were found to behave as fluoroionophores towards transition metal cations in buffered solution [4].

This result prompted us to synthesize the anthracene crown ethers of general structure **4** shown in Scheme 1. Whole strategy is based on functionalization of compound **1** *via* already known Richman-Atkins procedure [5].

For the first reaction, ethylenediamine-N,N'-di-p-toluenesulphonylamide was used as a nucleophile. This reaction was performed in DMF solution, according to already known procedure [5] and one product of structure **2** was obtained in 21% yield. The same procedure was applied for 1,10-di(p-toluenesulphonylamido)-4,7-dioxadecane, and the desired compound **3** was obtained in 62% yield.

A. W. Coleman (ed.), Molecular Recognition and Inclusion, 443–446.
© 1998 *Kluwer Academic Publishers. Printed in the Netherlands.*

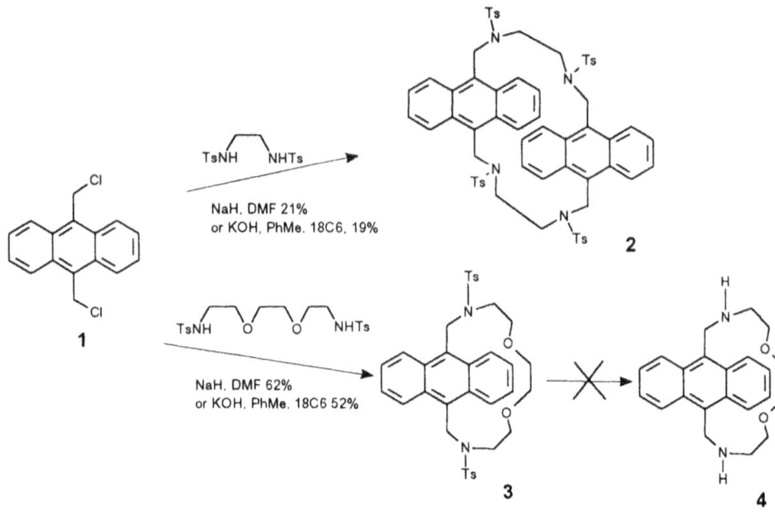

Scheme 1

Several attempts were made to obtain compound **4**. None of the reported methods of deprotection leads to formation of the desired compound in reasonable yield.

Since we were interested in a simple, efficient synthesis of anthracene crown ethers we turned our attention to a sulfuric acid-promoted reaction between veratrol and aliphatic aldehydes [6]. When benzo-15-crown-5 and aliphatic aldehydes were used, the reaction proceeded smoothly [7] and the desired crown ethers **6** were obtained in the yield of 21 to 53%, according to Scheme 2.

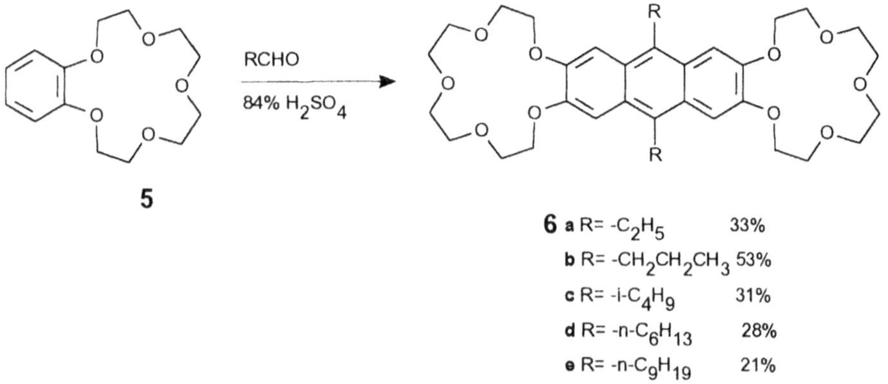

Scheme 2

The prepared crown ethers are able to form several types of complexes. Incorporation of the compound in a solvent polymeric membrane electrodes gives general information about selectivity of the complexes formed. The logarithmic values of the selectivity coefficient (expressed as log $K_{K,M}$) of the membranes containing ligands were determined in a usual way [8]. Increasing selectivity towards potassium cation was observed while the number of carbon atoms in aliphatic groups was increased.

UV titration experiments for crown **6c** were performed in a chloroform - methanol mixture. Upon addition of KSCN, formation of a complex was clearly observed. The first step in complexation consists of formation of sandwich type complex (ratio cation to crown 1:2) with stability constant log $K_{1,2}$ = 8.6, then upon addition of cation formation of a 2:2 complex is observed with log $K_{2,2}$ = 3.6. While the same experiment was performed for NaSCN, only formation of 1:2 complex was observed with stability constant log $K_{1,2}$ = 7.7.

Since for sodium cation only one crown ether was involved in complexation, the synthesis of mono-anthracene crown of general structure **8** was carried out according to Scheme 3.

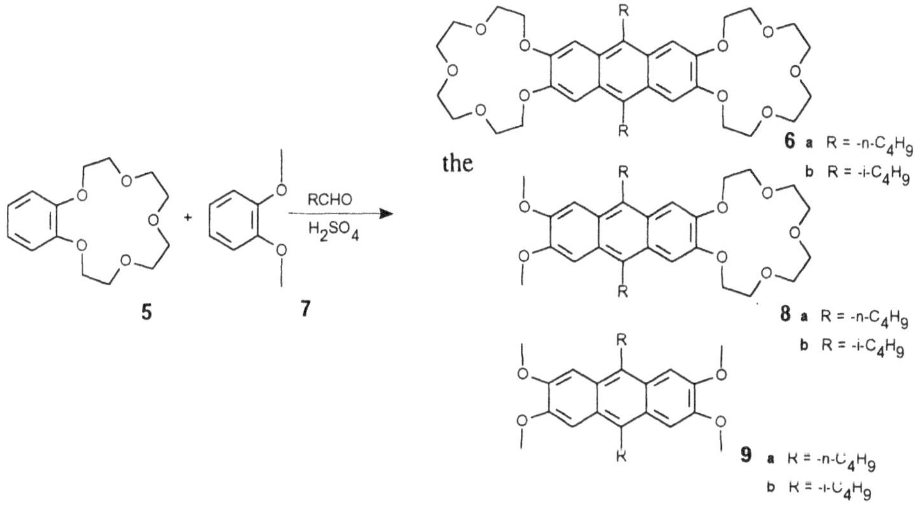

Scheme 3

For cross-condensation reaction benzo-16-crown-5 (**5**) and veratrol (**7**) were used. When reaction was performed in the presence of valeraldehyde, anthracene **9a** was obtained in 8% yield. Both anthracene crowns were obtained in 30% yield as almost unseparable mixture. The yield of single product was calculated from NMR spectra and anthracene crown **6a** was obtained in 18% yield while desired crown **8a** was obtained in 12% yield.

While isovaleraldehyde was used, only two products **9b** and **8b** were obtained. in 24% and in 6.3% yield, respectively.

It is evident that selectivity of reaction is modulated by steric hindrance of the alkyl groups. Larger isobutyl group favours formation of the desired compound, but simultaneously decreases yield of the reaction.

In conclusion, we have found that aliphatic chains which are present at position 9 and 10 of the anthracene moiety do regulate selectivity of the crown ethers studied. In a chloroform - methanol mixture predominate complexes of the crown to cation 2:1 ratio. Synthesis of model crown ethers was optimized and interesting substituent effect on the reaction course was observed. Bulky alkyl groups favoured formation of monoanthracene crown **8** while n-alkyl groups favoured formation of dianthracene crown **6**.

Acknowledgement

This work was supported by State Committee for Scientific Research grant no. 3 TO9A 050 011. We thank Prof. J. Jurczak for his kind interest and constant support.

References

1. Izatt, M.I., Pawlak, K., Bradshaw, J.S. (1995) Thermodynamic and Kinetic Data for Macrocycles Interaction with Cations, Anions, and Neutral Molecules, *Chem. Rev.* **95**, 2529-2586
2. Marquis, D., Desvergne, J.-P., Bouas-Laurent, H. (1995) Photoresponsive Supramolecular Systems: Synthesis and Photophysical and Photochemical Study of Bis-(9,10-anthracenediyl)coronands AAO_nO_n, *J. Org. Chem.* **60**, 7984-7996
3. Desvergne, J.-P., Laurent, J., Bouas-Laurent, H., Marsau, P., Lahrhar, N., Andrianatoandro, H., Cotrait, M., Synthesis, X-ray Structure, Spectroscopic and Cation Complexation Studies of Macrocyclic Ligands Incorporating the 9,9'-(Ethane-1,2-diyl)bis(anthracene) Photoactive Subunit, *Recl. Trav. Chim. Pays-Bas.* **114**, 504-513
4. Fabbrizzi, L., Licchelli, M., Pallavicini, P., Perotti, A., Taglietti, A., Sacchi, D. (1996) Fluorescent Sensors for Transition Metals Based on Electron-Transfer and Energy-Transfer Mechanisms, *Chem. Eur. J.* **2**,75-82
5. Richman, J., E., Atkins, T., J. (1974) Nitrogen Analogs of Crown Ethers, *J. Am. Chem. Soc.* **96**, 2268-2270
6. Muller, A.; Raltschewa, M.; Papp, M. (1942) Uber die Dimerisation des Isoeugenolmetylathers, *Chem. Ber.* **75**, 692-703
7. Ostaszewski, R. (1996) Synthesis of Anthracene Crown Ethers, *Synthesis*, submitted
8. Malinowska, E., Ostaszewski, R. (1996) The Influences of Alkyl Groups on Complexation of Anthracene Crown Ethers, to be published

CYCLO BIS-INTERCALAND RECEPTORS: STRUCTURE OF TWO INTERCALATIVE INCLUSION COMPLEXES. STABILITY OF THE HYDRATION NETWORK FACING TWO DIFFERENT SUBSTRATES

T. PARIS [a], J-P VIGNERON [a], J-M LEHN [a],
M. CESARIO [b], L. TCHERTANOV [b], J. GUILHEM.[b].
a) Laboratoire de chimie des interactions Moléculaires, Collège de France ,Paris, France.
b) Laboratoire de Cristallochimie , Institut de Chimie des Substances Naturelles, CNRS,Gif sur Yvette, France.

1. Introduction

Molecular recognition of flat substrates requires the design of receptor molecules containing complementary flat units. If two such units are incorporated into a macrocyclic framework, a **face to face inclusion** of a **planar substrate** may take place, leading to an **intercalative supramolecular structure**.
The water-soluble macrocyclic bis-intercaland receptor $[M]^{+}$, containing two naphtalene subunits ,linked by two positively charged oxy-bis-ethylamine binding sites is able to bind flat organic substrates. Crystal structures of terephtalate (1) and isophtalate (2) inclusion complexes of receptor $[M]^{+}$ are reported here :

(1)

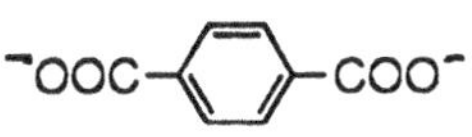

$[M]^{+}$

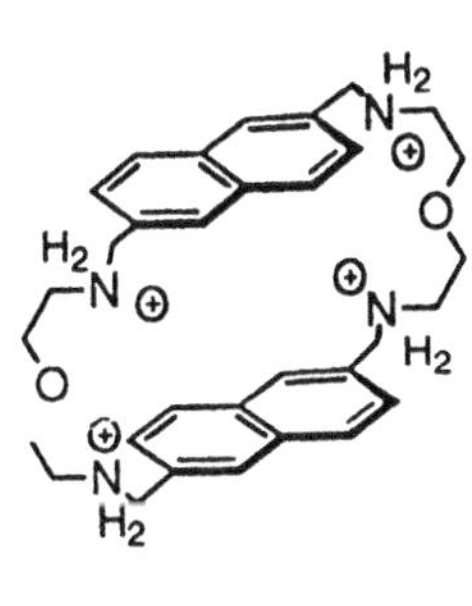

(2)

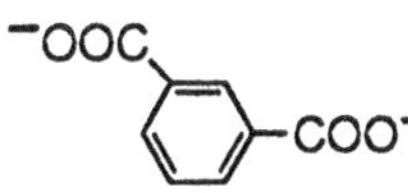

A. W. Coleman (ed.), Molecular Recognition and Inclusion, 447–450.
© 1998 *Kluwer Academic Publishers. Printed in the Netherlands.*

2. Crystal structures and fluorescence spectroscopy of complexes:

TEREPHTALATE

In the cell:

Ligand = [C32 N4 O2 H44]4+

Substrate = [C8 O4 H4]2-

Counter-ion = 2Cl-

Solvent = **10 water molecules**

(all H localized)

No disorder

R1 = 0.039 for 3168 Fo > 4σ(Fo)

R2 = 0.063 for all 4215 data(SHELX93)

ρmax = 0.25 e/Å^3

ρmin = -0.20 e/Å^3

Triclinic P-1 Z = 1

a = 7.717 (3) Å α = 99.00(7)°

b = 10.625(6) Å β = 99.70(6)°

c = 16.238(9) Å γ = 109.46(10)°

V = 1205 Å^3

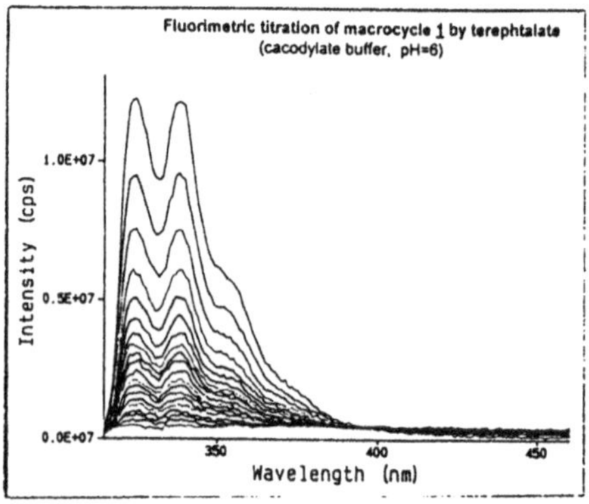

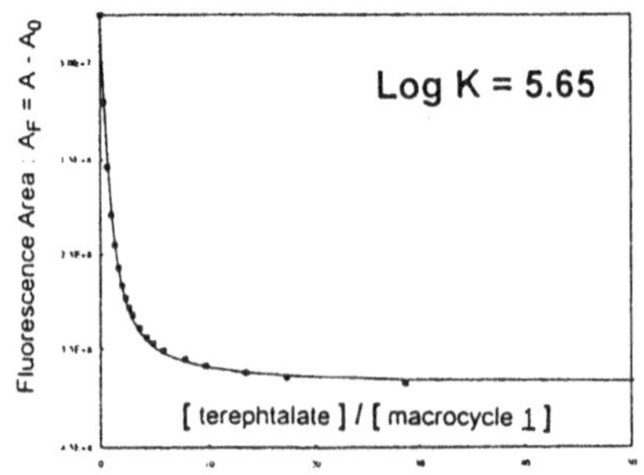

ISOPHTALATE

In the cell:

Ligand = [C32 N4 O2 H44]4+

Substrate = [C8 O4 H4]2-

Counter-ion = 2Cl-

Solvent = **10 water molecules**

(11/20 H localized)

No disorder

R1 = 0.064 for 2838 Fo > 4σ(Fo)

R2 = 0.12 for all 4269 data(SHELX93)

ρmax = 0.63 e/Å^3

ρmin = -0.33 e/Å^3

Triclinic P1 Z = 1

a = 7.513 (10)Å α = 98.81(5)°

b = 10.640(9) Å β = 99.70710)°

c = 16.164(10) Å γ = 109.36(12)°

V = 1170 Å^3

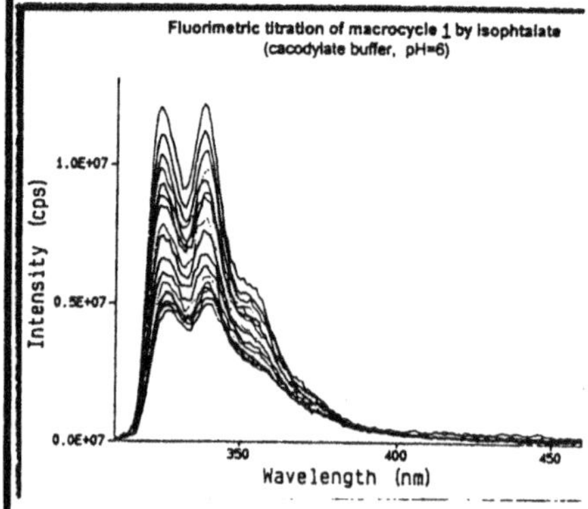

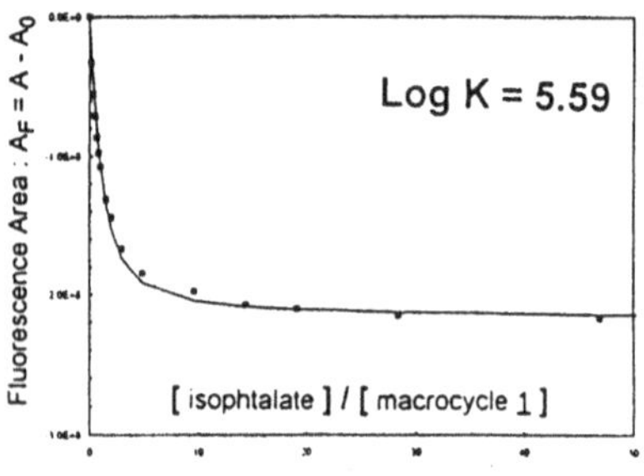

The binding of substrates (1) or (2) by the macrocyclic bis-intercaland receptor $[M]^+$ is controlled by fluorescence spectroscopy measurements. The similitude of the crystal structures of the two inclusion complexes are in accordance with the vary close values of stability constants.

3. Hydration network in the two inclusion complexes:

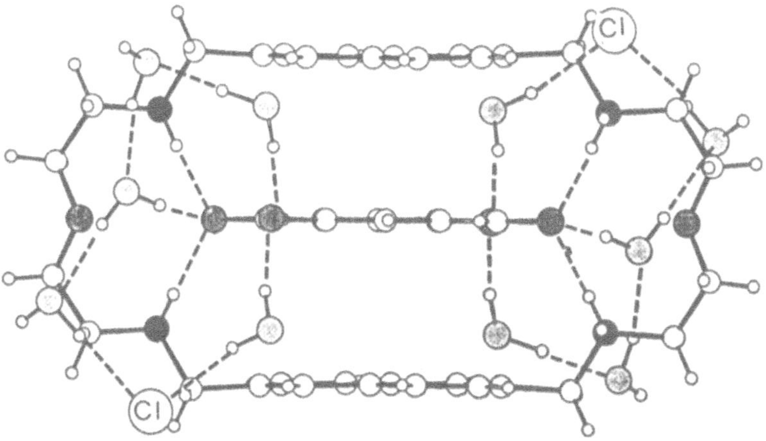

(1) Terephtalate substrate

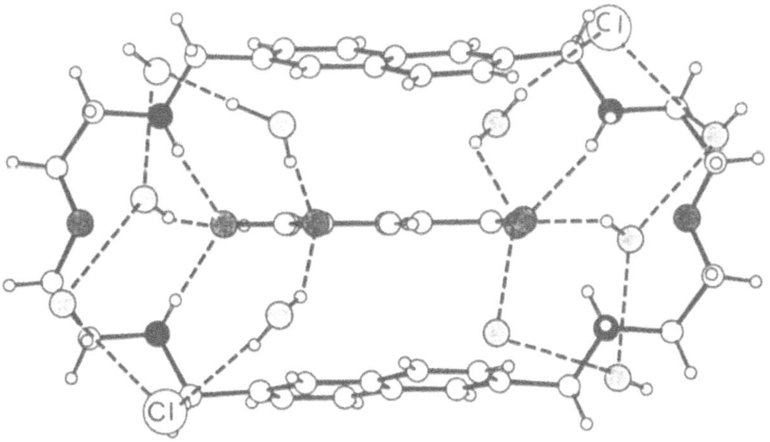

(2) Isophtalate substrate

Both complexes with included terephtalate and isophtalate anions, crystallize in the same triclinic cell. In each one, we find two chloride anions and ten water molecules, in spite of the different geometries of the polar H-bond acceptor groups of the substrates. In both crystal structures, complicated but **practically identical hydrogen bond networks** bind the different H-bonds donors ($NH2^+$), the acceptors (COO^-, Cl^-) and the donors / acceptors (water molecules):[All H bonds lengths 2.7Å< < 3.2Å].

Interactions between the **receptor** and the **substrate** are similar in the two complexes: identical parallel stacking between the aromatic planes of the host and that of the two different guests [only a slight angular deviation (9°) of the two naphtalene rings in the iso-structure].There is no observed disorder in the isophtalate complexe, neither by rotational motion of the substrate in its plane, nor by its statistical stacking in different orientations

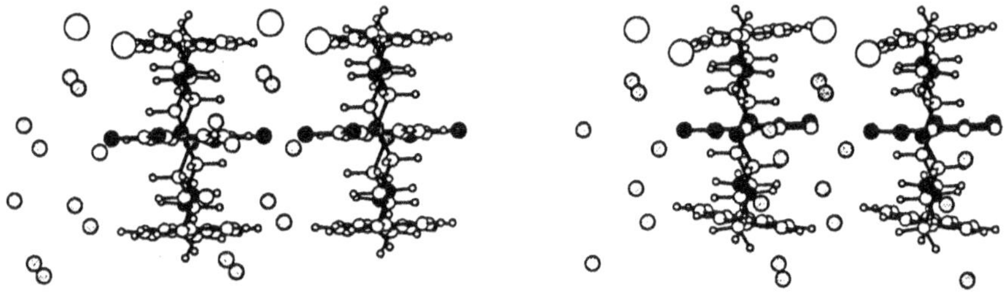

View of complexes (1) and (2) along the O...O axis of macrocycle bridges

4.Conclusion:

the

In **both** structures, the **water network** and the **intercalation process** are the **same**. From one structure to the other, the substrate aromatic ring glides in its plane to satisfy the H-bonding requirements of the medium, thus keeping the carboxylate groups nearly in the same position.

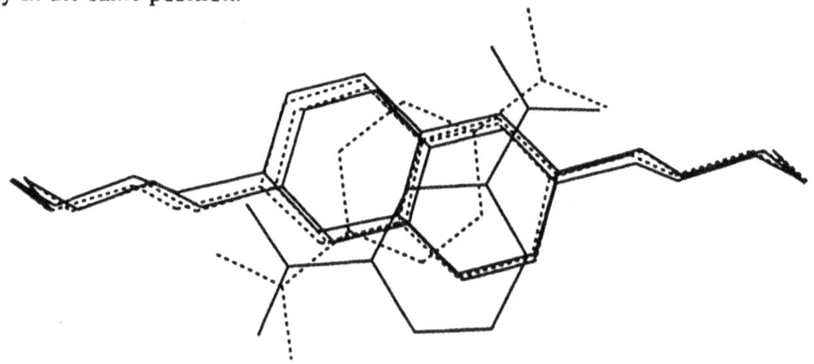

We then can consider the medium as a stable assembly : [water-/Choride anions /macrocycle]. This suggests the **pre-organization** of the water framework around and with the macrocycle, helping the planar anions to interact with the ligand, removing chloride anions. It would be interesting to isolate the structure of the free ligand neutralized with the 4 Cl^- anions to see if the macrocycle keeps its 3D shape or collapses.

ZEOLITES AS CATALYSTS: POROSITY OR ACIDITY? ALKYLATION OF BENZENE

G. Perez[a)], O. Ursini[b)], and E. Lilla[b)]

[a)]Istituto di Cromatografia del CNR and [b)]Istituto di Chimica Nucleare del CNR.Area della Ricerca, c.p. 10, 00016 Monterotondo Stazione (RM), Italy

Pore dimension and acidity play an important role in the zeolite catalytic activity. In the alkylation of benzene with ethanol over HZSM-5 zeolites the yield of ethylbenzene was found dependent on the acidity of the solid. However, more acidic zeolites, but with larger pore dimension, gave lower yields of alkylated products. When n-octanol is employed as an alkylating agent the effects of acidity and pore dimension can be distinguished.

INTRODUCTION

The catalytic activity and selectivity of zeolites are determined not only by the physicochemical properties of the active sites but also by the size, shape, and dimensions of the pores. The knowledge of the acidity and the understanding of the spatial distribution of the adsorbed molecules are therefore of fundamental importance. Nowadays, there is a general tendency to explain the selectivity features of zeolites on the bases of geometrical factors. While these factors can be dominant when the dimension of the activated complex is close to the size of the porous, they become less important when the dimensions of channels and cavities largely overcome those of reacting molecules and activated complex. Under these circumstances, and for acid catalysed reactions, the only parameter which is normally considered to explain zeolite catalyst behaviour is the acid strength.

H^+-forms of zeolites contain Brønsted and Lewis acid sites [1], being their acidity related to the Si/Al ratio. Even if many attempts have been done to establish an acceptable scale of zeolite acidity [2,3], its knowledge is still a real problem.

A. W. Coleman (ed.), Molecular Recognition and Inclusion, 451–454.
© 1998 *Kluwer Academic Publishers. Printed in the Netherlands.*

Since the rate on zeolites is limited by diffusion because of the microporosity and no relationship could be established between the acidity of different zeolites and the catalytic properties, the objective of this study was to elucidate the effects of pore dimension and acidity in the alkylation of benzene, using well-known mid-pore zeolites as ZSM-5, Mordenite and USY, and larger pore materials (Theta-1 zeolite and an alumina pillared montmorillonite, Al-PILC).

EXPERIMENTAL

In a typical experiment a mixture of 0.55 mmol of benzene and 0.11 mmol of alkylating agent was added to 6 mg of catalyst, activated at 450°C overnight, in a 6 mL Pyrex vial, which was then outgassed at liquid nitrogen temperature and then sealed. The reaction was carried out for 30 min. at 300°C. The identification of the products was performed by gas-chromatography and GC-MS.

RESULTS AND DISCUSSION

The effect of a decrease of the acidity of HZSM-5 zeolites (different Si/Al ratio) was investigated. The results, reported in Figure 1, indicate that acidity plays an important role in alkylating activity. For all the zeolites ethylbenzene was the principal product, while diethylbenzenes accounted for about 1/5 of ethanol conversion.

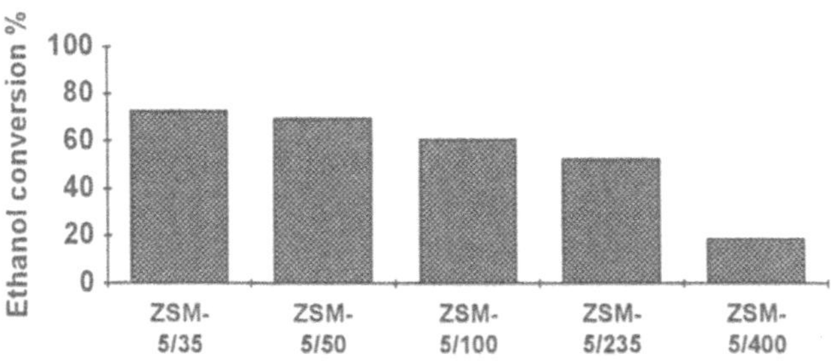

Figure 1. Ethanol conversion after 30 min at 300°C on different HZSM-5 zeolites.

A comparison between different pore dimensions is given in Figure 2, where the yield of the alkylated products have been calculated with respect to the amount of the alcohol. It can be easily observed that pore dimension more than acidity plays a role in benzene alkylation when different solids are compared. As a matter of fact, from previous studies [4,5], we found the following acidity scale:

USY ≥ Theta-1 > H-Mordenite > HZSM-5/50 > Al-PILC

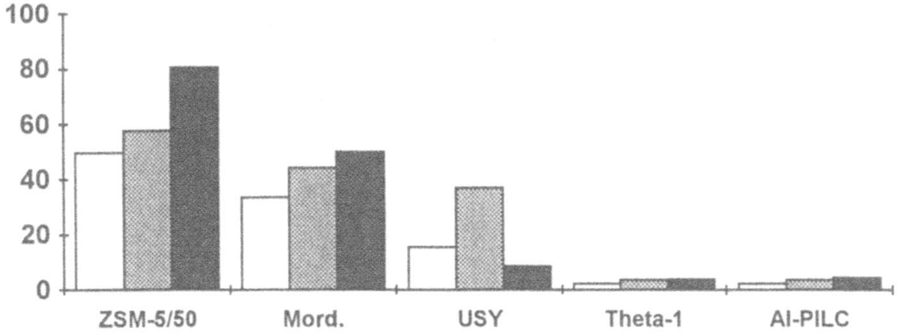

Figure 2. Yield (%) of alkylation products calculated with respect the amount of alcohols. (Ethanol, butanol and n-octanol from left to right).

From these results it seems that pore dimension more than acidity is responsible of alkylating efficiency, which seems to be in contrast with the results obtained on ZSM-5 zeolites.

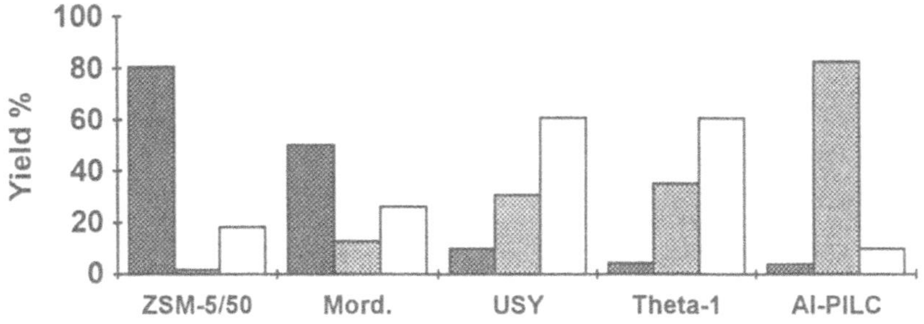

Figure 3. Fate of n-octanol on acidic solids. Benzene/n-octanol molar ratio =5. (Alkylation, dehydration, and C-C fission, from left to right.)

The effects of both parameters can be observed when n-octanol is employed as an alkylating agent. In this case three reaction pathways can be distinguished: alkylation of benzene with formation of alkylbenzenes (mainly octylbenzenes), dehydration of alcohol leading to octenes, and C-C cleavage with formation of lower molecular weight hydrocarbons. The extent of the three reactions, shown in Figure 3, can be explained on the basis of the following considerations.

1. The yield of alkylated products decreases at larger pore dimensions.
2. The extent of dehydration increases with the pore dimension.
3. Higher is the acidity, wider the fragmentation of n-octanol.

There is no doubt that the first step of the reaction is the protonation of the alcohol molecules by the zeolitic Brønsted sites [6]. This charged intermediate, an oxonium ion bound through the proton to an oxygen atom of the zeolite, can undergo reaction with benzene, via a carbenium ion, or loose a water molecule to form the correspondent alkene. These two reactions are competitive, being the first one favoured in the smaller cavities, where the probability of an impact between the two molecules increases. In larger pore it seems likely that benzene can escape the reaction.

As in the cracking processes the acidity plays a role in n-octanol C-C fission, as shown by the results obtained with the more acid USY and Theta-1, and with the less acidic Al-PILP. Such a fission has been shown to occur essentially on the starting material [7], as confirmed by our results which indicate that fragmentation of aliphatic chain of alkylaromatic occurs in a low extent.

In conclusion, both, acidity and pore dimension are responsible of benzene alkylation by alcohols, being the second parameter more determining since rate of the reaction is enhanced when the reactant concentration is increased in the zeolite micropore [8].

REFERENCES

1. Corma A., LLopis F., Viruela P., and Zicovich-Wilson C. (1994) Acid Softness and Hardness in Large-Pore Zeolites as a Determinant Parameter to Control Selectivity in Orbital-Controlled Reactions, *J. Am. Chem. Soc.* **116**, 134 and references therein.
2. Corma A. (1995) Inorganic Solid Acids and Their Use in Acid-Catalyzed Hydrocarbon Reactions, *Chem. Rev.* **95**, 559
3. Hölderich M, Hesse M., and Näumann F. (1988) Zeolites: Catalysis for Organic Syntheses, *Angew. Chem. Int. Ed. Engl.* **27**, 226
4. A. De Stefanis, G. Perez, and A. A. G. Tomlinson (1994) Pillared Layered Structures *vs.* Zeolites as Sorbents and Catalysts, *J. Mat. Chem.* **4**, 959
5. G. Perez, A. De Stefanis, and A. A. G. Tomlinson (1966) Pillared Layered Structures *vs* Zeolites as Sorbents and Catalysts. 3. Catalytic Activity of Alumina Pillared Smectites in Ethylbenzene to Diethylbenzene and Ethilbenzene to Styrene Conversions, *J. Mat. Chem.* (in press)
6. Izumi Y., Urabe K., and Onaka M. (1992) *Zeolite, Clay, and Heteropoly Acid in Organic Reactions*, Kondansha Ltd., Tokyo (Japan), Ch. 1.
7. Gonçalves de Almeida J. L., Dufaux M., Taarit Y. B., and Naccache C. (1994) Effect of Pore Size and Aluminium Content on the Production of Linear Alkylbenzenes over HY, H-ZSM-5 and H-ZSM-12 zeolites: Alkylation of Benzene with 1-Dodecene, *Appl. Catal. A: General* **114**, 141
8. van Santen R. A. and Kramer G. J. (1995) Reactivity Theory of Zeolitic Brønsted Acidic Sites, *Chem. Rev.* 95, 637

QUANTIFICATION OF SPECIFIC IMMUNOLOGICAL REACTIONS BY ATOMIC FORCE MICROSCOPY

A.PERRIN, A.THERETZ
Unité mixte CNRS/bioMérieux, ENS Lyon
46 Allée d'Italie, 69364 LYON cedex 07 FRANCE
Tel : (33 4) 04-72-72-83-63 Fax : (33 4) 04-72-72-85-33
E-mail : Alain.Theretz@ens-bma.cnrs.fr

In this paper we intend to demonstrate the ability of the Atomic Force Microscope (AFM) to become a tool for the sensitive quantitation of immunological reactions. Anti-ferritin antibodies were adsorbed on silane modified silicon oxyde wafers. The adsorption process was characterized and quantified by RIA, ELISA and AFM, using gold labelled antibodies. A linear correlation was found between the antibody concentration in solution and the number of gold particles, and this method was proved to be more sensitive than classical immunoassays. Finally, the binding reaction between anti-ferritin antibodies and ferritin antigens was studied by AFM. A linear relation was also found between the antigen concentration in bulk solution, and the number of ferritin molecules. The sensitivity threshold for the ferritin detection was 0.06 μg/ml.

1. Introduction

The atomic force microscope [1] is becoming a widely used tool in biosensors engineering, thanks to its ability to obtain excellent resolution on biomolecules under physiological conditions -ambient temperature and pressure, liquid environment-. One of the most promising application is the detection of immunological reactions. Some authors already obtained very encouraging results in this field [2,3]. Here, we aim at studying the ferritin/anti-ferritin system in order to get quantitative informations about the reaction, i.e. to relate the number of ferritin molecules, linked to anti-ferritin antibodies, to the initial antigen concentration in solution. The AFM will be used as a quantitative tool [4], allowing the direct detection and quantitation of ferritin molecules.

A. W. Coleman (ed.), Molecular Recognition and Inclusion, 455–458.
© 1998 *Kluwer Academic Publishers. Printed in the Netherlands.*

2. Experimental section

Silicon oxyde wafers were cleaned under acidic conditions and silanized by Methyl n Octadecyl DiEthoxy Silane (MODES) [5,6]. The silanation process was controlled by XPS, contact angle measurements, and AFM. Anti-ferritin antibodies were adsorbed onto these silane modified surfaces. The adsorption was quantified by RIA using ^{125}I radio labelled anti-ferritin antibodies, by ELISA using Alkaline Phosphatase (AKP) labelled anti-mouse antibodies, and by AFM (non contact mode in air) using gold labelled (diameter=40 nm) anti-mouse antibodies. Ferritin solutions of increasing concentrations were allowed to incubate on antibody coated (40 μg/ml) wafers, and antigens were detected by AFM in contact mode in air.

3. Results and discussion

3.1. SILANATION PROCESS

The silanation process was checked by XPS, confirming the presence of carbon atoms (atomic ratio: 7.9%), and by contact angle measurements, showing that hydrophobic surfaces were obtained (θ_w=90°). Furthermore, the surface roughness was measured by AFM, in order to check that wafers were sufficiently smooth (R_{rms}=0.7 nm).

3.2. ANTIBODY DETECTION BY RIA

the

The increase of the radioactivity signal as a function of the antibody concentration in solution was related to the amount of adsorbed antibodies (Figure 1). The detection threshold was calculated from the intial slope of the curve, and was found to be 11 pg/mm^2 of anti-ferritin antibodies.

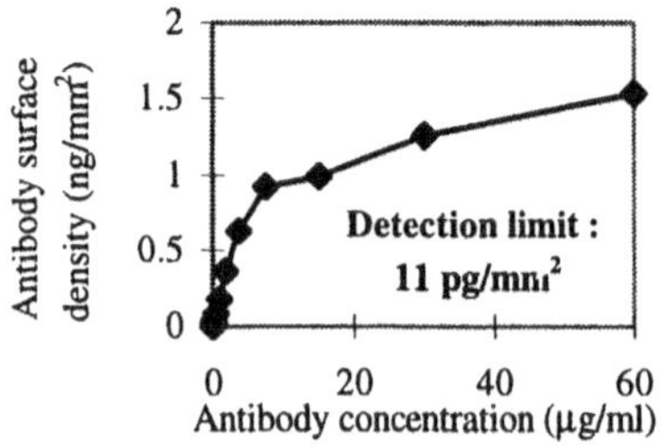

Figure 1. Antibody detection by RIA.

3.3. ANTIBODY DETECTION BY ELISA

The sensitivity threshold for the antibody detection by enzym labelled conjugates was estimated to 12 pg/mm^2, which is very close to the RIA result (Figure 2).

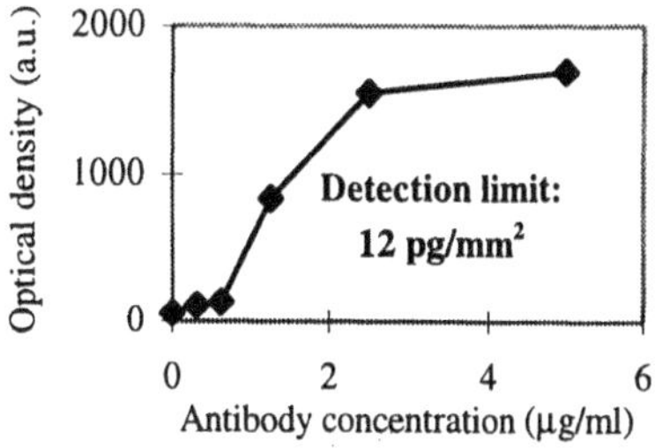

Figure 2. Antibody detection by ELISA.

3.4. ANTIBODY DETECTION BY AFM

No gold spheres were detected when no antibody were present on the surface (Figure 4), proving that no non specific adsorption of gold labelled conjugates occured. Gold particles were found to be evenly distributed, proving that the antibody layer was homogeneously adsorbed on the MODES modified surface. Counting the gold particles number on images as a function of the antibody bulk concentration was found to be the more sensitive method (Figure 3). As less as 0.3 pg/mm^2 of antibodies were detected by this method, which is 40 times better than classical immunoassays.

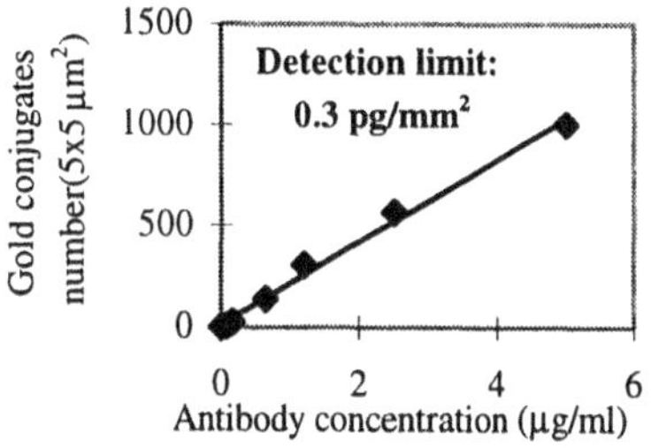

Figure 3. Antibody detection by AFM.

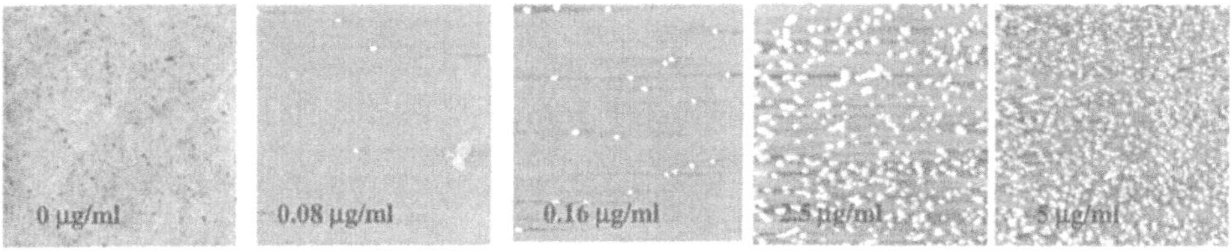

Figure 4 . Antibody detection by gold conjugates when increasing antibody concentration.

3.5 ANTIGEN DETECTION BY AFM

Ferritin molecules were found to be individually detected on a rather large scan area (3x3 µm^2) (Figure 6), although antibodies were not discrimated at this scale. We beleive that the surfactant (Tween 20) used in antigen buffer to avoid not specific adsorption filled holes between antibodies and smoothened the IgG layer. Some ferritin aggregates were sometimes seen, possibly due to the tip pushing some molecules. Nevertheless, ferritin distribution was found to be homogeneous. The sensitivity for ferritin detection

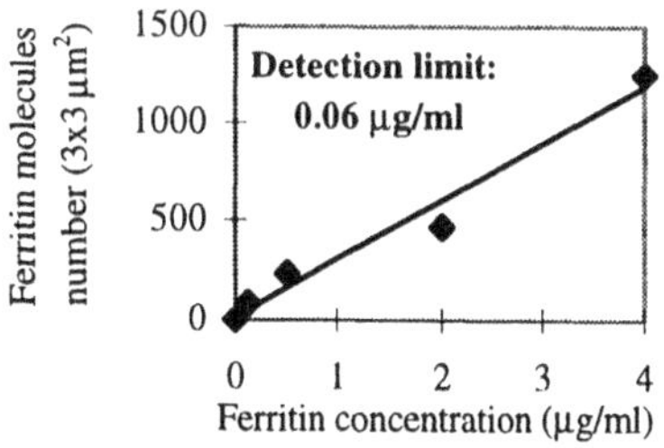

Figure 5. Antigen detection by AFM.

by AFM was 0.06 µg/ml (Figure 5). Up to now this is less sensitive than what is obtained on the Vidas automate (bioMérieux) (0.002 µg/ml), but this preliminary result is encouraging in a view to use the AFM as a new sensitive immunoassay tool.

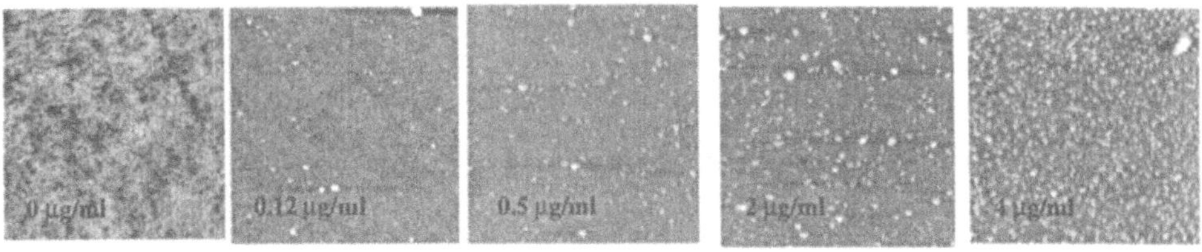

Figure 6. Ferritin direct detection when increasing antigen concentration.

4. Conclusion

The aim of this paper was to demonstrate the feasibility of a new immunoassay based on the AFM. First, we proved that the use of gold labelled antibodies allowed to reach a better sensitivity threshold than classical immunoassays (RIA and ELISA) for antibody detection. Then, the antibody-antigen reaction was quantified by AFM, without using any labels, as the large size of the ferritin molecule allowed its direct detection. It has been shown that antigens can be detected and quantified using the AFM.

(1) Binnig, G., Quate, C.F. and Gerber, C. (1986) Atomic Force Microscopy, *Phys.Rev.Lett.* **56**, 930.
(2) Masai, J. and Sorin, T. (1990) Scanning Tunneling Immunoassay: a preliminary experiment, *J.Vac.Sci.Technol.* **A8(1)**, 713.
(3) Davies, J., Dawkes, A.C., Haymes, A.G., Roberts, C.J., Sunderland, R.F., Wilkins, M.J., Davies, M.C., Tendler, S.J.B., Jackson, D.E. and Edwards, J.C. (1994) A STM comparison of passive antibody adsorption and biotinylated antibody linked to streptavidin on microtiter wells, *J.Immunol.Methods* **167**, 263.
(4) Perrin, A., Lanet, V. Theretz, A. (1996) Quantification of immunological reactions by atomic force microscopy, *Accepted in Langmuir.*
(5) Duvault, Y., Gagnaire, A., Gardies, F., Jaffrezic-Renault, N. and Martelet,C. (1990) Physicochemical characterization of covalently bound alkyl monolayers on silica wafers, *Thin Solid Films* **185**, 169.
(6) Bennett, D.R., Matisosn, J.G., Netting, A.K.O., Smart, R.C. and Swincer, A.G. (1992) Silicones on glass surfaces, *A.G. Polym.Int.* **27**, 147.

TRANSPORT STUDIES OF INORGANIC AND ORGANIC CATIONS ACROSS LIQUED MEMBRANES CONTAINING MANNICH-BASE CALIX[4]RESORCINARENES

OKSANA PIETRASZKIEWICZ, MAŁGORZATA KOŹBIAŁ, MAREK PIETRASZKIEWICZ*
Institute of Physical Chemistry, Polish Academy of Sciences
01224 Warsaw, Kasprzaka 44/52, Poland

1. Introduction

Calixarenes are very versatile macrocyclic molecular receptors comprising phenolic units. Their easy availability, complexing features and possibility for a variety of chemical modifications makes them particularly attractive to serve as target molecules for various studies in supramolecular chemistry. Three monograph books on calixarene chemistry appeared to date [1-3].

One of the features of calixarene chemistry is the possibility of complex formation with organic/inorganic cations, anions and neutral molecules, and we have been interested in recognition of inorganic and organic cations using chemically modified Mannich-type calix[4]resorcinarenes by means of transport experiments across bulk liquid membrane containing the carriers.

2. Experimental

Lipophilic calix[4]resorcinarene was prepared by a standard method [4] from resorcinol and lauryl aldehyde, and further functionalized by reaction with piperidine/N-methylpiperazine and paraformaldehyde [5]. The products were purified by a double cristallization from acetonitrile, and dried in high vacuum at $80^{o}C$ for several hours. Thansport experiments were performed in a custom-made apparatus in isooctane and toluene as bulk liquid membranes at $25^{o}C$. The concentration of the carriers in the membranes were 10^{-3}M. Concentration of the picrates in the receiving phase were measured spectroscopically. Alkali metal picrates and selected primary ammonium picrates (ethanoloamine picrate **A**, n-butylamine picrate **B**, and (S)(-)phenylethylamine picrate **C**) were used.

3. Results and Discussion

Two Mannich-base calix[4]resorcinarenes incorporationg piperidine (L2) and piperazine (L3) units were used, as well as the unsubstituted calix[4]resorcinarene (L1) for comparison. The L1 turned out to be ineffective in both transport and extraction

A. W. Coleman (ed.), Molecular Recognition and Inclusion, 459–462.
© 1998 *Kluwer Academic Publishers. Printed in the Netherlands.*

experiments, indicating insignificant complexing properties. In contrary, the substituted ligands L2 and L3 displayed pronounced selectivities that are collected in the Tables 1 and 2 containing the extraction and transport data, respectively.

Figure 1

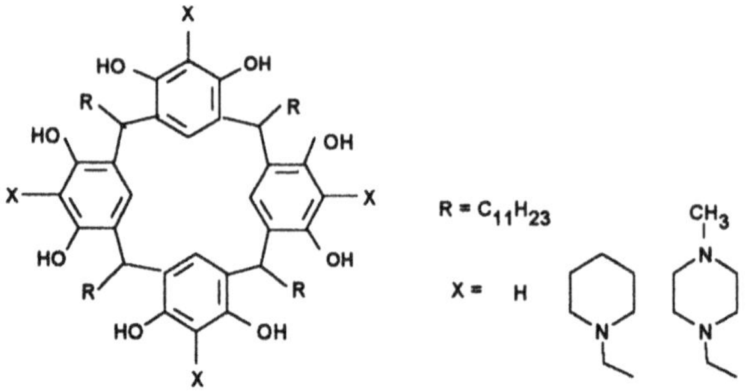

For inorganic and organic cations the extraction coefficients for isooctane phase were lower than for those in toluene. The extractability order in toluene for inorganic cation was the following: L3 > L2 >> L1. For the primary ammonium picrates the highest extraction coefficient was in case of phenylethyl ammonium picrate in both isooctane and toluene membranes (for L2 and L3). Several bar prolts show the pattern of ionoselectivities of the ligands towards inorganic and organic cations estimated in transport experiments

TABLE 1.Extraction coefficients for organic phase/wateı

Cation	Ligand	$k_{o/w}$ x 10^{-2} organic phase Isooctane	$k_{o/w}$ x 10^{-2} organic phase Toluene
Li	L2	1.08	4.30
Na	L2	1.83	4.53
K	L2	1.11	5.25
Rb	L2	1.23	4.51
Cs	L2	1.05	5.15
A	L2	3.10	9.17
B	L2	1.69	7.53
C	L2	10.13	17.23
Li	L3	1.22	10.93
Na	L3	2.21	11.09
K	L3	1.75	13.29
Rb	L3	1.35	12.20
Cs	L3	1.72	12.40
A	L3	6.50	13.64
B	L3	2.25	11.37
C	L3	15.71	21.43

TABLE 2. Transport efficiency of picrates through bulk liquid membrane

Cation	Ligand	Flux $J \times 10^{-12}(ns^{-1}cm^{-2})$ Isooctane membrane	Flux $J \times 10^{-12}(ns^{-1}cm^{-2})$ Toluene membrane
Li	L1	1.42	1.00
Na	L1	1.50	2.00
K	L1	1.92	1.30
Rb	L1	1.42	0.90
Cs	L1	1.00	1.17
A	L1	0.66	0.30
B	L1	1.08	1.71
C	L1	0.75	1.53
Li	L2	1.35	2.39
Na	L2	2.30	2.25
K	L2	2.45	6.47
Rb	L2	2.25	3.18
Cs	L2	2.19	5.56
A	L2	4.02	4.22
B	L2	2.85	3.32
C	L2	7.12	6,45
Li	L3	2.35	27.34
Na	L3	2.79	1.71
K	L3	3.38	1.83
Rb	L3	2.79	1.65
Cs	L3	3.98	1.83
A	L3	2.83	3.54
B	L3	3.36	5.83
C	L3	5.07	13.39

The phenylethyl ammonium picrate was transported with the highest rate in both isooctane and toluene in the presence of L2 and L3. Comparison of the transport rates is the following: isooctanc L2 > L3, toluene L3 > L2. Better recognition was observed in toluene. The preferential transport and extraction of phenylethyl ammonium picrate over other ammonium cations may reflect its higher lipophilicity, compared to less lipophilic cation A and B. In both isooctane and toluene the transport rates displayed the same pattern of selectivity.

The differences in inorganic cation selectivities estimated in transport experiments were not very pronounced. For L3 in isooctane cesium ion had the highest rate of transportation, whereas in toluene Li^+ selectivity was remarkable.

It has been observed that the UV-Vis spectra of the complexes of L2 and L3 with primary ammonium picrates were different from the spectra of initial substrates, which indicated an interaction between the picrate anion and the aromatic part of the calix[4]resorcinarene

Figure 2

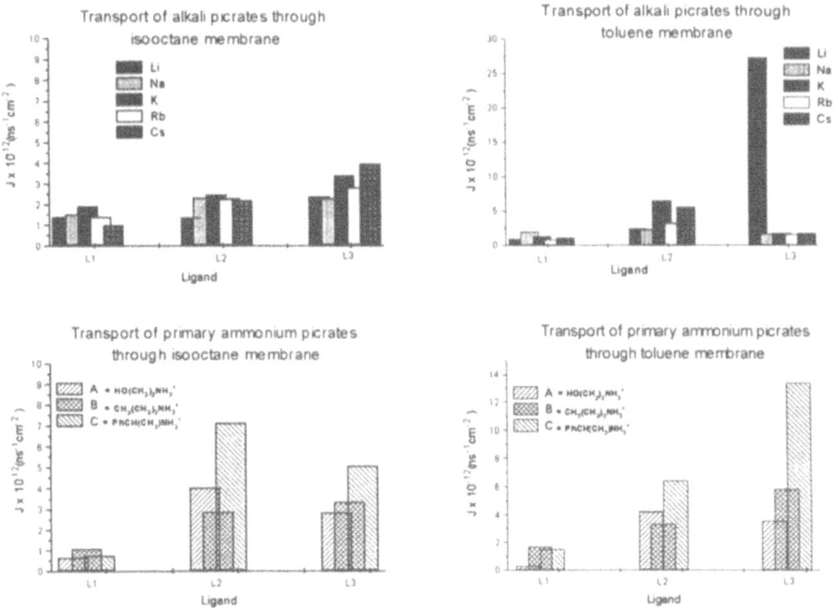

4. Conclusions

This work indicated that there are three factors influencing the cation selectivity of the modified calix[4]resorcinarenes: the substituent, the bulk membrane composition, and the counteranion present in the investigated salts. The interplay between these three variables may bring very specific selectivities from the point of view of separation science.

5. Acknowledgement

This work was supported by the National Committee for Scientific Research, Grant No. 3 T09A 025 011.

6. References

1. Gutsche, C. D. (1989) *Calixarenes, Monographs in Supramolecular Chemistry,* Royal Society of Chemistry, Cambridge.
2. Vicens, J. and Böhmer, V. (1991) *Calixarenes: a Versatile Class of Macrocyclic Compounds,* Kluwer Academic Publishers, Dordrecht.
3. Vicens, J., Asfari, Z., and Harrowfield, J. M. (1994) *Calixarenes 50th Anniversary: Commemorative Volume,* Kluwer Academic Publishers, Dordrecht
4. Aoyama, Y., Tanaka, Y., Toi, H., and Ogoshi, H. (1988) Polar host-guest interaction. Binding of nonionic polar compounds with a resorcinol-aldehyde cyclooligomer as a lipophilic polar host *J. Am. Chem. Soc.*, **110**, 634.
5. Matsushita, Y. and Matsui, T. (1993) Synthesis of aminomethylated calix[4]resorcinarenes, *Tetrahedron Lett.*, **34**, 7433-7436.

CHIRAL RECOGNITION STUDIES OF AMINO ACIDS BY CHIRAL CALIX[4]RESORCINARENES IN LANGMUIR FILMS

MAREK PIETRASZKIEWICZ[a*], PIOTR PRUS[a], WOJCIECH FABIANOWSKI[b]

a) Institute of Physical Chemistry, Polish Academy of Sciences
01224 Warsaw, Kasprzaka 44/52, Poland
b) Technical University of Warsaw, Department of Chemistry
00664 Warsaw, Noakowskiego 3, Poland

1. Introduction

Molecular monolayers ordered in Langmuir films may display molecular recognition properties, provided that molecular composition of the monolayer film is capable of forming molecular complexes with the molecules dissolved in the subphase [1]. One of the most interesting features is the possibility of chiral discrimination of optically active molecules present in the subphase by Langmuir films composed of amphiphilic and chiral molecular receptors.

Calixarenes are the most versatile host molecules capable of binding practically all conceivable organic and inorganic ions, and/or molecules. Their functionalizations provide a vast number of modified hosts designed for specific function on molecular level. We have been interested in lipophilic, chiral calix[4]resorcinarenes as hosts for chiral discrimination of organic species. The LB films formed by chiral calix[4]resorcinarenes may be of practical importance as sensory tools in detection, or discrimination of optically active molecules, or ions.

Figure 1

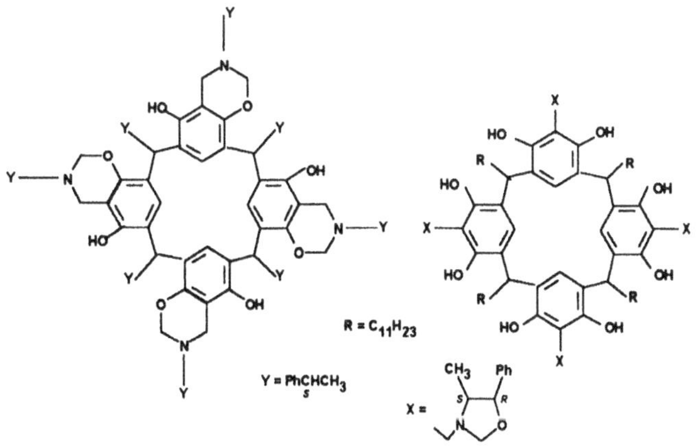

A. W. Coleman (ed.), Molecular Recognition and Inclusion, 463–466.
© 1998 *Kluwer Academic Publishers. Printed in the Netherlands.*

2. Experimental

The lipophilic calix[4]resorcinarene incorporating n-undecyl chains [2] was subjected to reaction with (S)(-)-phenylethylamine, or (1R,2S)(-)-ephedrine, and paraformaldehyde according to the literature [3]. The products, CAV-11PhEt and CAV-11EPH, respectively, after crystallization from acetonitrile, were dried in high vacuum for 10 h at 50°C. The solutions of the ligands were prepared in chloroform (10^{-3} M) for LB film formation. The concentrations of amino acids in deionized water were the same. LAUDA FW-2 apparatus was used to generate the films and to measure the surface tension isotherms. Each isotherm measurement was repeated three times. The selected amino acids were D- and L-forms of alanine, valine, leucine and tryptophane.

3. Results and Discussion

Two Mannich base-type calix[4]resorcinarenes formed Langmuir films on the water surface. The surface tension isotherms for the free ligands on the pure water and for the subphase containing various amino acids were not uniform, as shown in the *Figure 2* and *3*, indicating some conformational changes during the film compression, or release of the solvent molecules in which the calixarenes were dissolved. Another alternative would be a rearrangement of the calixarene molecules on water surface that account for an isotherm behaviour. An interesting point is that for enantiomeric pairs of amino acids in the subphase we have observed distinctly different isotherms that account for different diastereoisomeric species formed in the film. The Figure 2 shows the selected isotherms for different amino acids and the Table 1 contains the calculated average surface area per molecule, and for enantiomeric pairs these values are different.

Figure 2

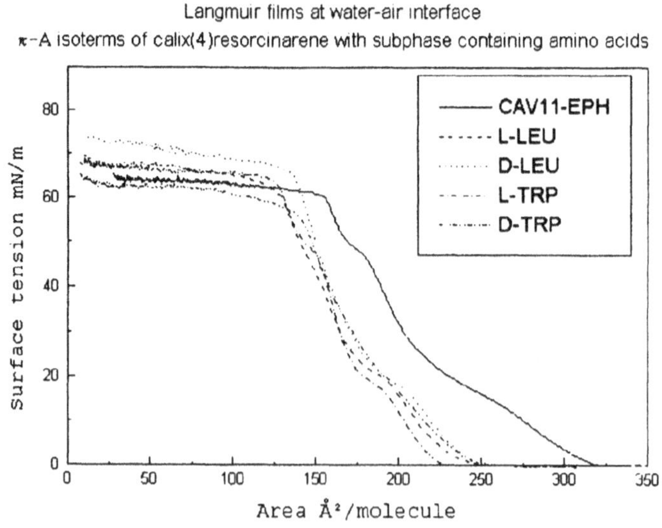

TABLE 1. Average surface area per molecule ($Å^2$/molecule)

Amino acid	Calix[4]resorcinarene	
	CAV11-EPH	CAV11-PhEt
-	239	165
L-Ala	183	144
D-Ala	194	156
L-Val	197	170
D-Val	191	180
L-Leu	195	156
D-Leu	186	144
DL-Leu	-	182
L-Trp	215	182
D-Trp	198	161
DL-Trp	-	163

The *Figure 3* illustrates the time-dependence of surface tension isotherms for CAV-11PhEt and DL-tryptophane in the subphase. The Table 2 contains the time-dependent data of surface area as the function of time. It is interesting to observe that the registration of isotherms in different time intervals shows different shape of these isotherms, and thus different surface areas the comparing the surface area for L- , and D-tryptophane in the subphase after 1 h we see that these two values are practically the same within experimental error. This means that after certain time all binding sites are saturated regardless the kinetics and thermodynamics of complex formation (for diastereomeric complexes they will be different), or difussion process. So, to make the qualitative test for chiral recognition the surface area should be assessed for the same period of time for both enantiomers before the saturation of all binding sites in the film.

Figure3

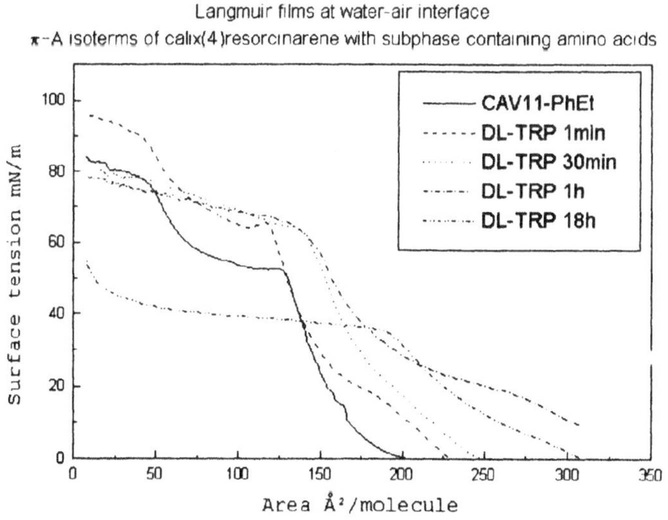

TABLE 2. Time dependent surface area for CAV11-PhEt

Amino acid	Time (h)	surface area ($Å^2$/molecule)
-	-	165
D-Trp	0	162
D-Trp	1/2	180
D-Trp	1	210
L-Trp	0	183
L-Trp	1/2	178
L-Trp	1	210
DL-Trp	0	162
DL-Trp	1/2	189
DL-Trp	1	206
DL-Trp	18	265

4. Conclusions

The Langmuir isotherme technique proved to be a fast method for the estimation of enantioselection by chiral macrocyclic ligands, provided they formed Langmuir film on the water surface. The test can be considered as qualitative only, but it gives an essential information about the magnitude of enantioselection. For the future investigations we plan to use other techniques based on LB films, such as static surface potential measurements. Thus it seems to be conceivable to develop a new analytical tool for enantiomeric purity assays, which is very essential in pharmaceutical, agricultural, and food industries.

5. Acknowledgement

This work was supported by the Polish Committee for Scientific Research, grant No. 3 T09A 025 011

6. References

1. Cha, X., Agira, K., Onda, M., and Kunitake, T. (1995) Molecular recognition of aqueous dipeptides by noncovalently aligned oligoglycine units at the air/water interface, *J. Am. Chem. Soc.*, **117**, 11833-11838.
2. Aoyama, Y., Tanaka, Y., Toi, H., and Ogoshi, H. (1988) Polar host-guest interaction. Binding of nonionic polar compounds with a resorcinol-aldehyde cyclooligomer as a lipophilic polar host. *J. Am. Chem. Soc.*, **110**, 634.
3. Matsushita, Y. and Matsui, T. (1993) Synthesis of aminomethylated calix[4]resorcinarenes, *Tetrahedron Lett.*, **34**, 7433-7436.

Preorganized Macrocyclic Dicarboxylic Receptors
Synthesis, Inclusion Behaviour and Structural Study

ROLF POLLEX, EDWIN WEBER and MÁTYÁS CZUGLER
Institut für Organische Chemie, Technische Universität Bergakademie Freiberg, Leipziger Str. 29, D-09596 Freiberg/ Sa., Germany.

1. Introduction

Biological systems often use the H-bonding properties of carboxylic acid functions in their active sites for binding and manipulation of substrates. On the other hand, synthetic molecular receptors containing preorganized carboxylic acid groups are rare [1] and macrocycles featuring *endo*-orientated carboxylic acid functions have not been reported recently [2].

Here we describe the synthesis of four new macrocyclic host derivatives (**1** - **4**) based on an unique framework which are modified in the bridging diphenylmethane elements. Molecular recognition of alcoholic guests and conformentional properties are presented including the X-ray crystal structures of three inclusion compounds with ethanol.

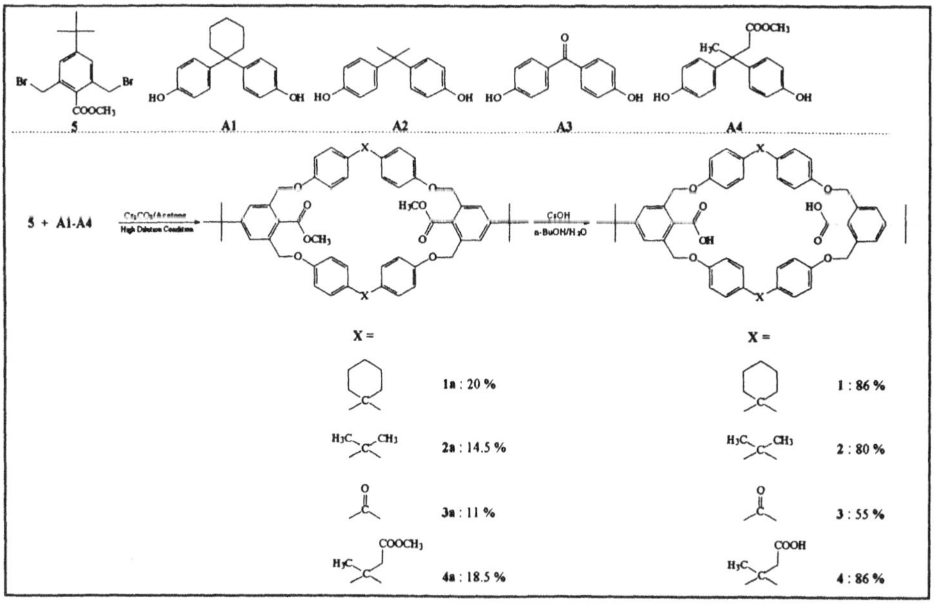

Figure 1. Ring closure reactions to the macrocyclic host compounds **1** - **4**

A. W. Coleman (ed.), Molecular Recognition and Inclusion, 467–470.
© 1998 *Kluwer Academic Publishers. Printed in the Netherlands.*

2. Synthesis

The preparation of the starting materials follows common routes and is well described for **5** [3], **A1** [4] and **A4** [5]. **A2** and **A3** are readily available. The synthetic key steps to the macrocyclic hosts are shown in Fig. 1 and the ring closure reaction was optimized for **1a** by a series of experiments under various conditions (*cf.* Tab. 1 and Tab. 2) resulting acetone as the solvent of choice and a towfold excess of cesium carbonate to secure complete deprotonation of the bisphenols without hydrolizing compound **5**. The hydrolyses of the macrocyclic di- or tetraesters **1a** - **4a** to their corresponding carboxylic acid compounds **1** - **4** succeeded in n-butanol/water with CsOH as base. All yields of the several rections are given in Fig. 1.

TABLE 1. Yields of ring closure reactions to compound **1a** (**5**, **A1** = 5 mmol, base = NaH, KOH, CsOH = 10 mmol, Cs_2CO_3, K_2CO_3 = 5 mmol, solvent volume = 500 ml, h. d. = high dilution conditions, x = not tested, - = no yield of **1a**)

Solvent/Base	NaH	KOH	CsOH	Cs_2CO_3	K_2CO_3
DMF (65 °C)	-	-	-	14.5%	7.3%
DMSO (65 °C)	-	-	-	9.5%	5.5%
Acetone (h.d.)	x	-	-	13.9%	4.8%
Cyclohexanone (h.d.)	x	x	-	4.8%	-
Methanol (h.d.)	x	-	-	-	-
Ethanol (h.d.)	x	-	-	-	-
Acetonitrile (h.d.)	x	-	-	5.7%	< 3%
Toluene (65 - 70 °C)	-	x	x	x	x

TABLE 2. Yields of the ring closure reaction fı the at different concentrations of **5**, **A1** and Cs_2CO_3 (DMF, DMSO: T = 65 °C, Acetone = high dilution conditions, solvent volume = 500 ml)

Cs_2CO_3 =	equimolar			twofold excess			threefold excess		
5,**A1**/Sol.	DMF	DMSO	Acet.	DMF	DMSO	Acet.	DMF	DMSO	Acet.
10 mmol	8.9%	8.7%	10.3%	7.8%	5.3%	15.8%	8.0%	3.3%	14.7%
5 mmol	14.5%	9.5%	13.9%	13.9%	8.8%	19.7%	13.8%	3.8%	19.8%
2.75 mmol	23.0%	11.4%	13.5%	22.1%	9.7%	20.5%	19.5%	5.6%	19.2%
1.6 mmol	22.4%	11.0%	13.1%	22.3%	9.5%	19.8%	18.7%	4.8%	17.4%

3. Inclusion behaviour and structural studies

All macrocyclic hosts form solid inclusion compounds with small alcohols such as ethanol or 2-propanol with respect to the polar groups attended to their skeletons and there is a rather good discrimation to bulkier guest molecules (*cf.* Tab. 3). Also cyclopentanol, cyclohexanol, cis/trans 1,4-cyclohexandiol or hydroquinone do not build crystalline inclusion complexes. The host/guest ratio has been determined by ^{1}H-NMR spectroscopy and the spectra of these solutions show different binding properties depending on the host molecules. Normal molecular acid-base-interactions occurs using hosts **3** or **4** as complexing agent and the signal of guest protons neighbouring the OH-group is shifted downfield (+0.05 - +0.2 ppm). On the other hand a significant upfield shift (-0.1 - -0.2 ppm) of this signal could be detected for all complexes of **1** and **2** which is probable caused by convergent binding of the guest molecules.

TABLE 3. Solid inclusion compounds of hosts **1** - **4** with alcohols (verified by ^{1}H-NMR spectroscopy)

Host/Guest	CH_3OH	C_2H_5OH	2-C_3H_7OH	1-C_3H_7OH	2-C_4H_9OH	1-C_4H_9OH	$HO(CH_2)_2OH$
1	1:2	1:2	1:2	+	-	-	-
2	1:2	1:2	1:2	1:2	-	-	-
3	1:3	1:3	1:3	1:3	-	-	-
4	1:4	1:4	1:4	+	-	-	-

Suitable colourless single crystals for X-ray analysis were obtained by slow concentration of solutions of **1**, **2** and **3** in ethanol at room temperature. For **1•**2 EtOH the complex decomposes so quickly that the guest molecules could not be determined and the structure was not refined to convergence (SG = $P\bar{1}$, a = 26.632 Å, b = 10.636 Å, c = 11.850 Å, α = 90°, β = 93.91°, γ = 90°). **2** yielded the described 1:2-complex as a colourless powder under absolute conditions. Surprisingly slow concentration in an open flask gave well figured crystals with a stochiometry of **2•**2 EtOH•2 H_2O.
The data sets for X-ray analyses were collected on a CAD4 diffractometer (ENRAF-NONIUS) at low temperature (T=130 K, CuK_α-radiation) and the first models were obtained by using SHELXS-86. The inclusion compounds **2•**2 EtOH•2H_2O (R_w= 0.125,

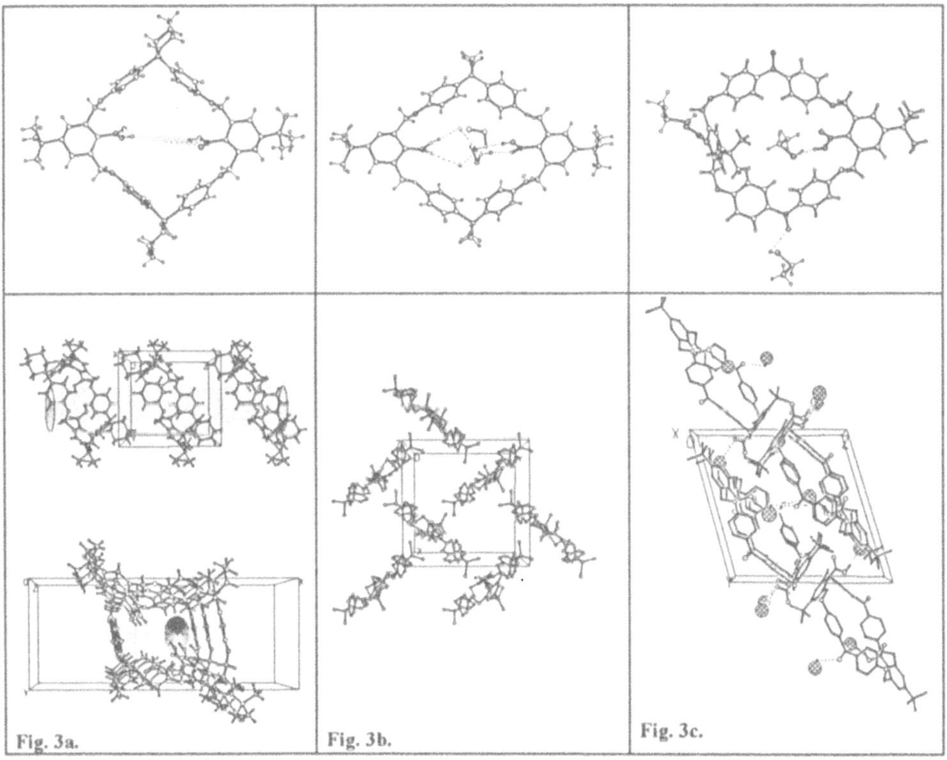

Figure 2. X-ray crystal structures of the inclusion compounds of (a) **1•**2 EtOH, (b) **2•**2 EtOH•2 H_2O and (c) **3•**3 EtOH

SG = $P2_1$, a = 14.639 Å, b = 15.227 Å, c = 25.127 Å, β = 91.13°) and **3**•3 EtOH (R_w= 0.143, SG = $P\bar{1}$, a = 11.466 Å, b = 15.875 Å, c = 16.028 Å, α = 73.14°, β = 86.60°, γ = 77.77°) were refined to convergence.

Fig. 2 presents the molecular structures with the host-guest relationships and packing diagrams without hydrogens and guest molecules (except for host **1**) of the inclusion compounds. The molecular representations of **1** and **2** show the convergent orientation of the two carboxylic groups which could be expected from the NMR spectra (*cf.* Fig. 2a and 2b). It is to be supposed that the not determined ethanol molecules bind directly to host **1** without water as bridging element (distance of the functional groups in Fig. 2a = 6.34 Å). In spite of the similar binding abilities of both hosts in the molecular view the packing diagrams illustrate large differences: while **1** stores the ethanol molecules in an endless channel host **2** forms layers of discrete complexes with its guest compounds.

Due to the two carbonyl groups the conformantional properties of **3** differs and an *endo/exo*-binding of ethanol occurs with the proper orientation of the carboxylic functions. Additionally a third guest forms a hydrogen bond to the oxygen atom of the carbonyl group (*cf.* Fig. 3c). The packing diagram shows the stacking of two host molecules on their flat site which are building layers and the guest molecules are located between them.

4. Conclusions

In summary the synthesis of the macrocyclic compounds offers an efficientroute to endo-functionalized molecular receptors. The first successful recognition experiments of these hosts with alcohols promise a wide range of application possibilities. For example cyclic or acyclic ketones and amines or heterocyclic compounds should be suitable guest components.

Acknowledgement

The authors acknowledge financial support from the Deutsche Forschungsgemeinschaft and from the Fonds der Chemischen Industrie. M. C. is on leave from Central Research Institute of Chemistry, Hungarian Academy of Science, H-1525 Budapest, PO.B. 17, Hungary.

References

1. Diederich, F.: Cyclophanes, RSC, Cambridge, 1991.
2. Gloe, K., Stephan, H., Heitzsch, O., Bukowsky, H., Uhlemann, E., Pollex, R. and Weber, E. (1994) *J. Chem. Soc., Chem. Comm.*, 1955 - 1956.
3. Urban, F.J., Chappel, L.R., Girard, A.E., Mylari, B.L. and Pimblett, I. J. (1990) *J. Med. Chem.* **33**, 765 - 771.
4. Niederl, J.B., Niederl, V. and Charney, J. (1940) *J. Am. Chem. Soc.* **62**, 322.
5. (a) Mauz, O. (1974) *Ann.* **472**, 1; (b) Yu, A.J. and Day, A.R. (1958) *J. Org. Chem.* **23**, 145; (c) Parris, C.L., Dowbenko, R., Smith, R.V., Jacobson, N.A., Pearce, J.W. and Christenson, R.M. (1962) *J. Org. Chem.* **27**, 455.

New *endo*-Functional Cryptophanes as Selective Complexants

CHRISTIAN E.O. ROESKY, MÁTYÁS CZUGLER and EDWIN WEBER
Institut für Organische Chemie, TU Bergakademie Freiberg, Leipziger Str. 29, D-09596 Freiberg/Sachs., Germany

TORSTEN KRÜGER, HOLGER STEPHAN and KARSTEN GLOE
Institut für Anorganische Chemie, TU Dresden, Mommsenstr. 6, D-01062 Dresden., Germany

1. Introduction

endo-Functional receptors having a preorganized cavity are rare, in particular if carboxylic groups are the *endo*-functions [1]. However, hosts with proton ionizable groups are promising complexants [2] and they are also interesting models in proton transfer and other studies.

Here we report a new host design based on an almost spherical and highly preorganized cryptophane-like macrocyle with three convergent carboxylic acid functions kept in a noninteractive distance.

Cryptophanes are generally defined as hollow molecules made of two cyclotriveratrylene caps assembled to one another by three bridging units [3]. According to their structure these compounds possess three windows allowing access to the central cavity by suitable guest molecules. Only two diastereomeric types of cryptophanes have hitherto been synthesized where the bridges normally constitute a building element O(Z)O (Z corresponding to $(CH_2)_{2-10}$, H_2C-CH=CH CH_2 or H_2C-C≡C-CH_2).

The series of cryptophanes (**1-3**) described here are novel types in two respects. They feature (1) a dibenzylic ether subunit as the bridging link and (2) polar endo-functional groups (COOH, $COOCH_3$) that are highly preorganized. The basic scaffold of this new cryptophane framework is realized by compound **4** lacking the polar functions.

2. Synthesis

The synthetic strategy of the present D_3- (chiral) and C_{3h}-symmetry (achiral) cryptophanes is based on Collet's "Two-step Method" [3]. This route has in addition of its simplicity and shortness the advantage of rendering superfluous high dilution conditions. In our synthesis methyl 2,6-bis(bromomethyl)-4-*tert*-butylbenzoate **5** [4] is used as the starting material. By treatment of **5** with vanillin and K_2CO_3 in acetone we obtained the bis(vanillyl) derivative **6** in 88 % yield. Reduction of the dialdehyde **6** with

A. W. Coleman (ed.), Molecular Recognition and Inclusion, 471–474.
© 1998 *Kluwer Academic Publishers. Printed in the Netherlands.*

$NaBH_4$ yields to the cyclization precursor **7** (quant.) which is converted directly under mild acidic conditions (formic acid) to a mixture of the cryptophanes **2** (C_{3h}, 14 %) and **3** (D_3, 21 %). Subsequent hydrolysis of the isolated ester **2** with KOH/18C6 gave the tricarboxylic host **1** in 77 % yield. A similar procedure was applied to synthesize **4** (C_{3h}, 5 %) using 1,5-bis(bromomethyl)-3-*tert*-butylbenzene [5] as starting material. Remarkably, here only the C_{3h} isomer **4** and no D_3 diastereomer was found.

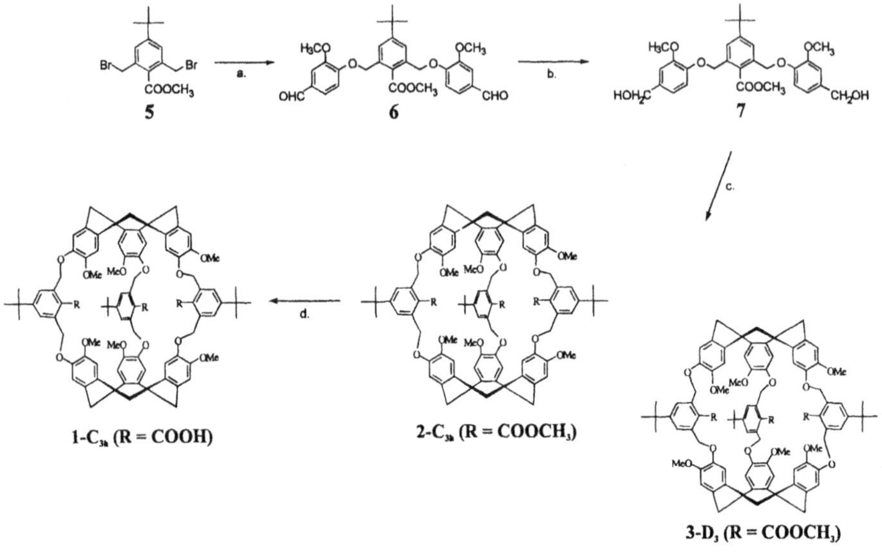

Fig 1: Synthesis of hosts **1-3**. a. vanillin, K_2CO_3; b. $NaBH_4$; c. HCOOH; d. KOH/ 18C6

3. Crystallographic and molecular modeling studies

Basic feature of the molecular structure of **2** (Fig. 2) is associated with the peculiar space group symmetry (pseudo $P\,\bar{6}$), *i.e.* the presence of a mirror plane through the central section of the molecule [6]. Chloroform and water solvent molecules entrapped in the crystal lattice obviously help for packing of the nearly spherical host constituents. These solvent particles occupy endless threefold symmetry channels formed between the lateral arms of the host molecules stacked at different heights in the crystal lattice (Fig 3). Conformation of the central placed ester groups is probably steered by electrostatically favorable C-H···O interactions such that they may adopt both symmetric ("up and down") orientations with respect to the mirror plane bisecting through the middle of the host. One calculated minimum geometry (PM3) of the thermodynamic most stable ligand conformation is in very good agreement with the X-ray data (r.m.s. deviation of $8.3\cdot10^{-2}$ Å).

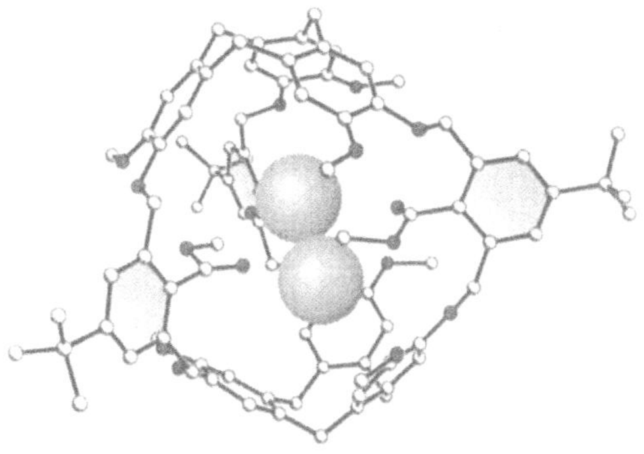

Fig 2: X-ray molecular structure of **2** with the possible intramolecular voids indicated by large spheres.

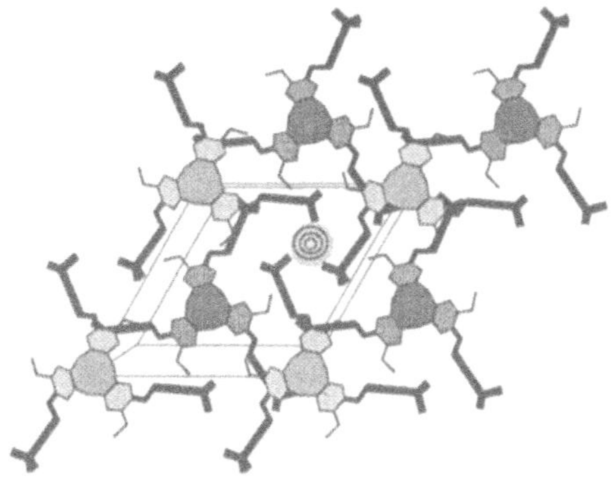

Fig 3: The packing of host 2 and the solvent channel in the crystal at (1/3, 2/3) indicated by the concentric circle.

4. Complexing properties of host 1

From molecular modelling studies it is apparent that the interior of host **1** is accessible for small ions or ball shaped molecules through the three inlets/openings between the methoxy side wings. Nmr titration experiments point out that complexation is only possible if the guest molecule matches the cavity's steric and electronic (H-bond D/A ability) requirements. This is met for only small molecules like methanol (K_{ass} = 7500 mol^{-1}) or ethanol (K_{ass} = 41 mol^{-1}) [7]. From the difference between the K_{ass} values it is apparent that host **1** has a high selectivity for methanol. By contrast, slim proton acceptor molecules such as acetonitrile and nitromethane failed to be complexed.
Apart from small molecules, metal ions are potential guests. A calculated structure model shows a preference for tight complexation of alkaline earth metal ions, e.g. Ca^{2+}, which was confirmed by single metal ion extraction experiments (Fig. 3) [8]. Metal ion selectivities as obtained from water/$CHCl_3$ interface extraction experiments are $Ca^{2+} > Sr^{2+} > Mg^{2+} > Ba^{2+} > Cs^{+} >> Na^{+}$. In all cases, under the experimental conditions, ligand **1** forms 1.1 (L:M) metal ion complexes, corresponding to a more or less tight coordination of the metal ions by the ligand.

5. Conclusion

Benzyl bridged cryptophanes have proved to be a new and fascinating family of receptor molecules having ability for the recognition of either metal ions or neutral guest molecules of suitable dimensions. The complexation selectivities are owing to the discrimination of both the steric and the electronic properties of the guest entities. These features, *e.g.* volume/ramification and simultaneous proton donor/acceptor behavior, can be quantified now on the basis of the X-ray models and molecular modeling studies to serve for the engineering of further derivatives. Considering the easy synthesis of this new type of cryptophanes it is obvious that the designing of other tailor-made cryptophane hosts that have different *endo*-functional groups such as OCH_3, F, NO_2 are promising.

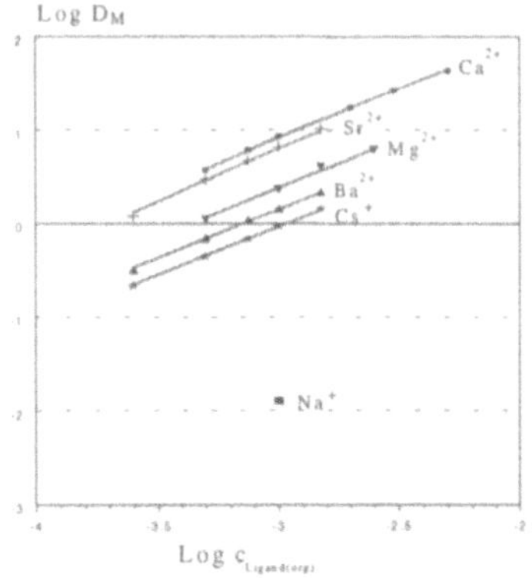

Fig 3: Extraction of alkali and alkaline earth metal ions as a function of ligand **1** concentration. [**1**] = 2.5 x 10^{-4} - 5 x 10^{-3} M in $CHCl_3$.

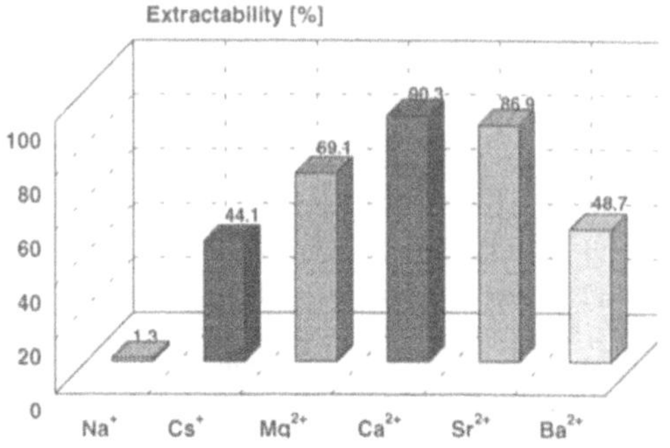

Fig 4: Extractability of alkali and alkaline earth metal ions using host **1**. $[M(NO_3)_2]$ = 1 x 10^{-4}; pH = 8,5 (TEA/HCl buffer); [**1**] = 1 x 10^{-3} M in $CHCl_3$.

Acknowledgement

We greatly thank the Deutsche Forschungsgemeinschaft and the Fonds der Chemischen Industrie for support of this work. M. C. is on leave from Central Research Institute of Chemistry, Hungarian Academy of Sciences, H-1525 Budapest, P.O.B. 17, Hungary.

References

1. Diederich, F.: Cyclophanes, RSC, Cambridge, 1991.
2. Gloe, K. Stephan, H. Heitzsch, O. Bukowsky, H. Uhlemann, E. Pollex, R. and Weber, E. *J. Chem. Soc., Chem. Comm.*(1994) 1955-1956.
3. (a) Review: Collet, A. in *Comprehensive Supramolecular Chemistry* Vol. 2 (Editor: F. Vögtle), Pergamon, Oxford, UK (1996) 325-366.
4. Urban, F.J. Chappel, L.R. Girard, A.E. Mylari, B.L. and Pimblett, I.J. *J. Med. Chem.* (1990) *33*, 765-771.
5. Fuson, R.C. and Freedman, B. *J. Org. Chem.* (1958) *23*, 1161-1166.
6. a. complete x-ray structure data will be published elsewhere b. Sheldrick, G.M.: SHELXS-86 Program for Crystal Structure Solutions. University of Göttingen, Germany, 1986 c. Sheldrick, G.M.: SHELXL-93 Program for Crystal Structure Refinement. University of Göttingen, Germany, 1993.
7. using HOSTEST (1994) developed by Wilcox, C.S.: Department of Chemistry, University of Pitsburgh, USA.
8. a. Gloe, K. and Mühl, P. *Isotopenpraxis* (1983) 19, 257. b. Gloe, K. Mühl, P. and Beger, J., *Z. Chem.* (1988) 28, 1-14.

IN SEARCH OF NEW TYROSINASE MIMETICS: ACYCLIC POLYAMINIC LIGANDS OF BENZO[g]PHTHALAZINE ABLE TO FORM DINUCLEAR COMPLEXES WITH Cu(II)

M. Rodríguez-Ciria,[a] A.M. Sanz,[a] F. Gómez-Contreras,[a] P. Navarro[b*], M. Pardo,[a] M.J.R. Yunta[a], and A. Castiñeiras[c]
[a] *Departamento de Química Orgánica I, Facultad de Química, Universidad Complutense, 28040 Madrid, Spain;* [b] *Instituto de Química Médica, C.N.Q.O. Manuel Lora-Tamayo, C.S.I.C., Juan de la Cierva 3, 28006 Madrid, Spain;* [c] *Departamento de Química Inorgánica, Facultad de Farmacia, Universidad de Santiago, Santiago de Compostela, Spain*

1. Introduction

The active center of tyrosinase plays an essential role in the synthesis of neurotransmitter catecholamines , since it catalyzes the hydroxilation of tyrosine to dopamine. [1] That activity is due to the presence of two Cu(II) ions able to experiment a reversible redox reaction *via* the formation of peroxo bridges.

It has been shown that dinuclear complexes of Cu(II) with ligands containing sp^2 heterocyclic nitrogens give oxygenated adducts able to transform phenol in catechol. [2] Furthermore, the optimal distance between the copper atoms (about 3-4 Å [3]) may be achieved by using rigid heteroaromatic spacers. [4] These facts prompted us to prepare ligands containing pyrazole units like **Ia** or **Ib**, that were able to form dinuclear complexes with Cu(II). [5] X-ray data show internuclear distances between the two copper atoms to be very close to those of hemocyanine (3.9 Å). [5a]

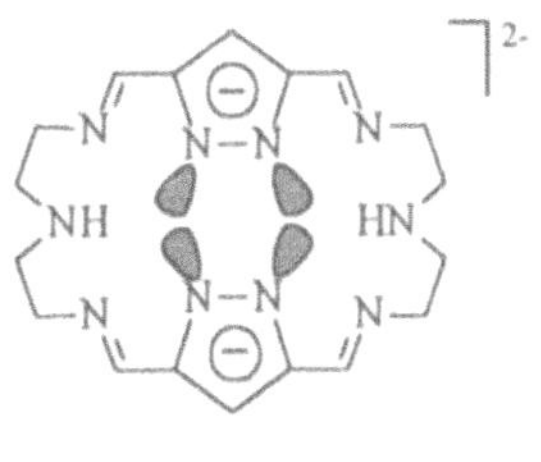

Ia

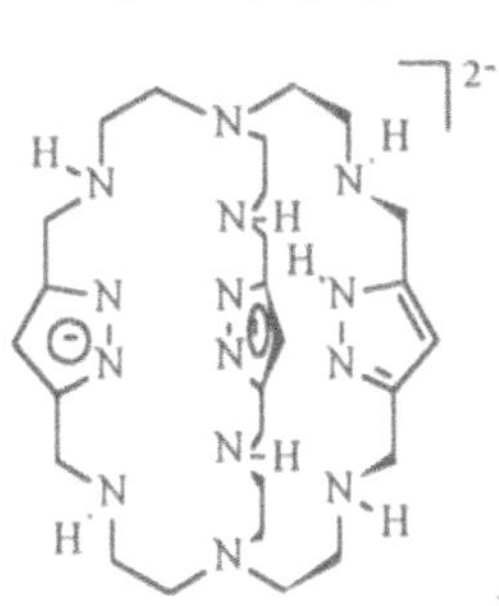

Ib

A. W. Coleman (ed.), Molecular Recognition and Inclusion, 475–478.
© 1998 *Kluwer Academic Publishers. Printed in the Netherlands.*

We are now trying to introduce the pyridazine ring as the heterocyclic spacer and sp^2 nitrogens supplier, as it has been stated by us that some simple 1,4-bis(alkylamino)-benzo[g]phthalazines form stable coloured complexes with Ni(II). [6] However, longer an more functionalized side chains are required for the formation of the desired dinuclear complexes, so that we have devised the synthesis of the polyaminic ligands **IIa**, containing four complexing heteroatoms in the side-chains, and **IIb**, in which the sp^3 primary amino groups of IVa have been substituted by pyridine rings.

2. Results and Discussion

The synthesis of IIa-b has been achieved in good yields by the nucleophilic substitution of 1,4-dichlorobenzo[g]phthalazine with 2,2'-oxibis(ethylamine) or 5-(2-pyridil)-4-oxapentylamine. The last amine was formed by alkylation of 2-hydroxymethyl-pyridine. In both cases, the monoaminoalkyl substitution products **IIIa** and **IIIb** were also formed in similar yields, and the mixtures of **II** and **III** were further separated by flash column chromatography on silica gel.

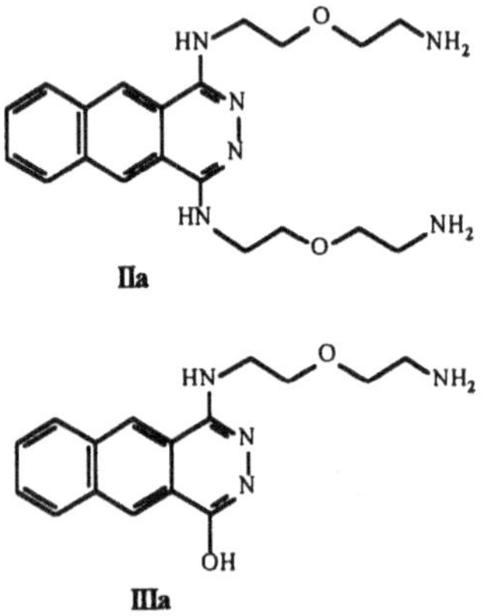

IIa

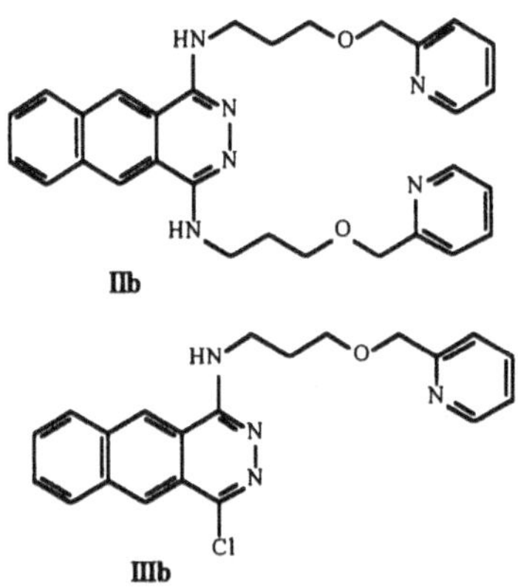

IIb

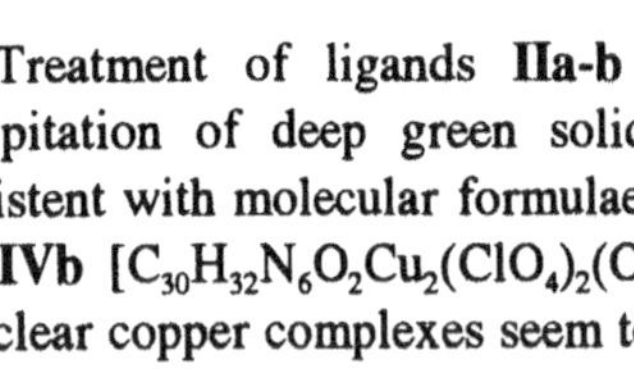

IIIa

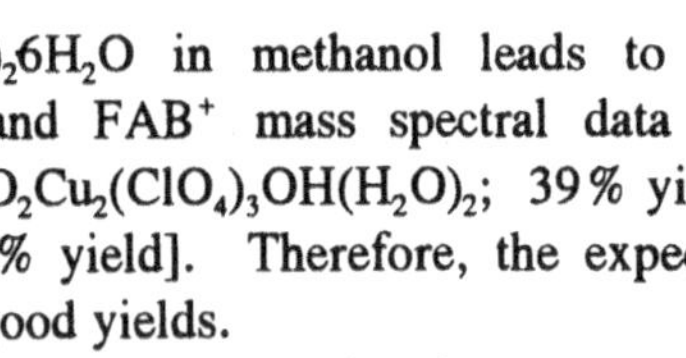

IIIb

Treatment of ligands **IIa-b** with $Cu(ClO_4)_2 6H_2O$ in methanol leads to the precipitation of deep green solids. Analytical and FAB^+ mass spectral data are consistent with molecular formulae **IVa** [$C_{20}H_{28}N_6O_2Cu_2(ClO_4)_3OH(H_2O)_2$; 39% yield] and **IVb** [$C_{30}H_{32}N_6O_2Cu_2(ClO_4)_2(OH)_2CH_3OH$; 75% yield]. Therefore, the expected dinuclear copper complexes seem to be formed in good yields.

The structure of both complexes has been studied on the basis of spectroscopic data (elemental analysis, FAB^+ high resolution MS, IR spectra, and magnetic susceptibility measures. Results obtained agree with structures **Va** and **Vb**, in which each copper ion is simultaneously bonded to two nitrogen atoms and the central oxygen and both ions are bridged by one hydroxy group and also by one perchlorate through the axial positions. Therefore, the metal ions would be pentacoordinated with a distorted square-pyramidal stereochemistry. This scheme is found in hemocyanin, [7] and also in a variety of synthetic complexes. [8]

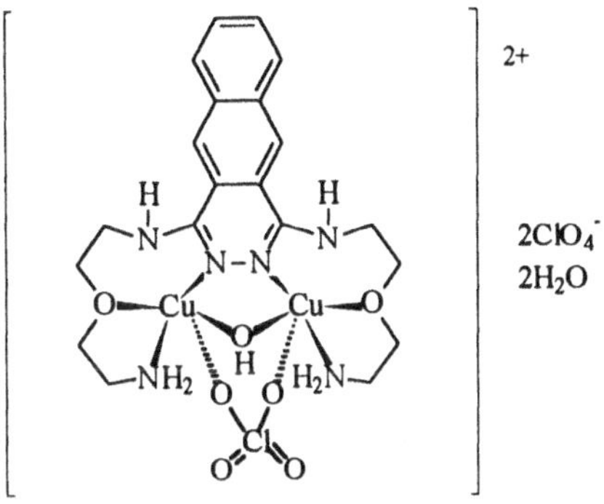

Va

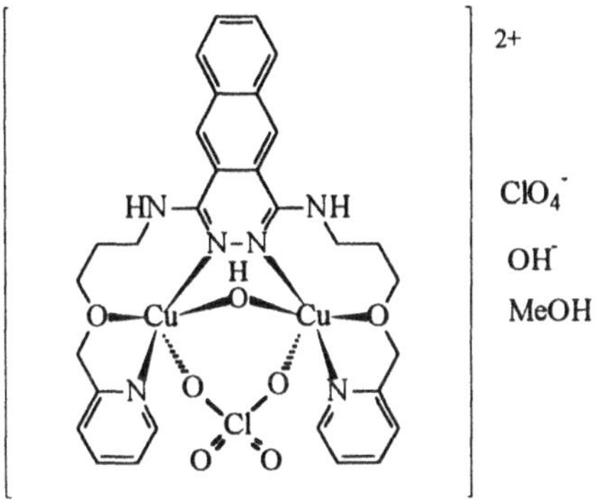

Vb

IR data are noteworthy in the assignment of structures Va-b. Absorptions around 325-350 and 400 cm^{-1} show that the copper ions are bonded both to nitrogen and oxygen atoms. Symmetric and antisymmetric C=C and C=N vibrations at 1626 and 1610 cm^{-1} indicate that the metal atoms are bonded to the pyridine nitrogen in **Vb,** whereas the NH signal associated with stretching vibrations is shifted to lower frequencies in **Va** with respect to **IIa,** demonstrating the formation of nitrogen-metal bonds. The N-N stretching vibrations of the benzo[g]phthalazine moieties are shifted to higher frequencies in **Va-b**, pointing to coordination through both nitrogen atoms.

On the other hand, the two complexes appear as hydroxyde bridged derivatives, as can be seen from absorptions in the range 3280-3560 cm^{-1}. Furthermore, compound **Vb** exhibits three different absorption bands associated with the perchlorate group between 1080 and 1150 cm^{-1}. This fact suggests the presence of both ionic and bridging bidentate perchlorates. A very broad perchlorate band is observed in the **Va** spectrum, pointing to a similar arrangement in both complexes.

From the fragmentation sequences found in the FAB^+ mass spectra it is also observed that the loss of the bridged hydroxy and perchlorate groups takes place in the final steps. Specially revealing is the case of **Va**, where one perchlorate remains after the loss of one of the copper atoms. This is clearly consistent with the structure proposed, since an ionic perchlorate in the coordination sphere should be much more easily released than the complexed copper ion.

Magnetic susceptibility measurements have been performed on powdered samples from both complexes in the temperature range of 5-300 K. [9] Plots of the susceptibility as a function of the temperature indicate the existence of antiferromagnetic exchange between the two Cu(II) centers in both cases. These results confirm the high interest of both dinuclear complexes as tyrosinase mimethics, and encourage to perform further studies concerning their redox properties.

The most energetically favored geometries for the stereochemistry proposed above have been also calculated for **Va** and **Vb** with the Hyperchem software using the MM^+ method. The preferred conformations are depicted in figures 1 (**Va**) and 2 (**Vb**) and suggest that distortion from the pure square pyramidal geometry around the copper atoms must very low, since dihedral angles N-O-N-O in the base of the pyramids oscillate between 7 and 14° in both complexes.

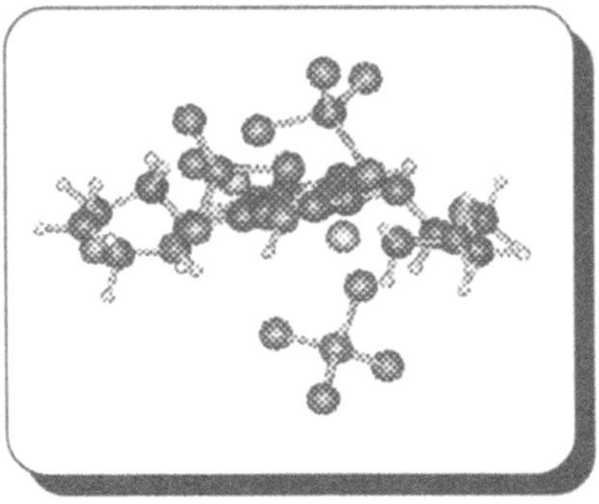

Figure 1

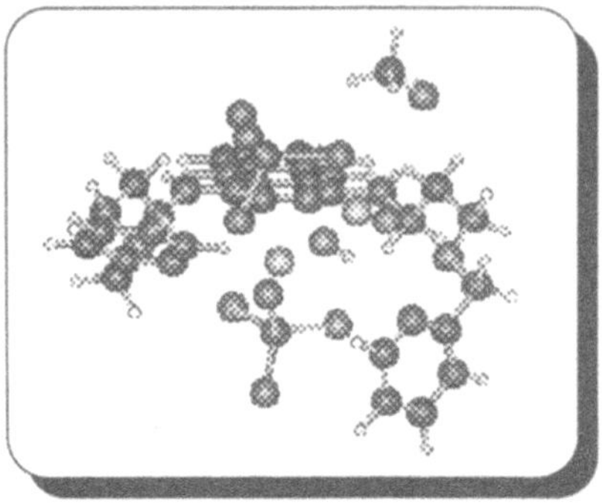

Figure 2

3. REFERENCES

1. Otsuka, S. and Yamanaka, T. (1988) *Metalloproteins: Chemical Properties and Biological Effects in Bioactive Molecules*; Ed. Elsevier 8, 311.
2. a) Reglier, M., Jorard, C., and Waegell, B. (1990) Binuclear Copper Complex Model of Tyrosinase. *J. Chem. Soc. Chem. Commun.*, **24**, 1752-1755; b) Simandi, L.I. (1992) *Catalytic Activation of Dioxygen by Metal Complexes;* Kleuwer Academic Publishers, London; Chapt. 5 and 6, pp. 196-296.
3. Gaykema, W.P.J., Hol, W.G.J., Verijken, J.M., Soeter, N.M., Bak, H.B., and Beintena, J.J. (1984) 3.2 A Structure of the Copper-containing , Oxygen-carrying Protein Panulirus Interruptus Haemocyanin. *Nature*: , 309, 23-29.
4. a) Karlin, K.D., Cruse, R.W., Guttmeh, Y., Faroo, A., and Zubieta, J. (1987) Dioxygen- Copper Reactivity. Reversible Binding of O_2 and CO to a Phenoxo-Bridged Dicopper (I) Complex. *J. Am. Chem. Soc.* **109**, 2668-2679; b) Thompson, L.K., Lee, F.L., and Gabe, E.J. (1988) Variable Antiferromagnetic Exchange in a Series of Binuclear Copper (II) Complexes of Tetradentate (N_4) and Hexadentate (N_6) Diazine Ligands. Magnetism versus Structure, Brindging and Terminal Ligands Groups and Chelate Ring Size. Crystal and Molecular Structure of [1μ- 1,4-Bis((6-methylpyrid-2-yl)amino)phthalazine](μ-hydroxo)(μ-nitrato-O-)-bis (nitratodicopper(II)-0.5-Water, $Cu_2C_{20}H_{19}N_9O_{10}-0.5H_2O$. *Inorg. Chem.* **27**, 39-46
5. a) Kumar, M., Arán, V.J., Navarro, P., Ramos-Gallardo, A., and Vega, A. (1994) Dinuclear Cu(II) Complexes with Two Pyrazolate Brindging Groups Formed from 26 Membered Oxaimine and Polyamine Macrocycles of 3,5-Disubstituted 1*H*-Pyrazole. *Tetrahedron Lett.* **35**, 5723-5726; b) Kumar, M., Arán, V.J., and Navarro, P. (1995) New Proton Ionizable 3,5-Disubstituted Pyrazole Criptands Able to form Tripyrazolate-Derived Di- and Tetra-Nuclear Complexes of Cu^{2+} or Zn^{2+}. *ibid.* **36**, 2161-2164.
6. Pons, M., Campayo, L., Martínez-Balbas, M.A., Azorlu, F., Navarro, P., and Giralt, E. (1991) A New Ionizable Chromophore of 1,4-Bis(alkylamino)benzo[g]phthalazine Which Interacts with DNA by Intercalation. *J. Med. Chem.* **34**, 82-86.
7. Coughlin, P. and Lippard, S.J. (1981) A Monohydroxo Bridged, Strongly Antiferromagnetically Coupled Dicopper (II) Center in a Binucleating Macrocycle. Comparisons with Binuclear Copper Sites in Biology. *J. Am. Chem. Soc.* **103**, 3228-3229; Coughlin, P. and Lippard, S.J. (1984) Copper Chemistry in Hexaaza Binucleating Macrocycles. Hydroxide and Acetate Derivatives. *ibid;* **106**, 2328-2336.
8. Sorrell, T.N. (1989) Synthetic Models for Binuclear Copper Proteins. *Tetrahedron* **45**, 3-68.
9. Kahn, O.; *Molecular Magnetism*; VCH, Weinheim, 1993.

The authors thank the Spanish Comisión Interministerial de Ciencia y Tecnología for the economical support given to this work (SAF93 - 0753 - C02 - 02).

Models Systems for Flavoenzyme Activity. Redox-Induced Modulation of Flavin-Receptor Hydrogen Bonding

VINCENT ROTELLO

Department of Chemistry, University of Massachusetts

Amherst, MA 01003 USA

Redox-active organic molecules are important species in both biological and man-made systems. In biological processes they serve as cofactors, performing redox transformations and serving as electron shuttles. In materials applications, they provide such diverse devices as light-harvesting devices, chemical sensors, and catalysts. While the properties and processes of these systems have been studied, one area that has been generally neglected is the interplay between intermolecular forces (such as hydrogen bonds and π-stacking) and redox behavior. To provide insight into these interactions, and direction for future materials research, we are currently investigating the properties of complexes of synthetic receptors with redox cofactors.

Flavoenzymes such as the flavin monooxygenases (FMO) and dehydrogenases (FDH), are proteins that use the $FADH_2$-FAD redox cycle (Scheme 1)(1) to catalyze a variety of biological redox transformations. These processes include the oxidation of amines to imines, the dehydrogenation of lipid esters and amides, hydroxylation of aromatic substrates, and the oxidation of a variety of other organic and inorganic substrates.(2)

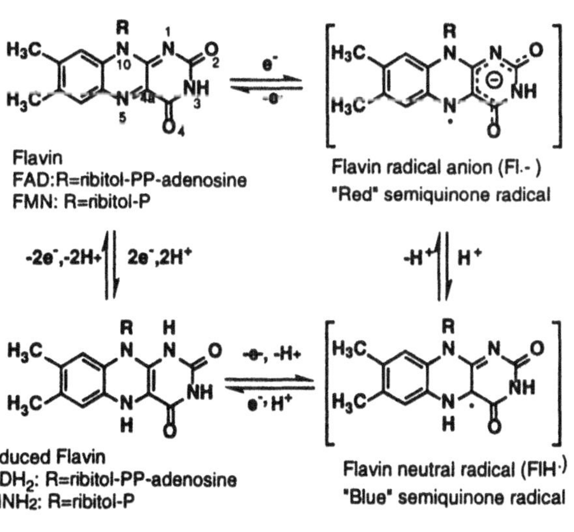

Scheme 1. Numbering scheme and common redox states of flavin derivatives.

A. W. Coleman (ed.), Molecular Recognition and Inclusion, 479–482.
© 1998 *Kluwer Academic Publishers. Printed in the Netherlands.*

In flavoenzymes, hydrogen bonding between the enzyme and cofactor is believed to be important in the regulation of flavin redox properties, and hence the biological activities of enzyme-bound flavin.(1 To provide insight into the ways that enzyme-cofactor interactions influence the behavior of flavins, we have designed receptors that allow us to isolate and quantify the effects of specific interactions on flavin spectroscopic and redox properties.(3)

The receptor **1**-flavin **2** complex (Fig. 1) provides hydrogen bonds to O(2), H(3), and O(4) of the flavin, analogous to a number of enzymes. The binding constant of receptor **1** with the organic-soluble flavin **2** in $CDCl_3$ was determined via NMR titration.(4)As shown in Fig. 2, addition of receptor **1** to flavin **2** resulted in a downfield shift of H(3), diagnostic for the formation of the hydrogen bound complex. Non-linear least-squares fitting of the resulting curve to a 1:1 binding isotherm provided a binding constant (K_a) of 540 M^{-1}.

With recognition established, we next examined the effects of complexation with receptor **1** on the electrochemistry of flavin **2**. Addition of receptor **1** to flavin **2** resulted in a movement of the flavin reduction potential for formation of the radical anion to a potential 155 mV less negative (Fig. 3)(3). This corresponds to a stabilization of the receptor **1**-bound flavin radical anion of 3.6 kcal/mol.

To provide insight into the origin of this stabilization, we have examined the effects of varying the concentration of receptor on the reduction potential of the flavin. As shown in Figure 4, increasing concentrations of receptor **1** result in steady movement of the flavin reduction potential (to the radical anion) to less negative values. This curve approached saturation at receptor **1** concentrations considerably lower than would be expected based on the K_a of the flavin **2**-receptor **1** complex. This results from redox-enhanced hydrogen bonding, observed first by us with flavins,(1) and later by Smith with *o*-quinones and imides.(5) While direct measurement of the binding constant is difficult due to electrode effects, the minimum binding constant fo the flavin radical anion-receptor complex is 20,000 M^{-1}, a *200-fold* increase in binding affinity. Preliminary *ab initio* computational studies indicate that this enhanced binding results from the increased hydrogen bond acceptor affinity of the anionic flavin radical anion. This arises from the concentration of negative charge on the O(2) and O(4) carbonyls of the flavin, resulting in increased electrostatic attraction of these functions to the protons of the receptor.

In summary, we have established that diaminopyridine **1** provides specific recognition of flavin derivative **2** through formation of a hydrogen-bound complex. Additionally, hydrogen bond complexation of flavin **2** with host **1** makes the potential required for reduction of **2** to the radical anion 155 mV less negative, corresponding to a biomimetic stabilization of the complexed radical anion of 3.6 Kcal/mol relative to the free radical anion. Concomitant with this stabilization is a 200-fold enhancement of recognition by receptor **1** of the flavin radical anion relative to binding of this receptor to the fully oxidized flavin. Applications of this redox-induced binding to the creation of sensors and other devices is currently being investigated, and will be reported in due course.

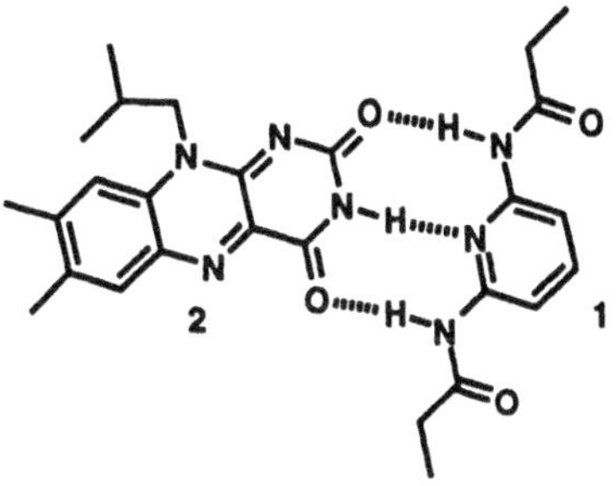

Figure. 1. Receptor **1**-flavin **2** complex receptor **2**

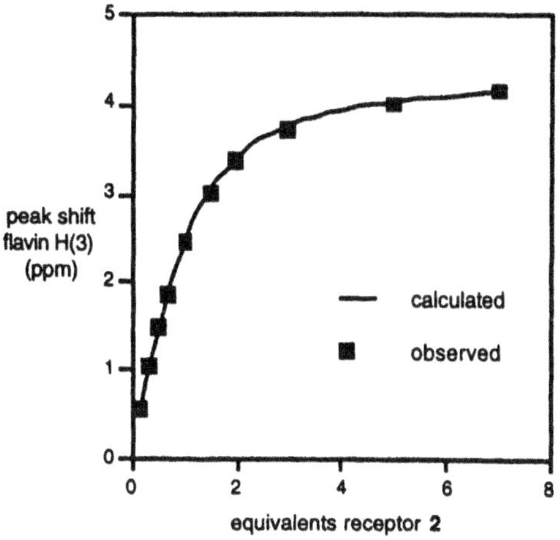

Figure 2. NMR Titration of flavin **1** with

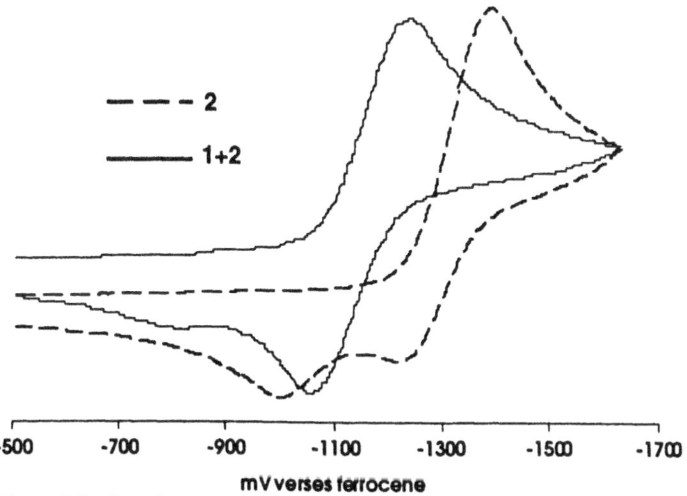

Figure 3 Cyclic voltammetry of flavin **2**, and the flavin **2**-receptor **1** complex in CH_2Cl_2, with tetrabutylammonium perchlorate as supporting electrolyte, [**1**]=5x10^{-3}, [**2**]=1x10^{-3}, scan rate 500 mV/s, T=23°C. Reduction potentials are relative to ferrocene as an external standard.(6)

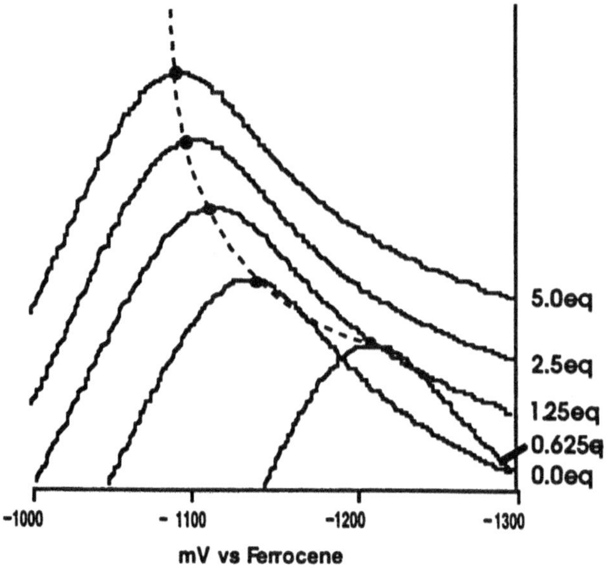

Figure 4. Cyclic voltammetry of flavin **2** in CH_2Cl_2 with varying concentrations of receptor **1**. Tetrabutylammonium perchlorate was used as supporting electrolyte, [**2**]=$1x10^{-3}$M.

Acknowledgments: This research was supported by the National Science Foundation (CHE-9528099) and the Petroleum Research Fund of The American Chemical Society (30199-G4). VR thanks Research Corporation for a Cottrell Fellowship.

---

1 Walsh, C. (1980) Flavin Coenzymes: At the Crossroads of Biological Redox Chemistry,*Acc. Chem. Res.* **13**, 148-155. Müller, F. (1983) The Flavin Redox-System and Its Biological Function, in *Topics in Current Chemistry 108*, Springer Verlag:New York, 1983, 71-109.

2 Ziegler, D.M. (1990) Flavin-Containing Monooxygenases: Enzymes Adapted for Multi-Substrate Specificity, "*Trends. Pharm. Sci.* 321-324.

3. Niemz, A., Rotello, V. M. Model Systems for Flavoenzyme Activity. The Effects of Specific Hydrogen Bonds on the ^{13}C and ^{1}H NMR of Flavins *J. Mol. Rec.*, in press. Deans, R., Rotello, V. (1996) Model Systems for Flavoenzyme Activity. 2-Aminopyridines as Spectroscopic Models for Flavoenzyme Active Sites, *Tetrahedron Lett.*, 4335.

4. Breinlinger, E, Niemz, A., Rotello, V.M. (1995) Model Systems for Flavoenzyme Activity. Stabilization of the Flavin Radical Anion Through Specific Hydrogen Bond Interactions, *J. Am. Chem. Soc.* **117**, 5739.

5 Ge, Y.; Lilienthal, R.; Smith, D. (1996) Electrochemically-Controlled Hydrogen Bonding Selective Recognition of Urea and Amide Derivatives by Simple Redox-Dependent Receptors ,*J. Am. Chem. Soc.* **118**, 3976.

6. Gagné, R., Koval, C. Lisensky, G.C. (1980) Ferrocene as an Internal Standeard for Electrochemical Measurements *Inorg. Chem.* **19**, 2854.

Selectivity in Thermodynamic Cyclisations of Cinchona Alkaloid Derivatives

S.J. ROWAN, P.A. BRADY and J.K.M. SANDERS*
Cambridge Centre for Molecular Recognition
University Chemical Laboratory
Lensfield Road, Cambridge CB2 1EW (U.K.)

Covalent organic structures have traditionally been synthesised using kinetically-controlled irreversible reactions, whereas the construction of non-covalent supramolecular assemblies generally uses thermodynamically-controlled reversible interactions.[1,2] Irreversible reactions lack the ability to proof-read and repair 'incorrect' bond formation, while supramolecular assemblies tend to lack the robust character associated with covalent bonds, so we have been exploring the synthetic potential of covalent chemistry under reversible conditions.[3] We now report the exceptionally efficient thermodynamic synthesis of macrocyclic trimers using new supramolecular building blocks derived from the cinchona alkaloids, quinine (**1a**) and cinchonidine (**1b**).

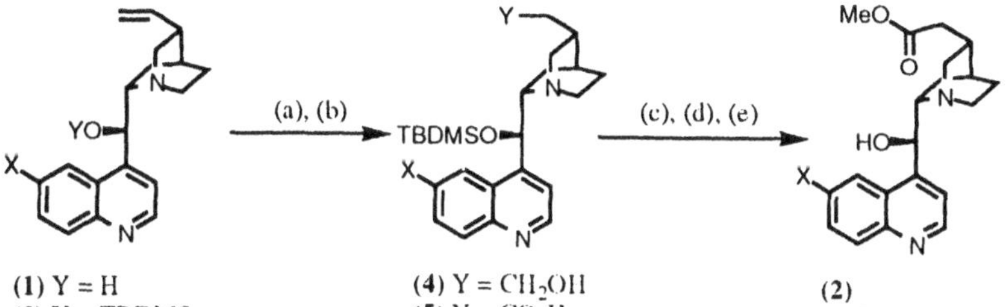

(**1**) Y = H
(**3**) Y = TBDMS

(**4**) Y = CH_2OH
(**5**) Y = CO_2H
(**6**) Y = CO_2Me

(**2**)
(**a**) X = OMe
(**b**) X = H

Reagents and Conditions
(a) TBDMSCl, Et_3N, DMAP, DMF, room temp. (b) 1. 5 equiv. BH_3.THF, diglyme, 0°C 2. Me_3NO, 100°C. (c) Jones Reagent, acetone, room temp. (d) MeOH, $HCl_{(conc.)}$, room temp. (e) TBAF, THF, room temp.

Scheme 1

The size and relatively concave shape of quinine, together with the functionality present in the natural product, make it attractive for macrocycle synthesis. As our thermodynamic cyclisations are transesterifications,[3] we require a hydroxyl group at one end of the molecule

A. W. Coleman (ed.), Molecular Recognition and Inclusion, 483–486.
© 1998 *Kluwer Academic Publishers. Printed in the Netherlands.*

and an ester elsewhere. There is already a hydroxyl group present at the 9-position of quinine but an ester group had to be introduced. This was achieved by modifying the vinyl group in five steps (Scheme 1)[4] to give the methyl ester **(2a)**. This was carried out by protection of the alcohol followed by hydroboration, oxidation and esterification to convert the olefin into the methyl ester. The alcohol was then deprotected to yield the required building block in an overall yield of 34% (Scheme 1). The cinchonidine monomer **(2b)** was prepared in an analogous manner starting from natural cinchonidine **(1b)**.

Thermodynamic cyclisations of **(2a)** were carried out utilising a procedure similar to that reported previously, (Scheme 2).[3,5] The catalyst (5% KOMe/18-Crown-6) was added to a refluxing solution (5mM) of **(2a)** in toluene, with azeotropic removal into molecular sieves of the methanol produced.

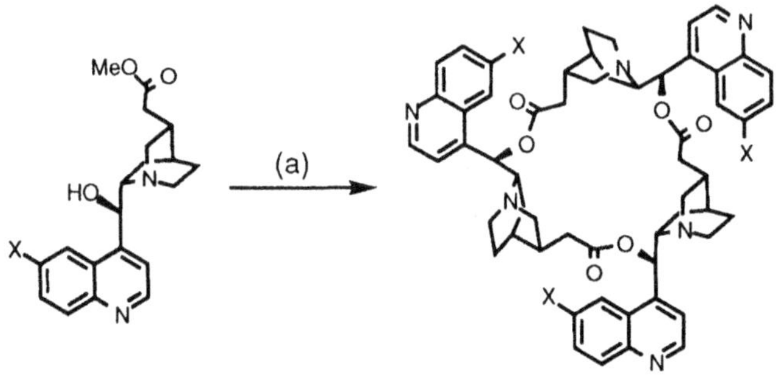

(2a) X = OMe
(2b) X = H

(7a) X = OMe
(7b) X = H

Reagents and Conditions
(a) KOMe, 18-crown-6, toluene, reflux

Scheme 2

The extent of the reaction was followed by electrospray mass spectrometry (ESMS). Remarkably, the cyclisation of **(2a)** gave virtually exclusively a single product: the cyclic trimer **(7a)** was formed in high yield (*ca.* 90% by NMR and 84% isolated yield). The reaction was complete after 10 minutes and no significant change in distribution was observed, even after one day. Cyclisation of **(2b)** results in the corresponding trimer **(7b)** in excellent yield. Reversibility of this reaction was demonstrated by formation of mixed cinchonidine/quinine cyclic trimers (as monitored by ESMS), when **(7a)** and **(7b)** were submitted to the thermodynamic conditions together.[5]

This led us to try the corresponding cyclisation under kinetic conditions[6] employing a modified Yamaguchi macrolactonisation (Scheme 3), using DMF to help solubilise the starting materials. The reactions were carried out at 5mM in order to ensure the formation of cyclic molecules, and were monitored by electrospray mass spectrometry and ^1H NMR.

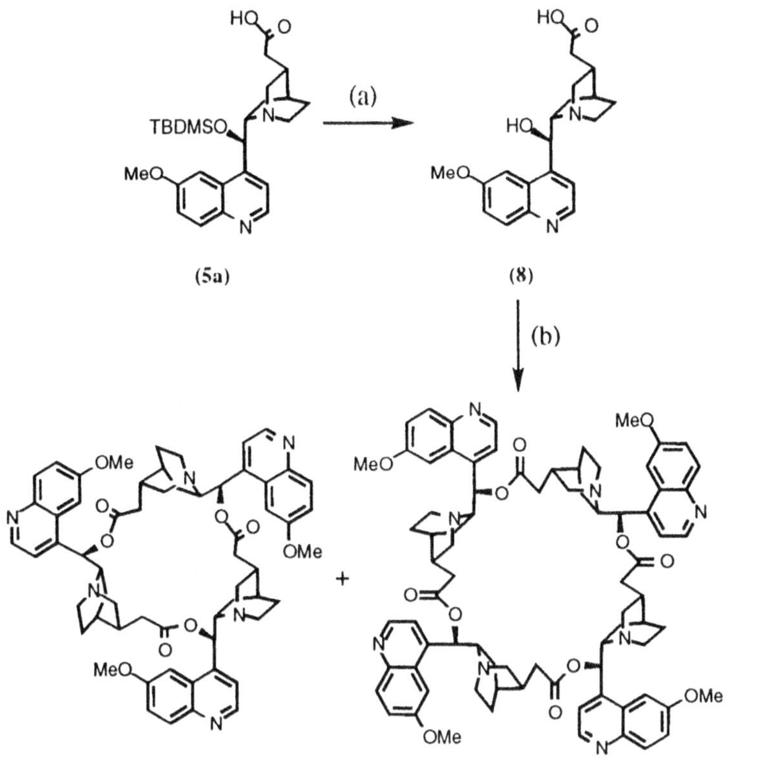

(7a) (9)

Mass Distribution: 37% 23%

+ 40% other higher oligomers

Reagents and Conditions
(a) TBAF, THF, room temp. (b) 1. 2,6-dichlorobenzoyl chloride, Et_3N, DMF.
2. DMAP, CH_2Cl_2, room temp., 18 hrs.

Scheme 3

The distribution of cyclic oligomers obtained from the kinetic cyclisation and identified by ^1H NMR (Only cyclic trimer **(7)** and tetramer **(9)** showed different NMR) is shown in Scheme 3. The interesting thing to note is that cyclic oligomers from trimer to octamer (by ESMS) are kinetically accessible. The kinetic cyclisation of monomer **(8)** gives a wide distribution of products, unlike the thermodynamic reaction.

In summary, we have demonstrated that cyclic oligomeric quinine derivatives can easily be obtained. We have shown that cyclic oligomers from trimer to octamer can be prepared kinetically, while under thermodynamic conditions only cyclic trimer **(7)** is formed. In theory, both kinetic and thermodynamic reactions should give a distribution of products. From this point of view, the kinetic reaction is well behaved, an ever decreasing amount of larger rings is obtained with trimer as the most abundant product (37% of monomer units). Kinetic cyclisations are dependent on the energy of the transition state and in our system the rate of cyclisation of the different linear species is broadly comparable, as demonstrated by the wide distribution of the cyclic products obtained. This confirms that all these cyclic products other than dimer are kinetically accessible and therefore that the absence of the larger oligomers in the thermodynamic reaction is not due to a kinetic effect. The high yield of the cyclic trimer (**7**) in the thermodynamic reaction (>90% of monomer units), which should be influenced by the energy of the ring system, suggests that this is a particularly stable molecule, relative to the other ring systems. We believe this is due to a degree of rigidity in our molecule which predisposes it to favour the conformation that it can adopt in the cyclic trimer (**7**). This selectivity suggests that, if a particular oligomer can be stabilised by an external agent (rather than by the internal predisposition, as here) the distribution should be shifted towards that oligomer. This would be thermodynamic templating, and is currently being investigated. We are also currently examining the possibility of mixing cinchona derived alkaloids with other building blocks already used in supramolecular chemistry (for example, cholic acids, porphyrins and xanthenes).

References

1 C. A. Hunter, *Angew. Chem.* **1995**, *107*, 1181; *Angew. Chem. Int. Ed. Engl.* **1995**, *34*, 1079-1081.

2 For examples see (a) P. Baxter, J. M. Lehn, A. DeCian and J. Fischer, *Angew. Chem.* **1993**, *105*, 92; *Angew. Chem. Int. Ed. Engl.*, **1993**, *32*, 69-72. (b) P. Baxter, J. M. Lehn, J. Fischer and M. T. Youinou, *Angew. Chem.* **1994**, *106*, 2432; *Angew. Chem. Int. Ed. Engl.*, **1994**, *33*, 2284-2287. (c) X. Chi, A. J. Guerin, R. A. Haycock, C. A. Hunter and L. D. Sarson, *J. Chem. Soc., Chem. Commun.*, **1995**, 2563-2566, 2567-2569. (d) M. Fujita, S. Nagao and K. Ogura, *J. Am. Chem. Soc.*, **1995**, *117*, 1649-1650. (e) P. J. Stang, K. Chen, and A. M. Arif, *J. Am. Chem. Soc.*, **1995**, *117*, 8793-8797.

3 P. A. Brady, R. P. Bonar-Law, S. J. Rowan, C. J. Suckling and J. K. M. Sanders, *Chem. Commun.*, **1996**, 319-320.

4 All new compounds were characterised by ^{1}H and ^{13}C NMR and mass spectrometry.

5 S.J. Rowan, P.A. Brady and J.K.M. Sanders, *Angew. Chem., Intl.. Ed. Engl.*, **1996**, *35*, 2143-2145.

6 S.J. Rowan, P.A. Brady and J.K.M. Sanders, *Tetrahedron lett.*, **1996**, *37*, 6013-6016.

A NEW BI-FUNCTIONAL RECEPTOR FOR ACETYLAMINO- FLUORENE MODIFIED GUANOSINE

M. Amélia Santos,[a] Adelino Afonso,[a] M. Matilde Marques[a] and Craig S. Wilcox[b]

a) Centro de Química Estrutural, Complexo I, Instituto Superior Técnico, Av.Rovisco Pais, 1096 Lisboa- Codex, Portugal; b) Department of Chemistry, University of Pittsburgh, Pittsburgh, P.A. 1526, USA.

Abstract: A new receptor based upon a carboxamidonaphthyridine unit derivatized with a polyaromatic pyrene segment has been synthesized and studied its binding interaction to the deoxyguanosine-C8 adduct with the carginogen 2-acetylaminofluorene. The high binding energy, in comparison with that of the non-derivatized analogue, is rationalized in terms of a bi-site interaction (hydrogen bonding and aromatic stacking).

1. Introduction

The modification of DNA by carcinogenic polyaromatic amines has attracted considerable attention, related to the formation of nucleoside covalent adducts with concomitant serious genotoxic lesions [1-3]. There has been much interest in the design of biomimetic receptors for nucleoside [4-6], in view of getting a better understanding of structural features behind the biological damage or of modeling enzyme repair activity.

We have been particularly interested in the design of abiotic receptors, able to recognize and selectively bind to the deoxyguanosine-C8 adduct with the carcinogen 2-acetylamino-fluorene [7]. Accordingly, a new receptor (**N2**) have been developed. Its architecture is based in a carboxamidonaphthyridine molecular unit of the **N1** type, providing a three hydrogen bond network that mimicks the cytidine interaction with guanosine; this unity is further functionalized with a pyrene unit for the sake of reinforcing the selectivity of the interaction to the polyaromatic adduct by π stacking effect. In addition the pyrene unity was thought as a possible probe of these effects due to its fluorescence properties.

In this paper, upon a brief reference to the synthetic scheme, we present ^{1}HNMR titration data, involving this new receptor as guest (titrant); deoxyguanosine and the corresponding *N*-acetylamino-fluorene C8 adduct were used as substrates (host). Accurate dimerization and subsequently host-guest binding constants have been calculated. An insight in the conformations the host-guest complex was also performed, using molecular simulation methods.

2. Results

The preparation of receptor **N2** 7-(4-pyrenyl)-butyrylamino-2-methyl-1,8-naphthyridine was accomplished in 40 % yield, by standard acylation of the amino group of the previously described 7-amino-2-methyl-1,8-naphthyridine with (4-pyrenyl)-butyryl-chloride. The preparation of the guanosine adduct, as well as the derivatization of all

A. W. Coleman (ed.), Molecular Recognition and Inclusion, 487–490.
© 1998 *Kluwer Academic Publishers. Printed in the Netherlands.*

nucleoside (for improving solubility and protecting sugar hydroxyl groups to reduce competing hydrogen bonding), were carried out as described previously [7], affording the following products: *N*-[2',3'-isopropylidene-5'-*O*-(-*tert*-butyldimethylsilyl)guanosin-8-yl]-2-acetyl-aminofluorene (**GAAF**), 2',3'-isopropyli-dene-5'-*O*-(*tert*-butyldimethyl silyl)guanosine (**G**) and 2',3'-isopropylidene-5'-*O*-(*tert*-butyldimethylsilyl)cytidine (**C**).

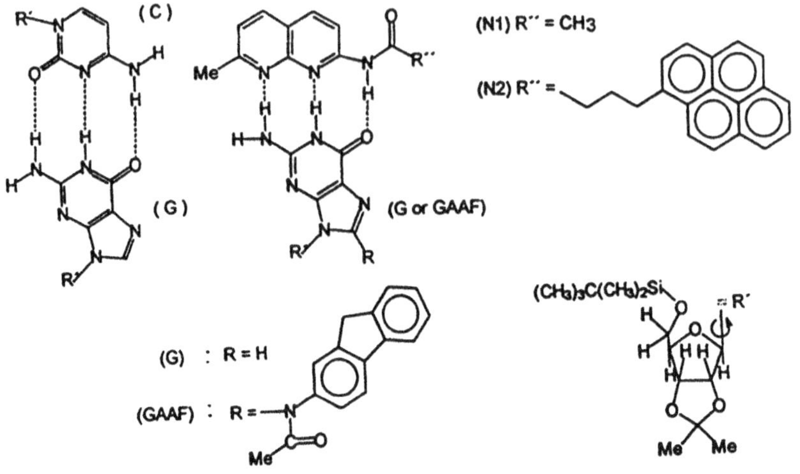

Quantitative evaluation of the binding interaction was based on the analysis of ^{1}HNMR titration curves. Each titration consisted of holding the substrate at constant concentration and varying the concentration of the receptor. Upon addition of the receptors it was observed downshift shift of proton resonances of the corresponding amide group as well as on the labile imino (N_1H) and amino(N^2H_2) protons of the both **G** and **GAAF**. Similar deshielding were observed on the same protons of **G** along the titration with **C** and this nucleobase pair has Watson Crick like interactions.

Titration curves were obtained by plotting the chemical shift of the substrate imino (N_1H) proton as a function of the total concentration of the receptor (Fig. 1a). All the host-guest titration curves present saturation behaviour with an inflexion at a host-

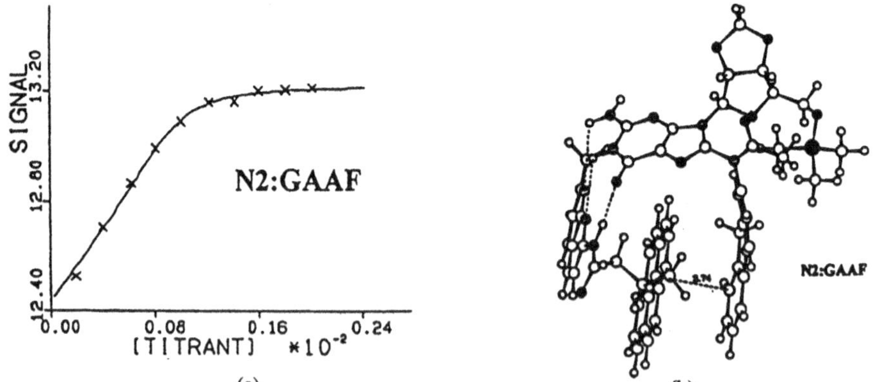

(a) (b)

Figure 1 (a) Plot of the δ signal for the N_1H protons of GAAF(1 mM) *vs* [N2]. (–) Calculated best fit; (x) experimental data points. (b) N2:GAAF molecular model structure calculated by MD/MM methods [10].

guest ratio of (1:1). Dimerization experiments were also performed for each component. Non-linear regression analysis of the dimerization and the titration curves gave the dimerization constants and the host-guest association constants, collected in Table 1.

Table 1: Binding constants (K_{bind}) and Gibbs free energies (ΔG^{295}) determined by ^{1}H NMR titrations*

Substrate	Receptor	$K_{bind}(M^{-1})$	ΔG^{295}(kcal/mol)
G	N1	$2.8\pm0.8 \times 10^3$	-4.6
GAAF	N1	$4.6\pm1.1 \times 10^3$	-4.9
G	C	$2.5\pm1.0 \times 10^3$	-4.6
GAAF	GAAF	2.5×10^2	-3.2
G	N2	$4.3\pm1.2 \times 10^4$	-6.2
GAAF	N2	$4.6\pm1.1 \times 10^4$	-6.3

*The dimerization constants for **G**, **N1** and **N2** are negligible(< 5 kcal/mol)

Since comparison of absolute values obtained by diferent methods is limited, a direct comparison between the complexation ability of the abiotic receptor **N1** and cytidine **C** for **G** was performed and it confirms they are comparable. The tenfold increase in the binding constants involving **N2**, as compared to those with **N1**, is consistent with the formation of **N2** complexes that combine both the aromatic stacking and three hydrogen interactions, while for the **N1** containing complexes only a triple hydrogen bond network should be involved.

Molecular modeling studies [10] suggested that for the **N2:GAAF** complex it is possible to have a quite strong intermolecular π-stacking interaction characterized by nearly parallel alignment of both polyaromatic rings (fluorene and pyrene) and by interplanar spacing in the vicinity of 3.75 Å (Fig. 1b). It is also suggest for this complex the existence of some internal π-stacking effect in **GAAF** itself that is able to induce some distortion on the hydrogen bond network. On the other hand the similarity of the binding constants obtained for **N1:G** and **N1:GAAF** suggest that a triple hydrogen bond net work should be involved in both the binding process. The strongest hydrogen bond in the **N1** or **N2** containing host-guest complexes is suggested to be the one involving the amide proton of the abiotic receptor. and the guanosine exocyclic carbonyl.

The enhanced k_{bind} of **N2** (or **N1**) to **GAAF**, relative to that for the same interaction to **G**, is outside the range attributable to experimental and may be due to some electronic effects. However, an opposite effect could also be expected, taking into account some steric due to predominancy of the glycosyl bond *syn* conformation for guanosinic adduct (**GAAF**) as compared with the normal *trans* conformation in guanosine [3]. Besides, previous solution studies indicate that the change in the glycosyl bond conformation in other monomeric nucleoside derivatives might not be determinant on the value of the host-guest binding constants [11]. Therefore, although in the *in vivo* oligonucleotides the DNA damage derived from the C8-guanosine adduct with the carcinogen acetylaminofluorene has been associated to a change in the glycosyl bond from the normal *trans* to the *cis* conformation, at the level of the monomers in solution, the effect on the Whatson-Crick hydrogen bond interaction is not so clear.

Concerning preliminar fluorescence results, the pyrene-naphthyridinic receptor **N2** showed fluorescence consistent with published for the polycyclic aromatic parent structures, while the **GAAF** adduct showed a strong broad emission band at 470 nm suggesting the existence of an excimer, λ_{em} being concentration independent. Upon addition of **N2** to **GAAF** (1:1), it was observed the disapearing of the that band with concomitant lowering of the pyrene peak intensity (by *ca.* 3-fold), thus suggesting considerable quenching by **N2**. Experiments directed to the fluorescence are underway.

It is of interest to note that the interaction order in solution involving monomeric nucleosides may be different from that in the oligonucleotides *in vivo.* However we hope the present results may shed some light on the problem of how the enzymes recognize the C8 guanosine adducts..

Conclusions: we have presented a new *bi-site* approach for the recognition of guanosine C8 polyaromatic adducts, involving simultaneously hydrogen bond and aromatic stacking interactions. Besides the good complementarity between these naphtyridinic receptors and guanine, thus mimicking the cytidine in its Watson-Crick interactions, the incorporation of the polyaromatic group in the naphthyridine skeleton leads to above 10-fold increase in the binding constant.

Theoretical modeling suggested there is a strong π-π interaction between **N2** and **GAAF**. Some internal π-π interaction in **GAAF** suggested also some distortion on the hydrogen bond network. The fluorescence spectrum of **GAAF** shows the existence of an "excimer" that seems to be quenched by binding interaction with **N2.**

This new receptor displays remarkable photophysical and complexing properties, namely with the guanosine aromatic adduct. This results could be helpful for the design of new fluorescent sensors for molecular recognition of nucleobases or corresponding carcinogenic aromatic adducts.

REFERENCES AND NOTES

1. Beland, F.A., and Kadlubar, F.F.(1990), *Handbook of Experimental Pharmacology. Carcinogenesis and Mutagenesis, vol. 94/1* (C.S. Cooper e P.L. Grover, eds.), Springer-Verlag, Heidelberg, 267-325.

2. Shapiro, R., Sidawi, D., Miao, Y.-S., Hingerty, B.E., Schmidt, K.E., Moscowitz, J., and Broyde, S. (1994) *Chem. Res. Toxicol.* **7**, 239-253.

3. Marques, M.M., Mourato, L.G., Santos, M.A., and Beland, F.A. (1996) *Chem. Res. Toxicol.* 99-108.

4. Hamilton, A.D. (1990) *J. Chem. Ed.* **67**, 821-828.

5. Conn, M.M., Deslongchamps, G., Mendonza, J., and Rebek, J. (1993) *J. Amer. Chem Soc.* **115**, 3548-3557.

6. Bell, T.W., Steven, Z.H., Zimmerman C., and Thiessen, P.A. (1995) *Angew. Chem. Int. Ed. Engl.* **34**, 2163.

7. Santos, M.A., Ferreira, T., Esteves, M.A., Drew, M.G., Beland, F.A., and Marques, M.M. (1995) *Supramol. Chem.* **5**, 239-249.

8. P.f. 260-262 °C; I.V. (KBr) 1720 cm^{-1} (C=O); m/z 429; ^{1}HMNR ($CDCl_3$) δ 2.37 (2 H, m, $CH_2\mathbf{CH_2}CH_2$), 2.68 (2 H, t, J = 6.5 Hz, $CH_2\mathbf{CH_2}CO$), 2.78 (3 H, s, $\mathbf{CH_3}$), 3.49 (2 H, t, J = 7.5 Hz, pyr$\mathbf{CH_2}$), 7.30 (1 H, d, J = 8.1 Hz, H3Naph), 7.89-8.17 (10 H, m, 2HNaph + 8H Pyr), 8.34 (1 H, d, J = 9.9 Hz, Hpyr), 8.42 (1 H, d, J = 8.7 Hz, H6Naph). Calc. for $C_{29}H_{23}N_3O$: C, 81.08; H, 5.40; N, 9.79; Found.: C, 80.74; H, 5.35; N, 9.74.

9. HOSTEST program; Wilcox, C.S. (1991) *Frontier in Supramolecular Organic Chemistry and Photochemistry*, Schneider, H., and Durr, H., Ed.; VCH: NY.

10. INSIGHT/DISCOVER: Insight II, version 2.3.5 (1995) BIOSYM. Inc., San Diego, C.A..

11. Gannet, P.M.,. Sura, T.P. (1993) *Chem. Res. Toxicol.*, **6**, 6907.

Acnowledgments

Junta Nacional de Investigação Científica e Tecnológica *(Proj. nº STRIDA/C /AMB/ 347/92)* , and Hoechst Portuguesa for finantial support; Prof. J.M. Gaspar Martinho for facilities in fluorescence measurements.

MACROCYCLIC POLYETHERS AS DITOPIC CO-RECEPTORS FOR DUAL-CENTRED SECONDARY DIALKYLAMMONIUM GUESTS: FROM DOUBLE-STRANDED TO SINGLE-STRANDED PSEUDOROTAXANES

CESARE SCHIAVO, J. FRASER STODDART
School of Chemistry, University of Birmingham
Edgbaston, Birmingham B15 2TT, UK

DAVID J. WILLIAMS
Department of Chemistry, Imperial College,
South Kensington, London SW7 2AY, UK

Recently, we have uncovered a new simple self-assembling system [1] based on the mutual recognition between secondary dialkylammonium dications ($R_2NH_2^+$) — such as dibenzylammonium and dibutylammonium cations — and differently-sized macrocyclic polyethers — such as bis-*para*-phenylene-34-crown-10 (**BPP34C10**) and dibenzo-24-crown-8 (**DB24C8**). These complexes are stabilised predominantly by electrostatic interactions (including hydrogen bonding) between the $R_2NH_2^+$ cations and the polyether loops within the cavity of the crown ether. When the macrocycle is sufficiently large, the $R_2NH_2^+$ cation threads itself through the macrocyclic cavity so as to generate a superstructure with a pseudorotaxane-like geometry in the solid state. We have found that only a single strand, incorporating either one or two [1] (—NH_2^+—) centres can be accommodated within the cavity of **DB24C8** to form the so-called single- and doubly-encircled pseudorotaxanes, respectively (*e.g.* a [2]- and [3]-pseudorotaxane). However, upon utilising the larger crown ether **BPP34C10**, two $R_2NH_2^+$-containing strands [1] are threaded through the centre of the cavity of either one or two macrocycles to generate a double-stranded [3]pseudorotaxane, and a double-stranded doubly-encircled [4]pseudorotaxane, respectively. These encouraging results have led us to employ [2] either larger macrocyclic polyethers (bearing more polyether loops) or cationic guests incorporating multiple (—NH_2^+—) centres to self-assemble multiply-stranded and multiply-encircled [n]pseudorotaxanes, respectively.

The X-ray crystal structure [1] of the 1:2 complex between **BPP34C10** and the $(PhCH_2)_2NH_2^+$ ion reveals that the two (—NH_2^+—) centres bind to two different polyether loops of the macrocycle. In the resulting inclusion complex, the macrocycle acts as a host incorporating two recognition sites — *i.e.* it is a *homotopic co-receptor.* [3] (Figure 1). Molecular modelling [4] suggests that the formation of single-stranded [2]pseudorotaxanes, rather than double-stranded [3]pseudorotaxanes, might be favoured by inserting an appropriate spacer between the two (—NH_2^+—) centres of a dual-centred secondary dialkylammonium dication (RNH_2^+—spacer—NH_2^+R) (Figure 1).

A. W. Coleman (ed.), Molecular Recognition and Inclusion, 491–494.
© 1998 *Kluwer Academic Publishers. Printed in the Netherlands.*

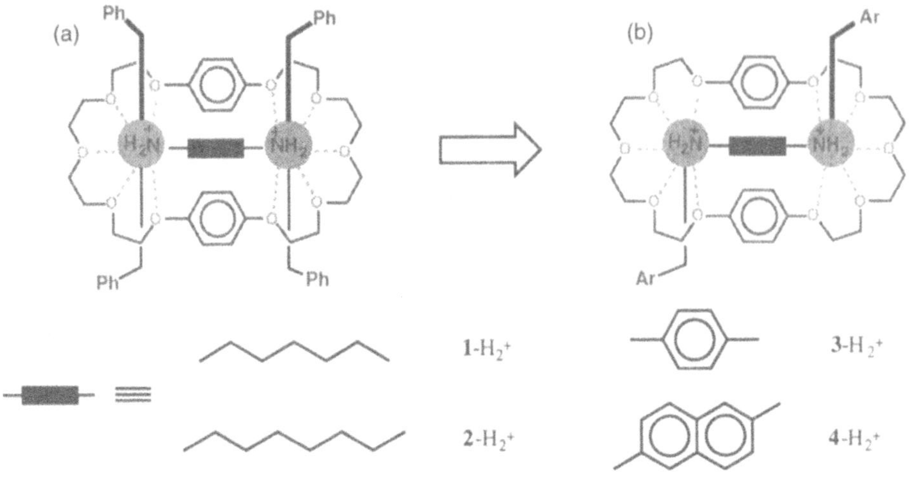

Figure 1. A schematic representation of (a) a double-stranded [3]pseudorotaxane and (b) a single-stranded [2]pseudorotaxane formed between **BPP34C10** and a single- or a dual-centred secondary dialkylammonium ion, respectively

Here, we report the experimental evidence for the existence of several complexes formed between **BPP34C10** and bis secondary dialkylammonium guests (RNH_2^+—spacer—NH_2^+R) of different structure, both in the solid state (X-ray crystallography) and in solution (^{1}H NMR spectroscopy), as well as in the 'gas phase' (FABMS). Initially, in our selection of suitable bisammonium guests for the macrocyclic host **BPP34C10**, we have considered flexible spacers of different lengths — such as an alkane chain bearing either five or six methylene units, *i.e.* as in **1**-H_2^{2+} and **2**-H_2^{2+}. In fact, a polymethylene chain should allow a favourable orientation of the guest inside the cavity of the macrocycle, to form optimal-strength [$NH^+\cdots O$] hydrogen bonds. Conversely, the complexation entropy should favour the binding of bis ammonium dications in which the two (—NH_2^+—) centres are separated by a rigid aromatic unit — such as a phenylene or a naphthalene ring, *i.e.* as in **3**-H_2^{2+} and **4**-H_2^{2+}. Furthermore, the presence of an aromatic moiety, ideally sandwiched between the hydroquinone residues, *i.e.* as in a 1:1 complex with pseudorotaxane-like geometry, should provide the system with additional enthalpic stabilisation as a result of π–π stacking interactions [5]. In addition, suitably-sized terminal groups (*e.g.* 3,5-di-*tert*-butylphenyl units) should prevent the insertion of a second strand through the macrocyclic cavity.

The complexes formed between the macrocycle **BPP34C10** and all the bis secondary dialkylammonium guests — as their bishexafluorophosphate salts — were investigated by ^{1}H NMR spectroscopy. All the spectra were recorded at room temperature in CD_2Cl_2 on equimolar mixtures between the components. In each case examined, the spectrum shows a time-averaged set of resonances, implying fast kinetics for complexation and decomplexation on the ^{1}H NMR (300 MHz) time-scale. This observation is not surprising, given the fact that CPK space-filling molecular models indicate that the 3,5-di-*t*-butylphenyl residues at the termini of the (RNH_2^+—spacer—

NH_2^+R) dication, have no difficulty in threading their way through the centre of the **BPP34C10** macrocycle. Generally, the resonances of the protons of the spacer unit experience significant upfield shifts [6] when the ^1H NMR spectrum is recorded in the presence of an equimolar amount of crown ether in solution. Interestingly, a (^1H-^1H) NMR NOESY experiment, recorded in CD_2Cl_2 at 245 °K, shows cross coupling peaks between the hydroquinone resonance and both the protons of the naphthalene ring and those of the terminal groups. These observations support the hypothesis of an inclusion complex in which the spacer unit is sandwiched in between the two hydroquinone residues within the macrocyclic cavity. FABMS reveals, for each complex, a peak at a *m/z* value corresponding to the "molecular weight" of the 1:1 complex with the loss of two PF_6^- counterions, confirming that the supramolecular entities persist in the 'gas phase'. The solid state structures of all the complexes reported herein were investigated by X-ray crystallography. Single crystals of the complex [**BPP34C10**·**1**-H_2]$^{2+}$ suitable for X-ray analysis, were grown from an equimolar mixture of the components in CH_2Cl_2 layered with hexane. The solid state structure reveals that threading of the dicationic unit through the macrocyclic ring does not occur, though the $CH_2NH_2^+(CH_2)_5^+NH_2CH_2$ chain is inserted into the convex face of the macrocycle. The terminal disubstituted benzyl groups protrude from the mean plane of the cavity, giving the complex a 'face-to-face' geometry [7]. The superstructure is sustained predominantly by hydrogen bonding interactions between the (—NH_2^+—) centres and some of their proximal polyether loops. Similarly, the complex [**BPP34C10**·**2**-H_2]$^{2+}$ was crystallised from a 1:1 mixture in dichloromethane layered with hexane and subjected to X-ray analysis. The solid state structure reveals that, on increasing the length of the polymethylene chain linking the (—NH_2^+—) centres from five to six carbon atoms, the dication now threads symmetrically through the centre of the macrocycle to form a single-stranded [2]pseudorotaxane (Figure 2). The host-guest stabilisation is achieved once again by [NH^+···O] hydrogen bonding between each cationic centre and the oxygen atoms of different polyether loops of the macrocycle. As a result, **BPP34C10** acts as a ditopic coreceptor for the dicationic strand. In addition, there is evidence for supplementary stabilisation *via* [CH···π] interactions between the hydrogen atoms attached to the two central carbon atoms of the hexamethylene chain of the dication and the π-system of the hydroquinone rings. The complexes [**BPP34C10**·**3**-H_2]$^{2+}$ and [**BPP34C10**·**4**-H_2]$^{2+}$ also possess a single-stranded [2]pseudorotaxane-like geometry in the solid state. These superstructures are sustained predominantly by a combination of [NH^+···O] and [N^+CH···O] hydrogen bonds involving each ammonium centre and a different recognition site (*i.e.* that is, a polyether loop) of the crown ether. Furthermore, some additional stabilisation is achieved as a result of π–π stacking interactions between the almost parallelly-aligned hydroquinone rings of the macrocycle and the central aromatic spacer of the dicationic guest, thus introducing an additional recognition property as predicted by our initial molecular modelling studies.

In summary, we have produced by self-assemby processes several single stranded [2]pseudorotaxane-like complexes between the macrocyclic polyether **BPP34C10** and

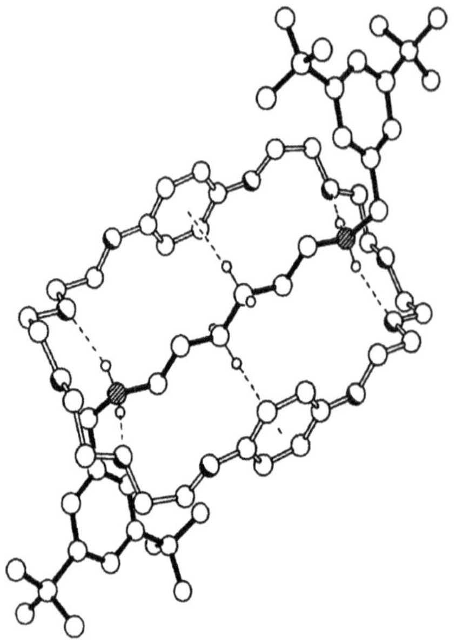

Figure 2. A perspective view of the centrosymmetric [2]pseudorotaxane in the solid state showing the intracomplex [$NH^{+}\cdots O$] hydrogen bonding, together with the [$CH\cdots\pi$] interactions

suitable dications (RNH_2^+—spacer—NH_2^+R) in the solid state. Strong supporting evidence confirms the integrity of these supramolecular entities in solution and in the 'gas phase'. The beauty of these systems rests on their great structural versatility which allows even simple changes in the nature and constitution of the spacer unit to be tailored to (i) specific recognition properties (*i.e.* interactions sustaining the pseudorotaxane complex) (ii) the geometry and (iii) stoichiometry of the resulting superstructure. These new pseudorotaxanes possess considerable potential as templates in the directed synthesis of a constitutionally-related novel class of rotaxanes bearing suitably-sized terminal groups.

References

1. Ashton, P.R., Chrystal, E.J.T., Glink, P.T., Menzer, S., Schiavo, C., Spencer, N., Stoddart, J.F., Tasker, P.A., White, A.J.P. and Williams, D.J. (1996) Pseudorotaxanes formed between secondary dialkylammonium salts and crown ethers, *Chem. Eur. J.* **2**, 709-728.

2. Ashton, P.R., Glink, P.T., Menzer, S., Stoddart, J.F., Tasker, P.A., White, A.J.P. and Williams, D.J. (1997) Triply- and quadruply-encircled and triple- and quadruple-stranded pseudorotaxanes, *J. Am. Chem. Soc.* In preparation.

3. (a) Lehn, J.-M. (1995) *Supramolecular Chemistry: Concepts and Perspectives*; VCH: Weinheim, and references cited therein.

4. The prospective guest species were assembled within the program Macromodel 5.0 and their suitability for inclusion into the crown ether cavity were assessed by either the AM1 semi-empirical Hamiltonian, as utilised in the program Spartan 4.1.1 or the MonteCarlo conformational searching routines resident within Macromodel 5.0.

5. (a) Hunter, C.A. (1994) The role of aromatic interactions in molecular recognition, *Chem. Soc. Rev.* **23**, 101-109; (b) Claessen, G.C. and Stoddart, J.F. π–π Interactions in self-assembly, *J. Phys. Org. Chem.* Submitted.

6. The Δδ values ($\Delta\delta = \delta_{uncomplexed} - \delta_{complexed}$) for the α-$CH_2$, β-$CH_2$ and γ-CH_2 (to the nitrogen atom) on the hexane chain of [**2**-H_2]·[PF_6] for the complex [**BPP34C10**·**2**-H_2]·[PF_6] in CD_2Cl_2 are: +0.29, +0.80 and +1.18 ppm, respectively.

7. Goldberg, I. (1984) Inclusion Compounds (Vol. 2) in *Complexes of Crown Ethers with Molecular Guests*, pp. 260-335, Atwood, J.L, Davis J. E. D., MacNicol D. D. (eds.), Academic Press, London.

SYNTHESIS OF A 20-CROWN-6 FROM D-GLUCOSE AND FIRST STUDY OF ITS ALKALI METAL CATIONS AFFINITY BY MALDI-FTMS

M.-F. SCHMITT-DUBESSY and J.-P. JOLY*
Laboratoire de Méthodologie et Synthèse Énantiospécifique de Biomolécules associé au CNRS, Université Henri Poincaré-NANCY I, BP 239, F-54506 Vandœuvre Cedex, France.
P.-J. CALBA, A. HACHIMI, and J.-F. MULLER
Laboratoire de Spectrométrie de Masse et de Chimie Laser, Université de Metz, IPC-1, Boulevard Arago, Technopôle 2000, F-57078 Metz Cedex 3, France.

1. Introduction

For many years now, some of us have been involved in synthesizing various crown ethers from suitably protected hexopyranosides [1,2] or D-mannitol frameworks [3,4]. We report here the synthesis of a novel macrocycle from D-glucose incorporating a flexible biphenyl unit and a primary amine function at C-6 on the sugar moiety and some of its complexing abilities towards various guests, examined for the first time by Matrix Assisted Laser Desorption Ionization with Fourier Transform Mass Spectrometry (MALDI-FTMS).

2. Synthesis of crown ethers

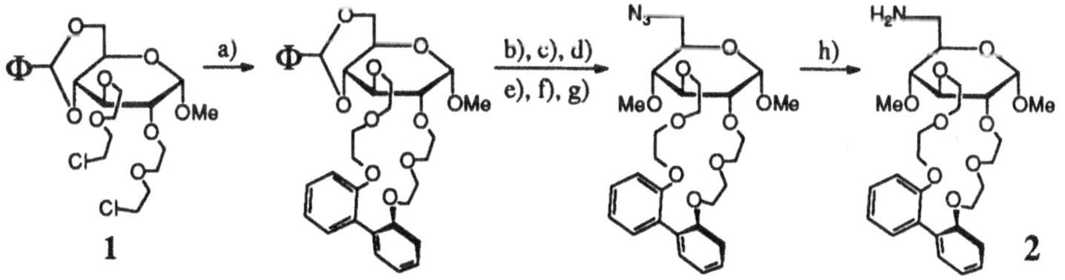

Figure 1. Synthesis of amino-crown ether **2** from half-crown **1** [1].
Reagents: a) 2,2'-biphenol, K_2CO_3, *n*BuOH, 55%; b) aq. AcOH, >95%; c) TrCl, pyr., >95%; d) Me_2SO_4, aq. KOH, tol., 80%; e) TsOH, MeOH, >95%; f) TsCl, pyr., 63%; g) NaN_3, DMF, >95%; h) $LiAlH_4$, THF, ~75%.

The 250 MHz ^{1}H-NMR spectrum of **2** exhibited two distinct doublets at δ~4.6 assigned to H-1 ($J_{1,2}$ = 3.5 Hz) on the anomeric carbone atom. This atropisomeric behavior, which is well recognised for small singly bridged biphenol derivatives [5,6], has never been described so far in macrocycles incorporating a non substituted 2,2'-biphenyl unit.

A. W. Coleman (ed.), Molecular Recognition and Inclusion, 495–498.
© 1998 *Kluwer Academic Publishers. Printed in the Netherlands.*

4. Study of alkali metal cation selectivity by means of MALDI-FTMS

An equimolar mixture of five alkali metal cations (Li^+, Na^+, K^+, Rb^+, Cs^+) and a two-fold excess of crown ether **2** were mixed with the matrix and evaporated before the analysis to produce comparable concentrations of 1:1 complexes in the probe [10, 11]. Under these conditions, it can be postulated that the mass spectrum presented below (Fig. 3) depends to the greatest extent on the relative cation affinities in the gas phase and only to a small extent on the ionization potential of the different cations as demonstrated by the right spectrum (Fig. 4):

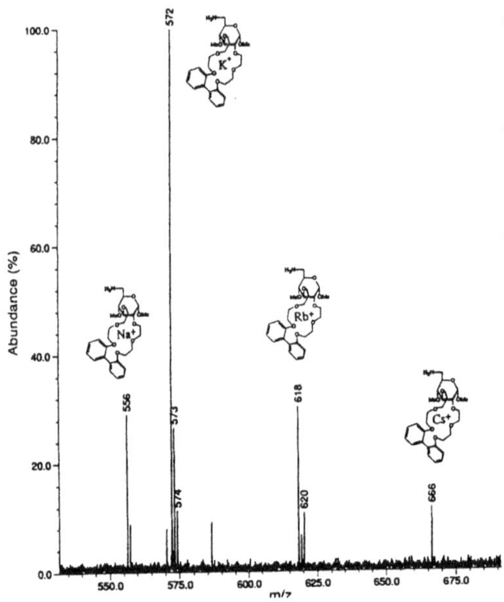

Figure 3. MALDI-FT mass spectrum of an equimolar mixture of NaCl, KCl, RbCl, CsCl, Li_3PO_4 (10^{-4} M), **2** (10^{-3} M) and matrix (10^{-1} M).

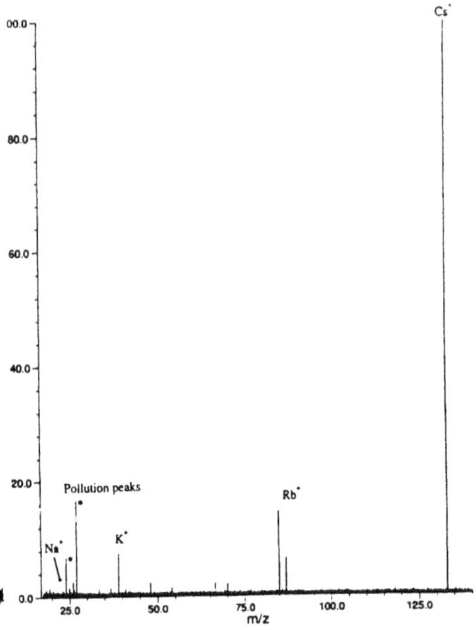

Figure 4. FT mass spectrum of an equimolar mixture of NaCl, KCl, RbCl, CsCl, Li_3PO_4 (10^{-4} M). Photonic energy at 355 nm=3.49 eV.

Without crown ether, the relative intensities of the peaks depend mainly on the ionization potential of the element: Li = 5.392, Na = 5.139, K = 4.341, Rb = 4.177, and Cs = 3.894 eV.

5. Conclusions and perspectives

The marriage of MALDI, as soft ionization method, and FTMS for obtaining high resolution of high-mass MALDI produced ions [12,13], gave us a powerful opportunity to get information on a complex (**2** + 2-phenylethylamine), which has been observed by ^1H NMR. Although the cation attachment process is still not well defined, some previous results suggested that the mass spectra may reflect the order of the stability observed in solution. The phenylethylammonium-attached molecular ion could not be detected owing to the relative higher affinity of the crown ether for lithium.

3. Extraction experiments of primary ammonium cations

Liquid-liquid extraction experiments of biogenic amines as guests (L-phenylalanine, tyramine, etc...) by crown ether **2**, as host, were performed at room temperature [2]. For instance, macrocycle **2** was able to extract 2-phenylethylamine hydrochloride, in a ~0.6 guest to host ratio, from $LiClO_4$ 1M medium into a $CDCl_3/CD_3CN$ mixture (74:26). As a complete interpretation of the ^{1}H-NMR spectrum after complexation was not possible, we decided to study the dried isolated complex [**2** + 2-phenylethylamine] by MALDI-FTMS. MALDI is a 'soft' ionization technique for introducing very delicate molecules such as macro-molecules into a mass spectrometer without fragmentation [7,8]. Our experiments were performed using a Laser-FTMS microprobe prototype equipped with a 3 T superconducting magnet [9]. For the sample desorption, a Nd-Yag laser (operating at 355 nm with a pulse of 6 ns) was used. Laser light was attenuated by a telescope to obtain an irradiance of $10^6 \rightarrow 10^7$ W/cm^2. The use of α-cyano-3-hydroxycinnamic acid (instead of 2,5-dihydroxybenzoic acid) as matrix produced a single ion signal assigned to $[M + Li]^+$ (*m/z* 540, 100%) for the postulated 'complex' [**2** + 2-phenylethylamine]:

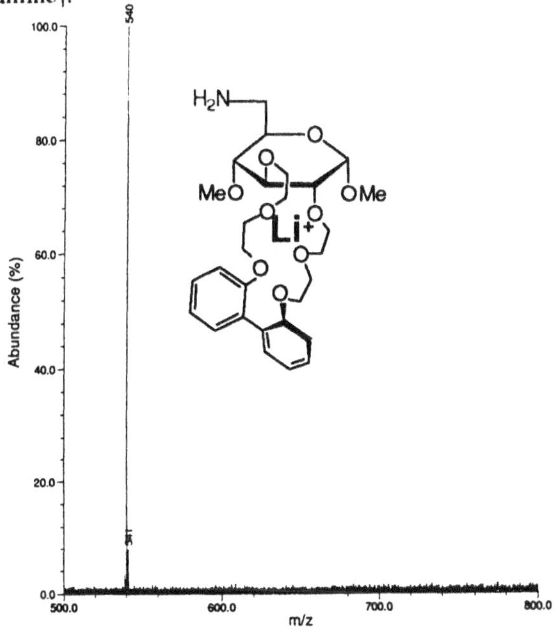

Figure 2. MALDI-FT mass spectrum of the 'complex' [2 + 2-phenylethylamine] (10^{-3} M) and matrix (10^{-1} M) in $H_2O/MeOH/CH_3CN$/0.1% TFA (1:1:1:1), irradiance = 5 10^6 W/cm^2.

None of these MALDI-FTMS data showed the presence of the complex (*m/z* 655) and of the organic cation (*m/z* 122) previously detected by ^{1}H NMR. A very close result was observed after the crown ether **2** was mixed with $LiClO_4$ (1 eq.) in $CHCl_3/CH_3CN$ (74:26) before the MS analysis. The non detection of the complex (**2** + 2-phenylethylamine) was ascribed to higher affinity of the crown ether for Li^+ than for 2-phenylethylamine *in the gas phase*. For that reason, we decided to study by the same method the alkali metal cations affinities of host **2** in the gas phase.

In a competition experiment with 5 alkali metals, the most intense ion signals observed were $[2 + K]^+$ and $[2 + Na]^+$ indicating a higher affinity for $K^+$and Na^+, and a lower affinity for Rb^+ and Cs^+ *in the gas phase*. Similar results have recently been claimed after the study of various 18-C-6 by electrospray-MS [14] and FAB-MS [15]. Furthermore, no protonated molecule of **2** $[M + H]^+$ could be observed even when the amino-crown ether **2** was treated with an excess of diluted $HClO_4$ before the analysis. Hence, the cationized molecules formed in high yield by exclusive binding of alkali metal cations seem to undergo the desorption process without significant fragmentation in the MALDI ionization process.

The possible effects of the counteranion and the matrix on the relative selectivity of host **2** and related compounds are currently under investigation.

6. References

1. Di Cesare, P. and Gross, B. (1979) Chiral Crown Ether Synthesis by Catalysis in a Two Phase System, *Synthesis*, 458-460.
2. Joly, J.-P., Nazhaoui, M., Dumont, B. (1994) Synthesis and complexation behavior of some crown ethers derived from D-hexopyranosides and D-mannitol towards racemic phenylglycine salts, *Bull. Soc. Chim. Fr.* **131**, 369-379.
3. Joly, J.-P. and Gross, B. (1989) Synthesis and Immobilization of New Crown Ethers Derived from D-mannitol For the Resolution of Free Amino Acids, *Tetrahedron Lett.* **30**, 4231-4234.
4. Nazhaoui, M., Joly, J.-P., Jean-Claude, B. J., Del Duca, V., Aubry, A. and (in part) Boubouh, M. (1995) *trans*-Fused crown ethers from 2,5-*O*-methylene-D-mannitol: synthesis, X-ray diffraction structure and full nuclear magnetic resonance spectroscopic data of 1,6-diazido-1,6-dideoxy-2,5-*O*-methylene-3,4-*O*-naphtalene-2,3-diylbis(oxyethyleneoxyethylene)-D-mannitol and 3,4-di-*O*-acetyl-1,6-diazido-1,6-dideoxy-2,5-*O*-methylene-D-mannitol, *J. Chem. Soc., Perkin Trans. 1*, 2919-2926.
5. Mislow, K., Hyden, S. and Schaefer, H. (1963) Stereochemistry of the 1,2,3,4-Dibenzcyclonona-1,3-diene System. A Note on the Racemization Barrier in Bridged Biphenyls, *J. Am. Chem. Soc.* **84**, 1449-1455.
6. Kurland, R. J., Rubin, M. B. and Wise, W. B. (1964) Inversion Barrier in Singly Bridged Biphenyls, *J. Chem. Phys.* **40**, 2426-2427.
7. Rosinke, B., Strupat., K., Hillenkamp, F., Rosenbusch, J., Dencher, N., Krüger, U. and Galla, H.-J. (1995) Matrix-Assisted aser Desorption/Ionization Mass Spectrometry (MALDI-MS) of Membrane Proteins and Non-covalent Complexes, *J. of Mass Spectrom.* **30**, 1462-1468.
8. Vincenti, M. (1995) Host-Guest Chemistry in the Mass Spectrometer, *J. of Mass Spectrom.* **30**, 925-939.
9. Muller, J.-F., Pelletier, M., Krier, G., Weil, D., Campana, J. (1989) A New Generation of Microprobe: Laser Ionization and FT/ICR Mass Spectrometry, in P. E. Russel (ed.), *Proceeding of the 24th Conference of the Microbeam Analysis Society-Ashville*, San Francisco Press Inc., pp. 311-316.
10. Dietrich, B., Viout, P. and Lehn, J.-M. (1991) *Aspects de la chimie des composés macrocycliques*, InterEditions/Editions du CNRS, Paris/Meudon.
11. Giraud, D., Scherrens, I., Lever, M.-L., Laprévote, O. and Das, B. C. (1996) Desorption of crown ether-alkali metal ion complexes in liquid secondary ion mass spectrometry (LSIMS), *J. Chem. Soc, Perkin Trans. 2*, 901-905.
12. Castro, J. A., Köster, C. and Wilkins, C. (1992) Matrix-assisted Laser Desorption/Ionization of High-mass Molecules by Fourier-transform Mass Spectrometry, *Rapid Comm. in Mass Spectrom.* **6**, 239-241.
13. Sheng, L.-S., Covey, J. E., Shew, S. L., Winger, B. E. and Campana, J. E. (1994) Matrix-assisted Laser Desorption Ionization Fourier-transform Mass Spectrometry, *Rapid Comm. in Mass Spectrom.* **8**, 498-500.
14. Leize, E., Jaffrezic, A. and Van Dorsselear, A. (1996) Correlation Between Solvation Energies and Electrospray Mass Spectrometric Reponse Factors. Study by Electrospray Mass Spectrometry of Supramolecular Complexes in Thermodynamic Equilibrium in Solution, *J. of Mass Spectrom.* **31**, 537-544.
15. Pulpoka, B., Asfari, Z., Vicens, J. (1996) Synthesis of Unsymmetrical Calix[4]arene Cryptand Crown-6 in 1,3-Alternate Conformation, *Tetrahedron Lett.* **37**, 6315-6318.

^{2}H NMR INVESTIGATIONS OF THE CYCLOHEXANE/TRI-*o*-THYMOTIDE INCLUSION COMPOUND

J. SCHUHMACHER AND K. MÜLLER
Institut für Physikalische Chemie, Universität Stuttgart
Pfaffenwaldring 55, D-70569 Stuttgart, Germany

Abstract

Variable temperature ^{2}H NMR studies are reported for the cyclohexane/tri-*o*-thymotide inclusion compound in a range between 100 K and 370 K. From a quantitative analysis of the available NMR data, comprising spin-lattice and lineshape experiments, the motional and structural features of the cyclohexane guests are evaluated.

1. Introduction

Tri-*o*-thymotide (TOT) is a well-known host component for inclusion compounds [1]. In the past TOT inclusion compounds with a variety of guest molecules have been studied. Detailed X-ray investigations have been reported which showed that the TOT molecules build up channel or cage structures with cavities large enough to accommodate small organic compounds like acetone or pyridine. More recently, dynamic NMR investigations have been perfomed on such systems which revealed a large degree of motional freedom of the guest molecules [2 - 4]. In this contribution we report on ^{2}H NMR studies of the TOT inclusion compound with perdeuterated cyclohexane. The main objective of this work is the evaluation of the motional and structural properties of the cyclohexane guests and the potential impact by the surrounding host matrix.

2. Experimental

The TOT inclusion compound was prepared according to literature procedures [4]. NMR measurements were performed on a Bruker CXP 300 spectrometer operating at a frequency of 46.07 MHz. The ^{2}H NMR spectra were obtained using a 5 mm homebuilt probe employing the quadrupole echo sequence with $\pi/2$-pulses of 2.0 - 2.7 μs width and a time interval between the pulses of 25 μs. Simulations have been performed using a FORTRAN simulation package developed to simulate the various types of dynamic NMR experiments.

A. W. Coleman (ed.), Molecular Recognition and Inclusion, 499–502.
© 1998 *Kluwer Academic Publishers. Printed in the Netherlands.*

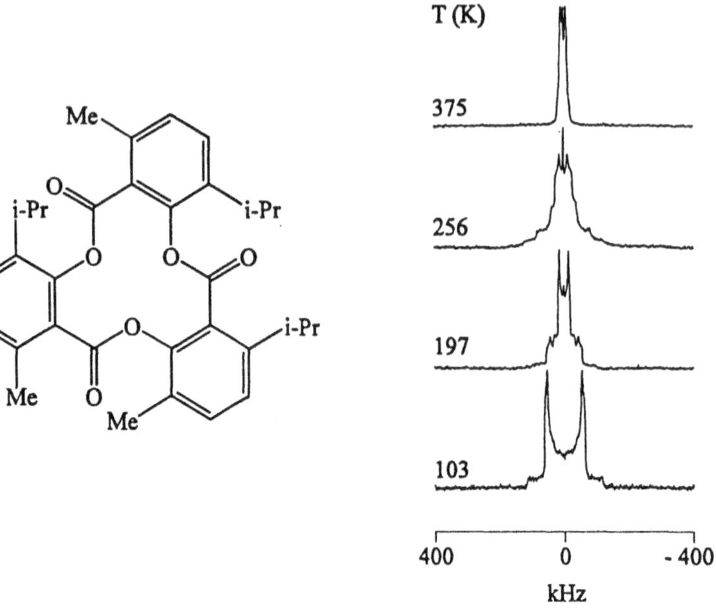

Figure 1: Structure of tri-*o*-thymotide (TOT) (left); experimental ^{2}H NMR spectra of cyclohexane in TOT in the temperature range between 103 and 375 K (right).

3. Results and discussion

An inclusion compound of TOT with perdeuterated cyclohexane has been prepared and studied by dynamic ^{2}H NMR techniques. Figure 1 displays representative ^{2}H NMR spectra obtained between 100 K and 380 K. In fact, they reflect a hierarchy of various motional contributions which determine the relaxation behaviour of the guest species within the temperature range covered here. To begin with, we focus on the low temperature range between 100 and 150 K (see Fig. 2). At 100 K a biaxial powder pattern (asymmetry parameter η = 0.3) is observed which is characteristic for an inherent fast wobbling or restricted reorientational motion of the molecular guests within the asymmetric potential of the host cavities. Above this temperature, a further spectral component starts to grow up in the central part of the ^{2}H NMR spectra until the spectrum at 150 K consists of a superposition of two subspectra. They can be attributed to the axial (broad component, quadrupolar splitting: 120 kHz) and equatorial deuterons (narrow component, quadrupolar splitting: 40 kHz) whose quadrupole coupling tensors are motionally averaged by the rotation around the C_3-axis of the molecules [4]. In principle, this motion can be modeled either by three-fold jumps between discrete positions or unrestricted free rotaions. Lineshape simulations have been performed which clearly show that for cyclohexane in the TOT inclusion compound the free rotation approach is adequate. The activation energy of 12 kJ/mol also is consistent with this motional process [5,6]. It is interesting to note that in this respect the TOT inclusion compound is different from the thiourea or cyclodextrin inclusion compound [5]. First of all, for the latter systems the cyclohexane guests are found to undergo three-fold jumps rather than free rotational motions. In addition, the absolute rate constants are about three orders of magnitude larger than for the system studied here. Although the reason for this behaviour is unknown, it is very likely that the differences of the rate constants are a consequence of the molecular constraints imposed by the rather narrow TOT cavities. Although a crystal structure is not available for the sample studied here, it is very likely that the TOT molecules build up a cage-like structure as found for the inclusion compounds with chlorocyclohexane or pyridine [1].

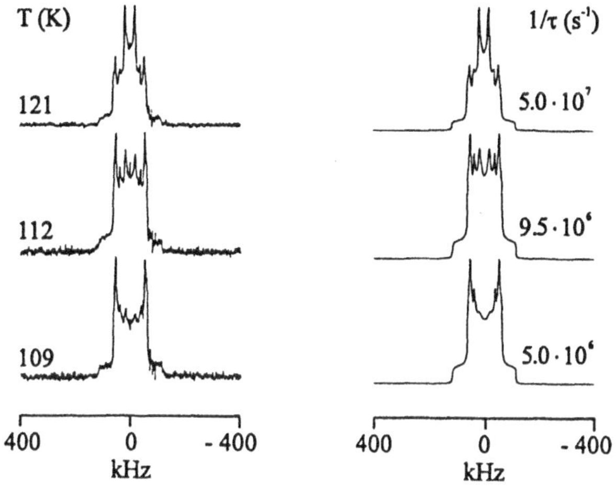

Figure 2: Experimental (left) and simulated (right) ^{2}H NMR spectra of cyclohexane in TOT in the temperature range between 109 and 121 K. The simulated spectra were calculated assuming a free rotation of the guest molecules with the rate constants given in the figure.

Between 220 K and 380 K, the experimental ^{2}H NMR lineshapes again are significantly altered (see Fig. 3). This observation primarily can be related to the ring interconversion process which has been detected in other inclusion compounds as well. Obviously, this type of motion is the slowest within the hierarchy of the motional modes found for cyclohexane in TOT. The kinetic parameters evaluated from the corresponding lineshape simulations are very similar to those reported earlier from solution NMR studies and investigations of other inclusion compounds. Thus, unlike the situation for the overall motions (see above), the internal ring interconversion process is practically unaffected by the TOT host lattice. Finally, a fourth motional mode is present at higher temperatures. This process is a very fast overall reorientational process (anisotropic tumbling) which leads to a significant narrowing of the ^{2}H NMR spectra in the high temperature range along with an increase in orientational disorder. Obviously, this type of motion replaces the wobble motion which was discussed earlier for the low temperature range. The analysis of the ^{2}H NMR spectra further yields the orientational order parameter which varies between 0.8 at 180 K and 0.3 at 380 K. This significant increase in orientational disorder might be a result of an increase of the host cavity size due to increasing thermal motions of the host molecules.

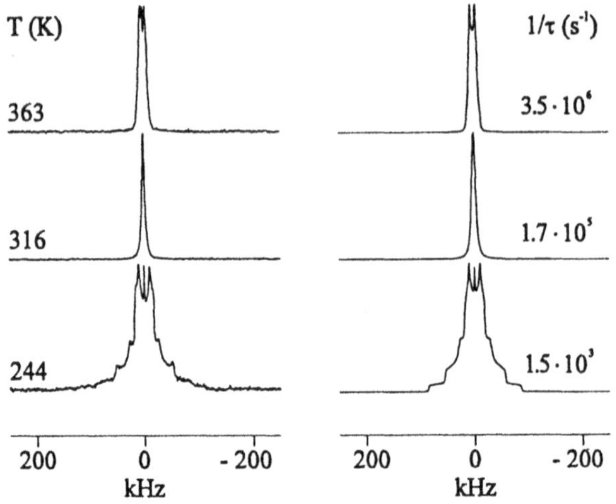

Figure 3: Experimental (left) and simulated (right) ^2H NMR spectra of cyclohexane in TOT in the temperature range between 244 and 363 K. The simulated spectra were calculated assuming exchange between the axial and equatorial deuterons with the rate constants given in the figure.

4. Conclusions

The inclusion compound of TOT with cyclohexane has been investigated using dynamic ^2H NMR methods. It is found that for the guest molecules four different types of motional contributions become relevant within the temperature range covered here: (i) wobbling motions, (ii) rotational motions around the C_3-symmetry axis, (iii) ring interconversions and (iv) anisotropic tumbling motions. It is found that the various types of overall motions are affected by the host lattice while the internal ring interconversion process remains unaltered.

Acknowledgment

We thank Mrs. H. Seidel for the synthesis of tri-*o*-thymotide. Financial support by the Deutsche Forschungsgemeinschaft and the Fonds der Chemischen Industrie is gratefully acknowledged.

References

1. Gerdil, R. (1987) Tri-*o*-Thymotide Clathrates, *Top. Curr. Chem.* 140, 71.
2. Facey, G. A., Ratcliffe, C. I., Hynes, R. and Ripmeester, J. A. (1992) Guest Dynamics and Crystal Structure of the Tri-*o*-Thymotide-Acetone Clathrate studied by Solid-State NMR, X-Ray Diffraction and Molecular Modelling, *J. Phys. Org. Chem.* **5**, 670.
3. Facey, G. A., Ratcliffe, C. I. and Ripmeester, J. A. (1995) Solid State NMR Studies of the Molecular Motion of Pyridine in its Tri-*o*-Thymotide Clathrate, *J. Phys. Chem.* **99**, 12249.
4. Aliev, A. E. and Harris, K. D. M. (1995) Dynamics of Benzene and Pyridine Guest Molecules in their Tri-*ortho*-Thymotide Inclusion Compounds, *J. Chem. Soc. Faraday Trans.* **91**, 2017.
5. Poupko, R., Furman, E., Müller, K. and Luz, Z. (1991) Reinvestigation of the Thiourea-Cyclohexane Inclusion Compound by Deuterium NMR Spectroscopy, *J. Phys. Chem.* **95**, 407.
6. Müller, K., to be published.

INCLUSION COMPLEXES OF SILICONHYDROFLUORIC ACID TRANSFORMATION PRODUCTS WITH THE CROWN ETHERS

Yu.A.SIMONOV[a], J.LIPKOWSKI[b], M.S.FONARI[a], V.Ch.KRAVTSOV[a], Ed.V.GANIN[c], V.O.GELMBOLDT[c], A.A.ENNAN[c]

[a] Institute of Applied Physics Academy of Sciences of Moldova, Kishinev 2028, Moldova; [b] Institute of Physical Chemistry Academy of Sciences of Poland, Warsaw, Poland; [c] Institute of Environment and Human Protection at Odessa State University, Odessa, Ukraine

Recently we have shown [1,2] that crown ethers and their aza-analogs are able to extract the distinctive species from the aqueous solution of siliconhydrofluoric acid which exists as an equilibrium mixture of the neutral and negatively charged compounds: $[SiF_4{\cdot}2H_2O]$, $[SiF_5{\cdot}H_2O]^-$ and SiF_6^{2-}. The macrocycles below have been used for recognition of these entities in solution.

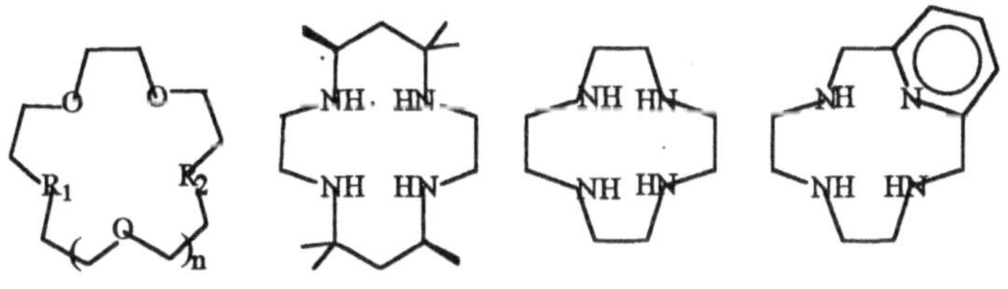

L1 $R_1 = R_2 = O; n = 2$ *L5* *L6* *L7*

L2 $R_1 = O, R_2 = NH; n = 2$

L3 $R_1 = R_2 = NH; n = 2$

L4 $R_1 = R_2 = NH; n = 1$

The final complexes belong to the host-guest type [3] and compose of the macrocycle as a host (*H*), inorganic entity as a guest (*G*) and water molecules. The following compounds have been isolated in the crystalline form and their structures have been studied by X-ray method.

A. W. Coleman (ed.), Molecular Recognition and Inclusion, 503–506.
© 1998 *Kluwer Academic Publishers. Printed in the Netherlands.*

TABLE 1. A list of complexes described by X-ray techniques.

Complex	Host	Complex composition	*H* : *G* : *W*	Crystal organization
I	*L1*	[*L1*(*trans*-$SiF_4 \cdot 2H_2O$)(H_2O)$_2$]	1:1:2	Chain
II	*L2*	[(*L2*H·H_2O)·($SiF_5 \cdot H_2O$)·H_2O]	1:1:2	Layer
III	*L3*	[(*L3*H_2)·($SiF_5 \cdot H_2O$)$_2$]	1:2	Chain
IV	*L4*	[(*L4*H_2)·(SiF_6)]	1:1	Ribbon
V	*L5*	[(*L5*H_4)·(SiF_6)$_2$·$4H_2O$]	1:2:4	Carcass
VI	*L6*	[(*L6*H_4)$_2$·(SiF_6)$_4$·$7H_2O$]	2:4:7	Carcass
VII	*L7*	[(*L7*H_3)$_2$·(SiF_6)$_3$·$4H_2O$]	2:3:4	Carcass

Only the 'all-O' containing host (*L1*) extracts neutral inorganic entity [$SiF_4 \cdot 2H_2O$] from solution. In (**I**) exclusively OH...O type of H-bonds stabilizes the structure. The components are alternately united in the infinitive chain. Mixed N,O-containing macrocycles are easily protonated on the N-atom and give the opportunities for the electrostatic interaction in complex formation. (**II**) and (**III**) are built of [*L2*H]$^+$ or [*L3*H_2]$^{2+}$ macrocyclic cations, [$SiF_5 \cdot H_2O$]$^-$ anions and water molecules. $NH^+ ... F^-$ short contacts appear in (**II**) and may be considered as hydrogen bonds. In (**III**) (Fig. 1) the [$SiF_5 \cdot H_2O$]$^-$ guest entities, ineterconnected *via* $OH ... F^-$ interactions, fulfill the double bridge function between two crown molecules joining to them through $NH^+ ... F^-$ contacts.

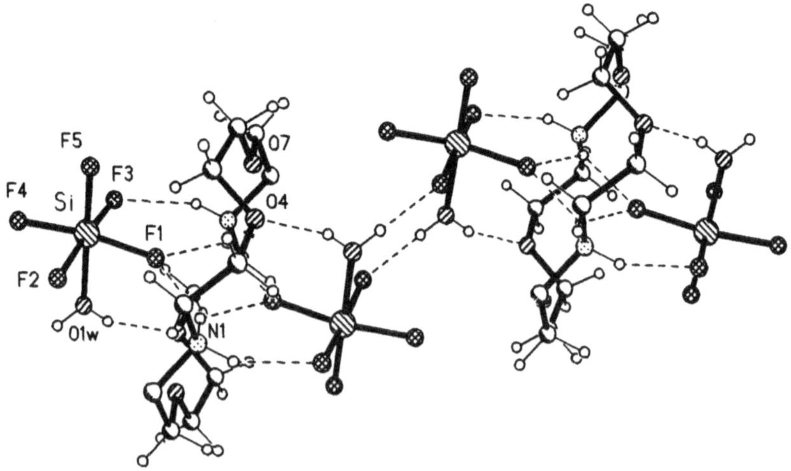

Figure 1. H-bonded polymeric chain in complex (**III**).

The number of the $NH^+...F^-$ interactions increases with the ring size diminution (*L4* in comparison with *L3*) and when the number of $>NH_2$-centers increases ('all-N' tetraazamacrocycles in comparison with mixed diaza-ones). As the number of the $>NH_2$ donor centers in *L2-L4* remains insufficient for saturation of all possible $NH^+...F^-$ contacts, the rather simple chain or layer motifs are shaped in the case of the N,O-mixed ligands. However, the rapprochement of $>NH_2$-functions in *L4* causes such an arrangement of host molecules around the inorganic SiF_6^{2-} anion which enhances the host-guest complementarity due to the number of fluorine atoms involved in interactions. For (**V-VII**) the X-ray analysis revealed complicated *3D*-networks comprised of quadri- (*L5, L6*) or tri- (*L7*) protonated macrocycles with $>H_2N^+$ interaction sites oriented outwardly, SiF_6^{2-} anions and water molecules (Fig.2).

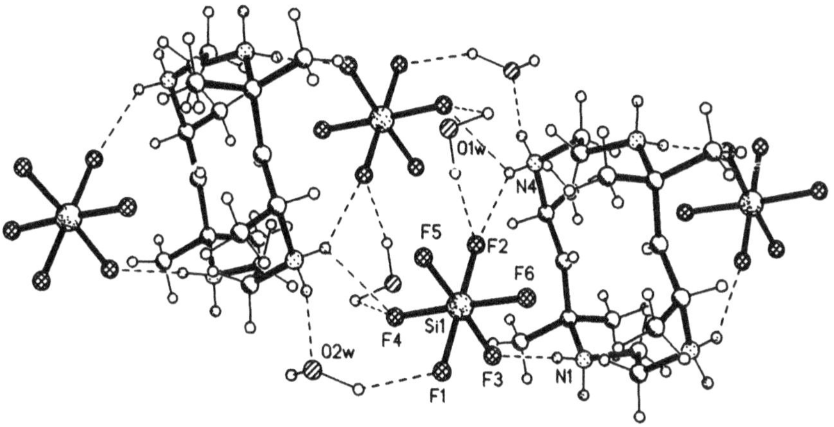

Figure. 2. The fragment of *3D*-carcass in complex (**V**).

All possible types of contacts mentioned above appear in (**V**)-(**VII**): $NH^+...F^-$; $NH^+...H_2O$; $F^-...H_2O$; $H_2O...H_2O$. The water molecules are distinguished by strong inequality of their binding in the structures. Only in complex (**IV**) water does not enter the crystal structure. In (**I**) external water molecule behaves as a go-between to bind an inorganic complex to the crown ether while coordinated to the inorganic moiety water molecule forms H-bond with the crown ether as well. Both in (**II**) and (**III**) it is coordinated to silicon atom with stabilizing of unstable $[SiF_5 \cdot H_2O]^-$ entities, in all the other cases (complexes **V** - **VII**) different number of solvent water molecules are incorporated in the structures. The $>NH_2$-binding sites are predominately attacked by fluorine-containing guest entities in the competition with the water lone pairs (Fig. 3).

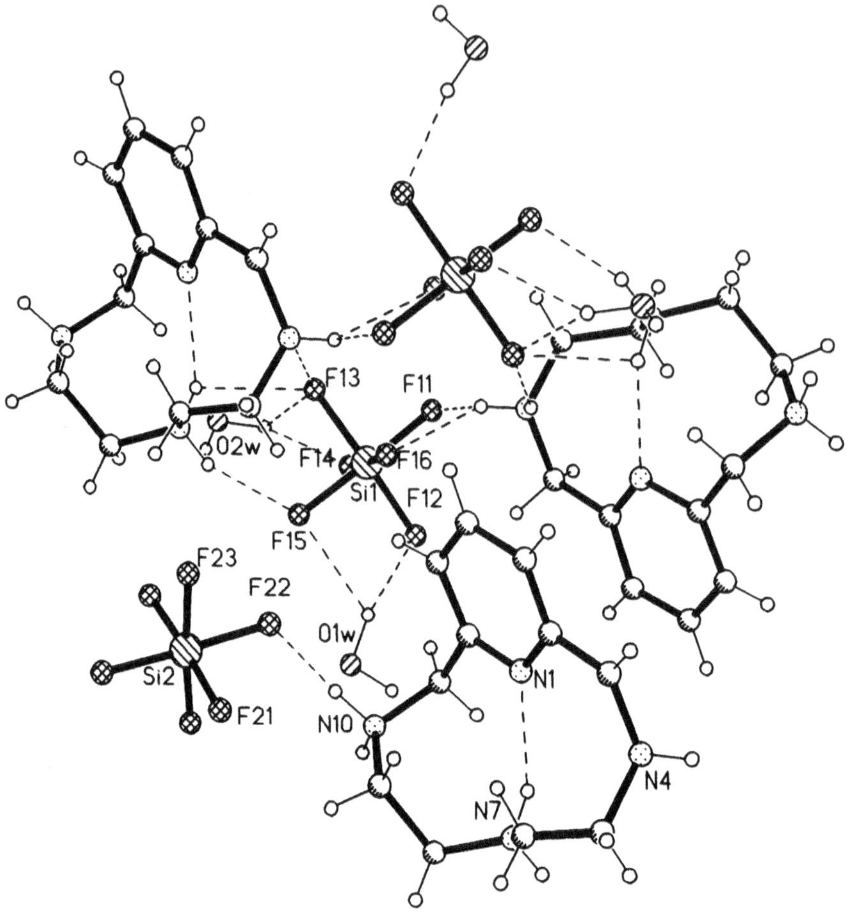

Figure 3. **The fragment of crystal structure in complex (VI).**

References

1. Simonov, Yu.A., Ganin, Ed.V., Dvorkin, A.A., Fonari, M.S., Gelmboldt, V.O. and Ennan, A.A. (1994) Stabilization of silicon tetrafluoride by crown ether complexation. Crystal and molecular structure of the host:guest complex 1:1:2 between *trans*-tetrafluorobis(aqua) silicon, 18-crown-6 and water, *Supramolecular Chemistry* **3,** 185-189.

2. Simonov, Yu.A., Fonari, M.S., Lipkowski, J., Gelmboldt, V.O. and Ganin, Ed.V. (1996) Inclusion compounds:
the products of the interaction of silicon, germanium and boron fluorides with crown ethers *J. Incl. Phenom.* **24,** 149-161.

3. *Host Guest Complex Chemistry Macrocycles. Synthesis, Structures, Applications* (1983). Ed. F.Vögtle, E E.Weber, Springer- Verlag, Berlin, Heidelberg, New York, Tokyo.

STUDY OF THE INTERACTION OF THE HOST:GUEST TYPE BETWEEN SnF_2: β-CYCLODEXTRIN

R.D. SINISTERRA[1]* C.A.L. FILGUEIRAS[1], C.A. ALVES DE CARVALHO[2], A. ABRAS[3], M. E. CORTÉS[4], AND C.A. MENEZES[1]

[1]Departamento de Química, Instituto de Ciências Exatas, Universidade Federal de Minas Gerais, C.P. 702, 30161-970, Belo Horizonte, MG, Brazil, e-mail ruben@juno.qui.ufmg.br . [2]Departamento de Química, Universidade Federal de Ouro Preto, MG, Brazil. [3]Departamento de Física, Universidade Federal de Minas Gerais, Belo Horizonte, MG, Brazil. [4]Departamento Odontologia restauradora, Universidade Federal de Minas Gerais, Belo Horizonte, MG, Brazil

INTRODUCTION

Cyclodextrins are oligosaccharides composed of at least six (1→4) linked α-D-glucosyl residues, which have the shape of a hollow, truncated cone. In the solid state as well as in solution they can form inclusion compounds with a variety of guest molecules. Cyclodextrins have been extensively used to modify the physicochemical properties of various drugs [1,2]. SnF_2 is a drug widely used in dentrifices and gels aiming at the remineralization of early carious lesions. However, this compound is very unstable to air and water and relatively little information is available to prove the efficacy of these self-applied fluoride gels and dentrifices.

In this study we have prepared the inclusion compound between SnF_2 and β-cyclodextrin in order to obtain higher stability to air and water and slow-release of the guest. The supramolecular compound was prepared by co-precipitation and dry-freezing. Similar studies with α-cyclodextrin were reported elsewhere [3].

EXPERIMENTAL

β-cyclodextrin hydrate (β-cd) was purchased from Aldrich and SnF_2 from Strem, and both were used without further purification.

The inclusion compounds were prepared by three methods. The first consisted of mixing an aqueous solution of β-cd and SnF_2 in a 1:1 molar ratio under N_2, and stirring for 15 minutes (preparation **P1**) and for 5 hours (preparation **P2**). In both cases the water was evaporated under vacuum and the host:guest compound obtained. The second method involved a reaction in the solid state between SnF_2 and β-cd in a 1:1 molar ratio under N_2 (preparation **P3**). The third preparation consisted of dry-freezing an aqueous solution of SnF_2 and β-cd, also in a 1:1 molar ratio (preparation **P4**). In order to compare the results a physical mixture of the reactants was also prepared with the same molar ratio as the host:guest compound (**P5**).

Several physical methods were used to characterize these products. Mössbauer spectra were obtained using a conventional constant acceleration spectrometer, moving a $CaSnO_3$ source kept at room temperature. The samples were measured at liquid nitrogen temperature. All spectra were fitted by means of Lorentzian line shapes. Isomer shift values are given with respect to the $CaSnO_3$ source.

X-ray powder diffraction (XRD) was carried out with a Rigaku diffractometer, fitted with a LiF monochromator and employing Cu Kα radiation with 5-40 ° (2θ) step

***** *Corresponding author***

A. W. Coleman (ed.), Molecular Recognition and Inclusion, 507–510.
© 1998 *Kluwer Academic Publishers. Printed in the Netherlands.*

scanning. TG and DTG curves were obtained from a Mettler-TA4000 system, using a dynamic N_2 atmosphere and a heating rate of 10°/min. The DSC curves were obtained from a Shimadzu DSC-50 apparatus with the same heating rate. I.r. spectra were recorded from a Perkin-Elmer spectrometer model 283B, using CsI pellets.

RESULTS AND DISCUSSION

Inclusion compounds are more stable to air compared to tin(II) fluoride. The XRD results suggest a highly amorphous character for the inclusion product, in stark contrast with either the physical mixture, SnF_2, or β-cyclodextrin (Figure 1). This points to the existence of a supramolecular host-guest compound. Lowering of the crystalline character upon inclusion is likely to result from loss of water molecules from the host [4]. Formation of the inclusion product is accompanied by a sharpening of all the bands in the i.r. spectrum, compared to those of free β-cyclodextrin and the physical mixture. In the OH stretching region this sharpening can be associated with a decrease in hydrogen-bonding, which is consistent with the XRD results. Likewise, sharpening of the C-O-C bands around 1050 cm^{-1} is also indicative of less hydrogen-bonding in the material upon inclusion.

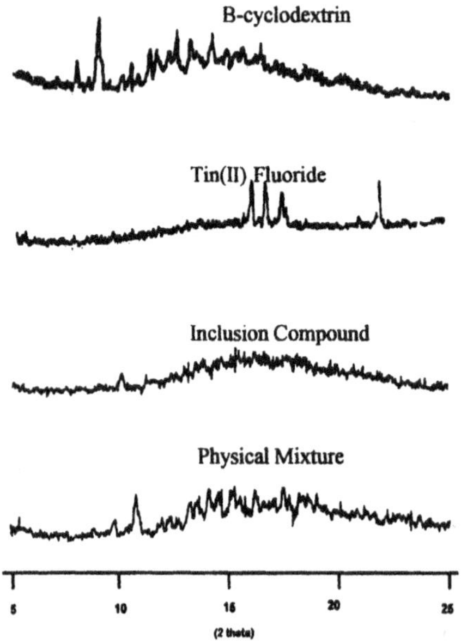

Figure 1. XRD patterns

The TG and DSC curves of the inclusion products show loss of water molecules in the range 28-105 °C, in accordance with the other data. The TG curve of SnF_2 shows its high thermal stability up to 506°C, suggesting intense association in the solid state. The TG curve of the supramolecular compound (dry-freezing), on the other hand, is markedly different from the physical mixture or free β-cyclodextrin and shows decomposition only at 270°C, which prevents any interpretation as a simple mixture.

Figure 2 shows the Mössbauer spectra of SnF_2 and their inclusion compounds prepared by different methods, and the corresponding parameters are given in Table 1, which includes literature data for comparison.

The SnF_2 used in this work (Fig. 2A and Table 1) presents 96% of the relative area as Sn(II), which is in excellent agreement with literature data for α-SnF_2 [5].

The remaining 4% relative area is due to impurities, chracteristic of Sn(IV) octahedrally coordinated by fluorine, as can be seen from the good agreement between our parameters and those reported for $(H_3O)_2SnF_6$ [6] and $[SnF_6]$ octahedra [7].

The Mössbauer parameters for the physical mixture (**P5**) are identical with those of free SnF_2 (Table1), showing that there is no host:guest interaction in this case. On the other hand, the only acceptable fitting for the Mössbauer spectra of the inclusion compounds were obtained assuming a combination of one Sn(IV) site with two Sn(II) sites, as shown in the Figure 2B/2C/2D and Table 1.

Isomer shifts and quadrupole splittings for site 1 in all products correspond to free SnF_2, whereas for site 2 those parameters have changed. This new Sn(II) site for which δ values decrease and Δ values increase were attributed to the inclusion products. This means that an interaction takes place between guest (SnF_2) and host (β-cd), so that the electronic configuration of the guest molecule is affected by the cavity of the cyclodextrin. The decrease in δ corresponds to a decrease in s electron density at the Sn nucleus, and the increase in Δ corresponds to an increase of the electric field gradient produced by an asymmetric distribution of charge about the nucleus. The variation in δ and Δ values can be attributed to the electrostatic field which exists in the cavity of the cyclodextrin[8]. Therefore, the guest molecules in the cavity are under a negative pressure, resulting in a decrease in the overlap between inner s shells of the Sn atom and the valence orbitals of the ligands, and consequently a decrease in δ is observed. We think that this electrostatic interaction might also be related to the increase in Δ values. Similar results have been observed for ferrocene-cyclodextrin and some ring-substituted ferrocene derivatives-cyclodextrins [8].

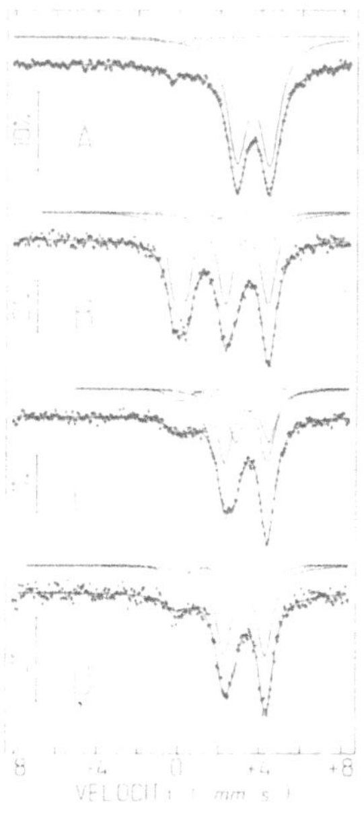

Figure 2. Mössbauer spectra at 85K of SnF_2 (A) and inclusion compounds: aqueous solution for 15 minutes (B), solid state(C) and dry-freezing(D)

It can be seen from Table 1 that except for free SnF_2 and compounds **P4** and **P5**, site 3 shows δ = 0 and Δ= 0.53-0.62 mm/s, which agree well with previous reported data for SnO_2 [9,10]. The amount fo Sn(II) from SnF_2 which oxidizes to SnO_2 depends on the method of preparation (Table1). In the method involving aqueous solutions, the SnO_2 relative area (RA) signals are respectively 34% (**P1**) and 79% (**P2**), when the reaction time changed from 15 minutes to 5 hours. Although the reaction in the solid state was performed under N_2, an SnO_2 RA signal of 6% (**P3**) was observed. We attributed this partial oxidation to the presence of water in the cyclodextrin cavity.

The relative intensity detected for SnO_2 in the Mössbauer spectrum through the RA parameter does not correspond to the relative amount of SnO_2 in the sample, because RA depends also on the recoil-free fraction (f), which is not the same for SnO_2 and SnF_2. It is known that $f(SnO_2)$= 0.44 and $f(SnF_2)$ = 0.34 at room temperature, i.e., $f(SnO_2) \cong 13f(SnF_2)$ [11].

As a result, oxidation of 1% of Sn(II) of SnF_2 to SnO_2 would give an SnO_2 signal of 13% of the total signal. Thus, the amount of SnO_2 formed in the various preparative methods are much smaller than the RA values indicated in the Table1.

On the other hand, it is quite reasonable to assume that the Mössbauer f-factor is the same for both Sn(II) sites. Therefore the RA parameter corresponds to the relative amount of site 2 in the product. From the RA values given in Table 1 one can conclude that the amount of inclusion compound (site 2) is large in the dry-freezing method (**P4**) and small in the solid state method (**P3**).

Although in the aqueous solution method (**P1** and **P2**) the amount of inclusion compounds depends on the time of reaction, P1 showed to be a method of inclusion as efficient as the solid state (**P4**). This result agrees with the importance of water for the inclusion phenomena with cyclodextrin [12].

Table 1. Mössbauer parameters at liquid N_2 temperature for SnF_2 and their inclusion compounds with β-cyclodextrin.

Compound	Site1(free SnF_2)			Site 2 (included SnF_2)			Site 3 (Sn(IV))		
	δ^b	Δ^c	RA^e	δ	Δ	RA	δ	Δ	RA
SnF_2	3.49(1)	1.62(1)	96%				-0.45(4)	0	4%
$\alpha SnF_2{}^a$	3.49(5)	1.62(1)							
P1	3.51(2)	1.65(4)	8%	3.18(1)	2.11(1)	50%	-0.05(1)	0.62(1)	34%
P2	3.49(3)	1.64(6)	4%	3.17 (4)	1.88(4)	17%	-0.07(1)	0.53(1)	79%
P3	3.49(2)	1.67(1)	39%	3.18(1)	2.09(1)	55%	-0.10(6)	0.56(3)	6%
P4	3.46(2)	1.65(3)	21%	3.17(1)	2.04(2)	71%	-0.15(5)	0	8%
P5	3.50(1)	1.63(1)	96%				-0.34(8)	0	4%

[a] ref 4, δ: isomer shift (mm/s) relative to $CaSnO_3$, Δ: quadrupole splitting (mm/s), and RA: relative area under resonance curve. Methods of preparations: aqueous solution, 15 minutes (**P1**) and 5 hours (**P2**), solid state (**P3**), dry-freezing (**P4**), and physical mixture (**P5**).

ACKNOWLEDGEMENTS

The authors would like to thank financial support by Brazilian agencies FAPEMIG, CNPq and FINEP. The Organizers of the 9th International Simposium on Molecular Recognition and Inclusion are also gratefully acknowledged.

REFERENCES

1.a. J. Szejtli: "Cyclodextrin Technology", Kluwer Academic Publishers (1989),
b. W. Saenger; Angew. Chem. Int. Ed. Engl. 19 (1980) 344
2. R.D. Sinisterra, R. Najjar, O.L. Alves, P.S. Santos, C.A. Alves de Carvalho and A.L. Conde da Silva, J. Incl. Phenom., 22 (1995) 91
3. R.D. Sinisterra, A. Abras, C.A.L. Filgueiras, C.A. Menezes, C.A. Alves de Carvalho, M. E. Cortés Segura and N. Tortamano, Procedings of Fith Latin American Conference on Mössbauer Effect, LACAME'96., (1996) pg P-05-02.
4. W. Saenger: "Inclusion Compounds", Ed. J.L. Atwood, J.E.D. Davies and D.D. MacNicol, Academic Press, London, Vol. 2, Ch. 8 (1984)
5. T. Birchall, P. A. W. Dean and R. J. Gillespie, J. Chem. Soc(A) (1971), 1777
6. N. N. Greenwood and T. C. Gibb, "Mössbauer Spectroscopy", Chapman and Hall Ltd, London 1971, pg. 391
7. G. Corrales and G. Denes, Hyp. Int., 90 (1994) 441
8. Y. Maeda, N. Ogawa and Y. Takashima, J. Chem. Soc. Dalton Trans., (1987) 627
9. F. Gauzzi, A. Maddalena, G. Principi and B. Verdini, "Applications of the Mössbauer Effect", ICAME'83, Vol.3 pg. 919, Ed. Yu. M. Kagan and I. S. Lyubutin.
10. J. K. Lees and P. A. Flinn, J. Chem. Phys., 48(02) (1968) 882
11. G. Deness, J. of Solid State Chem., 77 (1988) 54
12. A. Ueno, F. Moriwaki, T. Osa, F. Hamada and K. Murai, Tetrahedron Letters, Vol.26(7) (1985) 899

^{13}C CP/MAS STUDIES OF RHODIUM (II) 3-FLUOROBENZOATE AND THEIR INCLUSION COMPOUND IN β-CYCLODEXTRIN

R. D. SINISTERRA[1]*, R. NAJJAR[2], P. S. SANTOS[2], O.L. ALVES[3], C.A. ALVES DE CARVALHO[4], E. MUNSON[5] AND K. THAKUR[5]

[1]Departamento de Química da Universidade Federal de Minas Gerais, P.O. 13702, Belo Horizonte, MG, Brazil CEP 30161-970, e-mail ruben@juno.qui.ufmg.br. [2]Departamento de Química Fundamental, IQ-USP, São Paulo, SP, Brazil. [3]Laboratório de Química do Estado Sólido, IQ -UNICAMP, Campinas, SP, Brazil. [4]Departamento de Química-ICEB, UFOP, Ouro Preto, MG, Brazil.[5] Department of Chemistry of the University of Minnesota, USA.

INTRODUCTION

Bear et al. in 1972 first reported the antitumor activity of Rhodium acetate, after the discovery of cisplatin. We have sinthetized some new rhodium carboxylates aiming to get less toxic or more active complexes than those previously reported.
Rhodium(II) carboxylates are well known antitumor agents, but their extremely low solubility has prevent their use. Aiming to circunvenvent this problem we are systematically investigating their inclusion of such carboxylates in cyclodextrins. In fact, we have recently reported a study on th inclusion of rhodium(II) carboxylate in cyclodextrin that resulted in a substancial increase in the solubility [1]. In the present work the synthesis of 3-fluorobenzoate and its inclusion in β-cyclodextrin, β-cd, are reported.

EXPERIMENTAL

Materials and methods

β-cyclodextrin hydrate was supplied by Aldrich and used without further purification. Rhodium (II) 3-fluorobenzoate was prepared by a modification of the method reported in the literature [2].

Preparation of the inclusion compound

The inclusion compound was prepared by two methods, namely by co-precipitation and drying-freezing. In the first an aqueous solution of β-cyclodextrin was added dropwise under constant stirring to an ethanolic solution of rhodium (II) 3-fluorobenzoate. The system was kept at ca 40-50°C for 24 hours, the final volume being reduced to ca 10ml. The green solid produced was collected by filtration and washed with water and dried in vacuum at room temperature. In the drying-freezing process the solid rhodium (II) 3-fluorobenzoate was added to an aqueous solution of β-cyclodextrin and freeze-dried for 24 hours approximately.
The physical mixture of β-cd and rhodium (II) 3-fluorobenzoate in a 1:1 molar ratio was also prepared by gentle grinding of the components until a homogeneous powder was obtained.

Author for correspondance

A. W. Coleman (ed.), Molecular Recognition and Inclusion, 511–514.
© 1998 *Kluwer Academic Publishers. Printed in the Netherlands.*

Physical measurements
X-ray powder patterns were obtained using a Rigaku Geigerflex instrument, employing kα Cu, with 2θ between 5°-40°. The infrared spectra were obtained using a Perkin-Elmer 283B spectrometer. The TG curves were registered in a Mettler TA-4000 equipment in an N_2 atmosphere. The DSC curves were obtained with a Shimadzu apparatus. The ^{13}C CP/MAS NMR spectra were obtained in a Chemagnetics 300 MHz, equipment using 3-7 KMAS spinning speed.

RESULTS AND DISCUSSION

As previously reported [1] the first evidence for the inclusion of rhodium (II) 3-fluorobenzoate came from the solubilization of the complex in the β-cyclodextrin organic-aqueous solution, as can be easily observed by visual inspection. The β-cd/Rh(II) compound showed a higher solubility in water as compared with the virtual insolubility of the pure carboxylate.
The XRD patterns are shown in the Figure 1. The XRD pattern of rhodium (II) 3-fluorobenzoate shows high crystallinity. On the other hand, the XRD patterns of the inclusion compounds prepared by solution (B) and drying-freeezing (C) show a loss of crystallinity as compared to rhodium (II) 3-fluorobenzoate, β-cyclodextrin (D) and the physical mixture (E). This result is a strong evidence of the inclusion of the guest in β-cyclodextrin.

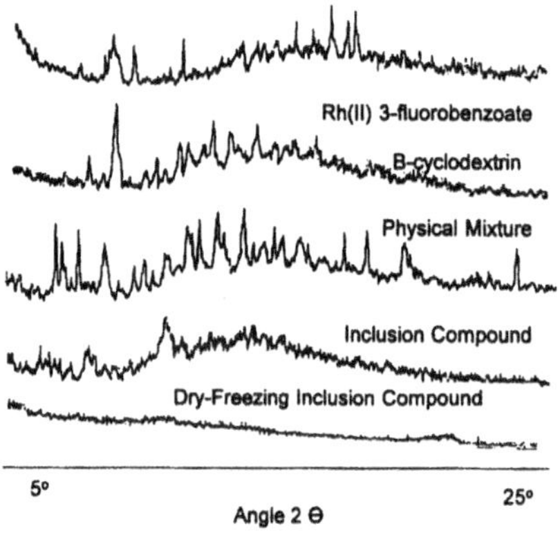

Figure 1. XRD patterns

The TG and DSC curves of rhodium (II) 3-fluorobenzoate, β-cyclodextrin, the inclusion compounds as well as the physical mixture are shown in Figure 2.
The TG curve of rhodium (II) 3-fluorobenzoate shows two steps of thermal decomposition. The first appears around 124°C, corresponding to melting with decomposition, as verified by the DSC curve, as a sharp peak and reinforced by the determination of the melting point. The second step lies around 337°C, possibly due to a breakdown of the cage. The TG curves of the inclusion compounds show higher thermal stability for this system compared to rhodium (II) 3-fluorobenzoate and to the physical mixture. These results are a strong evidence of the formation for the new crystalline phase when the inclusion compound is obtained.
In the infrared spectrum of rhodium (II) 3-fluorobenzoate one observes the typical ν_{as} and ν_{sim} vibrations of the COO^- group at 1550cm^{-1} and 1400cm^{-1} respectively and show up the aromatic modes at 1570cm^{-1} and 1450cm^{-1}.

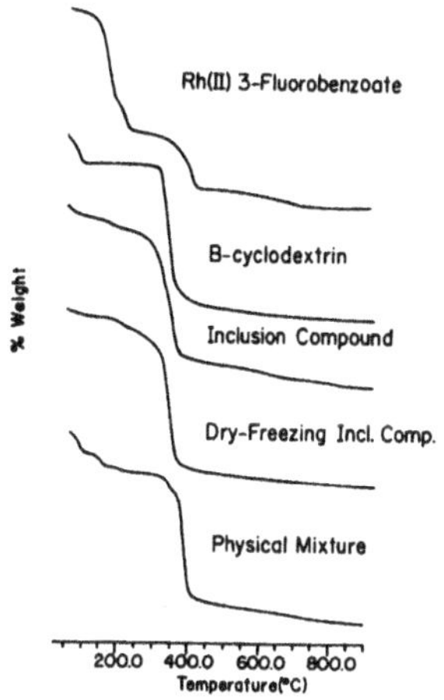

Figure 2. TG curves

Analysing the spectrum of the inclusion compound one a observes a substancial decrease in the intensity of the vibrational modes assigned to the aromatic moiety and to the carboxylates, suggesting the strong interaction between rhodium (II) 3-fluorobenzoate and β-cyclodextrin, as opposed to the spectrum of the physical mixture in which the characteristic modes of the guest and host were observed without any change.

^{13}C CP/MAS spectrum of rhodium (II) 3-fluorobenzoate, β-cyclodextrin, the inclusion compound and the physical mixture are shown in Figure 4.

The present work is the first report using ^{13}C NMR CP/MAS to rhodium (II) carboxylates as a characterization technique. Analysis of the spectrum of rhodium (II) 3-fluorobenzoate, in the region of the carboxylate group shows four peaks, i.e., one triplet (189.8, 185.9 and 184.4 ppm) and one singlet at 174.8 ppm. These four different peaks suggest four different carboxylate groups per unit cell and can be assigned to COO(1), COO(2), COO(3) and COO(4) respectively.

These four different carboxylates show the possibility of a stacking as was verified in the literature for single cristal XRD of similar compounds [3]. A doublet was observed at 160.8 ppm, which can be attribuited to C-F resonance, and also suggests the non-equivalent enviroment in the unit cell of this compound. On the other hand, a multiplet was observed in the region from 134-118.5 ppm, which can be attributed to the aromatic carbons (C-2, C-4, C-5, C-6).

The NMR spectrum of β-cyclodextrin is very similar to those reported in the literature [4]. Observation of the ^{13}C NMR CP/MAS of the physical mixture, proved to be a very effective techique to highlight the difference between a physical mixture and an inclusion compound in the solid state, since, is an exact summation of the spectra os the components, suggesting that there is no interaction between host and guest in the physical mixture. On the other hand, in the ^{13}C NMR CP/MAS spectrum of the inclusion compound, it was observed that all the peaks were broadened when compared with host and guest resonance in the free form.

In the region of the carboxylate moiety we observed three peaks: two peaks at 184.0 and 173.5 ppm,which can be attributed to carboxylates 1 and 2; the third peak in 167.3 ppm is possibly due to a shielded carboxylate carbon after its inclusion in β-cyclodextrin. A sharp peak was observed 160.5 ppm, which can be assigned to C-F, as discussed above

rhodium (II) 3-fluorobenzoate. This peak showed changes when compared with the doublet observed in the spectrum of the free guest. In the aromatic region (131.8-115.9 ppm) drastic changes were observed in the intensities of the peaks. Additionally a broadening of the multiplet was observed compared to the spectrum oof the free guest. These results strongly suggest the inclusion of the aromatic moiety and changes in the molecular mobility of the carbons after inclusion.
These results are in accordance with XRD data, which proved that the inclusion compound is more disordered, with less crystallinity when compared with the patterns of the physical mixture, as well as the host and guest. Changes were also observed in the region of ^{13}C ressonance of β-cyclodextrin. A broad peak appeared at 102.9 ppm, C-1, in contrast to what was observed in the spectra of β-cyclodextrin and the physical mixture. This result suggests possible changes in the torsional angle ψ_1 as reported in the literature for simple organic compounds [5]. Additionally a broadening of the peaks at 84.3 ppm (C-4), 72.4 ppm (C-2, 3-5) and 60.9 ppm (C-6), was also observed compared to the sextuplet, quintuplet and triplet, respectively, which were observed in the spectra of β-cyclodextrin and the physical mixture.

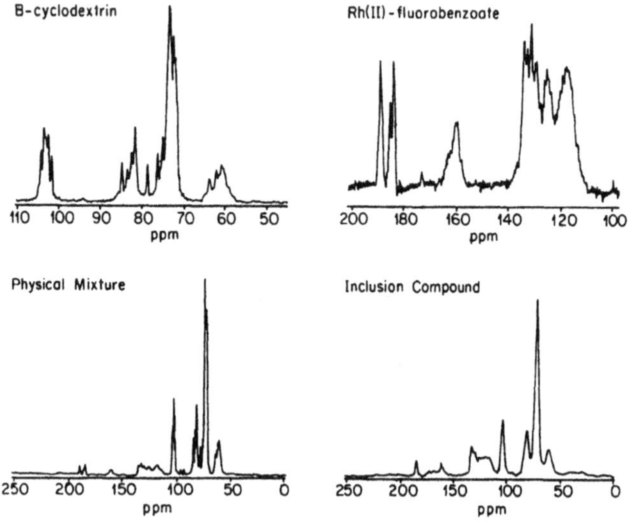

Figura 3. ^{13}C CP/MAS spectra

ACKNOWLEDGEMENTS

The authors would like too thank financial support by Brazilian agencies FAPEMIG, CNPq, FINEP, FAPESP, SBQ and ABQ and American Chemical Society (ACS) by the financial support to do the NMR solid state spectra. The Organizers of the 9th International Symposium on Molecular Recognition and Inclusion are also gratefully acknowledged.

REFERENCES

1. R. D. Sinisterra, R. Najjar, O. L. Alves, P. S. Santos, C.A. Alves de Carvalho and A.C. Conde da Silva, J. Incl. Phenom. 22, 91 (1995)
2. R. Najjar, Oliveira, W., Carducci, J.B.; Polyhedron 8, 1157 (1989)
3. T. Felthouse, Progr. Inorg. Chem. 29, 73 (1982)
4. J. A. Ripmeester, J. Incl. Phenom. 6, 31 (1988)
5. J. A. Ripmeester, J. Inc. Phenom. 4, 129 (1986)

ANION RECOGNITION USING BORONATE-UREAS

BRADLEY D. SMITH,* and MARTIN PATRICK HUGHES
Department of Chemistry and Biochemistry, University of Notre Dame, Notre Dame, IN 46556, USA

Neutral Acetate Receptor

Neutral receptors for anions have been recently reported by several research groups.[1] The primary inspiration for our acetate receptor came from a study by Etter and coworkers, who discovered that diarylurea derivatives with electron withdrawing groups in the *meta* positions were exceptionally good at forming 1:1 crystalline complexes with hydrogen bond acceptors.[2] Isomeric derivatives with *ortho* or *para* substituents did not form such co-crystals. This "meta effect" was attributed to a substituent induced intramolecular C-H···O hydrogen bonding interaction between the carbonyl oxygen and the acidified *ortho* hydrogens. Such an intramolecular interaction was postulated to weaken the hydrogen bond accepting ability of the urea carbonyl.

The Etter explanation now appears to be incomplete due to subsequent results obtained by Wilcox.[3] However, it provided us with an idea. We reasoned that a C-H group was a very poor Lewis acid, and that this putative effect could be dramatically increased by using a stronger Lewis acid. Thus, we initiated a study of receptors whose structures were related to the generic compound **1** (Figure 1). In line with our general interest in organoboron compounds it was an obvious step to examine the structural class **2**. The structure for receptor **2** can be represented by two limiting forms, **2a** or **2b** (Figure 1). Based on work by Groziak involving a similar boron heterocycle, we expected the structure to exist as the zwiterionic isomer **2b**, with sp^3 hybridized boron.[4] Thus, we synthesized receptors **5 - 8** (Figure 2). This entailed treating 2-aminophenylboronic acid, **3**, with various alkylisocyanates to give compounds **4** (structure of **4a** proven by X-ray crystallography). Condensation of **4** with pinacol gave the organic soluble

A. W. Coleman (ed.), Molecular Recognition and Inclusion, 515–518.
© 1998 Kluwer Academic Publishers. Printed in the Netherlands.

products **5** - **8**. Compound **5** proved to be highly water-sensitive and readily hydrolyzed to regenerate compound **4a**, thus giving unreliable binding data. The more sterically hindered compounds **6**, **7**, and **8** however, were significantly more stable and provided highly reproducible association behavior.

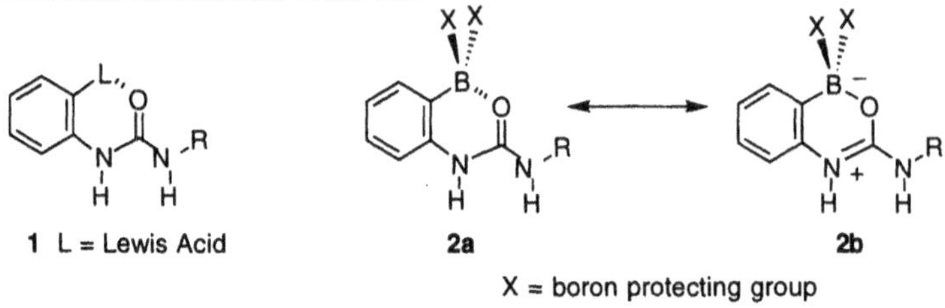

Figure 1.

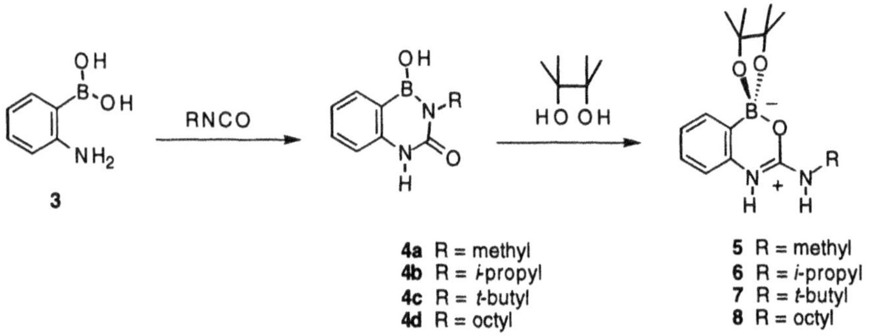

Figure 2.

As summarized in Table 1, the acetate binding ability of compounds **6**, **7**, and **8** have been determined in DMSO, and compared to the binding ability of control compounds **9** - **12**. Association constants were measured by ^{1}H NMR titration studies. Specifically, the downfield changes in host N-H resonances were monitored upon addition of aliquots of tetrabutylammonium acetate. In each case 1:1 binding was verified by a Job plot. Control titration experiments involving deprotonated analogues of the receptor compounds indicate that there is no proton transfer within the host:guest complex. In addition, control experiments showed that acetate anion has no affinity for the trigonal boron in **9**.

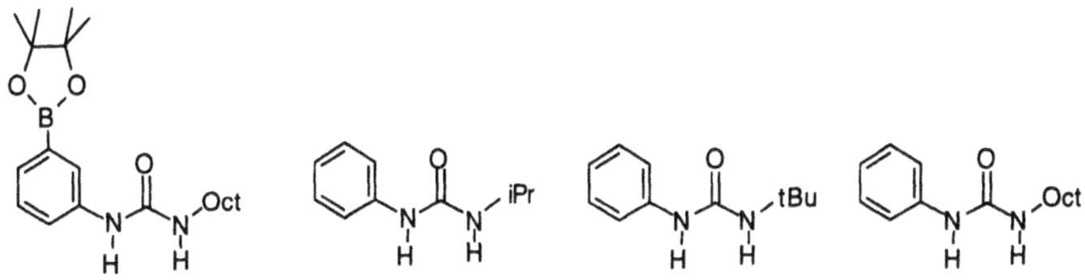

9 10 11 12

TABLE 1. Acetate association constants, and $\Delta\delta_{max}$ values from ^{1}H NMR titrations in DMSO at 295 K.

Host	K_{assn} $(M^{-1})^{a}$	ΔG_{295} (kcal/mol)	$\Delta\delta_{max}$ $(ppm)^{b}$
6	$(6.9 \pm 0.9) \times 10^{3}$	-5.2	3.83
7	$(7.1 \pm 0.3) \times 10^{3}$	-5.2	3.63
8	$(7.0 \pm 2.0) \times 10^{3}$	-5.2	3.63
9	$(3.9 \pm 0.4) \times 10^{2}$	-3.5	2.16
10	$(3.1 \pm 0.5) \times 10^{2}$	-3.4	2.39
11	$(2.6 \pm 0.6) \times 10^{2}$	-3.4	1.66
12	$(3.7 \pm 0.4) \times 10^{2}$	-3.5	2.14

[a]Average of three independent trials. [b]Average of $\Delta\delta_{max}$ for both N-H residues, derived for host at >90% saturation.

In agreement with our initial hypothesis, hosts **6**, **7**, and **8** bind acetate better than controls **9** - **12**. Thus, it appears that the strategically placed boron in **6**, **7**, and **8** is able to polarize the urea carbonyl, and increase acetate binding by acidifying the N-H residues and enhancing the ion-dipole interaction between host and acetate. The most likely structure for this host:actetate complex is shown in Figure 3.

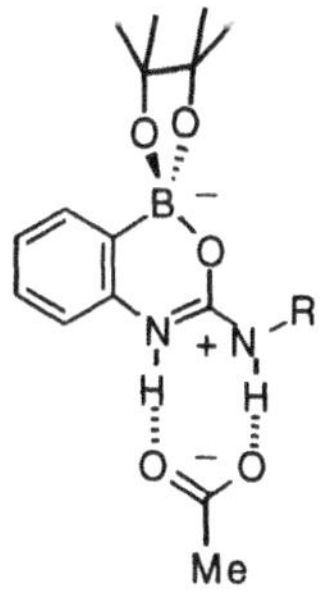

Figure 3. Likely structure for borate-urea host:acetate complex.

We also examined the enthalpic and entropic factors affecting this complexation: variable temperature experiments involving van't Hoff analysis plots were performed for compounds **6** and **10**. As shown in Table 2, the enhancement in the acetate binding free energy for the borate-urea **6** over the urea control **10** is due to an enthalpic difference.

TABLE 2. Binding free energy, enthalpy and entropy measurements for the host:acetate complexation of compounds **6** and **10**.

Host	ΔG_{295} (kcal/mol)	ΔH (kcal/mol)	ΔS (cal/mol*K)
6	-5.2	-5.6	-1.2
10	-3.4	-3.9	-1.5

Acknowledgments

This work was supported by a grant from the US National Science Foundation. M. P. H. acknowledges the support of the Schmidt Foundation in the form of a research fellowship.

References

1. Hughes, M. P., Shang, M. and Smith, B. D., (1996) High affinity carboxylate binding using neutral urea-based receptors with internal Lewis acid coordination, *J. Org. Chem Soc.* **61**, 4510-4511.
2. Etter, M. C., Urbanczyk-Lipkowska, Z., Zia-Ebrahimi, M., Panunto, T. W. (1990) Hydrogen bond directed cocrystallization and molecular recognition properties of diarylureas, *J. Am. Chem. Soc.* **112**, 8415-8426.
3. Wilcox, C. S., Kim, E. K., Romano, D., Kuo, L. H., Burt, A. L. and Curran, D. P. (1995) Experimental and theoretical studies of substituent effects in hydrogen bond based molecular recognition of a zwitterion by substituted arylureas, *Tetrahedron* **51**, 621-634.
4. Groziak, M. P., Ganguly, A. D. and Robinson, P. D. (1994) Boron heterocycles bearing a peripheral resemblance to naturally-occurring purines: design, syntheses, structures, and properties, *J. Am. Chem. Soc.* **116**, 7597-7605.
5. Conners, K. A. (1987) *Binding Constants, The Measurement of Molecular Complex Stability*, Wiley, New York.
6. Wilcox, C. S. (1991) Design, evaluation of an efficacious functional group dyad. Methods and limitations in the use of NMR for measuring host-guest interactions, in Schneider, H. -J., Dürr, H., (eds) *Frontiers in Supramolecular Organic Chemistry and Photochemistry*, VCH, Weinham, pp. 123-143.

SYNTHESIS AND SELF-ORGANISATION OF NEW CYCLODEXTRIN AMPHIPHILE

Takeshi Sukegawa[1], *Masao Matsuda*[1], *Shin-Ichiro Nishimura*[1*]
Masatsugu Shimomura[2], *Kunihiro Ijiro*[2], *Oraf Karthaus*[2]
[1]*Graduate School of Science, Hokkaido University,*
Kita-10, Nishi-8, Kita-ku,Sapporo 060, JAPAN
[2]*Research Institute of Electronic Science, Hokkaido University,*
Kita-12, Nishi-6, Kita-ku, Sapporo 060, JAPAN
Tel: 81-11-706-2772 Fax: 81-11-706-3435
e-mail: nishimura@indy.polymer.hokudai.ac.jp

Abstract

A novel amphiphilic cyclodextrin derivative having sialic acid residues was systematically synthesised. Hexakis (2,3-di-*O*-palmitoyl)-di-6-*O*-toluenesulfonyl-α-cyclodextrin was converted to hexakis (2,3-di-*O*-palmitoyl)-6-di-(2-aminoethyl)-amino-6-dideoxy-α-cyclodextrin by reaction with ethylenediamine as a reactive spacer. Coupling of a sialic acid intermediate and α-cyclodextrin derivative gave an amphiphilic cyclodextrin derivative having two sialic acid residues. This unique compound showed an excellent ability to form monolayer membrane.

Introduction

The cyclodextrins are cyclic oligosaccharides possessing six, seven or eight α-1,4-linked glucopyranose units and forming a cylindrical structure. The resultant hydrophobic cavity is capable of including a wide variety of organic molecules. The cyclodextrins have two hydrophilic face (the primary and secondary hydroxyl groups) surrounding a hydrophobic cavity. Therefore, regioselective chemical modification of either hydrophilic face with hydrophobic moieties can be induced the amphiphilic cyclodextrin derivative of tunnel form.[1)] On the other hand, the cyclodextrins have been capped with organic molecules to control the recognition and inclusion capacity.[2)] We now select carbohydrate as a specific cap of the cyclodextrin, since it exists on a surface of cell

A. W. Coleman (ed.), Molecular Recognition and Inclusion, 519–522.
© 1998 *Kluwer Academic Publishers. Printed in the Netherlands.*

membrane, and these molecules are specifically recognized by proteins on a viral membrane when viruses infect a targeting cell. The cyclodextrin capped with sugar is able to control the release of a guest-drug molecule, and makes possible to target virus and cells recognizing sugar. Therefore, self-organisation of amphiphilic cyclodextrin derivative would contribute to the developement of glycodrugs to inhibit the viral infection.

Results and Discussion

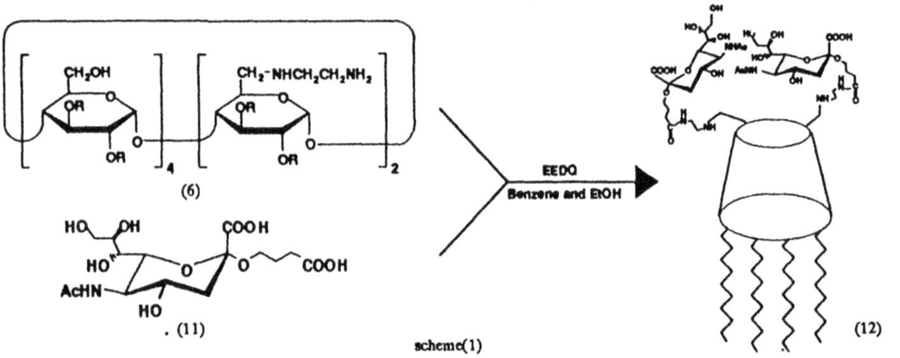

scheme(1)

Scheme 1 indicates a synthetic route of the cyclodextrin derivative designed in this study. The synthesis of compound (**12**) was performed by conjugation of a reactive cyclodextrin derivative (**6**) and sialic acid intermediate (**11**). Treatment of dried α-cyclodextrin (**1**) with *tert*-butyldimethylsilyl chloride in N,N-dimethylformamide (DMF)

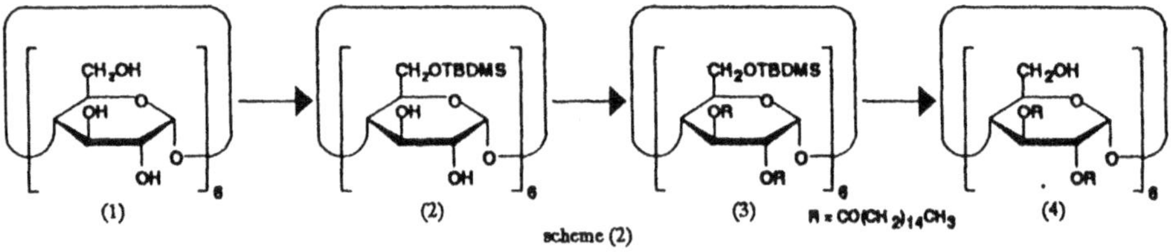

scheme (2)

in presence of imidazole for 1h at room temperature gave hexakis (6-*O*-*tert*-butyldimeth ylsilyl)-α-cyclodextrin (**2**) as the major product.[3] The 6-*O*-protected compound (**2**) was isolated with column chromatographic separation. Acylation of (**2**) with palmitoyl chloride in the presence of dimethylaminopyridine (DMAP) in dried pyridine afforded hexakis (2,3-di-*O*-palmitoyl-6-*O*-*tert*-butyldimethylsilyl)-α-cyclodextrin (**3**) in 65% yield. Desilylatioin of (**3**) with boron trifluoride etherate in dichloromethane proceeded smoothly and gave hexakis (2,3-di-*O*-palmitoyl)-α-cyclodextrin (**4**) in 67% yield. Partial *p*-toluenesulufonylation and silica-gel chromatographic separation gave hexakis (2,3-di-*O*-palmitoyl)-6-di-*O*-tolenesulufonyl-α-cyclodextrin (**5**) in about 20% yield. Finally, hexakis (2,3-di-*O*-palmitoyl)-6-di-(2-aminoehtyl)-amino-6-dideoxy-α-cyclodextrin (**6**) was obtained by nucleophilic displacement reaction of (**5**) with 1,2-diaminoethane.[4]

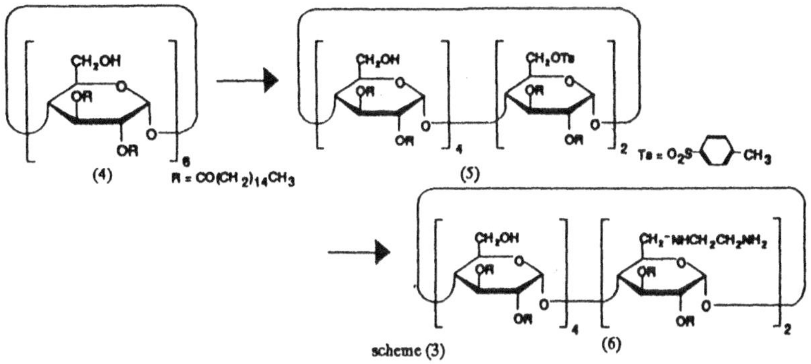

scheme (3)

On the other hand, a sialic acid intermediate (**11**) as a cap of the cyclodextrin amphiphile was synthesized from a sialic acid derivative (**7**). The replacement of acetoxy group of anomeric position in the derivative (**8**) with methylthio group by heating with (methylthio)trimethylsilane in dried 1,2-dichloroethane in presence of trimethylsilyle trifluoromethane sulfonate gave methyl (methyl 5-acetamide-4,7,8,9-tetra-*O*-acetyl-3,5-dideoxy-2-thio-*D-glycero-D-galacto*-nonuropyranosido)nate (**9**) as a 1:1

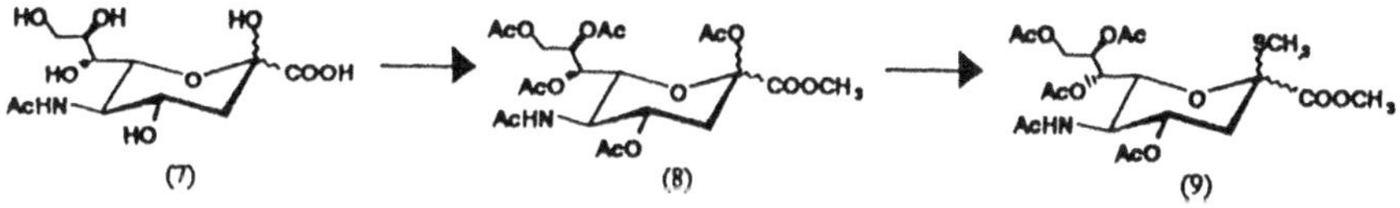

scheme(4)

anomeric mixture.[5] Methyl 4-hydroxybutylate as a spacer between cyclodextrin derivative and the sialic acid intermediate was synthesized by methylation of 4-hydroxybutyric acid, sodium salt with Dowex 50W-X8 in dried methanol. Regio- and α-steleoselective glycosylation of the derivative (**9**) with methyl 4-hydroxybutyrate in dried acetonitrile in presence of N-iodosuccinimide (NIS) and trifluoromethane sulfonic acid gave methyl [3'-(methoxycarbonyl)propyl 5-acetamide-4,7,8,9-tetra-*O*-acetyl-3,5-dideoxy-2-*D-glycero-D-galacto*-nonuropyranosido]nate (**10**) after column chromatography.[6] Deprotection of hydroxyl and carboxyl groups in the sialic acid intermediate (**10**) afforded 3'-carboxypropyl 5-acetamide-3,5-dideoxy-2-*D-glycero-D-galacto*-α-nonuropyranoside (**11**).

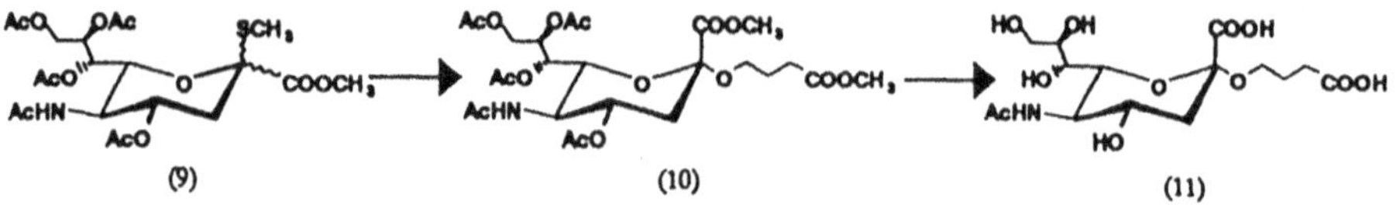

scheme (5)

New cyclodextrin amphiphile having the sialic acid residues (**12**) was synthesized by coupling of the cyclodextrin derivative (**6**) and the sialic acid intermediate (**11**) after Sephadex LH-20 chlomatogrphic separation.

Figure 1 gives the surface pressure versus molecular area curve for deposited layer of new cyclodextrin amphiphile (**12**). This isotherm showed an excellent ability to form monomolecular layer by the amphiphile (**12**).

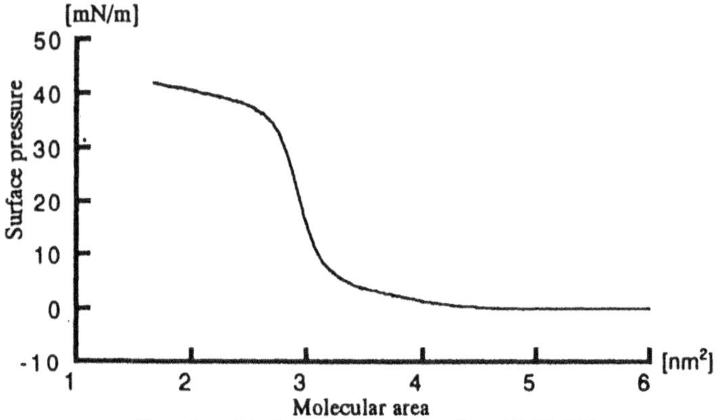

Figure 1: π-A isotherm of new cyclodextrin amphiphile (12)

References

1) Coleman,A.W., Zhang,P. Ling,C.-C. Parrot-Lopez,H. Galons,H. (1991) Formation of amphiphilic cyclodextrin via hydrophobic esterification at the secondary hydroxyl face, *Tetrahedron. Lett.*, **32**, 2769-2770

2) Tabushi,I. Shimokawa,K. Shimizu,N. Shirakata,H. Fujita, K. (1976) Capped cyclodextrin, *J.Am. Chem.Soc.*, **98**, 7855-7856, Tabushi,I. Shimokawa,K. Fujita,K. (1977) Specific bifunctionalization on cyclodextrin, *Tetrahedron.Lett.*, **18**, 1527-1530

3) Takeo,K. Uemura,K. Mitoh,H. (1988) Derivative of α-cyclodextrin the synthesis of 6-*O*-α-D-glucopyranosyl-α-cyclodextrin, *J.Carbohydr.Chem.*, **7(2)**, 293-308

4) Ikeda,T. Yoshida,K. Schneider,H.-J. (1995) Anthryl (alkylamino)cyclodextrin complexes as chemically swiched DNA intercalactors, *J.Am.Chem.Soc.*, **117**, 1453-1454

5) Hasegawa,A. Ohki,H. Nagahama,T. Ishida,H. Kiso,M. (1991) A facile, large-scale preparation of the methyl 2-thioglycoside of *N*-acetylneuraminic aicd, and its usefullness for the a-stereoselective synthesis of sialoglycoside, *Carbohydr.Res.*, **212**, 277-281

6) Hasegawa,A. Nagahama,T. Ohki,H. Hotta,K. Ishida,H. Kiso,M. (1991) Synthetic studies on sialoglycoconjugates 25:, *J.Carbohydr.Chem.*, **10(3)**, 493-498

MOLECULAR RECOGNITION OF ANIONIC SPECIES: HYDROGEN-BONDING PROPERTIES OF SULFATE ANDTHIOCYANATE

LUBA TCHERTANOV* AND CLAUDINE PASCARD

Laboratoire de Cristallochimie, Institut de chimie des Substances Naturelles, CNRS 91198 Gif-sur-Yvette, France

The study of hydrogen bonding and anion-cation interactions allows us to understand how molecules self-recognize and self- organize and to gain insight into the relationship between molecular and supramolecular crystal structures.

The coordination effects by the anions (anionic environment) which play a fundamental role in biological systems have been the subject of a growing interest in the recent years. The experimental data on the associations between macromolecules and anions being very rare and far from accurate, it appears that the best way to study the interactions with anions is the investigation of peculiarities of their H-binding through a survey of small-molecule crystal structures. The present work is in particular concerned with the **sulfate** and **thiocyanate** anions. Astudy of sulfate binding with proteins based on the limited PDB data was published [1]. Thiocyanate has been used lately as crystallizing agent in protein crystal growth, but the mode of its binding still remains unclear [2].

The present analysis is based on atomic coordinates retrieved from Cambridge Structural Database [3]; Brookhaven Protein Database [4] was used for comparison of the results with those on macromolecular structures.

Discrete **sulfate** dianion principal characteristic is to form as many H-bonds as possible, short and direct, up to 12 per dianion. There is no preference for H-bonding of sulfate with oxygen or nitrogen donors: the proportion of potential O-H and N-H donors corresponds to the proportion of observed O^-...H-O and O^-...H-N H-bond types. The most striking result is the high number of H-bonds to the sulfate which results in the sulfate being surrounded by a cloud of H-donors. This multiple binding (2 or 3) of each oxygen implies that the tetrahedral sulfate will gather around itself a maximum 3-dimensional environment of donors at close contact. The anionic oxygen exhibits a preferred pyramidal stereochemistry.

The environment of SO_4^{-2} in the structure BGDUSM10. Anion oxygen-acceptor function is characterized by multiple H-bonding. Sulfate oxygen participates in two (O2 and O3) and three (O1 and O4) H-bonds. Geometrical arrangement around oxygens: at O1 and O4 - pyramidal; at O2 -non-coplanar and O3 - coplanar. Water molecule binds with two oxygens of the same sulfate at one time. The H-bonds are shown by dashed lines.

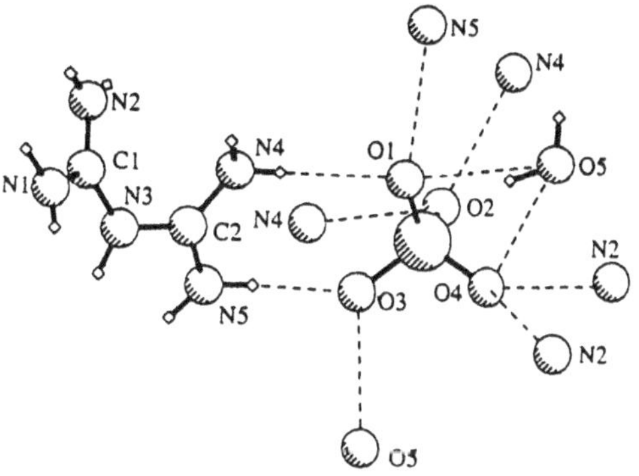

A. W. Coleman (ed.), Molecular Recognition and Inclusion, 523–525.
© 1998 *Kluwer Academic Publishers. Printed in the Netherlands.*

The high percentage of hydrates (78%) observed in the number of retrieved structures allows completion of the sulfate coordination. When the number of molecular H-donors is not sufficient (or the packing is subjected to steric problems), water molecules enter the binding sphere to complete each oxygen binding. Consequently, the sulfate anion will behave as a steric "tightener", placing the attracted partners in close contact

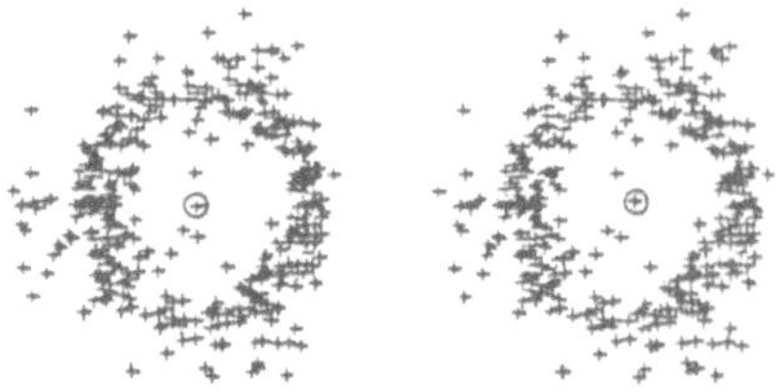

Stereodiagram of the distribution of donor groups around sulfate. The giant molecule(all fragment superimposed) is projected along the S-O bond (S and O superimposed)

SCN^- has a very unsual behaviour. Two main families of compounds were studied: (i) crystal structures containing discrete SCN^- anion, (ii) compounds containing SCN bonded to one metal atom only, either through S (thiocyanate) or through N (isothiocyanate). The search around the discrete anion shows that sulfur, nitrogen and π-system of the anion are H-bond acceptors.The preferred areas of the H-donors are determined: around sulfur, on a torus centered at the S-end, the axis of which is the linear anion; around nitrogen, on a spherical cap. In most cases (80%), thiocyanate binds by both terminal atoms (S and N) with multiple H-bonding. In a molecular assembly with a deficiency of H-bond donors, π-electrons of aromatic systems represent the next best partner for the thicyanate unit.

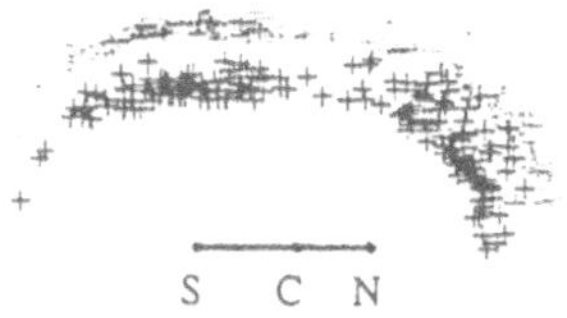

Distribution of the H-donors around SCN.

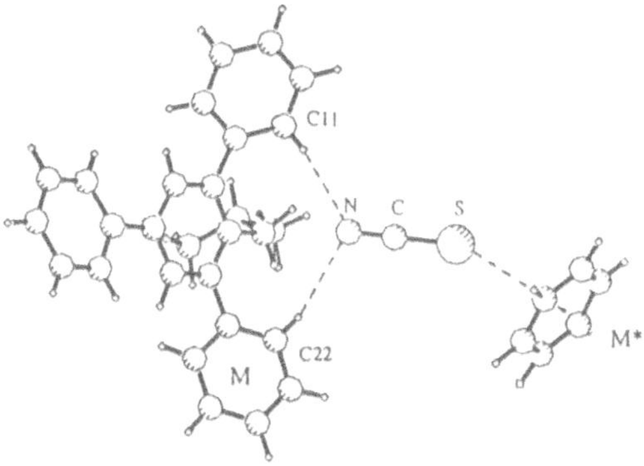

PLUTO diagram illustrating the C-H...N and S...π interactions.

In the metal complexes, the H-accepting function of SCN unit depends on the mode of its coordination to the metal (by S or N) and on the nature of metal. Complexing properties (metal-thiocyanate interactions) were analysed for different metal cations: class a - alkali and alkaline earth metals, class b - groups of Zn and Cu and transition metals (group VIII). Fine correlations between the nature of metal (radius, oxidation state, charge) and its position with respect to the thiocyanate unit, were found. 80% of the metals bind through the N-terminal atom, 9% through the sulfur and 11% as bidentate (by both end). SCN forms strong complexes with the metals: ionic with classe a and covalent with the metals of the other groups. In M-SCN fragment variations are observed only for metal bond distance in correlation with its covalent radius; the angular parameter is nearly the same for all thiocyanate complexes (about 100°). The coordination geometry of M-NCS fragment shows more variations as well as for distance M...N as for angular parameter at nitrogen. Whereas the bond length follows regularly the increasing cation radius (ionic or covalent), various metal positions with respect to the linear anion are observed: the largest cations interacting with the π-system of the anion.

Our results give an evidence of an important influence of covalent or ionic bonding on geometry of SCN unit in both modes of coordination, M-NCS and M-SCN. In first mode, the bond length S-C is shorter and in the second is larger whan in the discrete SCN anion. Such features would be consistent with a contribution from a structure such as S=C=N-M less important than $^{-}S\text{-}C{\equiv}N^{+}\text{-}M$ for the N-bonded thiocyanate and M-S-C≡N more important than $M\text{-}S^{+}{=}C{=}N^{-}$ for S-bonded. The cation provokes a redistribution of anionic charge in metal-bonded SCN and the resulting distribution of electron density controls the hydrogen-bonding properties of terminal acceptor atom. A preferential localization of negative charge on the terminal sulfur in M-NCS complexes leads to good hydrogen-bonding situation.

Bidentate character of SCN unit, possessing two different accepting centers (S and N), allows to predict the intermolecular interactions by the selection (as a function) of a third partner in the system: X-SCN-Y, where X or Y can be a combination of metal and H-donor groups.

Our study demonstrates the extraordinarily active behaviour of sulfate [5] and thiocyanate [6]. Both units are seen to act as powerful anionic bridges between the different molecular fragments.

References:

1. Chakrabarti, P., (1993). Anion Binding Sites in Protein Structures, *J.Mol.Biol.* **234**, 463-482.

2. Ducruix, A. & Giegé, R. (1992). Crystallization of proteins and nucleic acids, IRL/Oxford Press.

3. Allen, F.H., Bellard, S., Brice, M.D., Cartwright, B.A., Doubleday, A., Higgs, H., Hummelink, T., Hummelink-Peters, B.G., Kennard, O., Motherwell, W.D.S., Rodgers, J.R. & Watson, D. R. (1979). The Cambridge Crystallographic Data Centre: Computer-Based Search, Retrieval, Analysis and Display of Information. *Acta Cryst.*. **B35**, 2331-2339.

4. Bernstein, F. C., Koetzle, T. F., Williams, G. J. B., Meyer, E. F. Jr., Brice, M. D., Rodgers, J. R., Kennard, O., Shimanouchi, T. & Tasumi, M. (1977). The Protein Data Bank: A Computer-based Archival File for Macromolecular Structures *J. Mol. Biol.* **112**, pp. 535-542.

5. Tchertanov, L. & Pascard, C. (1996). Statistical Analysis of Noncovalent Interactions of Anion Groups in Crystal Structures. I. Hydrogen Bonding of Sulfate, *Acta Cryst.*. **B52**, 677-684.

6. Tchertanov, L. & Pascard, C. (1996). Statistical Analysis of Noncovalent Interactions of Anion Groups in Crystal Structures. II. Hydrogen Bonding of Thiocyanate, *Acta Cryst.*. **B52**, 685-690.

SYNTHESIS AND CHARACTERIZATION OF Na^+ AND Ba^{2+} COMPLEXES WITH SOME LIPOPHILIC DIAZA-18-CROWN-6 DERIVATIVES

Hamdi Temel*, Halil Hoşgören, Ümit Çakır ** and Mustafa Boybay*****
**) Department of Chemistry, Faculty of Education, University of Dicle, Diyarbakir*
***) Department of Chemistry, Faculty of Arts and Sciences, University of Dicle, Diyarbakir. ***) Department of Chemistry, Faculty of Arts and Sciences, University of Firat, Elazig, TURKEY.*

Abstract ; The solid and stable Na^+ and Ba^{2+} complexes of 7,16-Dibenzyl-4,7,10,13-tetraoxa-7,16-diazacyclooctadecane; 7,16-Dihexyl-4,7,10,13-tetraoxa-7,16-diazacyclo octadecane; 7,16-Dihep-tyl-4,7,10,13-tetraoxa-7,16-diazacyclootadecane, 7,16-Dioctyl-4,7,10,13-tetraoxa-7,16-diazacycloctadecane; 7,16-Dinon yl-4,7,10,13-tetraoxa-7,16-diazacyclooctadecane; 7,16-Didodecyl-4,7,10,13-tetraoxa-7,16-diazacyclooctadecane were prepared. The complexes were investigated using Elemental Analysis, IR, ^1H NMR, ^{13}C NMR spectroscopy. ^1H and ^{13}C chemical shifts observed in complexes are discussed in terms of conformational changes on the macrocycles.
Key words; Diazamacrocycles, Na^+, Ba^{2+}, cation complexes

1. Introduction

Hundreds of alkali cation complexes of macrocyclic polyethers and related macrocycles have been synthesized and investigated [1-3].

After metal-diazapolyoxa macrocycle complexes obtained by Lehn [4] the azacrown compounds have received considerable attention because of their ability to complex a variety of metal and organic cations and anions [5]. The preferential complexation results when cation radii and ligand cavities are matched [6,7]. Additionally, the azacrowns are important intermediates for the synthesis of the cryptands and the nitrogen pivot lariat crown ethers [8]. They are used as synthetic receptors in molecular recognition processes, as catalysts in nucleophilic substitution, oxidation [9, 10] and in design of chromogenic reagents [4,11]. They even have anion complex formation properties which are similar to those found in certain biological systems [12].

In the presented work some very stable (1:1) ratio complexes of diazacrown ethers with Na^+ and Ba^{2+} perchlorate salts were prepared..

A. W. Coleman (ed.), Molecular Recognition and Inclusion, 527–530.
© 1998 *Kluwer Academic Publishers. Printed in the Netherlands.*

2. Experimental

The chemicals were from FLUKA otherwise cited. Ligands were recently synthesized by Hoşgören [13]. A solution of $M^{+n}(ClO_4)_2$ (0,50 mmol) in ethyl alcohol (2 ml) was added drop wise to the solution of diazacrown (0,5 mmol) in ethyl alcohol (2 ml). The mixture was stirred at room temperature over 24 h then the precipitated solid complex was filtered, washed with ethyl alcohol and re crystallized from ethyl alcohol and dried in desiccator. Structural data of complexes with the elemental analyses performed on a Carlo-Erba model 1200 instrument are given in Table 1.

Table 1. Structural Data of Metal-dialkyldiazacrown Complexes.

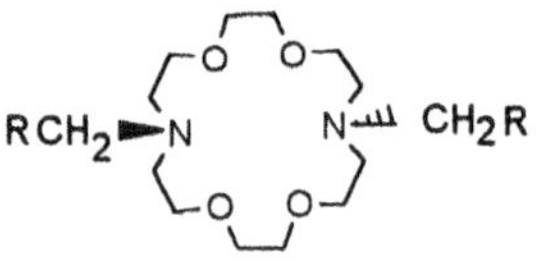

Comp	R
I	C_6H_5
II	C_5H_{11}
III	C_6H_{13}
IV	C_7H_{15}
V	C_8H_{17}
VI	$C_{11}H_{23}$

	Complex Formula	m.p. °C	Yield	N %	C %	H %	N %	C %	H %
I	$C_{26}H_{40}N_2O_9ClNNa$	103	88	4.80	53.56	6.91	4.83	53.85	6.89
I	$C_{26}H_{40}N_2O_{14}Cl_2Ba$	265-270	64	3.52	39.19	5.06	3.57	39.20	4.96
II	$C_{36}H_{76}N_2O_9ClNa$	88-89	87	3.77	58.67	10.31	3.72	58.64	10.34
II	$C_{36}H_{76}N_2O_{13}Cl_2Ba$	171-173	43	2.93	45.56	8.01	2.95	45.32	8.03
III	$C_{28}H_{60}N_2O_9ClNa$	26-28	24	4.47	53.61	9.64	4.57	53.66	9.60
III	$C_{28}H_{60}N_2O_{13}Cl_2Ba$	158-160	65	3.33	39.99	7.19	3.28	40.09	7.16
IV	$C_{24}H_{50}N_2O_{12}Cl_2Ba$	215-217	23	3.65	35.59	6.57	3.60	35.64	6.50
V	$C_{26}H_{56}N_2O_{13}Cl_2Ba$	162-163	62	3.45	38.41	6.94	3.44	38.48	6.90
VI	$C_{30}H_{64}N_2O_{13}Cl_2Ba$	104-106	24	3.22	41.48	7.42	3.20	41.54	7.40

[I]7,16-Dibenzyl-4,7,10,13-tetraoxa-7,16-diazacyclooctadecane, [II] 7,16-Dihexyl-4,7,10,13-tetraoxa -7,16-diazacyclooctadecane, [III] 7,16-Diheptyl-4,7,10,13-tetraoxa-7,16-diazacyclooctadecane. [IV] 7,16-Dioctyl-4,7,10,13-tetraoxa-7,16-diazacyclooctadecane, [V] 7,16-Dinonyl-4,7,10,13-tetra oxa-7,16-diazacyclooc-tadecane. [VI] 7,16-Didodecyl-7,10,13-tetraoxa-7,16-diazacyclooctadecane..

The melting points are uncorrected, IR spectra of the complexes were recorded with MIDAC 1700 instrument in KBr pellets. ^1H and ^{13}C NMR spectra at 200 MHz and 50.2 MHz respectively as well as the 2D spectra and decoupling experiments of the free and complexed macrocycles were recorded on a BRUKER-AC spectrometer in $CDCl_3$ and reported in ppm (δ) from internal TMS. The elemental analyses were conducted on Carlo Erba instrument.

Table 2. Stretching Vibrations of COC(asym) and CCO(sym) in Metal Complexes.

Crown	Salt	COC (cm^{-1})	CCO (cm^{-1})	COC(cm^{-1})	CCO(cm^{-1})
I		1121.5	1105.3	994.8	939.3
I	$NaClO_4.H_2O$	1102.2	1036.1	975.2	935.2
I	$Ba(ClO_4)_2.H_2O$	1093.5	1063.5	981.7	917.1
II		1120.8	1096.7	980.6	945.7
II	$NaClO_4.H_2O$	1099.3	1042.0	967.6	936.7
II	$Ba(ClO_4)_2.H_2O$	1099.1	1063.0	972.5	944.3
III		1127.8	1074.1	989.5	932.2
III	$Ba(ClO_4)_2.H_2O$	1109.9	1051.1	955.2	931.6
III	$NaClO_4.H_2O$	1106.3	1028.0	975.2	928.7
IV		1122.0	1090.6	990.4	936.4
IV	$Ba(ClO_4)_2$	1104.1	1027.0	960.6	926.7
V		1127.7	1078.6	995.2	921.9
V	$Ba(ClO_4)_2.H_2O$	1103.2	1035.7	961.5	903.3
VI		1127.8	1074.1	989.2	932.2
VI	$Ba(ClO_4)_2.H_2O$	1096.4	1000.0	960.5	894.2

Table 3. Proton Chemical Shifts (δ) of the free Macrocycles and with $NaClO_4.H_2O$ in ppm.

	Ligand		complex	
proton	I	VI	I	VI
H_C	2.84	2.76	2.71	2.67
H_B	3.64	3.60	3.60	3.58
H_A	3.62	3.61	3.50	3.68
H_D	3.70	2.47	3.62	2.53

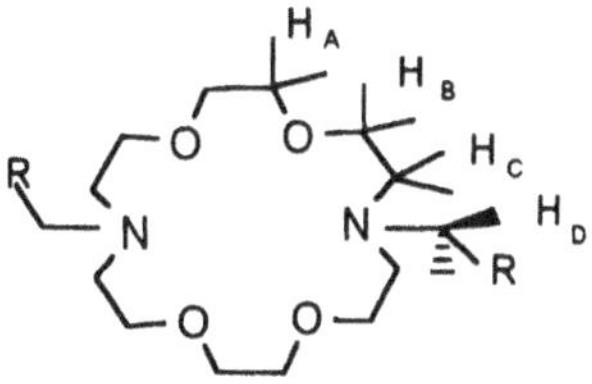

Table 4. ^{13}C Chemical Shifts (δ) of Macrocycles and Complexes in ppm in $CDCl_3$

	Ligand	Ligand	$NaClO_4$	, complex	Ligand	$(BaClO_4)_2.H_2O$	Complex	
	I	VI	I	VI	IV	II	III	IV
C-1	72.75	72.75	70.61	70.56	72.67	72.01	72.07	72.00
C-2	72.11	72.06	69.22	68.83	71.98	70.53	70.45	70.45
C-3	62.03	59.00	60.75	56.04	57.97	53.52	53.66	53.56
C-4	55.86	55.86	54.42	54.62	55.90	53.52	53.49	53.49

3. Results and Discussion

The IR spectra of the complexes clearly demonstrated that the COC and CCO stretching vibrations are altered compared to ligands due to conformational changes[14]. The ν(R-O-R) absorption is shifted to lower wave numbers in complexes providing another evidence of complexing [15]. IR spectra of the complexes of 3300-3400, 1650-1660 and 560-570 cm^{-1} regions are due to ν(OH), δ(HOH) and ρ(HOH), respectively, indicating the coordinated water molecules [16]. Accordingly, the IR bonds due to (ClO_4^-) is found at 620-630 cm^{-1}. The band at 1110 cm^{-1} due to $\nu(ClO_4^-)_{asym}$ overlaps the ν(COC) which is found at roughly the same frequency, Table 2 [17].

In the NMR spectroscopy of the macrocycles and N-substituted methylene protons in alkyl chains were assigned by means of selective heteronuclear decoupling experiments and 2D-HETCOR spectroscopy , Table 3,4.

^{1}H NMR of **I** and **II**, the methylenes moved 0,05 - 0,10 ppm in complexes, In **II**, H_A and H_D shifted to lower field, however, such roles are best interpreted in terms of the electric field effect which is observed not only on the methylene protons since the shifts are depended on the distance from the complexing site [18]. The ^{13}C chemical shifts involving the conformational changes due to complexing are moved to higher fields.

Acknowledgment,
For the structural discussions the authors would like to express their thanks to Prof. Dr. Çakıl Erk, from Istanbul Technical University,.

REFERENCES

1. C. J. Pedersen, *J. Am. Chem. Soc.*, **89,** 2495 (1967).
2. D. Dietrich, P. Viout, J. M. Lehn, *Macrocyclic Chemistry.*(1993).
3. V. G. Vladimir, B. Corinne and A. Howard, *Inorg. Chem.* **32**, 345 (1993)
4. B. Dierich, J. M. Lehn, J. P. Sauvage, *Tetrahedron Lett.* 2885 (1969).
5. J. S. Bradshaw, K. E. Krakowiak and R. M. Izatt, *J. Heterocycl. Chem.* **26**, 565 (1989).
6. B. Dietrich, J. M. Lehn, I. P. Sauvage and J. Blanzat, *Tetrahedron*, **29,**2629 (1973).
7 J. M. Lehn, *Acc. Chem. Res.* **11**, 49 (1978)
8. R. A. Schultz, B. D. White, D. M. Dishong, K.A. Arnold and G. W. Gokel, *J. Am. Chem. Soc.* **107**, 6659 (1985).
9. A. R. Kausai, *J. Chem. Soc. Perkin I*, **5,** 227 (1983)
10 E. Weber, *Kontakte* (Merck, Darmstad), **38** (1983)
11 E. Weber, *Kontakte* (Merck, Darmstad), **26** (1984).
12 M. W. Hosseini, J. M. Lehn, S. R. Duff, K. Gu and M. P. Mertes, *J. Org. Chem.* **52,** 1662 (1987).
13 H. Hoşgören, M. Karakaplan and M. Toğrul, *Collect. Czach. Chem. Commun.* ***61,****622 (1996).*
14. S. A. Kulyukhin, N.B. Mikheev, I. E. Veleshkocnd, L.N. Auerman, Translated from *Radiokhimiya,* **34**(1)1992, pp. 88-94.
15. T. Lu, X. Gan, N. Tang and M. Tan, *Polyhedron* **9**(19), 2371 (1990).
16. T. Lu, X. Gan, M. Tan, C. Lý, K. Yu, *Polyhedron* **12(13),** 1641 (1993).
17 E. Rüvarola, F. Saüano, A. Fontana, *J. Organometal. Chem.* **317**, 285 (1986).
18. E. Kleinpeter, S. Stoss, M. Gabler and W. Schroth, *Magnetic Resonance in Chemistry*, Vol. 27, 676-683, 1989.

The Effect of Macrocycles on the Reaction Rate

Part V. The Effect of [18]crown-6 on the Aromatic Nucleophilic Substitution Reactions in dioxane- water Solutions.

Hülya Tuncer* **and Çakıl Erk**

Istanbul Technical University, Science Faculty, Department of Chemistry, Maslak, 80626, Istanbul, Turkey.

Abstract ; The effect of 1,4,7,10,13,16-hexaoxacyclooctadecane, [18]crown-6 on the rate of reaction of 2,4-dinitrochlorobenzene and NaN_3 was investigated in water / dioxane (50/50) mixtures. The second order bimolecular reaction rates were estimated with a chloride selective electrode located in the solution The

Keywords ;

1. Introduction

The effect of macrocycles of cyclic oligomers of ethyleneoxide were so far recognised to alter the rates and the mechanism since they were tried in diplacemenr reactions [1]. The simple observation has been helpful to those concerned shortening reaction times, increasing yields, decreasing side reactions in the most common organic reactions.

Particular roles of macrocycles were really dramatic in non hydroxylic media due to formation of the ion pairs which accelerate the nucleophilic power and therefore the rate in a reaction [1]. Recently, we initiated a work of investigation of the effect of macrocycles on the aromatic nucleophilici substitution reactions in dioxane / water, [2,3] mixtures of which the composition of the solvent could be rearranged to see effects from hydroxylic to non hydroxylic media as described below.

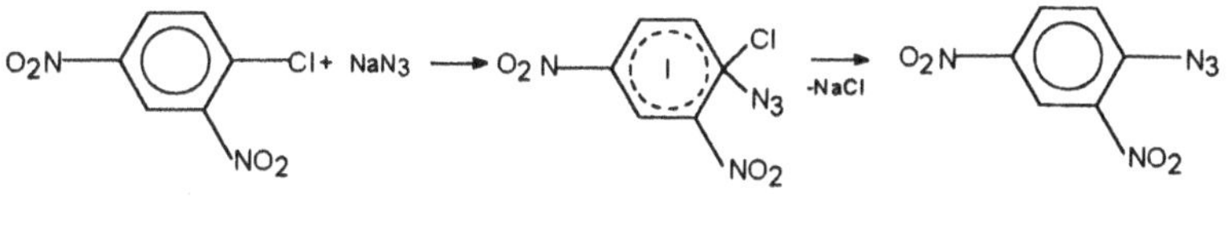

We also recently developed an analytical-electrochemical method for the determination of the mechanisms and reaction rate constants of the such reactions in a solution precisely to be able to see even the slight effects on the reaction rates which were simultaneously followed by potentiometry with a chloride selective ion electrode located in dioxane / water reaction solution.

*) Submitted for the partial fullfilment of Ph.D. degree.

A. W. Coleman (ed.), Molecular Recognition and Inclusion, 531–534.
© 1998 *Kluwer Academic Publishers. Printed in the Netherlands.*

2. Experimental

We designed a system of an electrochemical device consisting of the ionmeter (SCHOTT CG-804) connected to a PC via an interface (SCHOTT TI145). The millivolt changes due to chlorine formation during the reaction are detected simultaneously with an electrode (ORION 9617BN) inserted into the reaction flask together with a PT-100 thermo-element. Reactants of proper concentrations, Table 1, were placed in a cell thermostated with a circulating bath. The formation of the Cl^- anion was followed continuously being recorded every five minutes on the PC connected to the interface. Results were simulated for maximum correlation coefficients and optimised for the initial concentrations to estimate the rate constants according to Equations 1-3. Chloride ion concentrations were calibrated at similar reaction conditions. NaN_3 and the chemicals were from MERCK. Dioxane was repurified by distilling on the metallic sodium. 2,4-dinitrochlorobenzene was twice recrystallised from methanol.

3. Results

The second order kinetics and rate constants were determined according to following equations for S_N2 bimolecular reactions with equimolar amount of reactants,

$$A + B \rightarrow \text{products} \qquad (1)$$

$$dX/dt = k[A_o\text{-}X][B_o\text{-}X] \qquad (\text{since } dx/dt = k[A_o\text{-}X]^2) \qquad (2)$$

$$kt = 1/[A_o\text{-}X] - 1/A_o \qquad (\text{or } 1/[A_o\text{-}X] = kt + 1/A_o) \qquad (3)$$

$[A_0] = [B_0]$ in reaction mixtures. The $1/(A_o\text{-}X)$ values obtained via potentiometry are plotted versus time. However, we have simulated the data to fit the linear equations with the highest correlation coefficient and y-intercept and the slops obtained gave the rate constants, Figure 1.

Table -1 The Reaction Rates of 24DNCB with NaN_3 in dioxane / water (50/50)

Temp (°C)	Conc. (+18Cr6) a)	Intercept b)	Corr.	Rate, K x1000 c)	Ln K
5	0.02	49.99	0.993	0.922	-6.988
5	0.02 + 0.02 Cr	49.88	0.996	1.304	-6.642
25	0.02	47.16	0.995	17.183	-4.064
25	0.02 + 0.02 Cr	50.00	0.994	24.487	-3.709
25	0.02 + 0.1 Cr	50.00	0.997	34.732	-3.360
25	0.02 + PEGDM250 d)	50.00	0.993	34.096	-.3.378
40	0.02	50.00	0.993	26.737	-3.621
55	0.02	50.00	0.998	103.129	-2.272

a. Concentrations in M / l. b. intercept of plots, see Figs 1,2. c. Rates are (1 / M x min)
d. Polyethylene glycol dimethyl ether 250 from Merck.

We have tried the reactions in a mixture that it could be capable to dissolve the both of the salt and the aromatic substrate. Higher amounts of the dioxane caused also lesser sensitivity of the ion selective electrode as well as the poor solubility of the NaN_3.

It was particularly observed that the equimolar amounts of the macrocycle [18]crown-6 with NaN_3 has shown the increased displacement reaction rates with 2,4-dinitrochlorobenzene in dioxane / water (50:50) mixtures while the very excess of macrocycles has not much role in such a solvent mixture, Table 1 [5].

In fact, the reaction rates, therefore, the free energy of activations found were depended on the rate of addition of the nucleophile to the aromatic substrate [6,7]. Thermodynamics of the displacement reactions of anions, [3] with the uncharged species in the presence of macrocycles were also investigated at several temperatures.

The rates reported are the overall rates calculated. However, there is even a curvature reflected by correlation coefficients due to the ion pairs equilibrium of the [18]crown-6 between the NaN_3 and NaCl salts in the solution during a reaction [3-5,8]. The small effect on the cation solvation caused the small anion efficiency due to large role of water in the dioxane / macrocycle / water mixture where the hydrogen bonding may be still even effective in the presence of 50 % percent dioxane. However, the rate found at 25 °C is as fast as those found in DMF for similar reactions [9].

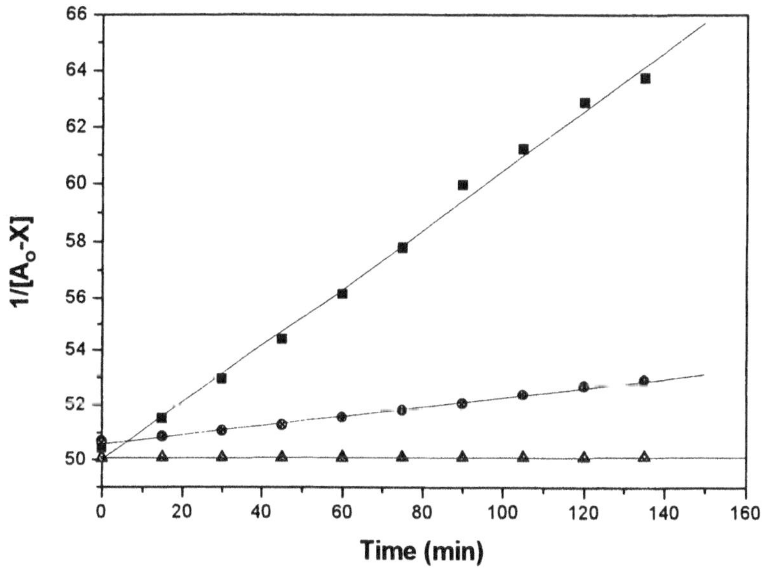

Figure 1. The plot of 1/(Ao-X) versus time at 5 °C (▲), 25 °C (•) and 55 °C (■) in the reactions of 24DNCB with NaN_3 in dioxane / water.

Acknowledgement ; This work has been kindly supported by the Research foundation of the Technical University of Istanbul.

References

1. a. D. J. Cram,. *Angew. Chem* Int. Ed. Engl. **21,**1039(1986).
   b. E de Boer, A. M. Grotens and J. Smid, *J. Am. Chem Soc.* **92**,4742(11970)
   c. M. Szwarc, in *Ions and Ion Pairs in Organic Reactions*, John Wiley, 1972, New York.
2. a. H. Çetişli and Ç. Erk, *Communic. Fac. Sci. Univ.*, Ankara. **B25**,9(1979).
   b. H. Çetişli and Ç. Erk, *Communic. Fac. Sci. Univ.*, Ankara. **B26**,29(1980).
3. F. P. Scmidtchen, *Angew. Chem.* Int. Ed. Engl. **20,**466(1981).
4. H. Tuncer and Ç. Erk, Abstracts, 9th IUPAC congress, 1995, Istanbul, TR.
5. H. Tuncer and Ç. Erk, 1994, Unpublished results,.
6. R. Alexander, E. C. F. Ko, A. J. Parker and T. J. Broxton, *J. Amer. Chem. Soc.* **90,**5049(1968).
7. A. J. Parker, *Chem. Rev.* **69**,1(1969).
8. B. G. Cox, and P. T. McTigue, *Australian J. Chem.* **20**,1815(1967).
9. H. L. Friedman, *J. Phys. Chem.* **71**,1723(1967).

Stereocontrol and Rate Enhancement of a Diels Alder Reaction Within an Unsymmetrical Porphyrin Host

Lance J. Twyman, Anton Vidal-Feran
Nick Bampos and Jeremy K. M. Sanders.
Cambridge Centre for Molecular Recognition
University Chemical Laboratory
Lensfield Road, Cambridge, CB2 1EW (UK)>

The synthesis of molecules capable of mimicking the action of enzymes provides us with a considerable intellectual challenge. In recent work we have described both the synthesis[1] and catalytic properties of a symmetrical porphyrin host.[2] Of particular note is the ability of this symmetrical 222 trimer[3] to stereoselectively accelerate the Diels Alder reaction shown in scheme 1, producing exclusively the thermodynamically *exo* adduct. This is one of a few known examples where a synthetic enzyme binds both reactive substrates.[4] Molecular modelling suggests that a smaller host cavity, of around 12 Å, should be capable of accelerating the formation of the thermodynamically disfavoured *endo* adduct To this end the 112 porphyrin host trimer 1 was synthesised.

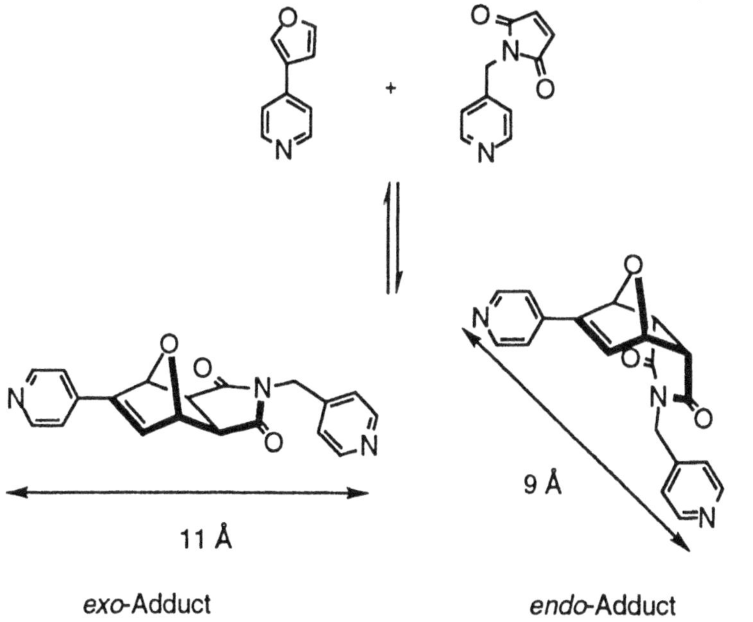

SCHEME 1: The Diels Alder reaction showing dimensions of the *endo* and *exo* adduct.

A. W. Coleman (ed.), Molecular Recognition and Inclusion, 535–538.
© 1998 *Kluwer Academic Publishers. Printed in the Netherlands.*

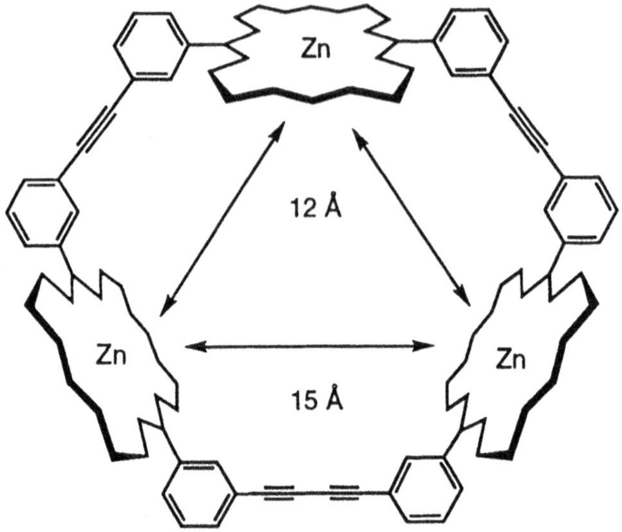

FIGURE 1: The 112 porphyrin trimer **1**, showing dimensions of the host cavity

This new unsymmetrical host contains two different Zn-Zn distances, firstly a longer 15 Ådistance, which favours formation of the *exo* adduct, and secondly, two shorter 12 Å distances, which it is hoped would favour formation of the smaller *endo* adduct Figure 1. At 60°C with 9.0mM dm^{-3} substrates, and in the absence of any catalyst, the forward rate constant for the bimolecular reaction is $2x10^{-4}$ dm^3 mol^{-1} s^{-1}. However, with 0.9mM dm^{-3} concentrations the reaction is much slower, with the equilibrium lying far on the side of the starting materials. The ratio of *exo* adduct to *endo* adduct is 4:1. At 60°C and in the presence of 1 equivalent of 112 trimer **1**, the initial forward rate (in tetrachloroethane solution at 0.9mM) is accelerated around 300 fold; now the *endo* and *exo* adducts are formed in essentially the same proportion Figure 2. Since the intrinsic *endo* rate is around one quarter that of the *exo* reaction,[4] the accelerations are 150 fold for the *exo* reaction, and 600 fold for the *endo* reaction. After about 30 hours the formation of *endo* adduct begins to slow, whilst the amount of *exo* adduct continues to increase, the reason for this arises from the reversibility of the reaction. The *endo* adduct can revert back to starting materials, which can then re-bind to form the thermodynamically favoured *exo* adduct.

FIGURE 2: Formation of endo and exo adduct in 112 trimer at 60°C.

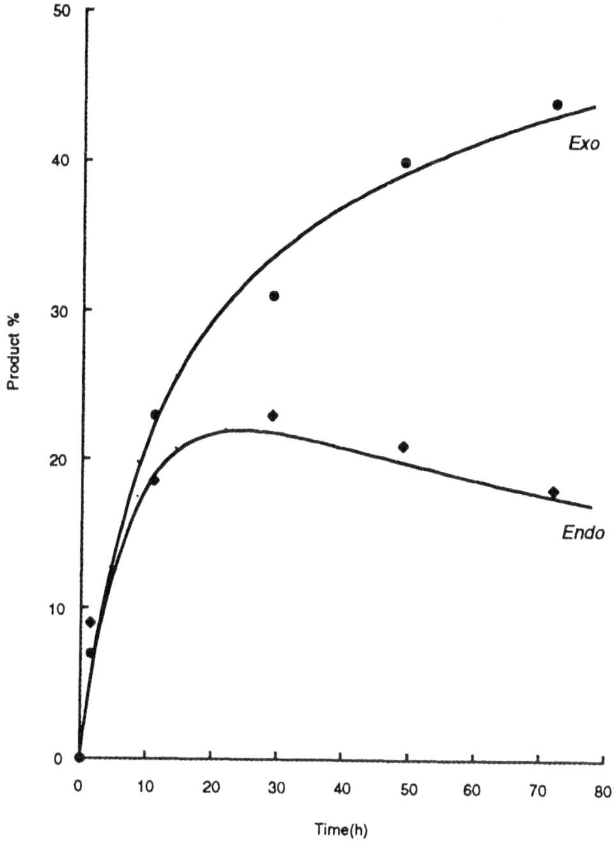

The measured binding constants are *ca* 1 x 10^6 and 2 x 10^6 M^{-1} for the *endo* and *exo* adducts respectively, Table 1, but these cannot be separated into the two seperate binding constants, 1) binding of *exo and endo* adducts across the shorter 12 distance, the K_{ethyne} component and 2) binding of *exo and endo* adducts across the longer 15 distance, the $K_{butadiyne}$ component. Quantitatively, there are too many binding events and rate constants (forward and back; k_1 *exo*, k_{-1} *exo*, k_1 *endo* and k_{-1} *endo*) in this system to allow meaningful simulation of the entire reaction: it has not been possible, therefore, to find a unique and reliable fit[5] to the experimental data.

TABLE 1: Binding constants of starting materials and products to 112 porphyrin **1**.

Ligand	Binding Constant M^{-1}
Diene	2×10^3
Dienophile	1×10^3
exo Adduct	2×10^6
endo Adduct	1×10^6

We have shown that subtle changes in host design can lead to controlled and predicted changes in the stereochemical outcome of a Diels-Alder reaction where there are two finely balanced pathways. We are currently synthesising similar host molecules with different binding properties as well as constructing host systems with smaller (*endo* selective) cavities.

Acknowledgements

We thank the European Union Human Capital and Mobility program and EPSRC for financial support.

References.

1. Anderson, H. L. and Sanders, J. K. M. (1989) *J. Chem. Soc., Chem. Commun.*, 1714
2. Walter, C. J. and Sanders, J. K. M. (1995) *Angew. Chem. Intl. edn.*, **34**, 217.
3. The numbers 222 and 112 refer to the number of acetylene spacers between the diaryl porphyrins. For example; the 112 trimer has two mono-acetylene spacers (***11***2) and one di-acetylene spacer (11***2***).
4. Diederich, F., Schurmann, G. and Chao, I. (1988) J. Org. Chem., **53** 2744: , Breslow, R., (1990)*Pure Appl, Chem.,* **62** 1859: Jubian, V., Dixon, R. P., and Hamilton, A. D. (1992) *J. Am. Chem. Soc.*, **114**, 1120.
5. Vidal-Ferran, A., Muller, C. M., and Sanders, J. K. M. (1994) *J. Chem. Soc., Chem. Commun.*, 2657.
6. Wylie, R. S., and Sanders, J. K. M, (1995) *Tetrahedron*, **51**, 513

PREORGANIZATION OF LINEAR POLYAMINES IN THE SOLID STATE

ZOFIA URBAŃCZYK-LIPKOWSKA[a)], AGNIESZKA PĘCAK[b)]
[a)]*Institute of Organic Chemistry, Polish Academy of Sciences, 01-224 Warsaw*
[b)]*Department of Chemistry, University of Warsaw, 02-093 Warsaw, Poland*

The supramolecular approach to the synthesis of structurally complex systems turned our attention toward polyamine compounds. Nitrogen function present in the molecule affords the possibility of constraction of such „three-dimmensional" molecules like cryptands or lariat crown ethers. On the other side, molecules containing amine groups like simple ethylenediamine or podands are known to form in solution highly stabile complexes with metal ions due to ability of conformational change upon complex formation.

Simultaneous reactions leading to an azamacrocyclic ring formation, undergoing between more than two reacting centers, require preorganization of the substrates in the preceding stages. In such an approach the knowledge of preferred conformations of substrate molecules is helpful in understanding the reaction mechanism and planning more efficient synthetic procedures.

Macrocyclization reaction between the acyclic amines and α,ω-dimethyl esters of appropriate dicarboxylic acids leading to di- and tetraazacrown ethers [Jurczak *et al.*, 1993] was a model in our studies. The reaction is very efficient for when both diamine and diester molecules contain oxygen function. In the case of 1,4-butyldiamine and amines with longer chain special techniques i.e. such as high pressure or ultrasound irradiation have to be used in order to achieve reasonable yield or selectivity [Jurczak *et al.*, 1992]. Therefore, conformational changes of the diamine molecules **1**, **2**, and **3**, and selectivity of their binding pattern have been investigated in different crystal cavities using X-ray diffraction techniques.

Preorganization of the amine molecules in the crystal was modulated by co-crystallization of liquid amines with maleic, fumaric, succinic, adipic, and picric acids [Urbańczyk-Lipkowska, 1996, Urbańczyk-Lipkowska *et al.*, 1996]. In all investigated co-crystals both amine groups were protonated. Thus the crystal lattice was formed by

A. W. Coleman (ed.), Molecular Recognition and Inclusion, 539–542.
© 1998 *Kluwer Academic Publishers. Printed in the Netherlands.*

ionic species, what was manifested by melting points higher than in the case of crystals of parent dicarboxylic acids. Acid molecules exhibited some flexibility (succinic and adipic acid) or rigidity (fumaric, maleic, and picric acid). The 1:1 or 1:2 ratio of amine to acid molecules was observed in the solid state, depending on character of the amine, protonation ability of acid molecules (fumaric *versus* forming intramolecular hydrogen bond maleic acid) and number of water molecules present in the crystal.

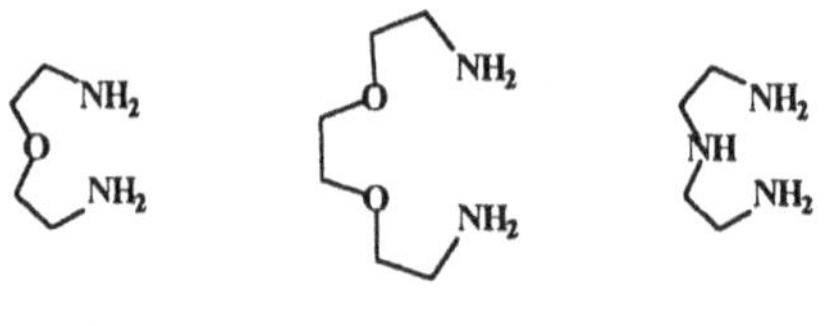

1 **2** **3**

The studies showed that molecules of diamines **1**, **2**, and **3** have folded conformations in the crystal lattice which, if retained in solution, would make ring closure reaction more efficient (Fig.1).

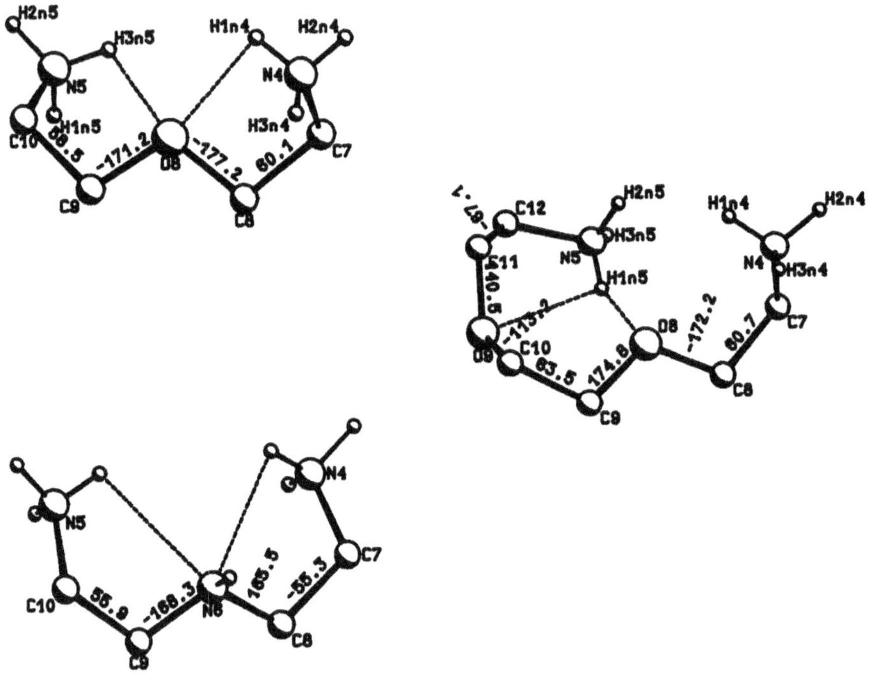

Fig. 1. Folded conformations of amine **1**, **2**, and **3** found in co-crystals with dicarboxylic acids

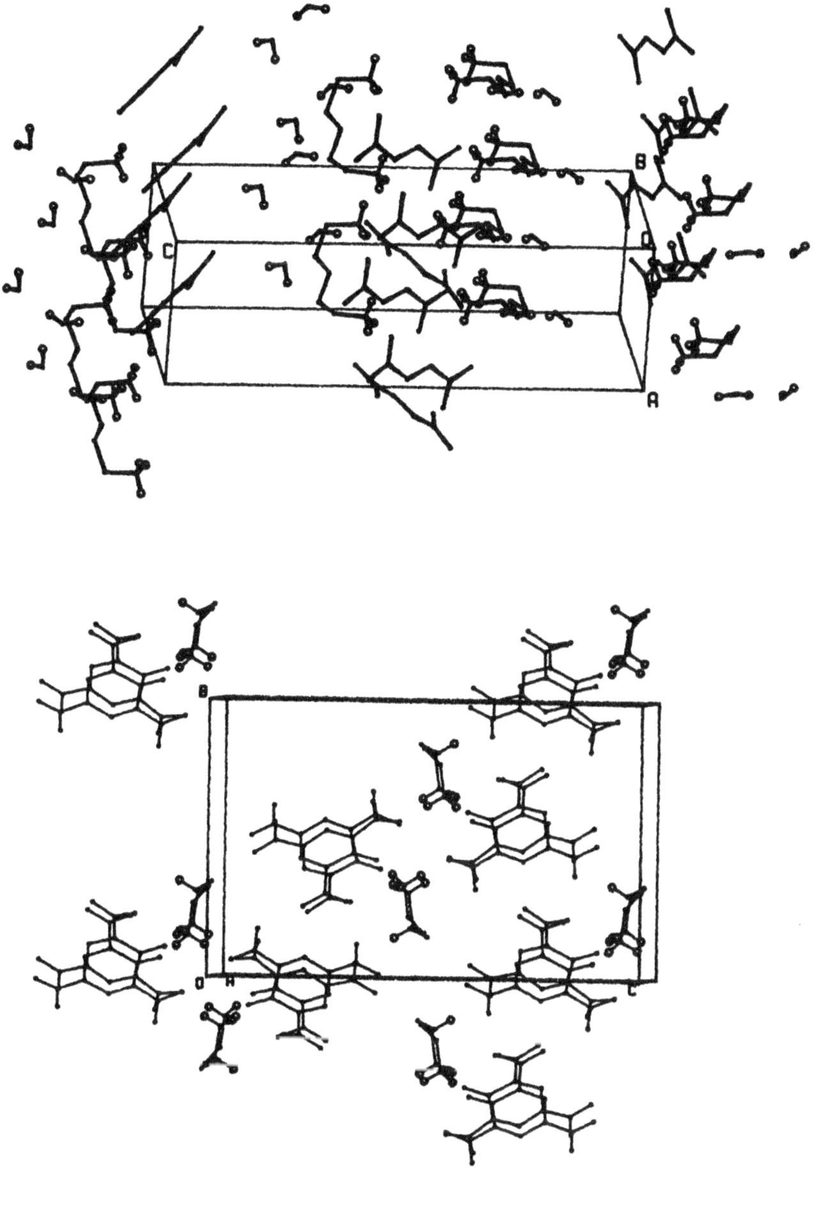

Fig. 2. Packing diagram of the systems amine **1** : fumaric acid : 2 H_2O (upper part) and amine **3** : 2 picric acids (lower part)

The folded conformation of the molecule is stabillized by two, usually weak, intramolecular hydrogen bonds between amine protons and the ethereal oxygen atom. As might be seen in Fig. 1, such conformation of the amine molecule affects its polarity: one side of the molecule containing ethylene bridges is hydrophobic, the opposite site containing two NH_3^+ groups is hydrophilic. Consequently, the molecular packing in the crystal is very characteristic for this group of compounds. The crystal lattice consists of layers of diprotonated amine molecules with their hydrophilic sides neighbouring the layer of dicarboxylate anions which are negatively charged at both ends. The less polar side of the amine layer is packed next to the layer of water molecules which play a role of interface between next layer of dicarboxylate anions. Such situation, shown in upper part of Fig. 2, has been found in the co-crystal of amine **1** with one molecule of fumaric acid and two water molecules [Urbańczyk-Lipkowska, 1996]. Another packing is observed in the case of amine **3** co-crystallized with two molecules of picric acid [Pęcak *et al.*, 1996]. View of the unit cell is shown in the lower part of Fig. 2. The alternating layers consist of molecules of diprotonated amines and picric anions. The layer of picrates contains antiparallelly located picric anions with O^- groups pointing to opposite directions.

In particular, intra- and intermolecular hydrogen bonding interactions have been investigated in order to find characteristic patterns wchich could explain reaction selectivity and yields.

X-ray diffraction techniques used for studying of molecular preorganization of polyamine molecules in the solid state showed that interaction with strong hydrogen bond acceptors force their folded conformations. Similar pattern of intramolecular hydrogen bonds was found to stabilize the molecular conformation of various diazacoronads.

References

1. Jurczak, J., Stankiewicz, T., Sałański, P. and Lipkowski, P. (1993) A New Method for the Synthesis of Diazacoronands *via* Double-amidation Reaction, *Tetrahedron* **49**, 1478-1488.
2. Jurczak, J. and Ostaszewski R. (1992) High Pressure Approach to the Synthesis of Diazacoronands and Cryptands, *J. Coord. Chem.* **27**, 201-214.
3. Urbańczyk-Lipkowska, Z. (1996) Preorganization of 2,2'-Diaminoethylether, 2,2'-(Ethylenedioxy)diethylamine and 1,4-Diaminobutane in the Solid State, *J. Mol. Struct.* **374**,305-311.
4. Urbańczyk-Lipkowska, Z. and Gluziński P. (1996) Molecular Recognition of Acyclic Amines I: Co-crystallization of 2,2'-(Eethylenedioxy)diethylamine with Dicarboxylic Acids, *Supramolecular Chem.* **7**, 113-118.
5. Pęcak, A. Urbańczyk-Lipkowska, Z., and Jurczak, J. (1996) in preparation.

Vibrational Spectroscopic Studies on the Inclusional Complexation by β-Cyclodextrin

Xuemei Wang and Hans-Jörg Schneider
Fachrichtung Organische Chemie, Universität des Saarlandes, D 66041 Saarbrücken, Germany

Abstract

Cyclodextrins are cyclic oligosaccharides whose structural feature gives the overall shape of a truncated cone with a hydrophilic exterior and a hydrophobic interior which possesses the interesting property of being able to recognize a wide variety of molecules by forming inclusion complexes. Although much effort has been expended in seeking for the special interaction characteristics of the cyclodextrin systems, yet the chemical affinity involved is still poorly understood. In view of the crucial importance to understand the molecular structure of supramolecular cyclodextrin complexes, in recent work we have tried to study the inclusional complexation by β-Cyclodextrin (β-CD) using vibrational spectroscopy and Surface Enhanced Raman Spectroscopy (SERS). The experimental results indicate that the SERS vibrational spectroscopic studies on the complex structure could afford some special structural information pertaining to the supramolecules.

In this report, we have investigated the particular structural features of the complex between barbitone (a pharmaceutically important barbituric acid) and β-Cyclodextrin. Our previous study indicates that a 1:1 β-CD/barbiturate complex would be formed in aqueous solution; therefore, we could obtain a white crystalline material of the 1:1 β-CD/barbitone complex through the coprecipitation method. In order to probe the molecular structure of this synthesized complex compound, Fourier transform methods (including both FTIR and FT Raman spectroscopy) were utilized to reveal its characterization. And the vibrational spectra of pure barbitone, pure β-Cyclodextrin, the 1:1 physical mixture of the two compounds, and the complex were obtained and compared.

Perhaps the most significant evidence for complexation indicated in Fourier Transform Raman spectroscopy (FT Raman) occurs in the C=O stretching lines of the barbitone spectrum at $1763cm^{-1}$, $1733cm^{-1}$ and $1703cm^{-1}$ where pure cyclodextrin exhibits no Raman intensity(Figure 1.). In the complex, the peak intensity of these lines decreases greatly and are

A. W. Coleman (ed.), Molecular Recognition and Inclusion, 543–546.
© 1998 *Kluwer Academic Publishers. Printed in the Netherlands.*

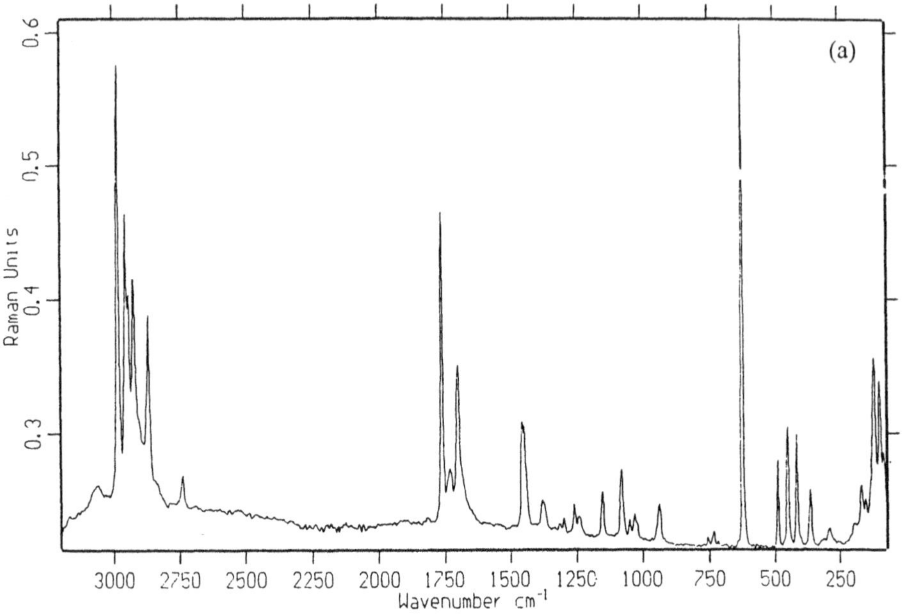
(a)
Raman Units
0.6
0.5
0.4
0.3
3000 2750 2500 2250 2000 1750 1500 1250 1000 750 500 250
Wavenumber cm⁻¹

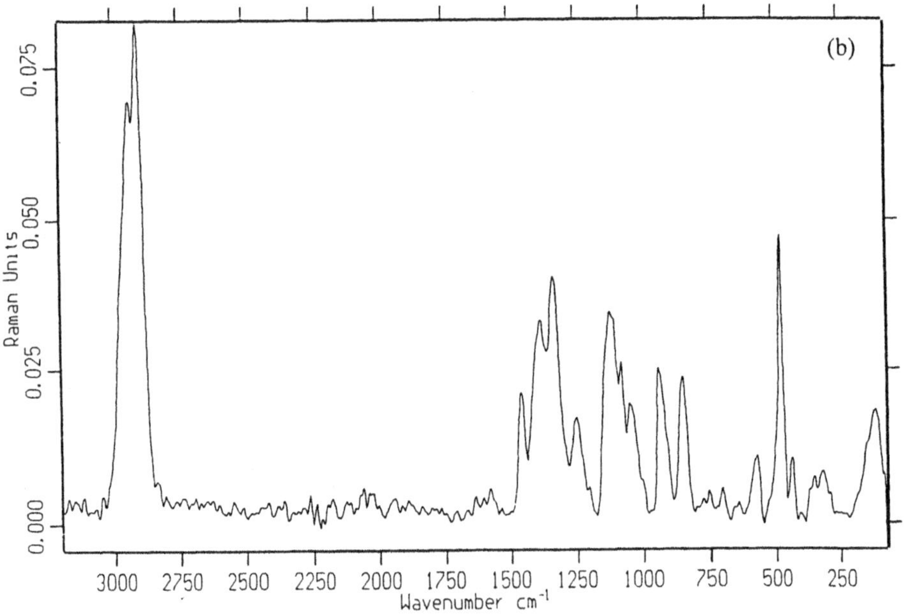
(b)
Raman Units
0.075
0.050
0.025
0.000
3000 2750 2500 2250 2000 1750 1500 1250 1000 750 500 250
Wavenumber cm⁻¹

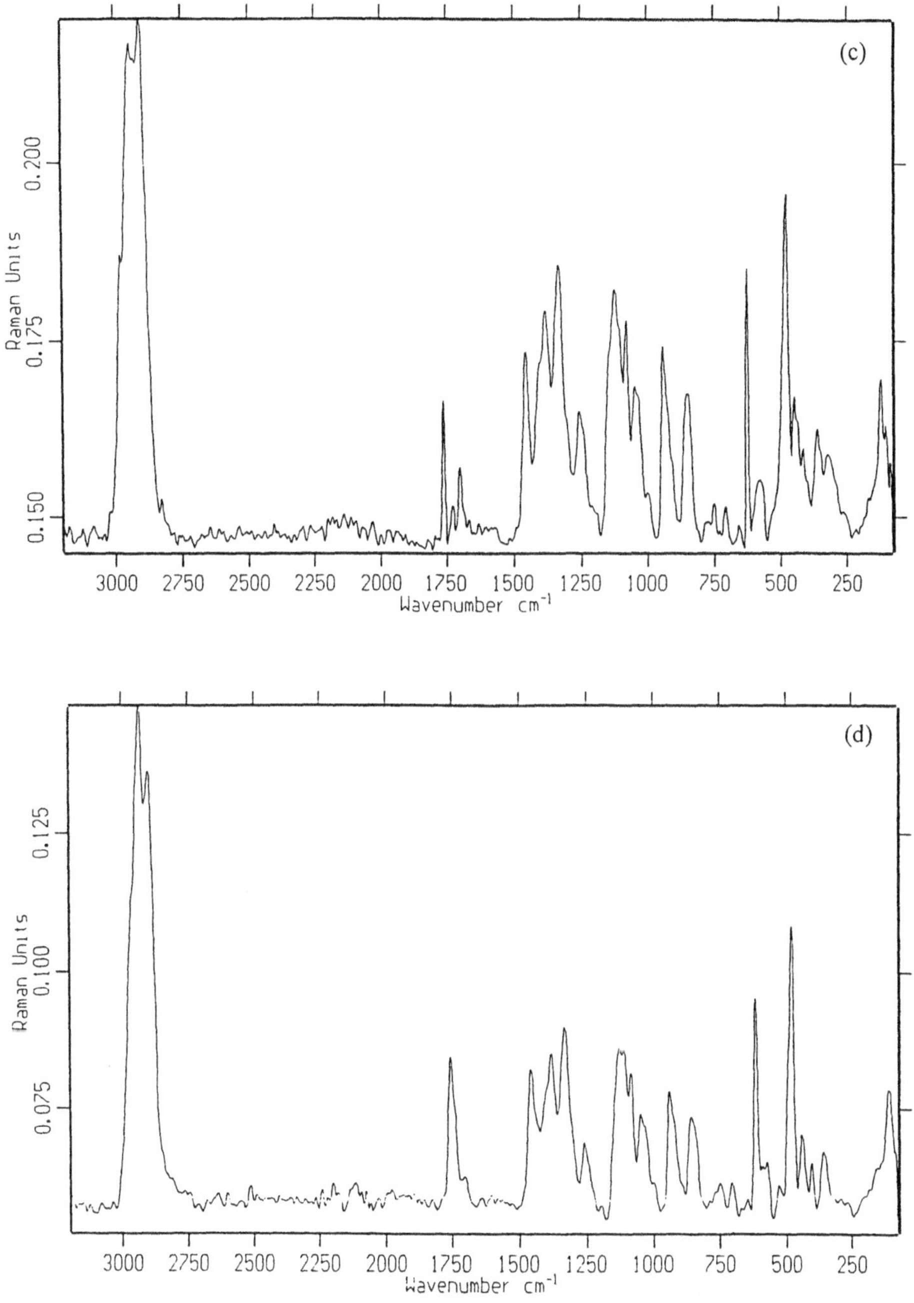

Figure 1. FT Raman Spectra of (a)barbitone, (b) β-CD, (c) β-CD/barbitone physical mixture, and (d) β-CD/barbitone compl

completely smeared out into one band peaking at $1756cm^{-1}$. On the other hand, the most intense band at $627cm^{-1}$ observed in the barbitone spectrum shifts to $617cm^{-1}$ upon complexation, accompanying a considerable decrease of the intensity in the complex. Since the vibrations near $627cm^{-1}$ are due to the various ring motions, these changes of the spectral features may suggest that hydrogen bonding is formed between the guest molecule and cyclodextrin. In addition, the peak of a strong band at $2985cm^{-1}$ in pure barbitone are found to disappear in the complex. This phenomenon may indicate that the side-chains of barbitone molecules are also included in the cavity of β-CD, which is supported by its particular structural information afforded by Fourier Transform Infrared (FTIR) spectroscopy.

Furthermore, another interesting phenomenon is remakable in the Surface Enhanced Raman Spectroscopic study which allows detailed molecular level analyses of the complex interactions between various species at the surfaces. On the Ag surface, a strong enhancement of the peak intensity in the $1230cm^{-1} \sim 1450cm^{-1}$ region of the spectra upon complexation clearly indicates that the side-chains of barbitone molecules are included in the cavity of β-CD and are perpendicularly orientated on the Ag surface. Our findings suggest that this approach may serve as a novel strategy to sensitively probe the microscopic environment of guest molecules and afford additional information for the molecular complexes. Therefore, as is mentioned above, the SERS and vibrational spectroscopic studies as a means of probing the complex structure are of inestimable value and have every prospect of contributing to the biological field.

Acknowledgement. An Alexander von Humboldt fellowship for W.X.M.

GPSR Compliance

The European Union's (EU) General Product Safety Regulation (GPSR) is a set of rules that requires consumer products to be safe and our obligations to ensure this.

If you have any concerns about our products, you can contact us on ProductSafety@springernature.com

In case Publisher is established outside the EU, the EU authorized representative is:

Springer Nature Customer Service Center GmbH
Europaplatz 3
69115 Heidelberg, Germany

Batch number: 10809436

Printed by Printforce, the Netherlands